LMPC 2013

Proceedings of the 2013 International Symposium on Liquid Metal Processing & Casting

LMPC 2013

Proceedings of the 2013 International Symposium on Liquid Metal Processing & Casting

Sponsored by

The Minerals, Metals & Materials Society (TMS)

Held

September 22–25, 2013

AT&T Executive Education & Conference Center

Austin, Texas

Edited by

Matthew J. M. Krane

Alain Jardy

Rodney L. Williamson

Joseph J. Beaman

Editors
Matthew J. M. Krane
Alain Jardy
Rodney L. Williamson
Joseph J. Beaman

ISBN 978-3-319-48587-4
ISBN 978-3-319-48102-9 (eBook)
DOI 10.1007/978-3-319-48102-9

Chemistry and Materials Science: Professional

Published by Springer International Publishers, Switzerland, 2016
Reprint of the original edition published by John Wiley & Sons, Inc., 2013, 978-1-118-00202-5

Printed on acid-free paper

This Springer imprint is published by Springer Nature
The registered company is Springer International Publishing AG
The registered company address is: Gewerbestrasse 11, 6330 Cham, Switzerland

TABLE OF CONTENTS
Proceedings of the 2013 International Symposium on Liquid Metal Processing and Casting

Electroslag Remelting

Defects

Vacuum Arc Remelting

Aluminum Processing

Miscellaneous

PREFACE

This proceedings volume represents the work presented at the tenth International Symposium on Liquid Metal Processing and Casting held in Austin, Texas, USA, from September 22-25, 2013. This symposium is held every other year to bring together an international group of researchers to discuss this important technical field and present their latest research results. The conference is the principal international forum for research and development in the fields of melting, refining and casting of special metals and alloys. Contributions cover a range of topics including vacuum arc and electroslag remelting, advances in process modeling, novel casting techniques, and characterization and modeling of inclusions, segregation, and structure in cast metals.

Papers from both the oral and poster sessions are included in this volume. The collection is divided into the following five areas: electroslag remelting (24 papers), defects (10 papers), vacuum arc remelting (4 papers), aluminum processing (3 papers), and a miscellaneous section with 11 papers on topics ranging from microsegregation to melt control. This volume contains a good mix of papers covering both experimental studies and mathematical modeling/simulation studies, with contributions from industry, academia, government laboratories, and teams from combinations of these institutions.

The organizers are pleased to welcome Mr. Rashmi Bhavsar, Global Materials Metier Manager for Schlumberger in Houston, Texas as the keynote speaker. His topic is

"Metal Alloy Challenges and Opportunities in the Oil and Gas Industry,"

which is in tune with the large demand for high quality metal products by this industry. Mr. Bhavsar has over three decades of experience working in the petroleum industry, with responsibility for material selection, alloy characterization, corrosion, welding, heat treating, surface treatment, and failure analysis for drilling and completion tools.

The Awards Committee would once again like to congratulate the authors of the LMPC 2011 Best Paper,

"Investigation of the Implications of the Current Conductive Mold Technology with Respect to the Internal and Surface Quality of ESR Ingots,"

authored by: H. Holzgruber, W. Holzgruber, A. Scheriau, M. Knabl, M. Kubin, J. Korp, R. Pierer.

The Conference Organizers are grateful to the members of the Scientific Committee who reviewed and sorted the abstracts and papers in a very short period of time. Ms. Rosalie Foster of the University of Texas was instrumental in setting up the logistics of the conference venue and Ms. Lisa Breese and the other TMS staff were very helpful in putting together the conference proceedings. The lead organizer would also like to thank Kyle Fezi and Alexander Plotkowski from Purdue University for help with paper reviews and copy editing.

The conference organizers are:

Matthew J. M. Krane	Purdue University, USA
Alain Jardy	Institut Jean Lamour, CNRS/Universite de Lorraine, France
Rodney L. Williamson	University of Texas at Austin, USA
Joseph J. Beaman	University of Texas at Austin, USA

ABOUT THE EDITORS

Matthew J. M. Krane

Dr. Matthew Krane is an Associate Professor of Materials Engineering at Purdue University and a member of the Purdue Center for Metal Casting Research. His research is on design, development, and modeling of materials processes, with a focus on transport phenomena and microstructural development in the solidification of metal alloys. He holds a Ph.D. (1996) from Purdue University in Mechanical Engineering, with a concentration on heat transfer and fluid flow in materials processing. His M.S. degree (1989) is from the University of Pennsylvania and his B.S. (1986) from Cornell University, both in mechanical engineering. In addition to consulting with the metals processing industry since the mid-1990s, he also worked on thermal packaging and manufacturing issues for Digital Equipment Corporation in Andover, Massachusetts. In 2006, he was a Visiting Research Fellow working in the Interdisciplinary Research Centre at the University of Birmingham (UK) and has a courtesy appointment in Purdue's School of Mechanical Engineering. He has served as Chair of the TMS Process Modeling and Control and Solidification Committees and is on the international technical committees for the Modeling of Casting, Welding and Advanced Solidification Processes (MCWASP) and the Liquid Metal Processing and Casting (LMPC) conference series. Professor Krane's teaching experience includes heat transfer, fluid mechanics, engineering design, materials processing, numerical modeling, and ethics in engineering practice.

Alain Jardy

After graduation at Nancy School of Mines and a Ph.D. (1984) from Institut National Polytechnique de Lorraine in Materials Science and Engineering, Dr. Alain Jardy joined the French CNRS (Centre National de la Recherche Scientifique). He is now a Senior Research Scientist at Institut Jean Lamour, a Laboratory of CNRS and the University of Lorraine, where he is the Head of the Research Group "Materials Processing." He is involved in studies related to liquid metal processing, vacuum metallurgy and melting/remelting processes, which usually require the combination of several scientific skills, including thermodynamics, fluid flow, electromagnetics, and heat and mass transfer. Most of his work is performed in close collaboration with the metallurgical industry (e.g., steel, aluminum, titanium, zirconium, Ni-based superalloys) and associates mathematical modelling and laboratory-scale, pilot-scale or full-scale experiments. Dr. Jardy is the Head of the East Division of SF2M (French Society of Materials and Metallurgy), and is a Key Reader for *Metallurgical and Materials Transactions B*; he has authored or co-authored 70 publications in refereed journals and 80 conference papers, and supervised 18 Ph.D. theses. He has been a member of the Technical Committee of the Liquid Metal Processing and Casting (LMPC) conference for more than 12 years, and co-organized the 2003, 2007, and 2011 conferences in Nancy.

Rodney L. Williamson

Dr. Rodney Williamson currently holds a part-time research faculty position with the Advanced Manufacturing Center at the University of Texas, Austin, Texas, and is sole proprietor of Remelting Technologies Consulting. He is retired from Sandia National Laboratories where he worked for 23 years. While at Sandia he was instrumental in helping to form the Specialty Metals Processing Consortium (SMPC), a consortium of U.S. specialty alloy producers dedicated to pre-competitive research in the field of process metallurgy relating to nickel-base superalloys and titanium alloys. He holds a B.S. in chemistry from Brigham Young University (1976) and a Ph.D. in chemistry from the University of Washington (1983). He is currently working with colleagues at the University of Texas and the SMPC to apply advanced modeling and simulation techniques to control and monitor the vacuum arc remelting and electroslag remelting processes. He is a co-inventor on several patents in this field and is internationally recognized for his work in process metallurgy controls and diagnostic techniques. He has also conducted and published scientific and engineering research in the areas of optical and magnetic resonance spectroscopy, chemical laser diagnostics, metal vapor arc physics, thermal spray diagnostics and controls, and chemical sensor technology based on field-structured composite materials.

Joseph J. Beaman

Dr. Joseph Beaman is the Earnest F. Gloyna Regents Chair in Engineering at the University of Texas. His career work has been in both manufacturing and control. His specific manufacturing research interest is in Solid Freeform Fabrication, a manufacturing technology that produces freeform solid objects directly from a computer model of the object without part-specific tooling or knowledge. Dr. Beaman coined this term in 1987. Professor Beaman initiated research in the area in 1985 and was the first academic researcher in the field. One of the most successful Solid Freeform Fabrication approaches, Selective Laser Sintering, was a process that was developed in his laboratory. Besides his work in Solid Freeform Fabrication, Professor Beaman has worked extensively with the special metals processing industry to develop next generation process control for remelting processes that are used to produce special metals alloys such as superalloys and titanium alloys. Dr. Beaman is a Fellow of the American Society of Mechanical Engineers and a recently elected member of the National Academy of Engineering.

Electroslag Remelting

Liquid Metal Processing & Casting 2013

Proceedings of the 2013 International Symposium on Liquid Metal Processing & Casting
Edited by: Matthew J. M. Krane, Alain Jardy, Rodney L. Williamson, and Joseph J. Beaman
TMS (The Minerals, Metals & Materials Society), 2013

COMPUTATIONAL MODELING OF ELECTROSLAG REMELTING (ESR) PROCESS USED FOR THE PRODUCTION OF HIGH-PERFORMANCE ALLOYS

Kanchan M. Kelkar[1], Suhas V. Patankar[1]
Shesh K. Srivatsa[2], Ramesh S. Minisandram[3], David G.Evans[4], John J. deBarbadillo[5], Richard H. Smith[6]
Randolph C. Helmink[7], Alec Mitchell[8], Howard A. Sizek[9]
[1]Innovative Research Inc., Plymouth, MN 55447, USA
[2]Srivatsa Consulting LLC, Cincinnati, OH 45249, USA
[3]ATI Allvac, Monroe, NC 28110, USA
[4]Special Metals Corporation, New Hartford, NY 13413, USA, [5]Special Metals Corporation, Huntington, WV 25705, USA
[6]Carpenter Technology Corporation, Reading, PA 19612, USA
[7]Rolls-Royce Corporation, Indianapolis, IN 46206, USA
[8]Department of Materials Engineering, University of British Columbia, Vancouver BC, Canada V6T1Z4
[9]Materials and Manufacturing Directorate, Air Force Research Laboratory, Wright-Patterson AFB, OH 45433

Electroslag Remelting, Computational Analysis, Unsteady Behavior, Macrosegregation

Abstract

This paper presents a comprehensive computational model for the prediction of the transient Electroslag Remelting (ESR) process for cylindrical ingots based on axisymmetric two-dimensional analysis. The model analyzes the behavior of the slag and growing ingot during the entire ESR process involving a hot-slag start with an initial transient, near-steady melting, hot-topping and subsequent solidification of the slag and ingot after melting ends. The results of model application for an illustrative ESR process for Alloy 718 and its validation using results from an industrial trial are presented. They demonstrate the comprehensive capabilities of the model in predicting the behavior of the ingot and slag during the entire process and properties of the final ingot produced. Such analysis can benefit the optimization of existing process schedules and design of new processes for different alloys and different ingot sizes.

Introduction

The Electroslag Remelting process (ESR) is a consumable electrode remelting process for producing high quality ingots used in aerospace and energy applications. Figure 1 shows a schematic diagram of the ESR process that uses a stationary water-cooled crucible containing the slag and ingot. Passage of an Alternating Current (AC) from the electrode to the crucible creates resistive heating in the slag that melts the electrode to build up the ingot in the crucible. A slag skin is formed at the slag-crucible interface that envelops the growing ingot. Further, as the ingot solidifies, it shrinks away from the crucible causing a loss of contact with the crucible. Thus, the ESR process involves complex interactions among the electromagnetic, flow and thermal processes which control the transfer of heat to the water-cooled crucible and, in turn, the ingot solidification process. The segregation of alloying elements into the molten pool during solidification also creates compositional nonuniformity in the ingot. Because of this complexity of the interacting physical phenomena, a trial and error approach can be inefficient for assessing the relationships between operational parameters and process performance. Mathematical modeling offers a scientific and cost-effective approach for developing a fundamental understanding of the governing physical processes and predicting the effect of various process variables on the quality of the ingot produced.

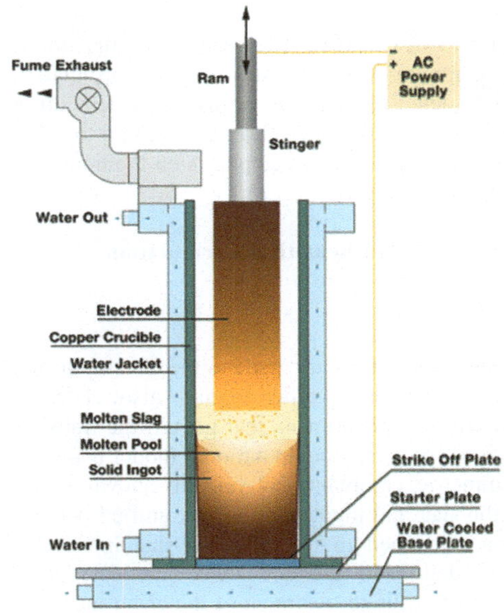

Figure 1. Schematic diagram of the Electroslag Remelting (ESR) process (Courtesy of ATI Allvac).

Computational analysis of ESR processes continues to be an area of active research and only a brief summary of the relevant studies is presented here. These studies involve computational analysis at various levels such as analysis of only the electromagnetic phenomena or a coupled analysis of all phenomena, analysis of a single region (slag or ingot) or both regions (slag and ingot together) and use of simple (Eddy Viscosity) or advanced (turbulent kinetic energy (k) and dissipation (ε)) models for turbulent mixing. An analytical model for the prediction of electromagnetic phenomena in ESR and Vacuum Arc Remelting (VAR) processes is presented by Patel [1]. Examples of segregated models for flow and heat transfer in the slag are presented by Dilawari and Szekely [2] and Jardy et al. [3]. A coupled analysis of the slag and ingot regions for steady conditions has been presented by Choudhury and Szekely [4]. More recently, a model for transient analysis of the ESR process that addresses ingot growth is presented by Weber et al. [5] while the study by Kharicha et al. [6] presents a three-dimensional analysis of the formation of metal droplets at the electrode-slag interface. Heat transfer from the slag and ingot to the water-cooled crucible is a critical factor that governs the solidification process. A detailed analysis of mechanisms of heat transfer to the crucible for various remelting processes is presented by Yu [7]. A complete model of the VAR process that addresses the effect of the progressive loss of contact due to the shrinkage of the ingot on the heat transfer to the crucible is presented by Bertram et al. [8].

The objective of this research effort is to develop an efficient and comprehensive computational model for the analysis of transient ESR processes that can be used routinely by process engineers. Since a majority of ESR ingots produced for high-performance applications are cylindrical, the effort in this study has focused on the development of a model for 2-D axisymmetric conditions and it constitutes an extension of the steady-state model that was developed by Kelkar et al. [9]. The sections below describe the mathematical formulations of the physical phenomena underlying the ESR process, their computational solution and the application of the transient model for the analysis of actual casting of an ingot of Alloy 718 performed by the Special Metals Processing Consortium (SMPC, [10]). The features of the transient model that are common with its steady-state formulation are not described in detail (e.g., the governing equations are not listed). This allows a concise and clear description of the complete computational model with an emphasis on the transient aspects of the ESR process analysis.

Mathematical Formulation

Assumptions

Appropriate assumptions are made in the development of a model that is computationally efficient and also addresses all the important physical phenomena that govern the transient behavior in the ESR process. The computational model is developed using an axisymmetric computational domain shown in Figure 2. It includes the slag and ingot regions in a unified manner. A frame of reference attached to the electrode-slag interface is utilized because the slag-metal interface is also stationary in this domain. This allows an accurate treatment of the electromagnetic, thermal and fluid dynamic interactions between the slag and metal phases for nominally planar slag-electrode and slag-metal interfaces. The growth of the ingot is addressed with a computational domain that expands at the instantaneous casting rate corresponding to the downward movement of the ingot-crucible interface in this frame of reference. The crucible and electrode regions are not included in the computational domain. Analysis is performed for a melt schedule that prescribes the transient variations of the total Root Mean Square (RMS) current and its frequency, power or slag resistivity, and melt rate to calculate the behavior of the slag and growing ingot during melting. A hot-slag start is assumed with the volume of the molten slag remaining constant during the melting process. Calculations are continued after the end of melting to determine the cooling of the slag and metal pools once the electrode has been removed. The model does not consider solidification shrinkage in the bulk of the ingot or slag. It assumes that the nominal densities of the molten and solidified states of the metal and slag are identical. The temperature-dependence of all other properties of the metal and slag is rigorously addressed.

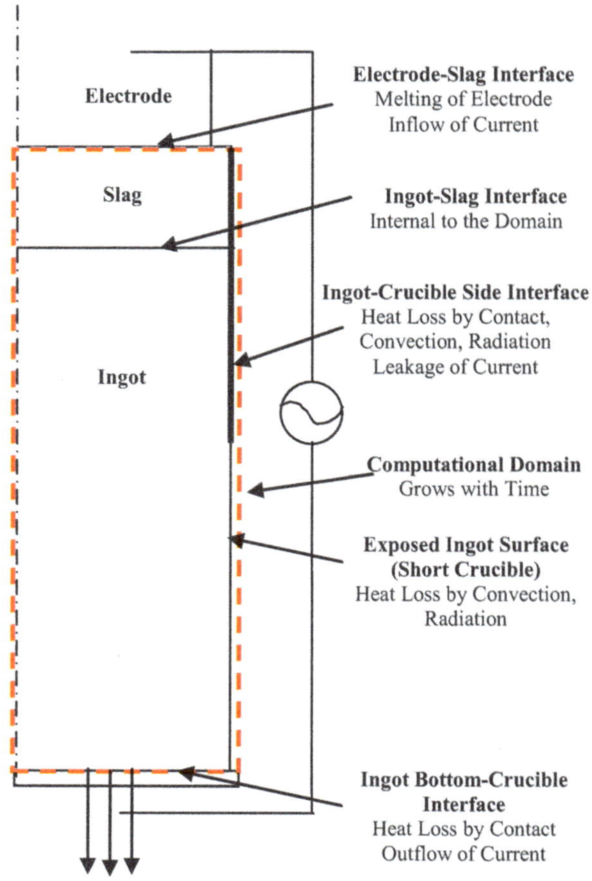

Figure 2. Computational Domain for the Analysis of the Unsteady ESR Process.

Electromagnetics

Practical ESR systems involve flow of Alternating Current in electrically conducting media so that magnetic field diffusion is the primary electromagnetic phenomenon. Further, the frequency of the AC power is sufficiently high so that the time scale for electromagnetic phenomena is orders of magnitude smaller than those for the flow and thermal phenomena. Hence, during transient analysis, electromagnetic fields are determined for the periodic steady state corresponding to the instantaneous process conditions and ingot length. This involves solution of the magnetoquasistatic form of Maxwell's equations to determine the variation of the induced magnetic field in the slag and partially-grown ingot for the prevailing sinusoidal current of a given RMS amplitude and frequency. The exposed portion of the top surface of the slag and also the circumferential surface of the ingot are electrically insulated. The formulation, however, allows analysis in presence of a leakage of current from the slag into the crucible. The large electrical conductivity of the electrode relative to the slag and the crucible relative to the ingot result in the condition of zero radial current at these interfaces.

The calculated distribution of the induced magnetic field is used to infer the current density field and, in turn, the distributions of Lorentz force and Joule heating that are needed in the analysis of the flow and temperature fields in the slag and ingot.

Fluid Motion

The unsteady macro-level flow in the molten slag and metal pools is driven by the buoyancy and Lorentz forces. This unsteadiness arises from the finite length of the ingot and time-dependent melt schedule. The unsteady Navier-Stokes equations govern the turbulent flow in the molten slag and metal pools. The two-equation k-ε turbulence model is used for the determination of the spatially nonuniform mixing in the molten pools. As a result, the computational model is very general and applicable for a wide range of ingot sizes and processing conditions. It should be noted that the pool sizes in the ingot during the entire process and, after the end of melting, in the solidifying slag are not known a priori. They are determined by the corresponding temperature fields which are strongly influenced by the instantaneous ingot length and melt rate. Thus, the flow and temperature fields are coupled not only through the buoyancy force but also through phase change in the ingot and slag during and after melting.

The interaction of the flow fields across the slag-metal interface is automatically addressed through the use of the unified computational domain. The velocity in the mushy regions of the ingot (during and after melting) and slag (after melting) below an assumed Immobilization Liquid Fraction is set to the instantaneous velocity of the corresponding solid phase (Bertram et al., [8]). Thus, the solidified portion of the ingot moves at the instantaneous casting velocity while melting is ongoing. After the end of melting, solidified portions of the slag and ingot are stationary. The flow in the molten pools is subject to no-slip boundary conditions on all solid boundaries.

Heat Transfer and Phase Change

The temperature field within the slag and metal pools is determined by fluid motion and turbulent diffusion. It is important to note that, unlike the optically-opaque molten alloy, the slag is semi-transparent. Therefore, radiative heat exchange is the dominant mechanism in the molten slag. This is handled by using the Rosseland approximation (Siegel and Howell, [11]). It involves a diffusion treatment of radiation using a temperature-dependent radiative diffusivity that augments the molecular and turbulent diffusivity in the molten slag. In the solid ingot region, the temperature field is determined by the local solid velocity and conduction. In order to achieve efficiency of calculation, the solidification of the slag during and after melting is handled through separate physically motivated treatments as follows:

- During melting, the bulk slag is assumed to be molten and enclosed in a thin layer of the slag skin slag formed on the crucible surface.
- After melting ends, the bulk slag solidifies. The computational technique transitions smoothly from "thin slag-skin" to the "bulk-slag solidification".

The enthalpy-porosity technique (Brent et al., [12]) is used for addressing the solidification of the metal (during and after melting) and the slag (after end of melting). It uses the specified variation of the solid fraction with temperature for the slag and alloy to deduce the solid/liquid fraction at all points in the domain. It is useful to note that the slag and metal pools solidify independently after the end of melting but the exchange of heat across the slag-metal interface influences their solidification progress.

Heat loss at the slag-electrode interface is calculated from a global heat balance on the electrode for the instantaneous melt rate. The molten metal droplets formed at the slag-electrode interface are assumed to fall through the slag under thermal equilibrium conditions and appear in the metal pool below the slag-metal interface. The exposed surfaces of the slag and the ingot (in a short-collared crucible) lose heat by radiation and convection to the environment. The energy conservation equation accounts for material addition during ingot growth. Finally, heat loss to the crucible from the ingot bottom is calculated using an empirically prescribed heat transfer coefficient that is representative of the good metal-to-metal contact created by the weight of the ingot.

Heat Loss to Crucible During Melting

As described earlier, during the melting process, a thin slag skin is formed at the slag-crucible interface due to loss of heat to the cooling water. Further, the ingot grows into the envelope of the slag skin which influences the heat loss from the ingot to the cooling water. Note that, during melting, the thin slag skin located at the interface of the slag/ingot and crucible is assumed to lie on the outer radial boundary of the computational domain. The thermal resistance it creates for heat loss to the crucible is calculated through the boundary treatments discussed below.

Heat Loss from the Slag - At any instant in the process, the thickness of the slag skin varies over the slag-crucible interface depending on the local rate of heat transfer from the bulk slag to the crucible. An uneven contact of the slag skin with the inside surface of the crucible results in transfer of heat to the crucible by conduction across the slag skin and radiation from its exposed surface as described by Yu [7].

Heat Transfer from the Ingot to the Crucible – The ingot grows into the slag skin formed at the slag-skin interface thereby also creating an axial variation of the skin thickness on it. The slag skin has a finite thermal resistance and its presence on the ingot surface influences the rate of heat transfer from the ingot to the crucible. Further, this heat loss is even more strongly affected by the shrinkage of the ingot as it solidifies. The extent of this loss of contact and its effect on heat transfer between the ingot and the crucible is determined using the method of Bertram et al. [8]. This heat loss occurs by conduction through the skin thickness where the ingot has not separated from the crucible and radiation from surface of the slag skin where ingot has shrunk away to cause a loss of contact with the crucible.

Heat Loss to Crucible After End of Melting

The bulk slag solidifies after the end of melting. Therefore, heat loss from the circumferential surface of the slag to the water flowing through the crucible is governed by exactly the same mechanisms as those for heat loss from the solidifying ingot to the cooling water. The only difference is that the temperature at which the solidified slag shrinks away from the ingot is expected to be much lower than that for the ingot.

Behavior of Inclusions

In ESR practice, inclusions may appear in the molten metal pool if they get embedded in the metal droplets that originate from the electrode. The slag and molten pools have very dissimilar transport properties (density and viscosity). Therefore, when an inclusion crosses the slag-metal interface, it experiences very different buoyancy (in magnitude) and drag (in magnitude and direction as exerted by the relative velocity of the surrounding fluid) forces. The relaxation time of an inclusion (time required to attain its terminal velocity), in turn, depends on this net external force and its inertia. The formulation of the equation governing the motion of an inclusion addresses these unique aspects of the ESR process. In addition, it also considers a temperature-dependent dissolution rate of an inclusion. The motion of an inclusion is tracked until it reaches the mushy zone or fully dissolves in the molten pool. It is assumed that the time scale for the change in the pool size is large relative to the residence time of the inclusions. Hence, the instantaneous flow and temperature fields in the ESR system at the time an inclusion enters the metal pool are used in calculating its behavior.

Thermal History and Defect Formation Probability

The structure of the cast ingot is determined by its thermal history during the solidification process. In an unsteady ESR process, the thermal history at any point in the solidified ingot depends on both its axial and radial locations. The evolving temperature field in the growing ingot is determined by the energy conservation equation. This calculation is coupled with a Lagrangian framework for tracking the temperature of each elemental metal volume to determine the thermal history of the growing ingot. The thermal history calculation is very comprehensive and includes determination not only of the local solidification time but also the cooling rate, temperature gradient, isotherm angle and speed of solidification at various solid fractions.

The goal of the remelting processes is to produce ingots that are free of solidification defects (freckles for superalloys and steels). Prediction of microsegregation by solving the species transport equations in the interdendritic region from first principles is computationally infeasible. Therefore, the method proposed by Auburtin et al. [13] is utilized in this study. It captures the underlying physics in an efficient manner by quantifying the probability of freckle formation through the Rayleigh number for thermosolutal convection in the interdendritic region. Thus, at each point in the solidifying ingot, alloy-specific correlations are used to relate the thermal history parameters to the corresponding metallurgical structure parameters that are needed in the calculation of the interdendritic Rayleigh number.

Macrosegregation of Alloying Elements

The properties of the cast ingot also depend on its macro-level composition. The preferential rejection or absorption of the alloying elements by the solid in the outer mushy region and subsequent imperfect mixing in the pool redistributes these elements in the final solidified ingot. Solution of the unsteady convection-diffusion equations for the concentrations of individual alloying elements over the ingot during and after melting is performed to determine their redistribution in the solidifying ingot. In this calculation, it is assumed that individual alloying elements are selectively rejected or absorbed by the solid in the outer mushy region up to a limiting Segregation Solid Fraction where they mix within the liquid metal pool due to convection and turbulent diffusion. This segregation flux for each alloying element at the boundary of the molten pool appears as a source term in the corresponding convection-diffusion equation.

Computational Solution

The equations governing the electromagnetic, flow, thermal and macrosegregation phenomena are solved using the framework of the control volume method (Patankar, [14]). It involves discretization of the axisymmetric domain into control volumes. The discretization equations are constructed by integrating the respective governing equations so that they represent exact conservation of the underlying physical quantities. An implicit time-stepping technique is used for discretizing the unsteady term. Due to the nonlinear and coupled nature of the equations, multiple iterations are needed to achieve convergence of the solution within each time step. Several enhancements of the basic method have been incorporated to address the special features of the transient ESR process as discussed below.

Ingot Growth - In the frame of reference attached to the slag-electrode interface, the growth of the ingot causes the interface between the bottom surface of the ingot and crucible to gradually uncover the computational grid. Special considerations are required for constructing the discretization equations for the "base-plate" control volume as it expands in size within each time step due to the growing ingot. Further, computational efficiency is achieved by limiting the computational solution to control volumes that contain the slag and only the already formed ingot at that time instant.

AC Electromagnetics - The solution of the magnetic diffusion equation for AC power is achieved through the use of complex variables. A staggered grid is utilized for calculating the current density and Lorentz force fields.

Inclusion Motion - The equation of motion of an inclusion is solved using a time-stepping technique that advances the inclusion along its trajectory. It involves an adaptive determination of the size of the time step to account for large differences in buoyancy and drag forces within the slag and metal pools experienced by the inclusion. This enables tracking the behavior of inclusions of a wide range of densities and sizes in an efficient manner.

Thermal History - A Lagrangian technique is developed for an accurate determination of the complete thermal history of each elemental metal volume in the solidifying ingot. It involves precise tracking of the location of each elemental metal volume in the growing ingot from the instant it starts solidifying. This enables determination of the time instants at which each elemental volume within the ingot reaches various specific temperatures and the corresponding cooling rates, isotherm angles and spatial thermal gradients. These thermal history parameters are then utilized for determining the probability of defect formation in each elemental metal volume that has fully solidified.

Macrosegregation – The solution of the equation for concentration of each alloying element accounts for convection and diffusion in the pool, absence of diffusion in the solidified ingot and presence of segregation flux in the control volumes over the pool boundary.

Illustrative Application of the Computational Model

The computational method described above has been incorporated in the commercial software program MeltFlow-ESR [15]. This program has been applied for the analysis of an ESR process of Alloy 718 for which detailed pool markings were made by

SMPC [10]. Details of the process conditions, and the model predictions and their comparison with marked pool shapes are now discussed.

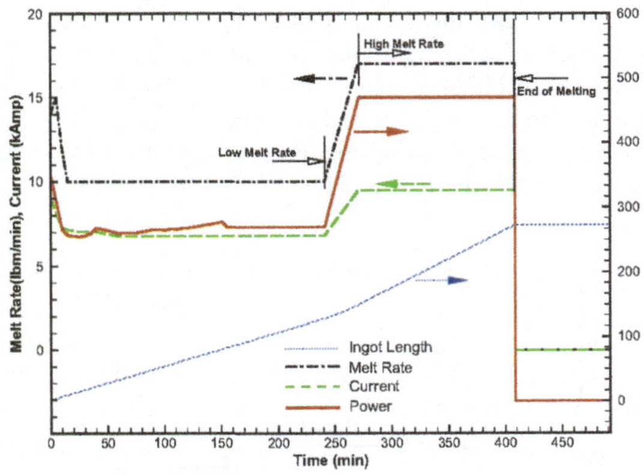

Figure 3. Melt Schedule of the ESR Process for Alloy 718. (Courtesy of Special Metals Processing Consortium (SMPC)).

Physical Situation

The ESR process studied by SMPC involved casting of a long ingot of Alloy 718. The melt schedule involved two melting stages – first stage with casting at a lower melt rate and second at a higher melt rate, with a short transition between the two stages. A sufficiently long ingot was cast in each stage by maintaining the corresponding melt rate so as to reach steady conditions. The pool profiles were marked throughout the process and are very useful for validating the model. Table 1 lists the geometric details of the ESR system while Figure 3 shows the corresponding melt schedule. The temperature-dependence of material properties, including the variation of fraction solid with temperature, of the slag (70%CaF_2-15%CaO-15%Al_2O_3) and metal are incorporated in the analysis. Due to space constraints, only their representative values are listed in the Table 1. Properties of the slag correspond to the reported values from prior studies (e.g., Mills and Keene, [16]). Properties of the alloy are taken from the studies by Mills [17], Auburtin et al. [18] and Knorovsky et al. [19]. Also, the trajectories of inclusions of a wide range of diameters (0.1 mm to 30 mm) and relative densities (ratio of the density of inclusion to that of molten metal of 0.5 to 2.0) have been calculated.

A computational grid of 253(axial) x 35(radial) control volumes is used for the analysis. Calculations are made for the entire period which includes the durations for melting of the electrode and subsequent cooling of the slag and ingot after the end of melting when the electrode has been removed. This analysis required a total of fifteen hours on a PC with a 3 GHz Intel Processor.

Results and Discussion

The computational model provides detailed information about the distributions of the electromagnetic, flow, temperature, liquid fraction and alloy concentration fields in the slag and ingot during and after the end of melting, and also predicts the behavior of inclusions in the molten metal and slag pools during melting. In the sections below, results of analysis are discussed in terms of the overall process behavior and detailed field variations.

Table 1. Physical properties and operating conditions used in the analysis of an ESR process for Alloy 718.

Metal (Alloy 718)	
Density, liquid	7500 kg/m^3
Viscosity, liquid	6x10^{-3}kg/m-s
Liquidus temperature	1623 K
Solidus temperature	1473 K
Latent heat of fusion	2.1x10^5 J/kg
Specific heat, liquid	720 J/kg-K
Specific heat, solid (773 K)	512 J/kg-K
Vol. thermal expansion coeff., liquid	1.5x10^{-4} K^{-1}
Thermal conductivity, liquid	30.52 W/m-K
Thermal conductivity, solid (773 K)	16.72 W/m-K
Electric conductivity	7.6x10^5 (Ω-m)$^{-1}$
% Concentrations (Segregation Coefficients) of Alloying Elements Al, Ti, C, Co, Cr, Fe, Nb, Mo	0.56(1.20), 0.95(0.55), 0.025(0.28), 0.25(1.05), 18.18(1.12), 18.10(1.16), 5.25(0.35), 3.12(0.90)
Slag	
Density, liquid	2800 kg/m^3
Viscosity, liquid	3.5x10^{-2} kg/m-s
Specific heat, liquid	1163 J/kg-K
Specific heat, solid (773)	1061 J/kg-K
Vol. thermal expansion coeff., liquid	9.0x10^{-5} K^{-1}
Thermal conductivity, liquid	0.5 W/m-K
Thermal conductivity, solid (773 K)	7.8 W/m-K
Liquidus temperature	1723 K
Solidus temperature	1618 K
Latent Heat	4.3x10^5 J/kg
Geometry	
Slag height	147 mm
Electrode diameter	330 mm
Ingot Diameter	384 mm
Final Ingot Length	2732 mm

Overall Process Behavior - The behavior of the overall process can be quantified in terms of the evolution of the volumes of the molten slag and metal, slag skin thickness and heat losses at the various interfaces in the ESR system as shown in Figures 4 and 5 respectively. In the first part of the melting process, the molten metal pool increases with time as the ingot grows and then reaches a steady state corresponding to the lower melt rate. The pool size then rapidly adjusts as the melt rate transitions to its higher value in the second stage of melting. The thickness of the slag skin formed at the slag-metal interface also responds to the change in the ingot length and melting power. In the early stages, the slag is colder due to lower power and higher heat loss through the ingot bottom resulting in a thick slag skin. After the ingot is sufficiently long, this thickness depends only on the melting conditions and reaches a constant value corresponding to the lower melt rate. As the melt rate transitions to the higher value, the skin thickness readjusts to a lower value as a result of a hotter slag. After the end of melting, the metal volume solidifies rapidly. The molten slag, however, takes longer to solidify due to its higher heat capacity and lower thermal conductivity. This behavior is consistent with observations of practical ESR processes.

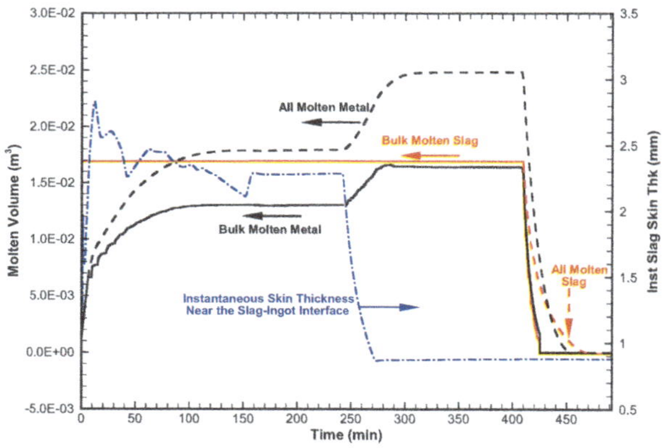

Figure 4. Evolution of the molten slag and metal volumes, and the instantaneous slag-skin thickness during the entire process.

A significant portion of the power used in the ESR process is lost to the cooling water through the slag-crucible interface. The heat loss flux to the crucible through the bottom face of the ingot is high in the initial stages and becomes insignificant once the ingot is sufficiently long. Finally, the radiative heat loss from the exposed surface of the slag shows a sudden increase after the end of melting because the electrode has been removed and the heat loss now occurs over the entire top surface of the slag.

Electromagnetic, Flow and Temperature Fields – Figures 6 and 7 show the electromagnetic, flow and temperature fields in the ingot. The electromagnetic fields show that the Joule heating is concentrated in the slag and the current is nearly uniformly distributed over its cross-section due to its low electrical conductivity. On the other hand, due to the high electrical conductivity of the ingot, the current distribution in the ingot shows the skin effect in which the current flow is concentrated in the region near the outer radius. The Lorentz force is in the direction perpendicular to the current density and is therefore mainly directed radially inward.

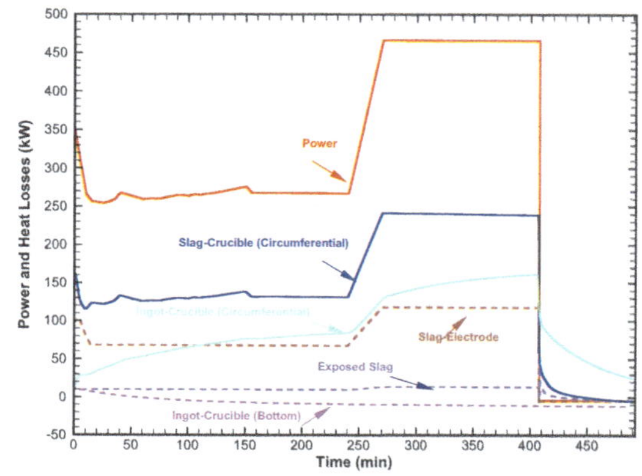

Figure 5. Evolution of heat losses at various interfaces in the ESR system during the entire process.

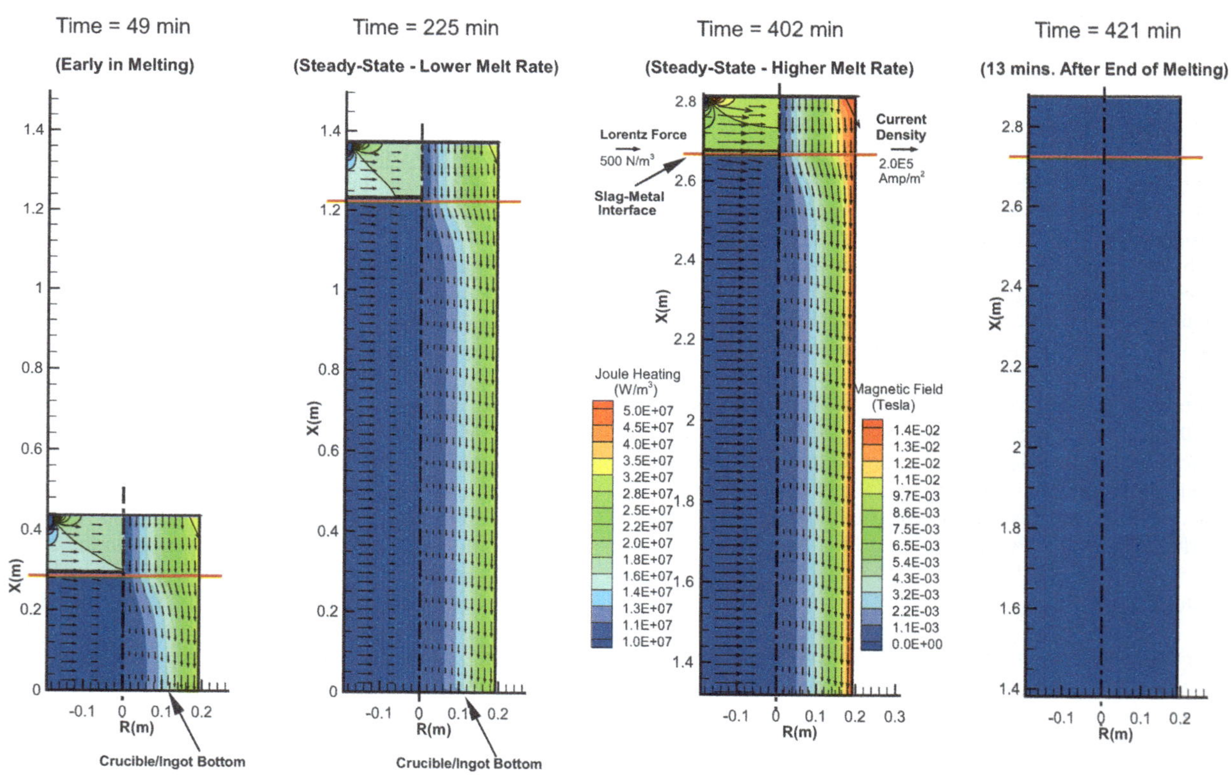

Figure 6. Electromagnetic fields and Joule heating in the slag and ingot at selected instants during and after melting.

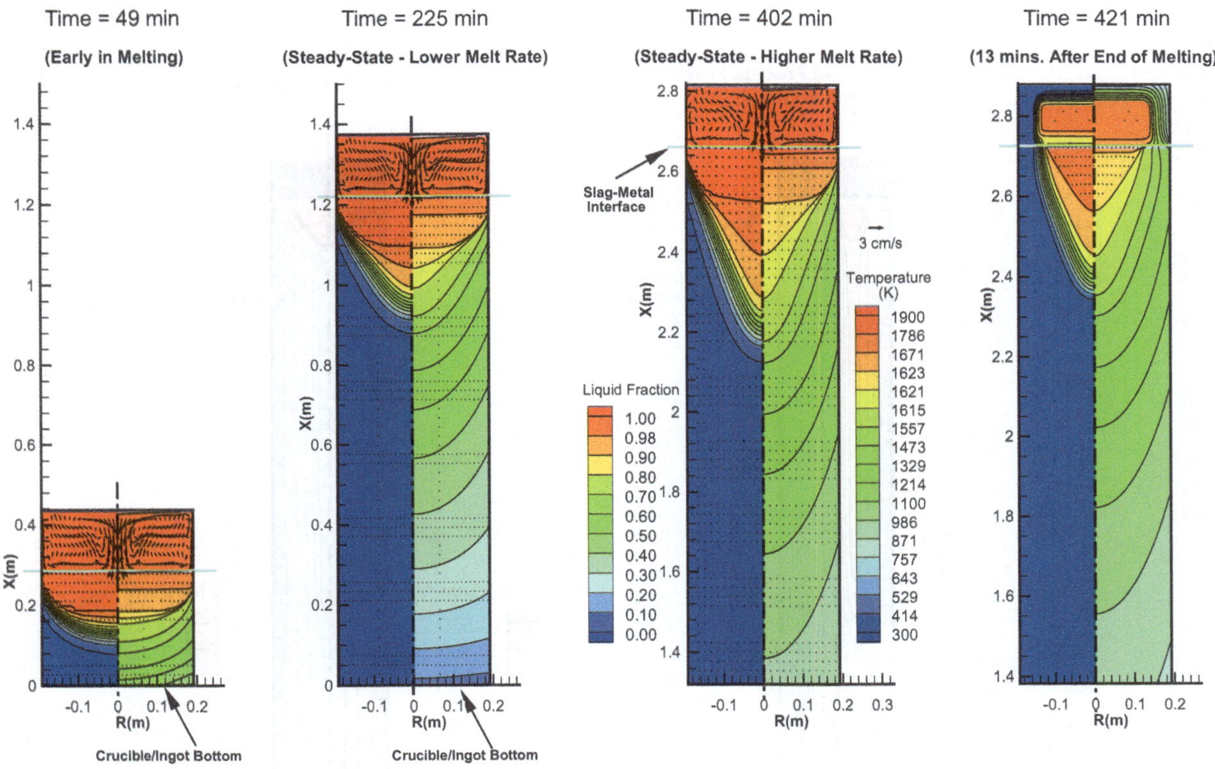

Figure 7. Velocity, temperature and liquid fraction fields in the slag and ingot at selected instants during and after melting.

The volume of the molten metal is small and the pool profile is flat in the initial stages when the ingot length is short and the heat loss flux through the bottom of the ingot is significant. After process conditions have been steady for a sufficiently long time, the metal pool reaches a steady volume corresponding to the low-melt rate condition. The metal pool is thermally stably stratified except over the pool boundary. At the higher melting rate in the second part of melting, the pool is seen to be significantly deeper and the mushy zone longer. Unlike the metal pool, the slag is thermally unstably stratified and shows strong mixing with velocities at least an order of magnitude larger than those in the molten pool. (Note that the slag-metal interface is assumed to be flat and the velocity vector at each point in the domain is drawn to reflect the magnitude and direction of the corresponding local velocity.) After the end of melting, the metal volume shrinks rapidly. The slag temperature also falls below its liquidus temperature relatively quickly but the core slag volume remains in the semi-solid state even after the metal pool has solidified. Thus, the hot slag creates a natural hot topping and helps to reduce the size of the solidification cavity in the ingot.

<u>Inclusion Motion</u> - Figure 8 shows the flow and liquid fraction fields in the ESR system and, for clarity, trajectories of only selected inclusions of small and large diameters (0.1 and 30 mm) and densities slightly higher and significantly lower than that of the metal (relative densities of 1.15 and 0.5), at an instant in the process when conditions have reached a steady-state at the higher melt rate. It should be noted that the inclusions of relative density of 0.5 are lighter than the molten metal but still heavier than the slag. Inclusions of various sizes and densities follow very different paths through the molten metal and slag pools. Smaller inclusions tend to move with the fluid while large inclusions are

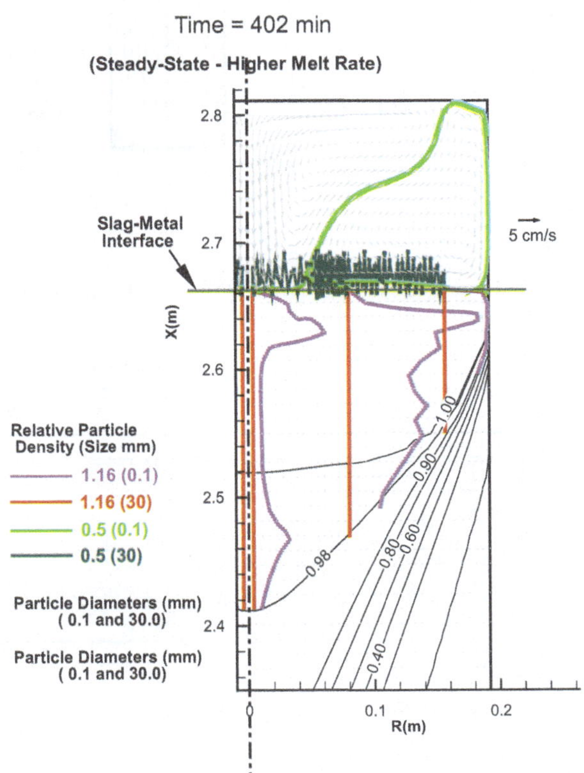

Figure 8. Flow and liquid fraction fields, and motion of inclusions of various sizes and densities in the slag and metal pools near the end of melting.

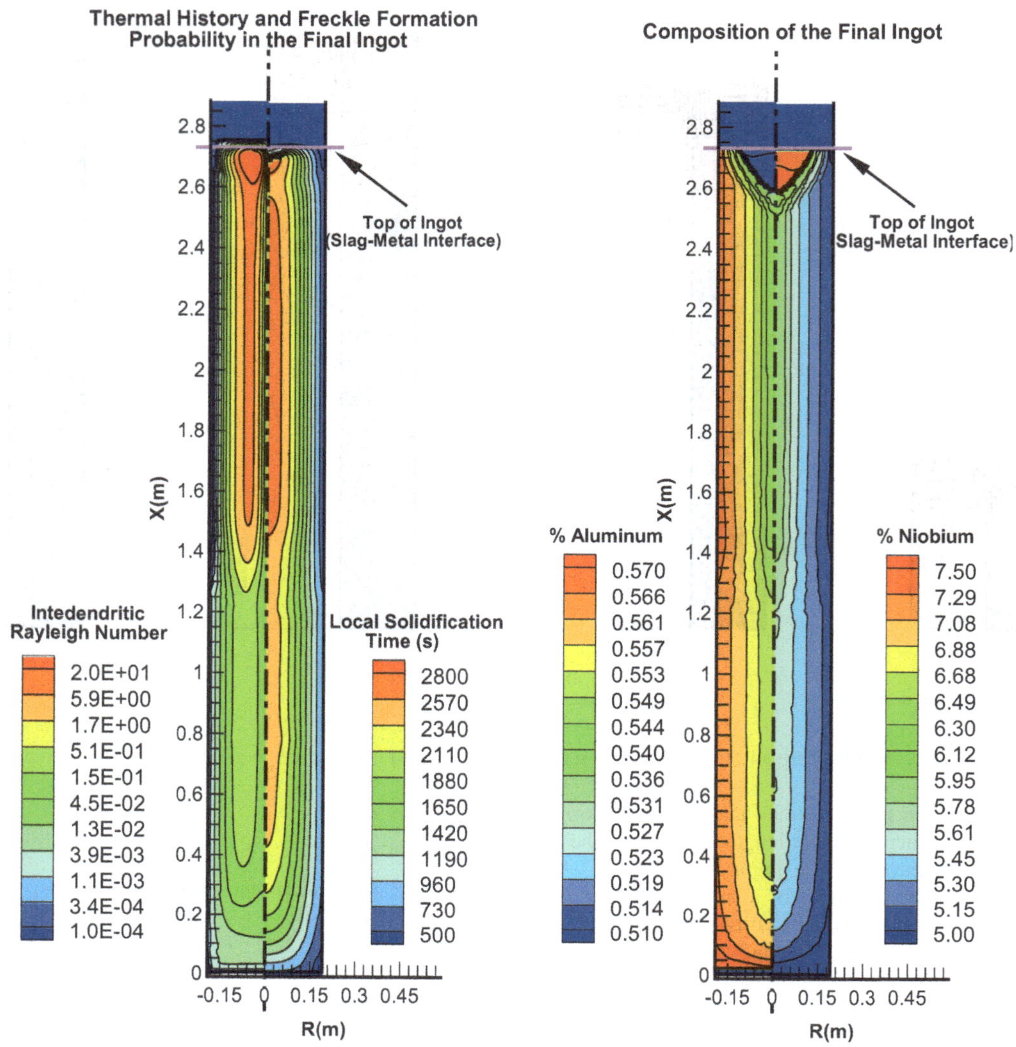

Figure 9. Thermal history and probability of freckle formation (left), and compositions of Niobium and Aluminum (right) in the final ingot.

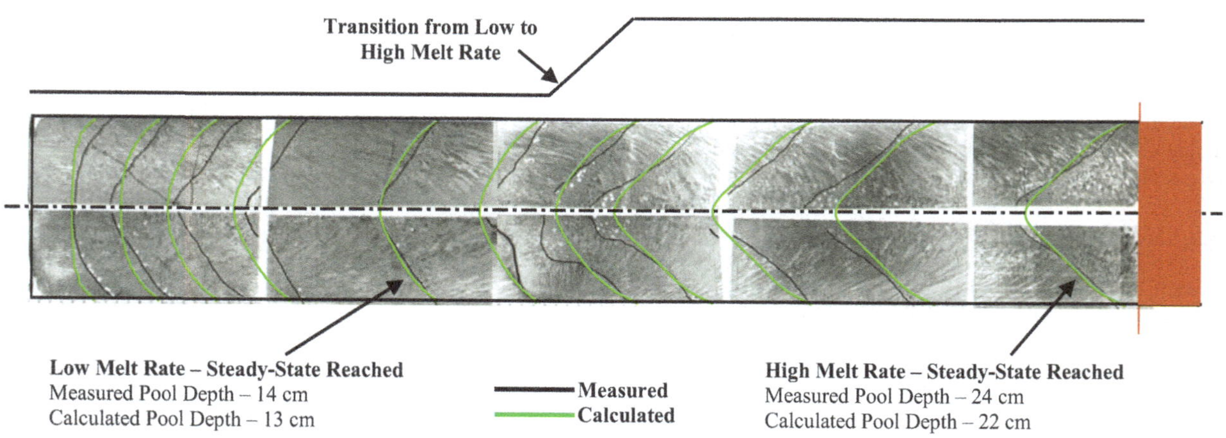

Figure 10. Comparison of the predicted and measured pool profiles for the SMPC ESR trial (Experimental Results Courtesy of Special Metals Processing Consortium (SMPC)).

strongly influenced by buoyancy. Thus, large heavy inclusions sink quickly through the metal pool while large light inclusions oscillate between the slag and the metal pools across the slag-metal interface and drift towards the center or the outer radius of the ingot depending upon their density. Very light and small inclusions enter the slag and continue to swirl around in it.

Thermal History and Probability of Defect Formation - Figure 9 shows the variations of the local solidification time and Rayleigh number for quantifying the probability of freckle formation in the solidified ingot produced in the ESR process. The cooling rate is high at the ingot bottom due to the high rate of heat transfer to the crucible bottom face in the very initial stages of the process when the ingot length is short. Therefore, the local solidification time is very low in the bottom portion of the ingot. As the ingot length increases and melting conditions reach a steady state at the lower melt rate, transfer of heat occurs primarily in the radial direction. As a result, local solidification time is high in the center and decreases radially outward. As the process conditions transition to the higher melt rate, the cooling rate decreases somewhat resulting in a higher local solidification time at the ingot center. After the end of melting, pool shrinks rapidly from the side but not as fast at the top due to the insulating effect of the hot slag. As a result, cooling rate is lower at the ingot top.

The probability of freckle formation scales in proportion to the magnitude of interdendritic Rayleigh number and its variation in the ingot is noteworthy. In the regions of ingot where the cooling rate is high (bottom and outer radius), dendrites are closely spaced and freckle formation is less likely. In the center region, the spacing between the dendrites is larger due to lower cooling rates, but the gravity-driven flow is parallel to the primary dendrites for which the permeability is lower relative to that for secondary dendrites. Therefore, in the portions of the ingot that correspond to steady melting, the probability of freckle formation is highest in the mid-radius region due to a combination of reasonably large spacing between the dendrites and the direction of the flow (pool angle) that involves a component along the secondary dendrites. When the melt rate is higher, this effect is more pronounced. As a result, both the maximum value of the interdendritic Rayleigh number is higher and nonuniformity in its radial variation is stronger in the ingot formed at the higher melt rate.

Ingot Composition - Variations in the concentrations of niobium and aluminum in the final ingot that result from macrosegregation are also shown in Figure 9. As the liquid metal solidifies, the solid formed preferentially rejects niobium and absorbs aluminum. The mass of the alloying element rejected into or depleted from the molten pool is redistributed into the pool due to the convective motion and diffusion. Since the motion of the molten metal is much weaker in the center of the pool, alloying elements from this region do not mix well with the bulk pool. As a result, the concentration of niobium varies in the radial direction with a higher concentration in the center of the ingot. Also, the pool is being enriched in niobium so that its radially-averaged concentration increases from the bottom of the ingot to the top of the ingot. The behavior for aluminum is exactly the reverse since it is preferentially absorbed by the solidifying alloy. The extent of segregation in the radial direction increases with the melt rate since the volume of the molten pool is higher and the mixing in the pool is weaker. Note that the redistributions of alloying elements during remelting depend on the assumptions made about the values of the limiting Segregation Solid Fraction, and element-specific effective segregation and molecular diffusion coefficients. Therefore, the results presented in Figure 9 are only for illustrating the capability of the model in predicting the compositional variation in the final ingot caused by macrosegregation.

Model Validation - A comparison of the predicted (green color) and the marked pool (dark color) profiles is shown in Figure 10. It is seen that the computational model reproduces the observed process behavior with good accuracy. Thus, in the initial stages of melting, the pool is flat and shallow. It transitions to a conventional pool shape and reaches a steady state after the ingot has grown to a sufficiently long length. It is important to note that this specific ESR melt showed an asymmetric ingot behavior when the melt rate transitioned from the low to high value. These three-dimensional effects cannot be resolved using the axisymmetric model discussed in this paper. But, the model correctly predicts the quick transition of the pool shape from its low-melt-rate profile with a smaller pool depth to its high-melt-rate profile that shows a deeper pool. Also, the predicted pool shapes and depths are consistent with their marked profiles. Measurements of ingot composition and investigations on the presence of freckles were not performed. However, the general aspects of their predicted behavior are deemed to be in agreement with corresponding observations in ESR practice.

The results of analysis illustrate the comprehensive capabilities of the model in predicting the important aspects of practical ESR processes for the casting of cylindrical ingots.

Summary and Conclusions

The present paper describes a comprehensive computational model for the analysis of electromagnetics, flow, heat transfer, macrosegregation and inclusion motion in the transient ESR processes for axisymmetric conditions. The mathematical formulation and computational model incorporate several physically motivated enhancements for accurate and efficient analysis of the unique aspects of the transient ESR process. Analysis is performed for the period during melting when the ingot grows and also after the end of melting when the slag and the ingot pools solidify. Thermal history of the entire ingot is determined in a comprehensive manner. Further, it is utilized for a quantitative prediction of the probability of freckle formation in the ingot based on the Rayleigh number for interdendritic flow. Use of the model is illustrated for the analysis of a transient ESR process for Alloy 718. Field variations of current density, volumetric heating, molten pool flow, liquid fraction, temperature and inclusion motion at various time instants during the process, and the evolution of pool volumes and boundary heat losses illustrate the insights the model can provide into the behavior of the ESR process. The predicted pool profiles compare well with the profiles marked in the final ingot. Thus, the computational model offers a cost-effective method for reducing trial-and-error during process design and for producing ingots of high quality.

Acknowledgments

The authors gratefully acknowledge the support for this work provided by the US Air Force through the Materials Affordability Initiative program under contract numbers FA8650-06-2-5211 and FA8650-11-2-5224.

References

1. Patel A.D., "Analytical Model for Electromagnetic Fields in ESR and VAR Processes," Proc. of Liquid Metal Proccessing and Casting, ed. Lee P. D et al., 205-214, 2003.

2. Dilawari A.H. and Szekely J., "Heat Transfer and Fluid Flow Phenomena in Electroslag Refining", *Metall. Trans. B, 9B,* 77-97, 1978.

3. Jardy J., Ablitzer D. and Wadier J.F., "Magnetohydrodynamic and Thermal Behavior of Electroslag Remelting Slags", *Metall. Trans. B*, 22B, 111-120, 1991.

4. Choudhury M. and Szekely J., "Modeling of Fluid Flow and Heat Transfer in Industrial-Scale ESR System," *Ironmaking and Steelmaking*, 5, 225-232, 1981.

5. Weber V., Jardy A., Dussoubs B., Ablitzer D., Ryberon S., Schmitt V., Hans S. and Poisson H., "A Comprehensive Model of the ElectroSlag Remelting Process: Description and Validation," LMPC 2007: Proc. 2007 Intl. Symposium on Liquid Metals Processing and Casting, Eds. Lee P., pp. 83-88, 2007.

6. Kharicha A., Ludwig A. and Wu M., "Droplet Formation in Small Electrslag Remelting Processes," Proc. of Liquid Metal Proccessing and Casting," Proc. of the 2011 Intl. Symposium on Liquid Metal Proccessing and Casting, ed. M. Krane, J. Bellot and A. Jardy, . R.L. Williamson and S. Ballantyne, pp. 113-119, 2011.

7. Yu Kuang-O, "Comparison of ESR-VAR Processes – Part 1, Heat Transfer Characteristics of Crucible," Proc. Vacuum Metallurgy Conference., ed. G.K Bhat and L.W. Lherbier, pp. 83-92, 1984.

8. Bertram L.A., Adasczik C.B, Evans D.G., Minisandram R.S., Sackinger P.A., Wegman D.D., Williamson R.L., "Quantitative Simulations of Superalloy VAR Ingot at the Macroscale," Liquid Metal Processing and Casting, ed. A. Mitchell and P. Aubertin, AVS, 110-132, 1997.

9. Kelkar K.M., Patankar S.V. and Mitchell A., "Computational Modeling of the Electroslag Remelting (ESR) Process for the Production of Ingots of High-Performance Alloys," Proc. of Liquid Metal Proccessing and Casting, ed. Lee P. D. et al., pp. 137-144, 2005.

10. Proprietary information provided by the Special Metals Processing Consortium, (SMPC), 2012.

11. Siegel R. and Howell J.R., *Thermal Radiation Heat Transfer*, Hemisphere Publishing Corporation, 1981.

12. Brent A.D., Voller V.R., Reid K.J., "Enthalpy-Porosity Technique for Modeling Convection-Diffusion Phase Change: Application to the Melting of a Pure metal," Numerical Heat Transfer, Vol. 13, pp. 297-318, 1988.

13. Auburtin P., Wang T., Cockroft S.L. and Mitchell A., "Freckle Formation and Freckle Criterion in Superalloy Castings," Metallurgical and Materials Transactions B, Vol. 31B, pp. 801-811, 2000.

14. Patankar S.V., *Numerical Heat Transfer and Fluid Flow*, Hemisphere Publishing Corporation, Washington, 1980.

15. MeltFlow-ESR Reference Manual, Innovative Research, Inc., 3025 Harbor Lane N., Suite 300, Plymouth, MN 55447, www.inres.com, 2013.

16. Mills, K.C. and Keene, B.J., "Physicochemical Properties of Molten CaF2-based Slags", *International Metals Reviews*, No. 1, pp. 21-69, , 1981

17. Mills, K.C., *Recommended values of Thermophysical Properties for Selected Commercial Alloys*, p. 190, NPL/ASM/Woodhead Publishing Limited, 2002.

18. Auburtin P., Cockroft S.L., Mitchell A., Schmalz A.J., "Center Segregation, Freckles and Development Directions for Niobium-Containing Superalloys," Proc Conf. Superalloys 718,625,706, ed. Loria E A et al., publ. TMS, pp. 47 – 54, 1997.

19. Knorovsky G.A., Cieslak M.J., Headley T.J., Romig Jr. A.D. and Hammetter W.F., "Inconel 718: Solidification Diagram", *Met. Trans. A.*, Vol. 20A, pp. 2148-2158, 1989.

Proceedings of the 2013 International Symposium on Liquid Metal Processing & Casting
Edited by: Matthew J. M. Krane, Alain Jardy, Rodney L. Williamson, and Joseph J. Beaman
TMS (The Minerals, Metals & Materials Society), 2013

A NUMERICAL STUDY ON THE INFLUENCE OF THE FREQUENCY OF THE APPLIED AC CURRENT ON THE ELECTROSLAG REMELTING PROCESS

E. Karimi Sibaki [1], A. Kharicha [1,2], M. Wu [1,2], A. Ludwig [2], H. Holzgruber [3]
B. Ofner [3], M. Ramprecht [3]
[1] Christian Doppler Laboratory for Advanced Process Simulation of Solidification and Melting, Univ. of Leoben, Austria
[2] Chair of Simulation and Modeling of Metallurgical Processes, Univ. of Leoben, Austria
[3] INTECO Special Melting Technologies GmbH, 8600 Bruck/Mur, Austria
Correspondent author: abdellah.kharicha@unileoben.ac.at

Keywords: Electroslag Remelting (ESR), Magnetohydrodynamics (MHD), Solidification, AC Current.

Abstract

Most conventional electroslag remelting (ESR) processes are operated with AC current, but the inductive loss of the power becomes a major problem when the process is operated at the high frequency for large scale ESR. Nowadays, the demand on large scale ingots has driven the people to think of operating the process with AC current at low frequency. Here the influence of the applied frequency of AC current on the large scale ESR process is numerically investigated. For this purpose, simulations with two operating AC frequencies (0.2 and 50 Hz) are performed. The main goal is to achieve some fundamental understanding of the two-phase flow and the formation of melt pool of the solidifying ingot under the influence of AC frequency. As we also know that the mold current (portion of electric current entering through solid slag skin into the mold) plays an important role in the process, calculations considering different current paths are also analyzed.

Introduction

The aim of the electroslag remelting (ESR) process is purifying and refining steel and other alloys such as Ni-based super alloys. Thermal energy is supplied to the process through the Joule heating that results in remelting the electrode and formation of droplets. The droplets then pass through the slag and reach the liquid pool. The melt pool solidifies directionally and builds the high-grade ingot in a water-cooled mold [1].

In the ESR process the electromagnetic field drives the flow of the molten slag and liquid melt pool, thus the electrical parameters of the process have a significant effect on the quality of the final product. Currently, the ESR process is mostly operating with AC current (50 - 60 Hz) in order to minimize the negative impacts of the electrochemical reactions occurring in the slag.

Comprehensive reviews of electrochemical reactions in the ESR process were presented by Peover and Mitchel [2-3]. In fact, the electric current is conducted by ions in the slag region [4]. Oxidation loss of elements such as Ti, Si, and Al is addressed as a big issue during DC remelting of the electrode in the ESR process [5-7]. The composition fluctuation of elements can significantly reduce the yield strength of the alloys. For instance, a major loss of Ti element (40-80 %) is reported by Etienne in a DC electrode remelting process [5]. Furthermore, the degree of sulphur removal by an ESR process operated with AC current is higher than by an ESR process operated with DC current [8]. Previously, Chang studied experimentally the effect of low frequency AC power supply on qualities of alloy steel [9]. It is generally recognized that the aforementioned alloying phenomena are due to electrochemical reactions taking place at the slag-metal interface when the process is operating with DC or quasi-DC current.

On the other hand, the inductive losses of the power for large scale ESR are the main disadvantage of operating the process at large frequencies. The demand on large scale ingots (diameter > 1 m) has been increased since the last decade. Unfortunately, the inductive losses become more significant as the ingot size increases which are the major drawbacks of applying large frequency for large scale ESR process. Therefore, the industry intends to operate the process at low frequency (0.1 - 5 Hz).

Over the last decades, some efforts were done to model ESR process using CFD [10-12]. Additionally, attempts have been made to investigate the effect of frequency of the applied AC current on the ESR process using numerical simulation. Kharicha studied the effect of the AC electromagnetic field on the flow and slag-pool interfaces [13]. In addition, Liang studied the quality of the ingot for producing GH4169 under different current frequency for a small scale ESR process [14]. Furthermore, Li developed a 3D finite element model (FEM) considering current frequency to simulate the current density, magnetic field, electromagnetic force and Joule heating for the process [15].

In the current work, the influence of the applied frequency of AC power supply for a large scale ESR process using a static mold is investigated. The main goal is to achieve some fundamental understanding of the two-phase flow and the formation of melt pool of the solidifying ingot under the influence of AC frequency.

Numerical Model

In the present study, the Finite Volume Method (FVM) is used for simulation of the fluid flow, heat transfer, and electromagnetic field in the process. The buoyancy and Lorentz forces as well as Joule heating due to the electrical resistance in the whole system are taken into account. The temperature field is obtained by solving the enthalpy conservation equation where Joule heating is added as a source term [16]. The momentum equation is solved to determine the velocity field. The drag resistance of the solidifying dendrites to the flow in the two-phase mushy zone is modeled according to the Black-Kozeny model [17]. Additionally, Lorentz force is added as a source to the momentum equation.

The origin of the Lorentz force is the interaction between the electric current and the self-induced magnetic field in the system. For the sinusoidal AC field, the magnetic field can be expressed using the phasor notation ($B_\theta = \widetilde{B}_\theta e^{i\omega t}$) where $\widetilde{B}_\theta$ is a function of position. The magnetic field has only azimuthal direction since there is no external magnetic field and the process conditions are axisymmetric. The equation for magnetic field that is derived from the Maxwell's system of equations must be solved in cylindrical coordinate [18]. The equation is given as:

$$\frac{\partial B_\theta}{\partial t}+\left[\frac{\partial}{\partial z}\left(\frac{1}{\sigma\mu_0}\frac{\partial B_\theta}{\partial z}\right)+\frac{\partial}{\partial r}\left(\frac{1}{r\sigma\mu_0}\frac{\partial(rB_\theta)}{\partial r}\right)\right]=0 \quad (1)$$

where μ_0 and σ denote the magnetic permeability and electric conductivity respectively.

After computing the real and imaginary components of the magnetic field, the electric current in the whole system can be obtained using the Ampere's law that is expressed as:

$$\tilde{\vec{j}}=\frac{1}{\mu_0}(\vec{\nabla}\times\tilde{B}_\theta) \quad (2)$$

Finally, the time average Lorentz force and Joule heating can be computed:

$$\vec{F}_L=\mathrm{Re}\left(\frac{1}{2}\vec{j}\times\tilde{B}_{\theta\ Conjugate}\right) \quad (3)$$

$$q_{Joule}=\mathrm{Re}\left(\frac{1}{2\sigma}\tilde{\vec{j}}\times\tilde{\vec{j}}_{\theta\ Conjugate}\right) \quad (4)$$

Configuration of the computational domain and boundaries are schematically shown in Figure 1. The computational domain includes the slag and ingot. A 2D axisymmetric calculation is performed. The slag-pool, electrode-slag, and slag-air interfaces are assumed to be flat. The electrode immersion depth is ignored.

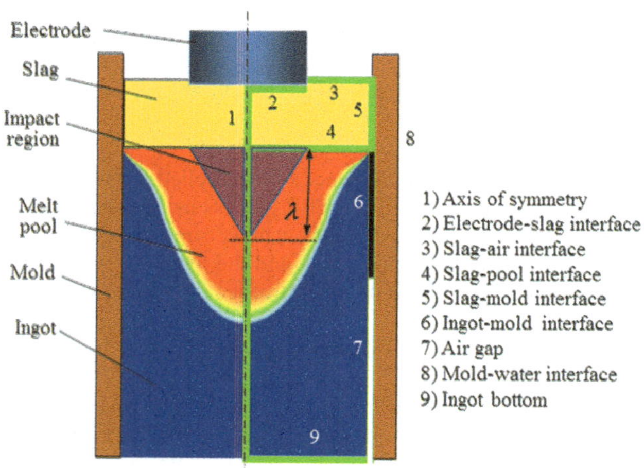

Figure 1: Schematic representation of the computational domain for the ESR process.

The classical way to consider the impact of falling droplets is to use a mass flux of liquid metal at the slag-pool interface [19-20]. Another way of modeling the droplets is prescribing a Gaussian distribution of velocity profile at the interface between the slag and melt pool [18]. In fact, the impact of droplets is not limited only at the surface but deeply inside the melt pool. Therefore, we modify the model considering the impact of droplets by introducing the parameter called impact depth of droplets (λ). With this approach, the droplets are considered to be as the mass, energy, and momentum source carriers which deeply penetrate into the melt pool. The amounts of the sources are assumed to be linearly decreased from the slag-pool interface till the impact depth. The impact depth can be obtained by experiment or direct numerical simulation of droplet–liquid surface splash phenomenon. For our simulations, we assumed that the droplet size is 1 cm and the corresponding impact depth is 15 cm. Furthermore, it is assumed that the droplets enter the pool under the shadow of the electrode. Non-slip boundary condition is applied for the flow at the electrode-slag interface and at the mold wall, whereas at the slag-pool interface the condition is free-slip. The heat is transferred by convection and radiation between electrode and air, and at the slag-air interface. A value of 0.8 is applied for the emissivity for the mentioned boundaries. The heat is conducted to the mold passing through the slag skin layer from the slag and ingot. In addition, the air gap between the slag skin layer and mold due to shrinkage of solidified ingot is taken into account. The gap alters the thermal boundary condition from the conduction condition to a combined radiation-convection condition. The tip of the electrode (slag-electrode interface) where the droplets form takes the liquidus temperature of the alloy.

In addition, the boundary conditions for the magnetic flux, Eq.(1), are obtained using the Ampere's law. The magnetic induction is prescribed at the slag-air and mould-water interfaces. The continuity of the magnetic induction is applied at the following interfaces: electrode-slag, slag-pool, slag-mold, and ingot-mold. Furthermore, an induction flux of zero is used for the ingot bottom. Details about the geometry and the average physical properties of the slag and steel are described in Table 1.

Table 1. Parameters used in the simulations.

Steel	
Density ($kg\cdot m^{-3}$)	7100
Viscosity ($Kg\cdot m^{-1}\cdot s^{-1}$)	0.006
Liquidus Temp. (K)	1779
Solidus Temp. (K)	1719
Specific heat, liquid ($J\cdot Kg^{-1}\cdot K^{-1}$)	800
Latent heat of fusion ($J\cdot Kg^{-1}$)	268000
Thermal Conductivity, liquid($W\cdot m^{-1}\cdot K^{-1}$)	40
Electric Conductivity, liquid($ohm^{-1}\cdot m^{-1}$)	880000
Slag	
Density ($kg\cdot m^{-3}$)	2800
Viscosity ($Kg\cdot m^{-1}\cdot s^{-1}$)	0.002
Specific heat, liquid ($J\cdot Kg^{-1}\cdot K^{-1}$)	1250
Thermal Conductivity, liquid($W\cdot m^{-1}\cdot K^{-1}$)	10
Electric Conductivity, liquid($ohm^{-1}\cdot m^{-1}$)	100
Geometry (Static mold)	
Mold radius (m)	0.9115
Electrode radius (m)	0.725
Slag height (m)	0.265

Several calculations under different operating conditions were performed using the commercial software FLUENT. Table 2 lists the operating conditions for the case studies.

Table 2. Operating conditions of the parameter studies

	RMS current (KA)	Frequency (Hz)	Mold current
Case I	36.5	0.2	No
Case II	36.5	50	No
Case III	36.5	50	Yes

Results

Case I

In Case I a low frequency (0.2 Hz) AC current is applied, and no current enters into mold. Figure 2 (a) shows the temperature field in the whole system and the isolines of solid fraction in the ingot region. The hottest area in the whole system is under the shadow of the slag-air interface.

In fact, an intense vortex forms under the edge of the electrode due to the Lorentz force and buoyancy force. The recirculation of the flow captures the released Joule heat in the slag region. The relative velocity between the melt and the ingot is illustrated in Figure 2 (b). The flow recirculation in the slag region corresponds to the maximum velocity in the whole system. The velocity is predicted to be much smaller in the melt pool than in the slag region. One point needs to be kept in mind is that the slag-pool interface is assumed to be stationary. This assumption would cause underestimation of the flow velocity in the melt pool especially in the vicinity of the slag.

The Lorentz force in the slag and melt pool is shown in Figure 2 (c). It is observed that the Lorentz force is quite homogeneously distributed. Exceptionally, the magnitude of the Lorentz force becomes large near the edge of the electrode. In fact, the variation in the Lorentz force and Joule heating in the slag region is consistent with the current density. Figure 2 (d) shows distributions of the current lines and Joule heating in the process. The maximum amount of Joule heating is released near the edge of the electrode where the current lines are denser. As a matter of fact, the amount of Joule heating is negligible in the electrode and ingot due to low electrical resistance of steel.

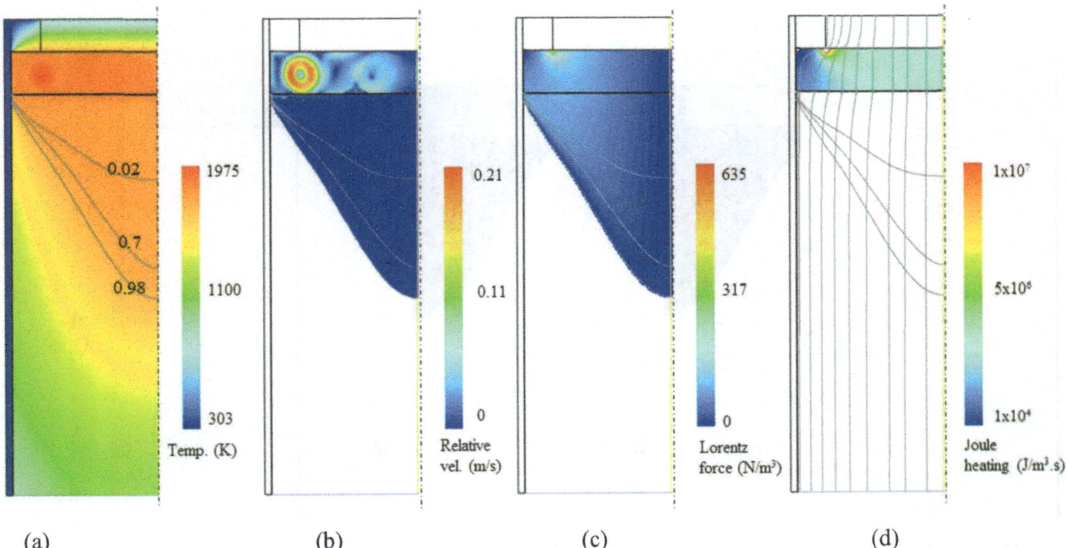

Figure 2. Contours of (a) the temperature field (b) magnitude of the relative velocity (c) Lorentz force and (d) Joule heating overlaid with the path of electric current for Case I. Isolines of fraction solid (0.02, 0.7 and 0.98) are plotted to indicate the mushy zone.

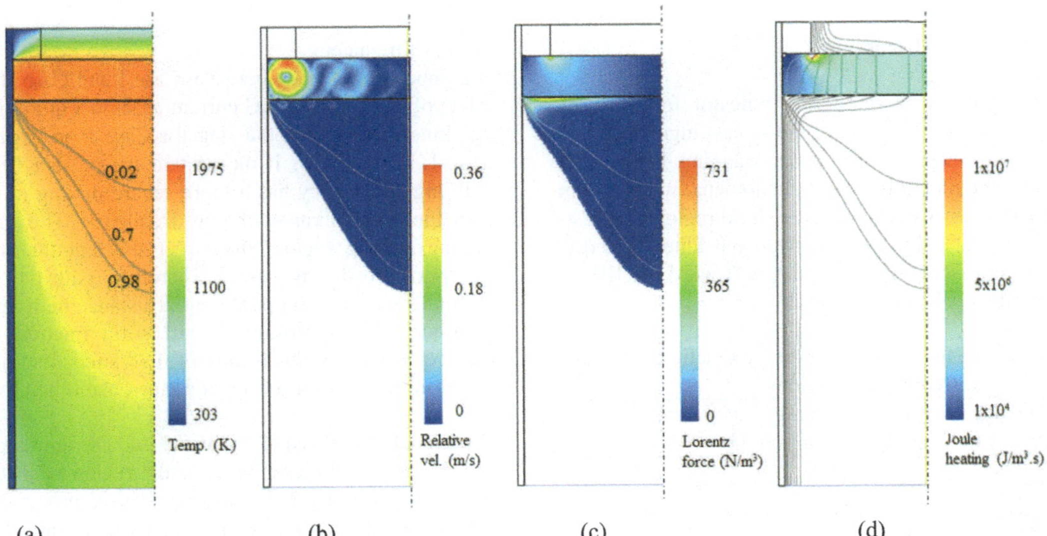

Figure 3. Contours of (a) the temperature field (b) magnitude of the relative velocity (c) Lorentz force and (d) Joule heating overlaid with the path of electric current for Case II. Isolines of fraction solid (0.02, 0.7 and 0.98) are plotted to indicate the mushy zone.

Case II

Figure 3 shows the contour of temperature, velocity, Lorentz force, and electric current together with Joule heating when the process is run under the frequency of 50 Hz. Again no current enters into mold since the skin layer is assumed to be an insulator. The region where the flow recirculates matches the hottest area in the slag zone as shown in Figure 3 (a). In the slag region, the Lorentz force is quite homogeneously distributed excluding the edge of the electrode where the force is the strongest. For this case, the force gradually increases from the bulk to the region near the mold wall in the melt pool as shown in Figure 3(c).This is due to change of electric current distribution once they enter the melt pool, as shown in Figure 3(d). The electric current flows radially towards the ingot surface as it passes the slag-pool interface. In fact, the electric current tends to flow along the ingot surface because of skin effect.

Case III

Figure 4 shows the modeling results of Case III. This case is same as Case II (50 Hz), but the electric current is allowed to cross the slag skin and enter into the mold.

The thickness of the slag skin layer is assumed to be 1 cm, and the contact length where the current can enter into the mold is 5 cm. Electric conductivities of the liquid slag and the solid slag skin are 100 and 48 respectively. In this case, around 90 % of total current is predicted to pass across the skin layer, as shown in Figure 4 (d). However, only 10 % of the total power is consumed in the skin layer. Due to the special current distribution in this case, strong Lorenz force is predicted in the region near the outer radius region of the slag layer and ingot, and only a very week Lorenz force acting in the bulk of the melt pool, as illustrated in Figure 4 (c).

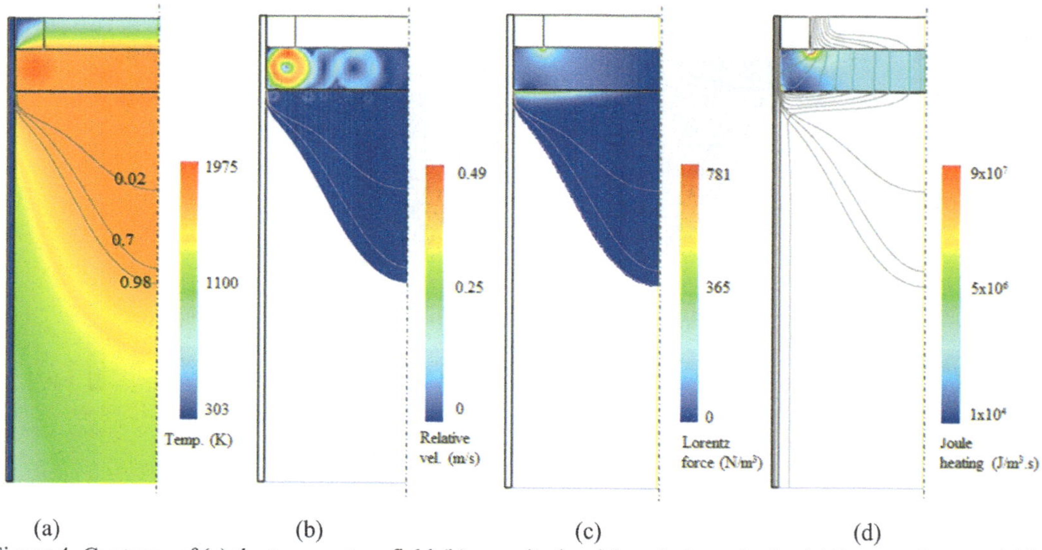

Figure 4. Contours of (a) the temperature field (b) magnitude of the relative velocity (c) Lorentz force and (d) Joule heating overlaid with the path of electric current for Case III. Isolines of fraction solid (0.02, 0.7 and 0.98) are plotted to indicate the mushy zone.

Discussion

The shape of melt pool is an important indicator for the ESR process, and it determines the quality of the as-cast ingot. Parameters of the melt pool (depth, mush zone, standing height) are mainly governed by the global transport phenomena, which are in turn related to the flow intensity of different fluid regions. Intensity of the flow is evaluated by the time-averaged kinetic energy. Influence of the process operating parameters (Case I, II, III) on the average kinetic energy is analyzed (Table 3).

Table 3. Summary of the computed average kinetic energy in different fluid regions.

	Average kinetic energy in slag (J)	Average kinetic energy in melt pool (J)
Case I	6.83	0.03
Case II	23.69	0.096
Case III	56.06	0.186

Effect of frequency

A comparison is made between Case I and Case II to analyze the effect of frequency of AC current as shown in Figure 5. The average kinetic energy in the slag for Case II is larger than that for Case I, see Table 3. It means that the present ESR process run with high frequency has a more severe mixing in the slag region than the process run with low frequency. The more severe the mixing in slag region, the larger the amount of energy being transport into the melt pool. Therefore, with the increase of the current frequency, the pool depth (distance between the slag-pool interface and the isoline of 0.02 solid fraction) is slightly increased. However, the isoline of 0.98 solid fraction moves down that results in thicker mushy zone for Case I than for Case II.

Additionally, the frequency of the AC current can affect the distribution of the Lorentz force at the region near to the slag-melt interface. As shown in Figure 5 (b), the electric current lines have similar distribution at the slag region independent of the applied frequency. Thus, the Lorentz force direction is identical for both cases, that is towards the axis of symmetry. However, at the slag-pool interface the direction is altered for Case II with 50 Hz frequency. It is horizontal and downward, as shown in Figure 6.

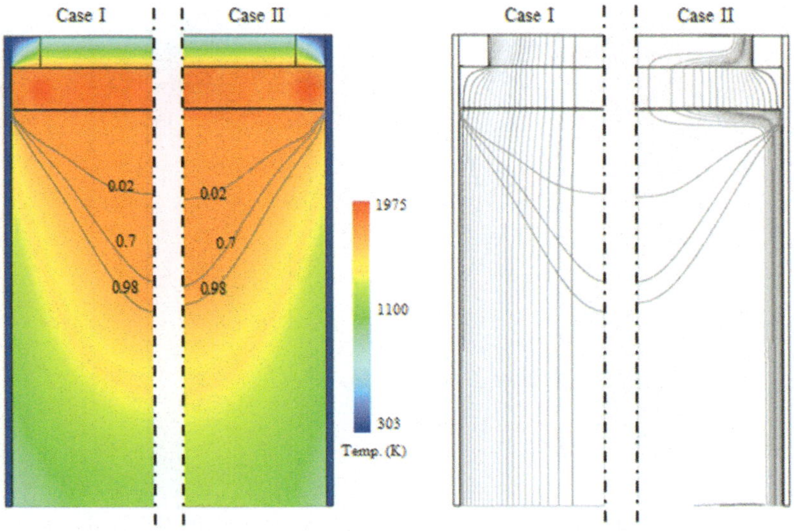

(a) (b)

Figure 5: Influence of the AC current frequency on (a) the temperature field and mushy zone (b) electric current distribution. Two cases are compared: Case I with frequency of 0.2 Hz: Case II with frequency of 50 Hz.

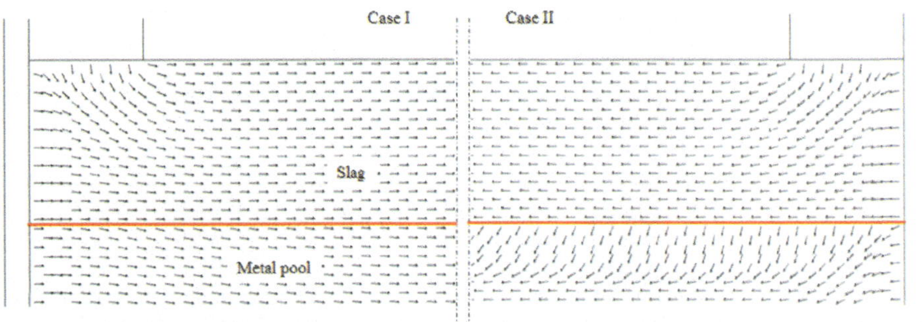

Figure 6: Influence of the AC current frequency on the direction of the Lorentz force, Two cases are compared: Case I with frequency of 0.2 Hz: Case II with frequency of 50 Hz.

Effect of mold current

Here, Case II and Case III are compered to study the influence of mold current on the pool shape, as illustrated in Figure 7. The average kinetic energy in the slag and melt pool is much larger for Case III than for Case II. Accordingly, the bulk of the slag is severely mixed in Case III with mold current. This mixing promotes the transfer of heat from slag to melt pool, hence leads to a raise of the melt pool temperature and a deeper pool.

The electric current lines are shown in Figure7 (b). For Case III a huge amount (92%) of the current enters into the mold in the region close to slag-pool interface where there is still contact between the metal and mold. This changes the pattern of the flow in this region, and results in larger standing height of the melt pool than for Case II without mold current. In addition, the skin effect in the copper mold is significant. Inside the mold the electric current only flows near the mold surfaces, as shown in Figure 7(b).

At high frequency, the current density and consequently the Lorentz force is week in the bulk of the melt pool where the flow is mainly driven by buoyancy. The direction of the Lorentz force is compared to study the effect of the mold current on the flow, as indicated in Figure 8. The direction of the Lorentz force is similar in the slag region and slag-pool interface in the central part of the system. However, for Case III with the mold current, the direction of the Lorentz force bents downwards near the mold wall both in the slag and melt pool regions. The Lorentz force and buoyancy act in the same direction and the flow is pushed downwards. The downward flow near the mold wall increases the standing height of the melt pool. The standing height is defined as the distance from the slag-pool interface to the start of solidification at the ingot surface (as shown in Figure 7a).

Joule heat is uniformly released under the electrode for both cases as shown in Figure 9. However, for Case II inhomogeneous distribution of Joule heat is released along the slag-mold interface. The minimum heat source is noticed near the slag-air interface for Case II where the slag skin layer acts as an electric insulator. In contrast, the magnitude of Joule heat source is significantly large at this region for Case III with mold current. Since this area is closer to the edge of the electrode, the electric resistance drops and more current flows through it. In the vicinity of the slag-pool interface Joule heating is pretty large for case III. This region is the most favorable path for the electric current to flow.

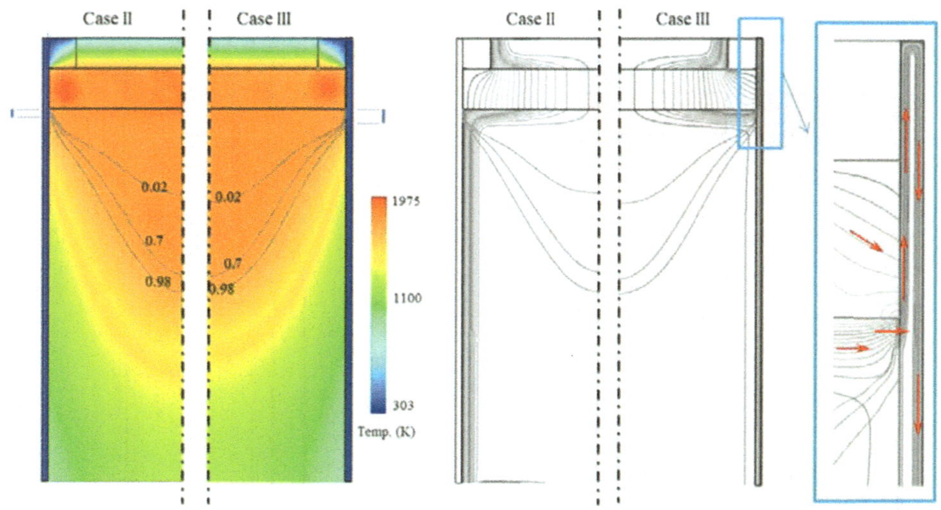

Figure 7: Influence of the mold current (frequency of 50 Hz) on (a) the temperature field and mushy zone (distance between dotted lines indicates the standing height) (b) electric current path (red arrows indicate the direction of electric current). Two cases are compared: Case II without mold current: Case III with mold current.

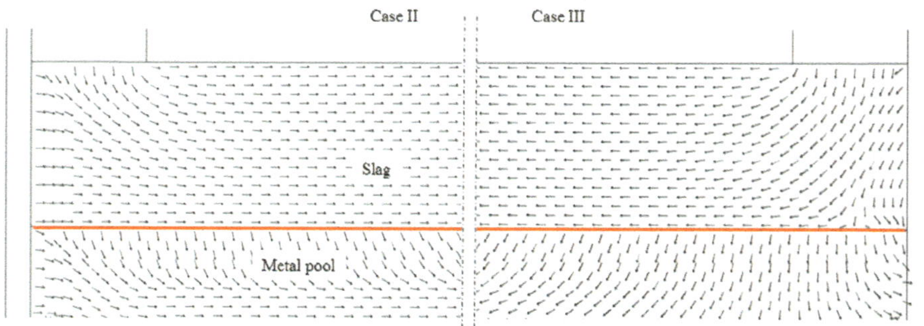

Figure 8: Influence of the mold current (frequency of 50 Hz) on the direction of the Lorentz force. Two cases are compared: Case II without mold current: Case III with mold current.

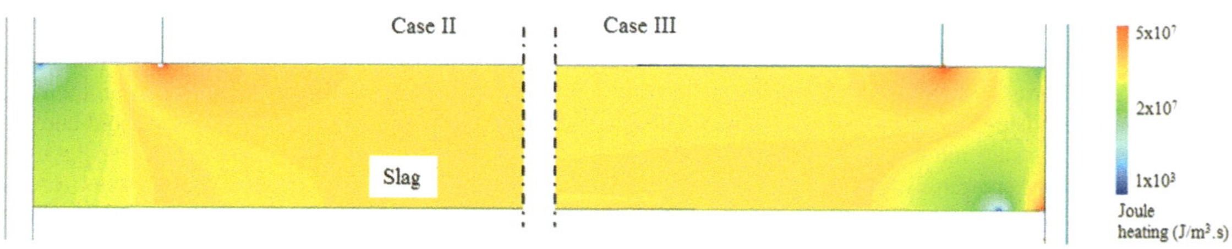

Figure 9: Influence of the Mold current (frequency of 50 Hz) on the distribution of the Joule heating, Two cases are compared: Case II without mold current: Case III with mold current.

Validity of the modeling result

A 2D axisymmetric model is used to solve the governing equations of electromagnetic field, heat transfer, and flow of the ESR process. The model is robust and computationally efficient to study the influence of electrical parameters on the solidification of the ingot. It is observed that the electrical parameters such as electric conductivity and frequency of the applied AC current have a huge influence on the shape of the melt pool. The electric conductivities of the slag (liquid and solid) are the main properties determining the electric current path, hence influencing other quantities of the process. As such properties are assumed, the current study is considered to be preliminary and qualitative. Some other factors influencing the quantitative accuracy of the modeling results should be kept in mind for improving the future model: the forming metal droplets under the electrode would adapt the electromagnetic field to a certain extend [21]; the motion of the slag-pool interface can affect the flow, temperature and electromagnetic fields of the system [22]; the nature of the flow in the real process is in 3D, but the current calculation can only be performed in 2D; the turbulence of the flow which is ignored would enhance the global energy transfer as well. In spite of the model simplifications, which are necessary at the current stage, the performed parameter studies in this paper can provide valuable information about the qualitative influence of the process parameters on some key solidification quantities of the ingot, such as the shape and depth of the melt pool, standing height, mushy zone, etc.

Summary

The influence of the applied frequency of AC current on the large scale ESR process with static mold has been numerically investigated. Main findings are summarized as follows.

- The melt pool shape of the solidifying ingot is slightly influenced by the applied frequency.
- In the case of high AC frequency, due to the skin effect, the current path changes its direction when the current passes the slag-pool interface, and so does the Lorenz force. As a consequence, it influences the hydrodynamic behavior of the slag and melt pool near the slag-pool interface.
- The distribution patterns of Joule heating and Lorentz force in the slag region are quite independent of the applied frequency in the slag region.

Additionally, the effect of the mold current (allow current entering into mold) is also considered. We found:

- Distribution patterns of the Lorentz force and Joule heating in the slag region are significantly influenced by the mold current, especially near the mold wall.
- Mold current promotes stirring in the slag region.
- Mold current increases the standing height of the melt pool.

Acknowledgements

The authors acknowledge the financial support by the Austrian Federal Ministry of Economy, Family and Youth and the National Foundation for Research, Technology and Development.

References

1. G. Hoyle, *Electroslag Processes* (London, Applied Science Publishers, 1983).
2. M.E. Peover, "Electroslag remelting, a review of electrical and electrochemical aspects," *J. Inst. Metals,* 100 (1972), 97-106.
3. A. Mitchell and G. Beynon, "Electrode polarization in the DC electroslag melting of pure iron," *Met. Trans B*, 10 (1971), 3333-3345.
4. Y. Kojima, M. Kato, T. Toyoda, and M. Inouye, "The formation of liquid droplet and behavior of oxygen in direct current electroslag remelting process," *ISIJ*, 15 (1975), 397-406.
5. M. Etienne, "The loss of reactive elements during electroslag processing of iron-base alloys," (Ph.D. Thesis, University of British Columbia, 1970).
6. G.K. Bhat and A. Mitchell, *Proc. Sec. Symp. On ESR technology*, Part I (1969).
7. B.I. Medovar, *Electroslag remelting* (Moscow, House of Literature on Ferrous and Nonferrous Metallurgy, 1963).
8. D.A. Whittaker, (Ph.D. Thesis, McMaster University, 1968).
9. L. Chang, X. Shi, H. Yang, Z. Li, "Effect of low-frequency AC power supply during electroslag remelting on qualities of alloy steel," *J. of Iron and Steel Research*, 16 (4) (2009), 7-11.
10. V. Weber, A. Jardy, B. Dussoubs, D. Ablitzer, S. Ryberon, V. Schmitt, S. Hans, H. Poisson, "A comprehensive model of the electroslag remelting process: description and validation," *Proc. of LMPC,* Nancy, 2007, 83-88.
11. Z. Jiang and Y. Dong, "Solidification model for electroslag remelting process," *Proc. of LMPC,* Nancy, 2007, 89-94.
12. Matthew J. M. Krane, Michael Fahrmann, Jeff Yanke, Enrique Escobar de Obaldia, Kyle Fezi, and Jonathan Busch, "A comparison of predictions of transport phenomena in electroslag remelting to industrial data," *Proc. of LMPC,* Nancy, 2011, 65-72.
13. A. Kharicha, M. Wu, A. Ludwig, B. Ofner, and H. Holzgruber, *CFD Modeling and simulation in materials processing* (US, Wiley publication, 2012), 139.
14. Q. Liang, X. Chen, H. Ren, C. Shi, and H. Guo, "Numerical simulation of electroslag remelting process for producing GH4169 under different current frequency," *Adv. Materials Research*, (482-484)(2012), 1556-1565.
15. B. Li, F. Wang, and F. Tsukihashi, "Current, magnetic field and joule heating in electroslag remelting processes," *ISIJ Int.*, 52 (7) (2012), 1289-1295.
16. A. Kharicha, W. Schützenhöfer, A. Ludwig, and R. Tanzer, "Numerical and experimental investigations on the ESR process of the hot work tool steel H11," *Proc. of LMPC,* Santa Fe, 2009, 235-242.
17. V. R. Voller, and C. Prakash, "A fixed grid numerical modeling methodology for convection-diffusion mushy region phase-change problems," *Int. J. Heat Mass Transfer*, 30(8) (1987), 1709-1719.
18. A. Kharicha, W. Schützenhöfer, A. Ludwig, and R. Tanzer, "Selected numerical investigations on ESR process," *Proc. of LMPC,* Nancy, 2007, 113-119.
19. A. H. Dilawari, and J. Szekely,"Heat transfer and fluid flow phenomena in electroslag refining," *Metall. Trans. B*, 9B (1) (1975), 77-87.
20. M. Choudhary, and J. Szekely, "Modeling of fluid flow and heat transfer in industrial-scale ESR system," *Ironmaking Steelmaking*, 8(5) (1981), 225-239.
21. A. Kharicha, A. Ludwig, M. Wu, "Droplet formation in small electroslag remelting processes," *Proc. of LMPC,* Nancy, 2011, 113-119.
22. A. Kharicha, A. Ludwig, M. Wu, "3D simulation of the melting during an industrial scale electro-slag remelting process," *Proc. of LMPC,* Nancy, 2011, 41-48.

Proceedings of the 2013 International Symposium on Liquid Metal Processing & Casting
Edited by: Matthew J. M. Krane, Alain Jardy, Rodney L. Williamson, and Joseph J. Beaman
TMS (The Minerals, Metals & Materials Society), 2013

A THERMODYNAMIC DATABASE FOR THE Al_2O_3-CaF_2-CaO SYSTEM

Lina Kjellqvist[1], Paul Mason[2], Qing Chen[1], Kaisheng Wu[2]
[1]Thermo-Calc Software, Norra Stationsgatan 93, SE-113 64 Stockholm, Sweden
[2]Thermo-Calc Inc, 4160 Washington Road, McMurray, PA 15317, USA

Keywords: Al_2O_3-CaF_2-CaO system, Calphad, Ionic two-sublattice liquid model

Abstract

Al_2O_3-CaF_2-CaO is the core system for modelling ESR slags. All ESR slags contain CaF_2 as the primary constituent. Oxides, such as CaO, Al_2O_3, MgO and other, may be added to the slag to raise the electrical resistivity or to modify the slag chemistry with regard to its reactivity with the metal being melted.

A good thermodynamic description of all phases in the system is essential in materials design. An efficient technique used to obtain a self-consistent thermodynamic database is called Calphad, where the Gibbs energy of each phase is described with a thermodynamic model. The pseudo-ternary Al_2O_3-CaF_2-CaO system and the bounding systems Al_2O_3-CaF_2 and CaF_2-CaO are assessed in this work using the Calphad approach. The liquid slag phase is described using the ionic two-sublattice model.

Several papers have been published on the phase equilibria in the Al_2O_3-CaF_2-CaO system and the bounding systems Al_2O_3-CaF_2 and CaF_2-CaO. However, wide discrepancies exist between the findings of the various authors. The present assessment gives a reasonably good description of available experimental data. A characteristic feature of the ternary phase diagram proposed by some authors is a rather extended egg-shaped miscibility gap in the high fluorite high alumina region. The calculated miscibility gap in the ternary reaches very close to the Al_2O_3-CaF_2 side, but it does not touch it.

Introduction

Electroslag remelting (ESR) is a process of remelting and refining steel and other alloys for different applications. ESR slags are CaF_2 based slags, with a mixture of primarily Al_2O_3, CaO or MgO, but could also encompass a wide range of additions that may include, but is not limited to, TiO_2, ZrO_2, SiO_2 and Cr_2O_3. The Al_2O_3-CaF_2-CaO system is thus the core system for modelling ESR slags.

This is the first part of a research program to construct a thermodynamic database for the oxy-fluoride system Al_2O_3-CaF_2-CaO-MgO-TiO_2-ZrO_2 using the Calphad (CALculation of PHAse Diagrams) approach [1, 2]. The thermodynamic database will be a basis for modelling other thermo physicochemical properties, such as density, surface tension, viscosity and thermal conductivity. The database developed within this project will improve the modelling capabilities of ESR slags.

The Al_2O_3-CaO system was assessed by Hallstedt [3]. Later, Mao et al. [4] developed a new model for liquid Al_2O_3 in order to better control the unwanted reciprocal miscibility gaps in the CaO-Al_2O_3-SiO_2 and MgO-Al_2O_3-SiO_2 ternary systems. The description of liquid Al_2O_3-CaO from Mao et al. [4] is used in this work together with the solid phases from Hallstedt [3].

The CaF_2-CaO system has been assessed recently by Kim et al. [5] using the Modified Quasichemical Model [6] to describe the molten slag phase. Their assessment is not considered in the present work since the Modified Quasichemical Model is incompatible with the ionic two-sublattice liquid model.

Both the Al_2O_3-CaF_2 and the CaF_2-CaO systems have simple eutectic phase diagrams with no intermediate phases. While the Al_2O_3-CaF_2-CaO system can be regarded as a ternary in sealed cells, Nikitin et al. [7-9] noted that under open cell conditions the system should be considered as the reciprocal system CaF_2-CaO-AlF_3-Al_2O_3. Two ternary compounds have been reported to exist in this system: $3CaO.3Al_2O_3.CaF_2$ (C_3A_3Fl) and $11CaO.7Al_2O_3.CaF_2$ ($C_{11}A_7Fl$). The recommended phase diagram [10] has a wide miscibility gap in the liquid phase.

Thermodynamic Models

The Calphad technique is used to describe the thermodynamics of materials. The energetic state of a closed system in equilibrium at fixed temperature, pressure and composition is described by its Gibbs energy. The main idea of the Calphad technique is to describe the Gibbs energy of all phases in the system, using appropriate thermodynamic models. All available experimental information, such as phase diagram data and thermochemical data, are collected and evaluated, and then used in an optimisation procedure to fit the model parameters. The determination of the coefficients is frequently called assessment or optimisation of a system. To get as good description as possible of the properties, it is necessary to use all experimental information simultaneously, both phase equilibria and thermochemical data. Each phase is usually modelled within the entire temperature and composition range, and not only in the range where it is stable. Some parameters will refer to metastable, or sometimes unstable, structures. As a consequence of this, a thermodynamic database provides the possibility to do calculations also of metastable states, where experimental information is hard to find.

A general formalism widely used in Calphad asssessments is the Compound Energy Formalism [11] (CEF). CEF was constructed in order to describe models of the thermodynamic properties of phases with two or more sublattices, which show a variation in composition, i.e. $(A,B)_i(C,D)_j$.

Liquid Slag

Within the framework of the CEF, the ionic two-sublattice liquid model [12, 13] was developed to be used when there is a tendency for ionization in the liquid, e.g. liquid oxides and sulphides. The same model can be used both for metallic and oxide melts. At low levels of oxygen, the model becomes equivalent to a substitutional solution model between metallic atoms. Two sublattices are assumed, one containing charged cations and one containing charged anions, neutrals and vacancies:

$$(C_i^{v_i})_P(A_j^{v_j}, Va^{-Q}, B_k^0)_Q \quad (1)$$

Where C represents cations, A anions, Va vacancies and B neutrals. The indices i, j and k denote specific constituents. Charged vacancies are introduced on the second sublattice to keep electroneutrality when the composition is approaching the metallic liquid. The site numbers, P and Q, on the sublattices vary so that electroneutrality is maintained:

$$P = \sum_j y_{A_j}(-v_j) + Qy_{Va} \quad (2)$$

$$Q = \sum_i y_{C_i}v_i \quad (3)$$

According to Mills and Keen [10], CaF_2 slags are generally considered to be ionic in nature and the Al_2O_3-CaF_2-CaO system should thus be modelled using $(Al^{+3},Ca^{+2})_P(AlO_2^{-1},F^{-1},O^{-2})_Q$. However, there will be problems when modelling the Al_2O_3-CaF_2 pseudo-binary system, since this is not a true pseudo-binary but part of the reciprocal system $CaF_2+Al_2O_3=CaO+AlF_3$. Al_2O_3-CaF_2 and Al_2O_3-CaF_2-CaO (and AlF_3-CaO) must then be evaluated simultaneously. This is in theory of course possible, but could be difficult in practice.

In the liquid state some oxides have a weaker tendency to form free O^{-2} ions whereas others are clearly ionic in behaviour. The former oxides are usually called acid whereas the latter ones with free O^{-2} are called basic. Within the frame of the ionic two-sublattice liquid model, it has long been accepted that basic oxide melts, such as Na_2O, CaO or MgO, are described with a cation in one sublattice and O^{-2} in the other. The acid behaviour of a slag can be taken into account by introducing the neutral oxide as a molecule in the anion sublattice. A typical acid slag is SiO_2. In general, the higher the valence of the cation, the more it behaves like an acid. Al_2O_3 is referred to as amphoteric, i.e. it behaves like an acid when mixed with a base and as a base when mixed with an acid. Liquid Al_2O_3 is described with both cation Al^{+3} in the first sublattice and anion AlO_2^{-1} in the second one. Selections of species in the modelling of acid, basic and amphoteric slags are extensively discussed in many studies [14-17]. If we introduce the neutral species CaF_2, the liquid model will read

$$(Al^{+3}, Ca^{+2})_P(AlO_2^{-1}, O^{-2}, CaF_2)_Q \quad (4)$$

In this new model, CaF_2 is described as an acid component. An equivalent model is used to model liquid Fe_2O_3. Fe_2O_3 is weakly basic, first modelled as $(Fe^{+3})_2(O^{-2})_3$ [18]. Later, Fe_2O_3 was treated as an acid oxide instead, using the neutral species $FeO_{1.5}$ [19]. This means that one assumes that liquid Fe_2O_3 do not readily give up their oxygen to form O^{-2} ions. Thus, the weakly basic Fe_2O_3 is now treated as an acid oxide with great success.

This present model (4) may seem less realistic as it does not make use of the experimental evidence on the existence of the F^{-1} ion. However, real oxide/fluoride melts have many microscopic features that are impossible to model accurately due to their complexity. The assumption of a neutral CaF_2 species will hopefully be closer to the macroscopic behaviour of the liquid.

Using this model, the Gibbs energy of the liquid is given by

$$\begin{aligned} G_m = {} & y_{Al^{+3}}y_{AlO_2^{-1}}G_{Al^{+3}:AlO_2^{-1}} + y_{Ca^{+2}}y_{AlO_2^{-1}}G_{Ca^{+2}:AlO_2^{-1}} + \\ & y_{Al^{+3}}y_{O^{-2}}G_{Al^{+3}:O^{-2}} + y_{Ca^{+2}}y_{O^{-2}}G_{Ca^{+2}:O^{-2}} + \\ & y_{CaF_2}G_{CaF_2} + \\ & PRT(y_{Al^{+3}}\ln(y_{Al^{+3}}) + y_{Ca^{+2}}\ln(y_{Ca^{+2}})) + \\ & QRT(y_{AlO_2^{-1}}\ln(y_{AlO_2^{-1}}) + y_{O^{-2}}\ln(y_{O^{-2}}) + \\ & y_{CaF_2}\ln(y_{CaF_2})) + {}^{E}G_m \end{aligned} \quad (5)$$

Where y is the site fraction and G represents the Gibbs energy of end-members. The excess Gibbs energy represents the deviation from the ideal behaviour and is described by ${}^{E}G_m$. The excess parameters used in this assessment are the following:

$$\begin{aligned} {}^{E}G_m = {} & y_{Al^{+3}}y_{Ca^{+2}}y_{AlO_2^{-1}}\,{}^{0}L_{Al^{+3},Ca^{+2}:AlO_2^{-1}} + \\ & y_{Al^{+3}}y_{AlO_2^{-1}}y_{CaF_2}(\,{}^{0}L_{Al^{+3}:AlO_2^{-1},CaF_2} + \\ & {}^{1}L_{Al^{+3}:AlO_2^{-1},CaF_2}(y_{AlO_2^{-1}} - y_{CaF_2})) + \\ & y_{Ca^{+2}}y_{O^{-2}}y_{CaF_2}(\,{}^{0}L_{Ca^{+2}:O^{-2},CaF_2} + \\ & {}^{1}L_{Ca^{+2}:O^{-2},CaF_2}(y_{O^{-2}} - y_{CaF_2})) + \\ & y_{Ca^{+2}}y_{AlO_2^{-1}}y_{CaF_2}(\,{}^{0}L_{Ca^{+2}:AlO_2^{-1},CaF_2} + \\ & {}^{1}L_{Ca^{+2}:AlO_2^{-1},CaF_2}(y_{AlO_2^{-1}} - y_{CaF_2})) + \\ & {}^{2}L_{Ca^{+2}:AlO_2^{-1},CaF_2}(y_{AlO_2^{-1}} - y_{CaF_2})^2) + \end{aligned} \quad (6)$$

A colon is used to separate species on different sublattices and a comma is used to separate species on the same sublattice.

α- and β-CaF_2

α- and β-CaF_2 are described using a model within the CEF with two sublattices. The solubility of CaO in CaF_2 is modelled by adding O^{-2} ions and vacancies on the second sublattice:

$$(Ca^{+2})_1(F^{-1}, O^{-2}, Va)_2 \quad (7)$$

The Gibbs energy of α- and β-CaF_2 are given by

$$\begin{aligned} G_m^{CaF_2} = {} & y_{Ca^{+2}}y_{F^{-1}}G_{Ca^{+2}:F^{-1}} + y_{Ca^{+2}}y_{O^{-2}}G_{Ca^{+2}:O^{-2}} + \\ & y_{Ca^{+2}}y_{Va}G_{Ca^{+2}:Va} + 2RT(y_{F^{-1}}\ln(y_{F^{-1}}) + \\ & y_{O^{-2}}\ln(y_{O^{-2}}) + y_{Va}\ln(y_{Va})) + \\ & y_{F^{-1}}y_{O^{-2}}L_{Ca^{+2}:F^{-1},O^{-2}} \end{aligned} \quad (8)$$

The parameter $G_{Ca^{+2}:Va}$ could appear in many systems and is therefore convenient to use as a reference; the simplest choice is to give it the value zero. CaO with the CaF_2 structure is described by the parameter $G_{Ca^{+2}:O^{-2}}$, but since the solubility of CaO in CaF_2 is fairly low it cannot be assessed with any accuracy and is therefore simply given a fixed large value compared to halite-CaO to assure that CaO with the CaF_2-structure does not become stable in the CaO-rich part of the system.

Other Solid Phases

CA, CA_2, CA_6, C_3A, Halite and Corundum were all accepted from earlier assessments and modelled as stoichiometric compounds. The two ternary compounds, C_3A_3Fl and $C_{11}A_7Fl$, are also assumed stoichiometric. The Gibbs energy is evaluated relative to the pure solid oxides as follows

$$G_m^{C_3A_3Fl} = 3\,{}^oG_{CaO}^{Halite} + 3\,{}^oG_{Al_2O_3}^{Corundum} + {}^oG_{CaF_2}^{\alpha-CaF_2} + A + BT \quad (9)$$

$$G_m^{C_{11}A_7Fl} = 11\,{}^oG_{CaO}^{Halite} + 7\,{}^oG_{Al_2O_3}^{Corundum} + {}^oG_{CaF_2}^{\alpha-CaF_2} + A + BT \quad (10)$$

Only enthalpy data were found for the two ternary phases. The variables A and B were related in such a way that C_3A_3Fl and $C_{11}A_7Fl$ are stable above 25°C (but not below). This is quite an arbitrary choice, but using only an enthalpy term would probably render C_3A_3Fl and $C_{11}A_7Fl$ much too stable at low temperature. The entropy term for C_3A_3Fl will by this choice be of about the same size as for the calcium aluminates, which seems quite reasonable. For $C_{11}A_7Fl$, the entropy term is slightly lower.

Experimental Data

The available experimental data mainly concern phase diagram data, such as liquidi and solidi.

CaF_2

A melting temperature of 1423°C has been adopted for pure CaF_2. Many lower values have been given for the melting point of CaF_2, but this is, according to Mills and Keen [10], a result of impurities (mainly CaO and SiO_2) in the materials.

There are two modifications of the solid CaF_2 phase. In the present work, the thermodynamic descriptions of the two phases were taken from Selleby and Hillert [20]. The solid-state transition temperature is 1151°C.

Al_2O_3-CaF_2

Experiments with CaF_2 containing slags are associated with difficulties. For example, the volatilization of CaF_2 and the tendency of CaF_2 to react with alumina and water vapour to form volatile AlF_3 or HF may cause changes in composition. Under open conditions AlF_3 can evaporate from Al_2O_3-CaF_2 melts and from ternary melts with low initial CaO contents causing an increase in CaO and decreases of the Al_2O_3 and CaF_2 contents. Therefore an investigation of the binary Al_2O_3-CaF_2 system under open conditions is impossible. Uncontrolled changes of sample composition may well be the reason for some of the disagreement between the reported data below.

The Al_2O_3-CaF_2 system is characterized by a simple eutectic phase diagram. The liquidus is S-shaped, indicating a tendency towards formation of a miscibility gap. The liquidus line measured by Ries and Schwerdtfeger [21] is in very good agreement with that reported by Zhmoidin and Chatterjee [22]. However, the liquidus lines determined by Kun and Sheng [23] as well as by Mitchell and Burel [24] deviate very much from the other results. Their liquidus lines are at considerably lower temperatures in the range below 30 mass-% Al_2O_3 and do not show inflection points. These authors conducted their experiments in open crucibles and it is not known what the final composition was under such conditions. Another phase diagram of the Al_2O_3-CaF_2 binary has been proposed by Hillert [25], who found a miscibility gap in sealed capsule experiments. No evidence for a miscibility gap was found in the binary system in the other investigations. However, when Ries and Schwerdtfeger added 1 mass-% CaO to the 80% CaF_2 – 20% Al_2O_3 mixtures, a second liquid was formed. This indicates that the miscibility gap in the ternary system extends very close to the Al_2O_3-CaF_2 join.

CaF_2-CaO

The binary CaF_2-CaO system has been subject of many investigations [21, 26-50]. It is a simple eutectic diagram with the eutectic at 1360°C. The composition of the eutectic melt varies between 16-19.6 mass-% CaO depending on the author. Baak [28] reported a miscibility gap on the CaF_2 rich side of the system, which is compatible with the activity measurements carried out by Hsu et al. [31]. But according to [32, 36, 39] there is no evidence for a miscibility gap. The CaF_2-liquidus lines given by the different investigators are in close agreement. The values for CaO saturation of the work by Ries and Schwerdtfeger [21] agree well with the data obtained by Kor and Richardson [38] and Zhmoidin and Chatterjee [42]. Data of the CaO-liquidus line were also published by Mukerjee [32] and Schlegel [36]. The liquidus lines proposed by these authors emerge from the eutectic point located similarly as the other investigators, but then swing rapidly over to much higher CaO contents than the others. Activity measurements have been performed in the CaF_2-CaO binary by several investigators. The gas equilibrium technique was used, which unfortunately can cause a significant solubility of gas species in liquid CaF_2-CaO which can change the composition of the liquid during the experiments. However, most of the reported activity curves increase more or less linearly with increasing CaO concentration in the liquid. Activity data have not been used in the optimization of the current study.

In a recent investigation by Kim et al. [50], noticeable solubility of CaO in α- and β-CaF_2 solids is observed above 1000°C. The maximum solubility of CaO in solid CaF_2 is about 5 mol-% at the eutectic temperature (about 1363°C). No solubility of CaF_2 in CaO was found.

Al_2O_3-CaF_2-CaO

In the first phase diagram of the subsystem CaF_2-CaO-5CaO.3Al_2O_3, Eitel [51] proposed a simple ternary eutectic system. A drastically modified phase diagram was presented by Chatterjee and Zhmoidin [52], who investigated several subsystems with the quenching technique. A characteristic feature of the phase diagram proposed by Chatterjee and Zhmoidin is a rather extended egg-shaped miscibility gap in the high fluoride high alumina region. They also found two congruently melting ternary compounds: C_3A_3Fl and $C_{11}A_7Fl$. They determined the melting points to 1507±1.5°C and 1577±2.5°C. A rather different phase diagram has been worked out by Nafziger [53], who conducted his experiments under open conditions. The phase diagram has no miscibility gap and only one ternary compound $C_{11}A_7Fl$. There is indication for a miscibility gap from Mitchell and Cameron [54], Ries [55] and Ries and Schwerdtfeger [21]. In addition to the investigation of the miscibility gap, Ries and Schwerdtfeger performed other experiments to determine the liquidus isotherms of CaO, Al_2O_3, CaO.6Al_2O_3 (CA_6) and CaO.2Al_2O_3 (CA_2). Gohil and Mills [56] determined values of (H_T-H_{298}) for C_3A_3Fl with a differential scanning calorimeter for the temperature rage -4 to 402°C and by drop calorimetry for the range 800 to 1547°C. Their data suggests a value of 1467°C for the melting temperature. They also obtained enthalpies for $C_{11}A_7Fl$. Kewu et al. [57] studied the liquidus temperature of

Al_2O_3-CaF_2-CaO slag with three different CaF_2 contents (10%, 15%, 20%) in a Al_2O_3 (68.4%)–CaO (31.6%) system.

Optimization and Results

The optimization of the parameters was performed using the PARROT module of the Thermo-Calc software [58]. The optimization was conducted by minimizing the sum of squares. By adjusting the weight of each experiment the optimization was repeated until a satisfactory description of most experimental data points was reached.

Al_2O_3-CaF_2

(H_T-H_{298}) values were determined by Gohil and Mills [56] for CaF_2 + 30 mass-% Al_2O_3. The result is shown in Figure 1. The calculated phase diagram for the Al_2O_3-CaF_2 binary is shown in Figure 2. The curve is very flat in the range 10 to 30 mass-% Al_2O_3, indicating a tendency towards formation of a miscibility gap. Experimental data on the liquidus close to CaF_2 deviate some from the calculated curve. It could be hard to fit experimental data close to an end-member if the data deviate from ideality. In this system, if trying to fit the experimental data close to CaF_2, a miscibility gap is formed in the liquid phase.

CaF_2-CaO

(H_T-H_{298}) values were determined by Gohil and Mills [56] CaF_2 + 16 mass-% CaO. The result is shown in Figure 3. The calculated phase diagram for the CaF_2-CaO binary is shown in Figure 4. The calculated diagrams together with experimental data are shown in Figures 5 and 6. The calculated liquidus of β-CaF_2 solid solution shows a maximum at about 2 mass-% CaO. The experimental data from Schlegel [36] shows this trend. This is due to the defect structure of CaF_2. When small amounts of CaO is added to CaF_2, the solid CaF_2 solution can be more stable than the liquid CaF_2-CaO solution at the same composition because the addition of CaO induces more entropy in solid CaF_2 (one O^{-2} replaces two F^{-1} with generation of a vacancy) than in liquid CaF_2 (one O^{-2} replaces two F^{-1} without generation of a vacancy). The calculated solubility of CaO in CaF_2 at the eutectic is 5.4 mol-%.

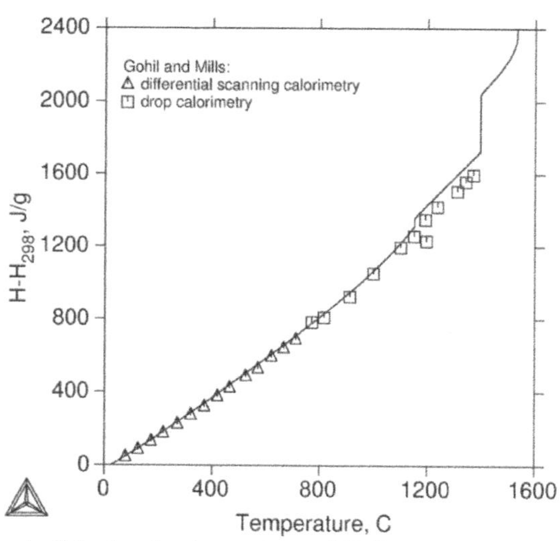

Figure 1: Calculated and experimental [56] enthalpy of CaF_2+30 mass-% Al_2O_3.

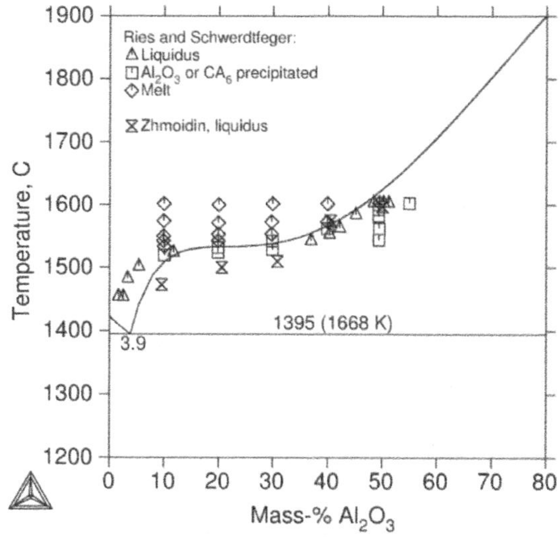

Figure 2: Calculated and experimental [21, 22] Al_2O_3-CaF_2 phase diagram.

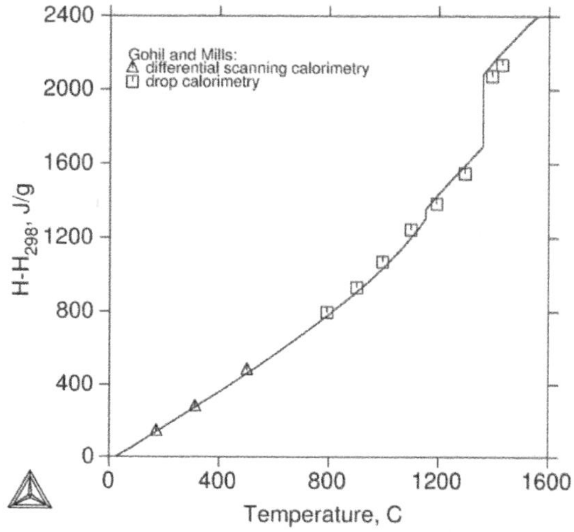

Figure 3: Calculated and experimental [56] enthalpy of CaF_2+16 mass-% CaO.

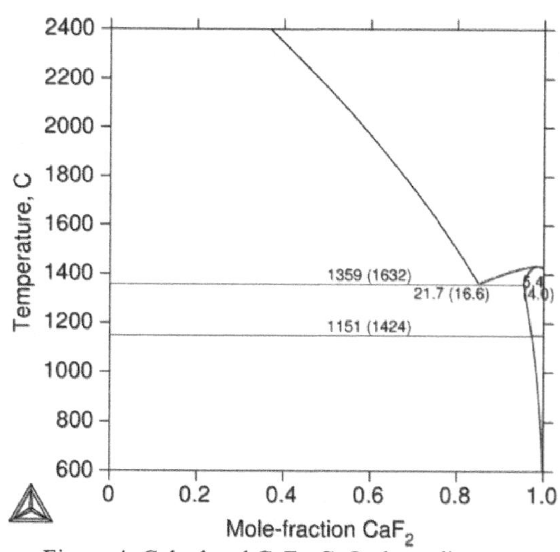

Figure 4: Calculated CaF_2-CaO phase diagram.

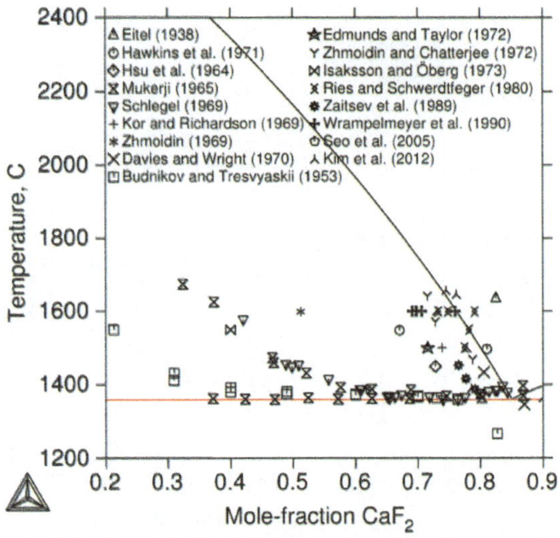

Figure 5: Calculated CaF_2-CaO phase diagram with experimental data [21, 26-27, 31, 33, 35, 37-38, 40-44, 47-50].

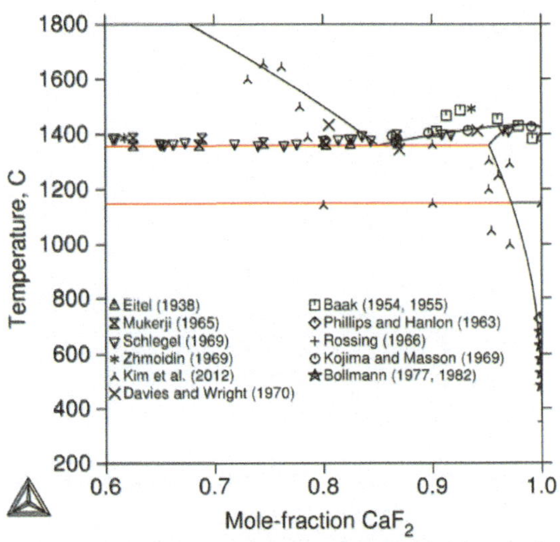

Figure 6: Calculated CaF_2-CaO phase diagram with experimental data [26, 28-30, 33-35, 37, 39-40, 45-46, 50].

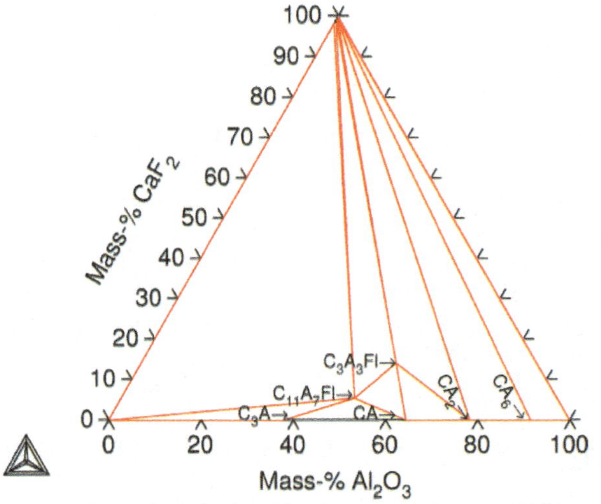

Figure 7: Calculated isothermal section at 1100°C.

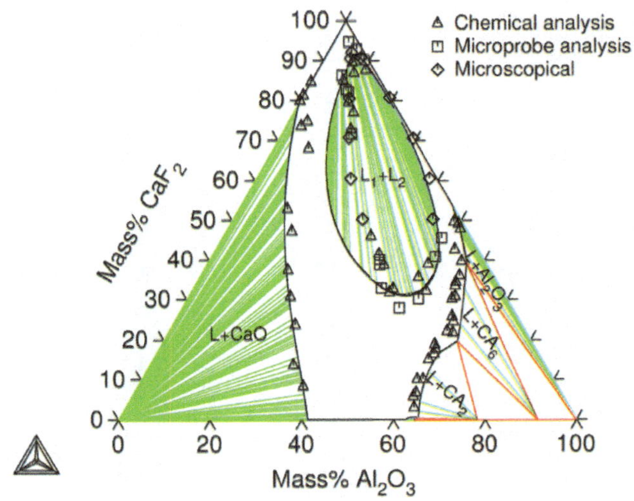

Figure 8: Calculated isothermal section at 1600°C, with experimental data from Ries and Schwerdtfeger [21].

Al_2O_3-CaF_2-CaO

Calculated isothermal sections for the Al_2O_3-CaF_2-CaO ternary system at different temperatures are shown in Figure 7-9. The miscibility gap with a starting mixture with 60 mass-% CaF_2, 12 mass-% CaO and 28 mass-% Al_2O_3 as a function of temperature is shown in Figure 10. The liquidus temperature for a Al_2O_3 (68.4%)–CaO (31.6%) system is shown in Figure 11. Calculated isoplethal section CaO·Al_2O_3-CaF_2 is shown in Figure 12 with experimental data from Gutt et al. [59].

The calculated melting points of C_3A_3Fl and $C_{11}A_7Fl$ are 1507 and 1577°C, respectively, the same result as the experimentally determined temperatures [52]. The calculated enthalpy of C_3A_3Fl and $C_{11}A_7Fl$ together with experimental data [56] is shown in Figure 13 and 14.

The compositions of the double saturated melts at 1600°C from Ries and Schwerdtfeger [21] and calculated values are shown in Table 1.

Table 1: Composition of double saturated melts.

	$L/Al_2O_3/CA_6$	$L/CA_6/CA_2$
CaF_2 (exp/calc)	38/39.5	24/19.8
CaO (exp/calc)	6/4.1	15/16.0
Al_2O_3 (exp/calc)	56/56.4	61/64.2

According to Ries and Schwerdtfeger [21], the liquid/Al_2O_3/CA_6 equilibrium could be located very exactly experimentally. The other equilibrium is more uncertain because close to the double saturated point the saturation state could not be elucidated unambiguously in the microscopic investigation. The liquid/CA_6/CA_2 equilibrium was therefore given a lower weight during the assessment procedure.

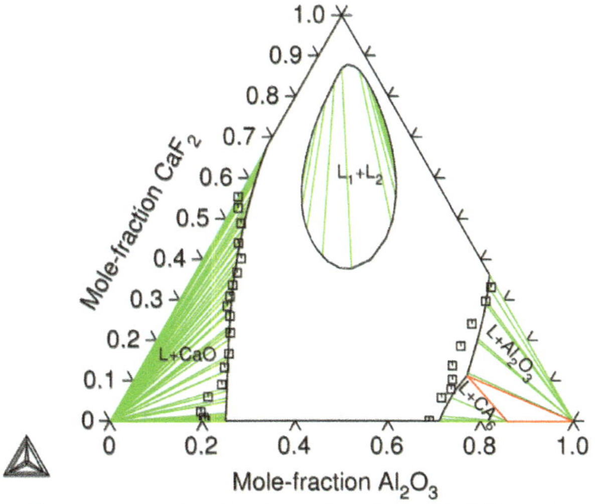

Figure 9: Calculated isothermal section at 1800°C, with experimental data from Zhmoidin and Chatterjee [42].

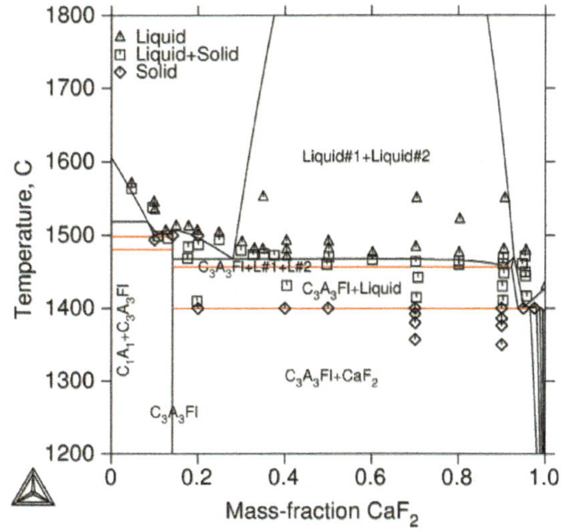

Figure 12: Calculated isoplethal section CA-CaF$_2$ with experimental data from Gutt et al. [59].

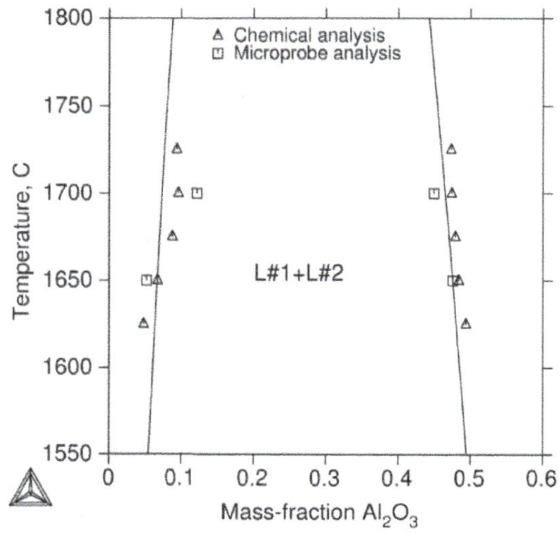

Figure 10: Calculated and experimental [21] miscibility gap as a function of temperature.

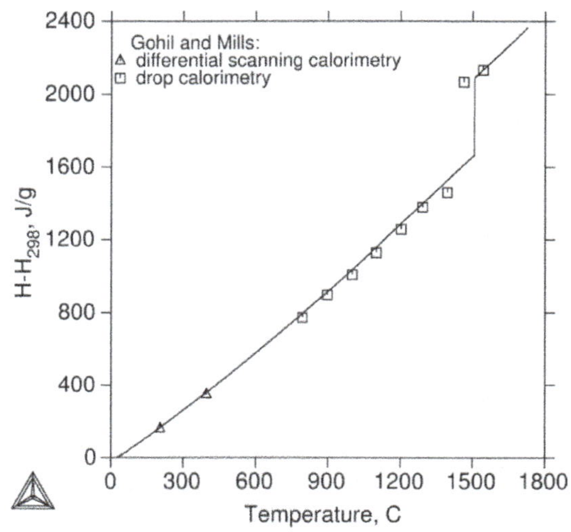

Figure 13: Calculated and experimental [56] enthalpy of C_3A_3Fl.

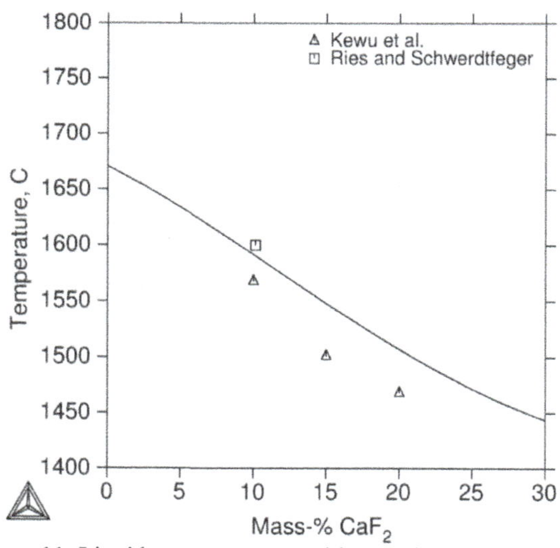

Figure 11: Liquidus temperature, with experimental data from Kewu et al. [57] and Ries and Schwerdtfeger [21].

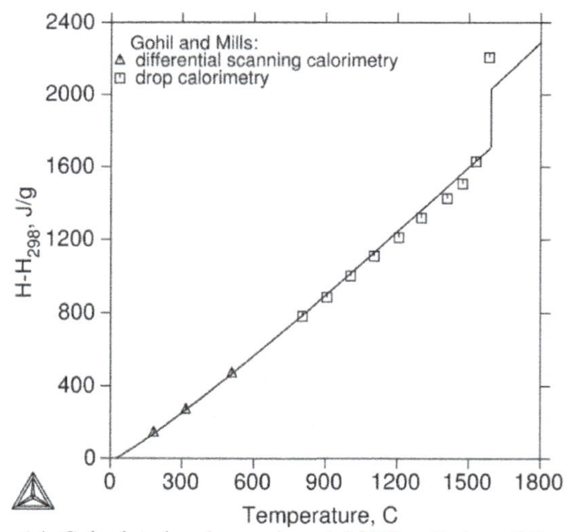

Figure 14: Calculated and experimental [56] enthalpy of $C_{11}A_7Fl$.

Summary

Al_2O_3-CaF_2-CaO is the core system for modelling ESR slags, and is assessed in this work. Several papers have been published on the phase equilibria in the ternary Al_2O_3-CaF_2-CaO system and the bounding systems Al_2O_3-CaF_2 and CaF_2-CaO. However, wide discrepancies exist among the findings of the various authors. The present assessment gives a reasonably good description of available experimental data. The ionic two-sublattice model with the neutral CaF_2 species was applied to the liquid. Isothermal and isoplethal sections were calculated and compared with experimental data. The calculated miscibility gap in the ternary reaches very close to the Al_2O_3-CaF_2 side, but it does not touch it. The Gibbs energy functions of C_3A_3Fl and $C_{11}A_7Fl$ were evaluated.

Acknowledgement

The authors wish to acknowledge the financial support of US Air Force under the auspices of the Materials Affordability Initiative through USAF Agreement Order FA8650-11-2-25224-B.

References

[1] H. Lukas, S. G. Fries and B. Sundman, Computational Thermodynamics, The Calphad Method, 2007.

[2] N. Saunders and A. Miodownik, Calphad (Calculation of Phase Diagrams): A Comprehensive Guide, 1998.

[3] B. Hallstedt, *J. Am. Ceram. Soc.,* vol. 73, pp. 15-23, 1990.

[4] H. Mao, M. Selleby and B. Sundman, *Calphad,* vol. 28, pp. 307-312, 2004.

[5] D.-G. Kim, M.-A. Van Ende, C. Van Hoek, C. Liebske, S. Van der Laan and I.-H. Jung, *Metall. Mater. Trans.,* vol. 43B, pp. 1315-1325, 2012.

[6] A. Pelton, S. Decterov, G. Eriksson, C. Robelin and Y. Dessureault, *Metall. Mater. Trans.,* vol. 31B, pp. 651-659, 2000.

[7] B. Nikitin, T. Litvinova and T. Raychenko, *Russ. Metall.,* vol. 6, p. 14, 1966.

[8] B. Nikitin, *Russ. Metall.,* vol. 6, p. 133, 1970.

[9] B. Nikitin, T. Litvinova and T. Raychenko, *Russ. Metall.,* vol. 3, p. 53, 1972.

[10] K. Mills and B. Keene, *International Metals Review,* vol. 2, pp. 21-67, 1981.

[11] M. Hillert, *J. Alloys Compd.,* vol. 320, pp. 161-176, 2001.

[12] M. Hillert, B. Jansson, B. Sundman and J. Ågren, *Metall. Trans.,* vol. 16A, pp. 261-266, 1985.

[13] B. Sundman, *Calphad,* vol. 15, pp. 109-119, 1991.

[14] M. Hillert, B. Sundman and X. Wang, *Metall. Trans.,* vol. 21B, pp. 303-312, 1990.

[15] B. Hallstedt, M. Hillert, M. Selleby and B. Sundman, *Calphad,* vol. 18, pp. 31-37, 1994.

[16] H. Mao, O. Fabrichnaya, M. Selleby and B. Sundman, *J. Mater. Res.,* vol. 20, pp. 975-986, 2005.

[17] H. Mao, M. Selleby and B. Sundman, *J. Am. Ceram. Soc.,* vol. 88, pp. 2544-2551, 2005.

[18] B. Sundman, *J. Phase Equilibria,* vol. 12, pp. 127-140, 1991.

[19] M. Selleby and B. Sundman, *Calphad,* vol. 20, pp. 381-392, 1996.

[20] M. Selleby and M. Hillert, *J. Phase Equilib.,* vol. 20, pp. 288-294, 1999.

[21] R. Ries and K. Schwerdfeger, *Arch. Eisenhuttenwes.,* vol. 51, pp. 123-129, 1980.

[22] G. Zhmoidin and A. Chatterjee, *Russian Metallurgy, Metally,* vol. 6, pp. 37-42, 1971.

[23] C. Kun and Y. Sheng, *Acta Chem. Sinica,* vol. 30, pp. 381-385, 1964.

[24] A. Mitchell and B. Burel, *J. Iron Steel Inst.,* vol. 208, p. 407, 1970.

[25] L. Hillert, *Acta Chem. Scand.,* vol. 19, pp. 2436-2438, 1965.

[26] W. Eitel, *Zement,* vol. 27, pp. 455-459, 1938.

[27] P. Budnikov and S. Tresvyakskii, *Dokl. Akad. Nauk SSSR,* vol. 89, pp. 479-482, 1953.

[28] T. Baak, *Acta Chem. Scand.,* vol. 8, p. 1727, 1954.

[29] T. Baak, *Acta Chem Scand.,* vol. 9, p. 1406, 1955.

[30] W. Phillips and J. Hanlon, *J. Am. Ceram. Soc.,* vol. 46, pp. 447-449, 1963.

[31] Y. Hsu, C. Hsi and L. Fu-lien, *Acta Metallurg. Sinica,* vol. 7, pp. 24-31, 1964.

[32] J. Mukerji, *Mém. sci. Rev. Metallurg.,* vol. 60, pp. 785-796, 1963.

[33] J. Mukerji, *J. Am. Ceram. Soc.,* vol. 48, pp. 210-213, 1965.

[34] B. Rossing, "Ph. D. Thesis," MIT, 1966.

[35] G. Zhmoidin, *Izv. Akad. Nauk. SSSR,* vol. 6, pp. 9-16, 1969.

[36] E. Schlegel, *Cercetari Metallurgice,* vol. 9, pp. 785-791, 1967.

[37] E. Schlegel, *Silikattechnik,* vol. 20, pp. 93-95, 1969.

[38] G. Kor and F. Richardson, *Trans. Metallurg. Soc. AIME,* vol. 245, pp. 319-327, 1969.

[39] H. Kojima and C. Masson, *Canad. J. Metallurg.,* vol. 47, pp. 4221-4228, 1969.

[40] M. Davies and F. Wright, *Chem. Ind. (London),* vol. 14, pp. 359-363, 1970.

[41] R. Hawkins, S. Meherali and M. Davies, *J. Iron Steel Inst.,* vol. 209, pp. 646-657, 1971.

[42] G. Zhmoidin and A. Chatterjee, *Zavodsokaja Laboratorija,* vol. 38, pp. 186-189, 1972.

[43] D. Edmunds and J. Taylor, *J. Iron Steel Inst.,* vol. 210, pp. 280-283, 1972.

[44] I. Isaksson and K. Öberg, *Scand. J. Met.,* vol. 2, pp. 1-4, 1973.

[45] W. Bollmann, *Krist. Technol.,* vol. 12, pp. 941-944, 1977.

[46] W. Bollmann, *Cryst. Res. Technol.,* vol. 17, pp. K107-108, 1982.

[47] A. Zaitsev, N. Korolev and B. Mogutnov, *Akad. Nauk SSSR Rasplavy,* vol. 3, pp. 58-64, 1989.

[48] J. Wramplemeyer, S. Dimitrov and D. Janke, *Steel Res.,* vol. 61, pp. 1-7, 1990.

[49] W.-G. Seo, D. Zhou and F. Tsukihashi, *Mater. Trans.,* vol. 46, pp. 643-650, 2005.

[50] D.-G. Kim, C. van Hoek, C. Liebske, S. van der Laan, P. Hudon and I.-H. Jung, *ISIJ Int.,* vol. 52, pp. 1945-1950, 2012.

[51] W. Eitel, *Zement,* vol. 30, pp. 17-21, 1941.

[52] A. Chatterjee and G. Zhmoidin, *J. Mater. Sci.,* vol. 7, pp. 93-97, 1972.

[53] R. Nafziger, *High Temperature Sci.,* vol. 5, pp. 414-422, 1972.

[54] A. Mitchell and J. Cameron, *Metallurg. Trans.,* vol. 2, pp. 3361-3366, 1971.

[55] R. Ries, "Ph. D. Thesis," Max-Planck-Institute fur Eisenforschung, Clausthal, 1979.

[56] D. Gohil and K. Mills, *Arch. Eisenhuttenwes.,* vol. 52, pp. 335-340, 1981.

[57] P. Kewu, Z. Peng, X. Jiangguo and M. Heli, *Applied Mechanics and Materials,* Vols. 117-119, pp. 269-272, 2012.

[58] J.-O. Andersson, T. Helander, L. Höglund, P. Shi and B. Sundman, *Calphad,* vol. 26, pp. 273-312, 2002.

[59] W. Gutt, A. Chatterjee and G. Zhmoidin, *J. Mater. Sci.,* vol. 5, pp. 960-963, 1970.

Proceedings of the 2013 International Symposium on Liquid Metal Processing & Casting
Edited by: Matthew J. M. Krane, Alain Jardy, Rodney L. Williamson, and Joseph J. Beaman
TMS (The Minerals, Metals & Materials Society), 2013

INFLUENCE OF THE POLARITY ON THE CLEANLINESS LEVEL AND THE INCLUSION TYPES IN THE ESR PROCESS

Armin Paar[1], Reinhold Schneider[2], Peter Zeller[2], Gerhard Reiter[3], Stefan Paul[3], Ingo Siller[3], Paul Würzinger[3]

[1]Eisenwerk Sulzau-Werfen, R&E Weinberger AG; Bundesstraße 4, 5451 Tenneck, Austria

[2]University of Applied Sciences Upper Austria, School of Engineering and Environmental Sciences; Stelzhamerstraße 23, 4600 Wels, Austria

[3]Böhler Edelstahl GmbH & Co KG; Mariazeller Straße 25, 8605 Kapfenberg, Austria

Keywords: electroslag remelting, cleanliness level, nonmetallic inclusions, power supply

Abstract

This paper describes the effect of different electrical power modes on the cleanliness level and the inclusion types in the ESR material X38CrMoV5-1. Therefore selected ingots, which were remelted in an open laboratory ESR plant, were investigated with SEM (combined with EDX) and with a conventional light microscope according to ASTM E45, method D. For these investigations, ingots with two DC polarities (cathodic and anodic) and with low frequency (1 Hz) AC power supplies were remelted under (apart from the polarity) comparable remelting conditions.

The results show, that the cleaning effect of both DC polarities is lower than the one with a low frequency AC operation, whereas the cathodic DC polarity results in better cleanliness levels than the anodic one. In addition the oxygen content of the ingots shows a strong increase when remelting with the anodic DC polarity.

A strong relation between the type of the electric power supply and the inclusion types in the ESR ingots could be found. The most pronounced differences occur with the oxide-type inclusions. Thereby, the amount of the originally dominating (Al, Mg)-spinell-type-oxides is further increased when remelting with DC-, but rather eliminated in favor of Al_2O_3-type inclusions when DC+ polarity is used. Regarding the oxysulfide inclusions, the examined electrical set-ups lead to Al-oxysulfides as the main part, while Mg- and Ca-oxysulfides are decreased respectively eliminated. Furthermore, a significant reduction of sulfides, especially CaS was found.

A comparison with industrial ESR ingots indicates that the results gained with the laboratory ESR plant are in good agreement with or even slightly better concerning the cleanliness than larger scale 50Hz-AC remelted ingots.

Introduction

The electroslag remelting process (ESR-process) has developed into an integral part in the production of steels, especially of special steels (such as tool steels) and nickel-based alloys. As time passed on, the main challenge of the ESR-process changed from desulfurization in the 1960^{ies} to the improvement of the cleanliness level, while the desulfurization treatment was transferred more and more towards the secondary metallurgy nowadays.

Regarding the improvement of the cleanliness level, early tests already showed the great potential of the power-supply type (e.g. [1-6]). Newer research focused more and more on the specific influence of low frequency AC and DC power supply and found specific dependences of the cleaning effect from the frequency and polarity (e.g. [7, 8]).

The laboratory ESR plant installed at the School of Engineering and Environmental Sciences at the University of Applied Sciences Upper Austria provides, amongst others, the possibility to feed the remelting electrode with a rather flexible polarity. This can be DC-, DC+ or a low frequency AC current (rectangular form, up to < 5 Hz). The results presented here are based on a cooperation with the special steel producer Böhler Edelstahl GmbH & Co KG, in the course of which the cleanliness levels of three power supply types (DC-, DC+, AC 1 Hz) were investigated. The analysis of the inclusions was carried out by automated SEM, where the inclusion-size distribution and the inclusion types were measured. These analyses were supplemented by the standardized optical method according to ASTM E45, method D, for comparison of these analysis methods.

Overview of the Influences on the Cleanliness Level in the ESR Process

Summarizing literature reports from the last 60 years of ESR research relating to the cleanliness level, several influences on the cleanliness can be pointed out. Those are, for instance, the deoxidation of the electrode material [9], multiple ESR [10], pressure or protective gas electroslag remelting (PESR) (e.g. [3, 11-13]), the size of the inclusions in the electrode [9, 14, 15] and the composition and viscosity of the slag (e.g. [15-19]).

Additionally to these factors, the process parameters (melt rate, current, voltage) and the power supply (AC, DC, frequency) play a decisive role in the removal of nonmetallic inclusions and will be described in more detail.

Influence of the melt rate on the cleanliness level

According to [10, 15], operating the ESR process with a lower melt rate generally results in a better solidification structure and an improved removal of nonmetallic inclusions. This is stated with an easier ascending of nonmetallic inclusions and gases in the steel. Furthermore, a lower melt rate provides more time for metallurgical reactions between steel and slag and favors a vertical solidification, which, on its own, decreases the "freezing" of the nonmetallic inclusions during the solidification.

On the contrary, [20] reports a positive effect of a *higher* melt rate, based on the fact, that when increasing the electrode speed, the immersion depth in the slag increases as well. This means, the hottest zone of the electrode and the one of the slag bath is moved to the center of the slag bath, which reduces the oxidation of the electrode tip.

Influence of the polarity and the frequency on the cleanliness level

An ESR plant can be supplied by alternating or direct current. When using a DC power supply, both varieties can be used: the anodic (DC+, DC-SP) and the cathodic (DC-, DC-RP) polarity. [14, 15, 17, 21-23]

If the electrode is fed by an anodic current, anions such as S^{2-}, O^{2-}, N^-, OH^- are pulled towards the electrode. These anions are absorbed to the liquid metal film on the electrode tip and transported together with the metal droplet into the metal pool below the slag. This leads to a poor cleanliness level concerning the oxygen and the sulfur content. [1, 2, 8]

Supplying the electrode with a cathodic current leads to a good desulfurization and a high cleanliness level. On the other hand, the cathodic polarity is connected with a poor melt rate (a high specific power consumption), which makes an ESR process with cathodic current economically irrelevant. [1, 2, 8]

Concerning these mentioned effects (cleanliness level and melt rate), the supply with an alternating current represents a very good compromise with a high melt rate, but no electrochemical reactions can occur when using standard AC. [23]

It has to be mentioned that it is supposed that electrochemical reactions are dependent to the electrode and ingot diameter. So, without providing further explanations, [2] states that at diameters larger than 1000 mm at a filling ratio of 0,6, no electrochemical reactions are expected for DC superposed alternating current power supplies.

When using an alternating current, the frequency plays an important role concerning the cleanliness level. [7] reports for an open laboratory ESR plant that with an increase of the frequency from 2,5 to 50 Hz, the oxygen content of the ESR ingot decreases, while the nitrogen content seems to be stable. Also the cleanliness level decreases with an increasing frequency. These results are explained by reaction (1) which has more time to take place at low frequencies.

$$O^{2-} - 2e^- \leftrightarrow [O] \qquad (1)$$

Investigation parameters

Laboratory ESR plant

The ingots were remelted with an open laboratory ESR plant, shown in Figure 1. The installation and the startup of this plant with first results concerning the influence of the power supply were already published in [24] and [25].

Figure 1. Laboratory ESR plant of the University of Applied Sciences Upper Austria.

The power supply is realized with a rectifier, which offers the possibility to feed the electrode with direct current (both polarities) or a (almost rectangular) low frequency alternating current (maximal frequency up to 5 Hz). The maximal current is 5000 A, at a maximal voltage of 100 V.

The electrodes are remelted in a water-cooled copper mold, fixed on a likewise cooled copper baseplate. The water flow in the mold and the baseplate was about 125 l/min. The mold had an upper diameter of 159 mm, a bottom diameter of 171 mm (cone-shaped) and a height of 775 mm. All ingots were remelted without a protective atmosphere.

Feedstock

As electrode material, the hot work tool steel grade X38CrMoV5-1 (Böhler W300 IsoDisc) was used. All electrodes were from the same heat. The electrode diameter was 101,5 mm and the length 1200 mm. The chemical composition of the electrode material is shown in Table I.

Table I. Chemical composition of the electrode material.

C [%]	Si [%]	Mn [%]	P [%]	O [ppm]
0,40	1,09	0,44	0,02	9
Cr [%]	Mo [%]	V [%]	S [%]	
4,97	1,14	0,34	<0,01	

The slag used for these investigations was the Wacker Electroflux ESR 2015 with the following composition in mass-% (Table II).

Table II. Composition of the slag Wacker ESR 2015. [26]

% SiO_2	% Al_2O_3	% CaF_2	% CaO	% MgO
1,5 ± 0,5	33,5 ± 2,5	31,5 ± 2,5	29,5 ± 2,5	3,0 ± 1,0
% TiO_2	% FeO	% H_2O*	% C	% Pb
≤ 0,2	≤ 0,2	≤ 0,06	≤ 0,06	≤ 0,005
% P	% S			
≤ 0,005	≤ 0,04			

* at 650°C

The height of the slag bath was about 80 mm (when 5 kg slag were used) respectively 125 mm (for 7,5 kg). The remelting processes were started with granulated cold slag.

Experimental parameters

Table III gives an overview on the investigated ingots. It can be seen that the melt rate of the most ingots lies in the range of 40-55 kg/h (~ 50 kg/h) with one exception which was remelted at a higher melting rate (100 kg/h). The designated polarity for the DC experiments (DC+, DC-) refers to the polarity of the electrode. The ingots with the denominations AC1, AC2 and AC3 can be regarded as identical to demonstrate the repeatability of the results. For all other experimental parameters only single ingots were produced.

Table III. Overview of the investigated ingots.

Designation	Polarity	Melt rate [kg/h]	Slag amount [kg]
DC-	DC-	40	5
DC+	DC+	52	5
AC1	AC 1 Hz	55	5
AC2	AC 1 Hz	53	5
AC3	AC 1 Hz	53	5
AC-high	AC 1 Hz	100	7,5

The remelted ingots were forged and analyzed at Böhler Edelstahl GmbH & Co KG. The analyses were carried out with an automated SEM, where both the inclusion type and the inclusion size were determined at an area of about 168 mm². Their types were analyzed with an EDX system and the size was calculated with the equivalent circle diameter (ECD). Additionally the ingots were investigated by automated light microscopy (according to ASTM E45, method D). The investigations included also the determination of the oxygen content of the remelted ingots by hot-extraction-analysis (LECO) and a chemical analysis of the slag after remelting.

Results

Analysis of the inclusion-size distribution (ECD with SEM)

Figure 2 shows the distribution of the inclusion in the DC- ingot. For comparison the scatter band of the inclusions size distribution of all investigated electrodes is included in this figure. It can be noted, that smaller inclusions <5 µm are built up during remelting, while larger inclusions are eliminated. Inclusions larger than 10 µm could not be found in the remelted ingot.

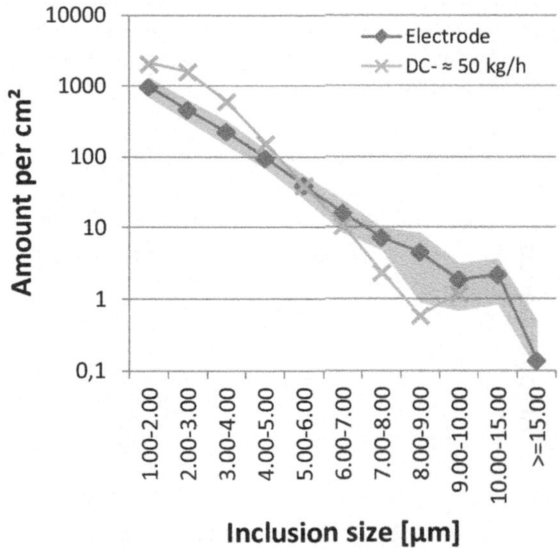

Figure 2. Inclusion-size distribution (ECD) of the cathodic DC feed (DC-, ≈50 kg/h).

In a similar way, Figure 3 shows the inclusion-size distribution of the anodic (DC+) remelted ingot. Compared with the DC- polarity it can be seen clearly, that a poorer cleanliness level was reached. Furthermore, the results illustrate that, in comparison with the electrode, more inclusions were found in all classes between 2 and 8 µm. No inclusions larger 15 µm were detected in the remelted ingot.

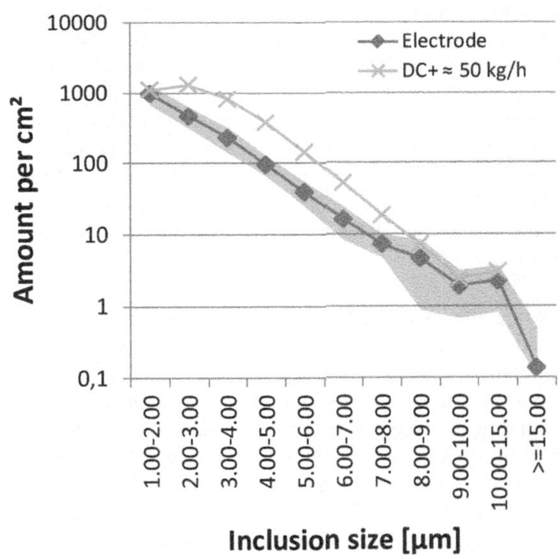

Figure 3. Inclusion-size distribution (ECD) of the anodic DC feed (DC+, ≈50 kg/h).

Regarding the inclusion-size distribution in Figure 4, it can be seen that the low frequency feed with alternating current result in the best cleanliness levels. Both the ingots with a melt rate of ≈ 50 kg/h and the one with ≈ 100 kg/h result in a strong improvement regarding the inclusion sizes. That means, that the amount of small inclusions is increased, while inclusions larger than 4 µm are decreased and no inclusions > 10 µm were found in the remelted ingots. The differences between the 2 melting rates (and slag heights) are largely within the scatter of the testing and measurement method.

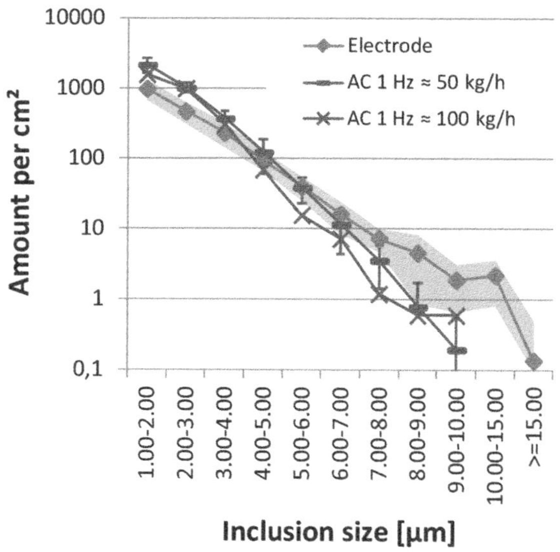

Figure 4. Inclusion-size distribution (ECD) of the low frequency AC feed (AC 1 Hz, ≈50 / ≈100 kg/h).

Analysis of the inclusion types (SEM+EDX)

The ingots have been analyzed with SEM+EDX to gain information about the inclusion composition respectively their types. The chart in Figure 5 shows the development of the main inclusion types "oxides", "oxysulfides" and "sulfides". Compared with the electrode material it can be noted, that the oxide-type as well as the oxygen content in the ingot experienced an increase at all polarities. The highest rise for both was detected for the anodic (DC+) polarity. In contrast, the amount of oxysulfidic but especially the sulfidic inclusions decreased in all cases. Overall, the best results could be achieved with the low frequency AC polarity.

Regarding the oxidic subtypes in Figure 6, one can see that the electrode material contained mainly (Al,Mg)-oxides (spinell-type) and a small amount of (Al,Ca,Mg)-oxides as well as some alumina inclusions. This composition changed during the remelting with different polarities. While the percentage of the oxidic subtypes remained quite the same at the DC- polarity, the (Al,Mg)-type oxides were nearly completely replaced by Al_2O_3-inclusions when remelting with DC+ polarity. Additionally the percentage of the (Al,Ca)-oxides increased slightly. Looking at the oxidic subtypes found after remelting with AC 1 Hz polarity, it can be noted, that both the share of (Al,Ca,Mg)- and of (Al,Ca)-type oxides increased. In consequence mainly the percentage of the (Al,Mg)-oxides decreased.

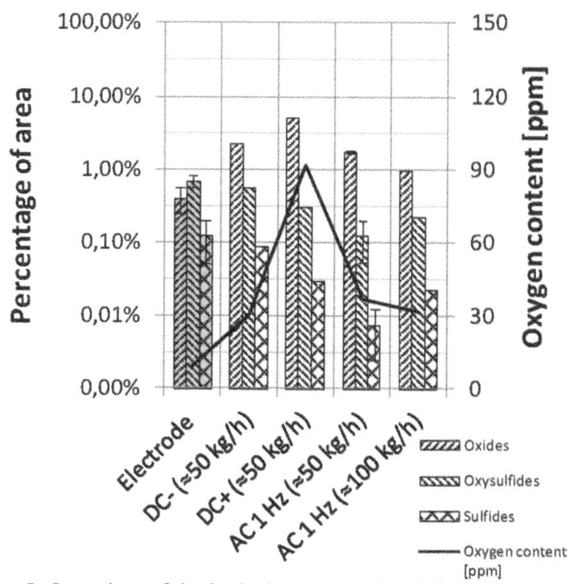

Figure 5. Overview of the inclusion types with different polarities.

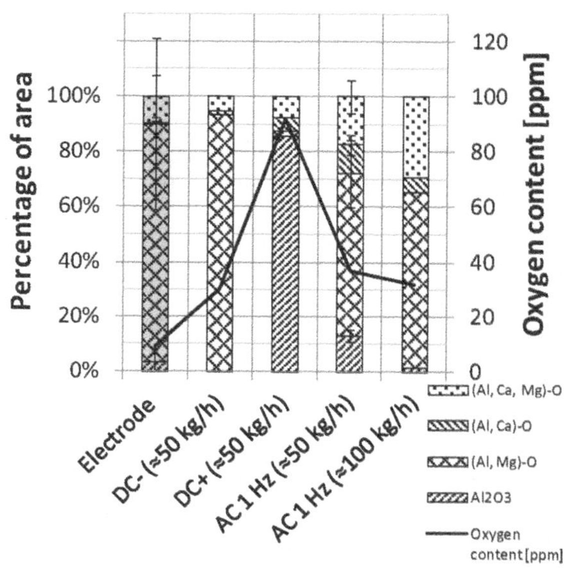

Figure 6. Oxidic inclusion subtypes according to the polarity.

The change of the oxysulfidic subtypes in the ingots, shown in Figure 7, is quite similar for all investigated polarities. While the electrodes contained about 35 % Al- and Ca-oxysulfides and the residual part Mg-oxysulfides, all ingots showed around 90% oxysulfides of the Al-type. The second type which increased were the Mn-oxysulfides. In consequence the percentage of Ca-oxysulfides decreased strongly and the Mg-oxysulfide type was almost completely eliminated.

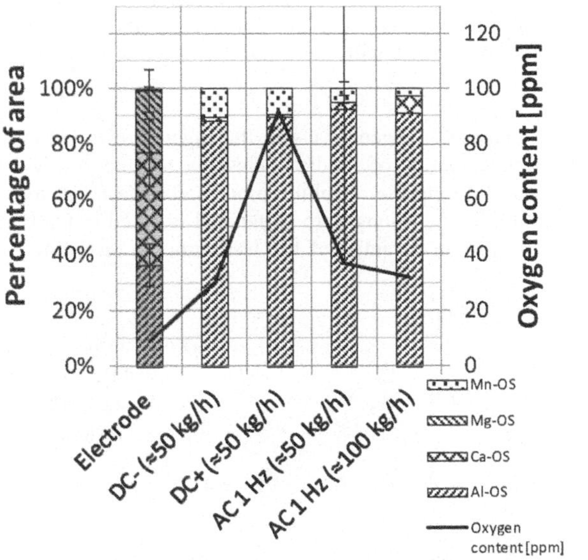

Figure 7. Oxysulfidic inclusion subtypes according to the polarity.

Regarding the sulfidic inclusions in Figure 8, the main type in the electrodes with an average content of about 70 % was CaS, the residual percentage is divided between MnS and (Ca,Mn)-sulfides. After remelting, the share of CaS decreased in all ingots, while the percentage of the MnS and the (Ca, Mn)-sulfides increased. This effect was stronger pronounced after remelting with a DC power supply.

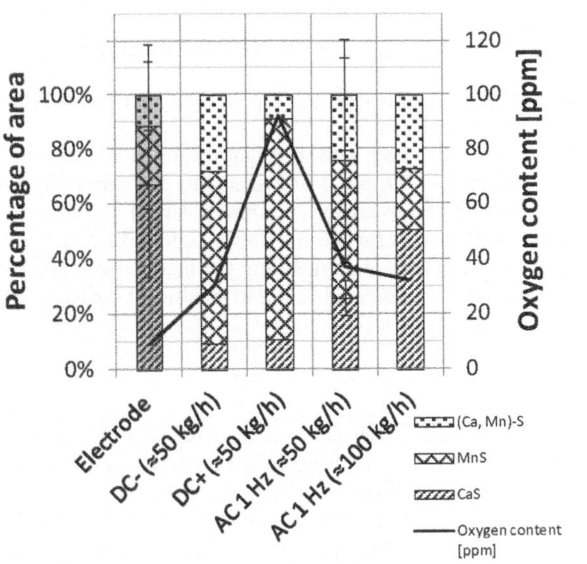

Figure 8. Sulfidic inclusion subtypes according to the polarity.

Impurity-levels

To simplify the comparison of the results based on the measurements according to ASTM E45, method D, the parameter "impurity-level" was established. This impurity-level is calculated with the following formulas (2) to (4). In the first step, a specific factor (size indication) is calculated with the first quartile of the upper and lower borders of the ASTM E45 severity levels. This factor is multiplied with the number of the fields found (Method D). The resulting "single-levels" are summed up to receive the impurity-levels for each inclusion type (A,B,C,D; each: thin and heavy).

$$\text{Factor}_{\text{Type B,D, X}} = 1^{st}\text{ quartile}(X,Y) \quad (2)$$

X... Severity Level lower border [µm]
Y... Severity Level upper border [µm]

$$\text{Single-levels}_{\text{Type B,D, X}} = \text{Factor}_{\text{Type B,D, X}} * \text{Amount}_{\text{Fields}} \quad (3)$$

$$\text{Impurity-level}_{\text{Type B,D}} = \sum \text{Single-levels}_{\text{Type B,D, X}} \quad (4)$$

This approach results in only one figure for each inclusion type. The inclusion types found in the ingots and electrodes were only from type B (alumina) and D (globular oxides).

The impurity-levels for the investigated polarities are shown in Figure 9. It's obvious that the electrode material contains many inclusions from the type Dt (D-thin) and several from the type Dh (D-heavy), while almost no alumina inclusions were found. Generally all investigated polarities result in better cleanliness levels in the ingots than in the electrodes. The best results were achieved with the AC varieties, with all inclusions from the type B being eliminated. The highest impurity-levels were found when remelting with the DC+ polarity, while DC- led to almost similar results as AC.

Comparing these results with measurements on material produced on an industrial scale (open mould, AC 50Hz), shows that both types of electric power supply, DC- and AC (1Hz) at the laboratory plant lead to similar or slightly better results.
Regarding the oxygen content in the ingots, an obvious relation between the oxygen content and the cleanliness level can be noted. The highest oxygen contents were found when remelting with the anodic (DC+) polarity, the same polarity with the poorest cleaning effect.

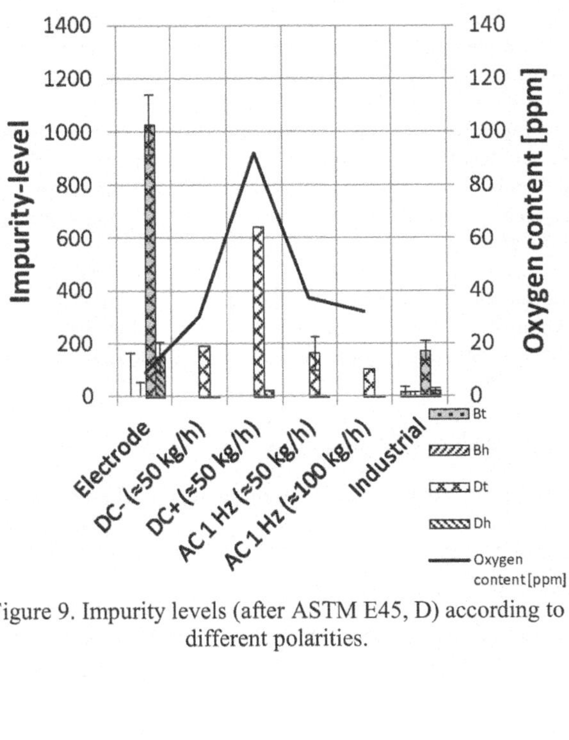

Figure 9. Impurity levels (after ASTM E45, D) according to different polarities.

Slag composition

The chemical composition of the slag after remelting is summarized in Table IV. It can be seen, that the contents of CaF , Al_2O_3 and MgO seem to be quite stable and independent from the type of the power supply. They are also in the same range as the initial slag composition. Regarding the contents of SiO_2 and FeO, it's obvious that there's a rather strong increase in amount. Both of the DC power supplies favor the highest contents of SiO_2 while the increase when using the AC power supply is lower especially for the higher melt rate. In contrast the percentage of CaO decreases during remelting. The highest and most obvious FeO pickup can be noted when remelting with DC+ (anodic) power supply.

Table IV. Slag composition after remelting with different power supplies.

Designation	CaF_2 Mass-[%]	Al_2O_3 Mass-[%]	CaO Mass-[%]	MgO Mass-[%]	SiO_2 Mass-[%]	FeO Mass-[%]
Original slag	31,5 ± 2,5	33,5 ± 2,5	29,5 ± 2,5	3,0 ± 1,0	1,5 ± 0,5	≤ 0,2
DC-	31,1	31,4	21,7	2,7	12,8	< 0,2
DC+	31,9	29,4	22,9	2,9	11,4	0,5
AC1	32,8	32,5	22,8	3,0	8,6	< 0,2
AC2	32,1	35,3	21,6	2,8	8,1	< 0,2
AC3	28,8	33,8	27,0	2,4	7,8	< 0,2
AC-high	29,2	33,1	29,1	2,8	4,7	< 0,2

Discussion

The size-distribution of nonmetallic inclusions

The determined size-distribution with the SEM (ECD) shows the expected tendency, which means, the amount of small inclusions is the highest and decreases with increasing inclusion diameter almost linearly on the logarithmic scale as long as classes of 1 µm are used (Figure 10). This is in good agreement with earlier investigations such as [27-29] and also with new ones with modern analysis methods [30, 31], whereby in the latter publication the maximum amount of inclusions was found between 3-5 µm which may be related to the detailed parameters of the analysis set-up.

Additionally the increase of the amount of smaller inclusions, combined with an elimination of larger inclusions, which was reported, amongst others in [29, 32], could be observed and confirmed.

Oxygen content and slag composition

The relation between the oxygen content in the ingots and the polarity was already investigated in [1, 2, 33]. The determined oxygen contents, with highest values for the CD+ polarity, are in perfect agreement with these results.

The fact that the oxygen level in the remelted ingots is higher than in the electrodes can be explained by the operation of the laboratory ESR plant without protective atmosphere, which means, that the slag and the hot part of the electrode are in permanent contact with air. This corresponds well with results in [17].

The increased oxygen levels in the ingots are in direct relation with the amount of oxidic inclusions (Figure 5), which also increased during the remelting process. The fact that oxysulfidic and sulfidic inclusions were eliminated is based in the very good desulfurization potential of the ESR process according to [1, 2, 16, 17].

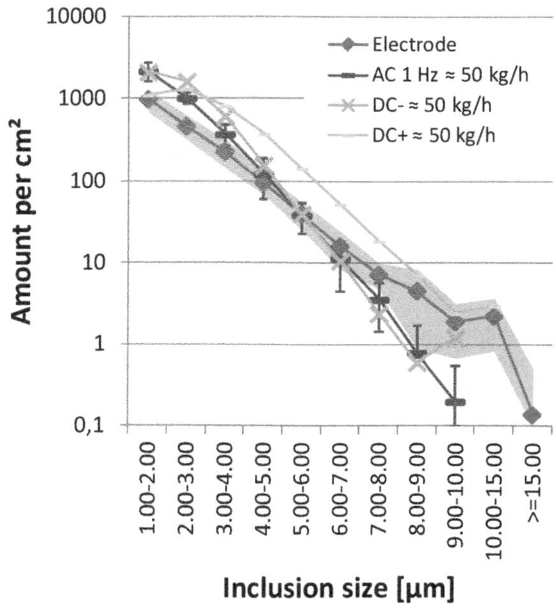

Figure 10. Inclusion-size distribution of all investigated materials.

The pickup of oxygen into the ingot results in a loss of Si, which is transferred to the Slag. SiO_2-contents are especially high for DC- polarity (followed by the DC+ polarity) and decrease with rising melting rates for the AC power supply. In case of the DC+ polarity, the uptake of oxygen into the ingot is so high, that even iron is partially oxidized, leading to significant contents of FeO in the slag.

Inclusion types

According to [2, 33] electrolytic reactions can be expected when working with ESR-plants under certain electrical parameters and geometrical proportions similar to those of the laboratory ESR-plant used for these experiments. When remelting with direct current, the following electrolytic reactions can take place at the polarized surfaces (see equations (5) to (10) according to [7, 8]).

$$O^{2-} - 2e^- \leftrightarrow [O] \quad (5)$$
$$(Si^{4+}) + 4e^- \leftrightarrow [Si] \quad (6)$$
$$(Mn^{2+}) + 2e^- \leftrightarrow [Mn] \quad (7)$$
$$(Fe^{2+}) + 2e^- \leftrightarrow [Fe] \quad (8)$$
$$(Ca^{2+}) + 2e^- \leftrightarrow [Ca] \quad (9)$$
$$(Al^{3+}) + 3e^- \leftrightarrow [Al] \quad (10)$$

These reactions form the basis for the changes in the inclusion types. When the electrode is polarized negatively (cathodic, DC-), predominantly equation (10) takes place at the electrode, which is the surface with the higher current density. Thereby the increased aluminum content is obvious in the analysis of the ingots, which does have no significant effect on the main inclusion type.

When the electrode is polarized positively (anodic, DC+), predominately equation (5) takes place, leading to an increase of the oxygen-content and an oxidation of Al and therefore higher Al_2O_3 content in the ingot, which was confirmed by chemical analysis. Table V summarizes these changes in the chemical composition and type of inclusions.

Table V: Changes of the oxygen and aluminum contents in the ingots according to the polarity.

	DC+	AC 1 Hz	DC-
Main type of inclusion	Al_2O_3 ↑↑	Al_2O_3 ↑ (Al,Ca)-O ↑ (Al,Ca,Mg)-O ↑	(Al,Mg)-O ↑
Oxygen content	[O] ↑↑	[O] ↑	[O] ↑
Aluminum content	[Al] ↑	[Al] ↑↑	[Al] ↑↑↑

According to [34], the high-MgO inclusions are eliminated or react with Al_2O_3 to spinal inclusions. Therefore, no pure MgO and Al_2O_3 inclusions should be found in the ingot after an ESR treatment. According to the SEM+EDX analysis, in the case of Mg-Oxysulfides, this could be confirmed in this paper (Figure 7).

Comparison between the SEM and the ASTM analyses

When comparing the analysis methods "SEM" and "light optical microscope" (ASTM), it's obvious that both methods prove, that larger inclusions are eliminated. Regarding the smaller (thinner) inclusions the results of the SEM-analysis show an increase of smaller inclusions, while measurements according to ASTM-E45 lead to lower contents of all types of inclusions. This fact may be explained by the specification of the ASTM-method. Thereby inclusions of the type B with a length < 17 μm and a width of < 2 μm (≈ ECD 6,5 μm) and inclusions of type D < 2 μm are neglected. [35] Furthermore, it has to be taken into consideration that inclusions of 2, 3 or 4 μm are generally difficult to distinguish at the typical magnification applied for the light optical analysis (see also [30]).

The inclusion types which were found with the two analysis methods are shown in

Table VI. It can be noted that with the optical microscope (ASTM), mainly globular oxides (type D) have been detected. This correlates well the SEM analyses and the dominance of (AlMg)-O-spinal-type inclusions. The only exceptions from this accordance are the alumina-type inclusions in the DC+ ingot. These inclusions should have been characterized as "type B" in the analyses according the ASTM.

Table VI: Comparison of the inclusion types found with SEM and ASTM.

	SEM	SEM → ASTM	ASTM E45
Electrode	(Al, Mg)-O	Type D	Type D
DC+	Al_2O_3	Type B	*Type D*
DC-	(Al, Mg)-O	Type D	Type D
AC 1 Hz	(Al, Ca)-O (Al, Ca, Mg)-O	~Type D	Type D

The explanation for this difference may be found in the small size of the alumina inclusions formed during the remelting process. Such small inclusions show no tendency to crack during plastic deformation of the ingot, thereby keeping their globular shape.

Comparison between the laboratory and the industrial ingots

The comparison between the laboratory and the industrial ingots (Böhler W300 IsoBloc # 390 mm) according to ASTM E45, indicates in excellent agreement for all types of inclusions, indicating that results from the laboratory plant can be well transferred to industrial production plants.

Acknowledgements

The authors want to thank Böhler Edelstahl GmbH & Co KG for the financial, personnel and intellectual support. Likewise, the financial support of the Austrian Forschungsförderungsgesellschaft mbH (FFG) within the Frame of the COMET-program and the "Competence Center for Excellent Technologies in Advanced Metallurgical and Environmental Process Development" (K1-Met) is gratefully acknowledged.

References

1. Jäger, H. *Schmelzflußelektrolyse beim Elektroschlacke-Umschmelz-Verfahren.* BHM 119 (1974). 11, p. 439-446.
2. Jäger, H. *Steuerung des Reaktionsablaufes beim Elektroschlacke-Umschmelzen durch Anwendung der Schmelzflussanalyse.* Dissertation, Montanistische Hochschlue Leoben, 1973.
3. Dewsnap, P. and R. Schlatter. *Process and Product characteristics of DC Electroslag Remelting of Alloy Steels.* Proceedings of Fifth International Symposium on Electroslag and other Special Melting Technologies, Oct. 16-18, 1974, Carnegie-Mellon Institute, Pittsburgh/USA p. 91ff.
4. Schlatter, R. *Application of the electroflux remelting process for high speed tool steels.* Iron and Steel International (1974), p. 197-204.
5. Schlatter, R. *Electroflux Remelting of Alloy tool Steels.* Proceedings of the 3rd International Symposium on Electroslag and other Special Melting Technologies. June 8-10 1971, Carnegie-Mellon-Institute, Pittsburgh/USA.
6. Schlatter, R. *Electroflux Remelting of Tool Steels.* Metals Engineering Quarterly 12 (1972), 1, p. 48-60.
7. Chang, L., X. Shi, H. Yang and Z. Li. *Effect of Low-Frequency AC Power Supply During Electroslag Remelting on Qualities of*

Alloy Steel. Journal of Iron and Steel Research, International. (2009), 16, p. 7-11

8. Kawakami, M, T. Takenaka and M. Ishikawa. *Electrode reactions in dc electroslag remelting of steel rod.* Ironmaking and Steelmaking 29, (2002), p. 287-292.

9. Li, Z. and Q. Wang. *The effect of electroslag refining on inclusioni compositions of ball-bearing steels.* Proceedings of the 1986 Vacuum Metallurgy Conference on Speciality Metals, Melting and Processing, June 09.-11. 1986; Pittsburgh/USA, Published 1987; ISBN 09322897088, p. 147-153.

10. Vacugov, G.A. und G.A. Antropova. *Zusammensetzung und Verteilung nichtmetallischer Einschlüsse im Block aus elektroschlacken-umgeschmolzenem Stahl SCh15.* Stal in Deutsch, (1966), 6, p. 553ff.

11. Boucher, A. *Electroslag Remelting trials in an inert atmosphere influence on the purity of a low alloy steel.* Proceedings of the Conference on Special Electrometallurgy, Kiev, 1972, Part 2. Publisher: Naukova Dumka, p. 47-62.

12. Schumann, R., U. Biebricher, H. Scholz. *Inert Gas ESR Produces Clean, Homogeneous Large Forging Ingots.* Industrial Heating. (2001), p. 39-42.

13. Koch, F., P. Würzinger and R. Schneider. *Advanced Equipment for the Economic Production of Speciality Steels and Alloys.* Compendium of the 4th Symposium on "Advanced Technologies and Processes for Metals and Alloys", June 16.-17. 1999, Frankfurt/Germany, p. 47-50.

14. Hoyle, G. *Electroslag Processes - Principles and Practice.* Essex/UK: Applied Science Publishers, 1983. ISBN 0-85334-164-8, p. 15-19, 99-107.

15. Duckworth, W.E. and G. Hoyle. *Electroslag Refining.* London/UK: Chapman and Hall, 1969.

16. Holzgruber, W. und E. Plöckinger. *Das Elektroschlacke-Umschmelzen - ein neues Verfahren zur Verbesserung der Qualität von Edelstählen.* Berg- und Hüttenmännische Monatshefte 113 (1968), 3, p. 83-93.

17. Holzgruber, W. und E. Plöckinger. *Metallurgische und verfahrenstechnische Grundlagen des Elektroschlacke-Umschmelzens von Stahl.* Stahl und Eisen 88 (1968), 12, p. 638-648.

18. Ayman, F., A. Azza, E.F. Hoda and E. Mamdouh. *Behaviour of precipitates and inclusions during ESR of nitrogen alloyed and conventional AISI M41 high speed steels.* Steel Grips 4 (2006), 4, p. 298-304.

19. Klujev, M.M. und V.M. Spicberg. *Abscheidung und Bildung nichtmetallischer Einschlüsse im Metall beim Elektroschlacke-Umschmelzverfahren.* Stal in Deutsch, (1969), 6, p. 590-594.

20. Paton, B.E., B.I. Medovar, Ju.V. Latas, B.I. Maksimovic und L.M. Stupak. *Gegenwärtiger Stand und weitere Perspektiven des Elektroschlacke-Umschmelzens.* Stal in Deutsch, (1963), 5, p. 487-492.

21. Kay, D.A.R. and J.R. Pomfret. *Removal of oxide inclusions during AC electroslag remelting.* Journal of The Iron and Steel Institute. (1971), p. 962-965.

22. National Materials Advisory Board - Commission on Sociotechnical Systems, National Research Council. *Publication NM AB-324: Electroslag Remelting and Plasma Arc Melting.* Washington, D.C./USA : National Academy of Sciences, 1976.

23. Peover, M.E. Electroslag Remelting: *A Review of Electrical and Electrochemicasl Aspects.* Journal of the Institute of Metals 100 (1972), p. 97-106.

24. Schneider, R., A. Paar, P. Zeller, G. Reiter, W. Schützenhöfer and P. Würzinger. *Aufbau und Inbetriebnahme einer Versuchs-ESU-Anlage.* Berg- und Hüttenmännische Monatshefte 156 (2011), 3, p. 112-118.

25. Paar, A., R. Schneider, P. Zeller, G. Reiter, P. Würzinger, M. Wöls and S. Paul. *Einfluss ausgewählter elektrischer Parameter auf den Elektroschlacke-Umschmelzprozess.* Proceedings of the 19th International Student's Day of Metallurgy, March 15-17 2012, Freiberg/Germany, p. 49-56.

26. Wacker Chemie AG. *Wacker Silicones - Electroflux.* Munich/Germany : Wacker Chemie AG.

27. Witek, C., E. Kaliszewski, J. Gepfert und G. Hansel. *Einfluß des Elektroschlacke- und Vakuum-Umschmelzens auf die Qualität von Wälzlagerstählen.* Stahl und Eisen, (1980), p. 32-48.

28. Jäger, H., P. Machner, O. Daghofer and P. Steinbach. *Einsatz des ESU-Verfahrens zur wirtschaftlichen Herstellung von Kaltwalzen mit gesteigerter Güte.* Berg- und Hüttenmännische Monatshefte 123 (1978), 9, p. 290-301.

29. Randak, A., A. Stanz and W. Verderber. *Eigenschaften von nach Sonderschmelzverfahren hergestellten Werkzeug- und Wälzlagerstählen.* Stahl und Eisen 92 (1972), 20, p. 981-993.

30. Schneider, R., F. Koch and P. Würzinger. *Metallurgical advances in Pressure-Electro-Slag-Remelting (PESR).* Proceedings of the International Symposium on Liquid Metal Processing and Casting 2001, Sept. 23-26 2001, Santa Fe/USA, p. 105-117.

31. Korp, J. *Einfluss der Schmelzrate auf die Charakteristiken nichtmetallischer Einschlüsse beim Elektroschlacke-Umschmelzen unter Schutzgas.* Berg- und Hüttenmännische Monatshefte 157 (2012), 5, p. 174-180.

32. Liddle, J. F. *Removal of Inclusions during Electroslag Refining.* Proceedings of the International Symposium on Chemical Metallurgy of Iron and Steel, (1973), p. 66-74.

33. Trenkler, H. und W. Krieger. *Gmelin-Durrer "Metallurgy of Iron" - Volume 9a/b "Practice of Steelmaking 3".* Berlin, Heidelberg, New York, Tokyo : Springer-Verlag, 4th Edition (1988), p. 202a-206a, 220a-258a, 178b-208b.

34. Beiler, C., M. Schwarz, H.-P. Jung and O. Ziegelmayer. *Einfluss- und Einstellgrößen auf den Reinheitsgrad bei hochlegierten Stahlwerkstoffen.* Stahl und Eisen 132 (2012), 11, p. 83-89.

35. ASTM E45: 2010 - *Standard Test Methods for Determining the Inclusion Content of Steel.*

36. Knüppel, H. *Desoxydation und Vakuumbehandlung von Stahlschmelzen.* Düsseldorf/Germany: Verlag Stahleisen GmbH, 1970.

Proceedings of the 2013 International Symposium on Liquid Metal Processing & Casting
Edited by: Matthew J. M. Krane, Alain Jardy, Rodney L. Williamson, and Joseph J. Beaman
TMS (The Minerals, Metals & Materials Society), 2013

HOT TEST AND SIMULATION OF ESR HOLLOW INGOTS FORMATION IN CURRENT SUPPLYING MOULD WITH ELECTRODES CHANGE

Xu Chen[1], Zhouhua Jiang[1], Lev Medovar[2], Fubin Liu[1], Ganna Stovpchenko[2], Borys Fedorovskii[3], Lebid Vitalii[2], Ximin Zang[4], Xin Deng[1]
[1]School of Materials and Metallurgy, Northeastern University, Wenhua Rd., Heping district, Shenyang, Liaoning, 110004, PR China
[2]E.O. Paton Electric Welding Institute of NAS of Ukraine, 11 Bozhenko street, Kyiv, 03150, Ukraine
[3]Elmet-Roll private company, P.o. Box 259, Kyiv, 03150, Ukraine
[4]University of Science and Technology Liaoning, Anshan, Liaoning, 114051, PR China

Keywords: Electroslag remelting, Electroslag casting, Hollow ingot, High Pressure Vessels, Thick wall tube, Shells, Current supplying mould, Two circuit power supply, Solidification, Mathematical modeling.

Abstract

A new electroslag remelting (ESR) technology for hollow ingot manufacture was recently commercialized due to the cooperation between Northeastern University of China and Elmet-Roll engineering company (Ukraine). The usage of T-shape current supplying mould (CSM) and ingot withdrawing system as well as the change of consumable multi electrode (electrodes) are the main features of performed process. Using the new process the 20t ESR hollow ingots (hollows) with perfect quality of surfaces and defect-free macrostructure were manufactured. Any visible defects in the external and internal surface of hollow ingot including all areas of electrodes changes were not found. It becomes to be possible due to non-stop supply of electric current via CSM from second transformer. Both the high cleanness and fine as-cast microstructure of hollow ingots were proved. Simulation of slag flow and heat transfer at ESR process was developed by using the ANSYS software. The model has been verified by measuring of geometry of liquid metal pool was revealed in macrostructure of 650/450mm hollow ingots.

Introduction

Application of hollow ingots instead of solid ones is known in the forging industry within many years [1-3]. Advantages of hollows usage are wide enough, but casting technology of such ingot is relatively more complicated. Now majority of other experts have accepted hollow ingot casting as the best way for manufacturing of heavy shells, high pressure vessels and thick wall pipe for various critical applications [4-7].

Positive effect of hollow ingot usage instead of solid one on the efficiency of final product manufacture arises due to less number of technological stages:

- In all cases both heating for piercing and piercing itself are excluded;
- An upsetting stage can be eliminated in many cases as well;
- For thick wall pipe production the usage of expensive extrusion presses can be not necessary due to the possibility to apply just common rolling operations.

The quality of products made from hollows is also better because of less heating stages (that is especially important for oxidation sensitive alloys) and of specific primary dendrite structure with low segregation.

Keeping in the mind that shells and ferrules make about 50% of production program of average open die forging shop, the urgency of technology for hollows manufacture becomes absolutely clear. It is vital problem because now a lot of heavy shells and pipes have great importance for high pressure vessels, steam generators, petrochemical and nuclear reactors [7, 8].

ESR technology for hollow ingots manufacturing (named Electroslag Casting - ESC) was developed 50 years ago by team of E.O. Paton Electric Welding Institute (PWI), Ukraine [1, 2]. Detailed overview of the ESR hollow technology developments and usage are given in the paper [9] will be presented within current symposium.

Two main reasons why the ESR hollow technology had quite limited spreading so far can be considered. First – both design and operation of furnace for hollows are more complicate. Second – the ESR furnace using traditional process to produce hollow ingot of the suitable length is too high.

Multi electrodes "candelabrum" location in narrow (comparing with ingot outer diameter) gap between inner and outer mould doesn't allow remelting with high filling ratio (ratio between cross sections of consumable electrode(s) and ingot). Very rare this ratio value is higher than 0.5. Up to now without electrodes change practice a furnace length was extremely high. For example, for manufacture of 10-12m heavy wall pipes the length of consumable electrodes is longer than 20-24m. Total furnace height makes 30m at least.

It is well known that problem of the electrodes length and furnace height limitation at ESR of solid ingots had solved due to consumable electrode change during ingot melting. But at standard ESR technology for hollow it is impossible to realize the same way as done because the high risk of inner mould pressing and jam/bite by solidifying ingot.

The usage of CSM makes it possible to change electrodes while hollow ingot melting. The essence of method consists of permanent input of power into a slag pool that is provided by the CSM within time of electrodes change operation. So, while electrodes change (no current goes from electrodes) the slag pool is heated by electric current that is going from the second power source via CSM wall to slag.

Principal diagram of new ESR process for hollow ingot manufacture with consumable electrode change is presented on Figure 1.

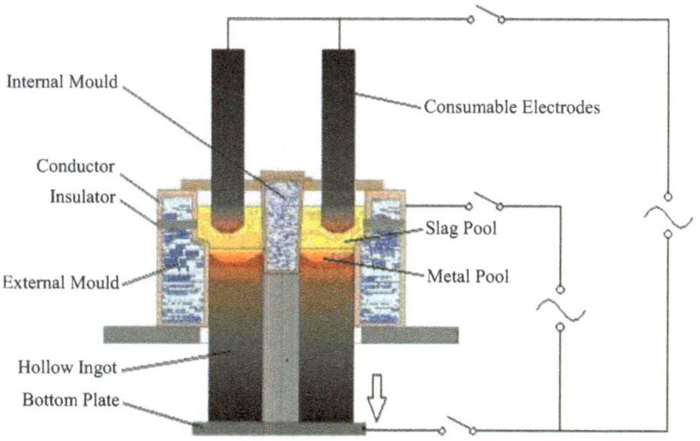

Figure 1. Principal diagrams of hollow ingots remelting with several consumable electrodes and electrodes change.

A new electroslag remelting technology and equipment for hollow ingots manufacture have mastered in industry by the cooperation between NorthEastern University of China and Ukrainian engineering company Elmet-Roll. The change of electrode, short collar mould with ingot withdrawal were applied in this ESR furnace in order to produce the solid and hollow ingots with maximum diameter 1100 mm and maximum length 6000 mm. A lot of new approaches, such as double power supply, T-shape current supplying mould, gantry type structure for furnace head and electrode movement, automatic control system based on eddy current type metal level sensor and load cell were used. This allow to guarantee a stable level of liquid metal while melting and ensure process from leakages of slag or liquid steel and internal mould jams. Due to double power supply, mould supplies heat from side of ingot during electrode change and both internal and surfaces of hollow ingot in the junction area have good quality, which was adopted for the first time in the world.

In the present paper, the preliminary results of investigation of quality of hollow ingots were produced during the hot tests of new ESR process are presented. The commercial software ANSYS was applied to calculate the fluid flow and heat transfer in the slag and metal pool in mould 650/450 mm in dia. The characteristic of heat and current transfer in the slag pool as well as some features of hollow ingot solidification were compared for two cases: with and without CSM usage while electrode change procedure.

Description of New ESR Process for Hollows

Electric diagram of new ESR method for hollow ingot manufacture consists of two power sources. One power source is connected to the electrodes are located uniformly in an annular gap between two cooling moulds. Second one is connected to upper section of current supplying mould.

The operation process for hollow ESR ingot manufacture is the following. A certain volume of liquid slag should be poured from a flux-melting furnace into the annular space between internal mould and external mould supported by the bottom plate. Then the consumable electrodes are immersed into the slag. Electric current coming from transformers goes through the multi-parallel consumable electrodes, slag bath, bottom plate, and high current loop. The resistance heat generated in slag bath stats to melt the tip of consumable electrode gradually. The molten metal drops fall down through the slag layer and form liquid metal pool, which depth is quite small because of strong cooling from internal and external moulds. The ingot withdrawal starts when the hollow ingot has reached a certain height.

In the stationary stage of hollow ingot melting the metal level should be stabilized by control of both melting rate of electrodes and speed of ingot withdrawing, based on the level sensor measurement. It is very important to avoid as a bite (jam) of internal mould as well as any leakage of slag or metal from any moulds side at melting. Therefore, it is key point to have possibilities to predict an actual situation in such complicated system that consists of a slag bath, metal pool and two water cooled moulds. This purpose new developed control system analyses a lot of data: figures of power input (voltage and current); measurements of withdrawing force by load cell; indications of metal level sensors, indicators of cooling water temperature in moulds, actual voltage between CSM sections etc. Electrode changes at hollow ingot melting are realized according prescribed procedure and takes less than 2 min.

Hot Tests and Results

Main Operation Parameters

Two sets of electrodes consisting of ten pieces of billets with the diameter of 160mm and length of 3000mm were used for hot tests to produce the hollow ingot with outer/inner diameters of 650mm/450mm (wall thickness of 100mm) and 6000 mm in length. The steel grade was 35CrMo. Multicomponent slag of CaF_2-CaO-MgO-Al_2O_3-SiO_2 system similar to ANF-29 grade was used. Addition of SiO_2 up to 10-15% improves slag's forming ability, reduces its melting point as well as increases of viscosity in solid state. At the first stage of hot tests, second power source (connected to CSM) was used while electrodes change operation only. Main power source (connected to the electrodes) worked between changes. The power input into the slag bath was approximately 1000-1100 kVA that corresponds to melting rate equal 750-850 kg/h. At electrode change operation the power input into the slag bath makes approximately 400-500 kVA.

Hot Tests Results

Figure 2 shows melting of electrodes and a hollow ingot withdrawing, respectively.

Figure 2. Hollow ESR ingot manufacturing in the shop.

Figure 3 shows general views of hollow ingot 650/450mm×6000mm, which was produced by CSM using from two sets of electrodes by the electrode change operation.

Figure 3. Appearance of hollow ingot of Φ650/450mm×6000mm.

It is obvious that the quality of both external surface and internal surface was quite good and acceptable for subsequent forging without grinding. Thus, the usage of second transformer allows us to supply power input into slag bath through CSM and to keep slag bath at high temperature in order to provide good quality of both surface and body of hollow ingot in the zone of electrode change where this operation is performed.

Figure 4 gives the comparison of appearance of hollow ingot surface quality in the cases with CSM working and without CSM working. There was a deep slag pit around the circle of ingot surface in the case of CSM power off and but no obvious defect in the junction where the operation of electrode exchange was performed in the case of CSM power on. In addition, preheating of electrode tip before electrode exchange also helped the temperature of slag bath not to decrease rapidly due to absorbing the heat by electrodes.

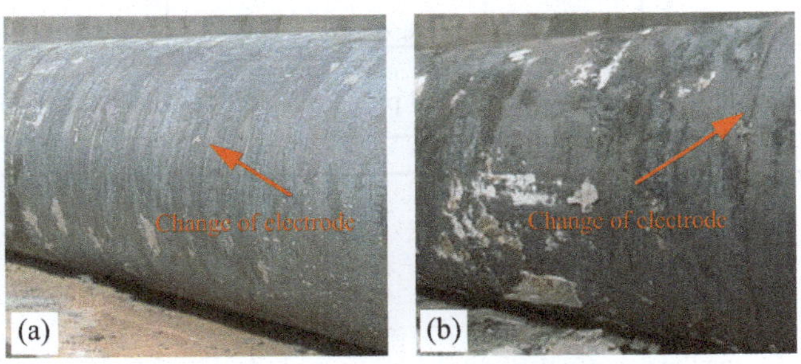

Figure 4. Surface quality of ingot with electrode exchange
(a) with CSM; (b) without CSM.

Investigation of Macrostructure and Cleanness of Hollow Ingot

The longitudinal and cross sectional templates of hollow ESR ingot Φ650/450mm×6000 mm were prepared by machine cutting, grinding, polishing and acid etching for macrostructure (Figure 5) investigation.

The macrostructure was very dense, fine grained and free of macro segregation. The absence of any visible defects either in the cross section or in the longitudinal section was proved. The direction of dendrite growth from outer mould (that was good visible in the longitudinal section) has approximately 45° angle to horizon, which was benefit to solidification quality of ESR ingot in general. The angle of direction of dendrite growth from inner mould side was a little bit less than for ones from the main mould side.

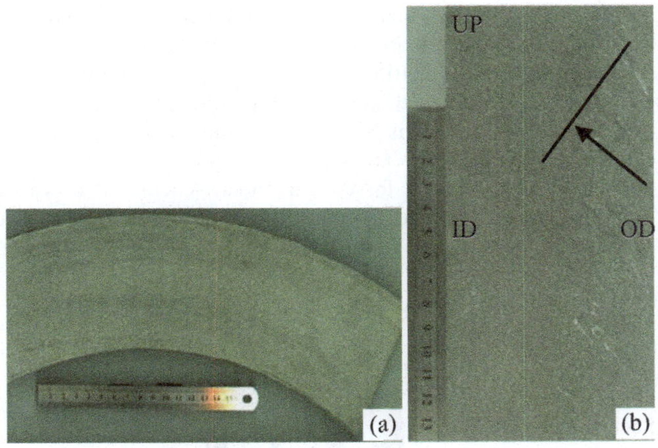

Figure 5. Macrostructure of hollow ESR ingot of Φ650/450mm×6000mm
(a) cross section; (b) longitudinal section.

Figure 6. Nonmetallic inclusions distribution in ESR hollow ingot, 50×.

Figure 6 shows the typical picture of nonmetal inclusion distribution in the steel sample was taken in the wall center of hollow ingot melted under stable stage of melting. Revealed nonmetallic inclusions were very fine and distribute uniformly.

ESR melting regimes on the stationary stage of process were in the range from 750 kg/h to 850 kg/h that corresponds to speed of ingot withdrawing 9-11 mm/min. So, hollow ingot can be formed much faster than for solid ingot with the same area of cross section (the usual melting rate is 5-7 mm/min).

Mathematical Simulation of Hollow Ingot Formation at ESR Process

Mathematic Modeling

The description of ESR process for hollow ingot formation in CSM should be formulated as conjugate problem of heat and mass transfer under strong interference of thermal, electrical and magnetic fields. First of all, 3D geometric model of actual mould is built. The whole area mould is divided by finite element grid using mapping principle, as shown in Figure 7. Boundary and initial conditions is formulated as well as simulated medias are defined by assigning of proper physical properties of slag and steel etc. The current densities are calculated according to actual current at ESR process and these data are used for determination of magnetic field and joule heat distribution in slag pool. Finally, the calculation result of magnetic force and joule heat as a volume force and volume heat are written for flow field definition by coupled calculate the distribution of speed and temperature to the convergence. The k-ε method is used to represent turbulence. Heat flow coupling algorithm is SIMPLE algorithm and the governing equations are solved by TDMA (tridiagonal matrix algorithm). Simplified flow diagram for computational scheme is shown in Figure 8. The method and procedure of establishing of this model is similar to our previous works [10-12].

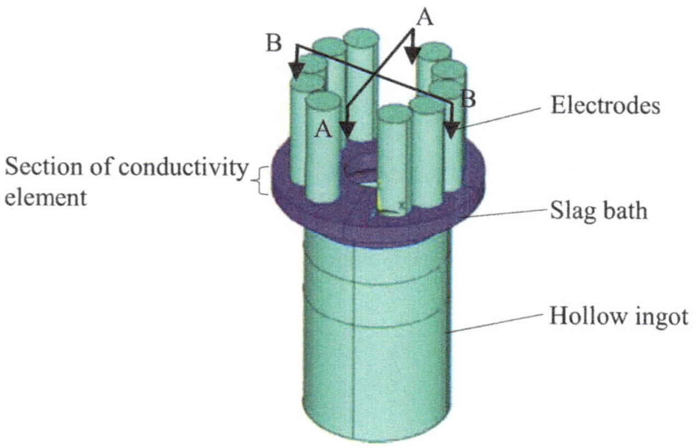

Figure 7. The 3D geometric model of hollow ESR.

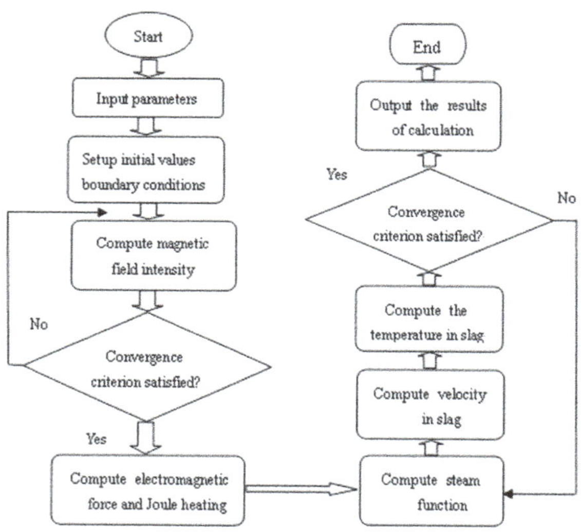

Figure 8. Simplified computational flow diagram.

The work to be described is based on the following assumptions:
(1) Quasi-steady state.
(2) The effect of metal drops and flow field of metal bath on the motion of the slag is neglected.
(3) Voltage losses are in the molten slag pool only.
(4) Continuity of heat flux at all the external surfaces and at the slag-metal interface.
(5)The tip of the electrode has the liquids temperature.
All governing equations used in the mathematic model are listed in the following Table I.

The main results obtained by this model are presented for the remelting of 10 electrodes into Φ650/450mm ingot of 35CrMo steel, with CaF_2-CaO-Al_2O_3-SiO_2-MgO slag. Slag and steel properties have been measured and calculated by laboratory apparatus and are listed in Table II-IV.

Table I. Governing equations used in the model

Governing equation's name	Model governing equation
Ampere's Law	$\nabla \times E = \frac{\partial B}{\partial t}$
Faraday's Law	$\nabla \times B = \sigma E$
Gauss's Law of electric field	$\nabla \cdot E = 0$
Gauss's Law of magnetic field	$\nabla \cdot B = 0$
Navier-Stokes Equations	$\rho(v \cdot \nabla)v = -\nabla P + \nabla \cdot [(\mu + \mu_t)\nabla v] + F$ $F = \mu_0 J \times H + \rho g$ $\rho = \rho_0[1 - \beta(T - T_0)]$
Turbulence Model	$\mu_t = 0.09\rho \frac{k^2}{\varepsilon}$
Heat Balance Equations	$\rho C_p v \cdot \nabla T = \nabla \cdot [(\lambda + \lambda_t)\nabla T] + q$

Table II. Values of all parameters in the model

Symbol	Meaning	Value
Re	Electrode radius, m	0.08
Num	The number of electrode, pcs	10
Ro	Outside radius of ingot, m	0.65
Ri	Inside radius of ingot, m	0.45
Ms	Mass of slag bath, kg	330
hI	Immersed depth of electrode in slag bath, m	0.02
I	Remelting current, A	15500

Table III. Slag thermo-physical properties used for the simulation

Meaning	Value
Density of slag, kg/m^3	2554
Density of ingot, kg/m^3	7800
Slag conductivity , $(\Omega\cdot m)^{-1}$	400
Metal conductivity , $(\Omega\cdot m)^{-1}$	7.1×10^{-5}
Specific heat, J/(kg·K)	1255
Thermal expan. Coefficient, K^{-1}	2.0×10^{-4}
Thermal conductivity, W/(m·K)	55

Table IV. Slag viscosity value at different temperature

Temperature, K	1614	1623	1683	1743	1773
Value, Pa·s	0.8536	0.2483	0.0697	0.0564	0.0537

The model has been verified by measuring the geometrical dimensions of metal pool of hollow ingot of 650/450mm (outer and inner diameters), as shown in Figure 9.

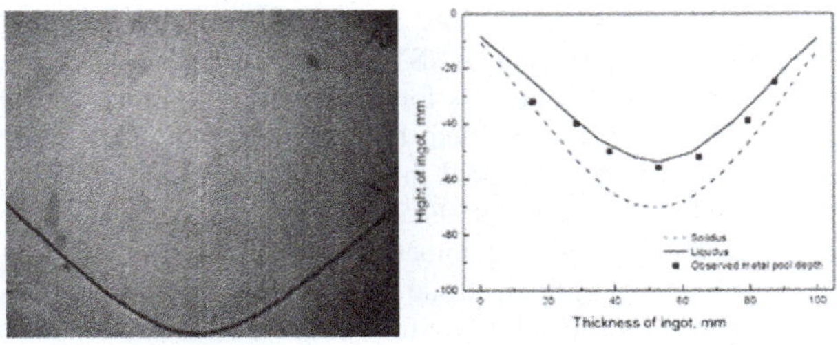

Figure 9. Observed and predicted metal pool profiles in ESR hollow ingot.

The observed metal pool depth makes 55.8mm that is in good agreement with the calculated data. At remelting rate 808 kg/h (10mm/min) the depth of metal pool, maximal width of mushy zone and maximal local solidification time (LST) are 53.8 m, 17 mm and 902 s, respectively.

The distribution of electric potential in simulated slag bath is presented on Figure 10.

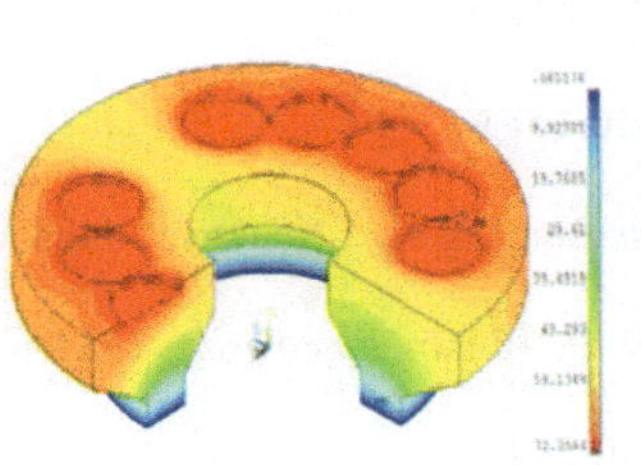

Figure 10. The distribution of electric potential.

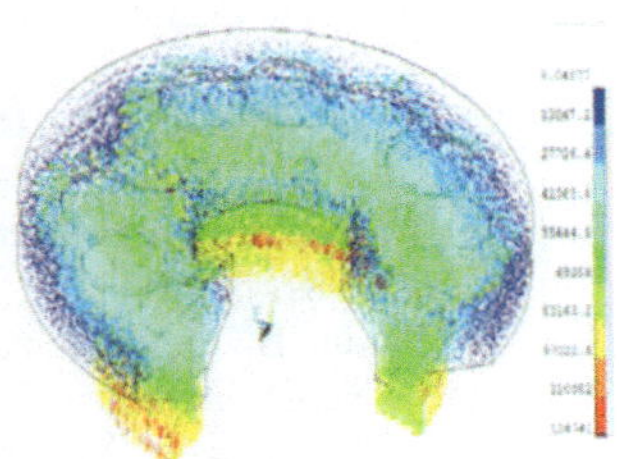

Figure 11. The distribution of current density.

As shown in the Figure 10, the greater electric potential gradient belongs to electrodes tips that illustrates of largest values of both current density and electric field in this region. There is a small electric potential near the interface of slag-metal. Its value is close to 0V and the same is in the hollow ingot itself.

Figure 11 presents the current density distribution in a slag bath. The maximum value of current density is $1.24x10^5$ A/m^2 on the electrodes tips. In other regions the current densities is much lower ($0.41x10^5$ A/m^2). Thus, the slag pool near the edge of electrodes is the place of heat releasing at melting of consumable electrode. At the same time, in T-shape mould, when the dimension of the cross section decreases, the current density will increase. As a result, the high temperature zone is removed down that allow us to provide good thermal conditions for obtaining a perfect surface quality of steel ingot.

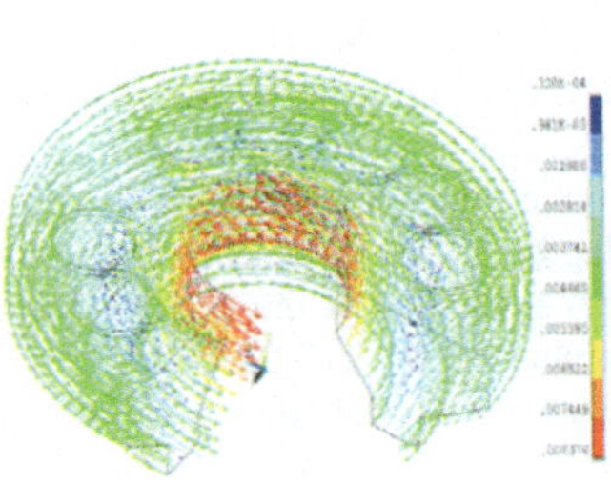

Figure 12. The distribution of magnetic induction intensity.

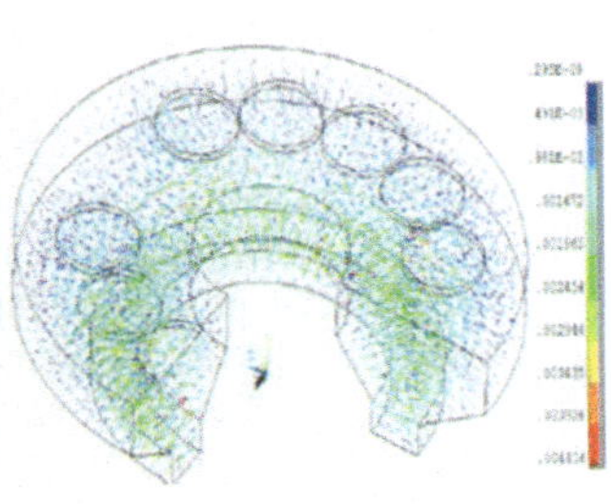

Figure 13. The distribution of electromagnetic force.

As shown on Figure 12, direction of magnetic induction intensity is defined by Ampere's right hand rule. At the side of outer mould the direction of magnetic induction intensity is clockwise. But at the side of inner mould it has counter clockwise direction. The value of first one is bigger than second one because of smaller distance between electrodes and this mould. The Figure 13 shows the calculated vectors of Lorentz force. Their direction in slag bath is as inwards as upward due to the usage of T-shape mould. Lorentz force, current density and magnetic induction intensity are consistent with Fleming's left hand rule.

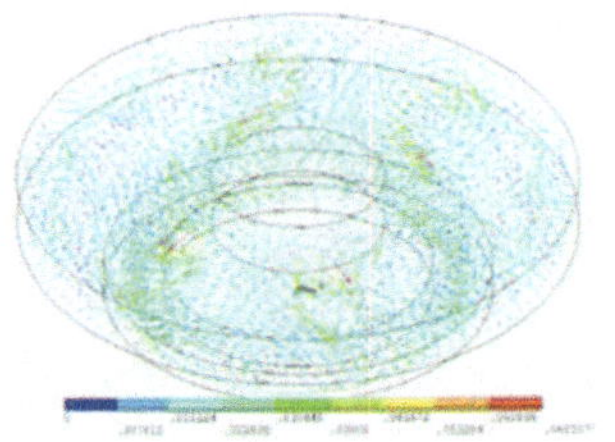

(a) Velocity distribution in the slag bath

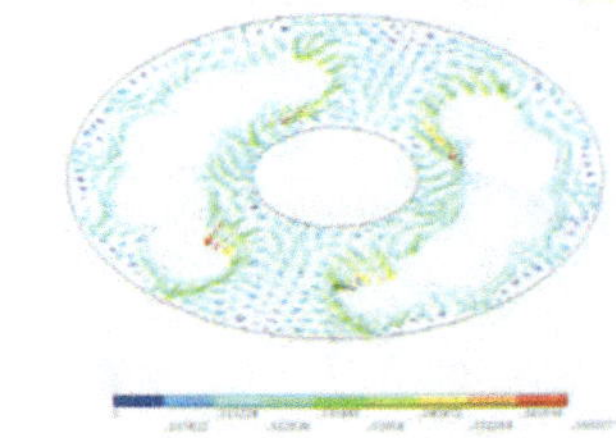

(b) Velocity distribution on the slag bath free surface

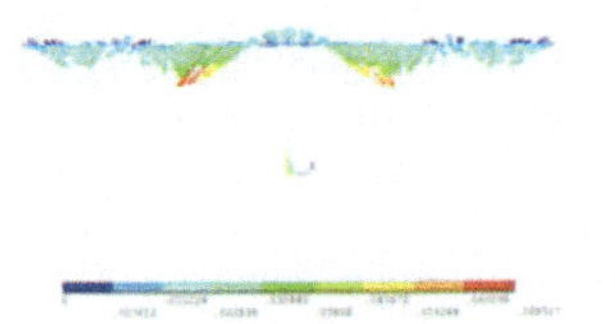

(c) Velocity distribution in the slag bath (horizontal)

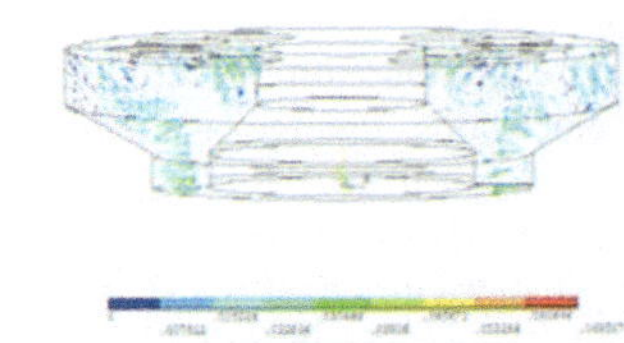

(d) Velocity distribution in the slag bath (vertical)

Figure 14. Calculated values of fluid flow velocities in the slag bath at hollow ESR melting.

It can be seen from Figure 14(a) and Figure 14(b) that the flow direction on the slag bath free surface is downwards at the end of electrodes and far away the A-A symmetrical plane. The calculation characteristic of velocity distribution on the slag bath free surface well agrees with phenomena observed at hot test. As in the Figure 14(d) there are two vortexes in the slag bath B-B plane: one forms near the internal mould and rotates clockwise, the other one forms near the external mould and moves counterclockwise.

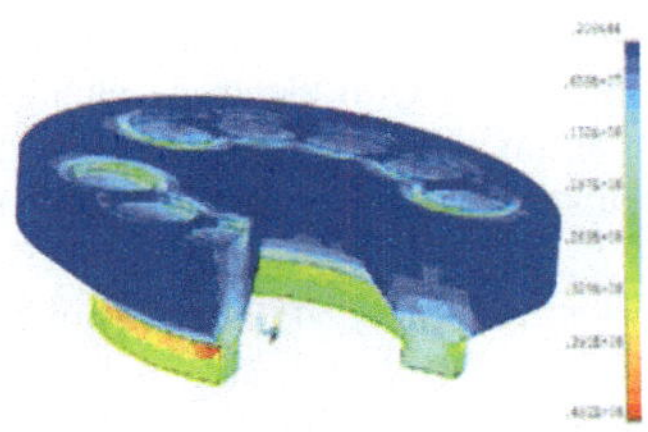

Figure 15. Distribution of Joule heat in hollow ESR slag bath.

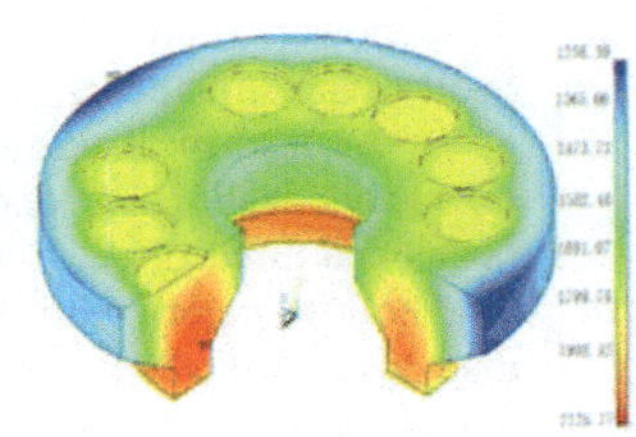

Figure 16. Temperature distribution in hollow ESR slag bath.

Figure 15 shows the distribution of Joule heat generated in the slag bath. The region of intensive heat generation is predicted in the near-electrode zone and in the T-shape mould region with minimal cross-section. Simulated picture is quite different from Joule heat generation in slag bath at conventional ESR process without T-shape mould. We will consider found features and verify received distribution of Joule density in order to be sure of made prognosis.

The same will be done for prognosis of high-temperature zone location (Figure 16) field is found far under the electrode in the slag bath. The maximum predicted temperature is about 2126 K. Because of electrodes arrangement in two symmetric groups both the current density and Joule heat generation field are symmetric but not uniform. As a result, the temperature distribution in circumferential direction on slag-metal interface is not uniform too. The depth of metal molten pool under an electrode is biggest.

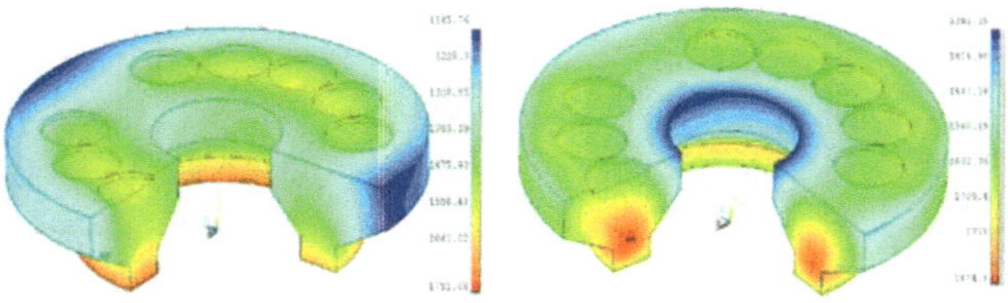

Figure 17. Temperature distribution after electrode exchange 120s (a) without CSM; (b) with CSM (400KVA).

The comparison of temperature distribution in slag bath while electrodes change is made. During this time electrodes are removed from slag. Figure 17 shows the temperature distribution in the slag bath after 120s from electrode removal without CSM (a) and with CSM (b) at power supply 400 KVA. The temperature of slag bath without CSM decreases rapidly, especially in the area near external mould, which makes liquid slag solidified. In contrary with the CSM the temperature of slag bath decreases very slowly during period of the electrode change. The place of heat releasing is not defined exactly so numerical experiments will be continues. But it is shown clearly by simulation that power supply from CSM is very important for hollow ESR manufacture with electrode change, especially in the case of thin wall hollow ingot. Macrostructure investigation of actual hollow ingots of industrial hot tests proved that this technique allows us to avoid surface and internal defects in hollow ingots.

Conclusions

1. A new ESR technology for hollow ingot manufacture has been successfully commercialized by the cooperation between China and Ukraine.
2. Two power sources, T-shape current supplying mould, gantry type structure of furnace head with electrode change and movement mechanisms, automatic control system based on metal level sensor and load cells are main features of developed furnace.
3. The hollow ingots with 650 mm (outer)/ 450 mm (inner) diameters (wall thickness 100 mm) and 6000mm in length have been successfully produced by using electrode change (two sets of electrodes).
4. The ESR hollow ingots with perfect surface quality, non-defect macrostructure, fine microstructure and high steel cleanness were manufactures by the new process.
5. The second power source and CSM usage are two key points for long hollow ESR manufacture because their usage guarantee permanent heating while electrode change that allow to avoid the defects in ingot body and surface.
6. Simulation of slag flow and heat transfer made by commercial software ANSYS allow us to predict main processes at ESR process for hollow ingot manufacture. The model has been verified by measuring the geometrical dimensions of metal pool in macrostructure of templates of actual hollow ingot of 650/450 mm.
7. Technology and equipment for sound hollow ingot manufacture with electrode change have passed hot tests program and shown ability to produce ESR hollow ingots of big length and mass with electrode change.
8. Developed technology and equipment is very promising for production of hollow ingots and heavy wall pipes for challenging application in most critical area of engineering.

Acknowledgement

The Chinese team of this project expresses the gratitude to the National Nature Science Foundation of China (grant #51204041) and the National High Technology Research and Development Program of China (863 Program) (grant #2012AA03A502) for support as well as to Program for Liaoning Innovative Research Team in University (grant # LT20120008).

References

1. B. Medovar et al., *Electroslag Metal*, Naukova dumka, Kiev, 1981, 677 P.
2. B. Paton, B. Medovar and G. Boiko, *Electroslag Casting,* Kiev, Naukova dumka, 1982, 282.
3. B. Fedorovskii, L. Medovar, and G. Stovpchenko, "ESR of hollow ingots: new approaches to a traditional problem" (Paper presented at the LMPC, Nancy, France, 25 September 2011), 97-104.
4. Rupert Wieser, "Forging ahead GFM radialforging," (Paper presented at the Proceedings of the INTECO Remelting and Forging Symposium, Shanghai, 2010), 186-207.
5. C. C. Zhao et al., "The development of the large hollow ingot," *Large castings and forgings*, 4(1999), 19-22.
6. T. H. Xu, "Hollow ingot manufacturing technology," *Technology of CFHI*, 2(2004), 28-31.
7. G. Girardin, "Hollow Ingots: Thirty Years of Use to Control Segregation and Quality for Nuclear and Petrochemical Large Shells" (Proceedings of IFM-18, Pittsburgh, PA, USA, 12-15 September 2011), 170-174.
8. L. Medovar et al., "ESR of heavy hollow ingots: New approaches to a traditional problem" (Proceedings of 18th Int .Forgemasters Meet, USA, 2001), 183-188.
9. L. Medovar et al., "Evolution of ESR technology and Equipment for Long Holow Ingots Manufacture" (Proceedings of LMPC, 2013 (CD)).
10. Yanwu DONG et al., "Simulation of Multi-electrode ESR Process for Manufacturing Large Ingot." *ISIJ International*, 52(12) (2012), 2226-2234.
11. Fubin Liu et al., "Couple modeling of electromagnetic, fluid flow and heat transfer in slag bath during current-conductive mould ESR process", (Paper presented at the LMPC, Santa Fe, New Mexico, 20 September 2009), 213-223.
12. Fubin Liu et al., "Comprehensive model for a slag bath in electroslag remelting process with a current-conductive mould," *International Journal of Minerals, Metallurgy, and Materials*, 19(4) (2012), 303-311.

Proceedings of the 2013 International Symposium on Liquid Metal Processing & Casting
Edited by: Matthew J. M. Krane, Alain Jardy, Rodney L. Williamson, and Joseph J. Beaman
TMS (The Minerals, Metals & Materials Society), 2013

DESIGN OF ESR SLAGS ACCORDING TO REQUESTED PHYSICAL PROPERTIES; PART 2: DENSITY AND VISCOSITY

Krzysztof Wroblewski[1], Brent DiBiaso[1,2], James Fraley[1], Jerry Fields[1], Stuart Rudoler[1]

[1] American Flux & Metal, 352 East Fleming Pike, Winslow, NJ 08095, USA
[2] Rowan University, Department of Chemistry and Biochemistry, 201 Mullica Rd., Glassboro, NJ 08028, USA

Keywords: ESR, slag, flux, viscosity, density, modeling, multifunction, topology

Abstract

Density and viscosity components of multifunction $\Gamma(T, \rho, \eta, \kappa)$ which for requested working temperature (T), density (ρ), viscosity (η) and electrical conductivity (κ) defines all six-component (CaF_2, CaO, MgO, Al_2O_3, TiO_2, SiO_2) slags of requested properties have been defined. The seven parameter correlation parameters describing molten flux density were calculated using Gauss-Jordan multivariable regression analysis approach and literature data. Mills and Sridhar method for estimating molten flux viscosity, has been used. The brute force algorithm for solving Γ has been updated and tested. Our results showed, despite of the exploratory nature of our work, that the calculated compositions included slags manufactured at American Flux & Metal, both: containing fluorspar and those composed only of oxides. The developed numerical algorithm for solving multifunction Γ is powerful enough to give us solutions in reasonable time so in the future additional components of ESR slags (MgF_2, MnO_2, LaF_3, La_2O_3, and ZrO_2) will be also included in the set of solutions.

Introduction

Electro Slag Remelting (ESR) is a process used for remelting and refining of steels and special alloys which are used for critical applications in aircraft, thermal and nuclear power plants, defense hardware, etc.[1] The quality of ESR ingots is significantly higher than obtained conventionally, but at the same time, the ESR process is more demanding, and selection of a proper slag is crucial for the success of the remelting operation. The old saying: "Look after the slag and the metal will look after itself" is still very appropriate. During ESR process an electrode made of an alloy that needs purification slowly melts in contact with hot slag, and droplets of molten metal travel through the pool of molten slag. During the period of residence in the slag the droplet of a metal is superheated by the heat transfer from surrounding slag and this is the main source of heat input to the molten metal pool. The residence time is defined by the following equation:

$$t = \frac{k\eta}{r^2(\rho_{metal} - \rho_{flux})} \tag{1}$$

Where: t – residence time, k – constant, η- viscosity, r – droplet radius, ρ - density

Viscosity and density of molten slag are very important properties in case of metallurgical melts because of their direct effect on the the kinetic conditions of the processes. In high temperature a successful separation of slag and metal is required, in which the difference of density between them plays an important role. However the experimental data are scarce due to the high reactivity of molten slag and the technical difficulty of taking precise measurements at such high temperatures. Density depends on relative affinities between the neighbor particles in the system. The bigger the affinities are, the larger the density. With the increasing of temperature the affinities will always become weaker. Therefore it is anticipated that the density will decrease with temperature. Arrhenius law describes the changes well enough.[2,3,4]

Viscosity is a measure of the resistance encountered when moving one set of atoms over a lower layer of atoms. Thus as the network structure becomes more polymerized the resistance to viscous flow will increase, and thus the viscosity will increase. Increased temperature and/or addition of molecules that loosens up the structure and decrease viscosity. It makes it practical to use viscosity as a measure of slag's structure. In most of ESR processes it is of considerable importance to keep the slag viscosity low to ensure fast heat and mass transfer. One of the most common measures to lower the slag viscosity is to add certain amount of calcium fluoride (fluorspar) to the slag.[5] In fact, these days practically all ESR slags contain fluorspar, and in most of them calcium fluoride is their main component. These days modeling software capable of predicting electromagnetic field, flow, temperature distribution, and other metallurgical parameters that are of interest to process engineers is commonly used in the industry[6,7] and in most of cases it requires physical parameters of molten slag.

The research at American Flux & Metal is aiming to design and manufacture custom ESR slags/fluxes, optimal for any particular metallurgical application. We adopted multifunction approach in searching for a solution of the optimal flux problem. A multifunction also known as a multiple-valued function or multivalued function, is a "function" that assumes one or more distinct values in its range for at least one point in its domain. Strictly speaking, a "well-defined" function associates one, and only one, output to any particular input. The term "multivalued function" is, therefore, a misnomer because functions are single-valued. Multivalued functions often arise from functions which are not injective. Such functions do not have an inverse function, but they do have an inverse relation. The multivalued function corresponds to this inverse relation. For example: Inverse trigonometric functions are multiple-valued because trigonometric functions are periodic. We have: $\tan(\pi/4) = \tan(5\pi/4) = \tan(-3\pi/4)$

=tan ((2n+1) π/4) = 1. Consequently arctan(1) is intuitively related to several values: π/4, 5π/4, −3π/4, and so on. We can treat arctan as a single-valued function by restricting the domain of tan(x) to $-\pi/2 < x < \pi/2$ – a domain over which tan(x) is monotonically increasing. Thus, the range of arctan(x) becomes $-\pi/2 < y < \pi/2$. These values from a restricted domain are called principal values. Multifunctions arise in topology and optimal control theory, and became more and more popular in different branches of physics and engineering. We are in a process of defining multifunction[8,9] Γ(T, ρ,η,κ) which for each set of T, ρ,η,κ, where: T - temperature (T>liquidus), ρ - density at the temperature T, η - viscosity at the temperature T, κ – electrical conductivity at the temperature T, assigns all possible sets of flux components. Fluorspar (CaF_2), lime (CaO), magnesia (MgO), alumina (Al_2O_3), silica (SiO_2) and titania (TiO_2) are the most common components of the ESR flux and presently vast majority of all manufactured ESR flux is based only on them. In fact their concentrations serve as coordinates of a single point in the n-dimensional space defined by the composition of flux (where n is a number of components). In this approach we limited n to 6. In the future less common components of ESR slag: MgF_2, MnO_2, LaF_3, La_2O_3, and ZrO_2 will be also added.

Materials and Methods

Densities of most of the components of the slag tend to be similar, and strongly depend on temperature. Seven parameter (including temperature) multilinear correlation parameters were calculated using multivariable regression analysis algorithm based on the Gauss-Jordan method[10]. 590 literature data points were used to obtain parameters (a1 - a7 and b) of the following equation:

$$\rho = a1T + a2CaF_2 + a3CaO + a4MgO + a5Al_2O_3 + a6SiO_2 + a7TiO_2 + b \quad (2)$$

Where: ρ – density (Poise), T – temperature (°C), CaF_2 CaO MgO Al_2O_3 SiO_2 TiO_2 are mass percent of respective components.

Correlations with log(ρ), log(T), log(1/T), 1/T were also obtained, but the total correlation coefficients were lower than obtained for T.

Viscosity of slag was calculated using NMP model by Mills and Sridhar.[11] This model provides reasonable estimates of viscosity for a wide range of industrial slags, including those containing calcium fluoride. We found it a little puzzling that two years after the NMP method was published, Mills in his paper "Estimating the Physical Properties of Slags"[12] uses Ribaud, Urbain, and Iidia models, without even mentioning the NMP one in this context, but we applied it anyway.

Generally it is much easier to correlate density data than viscosity ones, mainly because of the large experimental error which is associated with published viscosity data. Authors vary in their estimations but it is commonly accepted that this error lies between 30% and 50% Even viscosity data for fluorspar itself, which can be found in the Slag Atlas[13], and shown in Figure 1, are scattered all over the graph.

A numerical method using brute force algorithm[14] presented in Part 1 of this project[15] with a general flowchart presented in Figure 2, was extended for finding solutions for density and viscosity. Brute force is a straightforward approach to solving a problem, usually directly based on the problem's statement and definitions of the concepts involved. Generally it involves iterating through all possible solutions until a valid one is found. This algorithm given enough time always works. Unfortunately, "enough time" means relatively small number of dimensions that can be handled.

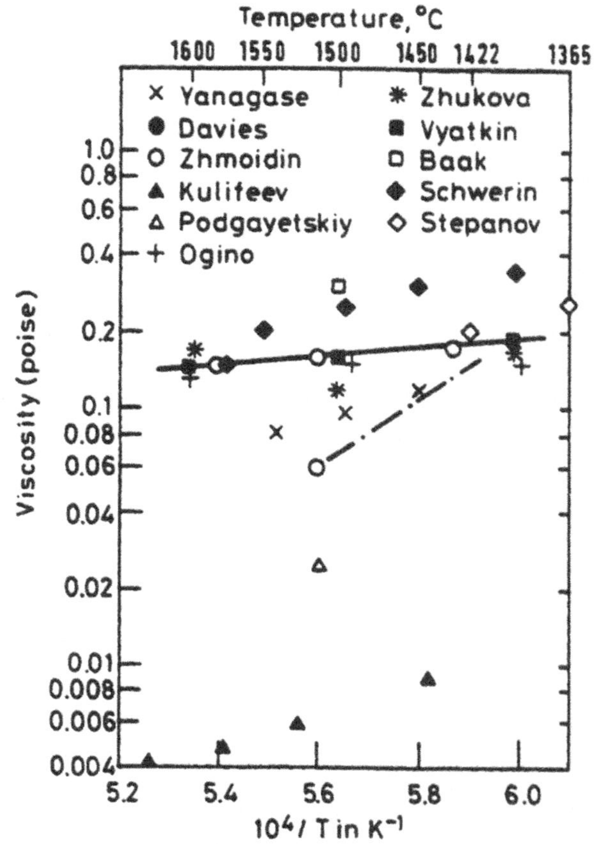

Figure 1. Viscosity of CaF_2 as a function of reciprocal temperature. (reprinted with permission of Verlag Stahleisen GmbH)

Luckily the number of components used for manufacturing ESR slags is relatively small so, as we have shown here, it is practical to use it. In six dimensional space (each dimension corresponds to the concentration of one flux component) all nodes corresponding to possible combinations of components were generated (ca. 10^8). The step in each dimension as well as the boundary conditions for each component i.e. starting and ending concentration, were set individually and are presented in Table 1. The nodes with physical meaning (concentrations adding up to 100% and temperature over liquidus) were tested against requested criteria (density, viscosity, or both). In this work in progress we present slag composition calculations based only on the density and viscosity. The developed algorithm is capable of handling in reasonable time (less than 20 min per one data set on a higher class PC) also additional flux properties as well as more complicated flux compositions.

All programs were written in Fortran 90.

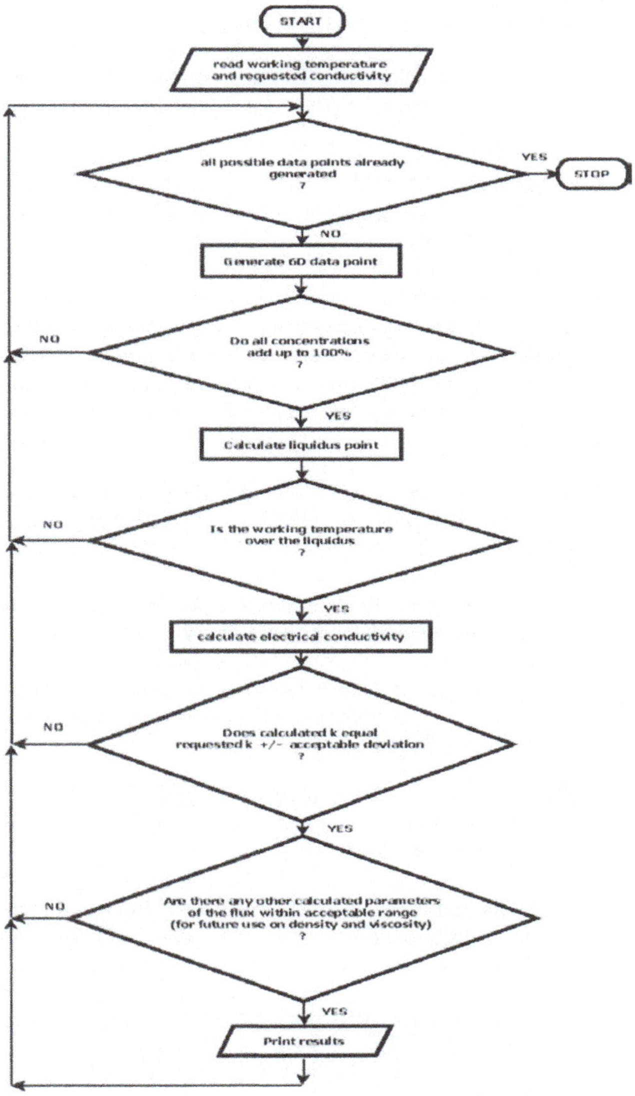

Figure 2. Flowchart of the algorithm for solving multifunction $\Gamma(T, \rho, \eta, \kappa)$

Results and Discussion

The seven parameter correlation parameters defining the correlation of molten slag density ρ with the temperature and slag composition are as follows: a1=4.11E-5, a2=-1.86E-3, a3=2.05E-3, a4=1.86E-3, a5=3.05E-4, a6=-4.13E-3, a7=1.01E-, b=2.632 The multilinear correlation coefficient c = 0.89

Table 1. The boundary conditions for the calculations: Min – minimal concentration of a component, Max – maximal concentration of a component step – distance between nodes in the particular dimension.

	CaF_2	CaO	MgO	Al_2O3	SiO_2	TiO_2
Min	0.0	0.0	0.0	0.0	0.0	0.0
Max	100.0	100.0	100.0	100.0	50.0	50.0
step	1.0	1.0	1.0	1.0	1.0	1.0

Calculations were carried out covering temperature range 1100-2000 °C every 100 °C, densities: 2.5, 2.57, 3.0, and viscosities 0.01, 0.05, 0.1, 0.5, 1.0, 1.5, 2.0
The number of solutions ranged between 0 and 3000. As shown in Figure 3 most of the solutions can be found for low viscosity melts (0.01-0.5) with additional local maximum at $\eta \sim 1$.

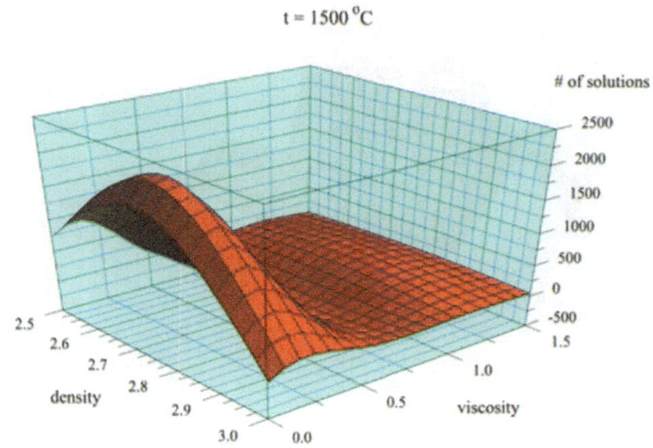

Figure 3. Number of slag compositions as a function of viscosity and density at T=1500°C

This is not a surprise. Typical ESR slags contain significant amount of fluorspar and only small amounts, if not only traces of silica. This is also why most of the viscosity models based on structure of silicate built up of Si cations which are surrounded with four oxygen anions arranged in the form of tetrahedrons joined together in chains or rings by bridging oxygens[13] may not be the best for describing ESR slags. The NPL (National Physical Laboratory) model used for calculations relates the viscosity of slags to the structure through the optical basicity[16] corrected for charge balancing (Λ^{corr}). $\Lambda(CaF_2)$= 1.2 was used for all calculations. The method assumes Arrhenius behavior of viscosity, and uses the following equation for the temperature dependence: $\ln(\eta) = \ln A + B/T$ where T-thermodynamic temperature, A,B – constants with respect to temperature. A and B are functions of Λ^{corr}.

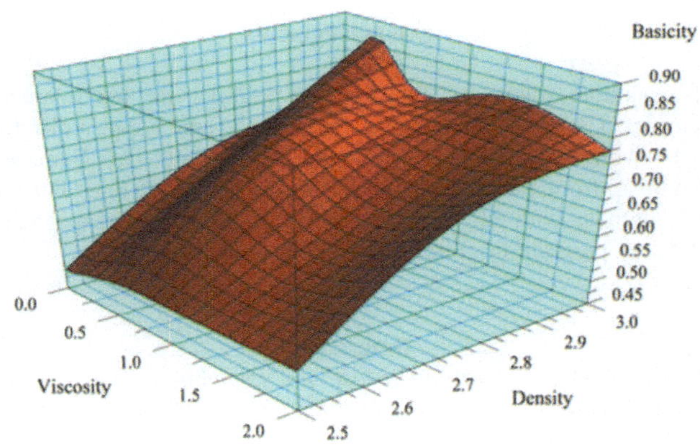

Figure 4. Optical basicity as a function of viscosity and density at T=1500°C.

As shown in Figure 4, calculated optical basicity for each data point at T=1500°C does not show any strong dependence from viscosity, especially for low densities. At higher densities one can observe increased role of Al^{3+} and Ti^{4+} which also form tetrahedra and can be incorporated into existing 3D structure.

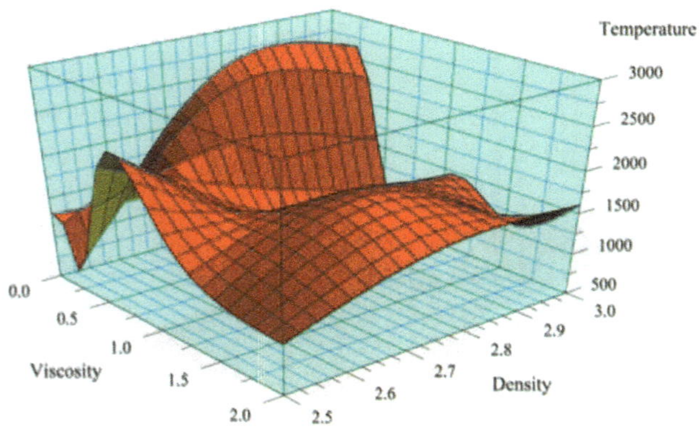

Figure 5. Temperature of melt as a function of viscosity and density.

Both density and viscosity strongly depend on temperature. In Figure 5 one can admire the inverse dependence which, at this moment, is to complicated to have any practical application.

Conclusions

The partially defined multifunction can be easily, and in reasonable time solved using numerical approach presented in this work. Our results show, despite of the exploratory nature of our work, that the calculated compositions include also slags manufactured at American Flux & Metal, both: containing fluorspar and those composed only of oxides, which means they are in a good agreement with properties of slags needed by particular ESR applications. The further work will be done on the remaining components of the multifunction $\Gamma(T, \rho,\eta,\kappa)$ as well on obtaining correlations with additional, less common, components of the ESR slags. The optical basicity approach even, though it appears less useful in silicate free slags,

Acknowledgments

The authors want to thank Joachim Rudoler for his financial support, Cathy Sullivan, Carol Steinhauser and Kipp Wolff for help with preparation of this manuscript, and Dr. Hamzech Roumani from York University for making available to us Fortran 90 compiler with SLATEC libraries. Graphics was created with PSI-Plot program from Poly Software International.

References

1. G. Hoyle, *Electroslag Processes Principles and Practice* (London and New York, NY: Applied Science Publishers, 1983).

2. G. Zhang, K. Chou, "Model for Evaluation Density of Molten Slag With Optical Basicity", "Journal of Iron and Steel Research, International 17 (1010) 104.

3. S. Akbari, J.Reitz, B.Friedrich, "*Thermophysical Properties of ESR-Electrofluxes (Part 1: Density)*", (Proceedings of the 2009 International Symposium on Liquid Metal Processing and Casting: Santa Fe, New Mexico, NM, 2009), 1-8.

4. P. Courtail, D.B. Dingwell, "Densities of Melts in the CaO-MgO-Al_2O_3-SiO_2 system", American Mineralogist 84 (1999) 465-476.

5. Wu, Jimmy Gran, Du Sichen, "The Effect of Calcium Fluoride on Slag Viscosity," *Metallurgical and Materials Transactions B* 42B (2011), 928-931.

6. K.M. Kelkar, S.V. Patankar and A. Mitchell, "*Computational Modelling of the Electroslag Remelting (ESR) Process Used for the Production of Ingots of High-Performance Alloys,*" (Proceedings of the 2005 International Symposium on Liquid Metal Processing and Casting: Santa Fe, New Mexico, NM, 2005), 1-8.

7. J. Jardy, D. Ablitzer, and J.F. Wadier, "Magnetohydrodynamic and Thermal Behavior of Electroslag Remelting Slags, Metall.Trans". B, 22B (1991), 111-120.

8. Hardy G.H., *A Course of Pure Mathematic- Centenial Edition*, (Cambridge: Cambridge Mathematical Library 2000).

9. C. Berge, *Topological Spaces Including a Treatment of Multi-Valued Functions, Vector Spaces, and Convexity,* (Mineola NY: Dover Publications, 2006).

10. W.H. Press et al., *Numerical Recipes; The Art of Scientific Computing (FORTRAN Version)* (Cambridge, Cambridge University Press, 1989) 24-26.

11. K.C. Mills and S. Sridhar, "Viscosities of Ironmaking and Steelmaking Slags," *Ironmaking and Steelmaking,* 26 (1999), 262-268.

12. K.C. Mills, L.Yuan, R.T. Jones, "Estimating the Physical Properties of Slags", *The Journal of The Southern African Institute of Mining and Metallurgy* 111 (2011) 649-658.

13. Verein Deutscher Eisenhuettenloute, ed., *Slag Atlas*, (Duesseldorf, Germany: Verlag Stahleisen GmbH, 1995).

14. M. Abaneth et al, "Occurences Algorithm for string Searching Based on Brute-force Agorithm," *Journal of Computer Science,* 2 (2006), 82-85.

15. K. Wroblewski, J. Fraley, J. Fields, R. Werner, S. Rudoler, "*Design of ESR Slags According to Requested Physical Properties; Part 1: Electrical Conductivity*" (Proceedings of the 2011 International Symposium on Liquid Metal Processing and Casting, Nancy, France 2011), 121-126.

16. J.A. Duffy "Optical Basicity: A Practical Acid-Base Theory for Oxides and Oxyanions," *Journal of Chemical Education.* 73 (1996) 1138-1142.

Proceedings of the 2013 International Symposium on Liquid Metal Processing & Casting
Edited by: Matthew J. M. Krane, Alain Jardy, Rodney L. Williamson, and Joseph J. Beaman
TMS (The Minerals, Metals & Materials Society), 2013

PREDICTING MELTING BEHAVIOR OF AN INDUSTRIAL ELECTROSLAG REMELTING INGOT

Jeff Yanke[1,2], Kyle Fezi[1], Mike Fahrmann[3], Matthew John M. Krane[1]
[1]Purdue Center for Metal Casting Research, School of Materials Engineering, Purdue University, West Lafayette, Indiana USA;
[2]currently Carpenter Technology Corporation, Reading, Pennsylvania USA; [3]Haynes International, Inc., Kokomo, Indiana USA

Keywords: Electroslag Remelting, Simulation, Volume of Fluid

Abstract

Electroslag remelting (ESR) is a secondary melting process used to cast stainless steel and superalloy ingots. In this process, current flows through a consumable electrode immersed in an electrically resistive slag, providing the heat to melt the electrode. Droplets from the electrode sink through the slag, pooling at the bottom of the mold and forming the final ingot. The electrode melt rate is a key parameter, affecting the probability of surface and macrosegregation defects. This work uses an axisymmetric model to simulate flow, heat and mass transfer, solidification, and electromagnetics in the production of industrial scale ESR ingots. The simulated melt rate, sump shape, and surface defects are qualitatively similar to measured data. However, quantitative comparisons are difficult to obtain due to large uncertainty in slag properties and lack of electrode motion in the present model.

Introduction

ESR is a secondary ingot production process used to manufacture high quality steels and nickel based superalloys[1, 2]. The process requires the manufacture of a consumable electrode of the desired alloy, which is hung in a water-cooled copper mold from a ram that serves to control the vertical position of the electrode as well as monitor the weight of the electrode throughout the process. AC current is passed through the electrode and slag, and then out the mold bottom. The electrically resistive slag generates Joule heating sufficient to melt the electrode. Liquid droplets of electrode material sink through the less dense slag to form a pool of metal at the bottom of the mold. Heat loss to the water-cooled copper mold solidifies the metal, forming a dendritic structure and maintaining a liquid pool that is 1-2 diameters deep throughout the process. The slag in contact with the mold wall freezes and forms a thin solid skin around the perimeter of the mold causing the volume of the slag cap to decrease throughout the process. As the metal fills the mold, the less dense slag floats on top (referred to as the slag cap) leaving behind this thin solid skin, on which the metal freezes.

Given the complicated phenomena involved in ESR and the difficulty and expense of running experiments on a real process, computational simulation provides an attractive method of understanding how process conditions affect ingot structure and defect formation. Numerical simulation of ESR presents challenges as it incorporates coupled fluid flow, heat and mass transfer, solidification, and electromagnetics in the interacting metal and slag. A comprehensive model must include a method of tracking the moving interface between the slag and metal. Several recent models have shown how processing variations affect ESR while simulating both the metal and slag regions

Kelkar et al [3] developed an axisymmetric, steady state model including both metal and slag in the computational domain by assuming a flat interface between them. This model was used to simulate industrial ESR processes at various input currents (and corresponding melt rates) and to compare computed sump shapes to pool markings in an experimental ingot. Kelkar et al have continued to use this model in [4, 5], simulating the effects of different alloys and process conditions on local solidification time and dendrite arm spacing.

An axisymmetric transient model assuming a flat slag/metal interface was developed by Weber et al [6]. This model was used to simulate a full experimental ESR run with various current and melt rates (used as model inputs). The predicted sump shapes matched well with experimental pool markings. By extrapolating melt rate information from the experiment, they showed the melt rate increased with fill ratio (the ratio of electrode to mold diameter), due to changes in the Joule heating distribution in the slag. This model was also used to predict slag skin thickness [7] in the slag cap as a function of process parameters and material properties. The frozen slag skin in the real system results in a thinner slag cap as the process progresses, but Weber's model assumed a constant cap thickness.

A method to track a non-flat slag/metal interface was developed by Kharicha et al [8], showing the effects of interface motion on the electromagnetic fields in ESR. Their results show that a curved interface has a significant effect on the distribution of Joule heating in the slag as well as the sump shape in the metal. They [9] also showed details of an immersed electrode melting with a 3D model of just the slag layer. It was shown that the formation locations of liquid metal droplets on the bottom of the electrode were not axisymmetric and that these droplets cause 0.01-0.02 m high waves as they impact the slag/metal interface.

This work presents a comparison of two industrial scale ESR trials to a transient axisymmetric model of fluid flow, heat and mass transfer, solidification, and electromagnetics in the ESR process, using a volume of fluid [10] method to track the slag/metal interface. The model predicts the melt rate, sump shape, and surface defects related to extra slag solidification present in the industrial scale ESR ingots.

Model Description

Transport Equations

Only a brief description of the numerical model will be provided here; details are available in [11]. A continuum mixture model for mass, momentum, energy, species, and magnetic field is used, as described by Bennon and Incropera [12], but the definition of mixture quantities is modified to account for the presence of two immiscible fluids.

The momentum and mass conservation equations are solved on a staggered grid using the SIMPLER algorithm[13].

$$\frac{\partial(\rho u)}{\partial t}+\nabla\bullet(\rho\vec{V}u)=\nabla\bullet(\mu\nabla u)-\frac{\partial P}{\partial x}+S_u \quad (1)$$

$$\frac{\partial(\rho v)}{\partial t}+\nabla\bullet(\rho\vec{V}v)=\nabla\bullet(\mu\nabla v)-\frac{\partial P}{\partial y}+S_v \quad (2)$$

$$\frac{\partial\rho}{\partial t}+\frac{\partial(\rho u)}{\partial x}+\frac{\partial(\rho v)}{\partial y}=0 \quad (3)$$

S_u and S_v are the source terms representing Darcy drag, thermo-solutal buoyancy, and Lorentz forces.

The enthalpy formulation of the energy equation is used in conjunction with multiple species equations:

$$\frac{\partial(\rho h)}{\partial t}+\nabla\bullet(\rho\vec{V}h)=\nabla\bullet\left(\frac{k}{c}\nabla h\right)+\mathrm{Re}\left(\frac{1}{2\sigma}\vec{J}\bullet\vec{J}\right)+S_h \quad (4)$$

$$\frac{\partial(\rho C^i)}{\partial t}+\nabla\bullet\left(\rho\vec{V}C^i\right)=\nabla\bullet\left(D\nabla C^i\right)+S_C \quad (5)$$

where S_h and S_C represent the advection and diffusion-like source terms [12].

The AC electromagnetics are solved by expressing time and frequency dependence of magnetic field as

$$\vec{H}=\mathrm{Re}(\vec{H}_\theta e^{j\omega t}). \quad (6)$$

Additionally, assuming that the magnetic diffusion is fast enough that a quasistatic magnetic field exists leads to the equation of magnetic diffusion which from Faraday's law is:

$$\nabla\bullet\left(\frac{1}{\sigma}\nabla\vec{H}_\theta\right)=i2\pi f_{AC}\mu_o\vec{H}_\theta \quad (7)$$

Equation (7) is divided into two equations representing the real and imaginary components. These two equations are solved for independently, but are linked together through the source terms resulting from the right hand side of equation (7). Current density is calculated from the magnetic diffusion:

$$\nabla\times\vec{H}_\theta=\vec{J}. \quad (8)$$

All thermophysical property data can be found in Table 1.

Table 1: Thermophysical properties used in simulations.

	718	Waspaloy	Slag
Density [kg/m3] [1, 3]	7500	8190	2490
Specific heat [J/kgK] [3, 14]	720	520	1260
Thermal Conductivity [W/mK] [3]	30.52	15.0	6
Viscosity [Kg/ms] [3]	0.0033	0.0023	0.025
Thermal expansion [1/K] [3]	0.00015	0.00015	0.00025
Emissivity [6]	-	-	0.87
Latent Heat [J/kg] [3]	2.1E+05	2.1E+05	4.75E+05
Eutectic Temperature [K]	1411	1474	1390
Electrical conductivity [1/Ωm]	7.1E+05	7.1E+05	variable

The heat transfer from the slag to the electrode is used to calculate the melt rate by assuming a constant thickness (8 mm) liquid metal film on the bottom of the electrode between the top of the domain and the solidus temperature of the electrode. A 1D conduction heat conduction equation is solved in the electrode to determine how much of the heat leaving the slag goes to heating up the electrode and how much goes to melting.

Boundary Conditions

The boundary conditions for the energy and electromagnetics equations, and the solution domain are shown in Figure 1. The heat transfer at the mold wall is determined by assuming a heat transfer coefficient that includes a fixed thickness slag skin and heat transfer through the copper mold. This heat transfer coefficient (along with the electrical conductivity of the slag) is varied for each alloy, as seen in Table 2. As the metal starts to freeze and a gap forms between it and the mold, the heat transfer coefficient is altered to account for the conduction through the air and radiation.

Zero flux conditions are enforced everywhere for the species equations. No slip conditions are applied to the momentum equations, except in the annulus between the mold and electrode and at the centerline, where no shear conditions are applied. Metal is added to the system with a mass source term in the mass, energy, and species equations at the slag/metal interface.

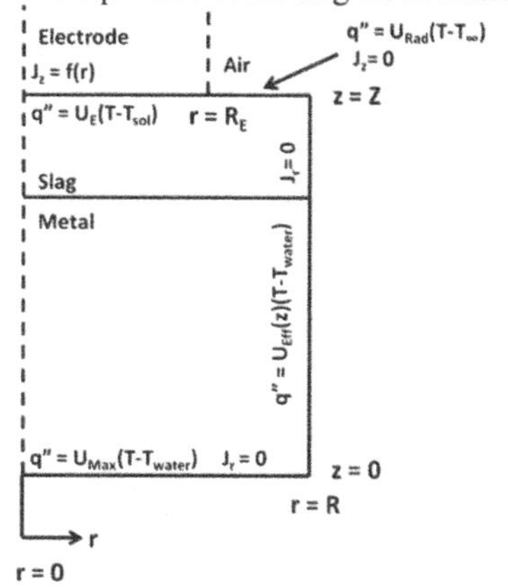

Figure 1: Schematic of the ESR domain showing the boundary conditions for the energy and magnetic field equations.

Table 2: Heat transfer coefficient at mold wall and slag electrical conductivity used in simulations.

Case	U_{eff} [W/m^{2}K]	σ [1/Ωm]
718-1	400	350
718-2	500	300
Wasp-1	400	325
Wasp-2	500	275

Experimental Procedure

Two industrial scale ESR ingots of alloys Inconel 718 and Waspaloy were cast by Haynes International, Inc., in 1981 as part of a previous NSF grant for the purpose of studying the relationship between process conditions and resulting ingot microstructure and segregation levels. Process data and samples (macro-etched slices) of both alloys were saved and are used as a basis for comparing the behavior of numerical simulation to industrial scale ESR. However, it should be noted that the conditions used to cast both ingots are not representative of standard ESR process conditions, which strive for near steady state conditions throughout the entire process to achieve a homogeneous ingot. These ingots were subjected to more transient conditions in order to learn as much as possible from each experiment.

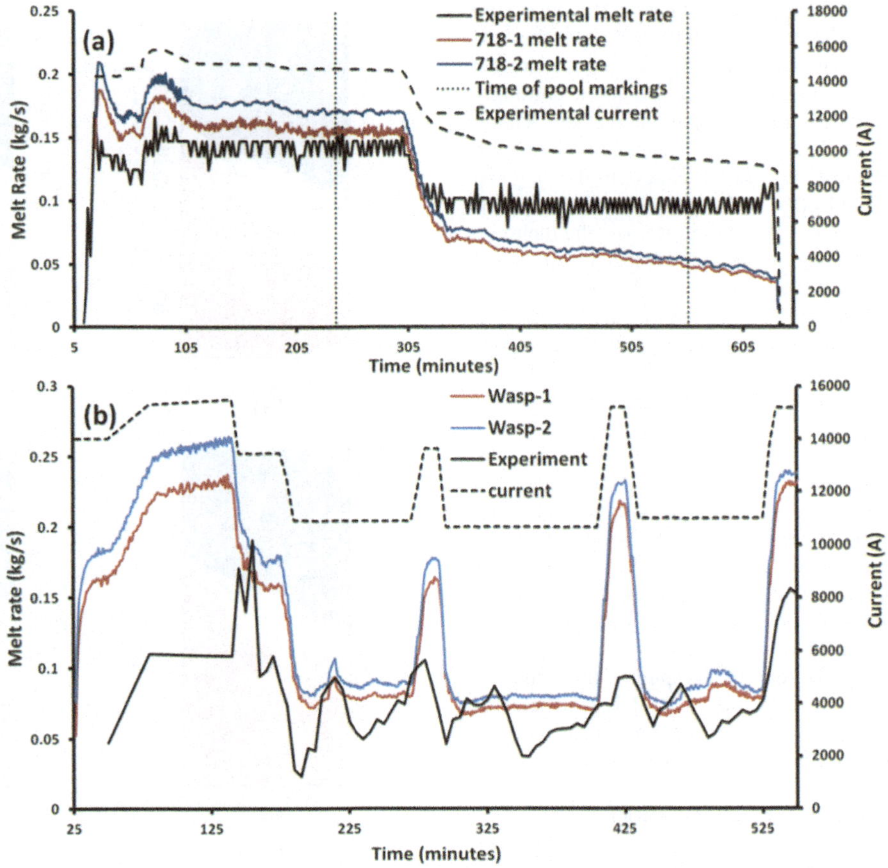

Figure 2: Process conditions and predicted melt rates for (a) 718 and (b) Waspaloy.

Results

718 Melt Rate

The comparison of melt rate and current input as a function of time for both numerical simulations and experimental ingot are shown in Figure 2a. The large spikes in melt rate at the beginning of the process (30 minutes) are an effect of the initial conditions used to start the simulations. A uniform slag temperature of 1700 K is used to begin melting so that the freezing metal front will not catch up to the slag/metal interface which can often lead to excessive (skin thicknesses half the radius of the mold) slag freezing to the wall. The initial spike in melt rate seen in the experimental case is likely due to the start-up process also, but is not related to the simulation start up process. The effects of the fluid flow in the slag are responsible for the high frequency oscillations in the melt rate seen in Figure 2.

In the first 300 minutes of the process it can be seen that both numerical cases overestimate the experimentally determined melt rate. However, at approximately 180 minutes a relatively small decrease in current (300 Amps) effects a noticeable decrease in the predicted melt rate of both simulations, while the experimentally determined melt rate stays within the bounds of the noise in the melt rate data. After this small current decrease, both numerically predicted melt rates decrease. Case 718-1 begins to approach the upper bounds of the experimental melt rate, while case 718-2 remains ~15% higher than the predicted melt rate. This suggests that the simulation is overresponsive or that the noise in the experimental melt rate data is too great to show responses to small changes in current.

As the current continuously drops during the low melt rate stage of the process (340 to 630 minutes), the experimental melt rate data apparently is not affected by the approximately 2000 amp reduction, keeping a steady melt rate within the experimental noise. The simulation results show a 50% reduction in melt rate over the same period of time. Part of the reason for the simulations' decreasing melt rate in the low current stage of the process is due to the formation of extra slag skin at the lower melt rates. Figure 3 shows the portion of the simulation domain including the slag cap and sump for both simulation cases at 380 and 484 minutes. These figures show the fraction solid along with the position of the slag/metal interface. In the high current region of the process, no additional slag was freezing to the side of the mold beyond the thickness assumed in the thermal boundary condition. Once the process enters the low current region, the temperature in the bottom of the slag cap is sufficiently low to freeze extra slag to the mold wall. This extra slag freezing to the wall depletes the volume of slag in the cap. Less slag in the cap will result in a lower melt rate.

A possible explanation for the difference in behavior between the simulation and experiment is the treatment of the thermal boundary condition at the mold wall in the slag cap. For the 718 cases, the slag skin thickness is assumed to be bounded by a minimum value at all times. While more solid slag may form under certain conditions, that minimum thickness sets a maximum value on the mold wall heat transfer coefficient. In the real process, at high currents and temperatures, the slag skin in the cap should become thinner than this minimum value and accordingly at lower currents and temperatures the skin in the cap should become thicker. However, Figure 3 shows that the simulation does not predict any change in the skin thickness in the slag cap itself as all the extra slag skin that is forming is at the bottom of the cap. If the skin thickness in the cap did increase, the melt rate should be higher due the insulating qualities of frozen slag. Also, at higher currents the slag skin may decrease in thickness, leading to a drop in melt rate. More realistic behavior of the slag skin in the cap may cause the correct melt rate behavior, that is, a lower melt rate than calculated at higher current and higher melt rate at lower current, which could potentially bring the simulation closer to the experimental results.

<u>718 Sump Shape</u>

The shape and depth of experimental liquid sumps have often been compared to ESR model predictions [3, 6] because they are an indication of the relative influence of melt rate and heat transfer to the mold. The details of the sump shape also affect the liquid flow and so macrosegregation, which is often a concern when casting larger ingots of specialty steels and superalloys. Perhaps the most attractive feature of comparing the sump to predictions is that in-situ measurements in ESR are difficult to accomplish, but by marking the sump profile with metal powder, the shape is preserved and may be examined post mortem.

At times indicated in Figure 2a, tungsten powder was used to mark the sump in the 718 process at Haynes at 240 and 556 minutes. These times are representative of the steady state sump shapes for the two melt rates used in the process. Comparisons of the predicted sump shape to both simulated cases are shown in the macrographs containing both pool markings (Figure 4). The Z axis is not representative of the position in the experimental ingot, as the differences in melt rate predicted by the model lead to differences in weight of the ingot (and Z) at the times that the pool was marked.

Figure 4a and 4b show that the 0.01 and 0.05 fraction solid contours for both simulations agree well with the actual marked pool shape and depth in the high current stage. The simulations show some disagreement within 3-4 centimeters of the centerline, which can be explained by the formation of equiaxed grains in the experimental ingot in this region. The simulation assumes a rigid, columnar mushy zone with no free-floating equiaxed grains. The equiaxed region has very different flow characteristics and mechanics of mushy zone front progression and these differences are not predicted with the present model. Of the two fraction solid contour lines plotted, the 0.01 contour seems to more accurately capture the shape of the sump, which is in agreement with the observations that that the tungsten powder came to rest on the dendrite tips[15].

The second pool marking at 556 minutes (in the low current stage) shows less agreement between simulation and experiment. The model predicts the sump to be shallower and more bowl shaped than the experiment. This difference is expected when the predicted melt rate from the low current melting stage is taken into consideration. Both simulations underpredict the experimental melt rate at 556 minutes by 50%, which will lead to the shallower and more bowl shaped sump.

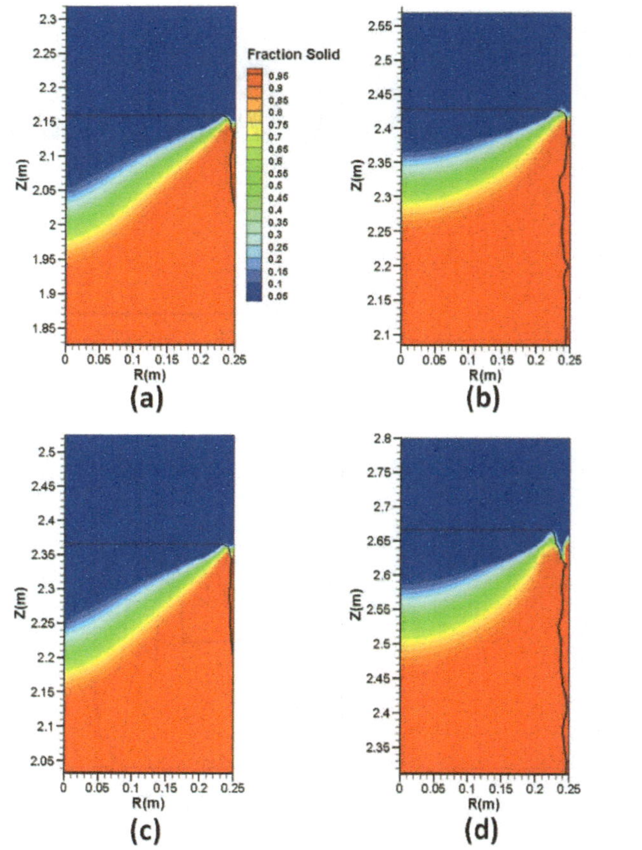

Figure 3: Contour plots of mass fraction solid (f_s). The black lines are the slag/metal interfaces, showing the freezing of slag at the wall. (a) and (b) show case 718-1 at times 380 and 484 respectively. (c) and (d) show case 718-2 at times 380 and 484 respectively

<u>Waspaloy Melt Rate</u>

From the results reported for the melt rate comparisons for the 718 process data, it is expected that the melt rate predictions for Waspaloy will be too large at higher currents and will be underpredicted at lower currents. Figure 2b shows the comparison of Waspaloy experimental and predicted melt rates from both simulations. As expected, the melt rates at larger currents are overpredicted, but they are also overpredicted at lower current values. However, unlike the 718 process where the simulation was overpredicting melt rates by 10 -15 % at higher currents, the Waspaloy simulations are predicting melt rates in the excess of twice the measured values. This discrepancy is due to the lower electrical conductivity the simulations required to produce a high enough melt rate in the low current regions to prevent the simulation from predicting a physically unreasonable frozen slag/metal interface. Also, melt rates in the Waspaloy process are frequently less than 0.05 kg/s, whereas the 718 process does not reach this low a level until the end of the process. By decreasing the electrical conductivity, the melt rate curve has been effectively shifted upward over the entire process.

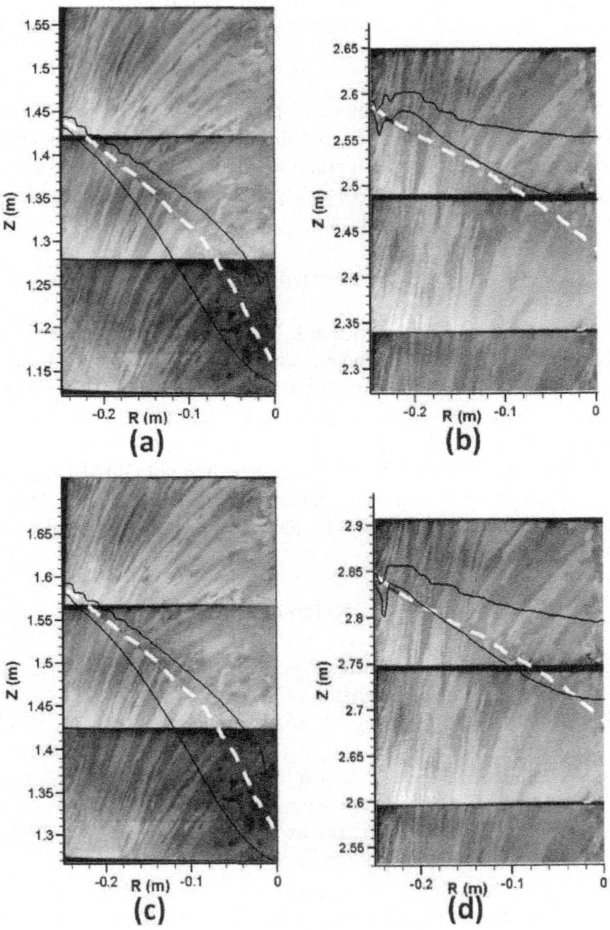

Figure 4: Fraction solid contours (black lines at f_{Sol} =0.01 and 0.05) overlaid on slices of 718 ingot with pool markings. (a)-(b) 240 minutes and (c)-(d) 556 minutes. Case 718-1 is shown in (a) and (c), case 718-2 in (b) and (d). White dashed lines drawn to clarify position of marking powder.

Another interesting difference between simulation and experiment is the effect of increasing the current for 30 minute intervals, which occur at 268-296 minutes and 404-437 minutes. The first current spike causes a 100% increase in the predicted melt rate from the simulations while only causing a 25-30% increase in the experiment. There is an even larger discrepancy in the second current spike where the predicted melt rate increases by 200% while the experimental melt rate only increases by 50%. This second spike is interesting as the current is spiked to the steady state value used later at 525-670 minutes. The simulations both predict that the melt rate during the spike nearly reaches the steady state value, while the experiment only reaches approximately 60% of the steady state value. A possible explanation for the difference in transient behavior is the control system used in the experiment. Unfortunately, no information was provided on the exact control mechanism used in the experiment. However, in most ESR control systems the data that is used to control the furnace is averaged over a length of time. The furnace control system information provided shows that the information was recorded every 5 minutes, so, depending on the particulars of the control system, the data used would have been at least 10 minutes old. This delay in the control system could account for the differences between the simulation and experiment during these 30 minutes current spikes. It is also likely that these current spikes effected the immersion depth of the electrode, which is not taken into account in the simulation.

Waspaloy Sump Shape

During the Waspaloy process, tungsten powder was added to mark the pool shape a total of 7 times at 78, 259, 372, 508, 650, 720, and 780 minutes. The process of adding the powder and the powder itself are assumed to be the same as in the 718 process, but the first 4 pool markings are indistinguishable in the bottom half of the macroetched ingot. However, at approximately 330 minutes of melting the entire metal pool nucleated equiaxed grains. This transition from columnar to equiaxed grains serves as the only suitable pool marking during the first half of the process.

No event in the process data is linked to the transition from columnar to equiaxed grains. There is an unaccounted-for drop in the experimental melt rate at this time. Figure 5 shows the 0.01 fraction solid contour overlaid on the columnar to equiaxed transition for both cases. Both cases predict the shape of the sump at this time as being fairly close to the transition region in the ingot, which is expected due to the melt rates of the simulations being within 10% of the experimentally observed value. This particular sump shape is qualitatively different from sumps observed in the 718 process due to the marking taking place in a region in the process where the melt rate is not at steady state.

In order to understand how this shape comes about, Figure 6 shows a montage of predicted sump shapes around the time that the columnar to equiaxed transition was observed. Before the current spike at 270 minutes the sump exhibits in Figure 6a a familiar V shaped pool. The sudden increase in current immediately increases the melt rate, causing the slag/metal interface to advance much quicker than the solidifying metal sump. This leads to a steeper sump near the wall where the close proximity to the mold wall allows for a fast response time to changes in the melting conditions as shown in Figure 7. As the current is lowered, the sump near the wall is again the first to react. However, the sump near the midradius advances faster than at the mold wall as it has adjusted to the higher melt rate and the depth is still increasing after the current has been lowered. This radial variation in transient sump velocity leads to the distinctive sump shape seen at the time of the columnar to equiaxed transition (Figure 6c). After being held at steady state for 60 minutes, all traces of the transient sump shapes are gone and the profile once again reverts back to the V shape.

Waspaloy Surface Defects

As shown in Figure 8, the surface of the Waspaloy ingot becomes very rough from 315 to 465 minutes in the process, The onset of this increased roughness coincides with the columnar to equiaxed transition discussed above, however it is unclear whether the increased thickness of slag skin frozen to the mold wall caused the transition. Both simulations also predict extra slag skin formation during this phase of the process as shown in Figure 9 for case wasp-2. The simulation correctly predicts additional slag freezing at the mold wall at 315 minutes. Over the next 85 minutes, the model predicts relatively smooth slag skin formation under one centimeter in thickness (shown in Figure 9a and 9b at times 319 and 374 minutes, respectively). At 400 minutes, the current is spiked for 30 minutes from 10.7 kA to 15.2 kA and the model predicts this short increase will stop extra slag skin formation (shown at 418 minutes in Figure 9c), as expected when considering the model severely overpredicts melt rate in this region. Once the low current level is restored at 430 minutes, extra slag skin formation is predicted to form again (shown in Figure 9d at 468 minutes). In the experimental images there is no gap in defect formation.

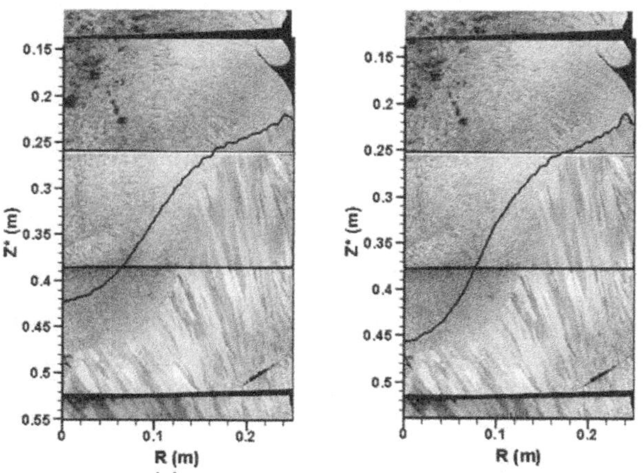

Figure 5: Comparison of 0.01 fraction solid contour for (a) Wasp-1 and (b) Wasp-2 to the columnar to equiaxed transition in the Waspaloy ingot at 330 minutes. It should be noted that, due to differences in melt rate between simulation and experiment, the Z* used is from the simulation and overlaid on the experimental ingot during the columnar to equiaxed transition.

Estimating the thickness of these defects is difficult. From the macro-etched slices, the variations in ingot radius are approximately 0.02-0.03 m, which is 3-4 times the extra slag skin thickness predicted by both simulated cases. Property values of slag skin are not well documented in the literature and so it is reasonable that quantitative predictions of slag skin thickness from the model will not agree with the experimental observations, however qualitatively the model is in agreement with the experiment showing extra slag skin formation during low current regions in the process.

Conclusions

The ESR model was used to simulate experimental work carried out Haynes International, Inc for two alloys remelted under different conditions. The melt rate predicted varied in accuracy in both processes. In alloy 718, the trend was that the simulations slightly overpredict melt rates at high currents, while more severely underpredicting melt rates at lower currents. In the Waspaloy process, the melt rate was severely overpredicted at high currents and only slightly overpredicted at lower currents. In both processes, the two simulations run for each alloy had very similar melt rates despite varying the slag electrical conductivity and heat transfer to the mold.

The sump shapes predicted by the model were compared to steady state as well as transient experimental shapes, showing generally good agreement, taking into account the closeness of the predicted melt rate to the experiment. The evolution of sump shapes was found to be primarily controlled by heat conduction through the solid alloy to the mold, with the distance from the mold wall determining how quickly the sump responds to changes in the melt rate. The Waspaloy process offered a look at the transient behavior of the sump during a non-standard ESR practice and the model was found to qualitatively agree with the experimental shapes.

Lastly, the model was found to predict extra slag formation in both processes. In the 718 case, this surface defect did not appear in the ingot because the simulations underpredicted the melt rate at lower currents. In the Waspaloy process, slag skin defects were correctly predicted to occur during low current levels.

Acknowledgements

The authors would like to thank Mr. Chris Jeanfils, formerly of Haynes International, who supervised the original ingot production, an effort supported by Haynes and the National Science Foundation. Major portions of the current research were funded by the National Science Foundation via grants CMMI-0456534 and CMMI-0900624, the latter being a GOALI award in partnership with Special Metals Corporation and Haynes International. Funding for Mr. Fezi was provided by a gift from PCC.

References

1. Hoyle, G., *Electroslag Processes*. 1983. New York, NY. Elsevier Science Publishing Company.
2. *Electroslag Remelting*. in *ASM Handbook*, 2008. **15**. p. 124-131.
3. Kelkar, K.M., S.V. Patankar, and A. Mitchell. *Computational Modeling of the Electroslag Remelting (ESR) Process Used for the Production of Ingots of high-Performance Alloys* in *LMPC 2005*. 2005. Santa Fe, NM, USA.
4. Patel, A.D. and K.M. Kelkar. *New Insights into the Electro-Slag Remelting Process using Mathematical Modeling* in *MCWASP XII*. 2009. Vancouver, Canada. pp. 69-76.
5. Kelkar, K.M., S.V. Patankar, A. Mitchell, R.S. Minisandram, and A.D. Patel. *Computational Analysis of the Vacuum Arc Remelting (VAR) and Electroslag Remelting (ESR) Processes* in *ASM Handbook*. 2010. **22B**. pp. 196-213.
6. Weber, V., A. Jardy, B. Dussoubs, D. Ablitzer, S. Rybéron, V. Schmitt, S. Hans, and H. Poisson. *A Comprehensive Model of the Electroslag Remelting Process: Description and Validation*. Metallurgical and Materials Transactions B. 2009. **40**(3). p. 271-280.
7. Weber, V., *Simulation numérique du procédé derefusion sous laitier électroconducteur*. Doctoral Thesis. 2008. Institut National Polytechnique de Lorraine. Nancy, France.
8. Kharicha, A., W. Schützenhöfer, A. Ludwig, and G. Reiter. *Influence of the Slag/Pool Interface on the Solidification in an Electro-Slag Remelting Process*. Materials Science Forum. 2010. **649**. pp. 229-236.
9. Kharicha, A., A. Ludwig, and M. Wu. *3D Simulation of the Melting During an Industrial Scale Electro-Slag Remelting Process*. in *LMPC 2011*. 2011. Nancy, France. pp.41-48.
10. Hirt, C.W. and B.D. Nichols, *Volume of fluid (VOF) method for the dynamics of free boundaries*. Journal of Computational Physics, 1981. **39**(1). pp. 201-225.
11. Yanke, J., *Numerical Modeling of Materials Processes with Fluid-Fluid Interfaces*. PhD dissertation. 2013. School of Materials Engineering, Purdue University. West Lafayette, IN, USA.
12. Bennon, W.D. and F.P. Incropera. *A continuum model for momentum, heat and species transport in binary solid-liquid*

phase change systems-I. Model formulation. International Journal of Heat and Mass Transfer. 1987. **30**(10). p. 2161-2170.

13. Patankar, S.V., *Numerical Heat Transfer and Fluid Flow*. 1980. Washington, DC. Hemisphere Publishing.
14. Weber, V., A. Jardy, B. Dussoubs, D. Ablitzer, S. Rybéron, V. Schmitt, S. Hans, and H. Poisson. *A Comprehensive Model of the ElectroSlag Remelting Process: Description and Validation* in *LMPC 2007*. 2007. Nancy, France. pp. 83-88.
15. Krane, M., M. Fahrmann, J. Yanke, E. Escobar, K. Fezi, and J. Busch. *A Comparison of Predictions of Transport Phenomena in Electroslag Remelting To Industrial Data* in *LMPC 2011*. 2011. Nancy, France. pp. 65-72.

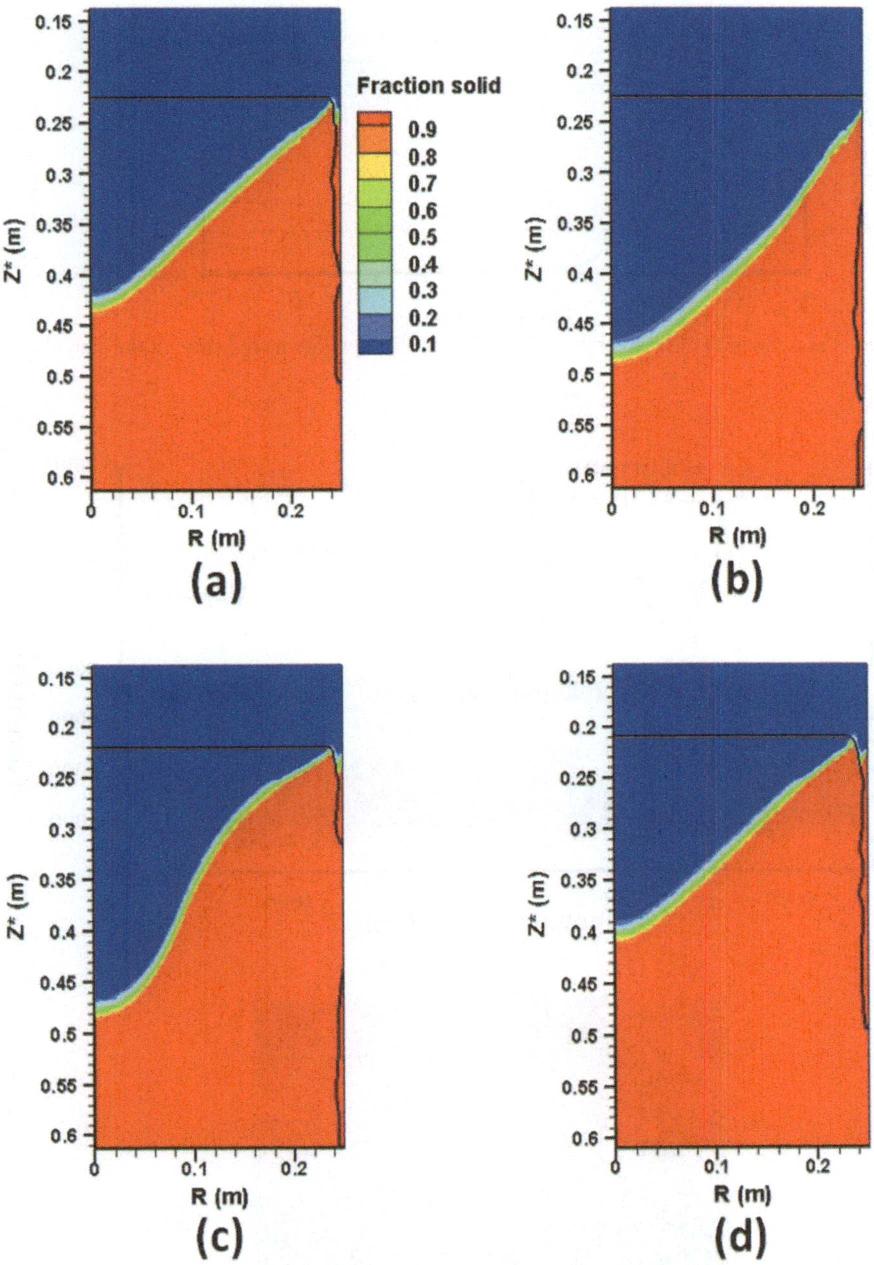

Figure 6: : Predicted sump shapes and slag skin formation (line contour of 0.5) for case Wasp-2 at times (a)-(d) of 270,294,330, and 390 minutes respectively showing the evolution of the sump shape due to a 30 minutes increase in current. Note that the Z* axis represents distance from the electrode.

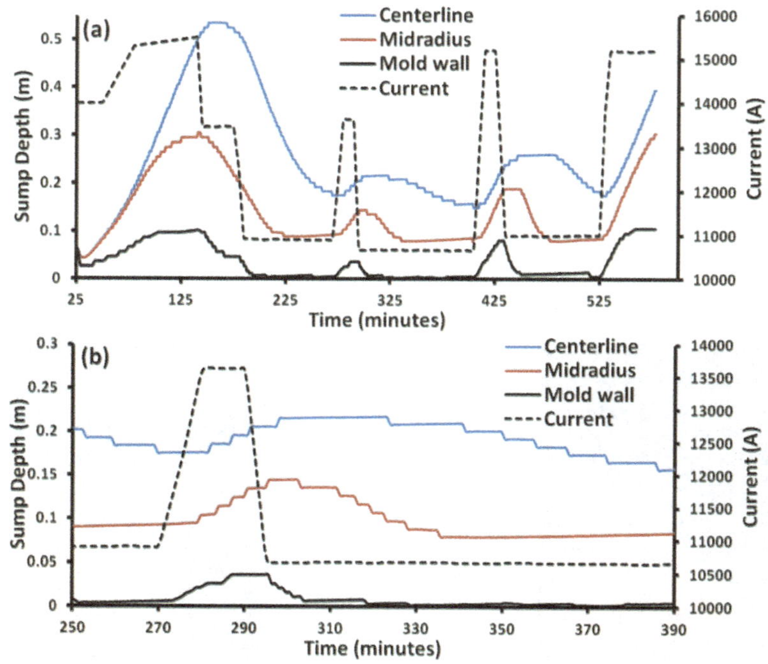

Figure 7: Sump depth and current as a function of time for case Wasp-2 with (b) showing the response of the sump to a 30 minute spike in current.

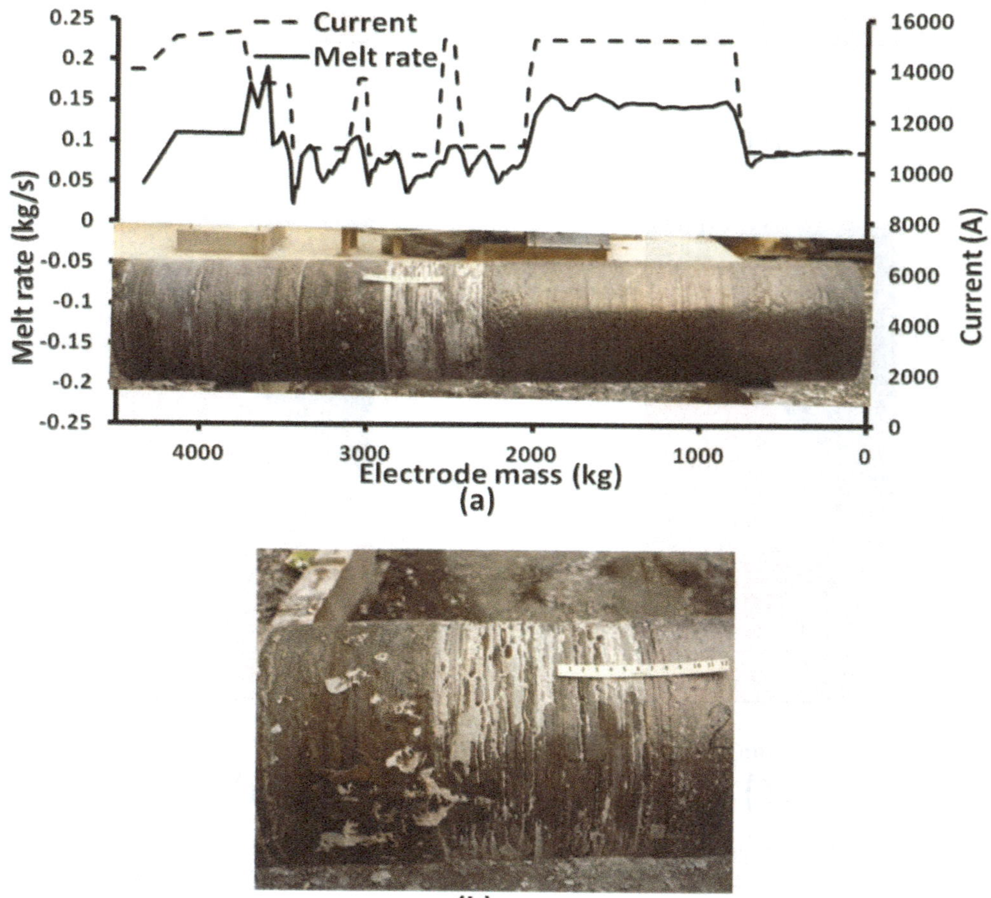

Figure 8: Waspaloy ingot surface (a) overlaid on a graph of current as a function of ingot weight to show the effect of the current on the surface quality of the ingot. (b) A close up of the roughest part of the ingot surface.

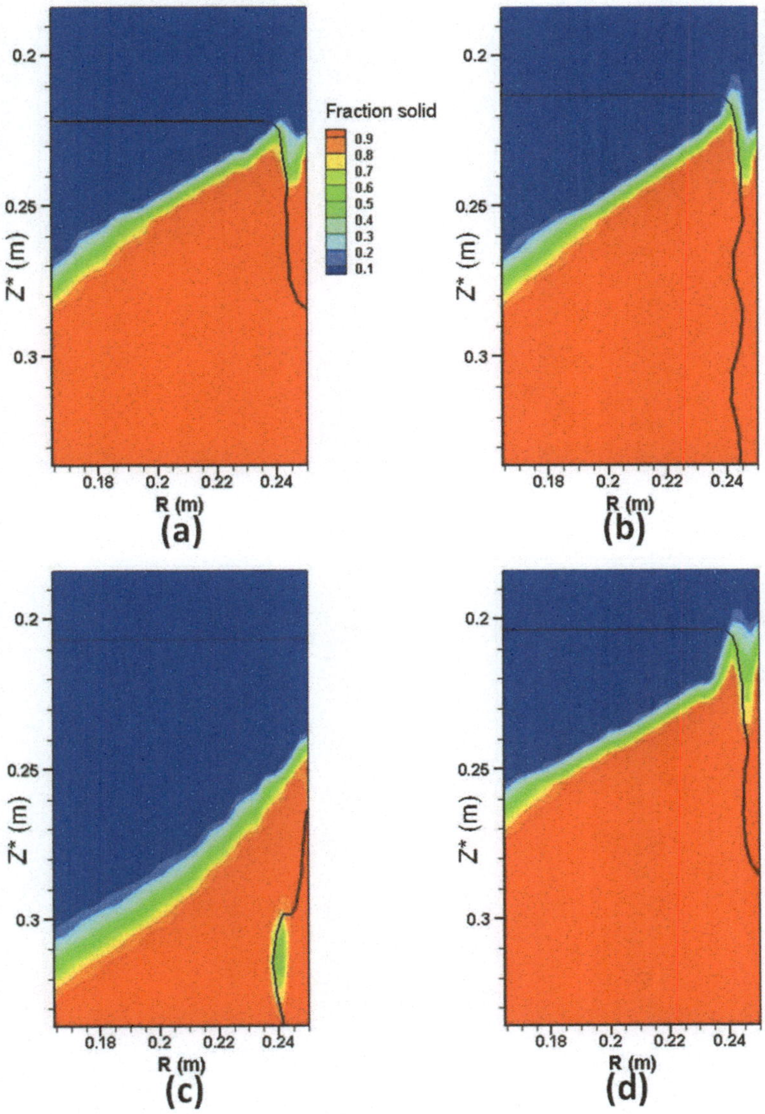

Figure 9: Fraction solid plots for case Wasp-2 with the black line representing the slag/metal interface. at times (a) 319, (b) 374, (c) 418, and (d) 468 minutes.

Proceedings of the 2013 International Symposium on Liquid Metal Processing & Casting
Edited by: Matthew J. M. Krane, Alain Jardy, Rodney L. Williamson, and Joseph J. Beaman
TMS (The Minerals, Metals & Materials Society), 2013

OPERATIONAL EXPERIENCE OF LARGE SIZED ESR PLANTS AND ATTAINABLE QUALITY OF ESR INGOTS WITH A DIAMETER OF UP TO 2600mm

Michael Kubin[1], Alexander Scheriau[1], Matthias Knabl[1], Harald Holzgruber[1] and Hiroshi Kawakami[2]
[1]INTECO Special Melting Technologies GmbH; Wienerstrasse 25; Bruck an der Mur, 8600, Austria
[2]Japan Casting & Forging Corporation; 46-59 Sakinohama, Nakabaru, Tobata-ku; Kitakyushu, 804-8555, Japan

Keywords: big ESR plants, start phase, hot topping, electrode change, inner quality, surface quality

Abstract

In the last years the field of remelting was mainly characterized by the commissioning of large sized ESR plants for the production of ESR ingots with a diameter up to 2.600 mm and weights up to 250 tons. These ESR plants are designed for short collar mold as well as static mold operation with the possibility of electrode change for both concepts. The steel grades of such big ESR ingots vary from low alloyed carbon steels for the nuclear power industry to complex chromium martensitic stainless steels for the power generation industry.

The present paper deals with the operational experiences of large sized ESR plants especially with the challenge of a well controlled starting and hot topping process as well as the electrode change procedure. Results will be a lso given regarding the attainable quality of ESR ingots produced in a 250 tons static mold as well as a 145 tons short collar mold ESR plant.

Introduction

The electroslag remelting (ESR) process is nowadays widely used for the production of heavy forging ingots. Especially the advantages of remelted products for end use in the power generation industry, chemical industry or oil and gas industry have gained increasing importance. The reason therefore is the better material quality, the higher yield as well as a better forgeability of ESR ingots as opposed to a conventional cast ingot. Hence, over the last years a huge demand of heavy forging ESR ingots has evolved up to an ingot weight of 250 tons. This development in ESR ingot dimensions poses a challenge for the plant engineering companies regarding the design of such large ESR plants, for instance the mould and furnace design enabling a well controlled electrode change technology in short collar mould as well as in static mould operation.

The plant concepts as well as the considerations which have to be taken into account in designing and engineering such big ESR plants were already published in [1]. Based on these considerations the validation of the calculated parameters with the measured ones was described in [2].

Up to now a number of big forging ingots have been produced so far during the commissioning phase of various large sized ESR plants. In this paper the operational experiences and most important considerations regarding each remelting phase (start phase, steady state phase including the electrode change procedure and hot topping phase) are described in detail. Furthermore selected results regarding the attainable ingot quality are highlighted.

Remelting Steps in the Production of Big Sized ESR Ingots

In principle the entire electroslag remelting process can be divided into three main process steps. These steps are the starting process, the steady state remelting phase including one or several electrode changes and the hot topping process. **Figure 1** shows a schematic illustration of these remelting steps in dependency of the power input level.

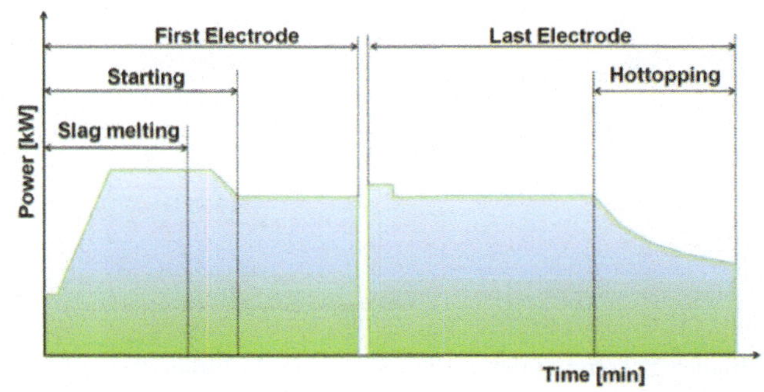

Figure 1: Schematic illustration of the remelting steps during the electroslag remelting process

In this chapter the most important parameters as well as remelting steps which need a precise control are described.

The Starting Process

For the production of big ESR ingots the start phase of the remelting process plays a very important role. It should be stated at this point that for these big dimensions the preferred start method is the cold start method with solid slag. The reason lies mainly in the easier handling of such big amounts of slag in cold conditions as well as the possibility to start with a defined start slag composition.

Therefore the following parameters must be precisely set and have a main impact on the starting behavior and subsequent steady state remelting phase:

- Starting parameters such as voltage, current and time to ensure the complete melting of the entire slag quantity
- Type and condition (dry) of the used slags especially to ensure no or limited hydrogen pick up
- Special additions in the start slag such as deoxidizing elements (FeSi, etc.)
- Right quantity of starting slag for a good usable ingot bottom
- Slag addition sequence for a close control of the chemical analysis
- Control of furnace atmosphere

All these parameters need to be precisely adjusted and controlled to get a smooth change over to the steady state remelting phase and to ensure a high yield of the ESR ingot as well as close control of the chemical analysis especially in the bottom part of the ingot.

In

Figure 2 the trend of the start phase of a 2.600mm dia. ESR ingot is shown. As can be seen the electrical parameter are set in a way that the melt rate slightly overswings the melt rate set point. This is important to compensate the bottom plate cooling effect. Furthermore the higher power input should ensure that the whole slag amount will be melted during the starting process.

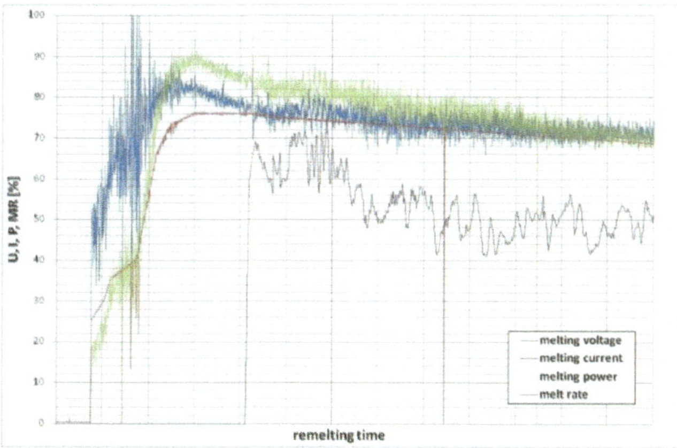

Figure 2: Trend of the starting phase of a 2600mm dia. ESR ingot. The blue line indicates the melting voltage (U), the red line the melting current (I), green line indicates the melting power (P) and the violet line is the melt rate (MR).

A further important point is that the aluminum and silicon content is kept within the required specification. Thus, certain issues have to be considered. Due to the nature of the chemical interaction of the liquid steel with the typically used slags in ESR a certain aluminum pick up as well as a silicon loss can be expected [3, 4]. Therefore the right selection of the type and composition of start slag and the subsequent dosing of the remaining slag amount during the start phase play a decisive role for a close control of the chemical composition in the ESR ingot.

Moreover, operational experiences showed that the furnace atmosphere has an important influence on the achievable chemical distribution of alloying elements in the ESR ingot such as the right mixture of the used protective gas or the right time of protective gas purging.

In **Table I** the aluminum pick up and silicon loss compared to the electrode values of ESR ingots which were produced in a short collar mold ESR plant are shown. It can be concluded that the maximum losses of silicon is very limited and practically no aluminum pick up can be achieved with the right start up and remelting technology.

Table I: Silicon loss and aluminum pick up of 1900mm dia. ESR ingots produced in a 145ton short collar mold ESR plant at Japan Casting & Forging Corporation.

Steel grade	**Analysis position**	**Si loss[%]**	**Al pick up [%]**
Low alloyed steel	Top	max. 0,02	0
	Middle	max. 0,03	max. 0,001
	Bottom	max. 0,02	max. 0,002
10% Cr steel-A	Top	0	0
	Middle	0	0
	Bottom	max. 0,01	0
10% Cr steel-B	Top	max. 0,01	0
	Middle	max. 0,01	0
	Bottom	max. 0,02	0
10% Cr steel-C	Top	max. 0,01	0
	Middle	max. 0,01	0
	Bottom	max. 0,01	0

If all these parameters mentioned above are correctly set and adjusted, the bottom part of the ESR ingot not only shows a close control of the chemical analysis but also a very smooth surface and no unmelted slag can be recognized (see **Figure 3**). This is the main precondition to get a good usable ingot bottom part which highly increases the yield respectively good usable ingot part.

Figure 3: Bottom Part of a 1.900mm dia. ESR ingot produced in a 145t short collar mold ESR plant at Japan Casting & Forging Corporation. The bottom part shows a rectangular appearance without any unmelted slag and the surface is smooth.

Steady State Remelting Phase and Electrode Change Procedure

In the steady state remelting phase the process is controlled by the melt rate and immersion depth. In this process step the power input is controlled in such a way that a predefined melt rate is kept constant. Also the electrode immersion depth which has a big influence on the surface appearance is controlled using a swing controller.

The special focus lies on the following parameters:

- Control of a predefined melt rate
- Control of the electrode immersion depth
- Control of a predefined slag bath level
- Well controlled electrode change procedure
- Well controlled furnace atmosphere

Figure 4 shows the melt trend of a 2600mm dia. ESR ingot of a low alloyed steel grade usually used for large vessels and rings. The blue line indicates the melting voltage, the red line the melting current and the green line the melting power. This ESR ingot has a weight of approximately 180 tons. The melt trend indicates a very stable remelting operation over the whole ingot length which is very important due to the fact that the remelting operation lasts for several days. Furthermore it can be seen that this ingot was produced out of three electrodes.

The electrode change sequence represents an interference to the steady state remelting phase as a power interruption is necessary to continue the process with a new preheated electrode. Hence, special attention has to be given to this process step. Several important parameters have to be considered and controlled:

- Power input before/after electrode change
- Control of the immersion depth of the new electrode
- Resistance increase after electrode change due to a chilled slag pool
- Preheating temperature of the new electrode
- Stop/Start of baseplate retraction before/after the electrode change (in case of short collar mold/retractable base plate operation)
- Time between power off/on during electrode change

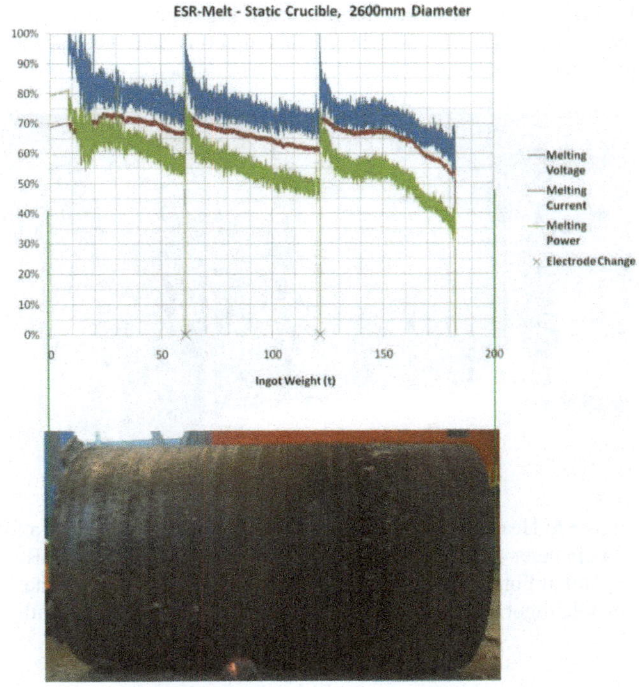

Figure 4: Multiple electrode melt trend of a 2600mm dia. ESR ingot produced in static mold operation with electrode change. The blue line indicates the melting voltage, the red line the melting current and the green line the melting power. The electrode change position is marked with an X.

Besides the physical interchange of the electrode also the pre-change-phase and the post-change-phase need to be controlled in an optimized way. The highly automatic and precisely adjusted electrode change operation ensures a very short interruption in the remelting operation as can be seen in **Figure 5**. The electrode change duration for such big ingots is in the range of 3 - 5 minutes. The whole ingot length shows a very good surface appearance and is smooth even at the position of the electrode change.

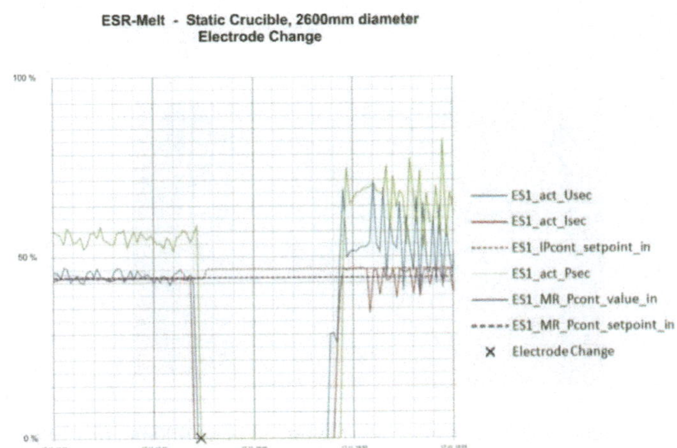

Figure 5: Melt trend of an electrode change in static mold operation. The duration of the electrode change is approximately five minutes. The blue line indicates the melting voltage, the red line the melting current and the green line the melting power.

Figure 6 shows a heavy forging ESR ingot produced in static mold operation. The weight of this ingot is 250 tons with a diameter of 2600mm.

Figure 6: Heavy forging ESR ingot with a weight of 250 tons and a diameter of 2600mm produced in a 250ton static mold ESR plant at Forgiatura A. Vienna. The surface is smooth over the whole ingot length, due to a very stable remelting process with well controlled electrode sequences.

In **Figure 7** a typical melt trend of a short collar mold ESR plant is shown on the example of a 1900mm dia. ESR ingot with a weight of approx. 60 ton. The steel grade of this ESR ingot is a 10% chromium steel which is commonly used in energy industry for big rotors and shafts.

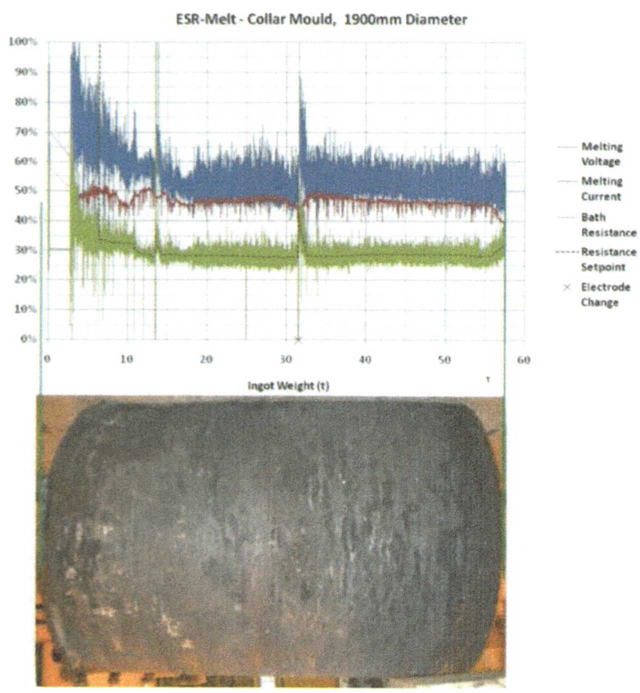

Figure 7: Multiple electrode melt trend of a 1900mm dia. ESR ingot produced in short collar mold operation with electrode change. The blue line indicates the melting voltage, the red line the melting current and the green line the bath resistance.

Figure 8 indicates the phase of the electrode change with a duration of the power interruption of approximately three minutes. This ingot also shows a very smooth surface and no distinctive grooves at the position of the electrode change.

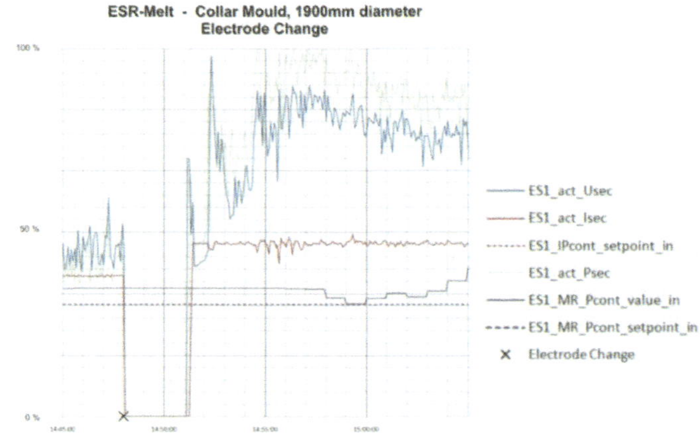

Figure 8: Melt trend of an electrode change in short collar mold operation. The power interruption lasts for about three minutes. The blue line indicates the melting voltage, the red line the melting current and the green line the melting power.

A further very important parameter in the steady state remelting phase is the freeboard setting respectively the liquid slag bath level.

For a precise detection and control of the liquid slag bath level INTECO special melting technologies has recently developed a radar slag bath level sensor, which already has reached industrial maturity.

The previous slag bath level determination method is based on calculations of the freeboard with assumed parameters, which results in some disadvantages and impreciseness of process control. Using a radar bath level control system it is possible to have continuous information of the actual slag bath level. This provides a more precise process control, e.g. automatic start and control of the ingot withdrawal mechanism and the operation at a preset slag bath level.

Figure 9 and **Figure 10** show the radar freeboard measurement of a 1900mm dia. short collar mould ESR plant indicating the actual freeboard (blue line) as well as the actual electrode carriage position (violet line) and actual baseplate position (green line) at the beginning of the remelting process. It can be seen that after the slag bath reaches a predefined level the baseplate retraction starts and the freeboard stays at a constant position.

Furthermore, in **Figure 10** it can be seen that as soon as the radar signal reaches a certain value the retraction of the base plate starts. These precise measurements and thus the steady-going base plate retraction have a positive impact on the process control and entire quality of ESR ingots.

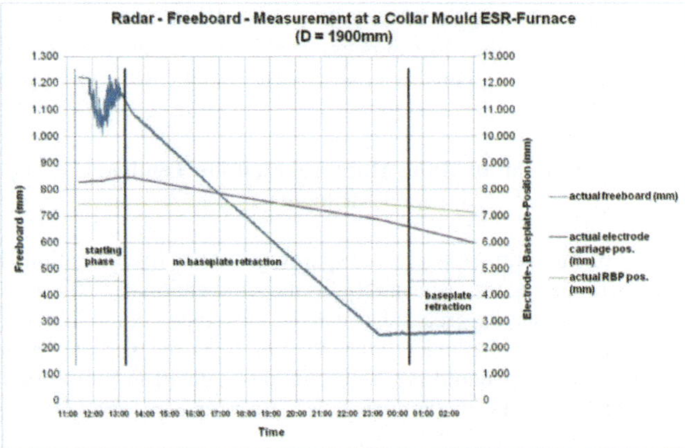

Figure 9: Trend of the radar freeboard measurement at a 1900mm dia. short collar mould ESR plant.

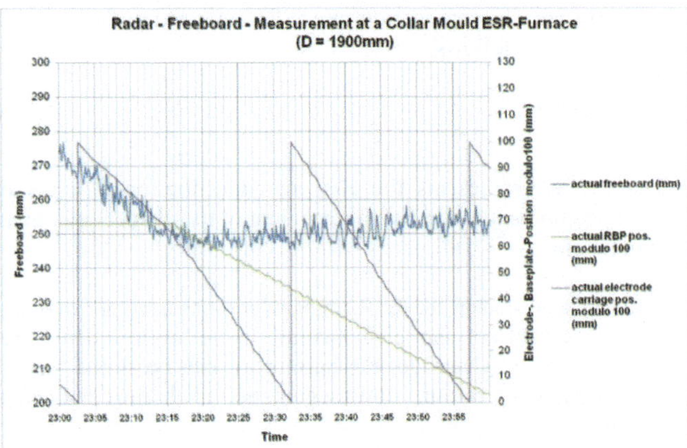

Figure 10: Radar freeboard measurement of the beginning of the baseplate retraction.

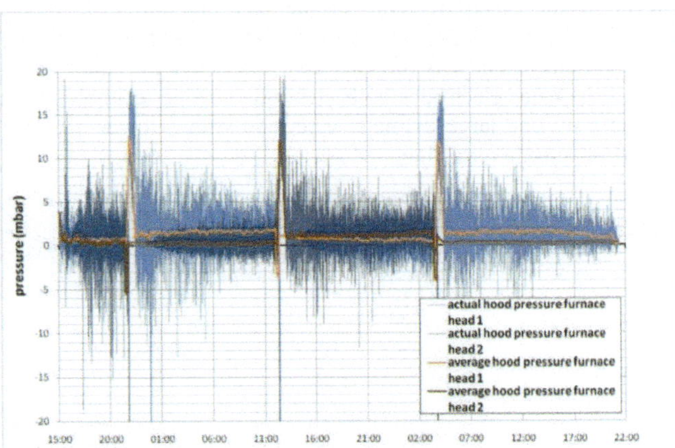

Figure 11: Hood pressure in dependency of the remelting time. The yellow and brown line indicate the average hood pressure. As can be seen a slight over pressure is ensured.

As already described in [5, 6] the atmosphere over the slag bath plays a decisive role for maintaining lowest gas contents of the remelted ingot. Therefore, protective gas which can be either argon, nitrogen or a mixture of both gases in conjunction with a slight overpressure (see **Figure 11**) is used to ensure no contact to the surrounding atmosphere or in other words no false air that can enter the furnace atmosphere.

For the same melt in Figure 12 the measured oxygen value within the protective gas hood can be seen. The oxygen content of the atmosphere was below 13ppm over the entire remelting time.

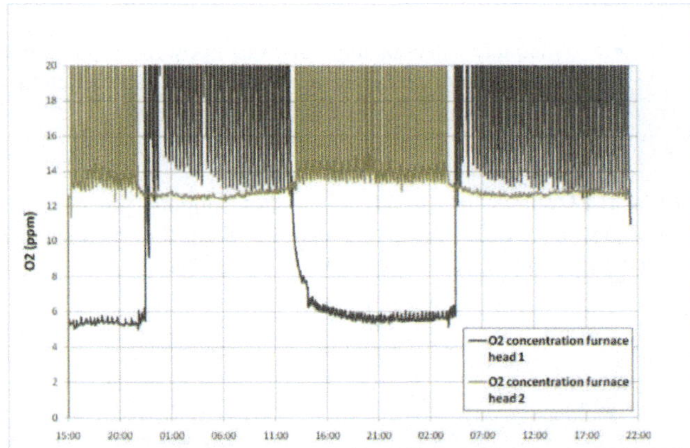

Figure 12: Oxygen content within the protective gas hood. The black line represents the oxygen content in furnace head 1 and the green line the oxygen content in furnace head 2.

This well controlled furnace atmosphere with respect to the hood pressure and oxygen content is also reflected in the obtained gas contents in the ESR ingots. In **Table II** the nitrogen and oxygen content of different ESR ingots are given.

Table II Oxygen and nitrogen pick up during the remelting process of ESR ingots produced in a 145ton short collar mold ESR plant at Japan Casting & Forging Corporation.

Steel grade	Analysis position	N pick-up [ppm]	O pick-up [ppm]
Low alloyed steel	Top	max. 13	0
	Middle	max. 15	0
	Bottom	max. 28	max. 6
10% Cr steel-A	Top	0	0
	Middle	max. 6	0
	Bottom	0	0
10% Cr steel-B	Top	0	0
	Middle	max. 14	0
	Bottom	max. 18	0
10% Cr steel-C	Top	max. 14	0
	Middle	max. 12	0
	Bottom	max. 12	0

The Hot Topping Process

An ideal hot topping process is mandatory to obtain a good usable ingot top part and has a big influence on the overall yield of the ESR process especially for big ESR ingots.

For a well controlled hot topping procedure, the following parameters have to be taken into account:

- Adequate decreasing of the power input level respectively right settings of the electrical parameters
- Sufficient time for the hot topping process
- Enough hot topping weight

Figure 13 shows the trend of the hot topping process of a 2.600mm dia. ESR ingot. As can be seen the melt rate is constantly decreased over the whole hot topping time.

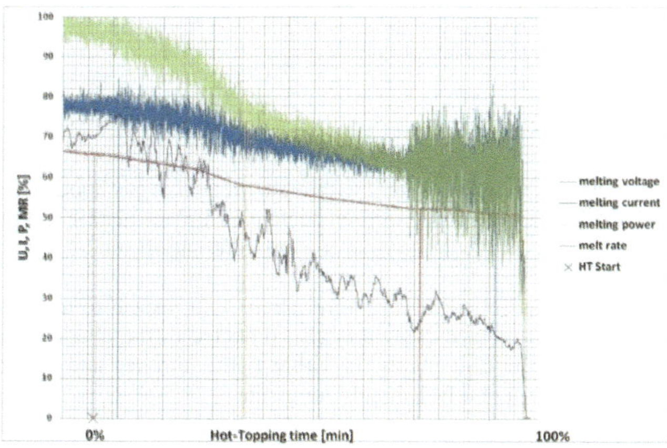

Figure 13: Trend of the hot topping process of a 2600mm dia. ESR ingot.

The main goal of the hot topping process is to get a dense top part of the ingot in conjunction with an acceptable surface quality.
If the hot topping curve is correctly adjusted a dense and sound top part of the ESR ingot can be expected. In **Figure 14** the cutted hot top part after the first forging step of a 2.600mm dia. 180tons ESR ingot is shown. The hot top part was cut longitudinally and analyzed with respect to the ingot structure. The results show that the hot topping was correctly performed and no porosities or slag inclusions were found.

Figure 14: Top part of a 2.600mm dia. ESR ingot after forging produced by Forgiatura A. Vienna. The structure is dense and no slag inclusions can be found.

The main advantages of a well executed hot topping process respectively of the ESR process is the higher obtainable yield compared to a conventionally cast ingot with the same dimensions.

First Evaluation of the Macrosegregation Tendency of big ESR ingots

A lot of literature is available regarding the macrosegregations in big forging ingots, especially for the conventionally cast ones. Generally, macrosegregations can be observed along the ingot axis and along the ingot cross section from the surface to the center. But this should not be discussed in detail in the scope of this paper. For further information the reader is referred to some selected literature [7, 8, 9].

To give a first idea about the chemical distribution, a comparison of the carbon segregation indices for the bottom, middle and top part of the ESR ingot with the corresponding electrodes is shown in **Figure 16**. The segregation rate (SR) indicated in the diagrams is calculated as follows:

$$SR = \frac{c_{position}}{c_0} \quad (1)$$

where $c_{position}$ represents the element concentration at a certain radial position and c_0 the average content of the element at a given ingot height.

It can be seen that in the ESR ingot which has a bigger diameter than the electrode, the elements are more evenly distributed compared to the electrode. Furthermore it can be stated that for these big ingots a certain center segregation is obvious and can be expected.
In
Figure 15 the carbon segregation rate of two electrodes with a different diameter is shown. It can be concluded that the bigger the electrode diameter the more pronounced is the segregation level of the electrode. Due to the fact that the ESR process generally improves the chemical distribution of the elements only to a certain extent it is obvious that it is better to remelt several small diameter electrodes with electrode change technology than using only one large diameter electrode with a lower quality in single electrode remelting.

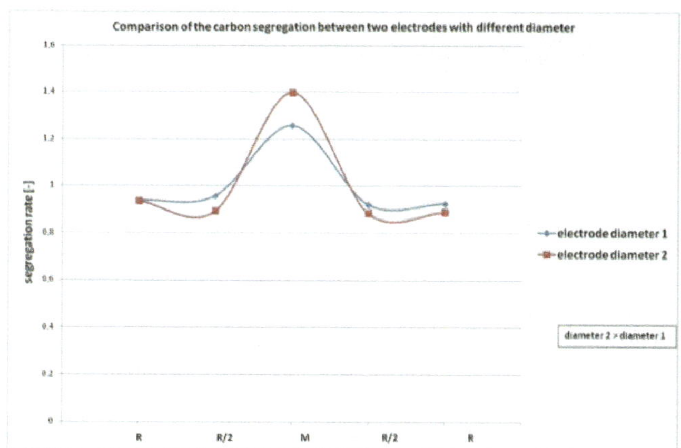

Figure 15: Comparison of the carbon segregation rate between two electrodes with a different diameter. As can be seen the

distribution of the carbon is much more uniform in the electrode with the smaller diameter.

Summary, Conclusion and Outlook

The results given in this article show that today it is absolutely possible to produce big forging ingots with a diameter up to 2.600mm in the ESR process in an excellent quality regarding the surface appearance as well as the chemical composition.

It is proven that ESR ingots with a weight exceeding 100tons show better results using smaller electrodes and applying electrode changes compared to single electrode remelted ESR ingots.

In order to achieve these results, certain issues have to be considered. In the starting phase the main influencing factor on the ingot quality is the used slag composition as well as the adequate slag addition sequence. This can ensure a well formed ESR bottom part with a uniform and well controlled chemical composition.

The most challenging factor in the steady state remelting phase is a well controlled electrode change. Furthermore the furnace atmosphere plays a very important role for controlling the gas content of the remelted product.

The hot topping process as the last main remelting step is most important to get the highest yield. This is achieved when the ESR ingot top part shows a dense and homogeneous structure.

At this point it should be also emphasized that all chemical analysis and metallurgical results which are presented in the scope of this paper are within customer's specifications. The results should show that also in these big dimensions the advantages of the ESR process are still valid compared to a conventionally cast ingot of the same dimensions.

The conclusions and selected results gained during the commissioning of the big ESR plants in close cooperation with INTECO's customers provide promising results for the remelting of ingots with a diameter up and exceeding even 2.600 mm.

Furthermore, our customer Forgiatura A. Vienna, in consideration of the obtained results, is currently evaluating an extension of the size of ESR ingot to a weight of 340 tons.

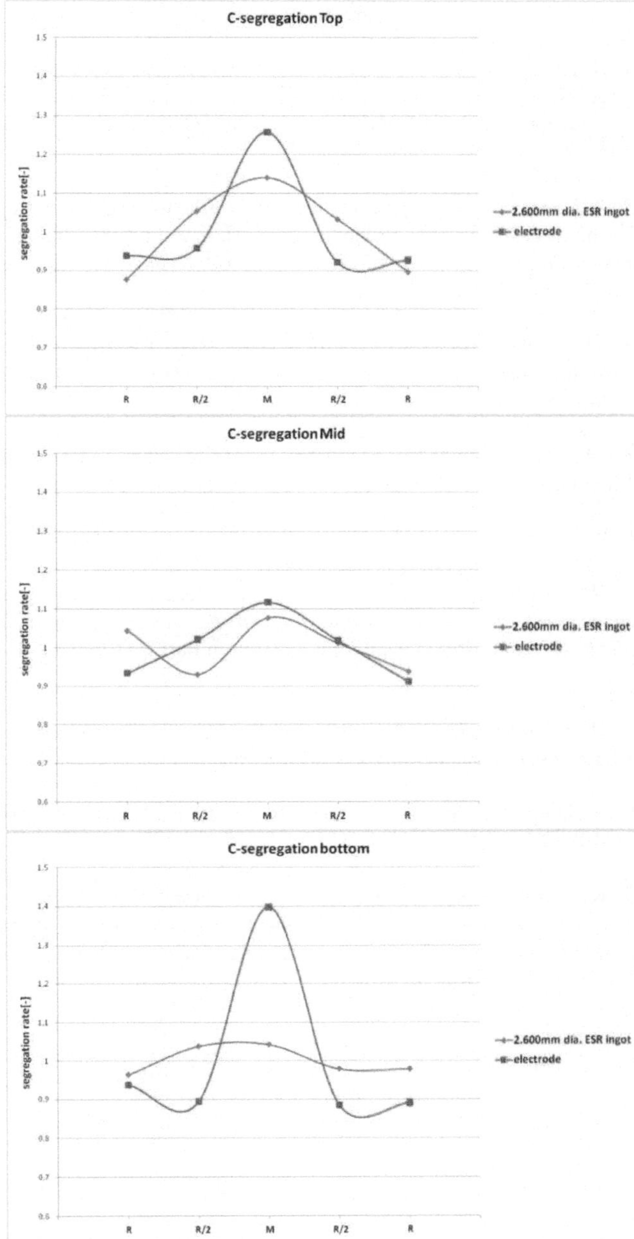

Figure 16: Carbon segregation rate along the cross section of the ESR ingot and the electrodes for the top, middle and bottom part. The carbon content is much more evenly distributed in the ESR ingot than in the electrode.

Acknowledgments

I gratefully acknowledge the extensive support of our customers Forgiatura A. Vienna and Japan Casting and Forging Cooperation in providing the data regarding the metallurgical results which are published in this work.

We, as INTECO special melting technologies, are very pleased to have such highly innovative and technologically experienced customers with who we have successfully commissioned the big ESR plants.

References

1. H. Holzgruber et al., "New plant concepts for the production of large sized ESR ingots", *Proceedings of the International Symposium on Liquid Metal Processing and Casting*, (2009), 273-285.

2. M. Kubin et al., "Future Trends in Electroslag Remelting Technologies for the Production of Heavy Forging Ingots", *Proceedings of the First International Conference on Casting, Rolling and Forging*, (2012),

3. W. Holzgruber and E. Plöckinger, "Das Elektroschlacke-Umschmelzen", *Berg- und Hüttenmännische Monatshefte*, 113 (3) (1968), 83-93.

4. A. Mitchell, "The chemistry of ESR slags", Canadian Metallurgical Quarterly 20 (1) (1981), 101-112.

5. W. Holzgruber et al., *Stahl und Eisen* 88, pp. 638/648 (1968)

6. W. Holzgruber, Oxygen control in the electroslag refining of steel, *First International Symposium on electroslag consumable electrode and casting technology*, Pittsburgh Part II (1967)

7. H. Trenkler and W. Krieger, *Gmellin-Durrer Metallurgie des Eisens. (10),* Springer Verlag, Berlin (1988)

8. B. Sang et al., Journal of Materials Processing Technology 210 (2010), 703–711

9. K. Kajikawa et al., "Development of 650ton Class Ingot for Turbine Rotor Shaft Forging Application", *Proceedings of the first International Conference on Ingot Casting, Rolling and Forging* 2012, Aachen.

Proceedings of the 2013 International Symposium on Liquid Metal Processing & Casting
Edited by: Matthew J. M. Krane, Alain Jardy, Rodney L. Williamson, and Joseph J. Beaman
TMS (The Minerals, Metals & Materials Society), 2013

EXPERIMENTAL RESEARCH ON THE ABSORPTION OF FLUORINE IN GAMMA-TiAl DURING ELECTROSLAG REMELTING

P. Spiess[1], B. Friedrich[1]
[1]IME Process Metallurgy and Metal Recycling, RWTH Aachen University, D-52056 Aachen, Germany

Keywords: TiAl, Recycling, PESR, Halogen Effect, Fluorine

Abstract

In literature, it has been reported that the so called "halogen effect" significantly improves the oxidation resistance of gamma-TiAl. This effect is based on the formation of gaseous Al-halides which are oxidized to Al_2O_3 on the surface resulting in a protective alumina layer. To investigate the absorption of fluorine in gamma-TiAl during electroslag remelting and to enhance the oxidation resistance of the produced alloy by bulk-fluoridation, a series of tests was performed at IME. In a 400 kW lab scale furnace multiple electrodes of Ti-45Al were remelted using an active Ca-CaF_2 slag. Parameters like melt rate and Ca content in the slag were systematically varied. The ingots were sectioned and characterized by GDOS and EPMA to demonstrate the influence of the melting parameters on the efficiency of fluoridation as well as their impact on solidification structure with special regard to the distribution of fluorine.

Introduction

The development of structural intermetallic alloys for industrial use has been a focus since the 1980s. Especially the group of titanium aluminides is characterized by its very good properties like high melting point (about 1450 °C), low density (3.9 – 4.1 g/cm³) high specific elasticity modulus, good oxidation resistance up to 700 – 800 °C and high specific creep resistance (39 – 46 GPa cm³ g⁻¹). Due to the steady increasing use of titanium in airframes (cf. Figure 1) and TiAl in combustion engines as an alternative to superalloys, fuel consumption as well as pollutant emissions can be reduced significantly. [1-3] Because of confidentiality no precise information on the TiAl content in combustion engines are available.

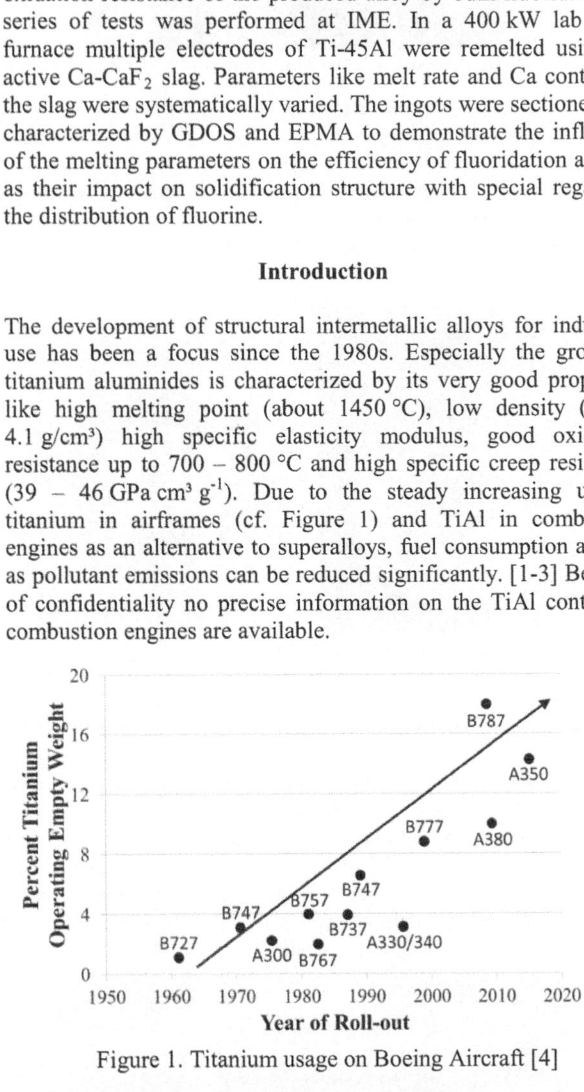

Figure 1. Titanium usage on Boeing Aircraft [4]

To further increase the efficiency of combustion engines it is appropriate to raise their operating temperature. However, operating temperatures above 800 °C lead to a considerable drop of the oxidation resistance of TiAl, due to the formation of an oxide layer consisting of Al_2O_3 and TiO_2. Both oxides have different rates of growth and expansion coefficients resulting in crack formation in the oxide layer. Although the oxidation protection can be increased by the addition of niobium since niobium lowers the Al activity. Niobium is attached to the Ti sub-lattice of α_2-Ti_3Al and γ-TiAl structure and influences the kinetics of flaking, whereby a finer grain structure is achieved. However, niobium contents above 2 wt.-% lead to a massive drop of Al activity which prevents the formation of Al_2O_3. Given that TiO_2 is a weak oxygen barrier the oxygen diffusion is accelerated. [5-6] Schütze [6] observed that the oxidation resistance can be further increased by microalloying of halogens like chlorine, iodine, bromine or fluorine, whereas Donchev [7] identified that only fluorine leads to a sustained increase of the oxidation resistance. This is based on the so called "halogen effect", wherein small amounts of a halogen are added into the alloy surface. At high temperatures the deposited fluorine in the metal edge zone selectively forms Al-fluorides, which shift through pores and microcracks to the metal surface. Due to the increasing oxygen partial pressure the fluorides decay and form gaseous F_2 and a dense Al_2O_3 layer.

The TiAl recycling process developed at IME consists of multiple industry approved processes and is explicitly explained by Reitz [8]. It is based on the deoxidization of via vacuum induction melting (VIM) consolidated casting scrap by pressure electroslag re-melting (PESR). To achieve a designated refining and deoxidization of the metal during PESR a reactive slag consisting of CaF_2 and metallic Ca is used. Due to its low vapour pressure, physical properties and good availability Nafziger [9] identified CaF_2 as the most suitable flux for ESR of titanium and titanium alloys. Because of the thermodynamic effects during ESR of titanium at high temperatures of around 1800 °C, CaF_2 decomposes (cf. eq. 1). By reaction of existant atomic fluorine radicals with the liquid metal, titanium fluorine TiF is formed (cf. eq. 2) and dissolves in the melt (cf. eq. 3).

$$(CaF_2) = \{CaF\} + \{F\} \quad \text{with } \Delta G_{R,1} \tag{1}$$

$$(Ti) + \{F\} = \{TiF\} \quad \text{with } \Delta G_{R,2} \tag{2}$$

$$\{TiF\} = [TiF]_{Ti} \quad \text{with } \Delta G_{R,3} \tag{3}$$

Scholz [10] reported that fluorine contents of 60 ppm after ESR of titanium are unavoidable. To obtain a positive effect of the F-effect on the oxidation resistance, an expected fluorine content of up to 500 ppm is needed. Since there is almost no thermodynamic data for equilibrium calculations between TiAl and CaF_2 as well as for evaporation tendencies, the conducted investigations engage on the experimental research on the absorption of fluorine in TiAl.

Experimental

In order to investigate the influence of the used process parameters on the fluorine absorption multiple remelting trials with a specific variation of single parameters were performed. The experiments were conducted at IME Process Metallurgy and Metal Recycling, Department and Chair of RWTH Aachen University.

Electroslag Re-melting Furnace (ESR)
Based on the method of electro-slag welding (ESW) in the 1880s Slavyanov developed the electroslag re-melting process. [11] An overview of the basic operating mode of ESR gives the subsequent Figure 2.

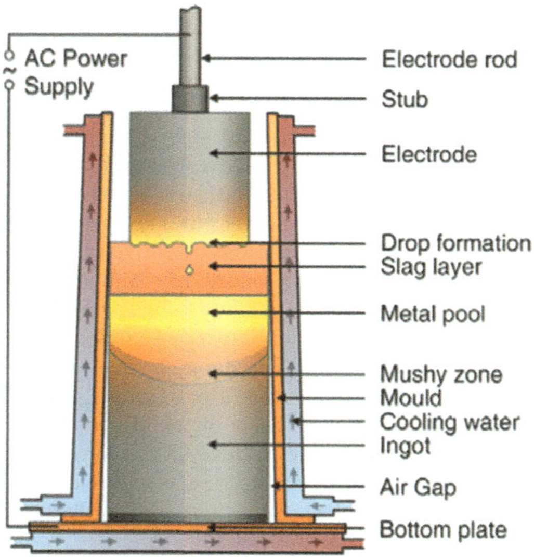

Figure 2. Principle design of an ESR furnace [12]

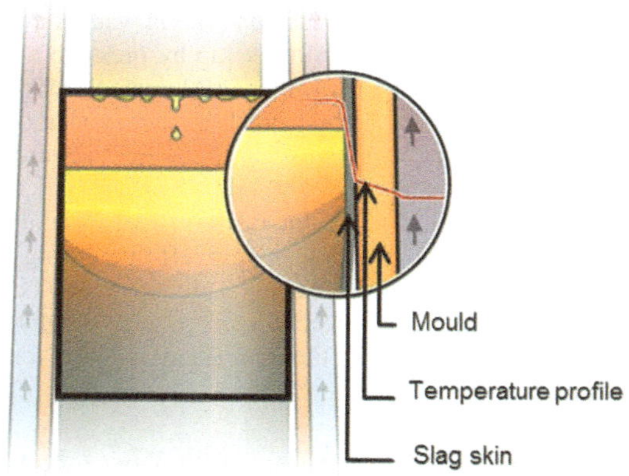

Figure 3. Heat removal during the ESR process [12]

During the electroslag re-melting process a self-consumable electrode is gradually molten by contact with a defined liquid slag. The dripping metal sinks through the slag while solid inclusions with a melting temperature higher than that of the base-metal will float into the slag. However, this float will only take place if the solid inclusions possess a lower density than the molten base-metal. Depending on the chemical properties and density of the inclusions they can be dissolved in the slag. The dripping metal solidifies partly directional in a water cooled copper crucible. [13]

Experimental Setup
The IME pressure electroslag remelting furnace used for the investigations (cf. Figure 4) is controlled by means of computer-aided software. As feedstock electrodes with a maximum length of 1340 mm and a diameter of up to 110 mm can be used. The available water-cooled molds have a height of up to 900 mm and an inner diameter of about 170 mm. The closed system can be at pressures up to 50 bar. The power supply of the furnace is carried out by a thyristor control, where an operating voltage of 80 V and a current of 5 kA, 66.6 V and 6 kA respectively can be tapped. Both cases result in a maximum power output of approximately 400 kW.

Figure 4. IME **P**ressure **E**lectro**S**lag **R**emelting furnace (PESR)

The experiments were performed in the PESR in closed state. An 890 mm high conical copper mold with a removable bottom plate was used. Its lower diameter is 170 mm while its upper diameter is 154 mm. A starting plate made of TiAl sputter targets (cf. Figure 5) is placed on the crucible bottom to ensure electrical contact during the starting phase of the process. Then the crucible is filled with slag.

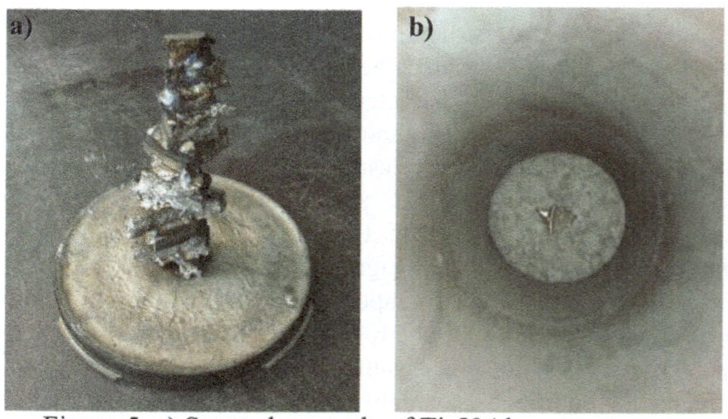

Figure 5. a) Starter box made of Ti-50Al sputter targets
b) Starter box in crucible surrounded by slag

As slag the technically pure CaF_2 slag Wacker Electroflux 2052 (> 97 wt.-% CaF_2) was used. According to the process requirements metallic calcium as deoxidization agent was added. For remelting, pressed Ti-45Al electrodes (cf. Figure 6) made of titanium sponge and aluminum rods with a length of 1200 mm and a circumference of 301 mm were used. The furnace pressure was chosen to 20 bar Ar. For this purpose and in order to avoid reactions with oxygen the vessel is evacuated to 1^{-10} mbar and bachfilled with Ar gas up to the desired pressure.

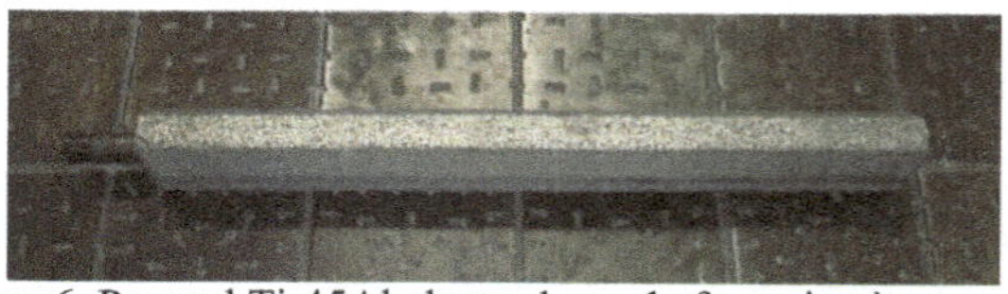

Figure 6. Pressed Ti-45Al electrode made from titanium sponge and aluminum rods

Experimental Procedure

After the setup of the furnace is completed, the process starts with an initial phase, whose process parameters are exemplified in the following

Table I. Thereby the control of the process at the beginning is done by controlling the current and the voltage. After the first 3 steps the regulation is changed to power and resistance.

Table I. Exemplary presentation of the stepwise change of process parameters during initial phase of each PESR experiment

Step	Current /kA	Voltage /V	Power /kW	Resistance /mOhm	Dwell /min
1	3.0	38.0	60.0	8.0	0.2
2	5.5	40.0	60.0	8.0	3.0
3	5.0	40.0	100.0	8.0	3.0
4	4.0	40.0	100.0	8.0	2.0
5	4.0	40.0	145.0	9.0	2.0

After reaching the melting phase a selective variation of the process parameters was carried out to examine their influence on the fluorine content of the TiAl alloy used. The varied process parameters are the remelting power as well as the amount of added deoxidization agent. An overview of the five PESR trials carried out with identical electrodes and the process parameter settings used are given in the following Table II.

Table II. Selective variation of the process parameters during remelting phase

Trial	Power /kW	Ca content in slag /wt.-%	Resistance /mOhm
A	125.0	0.0	9.2
B	100.0	0.0	9.2
C	125.0	2.5	9.2
D	125.0	5.0	9.2
E	145.0	2.5	9.2

Trial A serves as a reference experiment in which an average power of 125 kW and 4.5 kg Electroflux 2052 slag without any addition of deoxidization agent were used. For investigation of the influence of the remelting power in trial B the power has been reduced to 100 kW. Trials C and D are used to study the influence of the addition of deoxidization agent, where 2.5 wt.-% and 5.0 wt.-% respectively of the slag were replaced by calcium. Finally, in trial E the effect of an increased remelting power at medium calcium content in the slag was investigated.

After remelting the obtained ingots were sectioned and sampled in the mechanical workshop at IME and characterized by GDOS and EPMA. The detection limit for fluorine was 100 ppm.

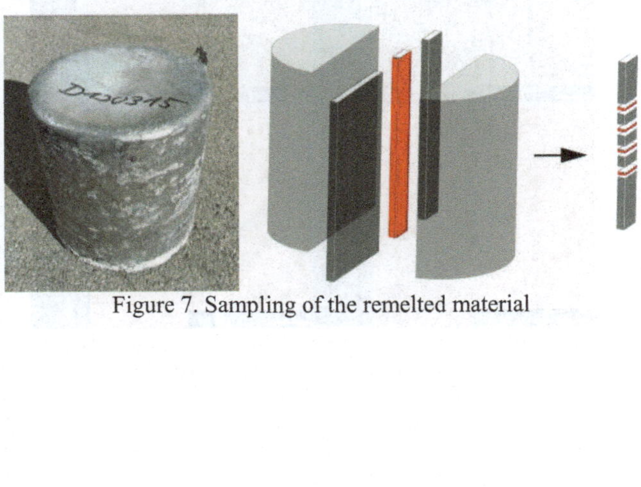

Figure 7. Sampling of the remelted material

Results

The prepared samples were analyzed both on the titanium and the aluminum content. The results are shown in the following Figure 8 and are compared with the composition of the used electrodes. In all experiments performed, a deviation between the targeted and the analyzed composition can be observed, whereas the deviation is illustrated by a raise of the aluminum content and a drop of the titanium content. The largest deviation with a value of 1.92 at.-% appears in Trial A, which was operated with a power of 125 kW (cf. Table II). Whereas in Trial B, the used remelting power of 100 kW results in a deviation of 1.19 at.-%. In both trials no calcium was added to the slag.

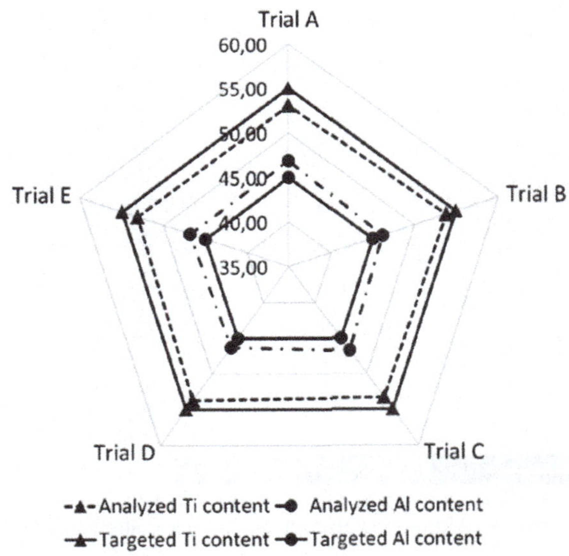

Figure 8. Analyzed and targeted aluminum and titanium content in remelted material

This effect is due to the nature of the used electrodes. As described in the experimental setup, the electrodes consist of pressed titanium sponge and aluminum rods. By reason of the lower melting temperature of aluminum compared to titanium, the aluminum rods melt down faster (cf. Figure 9) resulting in a slight raise of the aluminum content in the remelted ingot. This effect turns out the more, the higher the used remelting power.

Figure 9. Electrode tip after remelting with calcium addition

A similar effect can be observed in Trials C (1.69 at.-%), D (1.29 at.-%) and E (1.86 at.-%), where calcium was added to the slag. Due to the evaporation of calcium a thin vapor film is formed at the electrode tip and the appearance of arcs is advantaged. By the additional energy introduced more titanium is melted so that the deviation of the composition is lower than in melts without calcium addition. It can be observed that even a small addition of calcium to the slag results in a reduction of the deviation.

The following Figures 10 to 16 show the beam mapping of the prepared samples via electron beam back-scatter detector as well as the distribution of oxygen, fluorine, titanium and aluminum in the samples. The metallographic examination reveals only small differences in the solidification structure.

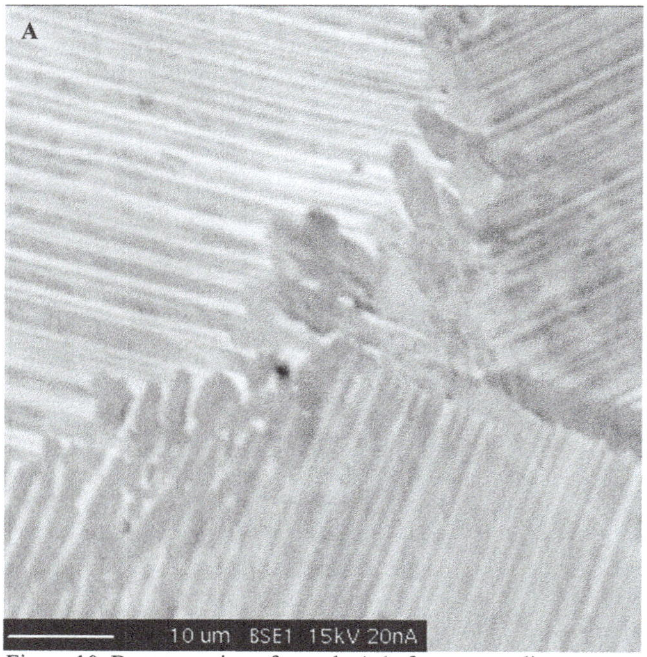

Figure 10. Beammapping of sample A (reference, medium power, no Ca addition) via electron beam back-scatter detector

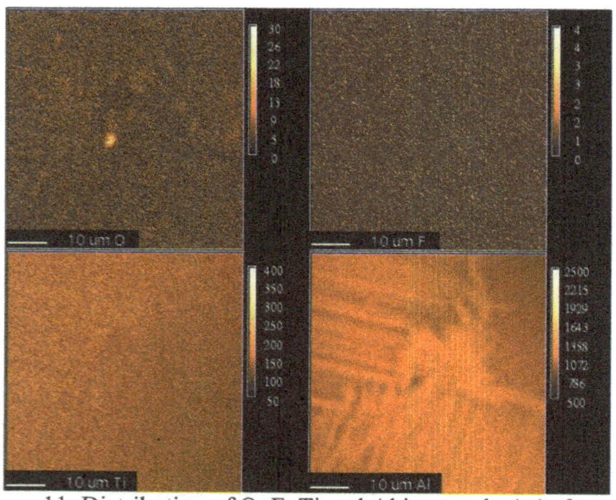

Figure 11. Distribution of O, F, Ti and Al in sample A (reference, medium power, no Ca addition); analyzed by EPMA

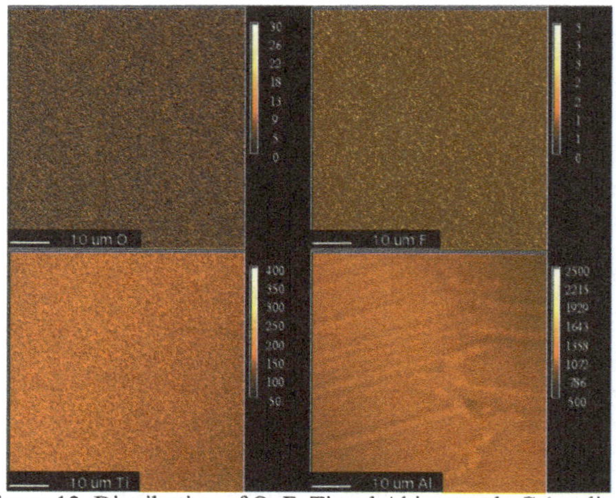

Figure 12. Distribution of O, F, Ti and Al in sample C (medium power, low Ca addition); analyzed by EPMA

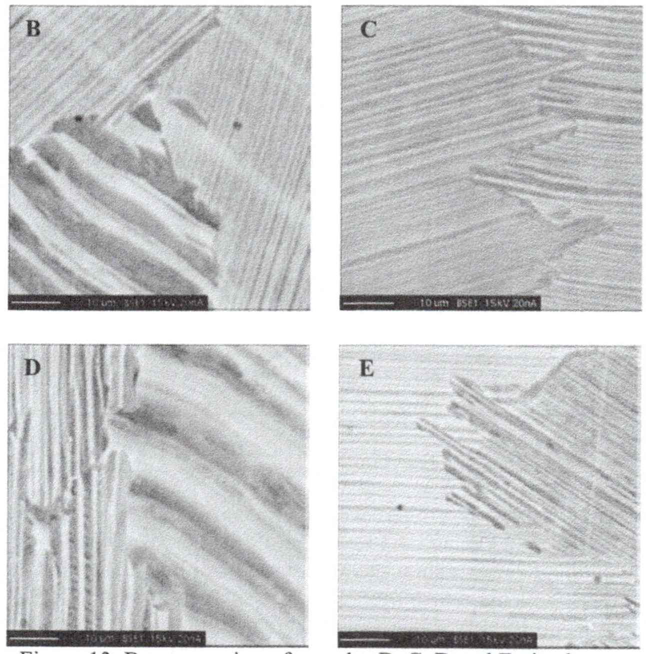

Figure 13. Beammapping of samples B, C, D and E via electron beam back-scatter detector

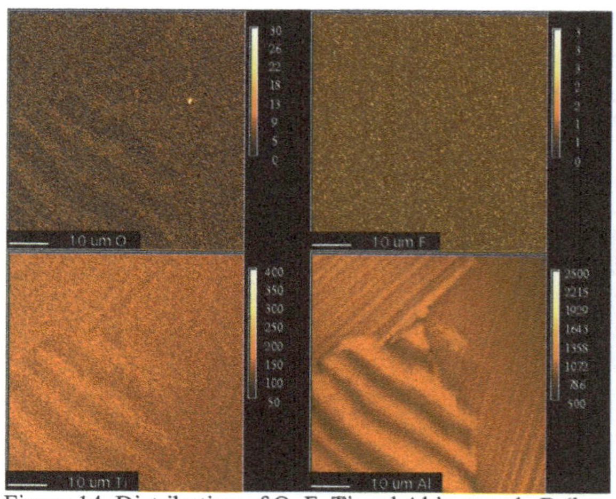

Figure 14. Distribution of O, F, Ti and Al in sample B (low power, no Ca addition); analyzed by EPMA

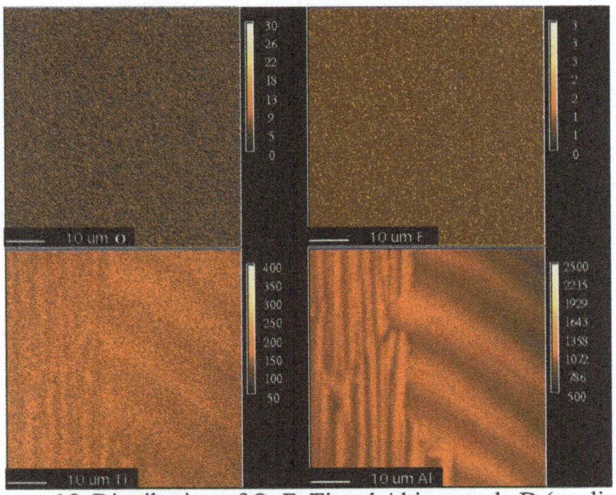

Figure 15. Distribution of O, F, Ti and Al in sample D (medium power, high Ca addition); analyzed by EPMA

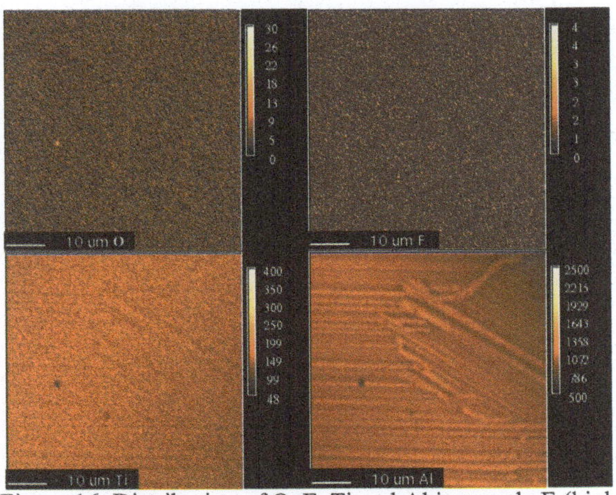

Figure 16. Distribution of O, F, Ti and Al in sample E (high power, low Ca addition); analyzed by EPMA

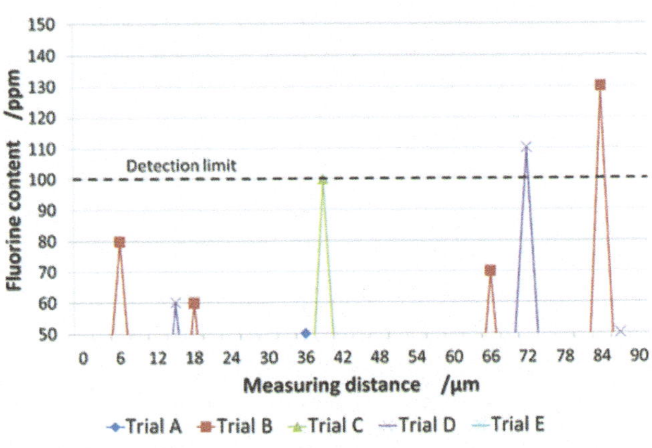

Figure 17. Results of the performed line scans for determination of fluorine

Upon closer examination of the oxygen distribution it can be seen that small nonmetallic inclusions (NMI) sporadically accumulate at the grain boundaries. These NMIs have a size of approximately 0.5 to 1 µm and occur primarily in the experiments without calcium addition (cf. Figure 11 and Figure 14). The background of adding calcium consists of the deoxidization of the used feedstock during remelting. In this case, by the reaction of the calcium dissolved in the slag with oxygen dissolved in the metal phase the development of calcium oxide is benefited. Furthermore, a reaction between the dissolved calcium and the nonmetallic inclusions takes places in which most of the NMIs get disbanded (cf. Figure 12 and Figure 15) or at least their size is reduced (cf. Figure 16). A complete removal of the nonmetallic inclusions is unlikely to be achieved.

A further effect of the calcium addition is the already mentioned decrease of the oxygen content. At detailed consideration of Figure 13-B and Figure 14, different oxygen contents between the titanium rich phase and the aluminum rich phase can be observed. Due to the large oxygen solubility in titanium, the oxygen content in the titanium rich phase is slightly higher than in the rest of the alloy. By deoxidization with metallic calcium the oxygen content is decreased what results in similar oxygen contents in both the titanium rich phase and the aluminum rich phase.

The determination of fluorine was carried out using electron probe micro-analysis with a fluorine detection limit of 100 ppm. As shown in Figures 11, 12, 14, 15 and 16 no fluorine in the bulk material above the detection limit could be determined. Hence, additional line scans were performed, whose results are shown in Figure 17. At rare intervals fluorine could be detected, whereas only in sample B and sample D the detection limit was exceeded. Therefore, it can be assumed that in case of an active CaF_2 slag the investigated process parameters (power and Ca addition) have a subordinated influence on the fluorine adsorption in γ-TiAl during pressure electroslag remelting. Higher fluorine contents than the reported value by Scholz (60 ppm) [10] could not be achieved in this investigation.

Nevertheless, even no fluorine could be detected at the grain boundaries. This suggests that there is no or at least a negligible deposition of CaF_2 on the grain boundaries what could have a negative influence on the mechanical properties.

Conclusions

(1) The electron probe micro-analysis has shown that the used electrode design caused increased aluminum contents in the remelted material. This effect can be reduced by the addition of even small amounts of calcium to the slag.

(2) Small nonmetallic inclusions with a size of 0.5 to 1 µm can occasionally be observed which primarily accumulate at the phase boundaries. By the addition of calcium to the slag both the NMIs frequency of occurrence as well as their size can be reduced. A complete removal of the nonmetallic inclusions is unlikely to be achieved.

(3) The usage of a pure and an active CaF_2 slag respectively leads to no significant fluorine enrichment and thereby no positive effect on the halogen effect. However, there is no deposition of CaF_2 on the grain boundaries.

Outlook

Due to the thermal stability of CaF_2 a partial substitution of the slag is considered in further investigations. Particular attention will be paid on the two compounds MgF_2 and NaF_2. Furthermore, there will be an exchange of the analysis methodology. By use of glow discharge mass spectrometry (GDMS) detection limits in the ppb range become feasible.

References

1. E.A. Loria, "Quo vadis gamma titanium aluminide," *Intermetallics*, 9 (2001), 997-1001
2. H. Clemens, " Intermetallische Werkstoffe für Anwendungen in Automobil- und Flugzeugtriebwerken," *BHM Berg- und Hüttenmännische Monatshefte*, 153 (2008), 337-341
3. S. Knippscheer, and G. Frommeyer, „Neu entwickelte TiAl-Basislegierungen für den Leichtbau von Triebwerks- und Motorenkomponenten – Eigenschaften, Herstellung, Anwendung," *Materialwissenschaft und Werkstofftechnik*, 37 (2006), 724-730
4. S. Seagle et al., *Titanium – The Ultimate Choice* (Broomfield, CO: International Titanium Association, 2007)
5. P.J. Masset, "Influence of alloy compositions on the halogen effect in TiAl alloys," *Materials and Corrosion*, 59 (2008)
6. M. Schütze, "The halogen effect in the oxidation of intermetallic titanium aluminides," *Corrosion Science*, 44 (2002)
7. A. Donchev, "Improvement of the oxidation behavior of TiAl-alloys by treatment with halogens," *Intermetallics*, 14 (2005), 1168-1174
8. J. Reitz, C. Lochbichler, and B. Friedrich, „Recycling of gamma titanium aluminide scrap from investment casting operations", Intermetallics, 19 (2011), 762-768
9. R.H. Nafziger, "Slag compositions for Titanium Electroslag Remelting and Effects of Selected Malting Parameters", *Proceedings of Second International Symposium on Electroslag Remelting Technology*, (1969)
10. H. Scholz et al., "An Advanced ESR Process for the Production of Ti-Slabs", *Ti-2003 Science and Technology*, ed. G. Lütjering and J. Albrecht (Weinheim, Wiley-VCH-Verlag, 2004), 205-212
11. B.I. Medovar, "Electroslag remelting of alloy steel," *Metallurg*, 4 (1970), 32-35
12. N. Giesselmann et al., "Numerical simulation of the electroslag remelting process in order to determine influencing parameters on ingot defects," *Proceedings of the 1st International Conference on Ingot Casting, Rolling and Forging*, (2012)
13. B. Friedrich, C. Lochbichler, and J. Reitz, „Closing The Material Cycle of Titanium – Thermochamical and Experimental Validation of a New Cycling Concept," *Proceedings of the 2007 International Symposium on Liquid Metal Processing and Casting*, (2007)

Proceedings of the 2013 International Symposium on Liquid Metal Processing & Casting
Edited by: Matthew J. M. Krane, Alain Jardy, Rodney L. Williamson, and Joseph J. Beaman
TMS (The Minerals, Metals & Materials Society), 2013

A PARAMETRIC STUDY OF SLAG SKIN FORMATION IN ELECTROSLAG REMELTING

Jeff Yanke[1,2] and Matthew John M. Krane[1]
[1]Purdue Center for Metal Casting Research, School of Materials Engineering, Purdue University, West Lafayette, Indiana USA
[2]currently Carpenter Technology Corporation, Reading, Pennsylvania USA

Keywords: Electroslag Remelting, Slag Skin, Simulation

Abstract

In electroslag remelting (ESR), the slag generates heat, chemically refines the melting electrode material, and forms frozen skin on the mold. An axisymmetric model is used to simulate fluid flow, heat transfer, solidification, and electromagnetics and their interaction with slag skin formation in ESR. A volume of fluid (VOF) method is used to track the slag/metal interface, allowing simulation of slag freezing to the mold. Mold diameter and applied current are varied to determine how these parameters affect melt rate and formation of slag skin during ESR. Variations in the slag skin thickness within the slag cap are found to have a significant impact on melt rate and depth of metal sump. Changes in slag cap volume resulted in small changes in melt rate.

Introduction

Electroslag remelting (ESR) is a secondary ingot production process used to manufacture high quality steels and nickel based superalloys. Ingots are formed by melting a consumable electrode by immersion in an electrically resistive slag, which also may react chemically with the molten metal as well as protecting it from the atmosphere. An electrical current is run through the system, and Joule heating in the slag provides sufficient energy to melt the electrode. Liquid droplets of electrode material sink through the less dense slag to form a pool of metal at the bottom of the crucible. Heat loss to the water-cooled copper mold solidifies the metal, maintaining a liquid pool that is only 1/2- 1 diameters deep throughout the process. Simultaneously, slag also freezes to the mold wall and forms a thin skin around the outer edge of the mold. If a thin layer is uniformly frozen, the ingot surface will have a smooth finish, but this slag skin, if thicker or of varying thickness, can be a source of surface defects resulting in severe grinding losses on the finished product [1, 2]. Understanding the mechanisms of skin formation is an important step in controlling these skin defects, as well as predicting the heat loss path from the metal to the mold in the ESR process. Given the complicated phenomena involved and the difficulty and expense of running experiments on a real process, computational simulation provides an attractive method of understanding how process conditions affect slag skin growth.

However, numerical simulation of ESR presents challenges as it incorporates coupled fluid flow, heat and mass transfer, solidification, and electromagnetics in the interacting metal and slag. A comprehensive model must include a method of tracking the moving interface between the slag and metal as well as modified discretization techniques for the proper calculation of transport phenomena in the presence of the thermophysical property differences between both fluids. Several recent models have shown how processing variations affect ESR while simulating both the metal and slag regions[3-5] .

The present work develops an axisymmetric transient model of ESR that simulates fluid flow, heat and mass transfer, solidification of both slag and metal, and electromagnetics. This model is used in conjunction with a simple 1D model of heat conduction and melting in the electrode to predict melt rates, and formation and possible remelting of slag skin as a function of input current and mold diameter. In order to predict the formation of solid slag skin and the interaction of the slag and metal, a method to track the interface between these two immiscible materials is used. A volume-of-fluid (VOF) method was adapted from the literature[6, 7] to fit the axisymmetric domain used in the current work.

Model Description

Transport Equations

The model is described in detail elsewhere[8] so only a brief overview will be given here. The simulation domain consists of the metal and slag, assuming the electrode is flat and rests directly on top of the slag cap. The SIMPLER algorithm is used to solve the equations of fluid motion on a staggered grid[9] using the continuum mixture model of Bennon et al[10] :

$$\frac{\partial(\rho u)}{\partial t}+\nabla\bullet\left(\rho\bar{V}u\right)=\nabla\bullet\left(\mu\nabla u\right)-\frac{\partial P}{\partial x}+S_u \qquad (1)$$

$$\frac{\partial(\rho v)}{\partial t}+\nabla\bullet\left(\rho\bar{V}v\right)=\nabla\bullet\left(\mu\nabla v\right)-\frac{\partial P}{\partial y}+S_v \qquad (2)$$

$$\frac{\partial\rho}{\partial t}+\frac{\partial(\rho u)}{\partial x}+\frac{\partial(\rho v)}{\partial y}=0 \qquad (3)$$

The source terms include Darcy drag in the mushy zone, thermo-solutal buoyancy, and Lorentz forces due to the interaction of the magnetic field and current density.

The energy and species conservation are represented by:

$$\frac{\partial(\rho h)}{\partial t}+\nabla\bullet\left(\rho\bar{V}h\right)=\nabla\bullet\left(\frac{k}{c}\nabla h\right)+S_h \qquad (4)$$

$$\frac{\partial(\rho C^i)}{\partial t}+\nabla\bullet\left(\rho\bar{V}C^i\right)=\nabla\bullet\left(D\nabla C^i\right)+S_{C^i} \qquad (5)$$

The source terms include the advection and diffusion-like source terms [10] and the Joule heating in the energy equation.

In ESR, the magnetic field intensity can be calculated from the diffusion equation:

$$\nabla\bullet\left(\frac{1}{\sigma}\nabla\bar{H}_\theta\right)=i2\pi f_{AC}\mu_o\bar{H}_\theta \qquad (6)$$

In order to calculate the Lorentz force and Joule heating, the current density is calculated from the magnetic field by:

$$\nabla\times\bar{H}_\theta=\vec{J} \qquad (8)$$

Boundary Conditions
Figure 1 shows a schematic of the solution domain and the boundary conditions applied to the energy and magnetic field equations. The present work is focused on the heat transfer boundary condition at the mold wall. Typically this boundary condition is expressed as an overall heat transfer coefficient that includes both the conduction through the slag skin and copper mold to the cooling water. In this work, the frozen slag skin is simulated directly so only the heat transfer through the copper mold is considered in the calculation of the uniform value of U_{eff}.

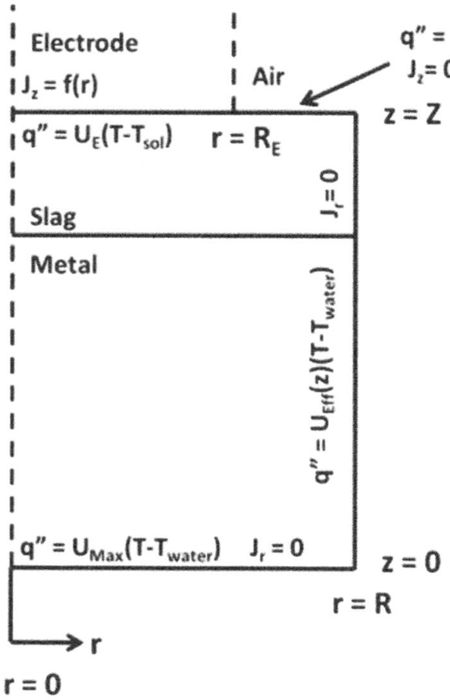

Figure 1: Schematic of the ESR domain showing the boundary conditions for the energy and magnetic field equations.

Melt Rate Calculation
In order to calculate the electrode melt rate, an iterative procedure was developed to split up the heat flux leaving the slag to the electrode into latent and sensible components. The actual heat flux from the slag is

$$q''_{slag} = k\frac{T_{slag} - T_{sol}}{\delta}, \quad (9)$$

where T_{slag} is the average temperature of the slag in contact with the electrode and T_{sol} is the solidus temperature of the alloy. The flux is assumed to travel across a liquid alloy layer δ thick, where δ is 8 mm in 718 and 4 mm in Waspaloy. The values are similar to those reported in [11], are were chosen based on trial and error. This heat flux is divided into two components:

$$q''_{Slag} = q''_{Latent} + q''_{Sensible}, \quad (10)$$

where the latent heat used is related to the melt rate,

$$q''_{Latent} = \frac{\dot{m}h_f}{A_{Electrode}}, \quad (11)$$

and the sensible part is found from the conduction in the solid electrode:

$$q''_{Sensible} = -k\left.\frac{\partial T}{\partial z}\right|_{Z=0,Electrode}. \quad (12)$$

An initial guess of $q''_{Latent} = 0.1\, q''_{Sensible}$ and the previous time steps' melt rate are used to solve a 1D conduction equation for the axial temperature profile in the electrode. The sensible heating in the electrode is calculated from equation (12) and the latent heating and melt rate are updated from equations (10) and (11). The temperature solution in the electrode is then solved with T_{sol} as a boundary condition. The updated melt rate and this procedure is repeated until the sensible heating changes less than 1% and the electrode temperature change at any given control volume is less than $1\text{x}10^{-4}$ % from the previous iteration's value.

Simulation Parameters

In order to demonstrate the effects of including slag skin in the ESR model described above, it was used to simulate the process under a variety of different conditions similar to industrial practice. For all cases, properties for alloy 718 and a 70-15-15 (CaF_3-CaO-Al_2O_3)slag were used, as listed in Table 1. The metal alloy properties are well characterized, but there is much more uncertainty in the slag data. Initial conditions are identical for all cases, with 0.07 m of liquid metal and a liquid slag cap, both at a uniform temperature of 1700 K. Also, a 0.03 m thick solid slag skin at 500 K was included at the mold wall. The computational mesh was refined near the mold wall with a control volume size of 1.5 mm for the 3.0 mm near the wall. The details of the 5 cases simulated are shown in Table 2, with U_{eff} representing the heat treansfer between the cooling water and surface of the copper mold.. Cases 1-3 include the standard case (Case 1) and the effects of increasing current density (Case 2) and ingot radius (Case 3) on slag skin thickness and macrosegregation. The larger diameter in Case 3 requires an increase in current to achieve stable melting conditions. Initially, the current densities were kept the same, scaling the current up from 13 kA to 28.9 kA. When Case 3 was run at 28.9 kA, the slag skin in the cap completely melted, and without the insulating slag skin, the heat lost to the mold was large enough to prevent any melting of the electrode. The current for Case 3 was determined through trial and error and 25 kA was found to provide a stable melt rate without completely melting the slag skin.

Cases 1-3 and 5 are run at the constant process currents shown in Table 2, but Case 4 predicts the effect of a sudden increase in power on the thickness of slag skin and transient reaction of liquid metal head and sump shape. Case 5 also shows how the model behaves differently if, instead of including slag skin in the model, a constant thickness of slag skin is assumed and used to calculate an overall thermal resistance at the mold wall.

Table 1: Thermophysical properties for alloy 718 and slag used in the simulations.

	718	Slag
Density [kg/m3] [3, 12]	7500	2490
Specific heat [J/kgK] [3, 13]	720	1260
Thermal Conductivity [W/mK] [3]	30.52	6
Viscosity [Kg/ms] [3]	0.0033	0.025
Thermal expansion [1/K] [3]	0.00015	0.00025
Emissivity [4]	-	0.87
Latent Heat [J/kg] [3]	2.1E+05	4.75E+05

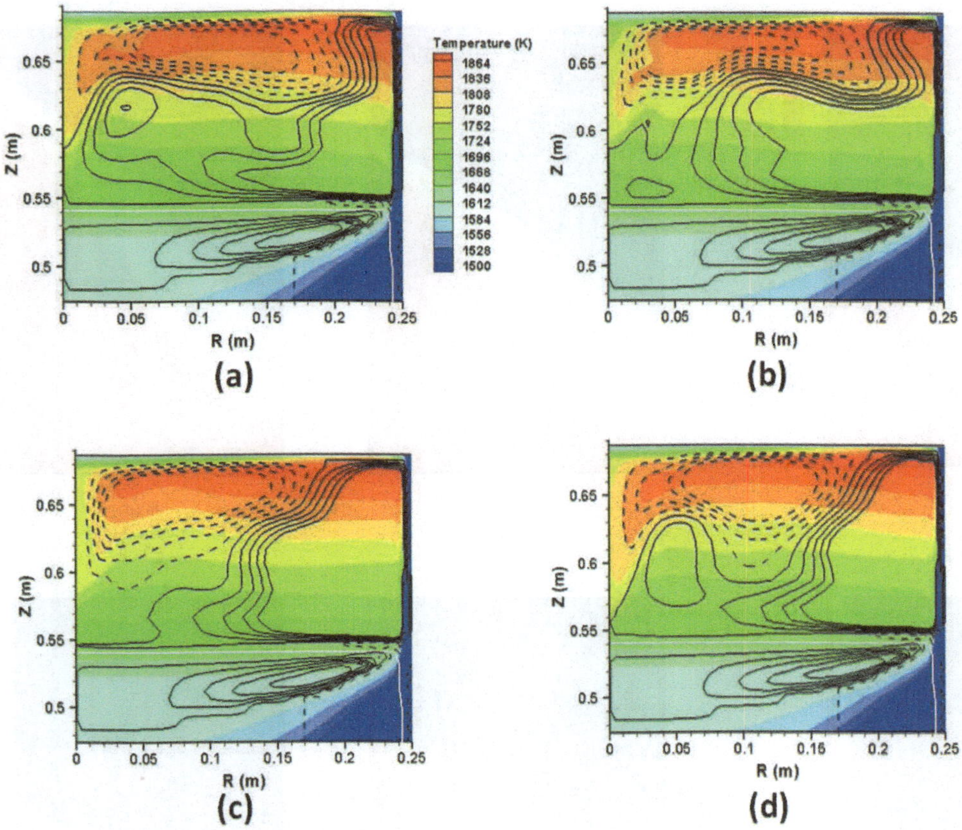

Figure 2: A series of temperature contour plots with velocity vectors showing an example of the unsteady fluid flow observed in all cases. The these snapshots are taken from Case 1 at (a) 159 minutes and 1.36 s, 2.06 s, and 3.18 s afterwards for b, c, and d respectively . The white line is the 0.5 VOF contour. The black lines are the stream function (dashed is negative, counterclockwise flow) ranging from -0.4 to 0.4 with a delta of 0.01 kg/s.

Eutectic Temperature [K]	1411	1390
Electrical conductivity [1/Ωm]	7.1E+05	variable

Table 2: Process parameters for slag skin thickness cases.

Case	I (kA)	R_{ingot} (m)	$R_{electrode}$ (m)	U_{eff} (W/m^2K)
1	13	0.25	0.23	2000
2	15	0.25	0.23	2000
3	25	0.381	0.343	2000
4	13 -> 15	0.25	0.23	2000
5	13	0.25	0.23	400 (no slag formation)

Results

Temperature Field and Fluid Flow in Slag Cap

Unsteady fluid flow patterns are observed in the slag cap even at a fairly steady melt rate. A sequence of representative snapshots of the flow patterns and temperature fields are shown in Figure 2. Figure 2a shows a typical flow pattern in the slag cap consisting of two thermally driven cells. At the mold wall, the heat extracted by the mold lowers the buoyancy of the slag there, driving a stable clockwise cell. The downward flow at the mold is the strongest in the slag, with a very thin boundary layer there, which turns the corner to form a radial jet along the top of the metal pool. The strength of this jet diminishes as it moves inwards and finally turns up, pulled by the positive buoyancy generated by the high Joule heating just under the outer radius of the electrode. At the centerline, a counterclockwise cell is caused by heat lost to the electrode. This colder slag plunges down the centerline, displacing the hotter slag closer to the slag/metal interface. Figure 2b shows that as this cell drags warmer fluid further down into the slag cap, it loses its relative buoyancy and turns towards the outer radius. Like the clockwise cell, this flow is driven the high heat generation under the electrode. The strength of this slag flow pattern is not enough to transition to turbulence, but it is strong enough to exhibit the periodic fluctuations seen in Figure 2, where the process repeats itself about every 4.8 seconds.

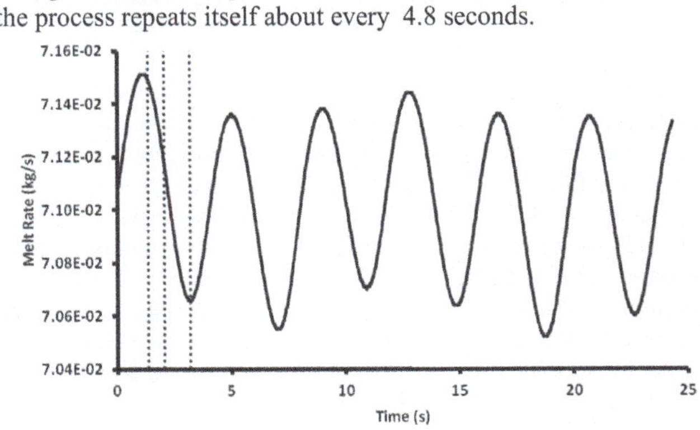

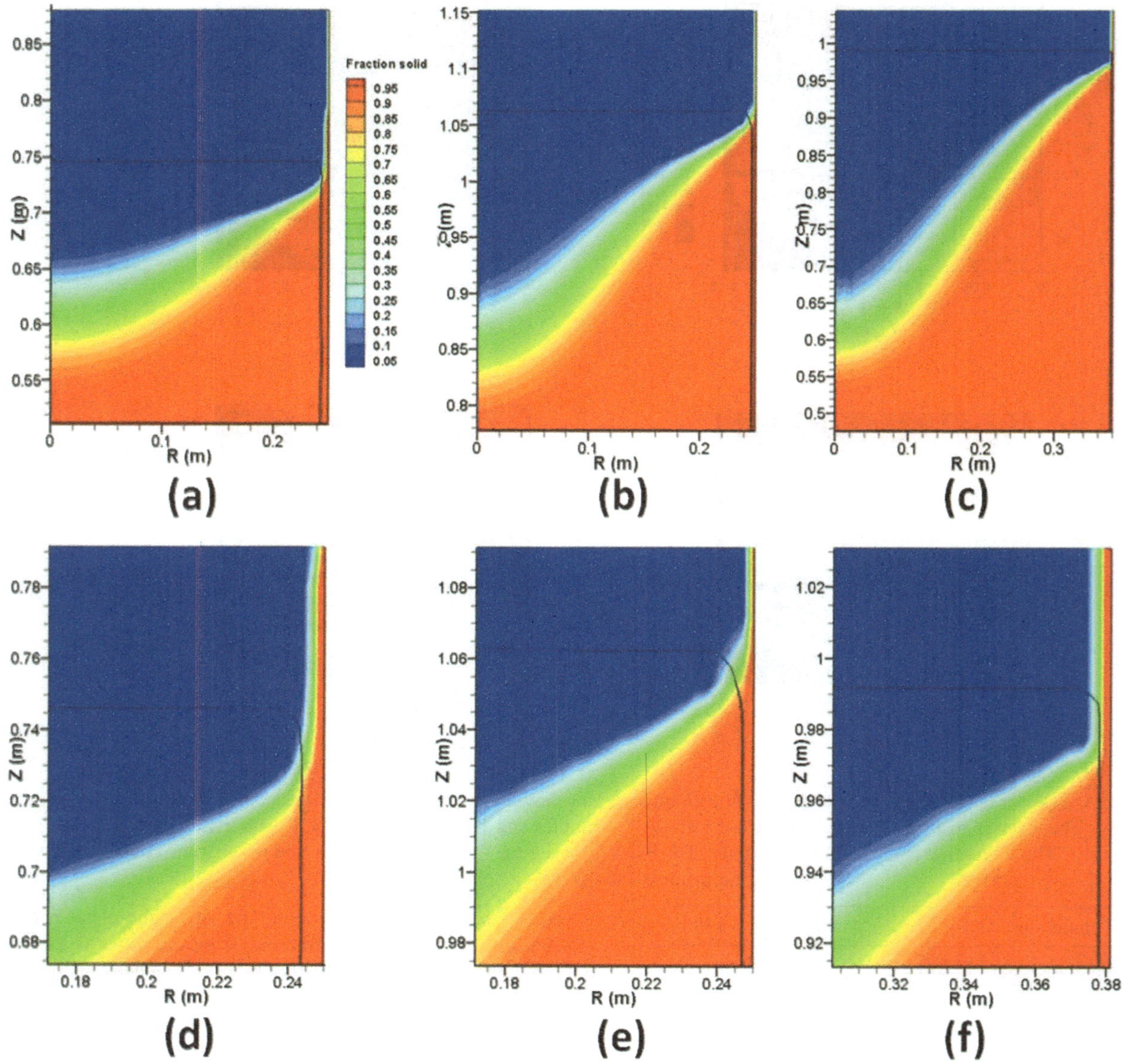

Figure 4: Sump shape near the mold wall at 200 minutes for (a) Case 1, (b) Case 2 (increased current density), and (c) Case 3 (increased ingot radius). A closer view of the mold wall shows the slag skin thickness for (d) Case 1, (e) Case 2, and (f) Case 3 at 200 minutes. The black lines indicate the slag-metal interface and the shading the fraction solid field.

Figure 3: A plot showing the melt rate for Case 1 as a function of time, starting at minute 159, showing the periodic changes caused by unsteady flow underneath the electrode. Dashed lines represent the times shown in Figures 2b-d.

The unsteady flow pattern also creates periodic fluctuations in the melt rate as the pattern repeats itself. Figure 3 shows the melt rate as a function of time for the case presented in Figure 2 and several seconds afterwards. The swing in the predicted melt rate is apparently +/- 0.5%. The changes in melt rate shown in Figure 3 can be explained by looking at the flow and temperature fields in Figure 2. When the counterclockwise flow cell running along the bottom of the electrode and down the centerline is strongest, Figure 2b, the melt rate reaches a maximum. As the centerline flow diminishes, Figure 2d, colder fluid is resident near the electrode longer, reducing the melt rate.

These unsteady flow conditions are observed in all cases presented here. In general, the higher melt rates increase the magnitude of the flow down the centerline and so the depth into the slag cap that the plunging flow will reach. However, even when this flow reaches the slag/metal interface, no significant deformation was observed. Instead the flow turns the corner and continues out into the midradius. The depth this flow plunges into the slag cap was unaffected by a larger slag cap except that, with a larger slag cap, in general a larger melt rate was observed.

Melt Rate and Sump Depth

Figure 4 shows a comparison of slag skin thickness and sump shape near the mold wall for Cases 1-3 at 200 minutes into the process, which was found to be enough time for the all the sump shapes to reach a steady state. For a constant diameter, an increase in current density of ~15% (Case 1 to Case 2) decreases the slag skin thickness from 0.0065 m to 0.003 m, a trend in accord with industrial observations. The sump shape at lower current is shallower and bowl shaped, whereas the higher power Case 2 shows a deeper V shaped sump.

Similar trends can be seen in Cases 1 and 3, shown in Figure 4. Case 3 shows the same slag skin thickness (0.003 m) and similar V shaped sump as Case 2, which is explained by their similar filling velocities. The filling velocity is:

$$u_f = \frac{\dot{m}}{\rho_{Metal} A_{Mold}} \tag{13}$$

Filling velocity is used to compare cases of different diameter because it represents the speed that the mold is being filled which, under stable operating conditions, will be similar between different diameters while the melt rates will be very different. Filling velocity histories for Cases 1-3 (Figure 5) show that Cases 2 and 3 are within ~10% of each other even though the current supplied to Case 3 is 67% higher than in Case 2. The larger melt rate produced by this higher current is offset by the larger volume (2.3 times the volume of Case 2) that Case 3 must fill to move the slag/metal interface. Both Cases 2 and 3 have filling velocities that are 60% higher than Case 1. This similarity in filling velocities between Case 2 and 3 explains the similarly shaped sumps shown in Figure 4.

In Figure 6, normalized steady state sump depths are plotted as a function of the normalized radius for Cases 1-3. The stair-stepped increases in sump depth are due to the coarseness of the numerical grid. In Case 1, the sump depth increases sharply moving in from the outer radius, steadying out near the midradius, which results in a shallow bowl shape sump. This shape is the result of the high heat transfer occurring at low filling velocities. As the filling velocity increases (Case 2 and Case 3), there is less time for heat transfer to the mold wall and so the shape of the sump becomes much more elongated, producing the "V"-shaped sump.

The filling velocities in Figure 5 also show a decrease over time of 10% for Case 1 and 5% for Cases 2 and 3. The formation of a frozen slag skin on the mold reduces the amount liquid slag available to generate heat in the slag cap. Case 1 exhibits a much thicker slag skin, which leads to a quicker reduction in slag cap thickness and so the filling velocity. However, the thickness of the slag cap is a secondary effect on the filling velocity. The short term variations in filling velocity in Figure 5 also show a trend that Cases 1 and 2 seem to have similar amplitude variations in filling velocity, where Case 3 shows higher frequency and amplitude variations. These variations are due to changing flow conditions in the slag cap even during constant current operation.

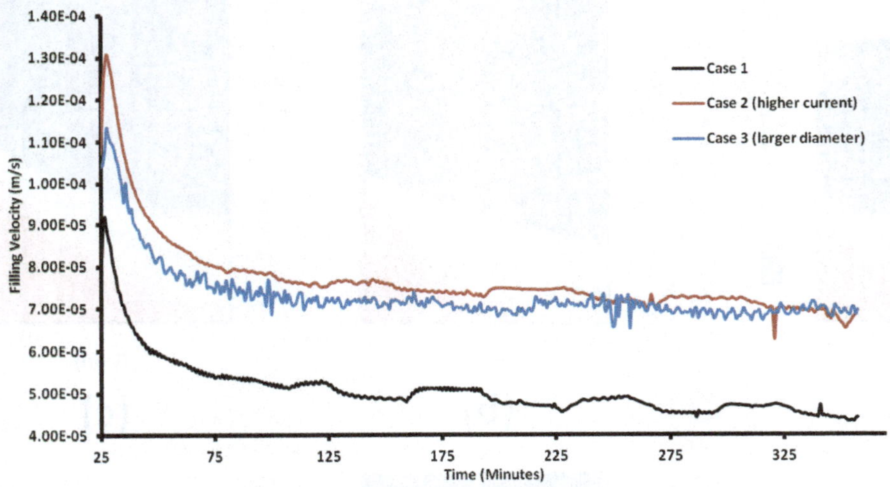

Figure 5: A comparison of filling velocity histories for Cases 1-3.

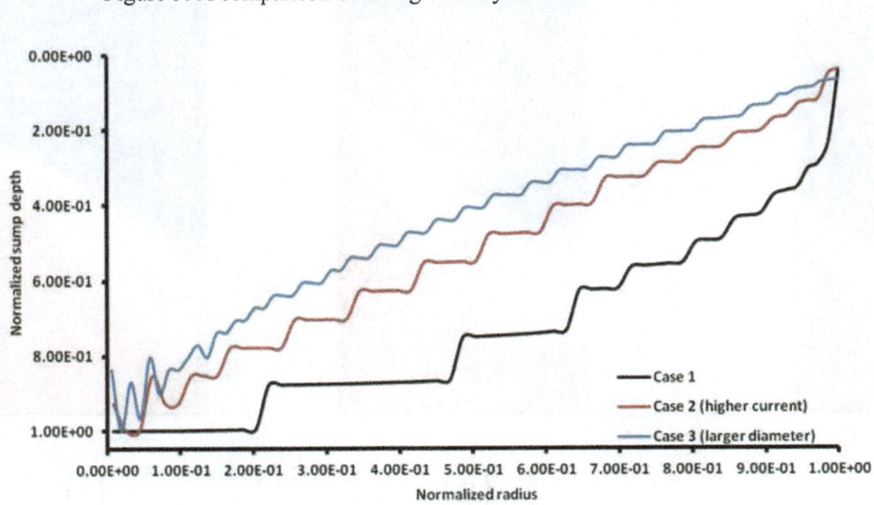

Figure 6: A comparison of normalized sump depth as a function of normalized radius for Cases 1-3 showing Case 2 and 3 having a more linear sump shape than Case 1.

Sudden Increase in Current

In Case 4, the current is increased instantaneously from 13 kA to 15 kA at 182 minutes into the process (in the steady state region), before which it is identical to Case 1. The subsequent changes in slag skin thickness and sump shape are shown in Figure 7. Immediately upon the increase in current, the slag heats up due to instantaneously increased Joule heating (in 2 minutes the slag under the electrode heats up from 1900 K to 2050 K), leading to an immediate increase in heat transferred to the electrode and melting of the insulating slag skin in the slag cap. The melt rate (and thus filling velocity) reacts quickly, causing a swift increase in the liquid metal head (the distance between the highest solid along the mold wall and the liquid metal/slag interface) and eventually exposing previously solidified slag skin to hot enough temperatures that remelting of the skin is observed as the metal head height increases (Figure 7d). The remelted slag flows up into the cap almost instantly and a thinner layer of solid slag skin is left behind.

As the advancing solid metal front passes this thinned slag skin, the increased heat transfer through the thinner layer leads to a rapid change in the shape of the sump as it transitions from the bowl shape typical at lower power to the V shape consistent with higher power melting. Figure 8 shows the sump depth at the centerline, midradius, and at r=0.24 m(chosen 1 control volume away from the slag skin) as a function of time for Case 4. The increase in sump depth and decrease back to a lower value takes 80 minutes, showing how sensitive ESR is to sudden changes in operating parameters. Figure 9 shows the filling velocity as a function of time for Cases 1,2, 4, and 5. It can be seen that the

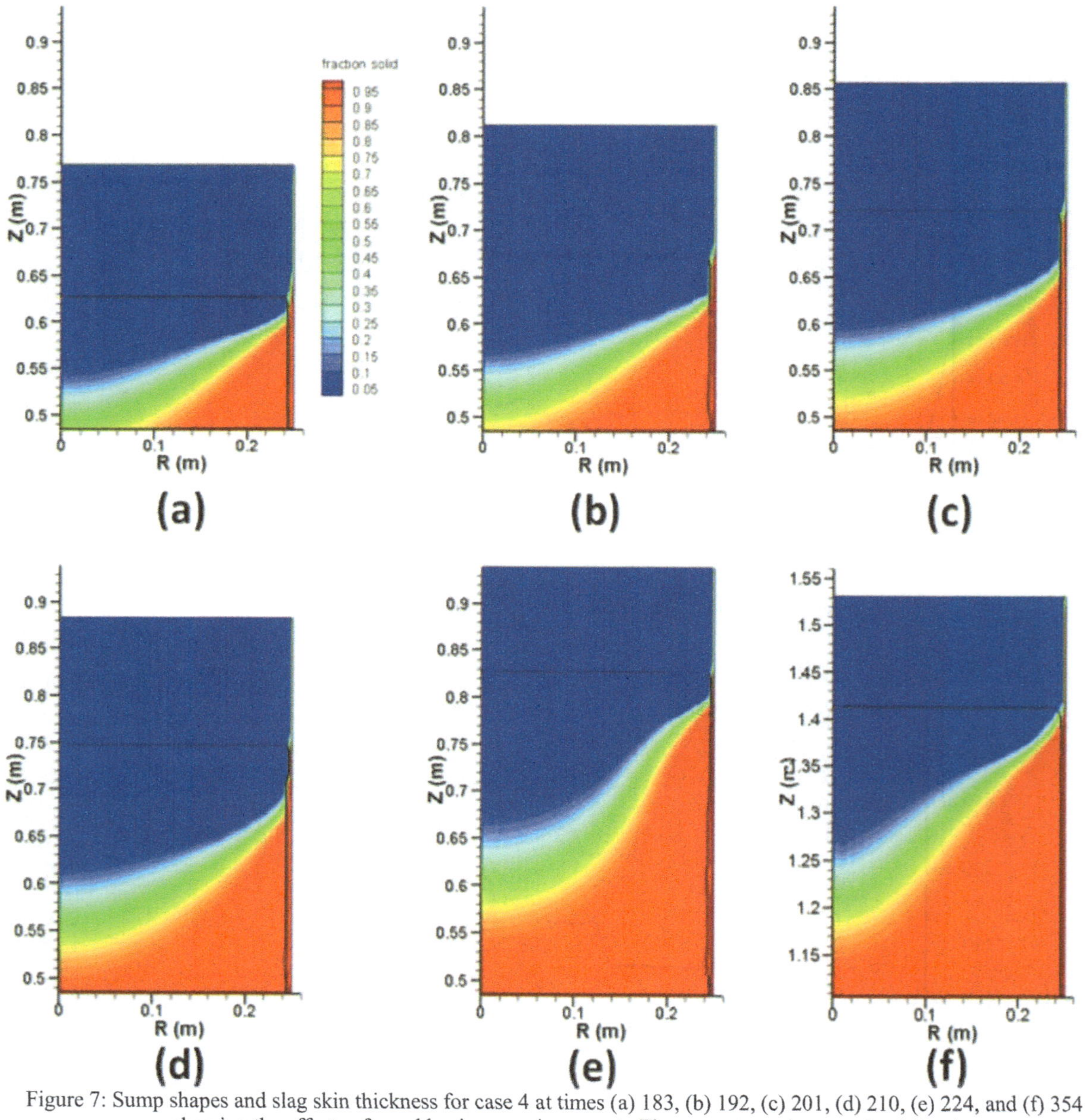

Figure 7: Sump shapes and slag skin thickness for case 4 at times (a) 183, (b) 192, (c) 201, (d) 210, (e) 224, and (f) 354 minutes showing the effects of a sudden increase in current. The current was increased at 182 minutes.

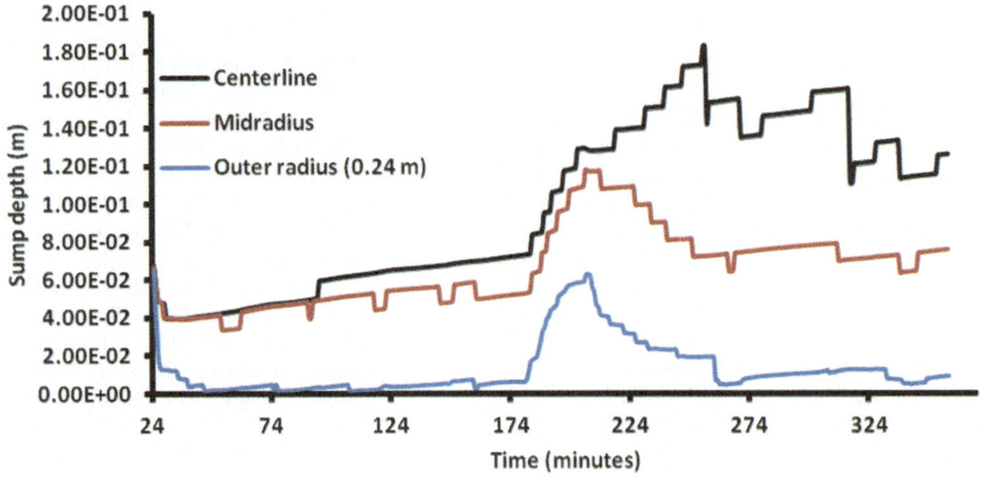

Figure 8: Sump depth as a function of time for Case 4 showing the increase and decrease in sump depth following the increase in current from 13 kA to 15 kA at 182 minutes.

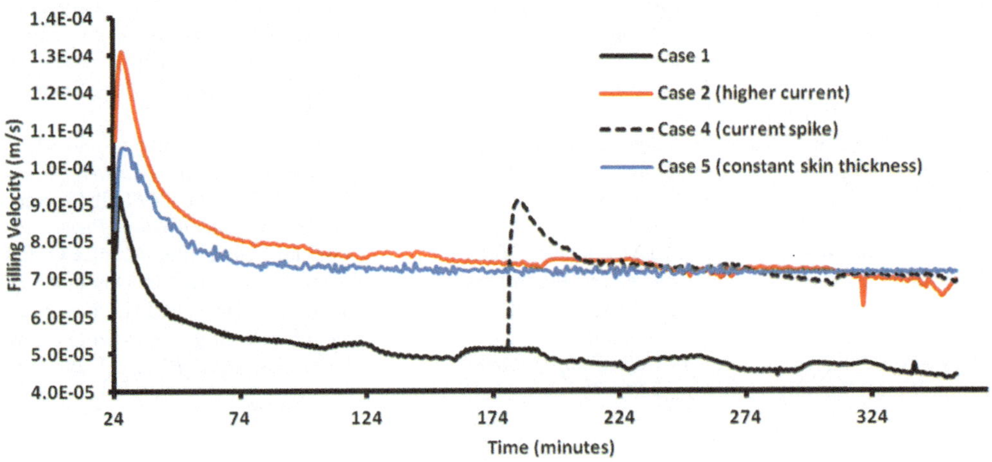

Figure 9: Filling velocity as a function of time for Cases 1, 2, 4, and 5.

response time of the filling velocity is nearly instantaneous, even overshooting the steady state filling velocity seen in Case 2. This temperature excursion can be explained by the transient response to the increased heat flux to the electrode and through the thinning slag skin in the cap. As the slag heats up when the current is increased, the increased heat loss to the electrode mostly goes to melting, as the temperature field in the electrode takes time to adjust to the increased melt rate. Simultaneously, the insulating frozen slag skin at the mold wall partially remelts, increasing heat flow to the mold and cooling the slag cap, causing the filling velocity to decrease to the steady state value of Case 2.

Comparison to Constant Thickness Slag Skin Model

In Case 5, the phenomenon of transient slag freezing is removed. Instead of forming a (possibly nonuniform) slag skin at the mold wall, which would give a thermal resistance as a local function of the process behavior, the thermal resistance of a fixed skin layer thickness is combined with the heat transfer coefficient between the cooling water and the mold wall. This effective heat transfer coefficient assumes a uniform 3 mm skin, while the other simulations predict a thickness variation in the cap. The assumed 3 mm is a few millimeters below the thickness predicted by Case 1 at the bottom of the slag cap, but about twice the thickness predicted in the top two thirds of the cap. The effect of this difference in treatment of the slag skin on filling velocity is shown in Figure 9. The change in thermal resistance due to different thicknesses of slag skin causes the filling velocity in Case 5 to be almost twice that for the variable skin in Case 1. Because Case 1 and 5 have the same electrode diameter, the melt rate of Case 5 is almost twice Case 1 also. Case 5 also exhibits the high frequency, low amplitude variations in filling velocity of Cases 1-4, but it does not show their gradual decrease in filling velocity because the slag cap remains at a constant thickness.

All of the results in this study assume that the only resistances to heat transfer from liquid slag to mold are the presence of slag skin and the heat transfer through the mold, ignoring any effect of a contact resistance between the solidifying slag and any air gap there due to solidification shrinkage. The results presented thereby amplify the sensitivity of the melt rate, sump shape, and macrosegregation in the ESR process to treatment of the slag skin. Another large influence on these results is the thermal conductivity of the slag skin, which is not well characterized, especially at the high temperatures in question. However, even given these uncertainties, the general trends that are presented should be valid.

Conclusions

The model has been used to simulate the effects of including the slag skin in the ESR simulation, allowing it to vary in thickness in response to the local thermal conditions for a variety of process conditions. This varying slag skin thickness decreases the predicted melt rate by 50% compared to assuming a constant thickness slag skin because partially melting the insulating skin causes a large increase in heat flow to the mold, and reduces the heat available to melt the electrode. Including slag skin into ESR process models is necessary. However, due to uncertainty in the slag skin thermal properties and solidification behavior, and heat transfer to the mold, as well as coarse grid resolution near the mold wall, an increase in slag skin thickness cannot be predicted by the current model.

Acknowledgements

This research were funded by the National Science Foundation via grant CMMI-0900624, a GOALI award in partnership with Special Metals Corporation and Haynes International.

References

1. Krane, M., M. Fahrmann, J. Yanke, E. Escobar, K. Fezi, and J. Busch. *A Comparison of Predictions of Transport Phenomena in Electroslag Remelting To Industrial Data*. in *LMPC 2011*. 2011. Nancy, France: pp. 65-72.
2. Busch, J., J. deBarbadillo, and M.J.M. Krane, *Flux entrapment and titanium nitride defects in electroslag remelting of Incoloy alloys 800 and 825*. Metallurgical and Materials Transactions A, 2013. published online.
3. Kelkar, K.M., S.V. Patankar, and A. Mitchell. *Computational Modeling of the Electroslag Remelting(ESR) Process Used for the Production of Ingots of high-Performance Alloys*. in *LMPC 2005*. 2005. Santa Fe NM, USA.
4. Weber, V., A. Jardy, B. Dussoubs, D. Ablitzer, S. Rybéron, V. Schmitt, S. Hans, and H. Poisson, *A Comprehensive Model of the Electroslag Remelting Process: Description and Validation*. Metallurgical and Materials Transactions B, 2009. **40**(3): p. 271-280.
5. Kharicha, A., W. Schützenhöfer, A. Ludwig, and G. Reiter, *Influence of the Slag/Pool Interface on the Solidification in an Electro-Slag Remelting Process*. Materials Science Forum, 2010. **649**: p. 229-236.
6. Rider, W.J. and D. Kothe, *Reconstructing volume tracking*. Journal of Computational Physics, 1998. **141**: p. 112-152.
7. Garrioch, S.H. and B.R. Baliga, *A PLIC volume tracking method for the simulation of two-fluid flows*. International Journal for Numerical Methods in Fluids, 2006. **52**: p. 1093-1134.
8. Yanke, J., *Numerical Modeling of Materials Processes with Fluid-Fluid Interfaces*. PhD dissertation, School of Materials Engineering, 2013, Purdue University: West Lafayette.
9. Patankar, S.V., *Numerical Heat Transfer and Fluid Flow*. 1980, Washington DC: Hemisphere Publishing.
10. Bennon, W.D. and F.P. Incropera, *A continuum model for momentum, heat and species transport in binary solid-liquid phase change systems-I. Model formulation*. International Journal of Heat and Mass Transfer, 1987. **30**(10): p. 2161-2170.
11. Kharicha, A., A. Ludwig, and M. Wu. *3D Simulation of the Melting During an Industrial Scale Electro-Slag Remelting Process*. in *LMPC 2011*. 2011. Nancy, France. pp.41-48.
12. Hoyle, G., *Electroslag Processes*. 1983, New York,NY: Elsevier Science Publishing CO. INC.
13. Weber, V., A. Jardy, B. Dussoubs, D. Ablitzer, S. Rybéron, V. Schmitt, S. Hans, and H. Poisson. *A Comprehensive Model of the ElectroSlag Remelting Process: Description and Validation*. in *LMPC 2007*. 2007. Nancy, France. pp. 83-88.

Proceedings of the 2013 International Symposium on Liquid Metal Processing & Casting
Edited by: Matthew J. M. Krane, Alain Jardy, Rodney L. Williamson, and Joseph J. Beaman
TMS (The Minerals, Metals & Materials Society), 2013

IMPACT OF THE SOLIDIFIED SLAG SKIN ON THE CURRENT DISTRIBUTION DURING ELECTROSLAG REMELTING

Mathilde Hugo[1], Bernard Dussoubs[1], Alain Jardy[1], Jessica Escaffre[2], Henri Poisson[2]
[1]Institut Jean Lamour - UMR 7198 CNRS/Université de Lorraine, Parc de Saurupt CS 50840, F-54011 Nancy Cedex, France
Laboratory of Excellence DAMAS
[2]Aubert & Duval, BP 1, F-63770 Les Ancizes, France

Keywords: ElectroSlag Remelting (ESR), numerical simulation, solidified slag skin, current path, Joule heating

Abstract

The ESR process is commonly used to produce defect-free ingots of high added value alloys such as special steels or Ni-based superalloys. Numerous simulation tools have been developed for the last 30 years to get better insight into the process and help its optimization. Most assume that no electrical current is able to cross the solidified slag skin and flow in the water-cooled mold. However, it has recently been claimed that the slag skin does not ensure perfect insulation, which is prone to modify the current distribution, hence some of the results previously assessed. This paper presents the assumptions made to simulate that phenomenon and some results in terms of current flow into the mold, depending on the thickness and electrical conductivity of the solidified slag skin. Results show that both parameters can have a great influence on the current distribution, hence the slag behaviour and final ingot quality.

Introduction

Critical applications such as aeronautics, energy or tooling, require high quality and defect-free metal, which cannot be achieved by primary melting. One way to fulfil these goals is the use of remelting processes such as ElectroSlag Remelting (ESR), which allows a good control of the alloy chemistry and solidification structure of the ingot. [1] The materials concerned are high value-added metallic alloys such as special steels or nickel-based superalloys.

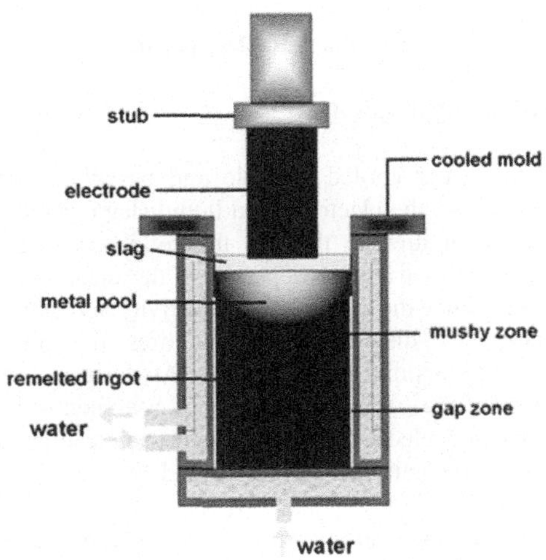

Figure 1: Schematic representation of an Electro Slag Remelting furnace

Remelting is caused by means of an electric current which goes from the primary ingot – also called electrode – whose lower part, as shown in Figure 1, is immerged into a highly resistive calcium fluoride based slag, to a water-cooled crucible, thus generating Joule heating in the slag. This causes the electrode to melt by contact, forming metal droplets which fall down through the slag to the liquid pool of the secondary ingot. Because of the water cooling, the latter solidifies continuously while growing.

It must be noticed that the slag is in contact with the water-cooled mould, which is responsible for the formation of a layer of solidified slag at the interface. As the secondary ingot rises, this layer is partially remelted and crushed between the metal and crucible, resulting in a slag skin which acts as a thermal insulator and provides ESR ingots with a smooth lateral surface. Besides, chemical reactions can occur at the electrode/slag, droplets/slag and pool/slag interfaces, thus influencing the final ingot chemistry and dissolving the non-metallic inclusions. [1]

The strategic importance of the remelted products makes it essential to acquire a detailed understanding of the process. Actual remelting experiments are very costly; so their repeated use cannot be considered as a suitable solution to get a better insight into the process and help its optimization. Apart being easy to use, mathematical modeling is a valuable tool to enhance fundamental understanding, since it allows linkage of operating parameters, such as the melting sequence, slag composition and properties, ingot diameter or cooling conditions, to local solidification conditions, hence the ingot final quality.

Therefore, several models of the ESR process have been proposed during the last 30 years, among which some of the most promising earlier ones were devoted to the electromagnetic and magnetohydrodynamic behaviour of the slag phase. [2-5]

More recently, especially during LMPC conferences, several researches were presented, aiming to simulate the whole process in a transient way, or discuss in more detail the electromagnetic fields in ESR. [6-11]

Among these models, the simulation software of the ElectroSlag Remelting process developed at Institut Jean Lamour (IJL) in Nancy, in close collaboration with Aubert & Duval, is a 2D axisymmetric transient model, similar in nature to the SOLAR model of Vacuum Arc Remelting. [12] It uses the finite volume method and a fully implicit time scheme to account for electromagnetic, fluid flow, heat transfer and phase change phenomena in a coupled way. Details of the coupled model are available in V. Weber's PhD dissertation. [13]

In the paper by Weber et al.,[14] it was stated that *"we assume that the solidified slag skin insulates electrically the slag and ingot from the mold. This assumption is particularly questionable and needs to be confirmed. Indeed, in some cases, the model predicts a discontinuous solid skin surrounding the slag cap, implying a possible electrical contact between liquid slag and mold"*. While this strong assumption is made classically in the literature devoted to ESR simulation[15-17], it was sometimes claimed that a certain amount of current is able to flow into the Cu crucible[18,19]. This phenomenon could modify the thermohydrodynamic behaviour of the slag and liquid pool, hence influencing the solidification process. Therefore, the goal of our study is to quantify this phenomenon, and determine the impact of the solid layer thickness and electric conductivity of the solidified slag on the current distribution during electroslag remelting.

Details Of The Electromagnetic Model

The principle of the ESR process is to melt the electrode, thanks to resistive heating generated by a current flowing through the slag. In most cases, AC is generally used with a frequency of 50 Hz. In that case, the magnetic and electric fields in the system are dependent on each other, hence phenomena are governed by the four Maxwell's equations. The system formed by these equations can be reduced to a single one under some assumptions[2]:

- Magnetization is negligible;
- Magnetic Reynolds number remains very low, which means that the flow of fluids does not affect the electromagnetic field. In such conditions, the simplified Ohm's law applies;
- Current displacement is much lower than the electric conduction, which is valid if the electrical conductivity is not too small.

Moreover, we use the phasor notation in order to account for the periodic time dependence. As an example, as the process conditions are axisymmetric and no external field is applied, the magnetic field has only one component in the azimuthal direction, which varies with time as:

$$H = H_{max} \cos(\omega t + \varphi) = H_{max} \cos(2\pi f t + \varphi)$$

where H_{max} is the maximum value, f is the frequency of the alternating current and the local phase angle. Both H_{max} and φ are dependent on the position. The phasor of the magnetic field is defined by:

$$H = \hat{H}e^{j\omega t} \qquad \text{with } j^2 = -1 \text{ and } \hat{H} = H_{max}\, e^{j\varphi}$$

Finally, the system of Maxwell's equations is reduced to a single one, with $\hat{H}$ as the main variable.

In a previous version of the model, the following equation was solved for both its real and imaginary parts:

$$\frac{1}{r}\frac{\partial}{\partial r}\left(r\frac{\partial \hat{H}_\theta}{\partial r}\right)+\frac{\partial^2(\hat{H}_\theta)}{\partial z^2}-j\omega\mu_0\sigma\hat{H}_\theta+\frac{\hat{J}_r}{\sigma}\frac{\partial\sigma}{\partial z}-\frac{\hat{J}_z}{\sigma}\frac{\partial\sigma}{\partial r}-\frac{\hat{H}_\theta}{r^2}=0 \qquad (1)$$

where $\hat{J}_r$ and $\hat{J}_z$ are the phasors of the radial and axial components of the current density. The latter are directly computed from the magnetic field by using the Maxwell-Ampere's law. The electromagnetic forces and Joule heating are then deduced from the averaged current density. In fact, the mathematical treatment of Maxwell's equations is such that the use of Eq. 1, along with appropriate boundary conditions (generalized Ampere theorem at the inner surface of the mould and the continuity of tangential electric field) is valid as far as the gradient of the electrical conductivity is relatively small.

However, because our aim is now to focus on the impact of a partially insulating solidified slag layer at the mould inner surface, it was found that actual computations had to be performed within this layer. In such a case, because of the very important difference between the electrical conductivity of liquid and solid slag, Eq. 1 is no more fully valid.

Therefore, a new formulation has been derived, using the electromagnetic stream function $r\hat{H}$ as the main variable and incorporating the electrical conductivity as a transfer coefficient. Therefore, all interfacial conditions at the boundaries between adjacent media are automatically accounted for. The resulting equation is:

$$\frac{1}{r}\frac{\partial}{\partial r}\left(r\frac{1}{r\sigma}\frac{\partial(r\hat{H}_\theta)}{\partial r}\right)+\frac{\partial}{\partial z}\left(\frac{1}{r\sigma}\frac{\partial(r\hat{H}_\theta)}{\partial z}\right)-\frac{1}{r^2\sigma}\frac{\partial(r\hat{H}_\theta)}{\partial r}-j\omega\mu_0\hat{H}_\theta=0 \qquad (2)$$

which is strictly identical to the one used, for example, by Kharicha et al. [10]

The real and imaginary parts of Eq. 2 were solved numerically using a finite volume method, on a non-uniform (r,z) grid composed of typically 160 x 295 cells in the liquid slag phase. In order to compute the current in the solidified slag zone with some accuracy, a minimum number of 4 nodes was required in that phase. It must be noted that, for all calculations presented in that paper, the solid slag thickness was kept as uniform on the slag/crucible and ingot crucible interfaces. In order to reach the convergence, around 10^6 iterations were found to be necessary.

As it will be shown later, an important parameter is the electrical conductivity of the solidified slag. In most cases, such a physical property is very poorly known, as few experiments have been carried out on fluorine-based materials at high temperatures. In the present study we used values in the range 1-15 $\Omega^{-1}.m^{-1}$, as proposed by Kharicha [20].

Results And Discussion

Input Data And Main Results

To reach the water cooled crucible and baseplate, the melting current supplied to the electrode and liquid slag can either flow in the ingot pool or directly through the solidified slag skin. The resulting current distribution depends on the electrical resistance of that phase, hence the solid slag conductivity and skin thickness. In this paper, we discuss the computation of electromagnetic phenomena with a simplified geometry. As discussed previously, the thickness of the solidified slag skin is assumed to be uniform and the assigned electrical conductivities are estimated values. The main input data are gathered in Table I.

The effects of the variations of two parameters (electrical conductivity and thickness of the solid slag layer, written with italic characters in Table I) on the current distribution and resulting Joule heating were studied. The electrical conductivity was allowed to vary in the range 10^{-3} - 400 $\Omega^{-1}.m^{-1}$ (10^{-3} $\Omega^{-1}.m^{-1}$ corresponds to a full insulation while 400 $\Omega^{-1}.m^{-1}$ is the conductivity of the liquid slag). The thickness of the solidified slag skin was set to 4 or 6 mm.

In the literature, the computed current distribution is most often represented by visualizing the magnitude of the current density phasor (i.e. the maximum value for each component of the current density), which classically leads to the observation of an important skin effect in the electrode and ingot.[2, 5,9,14...] Indeed, it is well known that the current distribution is related to the value of the skin depth into the different materials: if the latter is larger than the actual dimension of the domain, the current distribution is homogeneous, e.g. into the liquid slag.

However, within this study, we chose to represent the instantaneous current distribution at a precise moment in the alternating period t = 0. In addition to the visualization of the skin effect, such a representation also highlights the local variation in the phase angle caused by the variation in the induced magnetic field into the metallic conductors, as it was shown by Li et al.[11] A comparison between these two representations is presented on Figure 2.

Table I: Parameters used in the simulations

Melting current (maximum value)	10 kA
AC frequency	50 Hz
Electrode radius	26 cm
Mould external radius	30 cm
Mould thickness	2.5 cm
Electrode immersion depth	1 cm
Electrical conductivity of the metal	$10^6\ \Omega^{-1}.m^{-1}$
Electrical conductivity of the liquid slag	$400\ \Omega^{-1}.m^{-1}$
Electrical conductivity of the solid slag	*10^{-3} - $400\ \Omega^{-1}.m^{-1}$*
Thickness of the solidified slag skin	*4 / 6 mm*

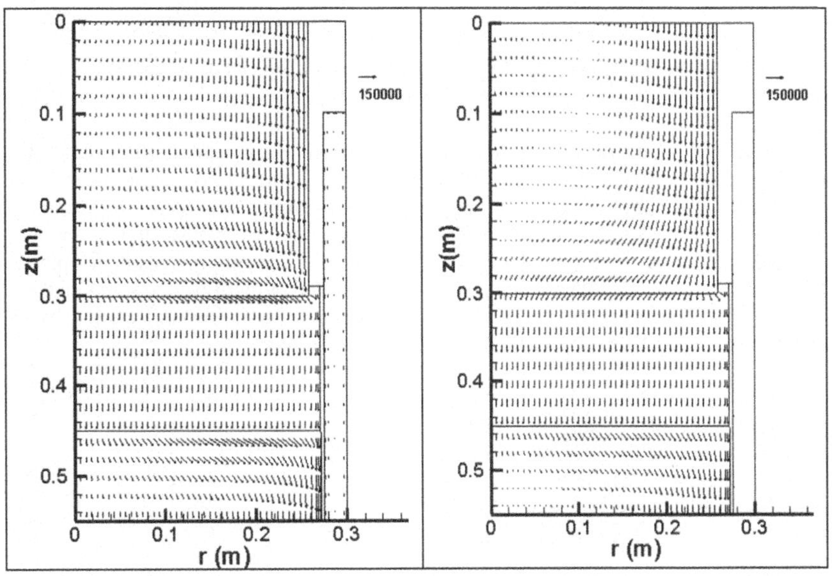

Figure 2: Comparison between the usual representation (maximum value) and the chosen representation (instantaneous value at t=0s) of the current density distribution ($A.m^{-2}$)

The first step of the study consists in confirming that part of the melting current is likely to flow through the solidified slag layer and directly enter the mould. Figure 3 presents the computed results obtained either when the solid layer behaves as a perfect insulator (such behaviour is reached as soon as the electrical conductivity is lower or equal to $10^{-3}\ \Omega^{-1}.m^{-1}$) or when the electrical conductivity is set to $15\ \Omega^{-1}.m^{-1}$. The solidified slag skin is supposed to be 4 mm thick.

Clearly, when the electrical conductivity of the solid slag is set to $15\ \Omega^{-1}.m^{-1}$, part of the current actually flows through the skin to reach the mould. The solidified slag layer does not act as a perfect electrical insulator and this modifies the current distribution in the system. Our result confirms some previous claims in the literature [18,19] and raises new questions regarding the consequences of such a loss of current on the process efficiency.

This conclusion is biased by the assumption made on the value of the electrical conductivity of the solid slag. However, supplementary simulations enabled us to conclude that this statement remains valid whatever the value of the conductivity (except if it is as low as $10^{-3}\ \Omega^{-1}.m^{-1}$, the reference case for the perfect insulation of the mould).

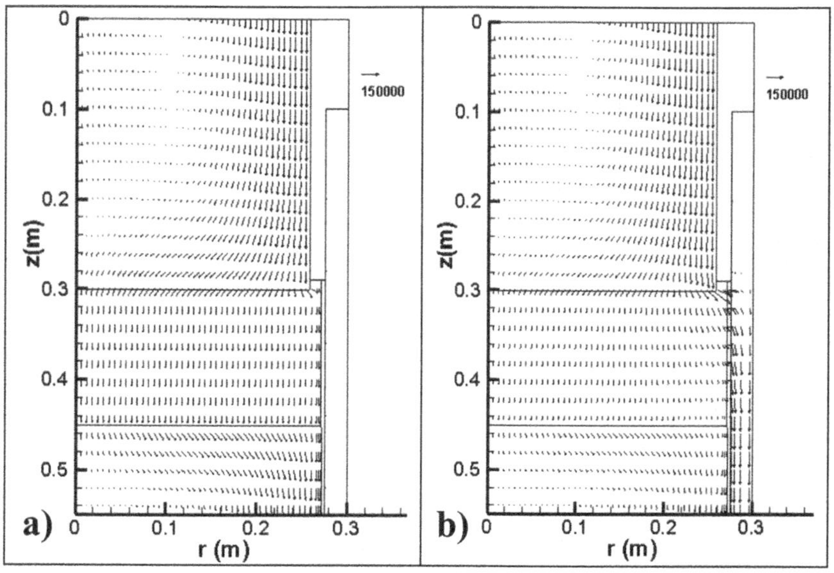

Figure 3: Current density distribution (A.m^{-2}) computed with two values for the electrical conductivity of the solidified slag skin : $10^{-3}\ \Omega^{-1}.m^{-1}$ (a) and $15\ \Omega^{-1}.m^{-1}$ (b)

It must be stated that the simulation whose results are reported in Figure 3-b voluntarily uses a wrong value for the electrical conductivity of the Cu mould, e.g. $10^6\ \Omega^{-1}.m^{-1}$. In the real case, since the conductivity of Cu is about $5.10^7\ \Omega^{-1}.m^{-1}$, the electrical skin depth in the mould is around:

$$\delta = \sqrt{\frac{2}{\omega\mu\sigma}} = \sqrt{\frac{1}{\pi f \mu\sigma}} = 1\ \text{cm}\cdot$$

Therefore, important eddy currents take place into the crucible, which prevents the observation of the melting current flowing from the slag. To avoid this phenomenon, we chose to attribute to Cu a lower electrical conductivity ($10^6\ \Omega^{-1}.m^{-1}$), for which the skin depth is larger than the mould width. All the results obtained remain unchanged in the electrode, slag and ingot, while the melting current flowing in the mould can be observed.

To illustrate that issue, Figure 4-a shows the results obtained when the real value of Cu conductivity is used: the resulting induced current prevents us from noticing the melting current path at the slag/mould interface. On the contrary, when the assigned conductivity is lower (Figure 4-b), this melting current is clearly discernible.

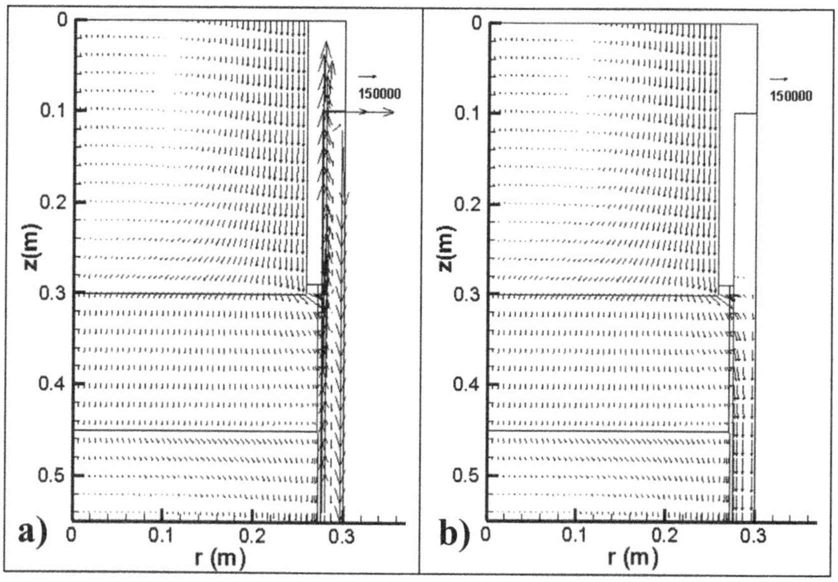

Figure 4: Current density distribution (A.m^{-2}) computed with two values for the electrical conductivity of the Cu mould : $5.10^7\ \Omega^{-1}.m^{-1}$ (a) and $10^6\ \Omega^{-1}.m^{-1}$ (b)

In the remaining part of this paper, that wrong value of Cu electrical conductivity will systematically be used, to allow visualizing the melting current distribution in the whole system, including the crucible.

Impact Of The Electrical Conductivity Of The Solid Slag

This parametric study aims to compute the impact of a variation of the conductivity - in the range 10^{-3}-400 $\Omega^{-1}.m^{-1}$ - on the current distribution, i.e. to observe the behaviour of the melting current when the slag/mould electrical contact varies from a total insulation to a perfect contact. Figures 5 and 6 present the current density distribution and the resistive heating computed if the conductivity of the 4 mm thick slag skin is set to 10^{-3} $\Omega^{-1}.m^{-1}$ (a), 1 $\Omega^{-1}.m^{-1}$ (b), 15 $\Omega^{-1}.m^{-1}$ (c) and 400 $\Omega^{-1}.m^{-1}$ (d).

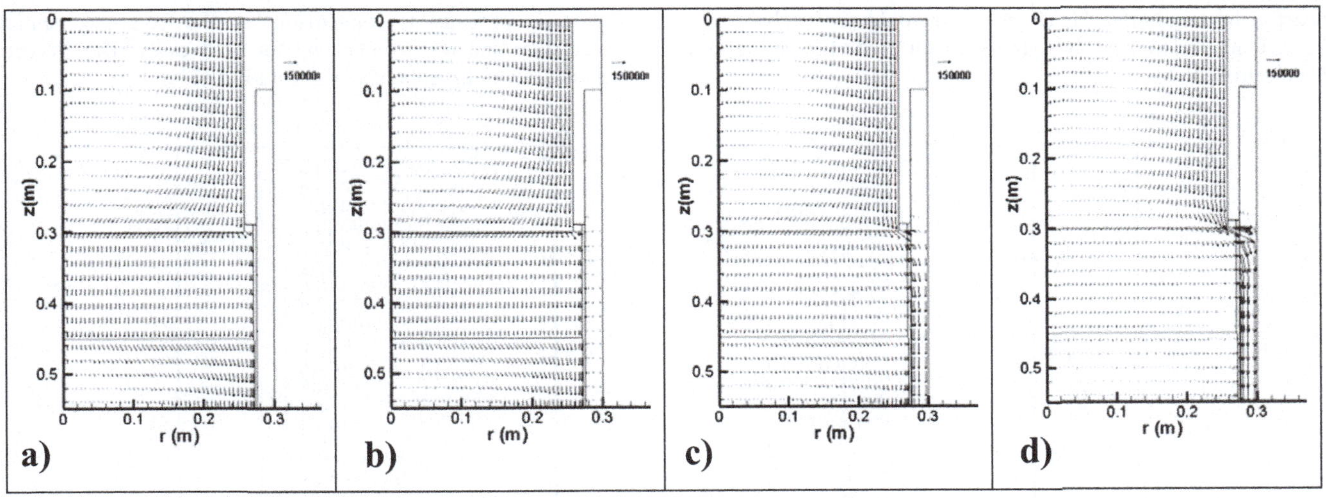

Figure 5: Current density distribution ($A.m^{-2}$) computed with different values for the electrical conductivity of the solidified slag : 10^{-3} $\Omega^{-1}.m^{-1}$ (a), 1 $\Omega^{-1}.m^{-1}$ (b), 15 $\Omega^{-1}.m^{-1}$ (c) and 400 $\Omega^{-1}.m^{-1}$ (d)

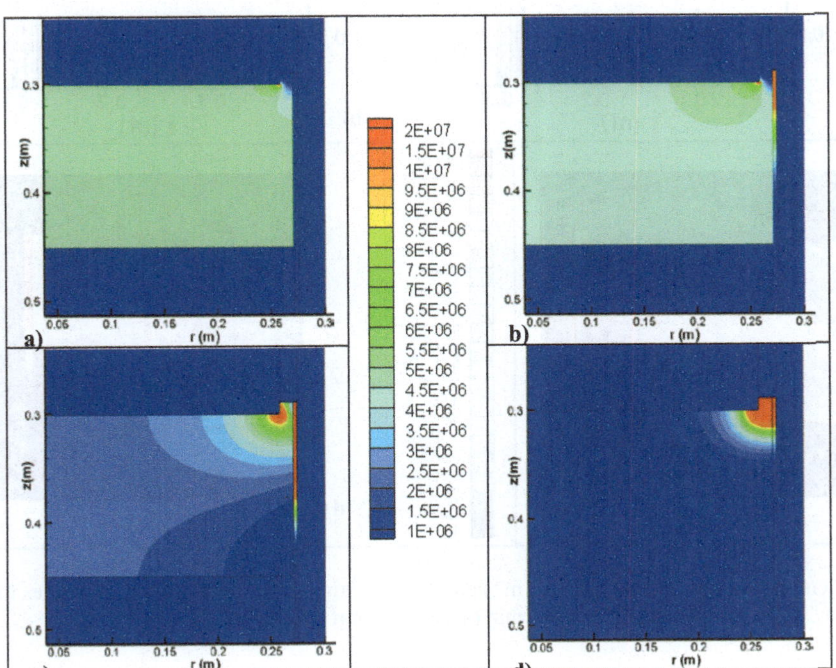

Figure 6: Computed Joule heating ($W.m^{-3}$) with different values for the electrical conductivity of the solidified slag : 10^{-3} $\Omega^{-1}.m^{-1}$ (a), 1 $\Omega^{-1}.m^{-1}$ (b), 15 $\Omega^{-1}.m^{-1}$ (c) and 400 $\Omega^{-1}.m^{-1}$ (d)

If the solid slag layer acts as a perfect insulator, the current distribution is homogeneous into the liquid slag (Figure 5-a), leading to a homogeneous Joule heating (Figure 6-a). Moreover, the amount of current into the insulating layer, hence the Joule effect, is negligible.

As the resistivity of the solidified slag decreases, more current flows through the solid skin: the remaining current in the liquid slag generates a lower Joule heating. On the contrary, an important Joule heating can be noticed in the solid phase, which could lead to the remelting of the skin. It can be noted that in absence of any partially insulating skin, the majority of the current would be attracted towards the mould (Figure 5-d). As a result, resistive heating in the slag (Figure 6-d) would very probably not be sufficient neither for the slag to remain in the liquid state nor the electrode to be melted.

Impact Of The Thickness Of The Slag Skin

As stated above, the thickness of the solidified slag layer is assumed as uniform along the height of the crucible. In a next step, a full coupling of the model described in the present paper with the overall thermo-hydrodynamic model (see the "Conclusion and Prospects" section) will allow us accounting for the calculated variation of that thickness with the axial position.

Based on industrial observations, we tested two values for this parameter: 4 mm and 6 mm. It must be noted that a slag skin thickness of 6 mm is in the upper part of the range of measured values.
Figure 7 presents a comparison of the computed current distribution and resistive heating for both cases. The electrical conductivity of the solid slag was set to 15 $\Omega^{-1}.m^{-1}$.

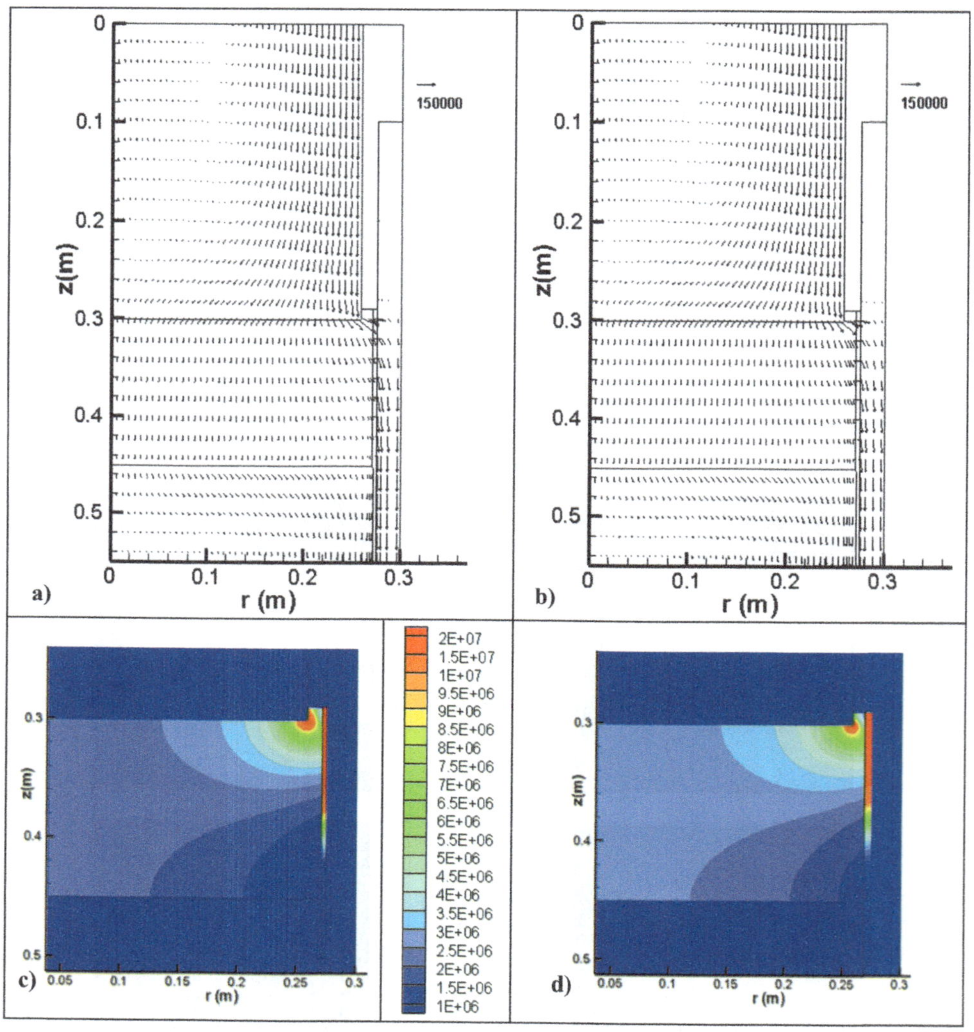

Figure 7: Computed current density distributions ($A.m^{-2}$) and Joule heating ($W.m^{-3}$) with different values for the skin thickness : 4 mm (a-c) and 6 mm (b-d)

As it can be deduced from the observation of Figure 7, the model predicts that even a thick (6 mm) slag skin cannot insulate completely the crucible as it was assumed until recently. Obviously, increasing the skin thickness has the same consequences as decreasing the electrical conductivity, since in both cases the electrical resistance of the solid layer is increased:

- The amount of current "lost" into the mould is slightly lower
- The total Joule heating in the liquid slag resulting from the current distribution is slightly higher, leading to a more efficient process
- On the contrary, the current within the skin, hence the Joule effect in the solid phase, is reduced

Figure 8 summarizes the main effects discussed in the previous sections: it represents the evolution of the total Joule heat generated according to both parameters.
The electrical conductivity of the solid slag appears to be a crucial parameter of the process. This observation emphasizes the necessity to have access to actual measurements. The thickness of the solidified slag layer also influences the current distribution in the system. However, in the range of tested values, the impact of this factor remains of secondary importance.

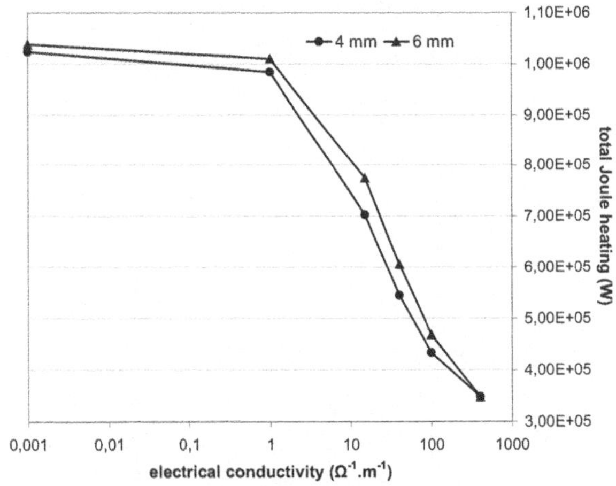

Figure 8: Evolution of the total resistive heating in the slag, according to the electrical conductivity and the thickness of the solidified slag skin

Conclusion and Prospects

The current distribution during an electroslag remelting is a major issue to reach the efficiency and the final quality of ingot required. The study presented in this article used a 2D axisymmetric model of the electromagnetic phenomena occurring during the process. Therefore, it is now possible to compute the current flowing into the electrode, slag, ingot and crucible, as well as the resulting Joule heating.

The model predicts that a fraction of the melting current flows directly from the slag within the mould. This amount depends on the electrical resistance of the solidified slag skin hence its electrical conductivity and thickness. The model definitely highlights that the solidified slag layer does not behave as a perfect electrical insulator unless the solid slag electrical conductivity is lower than 1 $\Omega^{-1}.m^{-1}$.

In the range of values tested, the electrical conductivity of the solidified slag appears to have a determining influence on the results, whereas the layer thickness plays a minor part. The need for accurate thermophysical data is put in evidence.

In its present state, this model considered a uniform layer thickness along the slag/crucible interface. However, this parameter is liable to vary from a negligible value to few millimetres during a real remelting. To take into account this variation, as well as to assess the influence of the electrical current distribution on the ingot solidification, the next step of our study will consist in a full coupling of the model with a numerical simulation of the whole ESR process. Results obtained will be compared to actual experimental observation.

Acknowledgments

As a part of the OPERAS (Optimizing Processes based on Electrode Remelting with Arc or Slag) Project, this work has been supported by the French National Research Agency (ANR-08-MAPR-0006-04).

References

[1] G.Hoyle, *Electroslag Processes – Principles and Practice*, Applied Science Publishers Ltd., 1983.
[2] A. H. Dilawari and J. Szekely, *Metall. Trans. B*, 8B, 1977, p. 227.
[3] J. Kreyenberg and K. Schwerdtfeger, *Archiv Eisenhüttenwesen*, 50, 1979, p. 1.
[4] M. Choudhary and J. Szekely, *Metall. Trans. B*, 11B, 1980, p. 439.
[5] A. Jardy, D. Ablitzer and J.F. Wadier, *Metall. Trans. B*, 22B, 1991, p. 111.
[6] K.M. Kelkar, S.V. Patankar and A. Mitchell, *Proceedings LMPC 2005*, ASM International, p. 137.
[7] A. Kharicha et al., *Proceedings LMPC 2009*, TMS, p. 235.
[8] A.D. Patel, *Proceedings LMPC 2011*, SF2M, p. 49.
[9] M.J.M. Krane et al., *Proceedings LMPC 2011*, SF2M, p. 65.
[10] A. Kharicha et al., *CFD Modeling and Simulation of Materials Processing*, Wiley/TMS, 2012, p. 139.
[11] B. Li, F. Wang and F. Tsukihashi, *ISIJ International*, 52, 2012, p. 1289.
[12] T. Quatravaux et al., *Journal of Materials Science*, 39, 2004, p. 7183.
[13] V. Weber, *Simulation numérique du procédé de refusion sous laitier électroconducteur*, Ph.D. Thesis, Nancy, 2008.
[14] W. Weber et al., *Metall. Trans. B*, 40B, 2009, p. 271.
[15] B. Hernandez-Morales and A. Mitchell, *Ironmaking and Steelmaking*, 26, 1999, p. 423.
[16] D.K. Melgaard and G.J. Shelmidine, *Proceedings LMPC 2001*, AVS, p. 148.
[17] K.M Kelkar et al., *Journal de Physique IV*, 120, (2004), p. 421
[18] A. Mitchell, *Materials Science and Engineering A*, 413-414, 2005, p. 10.
[19] A. Kharicha et al., *Proceedings LMPC 2007*, SF2M, p. 113.
[20] A. Kharicha, private communication

Proceedings of the 2013 International Symposium on Liquid Metal Processing & Casting
Edited by: Matthew J. M. Krane, Alain Jardy, Rodney L. Williamson, and Joseph J. Beaman
TMS (The Minerals, Metals & Materials Society), 2013

INVESTIGATION OF SLAG COMPOSITIONS AND PRESSURE RANGES SUITABLE FOR ELECTROSLAG REMELTING UNDER VACUUM CONDITIONS

S. Radwitz[1], H. Scholz[2], B. Friedrich[1]
[1]IME Process Metallurgy and Metal Recycling, RWTH Aachen University, D-52056 Aachen, Germany
[2]ALD Vacuum Technologies GmbH, Wilhelm-Rohn-Strasse 35, D-63450 Hanau, Germany

Keywords: ESR slags, oxygen content, refining of steel, vacuum conditions

Abstract

It is well known that high contents of oxygen and hydrogen in creep resistant structural steels like 21CrMoV5-7 have negative influence on a variety of material properties. To investigate the refining ability of various slag compositions under reduced pressure multiple experiments were performed in a 40 kW vacuum-induction furnace with the aim to ensure minimal oxygen and hydrogen contents. With regard to slag evaporation, different mixtures of fluorides and oxides as well as pure oxide systems were utilized. The pressure was varied in the range of 5 and 700 mbar.

The behavior of different slags under vacuum conditions as well as the results of various chemical analysis were used to evaluate the application of the investigated slag systems for ESR under vacuum conditions. Special attention was given to easily oxidizable alloy elements as aluminum and silicon, the amount of oxygen and hydrogen as well as on formation of non-metallic inclusions.

Introduction

The presence of non-metallic inclusions (NMIs) in metallic materials results in the occurrence of rolling and surface defects during further processing as well as in a decrease of the mechanical properties and corrosions resistance.

Thereby, the effect of NMIs on the material behavior depends strongly on their size, morphology, distribution and chemical composition. For instance, the fracture toughness and ductility are greatly reduced with an increasing content of oxides and sulfides. In addition, brittle oxide components (e.g. Al_2O_3 inclusions) could serve as potential crack initiators. [1]

Due to the fact that a dominating factor with regard to the amount of non-metallic inclusions consists in the content of dissolved oxygen in liquid steel, it is of interest to minimize the total oxygen (T.O.) content by various metallurgical treatments. According to that, Zhang and Thomas [2] have summarized typical steel cleanliness requirements, published by various authors, for common steel grades (cf. Table I).

In the ESR process, a consumable electrode is continuously melted in a water cooled cooper mold while being inserted into a liquid slag bath. On the electrode tip, liquid metal droplets are formed which sink through the slag bath after detaching from the electrode. After passing the slag, the droplets are collected in a typical U-shaped liquid metal pool before solidifying as the remelted ingot (cf. Figure 1).

Table I: Steel cleanliness requirements for various steel grades [2]

Product class	Maximum impurity fraction in ppm	Maximum inclusion size in µm
Tinplate	T.O. < 20, C < 30, N < 30	20
Tube seel	T.O. < 30, N < 35, S < 30	100
Ball bearing steels	T.O. < 10	15
Heavy plate steels	T.O. < 20, H < 2, N < 30-40	13 (Single inclusions) 200 (Cluster)
Tire cord	T.O. < 15, H < 2, N < 40	10-20
Wire rod	T.O. < 30, N < 60	20

Fundamentals

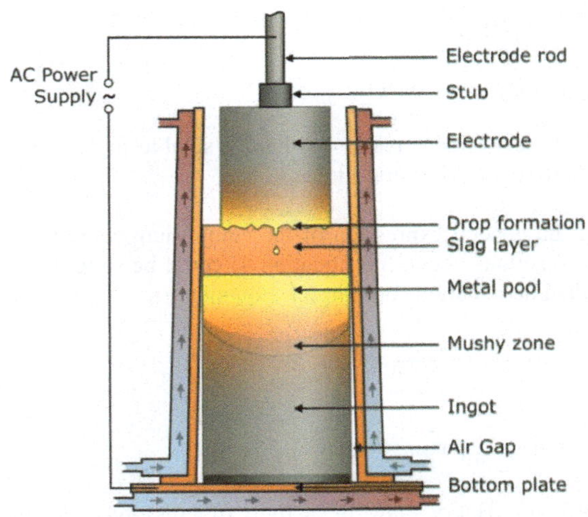

Figure 1: Schematic procedure of the ESR process [3]

In the ESR process various chemical and physical interdependencies between the remelted material (alloying elements, dissolved interstitial elements and non-metallic inclusions), the cat- and anions of the slag system and the gas atmosphere are present. An appropriate illustration of these interactions is given by Figure 2.

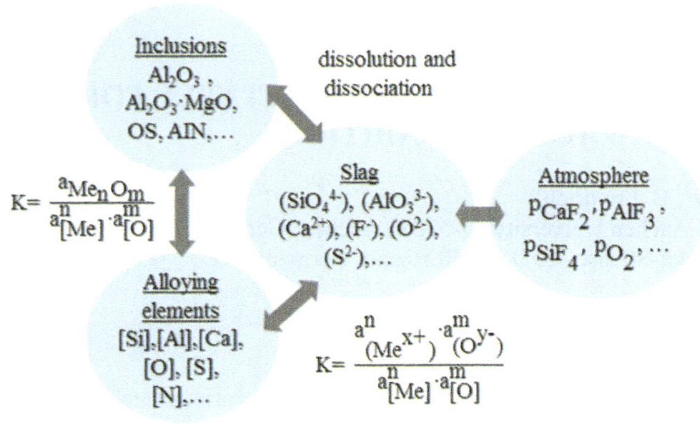

Figure 2: Chemical and physical interactions between alloy, slag and atmosphere during ESR [4]

It could be seen that one influencing factor on the oxygen content of the steel melt is given by the amount of dissolved elements with a high affinity for oxygen such as calcium, aluminum or silicon. On the one hand a reaction of these metals with oxygen would therefore result in a decreased oxygen content and oxygen activity respectively but on the other hand would also generate oxide inclusions which need to be removed from the liquid steel (cf. Equation 1-3).

$$2\,[Al]_{steel} + 3\,[O]_{steel} \leftrightarrow \langle Al_2O_3 \rangle \quad (1)$$

$$[Si]_{steel} + 2\,[O]_{steel} \leftrightarrow \langle SiO_2 \rangle \quad (2)$$

$$[Ca]_{steel} + [O]_{steel} \leftrightarrow \langle CaO \rangle \quad (3)$$

Since these forming oxides are hardly soluble in the steel melt their activity could be considered as one.

Thus, the general expression of the equilibrium constant for one of the reactions above (cf. Equation 4) could be simplified to the so-called deoxidation constant (cf. Equation 5).

$$K_{Me_xO_y} = \frac{a_{\langle Me_xO_y \rangle}}{a^x_{[Me]_{steel}} \cdot a^y_{[O]_{steel}}} \quad (4)$$

$$K'_{Me,O} = a^x_{[Me]_{steel}} \cdot a^y_{[O]_{steel}} \quad (5)$$

According to Henry's law, the activity of a dissolved element in a dilute solution is given by Equation 6.

$$a_i = f_i \cdot x_i \quad (6)$$

In order to calculate the according concentration equilibria, the method using interaction coefficients (e_i^j) presented by Wagener [5] could be utilized reliably for low alloyed steels. The Henrian activity coefficients could therefore be expressed by Equation 7.

$$\log f_i = \Sigma_i^n (e_i^j \cdot [\%j]) \quad (7)$$

As a result of such considerations, the correlation between dissolved oxygen, alloying elements and their corresponding oxides leads to the typical curve progression illustrated by Figure 3.

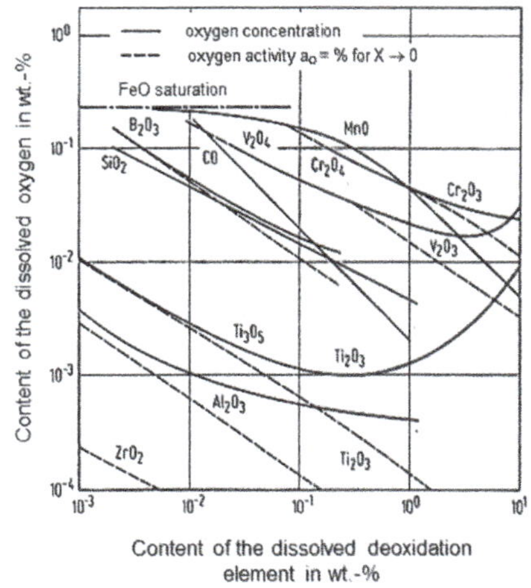

Figure 3: Equilibria between oxygen and various deoxidizing elements dissolved in liquid iron at 1600 °C [6]

At this point, it should be noted that in the ESR process such reactions mainly occur during solidification in the mushy zone (if no deoxidizing elements are added during remelting).

If this kind of new formed inclusions are able to reach the interface between the slag bath and the liquid metal pool by flotation (which is discussed controversial in literature [7]) they could be dissociated and absorbed by the slag. However, inclusions contained already in the electrode material could be removed more effectively at the various reaction zones between the liquid metal and the slag bath.

Another important factor determining the oxygen content of the steel is found in the interaction of the metallic melt and the liquid slag bath. Since in the case of electroslag remelting the common utilized slag systems are based on calcium fluoride with additions of comparatively stable oxides such as CaO, Al_2O_3 or MgO, the cleanliness of the metal is greatly affected by the oxygen activity as well as by the activity of inclusion-forming elements. Under these circumstances the expression for the equilibrium constant for the reaction between an alloying element and oxygen (cf. Equation 8) has to be modified as shown by Equation 9.

$$x\,[Me]_{steel} + y\,[O]_{steel} \leftrightarrow [Me_xO_y]_{slag} \quad (8)$$

$$K_{Me_xO_y} = \frac{a_{[Me_xO_y]_{slag}}}{a^x_{[Me]_{steel}} \cdot a^y_{[O]_{steel}}} \quad (9)$$

As could be seen the activities of slag components are of great importance for determining the appropriate equilibrium constants. With regard to the correlation given in Figure 3, a decrease of the oxide activity by dissolving it in the slag ($a_{[MeO]slag} < 1$) would lead to a shift of equilibrium towards smaller dissolved contents of oxygen and deoxidizing elements in the steel.

In this case, especially the availability of data for oxide activities in fluoride based slags is very limited in the current literature. Referring to pure oxide systems, consisting mainly of CaO and Al_2O_3, Riyahimalayeri [8] et al. have summarized different approaches for calculating the activity of Al_2O_3. Further suggestions of Mitchell [9] include the assumption for systems containing less than 70 wt.-% CaF_2 that calcium fluoride could be considered as an inert diluent.

In addition to the influence on the oxygen content of the metal various reactions with slag components could result in a loss of alloying elements as exemplified by Equation 10.

$$3\,[Si]_{steel} + 2\,[Al_2O_3]_{slag} \leftrightarrow 2\,[Al]_{steel} + 3\,[SiO_2]_{slag} \quad (10)$$

Besides to the contact with the liquid metal, the slag system is also driven to reach equilibrium with the gas phase.

Due to the high process temperatures fluoridic components are continuously evaporating as illustrated by Equation 11 and 12.

$$[CaF_2]_{slag} \leftrightarrow \{CaF_2\} \quad (11)$$

$$3\,[CaF_2]_{slag} + [Al_2O_3]_{slag} \leftrightarrow 3\,[CaO]_{slag} + 2\,\{AlF_3\} \quad (12)$$

The corresponding vapor pressure curves of selected components (cf. Figure 4) show that the evaporation of the present oxides is negligible compared to that of existing and forming fluorides.

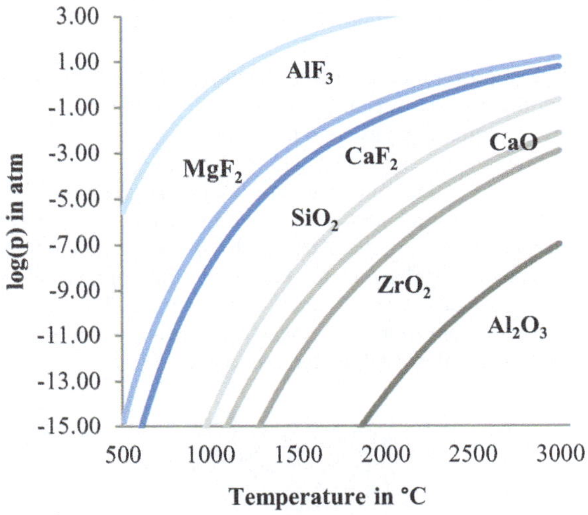

Figure 4: Correlation of vapor pressure and temperature for selected components (calculated with FactSage 6.3 [10])

Due to the fluoride evaporation the electroslag remelting is usually performed under normal or overpressure. To minimize the oxidative loss of alloying elements as well as the introduction of oxygen into the liquid metal a protective argon atmosphere could be applied. However, the findings of Saito et al. [11] and Sekiya et al. [12] indicate the possibility to achieve very low contents of oxygen (approximately 10 ppm) and non-metallic inclusions by carrying out the remelting under vacuum conditions in the range of 66 to 200 mbar.

Experimental Work

Since numerous factors are influencing the total oxygen content of a remelted steel the focus of the present work lies on the investigation of chemical interactions between metal and slag as well as on the slag behavior under reduced pressure. Therefore, a specific steel grade was melted under various slag compositions and pressure ranges in a lab scale vacuum induction furnace. Basing on these results subsequent remelting trials will be carried out as a next step to further examine factors determining the final oxygen content of the metal but are not part of the current work.

The performed investigation involves nine slag treatments of the creep resistant structural steel 21CrMoV5-7 as well as the sampling and analysis of metal and slag after each trial. Prior vacuum induction melting, the chemical composition of the used steel grade was analyzed by optical emission spectroscopy as well as by using combustion and inert gas fusion methods (cf. Table II and III).

Table II: Chemical composition of the utilized steel 21CrMoV5-7 in wt.-%

C	Si	Mn	Al	Cr
0.20	0.273	0.671	0.017	1.288

Mo	Ni	V	Zr
0.710	0.486	0.279	0.0015

Table III: Total content of interstitial elements in ppm

S	O	N	H
18	42	66	2

Vacuum Induction Melting

The melting was carried out in a 40 kW and 10 kHz lab scale induction furnace (cf. Figure 5) which is equipped with various devices for material charging, temperature measurement and sampling via a lock.

Figure 5: Utilized vacuum induction furnace at IME

To investigate the refining potential of different slag compositions with regard to the oxygen content of the steel, various mixtures of a premelted calcium-aluminate slag as well as of prefused fluorspar were blended. In this case, the oxidic slag contains equal amounts of alumina and lime. However, due to the conditions of manufacturing the slag also includes approximately five percent magnesia. The fluorspar has a purity of at least 97 wt.-% CaF_2. Moreover, the main impurities contained therein also consist of alumina and lime.

For all slag treatments, at least two different pressures inside the furnace chamber were applied. To minimize the influence of atmospheric oxygen on the total oxygen content of the metal the furnace chamber was evacuated at least to a pressure below 10^{-2} mbar before heating-up under a protective argon atmosphere with 700 mbar. An overview of the performed trials is given in Table IV.

Table IV: Performed VIM trials with variation of slag composition and process pressure

Trial	Utilized slag compositions	Pressure in mbar
T1	46.9 % CaO, 46.6 % Al_2O_3, 5.0 % MgO	700
T2	46.9 % CaO, 46.6 % Al_2O_3, 5.0 % MgO	200
T3	46.9 % CaO, 46.6 % Al_2O_3, 5.0 % MgO	7
T4	20 % CaF_2, 37.4 %CaO, 37.1 %Al_2O_3, 4.0 % MgO	700
T5	20 % CaF_2, 37.4 %CaO, 37.1 %Al_2O_3, 4.0 % MgO	5
T6	80.0 % CaF_2, 9.5 %CaO, 9.4 %Al_2O_3, 0.9 % MgO	700
T7	80.0 % CaF_2, 9.5 %CaO, 9.4 %Al_2O_3, 0.9 % MgO	5
T8	97.0 % CaF_2, 1.5 %CaO, 1.5 %Al_2O_3	700
T9	97.0 % CaF_2, 1.5 %CaO, 1.5 %Al_2O_3	5

As crucible material densely sintered zirconia (yttria-stabilized) was utilized due to the comparatively slow dissolution rate in the fluoride based slag. Furthermore, to minimize the reaction time between crucible and slag, the prior blended slag components were added on top of the metal melt via the lock. Once the slag was completely melted, the pressure was reduced in the corresponding trials until an extensive bubbling of the slag was observed. Independent of the fluorine content, this phenomenon constantly occurred when reaching a pressure of 5 mbar.

In contrast to the ESR process, where remelting is performed in a water-cooled cooper mold, the liquid metal is subjected to chemical interactions with the oxidic crucible material (cf. Equation 13).

$$\langle ZrO_2 \rangle \leftrightarrow [Zr]_{steel} + 2\,[O]_{steel} \quad (13)$$

However, according to the equilibrium position of the reaction above (cf. Figure 3), the equilibrium content of oxygen which is due to pick up from the crucible is only in the range of 2 ppm at 1600 °C.

By melting in an induction field heat is generated directly in the metal and only marginal in the slag. In order to prevent distinct temperature differences between metal and slag the crucible was placed in a tubular graphite susceptor (cf. Figure 6). Thereby, sticking of slag particles on the crucible wall was also avoided. With an inner diameter of 59 mm and a height of 93 mm, the utilized crucibles had a maximum filling volume of 220 ml. To keep the ratio between metal and slag constant in all trials approx. 790 g of metal and 103 g of slag were used. Hence, the crucible was half filled with liquid steel and to a quarter with liquid slag. In T1-T3 smaller crucibles had to be used with an inner diameter of 44 mm and a height of 73 mm but the ratio between metal and slag was equal to the other trials.

Figure 6: Arrangement of the experimental set-up

During the trials the temperature was measured using platinum-rhodium thermocouples by two different methods. One thermocouple was centrically placed below the bottom of the crucible. The second measurement was performed periodically by immersion measurements via the lock. Both measurements showed a metal temperature of approx. 1550 °C whereas the slag temperature was at least 50 °C lower.

In order to prevent an extensive mixing of slag and metal during casting, the material was allowed to solidify in the crucible. Thus, every trial required the use of a new crucible.

After melting, slag and vacuum treatment respectively, the obtained ingots were sectioned and sampled in the mechanical workshop at IME and characterized by the same methods as the input material. Referring to Figure 7, the optical emission spectroscopy was carried out by analyzing the blue illustrated parts. The combustion and inert gas fusion techniques were applied to the yellow highlighted sections.

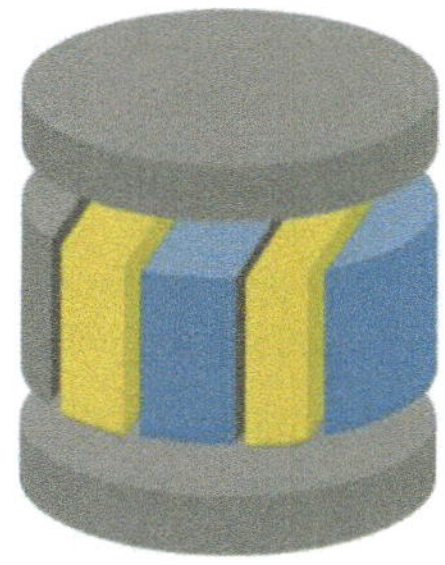

Figure 7: Sampling of the metal ingots

Results and Discussion

After each trial a strong adhesion of the utilized slag on the crucible wall was observed (cf. Figure 8, up). As expected the received metal ingots contained noticeable shrinkage cavities at the bottom and the side walls (cf. Figure 89, down).

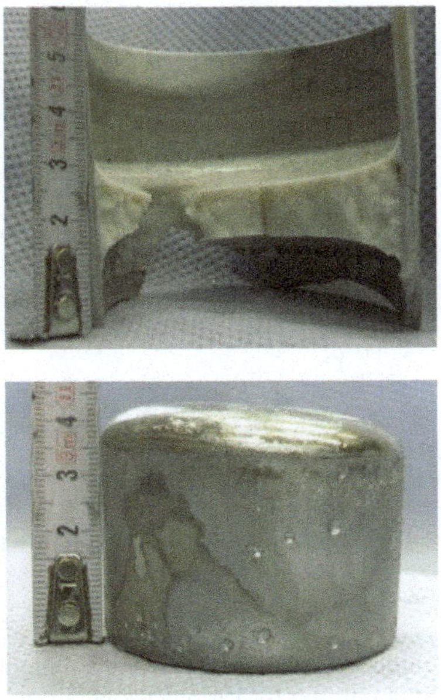

Figure 8: Representative illustration of the received slag layer attached to the crucible (up) and the obtained metal ingot (down)

A selection of analyzed elements via optical emission spectroscopy is given in Table V. For the additional elements (Cr, Ni, Mo and V) listed in Table II no significant differences to the contents of the input material were found.

With respect to the proportion of silicon a decrease was detected after treatment with all slags containing oxides in a notable quantity (T1-T7). Within these, the silicon loss was most distinct when the metal was interacting with a pure oxide slag. With an increasing amount of calcium fluoride the silicon loss could be reduced significantly. Furthermore, by utilizing a technically pure CaF_2-slag (T8, T9), the silicon content of the metal was fully retained.

Simultaneously to the decrease of silicon, an increase of the aluminum content was observed in all trials containing Al_2O_3 as slag component. These findings could be explained by the high alumina and low silica activity in the slag resulting in the reaction given in Equation 10. Additionally, it is noticeable that only the use of a pure fluoridic slag results in a comparatively strong decrease of the aluminum content. Referring to the deoxidation equilibrium of oxygen and aluminum, a possible explanation for that could be found in the greatly reduced alumina activity in the slag.

Additionally, with the utilization of Fact Sage 6.3 [10] a first approach of estimating the loss of silicon and the increase of the aluminum content was performed using FSstel and FToxid databases. Since under these circumstances only a reliable simulation of the interaction between steel and a solely oxidic slag is possible, the calculations were limited to T1-T3. In good agreement with the present findings a resulting silicon amount of 0.217 wt.-% was calculated for equilibrium conditions between steel and slag. However, the according aluminum content of 0.078 wt.-% was significantly higher than the measured values. On the one hand this amount is consistent with the trend observable in T4-T9 but on the other hand indicates to low aluminum contents in T1-T3.

The low manganese losses could either be the result of metal-slag reactions or slight evaporation effects.

By considering the amount of zirconium an increase was observed in all trials but much more intense when the slag treatment was performed under reduced pressure. However, a correlating variation of the resulting oxygen content in the metal was not detected. Since the dissociation of ZrO_2 is nearly not depending on pressure, the deviation of the measured values is suspected to be due to the inaccuracy of the analyzing method in the present concentration range. Additional calculations using FactSage 6.3 [10] analogously showed an equilibrium concentration of approx. 0.02 % Zr in the according steel melt while being in contact with solid ZrO_2 at 1550 °C. A simultaneous rise of the dissolved oxygen content was also not observable by that simulation.

At this point it should be noted the increased carbon content in T4 is due to an unintended immersion of the graphite tube in the steel melt which was used as protection cover for the submerged thermocouple.

Table V: Analysis of selected elements after slag treatment compared to the input material

	C	Si	Mn	Al	Zr
Input material	0.203	0.273	0.671	0.017	0.0015
T1	0.180	0.204	0.642	0.022	0.0023
T2	0.200	0.223	0.640	0.040	0.0022
T3	0.190	0.186	0.645	0.021	0.0017
T4	0.444	0.224	0.664	0.049	0.0082
T5	0.200	0.217	0.659	0.057	0.011
T6	0.208	0.248	0.667	0.037	0.0042
T7	0.198	0.248	0.665	0.038	0.011
T8	0.202	0.276	0.653	0.0067	0.0037
T9	0.203	0.281	0.651	0.0066	0.028

The according contents of the interstitial or inclusion forming elements are given in Table VI. Due to the content of calcium oxide in the slag it was possible to further reduce the sulfur fraction below 10 ppm, even in the trials T8 and T9. With regard to nitrogen a decreasing pressure resulted in reduced amounts in T1-T3. In the further trials T4-T9 no significant differences with respect to the resulting nitrogen content were found.

Similar observations were made when considering the hydrogen contents. For the trials T1-T3 it was not possible to influence the hydrogen content significantly by the combined vacuum-slag treatment. However, in T4-T9 the hydrogen content was further reduced to values below 1 ppm.

Furthermore, T1-T3 showed that it was possible to reduce the total oxygen content of the metal by decreasing the process pressure when melting under a pure oxide based slag (cf. Figure 9). In contrast to that, all trials including CaF_2 as slag component resulted in even lower oxygen contents between 5 and 8 ppm. A dependency on the process pressure could not be observed.

Table VI: Analysis of interstitial elements after slag treatment compared to the input material

	S	O	N	H
Input material	18	42	66	2
T1	< 10	19.2	73	1.78
T2	< 10	15.2	31	1.81
T3	< 10	9.8	16	1.7
T4	< 10	6	14	0.7
T5	< 10	6	12	0.6
T6	< 10	5	16	0.7
T7	< 10	8	16	0.7
T8	< 10	6	6	0.8
T9	< 10	7	12	0.8

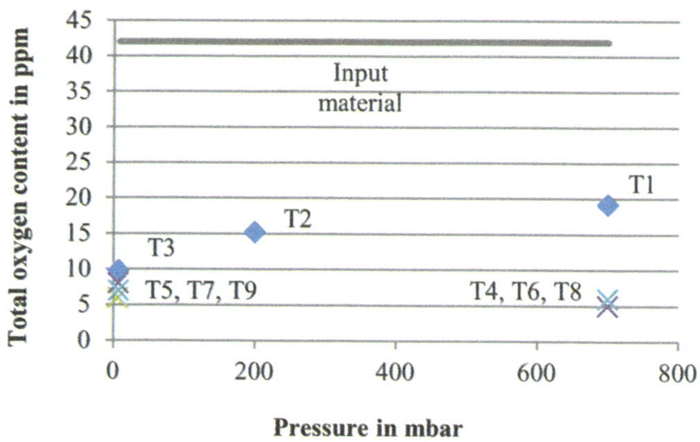

Figure 9: Correlation between the total oxygen content of the metal and the applied pressure compared to that of the input material

By considering the equilibria between oxygen and aluminum as well as between oxygen and silicon it could be concluded that the measured oxygen contents are clearly below those of the silicon deoxidation equilibrium. However, referring to the interaction of oxygen and aluminum, the analyzed contents noticeably exceed the corresponding equilibrium amounts at process temperature of 1550 °C even by assuming an alumina activity of one as shown in Figure 10.

Conclusions

Although the interaction between metal and the chosen slag systems leads to decreased contents of total oxygen in the steel, the main factor determining the final oxygen content is not clearly identifiable. The expected direct relation between the dissolved oxygen and aluminum contents in the metal and the alumina content of the slag are not sufficiently confirmed. Therfore, furhter detailled investigations are needed to identiefy the influence of fluorine on the oxygen activity of the slag and the corresponding equilibrium with the metal. Additionally the residual partial pressure of oxygen in the atmosphere needs to be considered with regard to the permeability of the various slag systems.

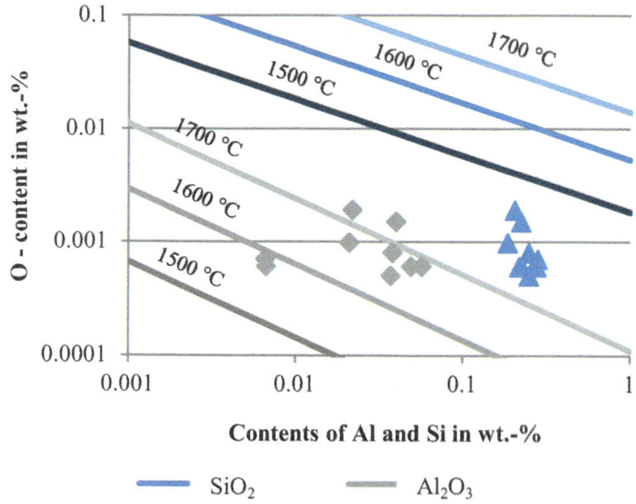

Figure 10: Equilibria between oxygen and aluminum and silicon respectively in liquid iron for various temperatures as well as the analyzed contents of the respective elements

Summary

(1) In order to investigate the influence of various mixtures within the system CaF_2-CaO-Al_2O_3-MgO during slag treatment of a 21CrMoV5-7 steel, nine trials were performed in a vacuum induction furnace. Additionally, the refining behavior was studied for at least two different pressures.

(2) The conducted treatments with a pure oxide slag as well as with various mixtures of oxides and fluorides resulted in significantly reduced total oxygen contents in contrast to the input material.

(3) A pressure dependence of the final oxygen content was only observed by melting under a pure oxide slag.

(4) Due to the interaction between metal and slag a noticeable loss of silicon accompanied by an increase of the aluminum content was detected for most of the trials. Thereby, the degree of that interaction strongly depends on the fluoride content of the slag.

Outlook

In order to validate the findings derived from the trials where solely oxides were used as slag components these will be repeated. Furthermore, analysis with respect to the distribution of non-metallic inclusions using SEM-EDX methods as well as additional slag investigations will be carried out.

Subsequently, all present results will be compared to the behavior of 21CrMoV5-7 steel after electroslag remelting with similar slags and atmospheric conditions.

References

1. G. Klösch et al., "Nichtmetallische Einschlüsse in Langprodukten Entstehung, Bestimmung, Auswirkungen - ein Überblick," *BHM Berg- und Hüttenmännische Monatshefte*, 154 (2009), 27-32

2. L. Zhang, and B.G. Thomas, "State of the Art in Evaluation and Control of Steel Cleanliness," ISIJ International, 43 (2003), 271-291
3. N. Giesselmann, "Numerical simulation of the electroslag remelting process in order to determine influencing parameters on ingot defects," *Proceedings of the 1st International Conference on Ingot Casting, Rolling and Forging*, (2012)
4. R. Schneider et al., "DESU-Prozessoptimierung zur Herstellung stickstofflegierter Stähle mit höchsten Reinheitsgraden" *BHM Berg- und Hüttenmännische Monatshefte*, 147 (2002), 1-5
5. H. Schenck, and E. Steinmetz, "Wirkungsparameter von Begleitelementen in Eisenlösungen und ihre gegenseitigen Beziehungen," *Stahleisen-Sonderberichte*, 7 (1968)
6. F. Oeters, "Metallurgie der Stahlherstellung," *Springer Verlag*, (1989), 57-75
7. A. Mitchell, "Oxide inclusion behavior during consumable electrode remelting," *Ironmaking and Steelmaking (Quarterly)*, 3 (1974), 172-179
8. K. Riyahimalayeri, P. Ölund, and M. Selleby, "Oxygen Activity Calculations of Molten Steel: Comparison With Measured Results," *steel research international*, 84 (2013), 136-145
9. A. Mitchell, "Slag Functions in the ESR Process," *Proceedings of the 2005 International Symposium in Liquid Metal Processing and Casting*, (2005)
10. GTT-Technologies, Herzogenrath, "FactSage 6.3"
11. K. Saito, R. Asano, and H. Matsuya "Enhancement of VSR Ingot Quality by Precise Melt Control," *Electric Furnace Steel (Denki-Seiko)*, 80 (2009), 151-156
12. A. Sekiya, S. Nakayama, and T. Taketsuru "Deoxidation of Stainless Steel on Vacuum Electroslag Remelting Process," *Electric Furnace Steel (Denki-Seiko)*, 66 (1995), 47-53

Proceedings of the 2013 International Symposium on Liquid Metal Processing & Casting
Edited by: Matthew J. M. Krane, Alain Jardy, Rodney L. Williamson, and Joseph J. Beaman
TMS (The Minerals, Metals & Materials Society), 2013

CONTRIBUTION OF THE MOULD CURRENT TO THE INGOT SURFACE QUALITY IN THE ELECTROSLAG REMELTING PROCESS

A. Kharicha [1,2], E. Karimi Sibaki [1], M. Wu [1,2], A. Ludwig [2]
[1] Christian Doppler Laboratory for Advanced Process Simulation of Solidification and Melting, Univ. of Leoben, Austria
[2] Chair of Simulation and Modeling of Metallurgical Processes, Univ. of Leoben, Austria
Correspondent author: abdellah.kharicha@unileoben.ac.at

Keywords: electroslag remelting (ESR), slag skin, mould current, slag electrical conductivity

Abstract

Thickness, uniformity and mechanical properties of slag skin are main factors controlling the surface quality of the electroslag remelting (ESR) ingot. Thin and homogeneous slag skin is a key to obtain a nice ingot surface quality. It is classically believed that surface quality is mainly influenced by the slag temperature at the vicinity of the mould. In the present paper considers the effect of the Joule heating on the slag skin energy balance. The slag skin between ingot and mould is expected to have a significantly different electrical resistivity than the bulk slag, complicating the electrical boundary conditions at the ingot/mould interface. The amount of electric current crossing the skin towards the mould depends on the electric slag skin conductivity. It is shown that the magnitude of Joule heating generated within the skin can control its thickness. Thus, it is suggested that the known sensibility of the ingot quality on the melt rate, the electrode immersion depth, fill ratio and the resistivity of the slag can be reduced to its dependence on the electric current path and intensity.

Introduction

The surface treatment of ingots produced by the ESR process is expensive. The optimum slag would enable the production of ingots with smooth surfaces, thus reducing the need for surface treatment prior to forming. Mitchel et al. [1] explain the mechanism of slag skin formation by using the phase diagrams of the corresponding phase. A slag skin is formed by the selective solidification of one or more phases on the water cooled copper wall [1]. Its formation plays an important role in controlling the heat transfer to the mould. The phases which solidify and their order of solidification are determined by the composition of the liquid slag and the governing phase relationships. Mitchell [2] has pointed out that the phases present in the slag skin have a direct influence on ingot surface quality. However Korousic and Osterc [3] performed mineralogical analysis of several sample of slag skin, and found that the crystallisation did not follow the path given by the phase diagram.

It has been observed that good quality surface is associated with the presence of a finite depth (few cm) of liquid-metal contact on the slag skin [4] also known as "liquid head" or "standing height". To achieve these requirements, the melting point of the slag must be higher than that of the metal. For this reason it was stated that a slag with higher Al_2O_3 content than the eutectic will produce good surfaces, while those with lower will not[2]. It was found that two slags with 25 and 10 wt-% gave uniform smooth slag skin. The slag with 1 wt-% Al_2O_3 was embedded within the ingot surface. The resistivity of the slag was also mentioned as an important parameter since it controls the heat generated within the slag [2]. Slags with lower fluoride contents (higher electric resistivity) are known to be operationally more efficient in achieving good ingot surface quality [5]. A resistive slag is believed to be able to generate a sufficiently high temperature at the slag/pool interface to sustain the existence of a sufficiently thick depth of liquid metal against the slag skin. However it can be argued that the temperature in a process is primarily determined by the targeted melt rate. A resistive slag needs a smaller amount of imposed current to reach the same melt condition than a more conductive slag. So the temperature might not be the only reason. In addition the explanations given, melting point of the slag versus electric resistivity, are clearly overlapping since slags with higher (Al_2O_3) content have generally higher electric resistivity and simultaneously higher melting point.

It was suggested by industry to run the process with warmed mould may encourage the formation of smooth skin [6]. But in most common industrial practice, the problem with poor surface has been dealt with increasing the power input (and so the melt rate), and by using the highest stable fill-ratio. An explanation based on slag temperature can again be given, a large fill ratio decreases the heat losses at the exposed slag surface, and thus increases the slag temperature near the mould. The thickness of the slag skin is also largely affected by immersion volume of the electrode. Immersing a large volume of electrode into the slag layer increases the amount of heat extracted by the electrode and by the mould, which lowers the temperature of the slag cap and thus increases the thickness of the slag skin. Therefore it is highly recommended to keep the immersion volume of the electrode as small as possible.

The slag viscosity was also mentioned as a parameter that can control the slag skin thickness. Due to similar molecular mechanism of the viscosity with the electric resistivity (ions movement), electrically resistive slags are more viscous than conducting ones. By using a slag with high viscosity the slag flow velocities are lower. Kusamichi et al.[7] reported larger radial temperature gradient in slags with low fluoride content than for larger one. The hottest region was located near the electrode, and the coldest near mould. Low fluoride slags are usually used together with lower applied electric current intensities, which together with a higher viscosity level mean lower magnitude of the electromagnetic forces, which in turn generate less stirring in the slag region. The fact that colder temperatures were found near the slag skin for low fluoride slags is an indication that the slag temperature is not the proper mechanism to explain why resistive slags are more efficient in producing thin slag skins.

In previous numerical works it was shown that the electric current flowing directly towards the mould has a considerable effect on the slag skin thickness [8-9]. For better prediction, the Joule heating generated by the electric current must be added to the heat balance of the slag skin. To our knowledge, Medina et al. [10] were the first to notice experimentally a possible relation between the electric current and slag skin thickness, these authors stated: "..a considerable increase in slag skin weight may be produced when remelting shows wide oscillation in current". A physical relation between the slag skin formation and voltage swing must exist. The amount of electric current flowing through the slag skin is controlled by the effective electric conductivity of the skin. This effective conductivity must also take into account the existence of a contact resistance at the slag skin/mould interface [11-12]. Just as for the heat transfer, this contact resistance might depend on the slag composition but also on the temperature. For some compositions or/and at high temperatures the slag skin possesses a "plasticity" which promotes a very good contact with the mould [12]. According to our own estimations, by taken into account the slag/mould contact resistance, the effective solid slag conductivity lies in the range of 0.1 to 50 Sm^{-1}.

In the case of "mould isolated" from the baseplate, J.Cameron et al. [13] estimated the amount of mould current to be less than 5% of the total current. However, the processes studied were small ESR units (moulds ~2.5-8 cm mould and 2.5-3 cm electrodes diameters) which in normal conditions operate with deep electrode immersion depth. The shape of the electrodes tip in such small size ESR are conical, the current path is thus mostly oriented towards the liquid pool. If the electrode tip would have been deepen close to the slag surface the amount of mould current would have probably reached 20 to 40 % of the total current (according to our own calculations). For other ESR scales the amount of mould current can reach up to 90% of the total applied current [8-9].

We propose the hypothesis that the current flowing through the slag/mould and the ingot/mould interfaces is a major key parameter to explain the empirical correlations gathered by industry on the quality of the ingot surface. After having presented some generalities about electric current path, the effect of mould current on the slag skin thickness is investigated.

Electric current path in ESR system

To correctly predict the electric current path, a numerical model based on the potential formulation A-ϕ of the electromagnetic field has been used. The mould and the electrode diameter are 0.75 and 0.5 m respectively. The slag height is 0.15 m, the liquid and solid slag conductivities are taken equal to 140 and 5 Sm^{-1}. The slag skin thickness is assumed constant along the height $\delta = 1\ mm$. The ingot is assumed to be in perfect electric contact over 3 cm under the slag/metal interface. An rms current of 14200 A is applied from the top electrode. The results of calculations show that different electric current paths can exist in the ESR process (Figures 1-2). The possible paths depend on whether the mould is connected or not to the baseplate.

In the case of a "live mould" the possible electric current circuits are:

a: Electrode->Slag->Mould
b: Electrode->Ingot->Mould
c: Electrode->Ingot ->Baseplate->Mould
In the case of "insulated mould":
d: Electrode->Slag->Mould->Ingot->Baseplate
e: Electrode->Ingot-> Baseplate

The magnitude of the current flowing in each circuit depends on the electric resistances of each path. In the isolated concept the current in the circuit (d) has to cross two times the skin. Thus if the slag skin covers the entire ingot, less current flows within the mould in the "isolated" than in the "live mould". However a distinction must be drawn between the current entering (or leaving) the mould at the slag and at the ingot level. At the slag level, for the present configuration, a small difference exists between the live and insulated mould concept. However at the mould level it can be seen that for the live concept, the slag skin has only a very limited power of insulation since almost 30% the current enters the mould. At 50 Hz for the insulated mould, eddy currents generated within the copper media propagate inside the steel media (Figure 2). In the case of live mould, the model predicts that a small amount of current is flowing from the ingot to the mould trough the liquid slag. If the slag skin is totally remelted at the liquid metal level, a good contact metal against the copper mould can increase considerably the amount of mould current. In the present configuration the mould and the ingot are assumed to be in contact over a height of 3 cm, in reality this height can be much larger especially in the presence of a "liquid head". So from a process to another this contact height can be very different, so the magnitude of current that can potentially flow through this boundary cannot be clearly defined. In opposite the amount of current flowing from the liquid slag to the mould, topic of the next section is much more universal.

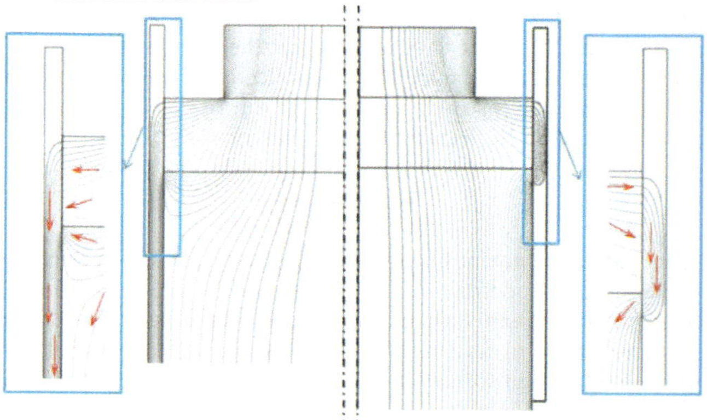

(a) Live Mould (b) Insulated Mould

Figure 1: Electric current path for a low frequency current (DC or < 5Hz)

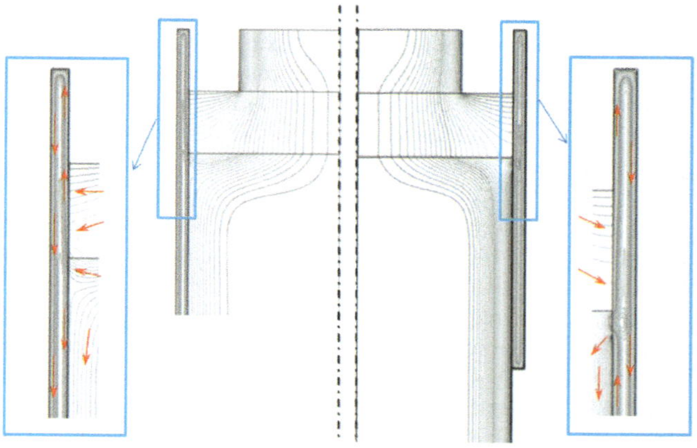

(a) Live Mould (b) Insulated Mould

Figure 2: Electric current path for a 50Hz frequency AC current

Effect of a finite slag skin conductivity on power generated

To illustrate the effect of the slag skin electric conductivity on the power generated a parametric study were performed on the case of a "live mould" ESR process. The slag skin thickness is still assumed to be constant along the height $\delta\ = 1\,mm$. Figures 3 shows the evolution of the total power generated in the slag (P_t) with the slag skin conductivity.

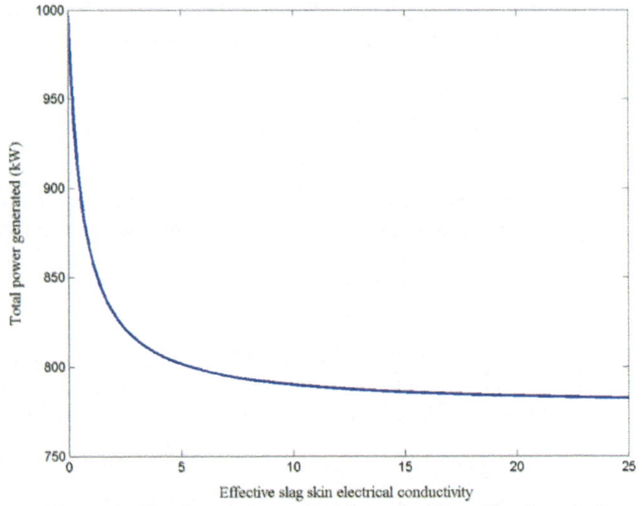

Figure 3: Total power generated versus slag skin electrical conductivity

The power decreases almost exponentially with the slag skin electric conductivity, which means that opening the path to the mould decreases the overall resistance. The power reaches 1000 kW for the case of perfectly insulating slag skin, for conductivities larger than 15 Ω/m the power almost reached its asymptotic level (~775kW).

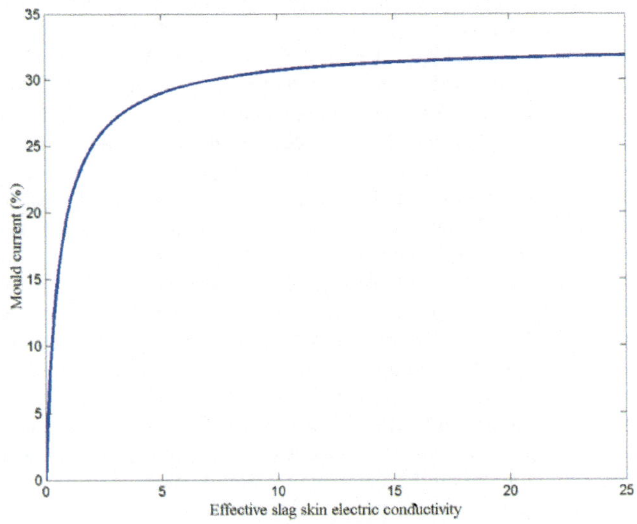

Figure 4: Proportion of electric current entering the mould at the slag level

The proportion of mould current I_m/I_t increases strongly for conductivities smaller than 3, for $\sigma > 15$ the proportion increases asymptotically towards a maximum of about 32 % (Figure4).

The proportion of power generated within the slag skin reaches a maximum of 27 % for slag skin electric conductivities $\sigma \sim 1 - 2\ \Omega$/m (Figure 5). The existence of a maximum can be explained by the fact that j_m^2 increases while $1/\sigma$ decreases with σ. Due to the lack of knowledge of the actual slag skin electric conductivity it is difficult to determine the actual state of the process. However in the real ESR plant the experimental power was found to fluctuate from 790 to 830 kW. This indicates that the actual slag skin conductivity might be around 2 to 3 Ω/m. If uncertainty about actual liquid slag conductivity is taken into account ($\sigma_l \sim 120 - 180\ \Omega$/m), the slag skin conductivity can be assumed to be within the range of 1 to 5 Ω/m.

Within this range the power generated within the slag skin (Ps) is high 80-220 kW, it represents 10 to 27 % of the total power. This power must be compared with the heat lost through the mould, which is in the order of 350 kW (assuming 10^6 Watt/m^2). Even in the largest case the Joule heating generated in the slag skin contributes for 2/3 of the heat received by the mould. So it can be stated that having 27% of the total power generated within the slag skin cannot be considered as unphysical. However this large amount of power generated in a so thin volume will definitively modify the actual skin thickness.

Effect of slag height and fill ratio on the mould current

The larger is the slag height, the higher is the resistance of the electrode/liquid pool path. For the case of slag skin conductivity of about $\sigma = 15$ and slag height of about 30 cm, 66% of the current flows to the mould in the live mould concept, and 50% in the insulated mould concept (Figure 6). Similarly the fill ratio controls the ratio between the electrode/mould and the electrode/pool distances. Smaller is the electrode radius, less current flows into the mould.

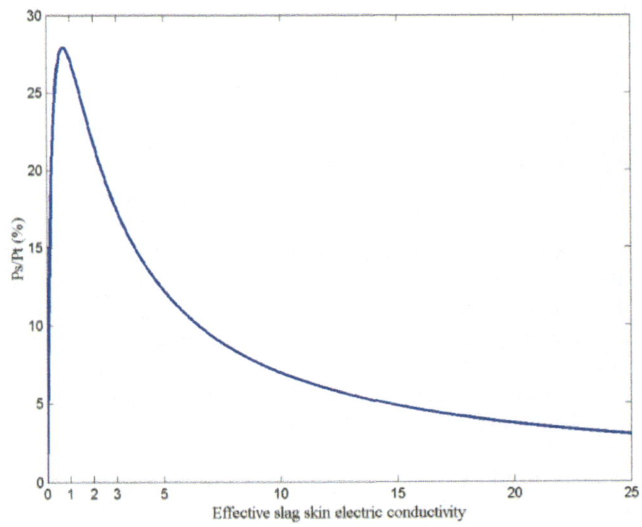

Figure 5: Proportion of the total power generated within the slag skin

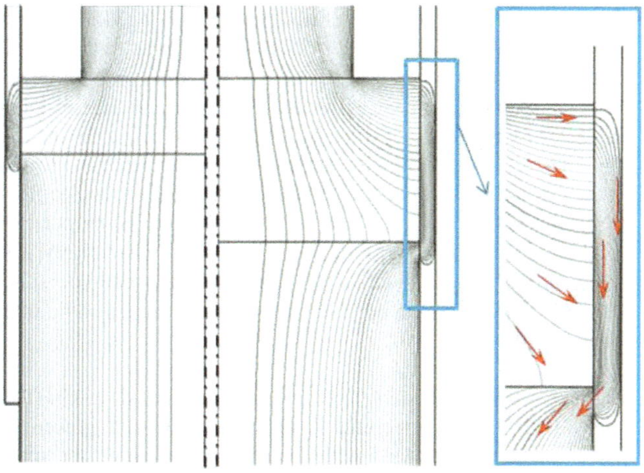

(a) 15 cm slag cap (b) 30 cm slag cap

Figure 6: Electric current path for two slag heights for the case of mould insulated from the baseplate. On the right the electric current vectors are shown in a the region near the mould

Effect of the mould current on the slag skin thickness

Let us consider the energy balance within the slag skin. The warmer liquid slag (or liquid metal) provides a heat flux Q_{Slag} wich enters the slag skin. At equilibrium the heat entering the mould Q_{Mould} consists in the sum of Q_{Slag} and the heat generated by the Joule heating within the slag skin Q_{Joule} :

$$Q_{Slag} + Q_{Joule} = Q_{Mould} \tag{1}$$

The thin attribute of the slag skin allows us to rewrite it in the following form:

$$\delta^2 \frac{j^2}{\sigma} + \delta Q_{Slag} - k_s \Delta T = 0 \tag{2}$$

Where j is the electric current crossing the skin, δ is the skin thickness, σ is the effective slag skin electric conductivity, and $\Delta T = T_{liq} - T_{mould}$ is the drop in temperature across the slag skin. In the present analysis consider the hot side of the skin to be the slag liquidus temperature. k_s is an average heat conductivity of the solid slag . The magnitude of the temperature drop across the slag skin depends on the quality of the heat contact with the copper mould. In the case of a perfect contact, $\Delta T \sim 1650 - 500 \sim 1000\,K$. In real process a jump in temperature exists at the interface with the mould, so the temperature drop across the skin is probably smaller $\Delta T \sim 400 - 900\,K$.

Without mould current (j=0) the thickness is simply inversely proportional to the heat flux:

$$\delta_0 = \frac{k_s \Delta T}{Q_{Slag}} \tag{3}$$

In the presence of mould current the solution is:

$$\delta(j) = Q_{Slag} \frac{\sqrt{1 + \frac{j^2}{j_c^2}} - 1}{2\frac{j^2}{\sigma}} \tag{4}$$

with j_c defined as:

$$j_c^2 = \frac{\sigma Q_{Slag}^2}{4 k_s \Delta T} = \frac{\sigma}{4\delta_0^2} k_s \Delta T \tag{5}$$

When $j \ll j_c$, the slag skin thickness decreases quadratically with the intensity of mould current :

$$\delta(j) \approx \delta_0 \left(1 - \frac{j^2}{2 j_c^2}\right) \tag{6}$$

For very large electric current density $j \gg j_c$, the thickness decreases with the inverse of the electric current density :

$$\delta(j) = Q_{Slag} \frac{\sqrt{1 + \frac{j^2}{j_c^2}} - 1}{2\frac{j^2}{\sigma}} \approx \frac{\sigma Q_{Slag}}{2 j_c\, j} \tag{7}$$

In the ESR process the current j is function of the difference in electric potential across the skin thickness is not independent of δ :

$$j \sim \sigma \frac{\varphi_{liq} - \varphi_{Mould}}{\delta} = \sigma \frac{\Delta\varphi}{\delta} \tag{8}$$

For the sake of simplicity we assume that the mould is connected with the baseplate, in this case we can assume φ_{Mould} constant and equal to zero. The solution of equation (Eq. 2) simplifies into:

$$\delta = \frac{k_s \Delta T - \sigma \Delta\varphi^2}{Q_{Slag}} \tag{9}$$

The slag skin thickness decreases linearly with σ and quadratically with the jump in potential across the skin. The maximum possible difference between the liquid slag and the mould electric potential is:

$$\Delta\varphi_{max} = \sqrt{\frac{k_s \Delta T}{\sigma}} \tag{10}$$

This represents a physical limit for the Joule heat released over which the skin layer cannot sustain its solid nature ($\delta > 0$). As example by using $k_s = 0.5$, $\sigma = 1$, we obtain $\Delta\varphi_{max} \sim 22$ Volt. The maximum difference in potential $\Delta\varphi$ is reached at the gas/slag interface. The question is whether the limit (Eq. 10) can be reached or not during the process. In both live and insulated mould cases, the potential of the mould φ_{Mould} can be considered as constant. The magnitude of $\Delta\varphi$ varies with the axial position along the mould and reaches its maximum at the top slag/gas interface. This maximum depends directly on the fill ratio and must be compared with $\Delta\varphi_{max}$. By increasing the size of the electrode, $\Delta\varphi$ becomes closer to the operating voltage of the process (typically 30-70 Volts). Therefore depending on the properties and the height of the slag, there exists a maximum fill ratio so that $\Delta\varphi < \Delta\varphi_{max}$. It is clear that more resistive is the slag skin, larger can be the fill/ratio. We believe that relation (Eq. 10) explains also why ESR processes are successfully run with large fill ratio only if the slag height is not too high.

Correlation between electric current distribution and ingot surface quality

The slag skin is first formed at the slag level, and then enters the liquid metal pool where it can either keep its size or melt. Good ingot surface quality is believed to be achieved if the slag skin melts completely or at least partly. For each empirically known correlation, a conventional explanation related to heat transferred to the slag skin (Q_{Slag} in Eq.1.) exists. The proposed explanation based on the Joule heat released within the slag skin Q_{Joule} is related to the magnitude I_m or proportion of the mould current (I_m/I_t). A summary of concurrent explanations is presented in table1.

The two cases (B-C) are highly related and cannot be separated, higher melt rate are only reached by increasing the magnitude of the applied current and so the power. The cases D and C decrease the heat losses by decreasing the area in contact with the electrode, the air and the mould. However these two cases, as mentioned in previous sections, are related to the geometrical characteristics which control the amount of mould current. The case A is better related to the electric explanation by the fact that if similar melt rates are obtained with two different slags, resistive slags although using lower electric intensity give usually better surface quality.

Table 1. Factors influencing the ingot surface quality and possible explanation

Cases	Empirical correlation towards better ingot surface quality	Explanation based on slag temperature	Explanation based on mould current
A	Higher slag resistance ↑	Higher slag temperature ↑	I_m/I_t ↑
B	Melt rate ↑	Higher slag temperature Stronger liquid pool stirring	I_m ↑
C	Power, applied current ↑	Higher slag temperature Stronger stirring in the slag and metal	I_m ↑
D	Fill ratio (Re/Rm) ↑	Heat lost at the exposed slag surface ↓	I_m/I_t ↑
E	Electrode penetration depth ↓	Slag/mould surface ↓ heat lost ↓	I_m/I_t↑

Summary

Despite its small electric conductivity, the solid slag skin can conduct a large amount of electric current. The proportion of electric current entering the mould is a function of the slag electrical conductivity, the slag height, electrode penetration depth and the fill ratio. These parameters are also known to influence the ingot surface quality. We suggest the hypothesis that the Joule heating generated within the slag skin contributes largely to the ingot surface quality. Results of simulations show that a large amount of Joule heating is generated within the slag skin volume. The electric current that crosses skin at the liquid metal level can easily reach high electric current density. Depending on the heat balance, this additional heat contributes certainly to the melting of the slag skin volume. It is obvious that through the slag skin thickness, the mould current can influence the liquid pool shape and thickness.

Acknowledgements

The authors acknowledge the financial support by the Austrian Federal Ministry of Economy, Family and Youth and the National Foundation for Research, Technology and Development.

References

1. A. Mitchell, M. Etienne, "The solidification of Electroslag fluxes," *Transactions of The Metallurgical Society of AIME,* 242 (1968), 1462-1464
2. M. Bell, A. Mitchell, "Some observations on the surface quality of Electroslag ingots," *Journal of the Iron and Steel Institute,* 209 (8) (1971), 658-670.
3. B. Korousic, V. Osterc, *Radex-Rundschau* (Österreichisch, Österreichisch - Amerikanische Magnesit A.G., 1976), 803-813.
4. J. Cameron, M. Etienne, A. Mitchell, "Some electrical characteristics of a DC electroslag unit," *Metallurgical Transactions*, 1 (7) (1970), 1839 – 1844.
5. J.D.W. Rawson, G. Jeszensky, A.W. Bryant, "Novel slags for ESR," *Proc. of 6th Int. Vac. Met. Conf. on Special Melting,* San Diego, 1979, 848-863.
6. P.O. Mellberg, "Temperature distributions in slag and metal during Electroslag remelting of ball-bearing steel," *ISIJ Proc. of 4th Int. Symp. on ESR Processes*, Tokyo, 1973, 13-25.
7. T.Kusamach, T.Ishii, T. Onoye, K. Narita, "Effects of composition of slags on heat transfer characteristics in electroslag remelting process," *Tetsu-to-Hagane*, 12 (1980), 1640-1649.
8. A. Kharicha , W. Schützenhöfer, A. Ludwig , G. Reiter, "Re formulation of the Joule heating in presence of turbulent fluctuation," *Int. J. Cast Metals Research*, 22 (2009), 155-9.
9. A. Kharicha , W. Schützenhöfer, A. Ludwig , G. Reiter, M. Wu, "On the importance of electric currents flowing directly into the mould during an ESR process," *Steel Research Int.*, 79 (2008) 632-36.
10. S.F. Medina and M.P. de Andres, "Electrical field in the resistivity medium (slag) of the ESR process: influence on ingot production and quality," *Institute of Metals,* 2 (1987), 110–121.
11. A.Mitchell and S.Joshi, "The thermal characteristics of the electroslag process," *Met. Trans.*, 4 (1973), 631-642.
12. A. Mitchell and S. Joshi, "Some observations on the electrical and thermal properties of the slag-skin region in the Eletroslag remelting process," *Met. Trans.*, 2 (1971), 449–455.
13. A. Mitchell and J. Cameron, "The electrical conductivity of some liquids in the system CaF_2+ CaO + Al_2O_3," *Met. Trans.*, 2 (1971), 3361-3366.

Proceedings of the 2013 International Symposium on Liquid Metal Processing & Casting
Edited by: Matthew J. M. Krane, Alain Jardy, Rodney L. Williamson, and Joseph J. Beaman
TMS (The Minerals, Metals & Materials Society), 2013

MECHANISMS OF CALCIUM OXIDE DISSOLUTION IN CaO-Al_2O_3-SiO_2-BASED SLAGS

Muxing Guo, Zhi Sun, Xiaoling Guo, Bart. Blanpain

Department of Metallurgy and Materials Engineering, KULeuven, 44 Kasteelpark Arenberg, Leuven, BE-3001 Belgium
Leuven, Belgium

Keywords: CaO dissolution, confocal scanning laser microscopy, thermodynamic calculation

Abstract

The dissolution of solid lime particles into liquid slags at high temperatures was investigated by means of confocal laser scanning microscopy (CLSM). An additional solid layer around the lime particle was observed at the intermediate stage of the dissolution into CaO-Al_2O_3-SiO_2 slags. The dissolution rate was decelerated due to the existence of the additional layer and the dissolution profile could be clearly distinguished into three stages, i.e. an early, intermediate and late stage. By adding 10 wt% MgO, this layer could be effectively eliminated and the slope of the whole dissolution profile kept relatively constant. The dissolution path and mechanisms were subsequently evaluated by incorporating thermodynamic calculations. Both direct and indirect dissolutions could be distinguished. It was realized that the decrease in composition range for solid precipitating after adding MgO could significantly reduce the interfacial reaction (IR) layer formation. Post-mortem analyses on quenched samples were further carried out to confirm the theoretical calculations. It was found that the solid layer in slags without MgO was $(CaO)_2 \cdot SiO_2$ and $(CaO)_3 \cdot SiO_2$ which is in line with the thermodynamic calculations. However, only $(CaO)_2 \cdot SiO_2$ was noticed in slags with MgO which both $(CaO)_2 \cdot SiO_2$ and MgO phases should be present according to the calculations. The non-equilibrium during dissolution may play an important role on phase transformation and MgO particles in much smaller quantity may have dissolved into $(CaO)_2 \cdot SiO_2$ phase.

Introduction

Dissolution, a process of dissolving solid or liquid into a solvent, is a ubiquitous phenomenon in the field of engineering. Decomposition of biomass [1], leaching of mineral ores [2], corrosion of refractory materials [3], and dissolution of solid additives for impurity removal in ladle metallurgy [4] are only a few examples where a breakdown of the crystal lattice occurs forming a homogeneous mixture with the solvent. There are mainly two categories of dissolution, i.e. direct and indirect [5]. During indirect dissolution, a firm product layer or non-flaking ash will be formed outside the particle while the whole solid particle size (including the product layer) remains unchanged. In this case, the process may include such steps as 1) diffusion of liquid reactant through the boundary layer (liquid film) surrounding the particle to the surface of the particle, 2) penetration and diffusion of liquid reactant through the layer of non-flaking ash, 3) chemical reaction at the interface between the non-flaking ash and the particle, 4) diffusion of the products through the product layer and the boundary layer [5]. The dissolution can be rate limited by any of the steps or a combination. It can be influenced by various factors, e.g., temperature, particle size, agitating conditions, etc. Although the apparent kinetics can be easily obtained by characterizing the samples after dissolution, the real-time dissolution mechanism is very complex and is difficult to be concluded [6]. Direct dissolution is relatively faster and no ash/residue is formed outside the particle or the ash is flaking [5]. In this research, lime is adopted to understand the dissolution mechanisms in CaO-based slags. The reason is that lime is very commonly used in many high temperature processes and its dissolution in slags is very representative. The dissolution nature is of broad interest and a better understanding of the lime dissolution in liquid slags will be helpful to chemical/metallurgical process design and optimization.

One process of using lime is the primary steelmaking e.g. basic oxygen furnace (BOF) and electric arc furnace (EAF) processes, where lime or dolomite is added as a fluxing additive to obtain a basic slag and help to remove harmful impurities such as sulfur and phosphorous [7, 8]. Another process involving lime dissolution is the entrained flow gasifier for power production where lime is used as flux to lower the melting point of the coal ash slag [9]. The ash that remains after coal combustion typically contains SiO_2, CaO, Al_2O_3 and/or FeO [9]. The slag properties, e.g. viscosity and refractory corrosion behavior, are associated with its evolution and slagging process where the lime dissolution is a key issue [7, 10-12]. By increasing the lime addition, the resulting slag becomes less viscous and can be tapped from the base of the gasifier. In both cases, the rapid and complete dissolution of lime into the molten slag is very important to ensure its effective utilization. Calcium fluoride (fluorspar) has been known as an efficient flux which can enhance the lime dissolution [13, 14] by increasing the diffusivity of CaO. However, owing to the emission of fluorine containing species into the atmosphere and potential contamination to the soil during slag disposal, the use of fluoride is limited in consideration of environmental protection [15]. Therefore, many efforts have been made to find alternative fluxes in place of fluorspar [15-18]. Some attempts have been performed to obtain alternative options in place of fluorspar. MnOx, FeO_x, and TiO_2 are found to increase the lime dissolution [17] where the viscosity of the slag and CaO diffusivity are modified. However, SiO_2 is found to decrease the lime dissolution by increasing the viscosity [19]. In a CaO-Al_2O_3-SiO_2 slag, a layer of Ca_2SiO_4 can be found and the diffusion in this solid product tends to be of vital importance in determining the dissolution [7]. If the slag contains a sufficient amount of Fe_2O_3 [20], the product layer becomes discontinuous and it will be beneficial for the lime dissolution. To detect the dissolution mechanism, the most common method is the so-called "finger test" technique where a "rotating or static lime cylinder" is dipped into a molten ash/slag and exposed for a certain period [11]. After the test, the cylinder is taken out and characterized. The dissolution kinetics as well as the mechanisms can be obtained by analyzing the distribution of the phases through the cross-section of the sample. However, this method suffers as intrinsically indirect determination and data may become doubtable sometimes especially when a discontinuous layer of solid phases are formed. In addition, the samples may be easily broken up and the local situations become different which make it hard to get the exact conclusions [9, 11, 12, 21, 22].

Recently, high temperature confocal scanning laser microscope (CSLM) has become available providing a concurrent observation of metallurgical phenomena at high temperatures. Highlighted work with respect to the CSLM technique has been performed on various topics, such as inclusion behavior in metal melts [23], phase transformations [24] and also dissolution of oxide particles such as magnesia, alumina, zirconia and spinel [16, 23, 25]. This technique not only provides an in-situ observation, but also makes it possible to control the dissolution process on-line by adjusting the temperature or heating rate. Nevertheless, one of the limitations of the CSLM so far is that on-line phase/composition analysis is still not available. Traditionally post-mortem analyses of quenched samples are combined with the in-situ observation [16, 23]. However, information about phase transformation may be lost, e.g. precipitates of very small size are neglected, since the dissolution process is far from equilibrium. On the other hand, instantaneous chemical reactions may still be possible during quenching and therefore some details of the dissolution can still be overlooked.

In this work, real-time thermodynamic calculations are performed in accordance with the in-situ observation and post-mortem analyses. This approach provides more reliable information on the dissolution nature of micrometer-sized lime particles into $CaO-Al_2O_3-SiO_2$ based slags. The information is obtained by steadily characterizing i) phase transformation/precipitate during dissolution, ii) effect of MgO additive on dissolution.

Materials and Experiments

Materials

The slags involved in the experiments consisted of Al_2O_3, SiO_2 and CaO with or without MgO. They were prepared by blending the pure oxide compounds in a molybdenum crucible. Before mixing, all the oxides were dried in a muffle furnace at 300 °C for 24 hours. After obtaining a homogeneous mixture, the crucible was placed in a tube furnace under purified argon atmosphere and held for at least 24 hours at a temperature of 1600 °C. As soon as the pre-melting finished, the slags were quenched against a steel plate and crushed into small particles that were then re-melted in the CSLM to release all the gases before running an experiment. The compositions of the pre-melted slags are listed in Table 1. The oxides used during the experiments were produced by Alfa Aesar, all with purities of 99.95 wt. %. The lime particles for dissolution were also produced by Alfa Aesar and prepared from coarse lime grains. Before dissolution, the particles were dried in a muffle furnace at 600 °C for 24 hours to remove water.

Experimental Apparatus and Procedures

The CSLM-IIF equipment (CSLM, Lasertec, 1LM21M-SVF17SP) was used to carry out in-situ observations of lime dissolution at high temperature. A detailed description of the CSLM can be found in our previous work [6, 26]. The furnace inside is gold-plated and ellipsoidal in shape. The crucible is placed in the upper focal point of the elliptical heating chamber and a halogen heating lamp is located in the lower focal point. The temperature is controlled by a B-type thermocouple welded at the bottom of the sample holder. HiTOS software contacting with a REX-P300 controller is used to program the temperature profiles. In order to ensure the accuracy of the measured temperature, thermocouple calibrations were performed using standard pure metal samples, e.g. copper, nickel and iron. A schematic view of the set-up is given in Figure 1.

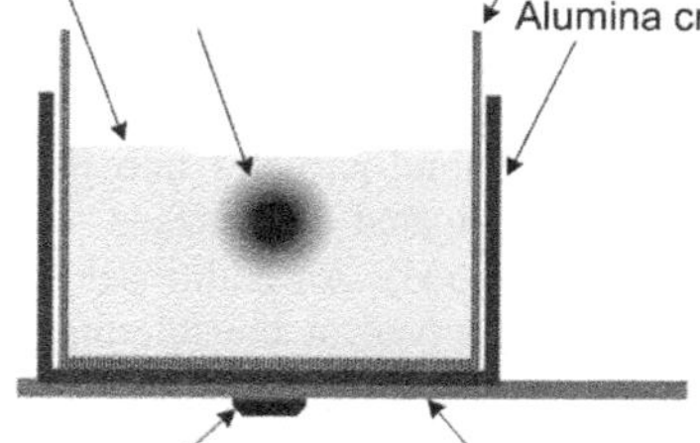

Figure 1. Schematic representation of the CSLM sample holder

Before the dissolution experiments, the heating chamber was evacuated to ca. 10 Pa and refilled with ultra-purified argon gas for at least three times. The Ar gas passed through a Restek triple filter and a Restek high capacity oxygen scrubber to remove oxygen, moisture and hydrocarbons before being filled into the chamber. Then the slag sample with a lime particle on the surface was heated to the desired temperature. The evaluation of the lime dissolution always started from the point where the set temperature was reached. In order to reduce the interaction of the slag and the particle before the slag melting, the sample was heated with a relatively high rate of around 200 ~ 350 °C/min. For each slag (Table 1), the dissolution experiments were performed at various temperatures.

In the present investigation, the lime particles were of irregular shape. In order to evaluate the changes of lime particle size during the dissolution, a border was extracted around the lime particle in the images obtained via HiTOS software. An example is given in Figure 2 for the lime dissolution in slag SM1 at 1480 °C. An equivalent radius was computed based on the measured area of the particle size. With the size evolution of the lime particle, the dissolution kinetics was derived.

Table 1. Compositions (wt. %) of the Slags Used in This Study

Slag Number	CaO	SiO_2	Al_2O_3	MgO	Basicity (CaO/SiO_2)
SM01	45.0	45.0	10.0	0	1.0
SM02	30.0	60.0	10.0	0	0.5
SM03	40.0	40.0	20.0	0	1.0
SM04	26.7	53.3	20.0	0	0.5
SM1	40.0	40.0	10.0	10.0	1.0
SM2	26.7	53.3	10.0	10.0	0.5
SM3	35.0	35.0	20.0	10.0	1.0
SM4	23.3	46.7	20.0	10.0	0.5

Results and Discussion

In-situ Observation of Lime Dissolution in CAS-Based Slags

Figure 2 shows the lime dissolution process in slag SM1 at 1480 °C. In each image, the boundary of the particle which is indicated by the dashed lines is extracted according to the image contrast. The particle which had an irregular shape becomes more spherical during dissolution. A rotation of the lime particle is also realized in Figure 2 which results from a composition build-up near the particle surface. During dissolution, the composition build-up originates from the local mass fluxes due to the change in the local radius of the particle when it is not spherical [6]. The rotation introduces errors in the measurement of the particle size and probably influences the shape of the radius evolution profile. The dissolution process may therefore be affected because the rotation can induce extra flow in the boundary layer between the particle and the slag bulk. The equivalent radius is calculated by assuming that the particle is a sphere and therefore the irregular shape of the particle is an important source of scatter in the dissolution profiles. However, the effect from these errors is not influencing its tendency, which is considered important in determination of the dissolution mechanisms.

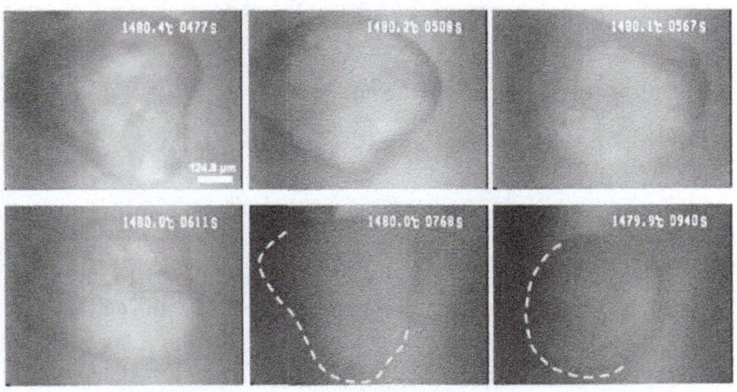

Figure 2. CSLM images of a slow CaO dissolution process in slag SM1 at 1480 °C.

Multi-Solid (MS) Layer Formation During Lime Dissolution In Slags Without MgO Addition

In some experiments, a clear MS-layer structure could be observed during CaO dissolution, as shown in Figure 3 (indicated by arrows). The in-situ observation showed that this layer formed after the CaO particle has been immersed into the molten slag for a certain time, suggesting an incubation period for its formation. The layer grew with time, reaching a maximum thickness. Afterwards, it dissolved into the slag, faded and finally disappeared. The remaining lime then dissolved with a rapid rate. It has been found that the MS layer is more easily observed in slags without MgO addition and at lower temperatures. The formation of the MS layer influences directly the dissolution mechanisms and a lower dissolution rate is normally realized.

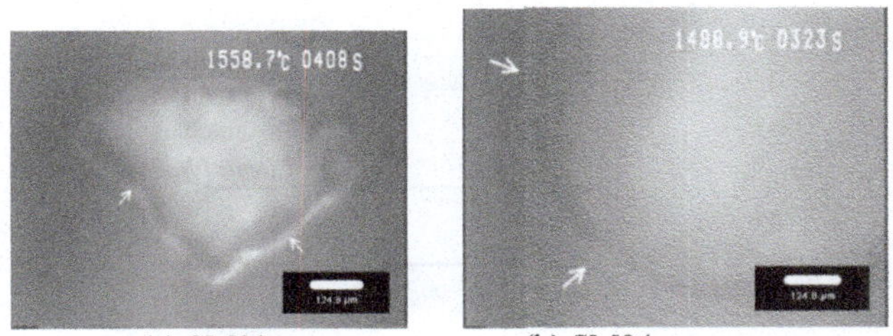

(a) SM02 (b) SM04

Figure 3. MS layer formed during CaO dissolved in slags without MgO: (a) slag SM02 (thinner layer); (b) slag SM04 (thicker layer)

Figure 4 gives the lime dissolution profile in slag SM04 at 1480 °C. It can be found that the dissolution rate becomes very low from around 315 s indicating the formation of a MS layer. At both the early and late dissolution stages where no MS layer forms, the dissolution rates are significantly faster. It gives a direct proof that the MS-layer structure prohibits lime dissolution. Therefore the current dissolution kinetics is intrinsically a multi-stage/multi-scale process and is influenced by a number of factors, e.g. temperature, slag compositions, diffusion nature of CaO and/or local flow field. Those factors are apparently affecting the formation of the MS-layer structure.

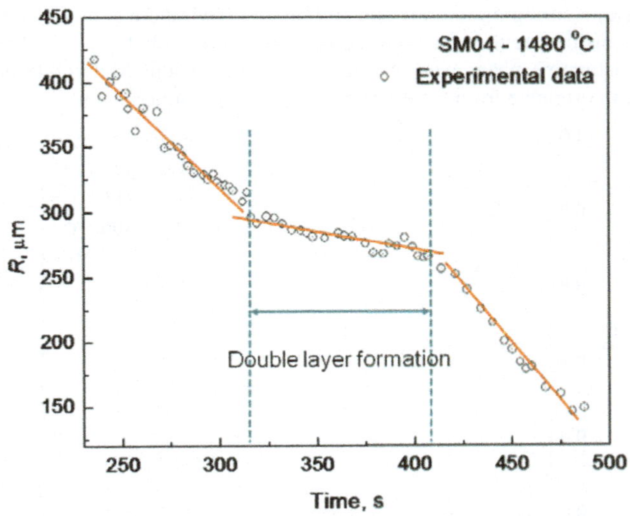

Figure 4. Lime dissolution profile in slag SM04 at 1480 °C (R: the equivalent particle radius). A clear platform is observed due to the MS layer formation during dissolution.

In slags without MgO addition, it has been detected that temperature influences significantly the observation of a MS layer structure with CSLM. This is reflected in the dissolution profiles. As shown in Figure 5, the MS layer formation range (with a lower dissolution rate) in the dissolution profile becomes much narrower with increasing the temperature. Since the lime particles were irregular and it was difficult to find two similar particles, the normalized particle size R/R_0 is provided in Figure 5 with R_0 the particle equivalent radius when the particle was just immerged in the slag which is corresponding to the starting time t_0 of the dissolution. At a sufficiently high temperature, e.g. 1600 °C, although the MS layer was still observed, it diminished rapidly.

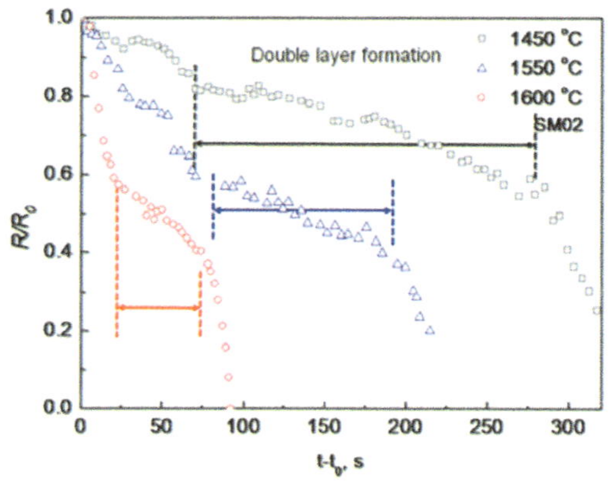

Figure 5. Lime dissolution profiles in slag SM02 without MgO addition at different temperatures. MS layer formation range becomes shorter with increasing temperature.

By adding MgO, the formation of MS layer structure is found to be effectively prohibited. As shown in Figure 2, only one border of liquid slag-solid in the images is distinguished during dissolution. Figure 6 shows the lime dissolution profiles in slag SM1 at different temperatures. The dissolution rate which is reflected by the slope of the dissolution profile increases dramatically with temperature. Although a minor change of the slope during dissolution can still be observed, there is no obvious transition region distinguishable comparing with Figure 5. The distinctive phenomena in slags with and without MgO addition can originate from the difference in their dissolution nature.

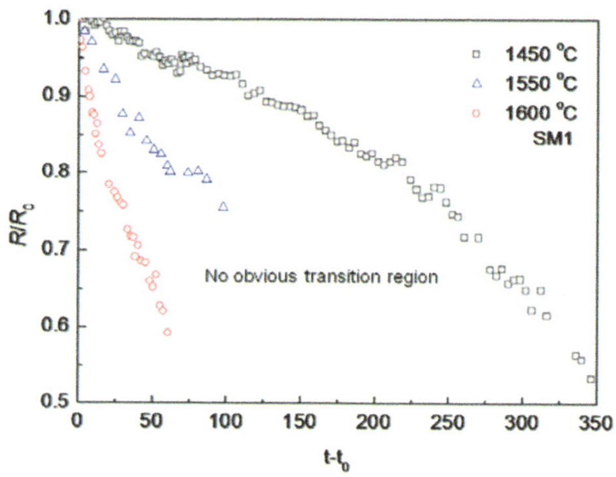

Figure 6. Lime dissolution profiles in slag SM1 with 10 wt.% MgO addition at different temperatures. No obvious transition region can be distinguished.

Dissolution Path and Phase Identification

Dissolution Path In order to understand the formation of the MS layer, the phase diagrams of the slag systems were calculated by using FactSage software with FToxid database [27]. During dissolution, a concentration gradient of CaO is built between the bulk (liquid slag) and the lime particle. The dissolution path of CaO at a certain temperature and in a given slag passes through different phase zones in the phase diagrams. Although the dissolution is a non-equilibrium process, the phases which may precipitate adjacent to the particle can still be predicted by the phase diagrams. Figure 7 shows the phase diagrams of slags with and without MgO addition at 1600°C. The dissolution should stop at the CaO content where CaO monoxide phase starts to precipitate in the phase diagram. In the case of slag SM02 (Figure 7 (a)), it happens at the composition of around 72 wt% CaO. With the accumulation of CaO content in the slag during lime dissolution, two solid phases i.e. $(CaO)_2{\cdot}SiO_2$ and $(CaO)_3{\cdot}SiO_2$ will precipitate adjacent to the particle according to Figure 7 (a). The precipitates subsequently form a continuous solid layer or a MS layer structure according to the in-situ observation (Figure 3). The CaO content for solid phase precipitate ranges from ~52 wt% to ~72 wt%. The accumulation of solid precipitates around the lime particle is considered to be the reason of a lower dissolution rate after the dissolution proceeds a certain time (Figure 5). At the early stage, CaO dissolves directly into the liquid slag inducing a local compositional accumulation and it continues to diffuse into the bulk slag. The direct dissolution reaction can be given as

$$CaO(s) \rightarrow (CaO) \quad (1)$$

where small bracket ( ) indicates species in liquid slag.
When the precipitate of solid phases is promoted and a solid layer is formed, lime dissolution becomes indirect as

$$3CaO(s) + (SiO_2) \rightarrow (CaO)_3 \cdot SiO_2(s) \quad (2)$$

$$2(CaO)_3 \cdot SiO_2(s) + (SiO_2) \rightarrow 3(CaO)_2 \cdot SiO_2(s) \quad (3)$$

and/or

$$2CaO(s) + (SiO_2) \rightarrow (CaO)_2 \cdot SiO_2(s) \quad (4)$$

$$(CaO)_2 \cdot SiO_2(s) \rightarrow 2(Ca^{2+}) + (Si^{4+}) + 4(O^{2-}) \quad (5)$$

When the slag contains MgO, e.g. SM2, the primary solid precipitate is $(CaO)_2{\cdot}SiO2$ from the liquid slag when CaO content is relatively lower (e.g. at ~56 wt%) according to Figure 7 (b). Solid MgO phase is also precipitating together with $(CaO)_2{\cdot}SiO2$ when the CaO is locally accumulating in the slag and has relatively smaller amount. The CaO content for $(CaO)_2{\cdot}SiO_2$ precipitate ranges from ~56 wt% to ~65 wt%. This small compositional range may dramatically decrease the amount of $(CaO)_2{\cdot}SiO_2$ formation comparing with slag SM02 (Figure 7 (a)). Therefore, no MS layer was clearly observed under CSLM and the dissolution rate kept nearly constant as shown in Figure 6. The dissolution is then following reactions (1), (4) and (5). If MgO crystals with large particle size are formed, the dissolution mechanisms may become more complex and spinel phase may be observed [6].

Microstructural Characteristics at Lime Particle/Liquid Slag Interface In order to verify the dissolution paths and reactions, the sample with MS layer formed during dissolution was quenched 1 second after the layer was clearly distinguished. It needs to be indicated that the formation of the MS layer is very rapid. The time is around 2~3 seconds from the first sight to a clear view of the layer. The sample with MgO addition,was quenched 1 second after the particle is fully immerged into the slag. Figure 8 shows the secondary electron images of lime dissolving into slags SM02 and SM2 as well as their energy-dispersed x-ray (EDX) mappings. A single sharp edge is observed between the lime particle and the slag bulk for slag SM2 (Figure 8 (a)), while an MS layer with CaO accumulation can be clearly distinguished for slag SM02 (Figure 8 (b)). The layer exhibits compositional gradient from the particle to the slag bulk. It gives a direct proof that a solid layer is formed during lime dissolution in slags without MgO addition.

The compositions of different phases formed during dissolution were further analyzed. The morphologies of the phases are shown in Figure 9. For slag SM02, the experimental results are well corresponding with the phase diagram predictions (Table 2). Both $(CaO)_2 \cdot SiO_2$ and $(CaO)_3 \cdot SiO_2$ were found and $(CaO)_3 \cdot SiO2$ generates first near to the lime particle. This is in agreement with the dissolution path given in Figure 7 (a) (the dash line indicates the phase evolution from the liquid slag to periphery of the lime particle). However, the layer of $(CaO)_3 \cdot SiO_2$ is very thin (<5 μm) and it is replaced immediately by $(CaO)_2 \cdot SiO_2$ when CaO content becomes lower in the surrounding liquid slag (or when the location is sufficiently far from the particle). The $(CaO)_2 \cdot SiO_2$ layer is found to be ~30 μm and continuous. This thickness was obtained when the solid layer started to be clearly distinguished. It had been realized that this layer formed in a very short time as mentioned above. Before the layer was formed, CaO diffusion/direct dissolution into the slag prevailed the dissolution process.

In slag SM2, a precipitate layer of ~5 μm is realized near to the lime particle (Figure 9 (a)). The layer is discontinuous and contains a set of $(CaO)_2 \cdot SiO_2$ crystals with a small amount of MgO (the compositions are shown in Table 2, C and C*). However, no pure MgO crystals were found in the sample which should precipitate together with $(CaO)_2 \cdot SiO_2$ according to the phase diagram (Figure 7 (b)). This may be due to two reasons, i.e. i) the fresh MgO crystals were too small to be detected with SEM; ii) MgO and $(CaO)_2 \cdot SiO_2$ precipitate at the same time; since MgO has relatively small amount, it may form solid solution with $(CaO)_2 \cdot SiO_2$ crystals (hypothetically).
According to the C and C*compositions in Table 2, the crystals can also be considered as distorted Merwinite phase. However, it is difficult to draw this conclusion since the process is far from equilibrium. The precipitation of new phases may cause a compositional divergence in the slag from its original composition and result in depleted zones of SiO_2, Al_2O_3 and/or MgO. The shortage of these elements may be suppressed quickly if the required mass transport condition is fulfilled. Otherwise, the formation of new precipitates is interrupted and distinct phases may also be generated. Therefore, a detailed experimental verification on phase equilibria is required to determine the exact phases.

In the present research, however, the kinetics and dissolution path are focused. Although the above analyses are not fully quantitative, they have already provided significant information to distinguish the effect of MgO on lime dissolution into liquid slags. On one aspect, the narrower CaO content range for solid precipitates in slags with MgO (Figure 7) addition initiates less solid precipitates and the solid layer becomes less visible under CSLM. On the other aspect, the layer of precipitates is discontinuous in slags with MgO (Figure 9 (a)) and it can be easily re-dissolved into the liquid slags with lower CaO content. It indicates that MgO plays important roles in accelerating lime dissolution into liquid slags.

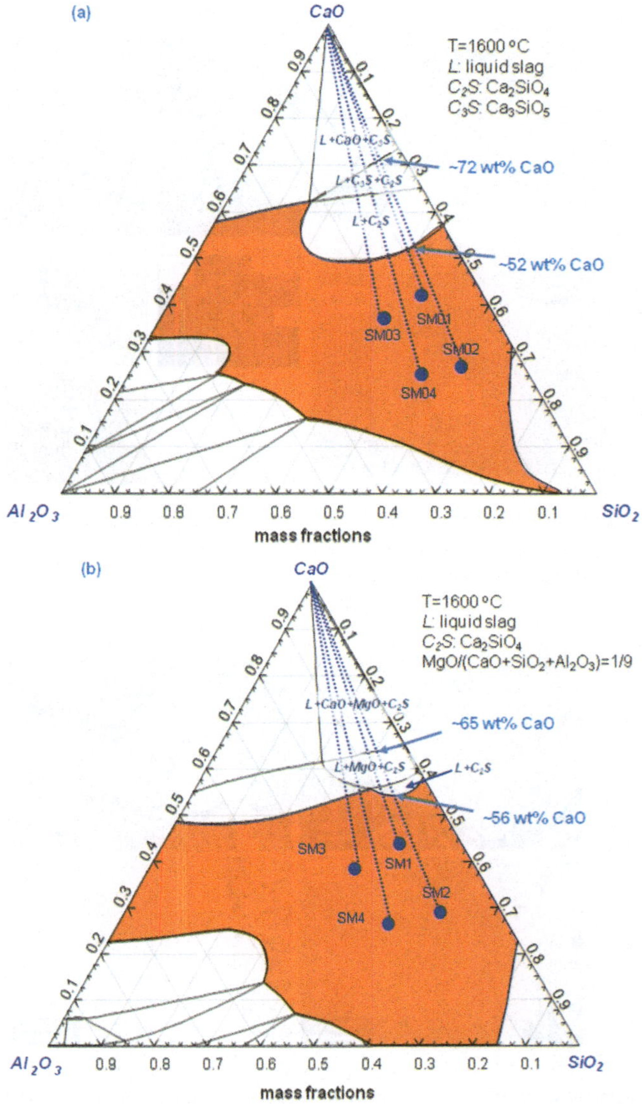

Figure 7. Phase diagrams of the slags in the present research. (a) slags without MgO addition; (b) slags with MgO addition. The direct dissolution paths for each slag are shown by the dash lines.

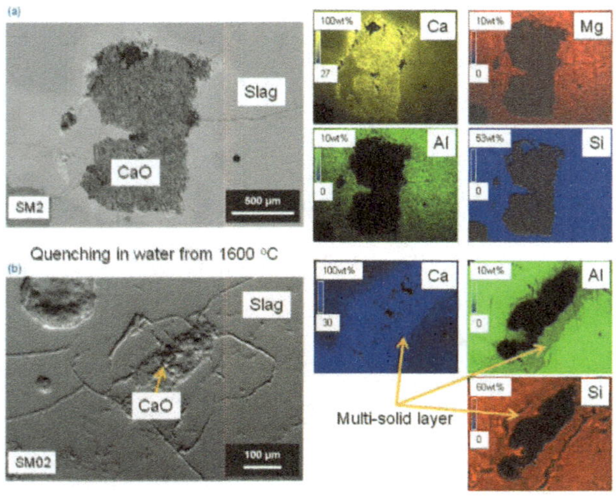

Figure 8. SEM images of lime particle dissolving into slags with (a) and without (b) MgO addition. Energy dispersed x-ray mappings indicates the distribution of the corresponding elements around the lime particles.

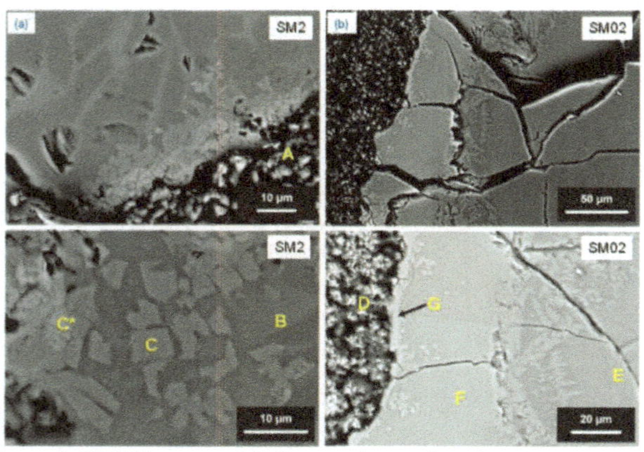

Figure 9. SEM images of different phases forming during lime particle dissolving into slags with (a)-SM2 and without (b)-SM02 MgO addition.

Table 2. Composition of Different Phases Measured with Energy Dispersed X-ray

	NO.	Ca at.%	Si at.%	Al at.%	Mg at.%	Phases
SM2	A	50.75	0.98	0.18	0.29	CaO
	B	12.31	21.99	4.61	3.69	Liquid slag SM2
	C	26.36	16.81	1.04	5.92	C_2S+MgO
	C*	32.01	17.84	0.84	4.74	C_2S+MgO
SM02	D	48.51	0.73	0.29	-	CaO
	E	14.35	25.19	4.71	-	Liquid slag SM02
	F	24.53	16.60	0.23	-	C_2S layer
	G	46.24	17.10	0.15	-	C_3S

Conclusions

The effect of MgO addition on lime dissolution has been discussed. Without MgO addition, the dissolution rate was decelerated due to the formation of an MS layer in slags and the dissolution profile could be clearly divided into three stages, i.e. the early, intermediate and late stages. The smallest dissolution rate i.e. slope of dissolution profile, is found at the intermediate stage of the dissolution. This implies that the MS layer forms when the local CaO accumulation becomes large enough. By adding 10 wt% MgO, this MS layer could be effectively gotten rid of and the slope of the whole dissolution profile then kept relatively constant.

Thermodynamic calculations are performed with FactSage software to evaluate the dissolution path and mechanisms. Both direct (CaO reacts with liquid slag) and indirect (through intermediate phase e.g. $(CaO)_2 \cdot SiO_2$) dissolutions could take place under the certain experimental conditions. It was realized that the narrower CaO content range for solid phases in slags with MgO (Figure 7) addition initiates less solid precipitates and the solid layer becomes less visible under CSLM. On the other aspect, the layer of precipitates is discontinuous in slags with MgO (Figure 9 (a)) and it can be easily re-dissolved into the liquid slags with lower CaO content.

Post-mortem analyses on quenched CSLM samples substantiate that the solid layer in slags without MgO are $(CaO)_2 \cdot SiO_2$ and $(CaO)_3 \cdot SiO_2$ which is in agreement with the thermodynamic calculations. Meanwhile, only $(CaO)_2 \cdot SiO_2$ is detected in slags with MgO addition whereas both $(CaO)_2 \cdot SiO_2$ and MgO phases should be found according to the calculations. This disagreement between thermodynamic calculations and post-mortem analyses is believed to be caused by the non-equilibrium between different phases during dissolution and MgO particles in small quantity may dissolve into $(CaO)_2 \cdot SiO_2$ phase forming a solid solution.

References

1. M. Becidan, G. Várhegyi, J.E. Hustad, Ø .Skreiberg, " Thermal Decomposition of Biomass Wastes. A Kinetic Study", *Ind. Eng. Chem. Res.*, 46 (2007), 2428.

2. Z. Sun, Y. Zhang, S.L. Zheng, Y. Zhang, "A new method of potassium chromate production from chromite and $KOH-KNO_3$-H2O binary submolten salt system", *AIChE J.*, 55(2009), 2646.

3. M. Guo, P.T. Jones, S. Parada, E. Boydens, J.V. Dyck, B. Blanpain, P. Wollants, "Degradation mechanisms of magnesia-chromite refractories by high-alumina stainless steel slags under vacuum conditions", *J. Eur. Ceram. Soc.*, 26(2006), 3831.

4. M. Valdez, K. Prapakorn, A.W. Cramb, S..Sridhar, " Dissolution of alumina particles in $CaO-Al_2O_3-SiO_2$-MgO slags", *Ironmak. Steelmak.* 29(2002), 47.

5. O. Levenspiel, *Chemical reaction engineering* (New York: John Wiley & Sons, 1999).

6. J Liu, M. Guo, P.T. Jones, F. Verhaeghe, B. Blanpain, P. Wollants, "In situ observation of the direct and indirect dissolution of MgO particles in $CaO-Al_2O_3-SiO_2$-based slags", *J. Eur. Ceram. Soc.*, 27(2007), 1961.

7. J. Yang, M. Kuwabara, T. Asano, A. Chuma, J. Du, "Effect of Lime Particle Size on Melting Behavior of Lime-containing Flux", *ISIJ Int.*, 47(2007), 1401.

8. R. Fruehan, "Desulfurization of liquid steel containing aluminum or silicon with lime", *Metall. Mater. Trans. B*, 9(1978), 287.

9. L.K. Elliott, J.A. Lucas, J. Happ, J. Patterson, H. Hurst, T.F. Wall, "Rate limitations of lime dissolution into coal ash slag", *Energy Fuels*, 22(2008), 3626.
10. W.E. Lee, S. Zhang, "Melt corrosion of oxide and oxide-carbon refractories", *Int. Mater. Rev.,* (44)1999, 77.
11. L. Elliott, S.M. Wang, T. Wall, F. Novak, J. Lucas, H. Hurst, J. Patterson, J. Happ, "Dissolution of lime into synthetic coal ash slags", *Fuel Process. Technol.* (56)1998, 45.
12 T. Deng, J. Gran, D. Sichen, "Dissolution of Lime in Synthetic $FeO-SiO_2$ and $CaO-FeO-SiO_2$ Slags", *Steel Res. Int.* 81(2010), 347.
13. G. Kor, "Effect of fluorspar and other fluxes on slag-metal equilibria involving phosphorus and sulfur", *Metall. Mater. Trans. B,* 8(1977), 107.
14. H. Wang, G. Li, Z. Ren, B. Li, X. Zhang, G. Shi, "LATS refining ladle slag modifying with $CaO-CaF_2$", *J. Univ. Sci. Technol. Beijing* 14(2007), 125.
15. A.B. Fox, K.C. Mills, D. Lever, C. Bezerra, C. Valadares, I. Unamuno, J.J. Laraudogoitia, J. Gisby, "Development of Fluoride-Free Fluxes for Billet Casting", *ISIJ Int.,* (45)2005, 1051.
16. A.B. Fox, M.E. Valdez, J. Gisby, R.C. Atwood, P.D. Lee, S. Sridhar, "Dissolution of ZrO_2, Al_2O_3, MgO and $MgAl_2O_4$ particles in a B_2O_3 containing commercial fluoride-free mould slag", *ISIJ Int.* (44)2004, 836.
17. S. Amini, M. Brungs, O. Ostrovski, S. Jahanshani, "Effects of additives and temperature on dissolution rate and diffusivity of lime in Al_2O_3-CaO-SiO_2 based slags", *Metall. Mater. Trans. B*, 37(2006), 773.
18. G.H. Wen, S. Sridhar, P. Tang, X. Qi, Y.Q. Liu, " Development of fluoride-free mold powders for peritectic steel slab casting", *ISIJ Int.* (47)2007, 1117.
19. A. Kondratiev, E. Jak, P. Hayes, "Predicting slag viscosities in metallurgical systems", *JOM Journal of the Minerals, Metals and Materials Society,* 54(2002), 41.
20 S.H. Amini, M.P. Brungs, S. Jahanshahi, O. Ostrovski, " Effects of additives and temperature on dissolution rate and diffusivity of lime in Al_2O_3-CaO-SiO_2 based slags", *Metall. Mater. Trans. B-Proc. Metall. Mater. Proc. Sci.* 37(2006), 773.
21. Y. Satyoko, W.E. Lee, E. Parry, P. Richards, I.G. Houldsworth,"Dissolution of iron oxide containing doloma in model basic oxygen furnace slag", *Ironmak. Steelmak.,* 30(2003), 203.
22. C. Orrling, S. Sridhar, A.W. Cramb, "In situ observation of the role of alumina particles on the crystallization behavior of slags", *ISIJ Int.,* 40(2000), 877.
23. S.H. Lee, C. Tse, K.W. Yi, P. Misra, V. Chevrier, C. Orrling, S. Sridhar, A.W. Cramb, "Separation and dissolution of Al_2O_3 inclusions at slag/metal interfaces", *J. Non-Cryst. Solids*, 282(2001), 41.
24. J.H. Kim, S.G. Kim, A. Inoue, "In situ observation of solidification behavior in undercooled Pd-Cu-Ni-P alloy by using a confocal scanning laser microscope", *Acta Mater.* 49(2001), 615.
25. J. Liu, F. Verhaeghe, M. Guo, B. Blanpain, P. Wollants, "In situ observation of the dissolution of spherical alumina particles in CaO-Al2O3-SiO2 melts", *J. Am. Ceram. Soc.*, 90(2007), 3818.
26 X. Guo, Z. Sun, J.V. Dyk, M. Guo, B. Blanpain, "Kinetic Consideration of Chemical Dissolution of Lime Particles in CaO-Al_2O_3-SiO_2-based Slags: an in-situ Observation Approach", Submitted.
27. http://www.factsage.com/.

Proceedings of the 2013 International Symposium on Liquid Metal Processing & Casting
Edited by: Matthew J. M. Krane, Alain Jardy, Rodney L. Williamson, and Joseph J. Beaman
TMS (The Minerals, Metals & Materials Society), 2013

EFFECT OF FLUORIDE CONTAINNING SLAG ON OXIDE INCLUSIONS IN ELECTROSLAG INGOT

Yanwu Dong, Zhouhua Jiang, Yulong Cao, Jinxi Fan, Ang Yu, Fubin Liu
School of Materials and Metallurgy, Northeastern University, Wenhua Rd., Heping District, Shenyang, Liaoning, 110004, PR China

Keywords: Electroslag, Slag, Fluoride, Non-metallic inclusions.

Abstract

Besides controlling homogeneous composition and compact solidification structure, removal of non-metallic inclusions is an important characteristic for electroslag remelting process. Many factors influence the non-metallic inclusions in steel including gas and inclusions original content in consumable electrode, atmosphere, slag amount and its composition, power input, melting rate, filling ratio and so on. Fluoride containing slag, which influences the non-metallic inclusions to a great extent, has been widely used for electroslag remelting process. The present paper focuses on the effect of fluoride containing slag on the inclusions in electroslag ingot based on the interaction of slag-metal interface. In this work, steel grade MC5 and several slags have been employed for investigating the effect of slag on inclusions. These experiments had been carried out in an electrical resistance furnace under argon atmosphere in order to eliminate the effect of ambient oxygen. Some specimens had been taken at different times for analyzing the content, dimensions, and type of non-metallic inclusions. Quantitative metallographic analysis method has been adopted for observing and examining the inclusions. SEM-EDS analysis has been used to investigate the composition of non-metallic inclusions of specimens at different time for investigating the modification behavior of inclusions. All the results obtained will be comparison to the original state inclusions in steel, which will be in favor of choose of slag for electroslag remelting process.

Introduction

It is a critical problem of controlling non-metallic inclusions in steel in steelmaking process, which has a great relationship to the performance of production. Inclusions will easily initiate micro-voids and cracks at interface between inclusions and steel, which might be the origins of fatigue fractures or other defects [1], and it is not exception for electroslag remelting process. As is well known that removal of inclusions are an important characteristic for electroslag remelting process [2]. The removal of non-metallic inclusions during electroslag remelting process is mainly at the tip of the electrode and it is mainly absorption and dissolution of liquid slag to the inclusions [3, 4]. However, cleanliness of conventional electroslag steel can not meet the requirement of customers with the development of economic and society. A lot of factors influence the cleanliness of electroslag steel including slag, melting rate, input power, and steel grade and so on, in which slag plays an important role for controlling of inclusions during the electroslag remelting process [5].

The present work focused on the behavior of non-metallic inclusions through interactions at the slag-metal interface. The development regulation has been investigated in this paper.

Experiments

Experimental Material

In this work, cold roll steel MC5 has been employed for the experiment and its chemical compositions is shown in Table I. The steel sample, which is manufactured by EAF-LF-VD-DC process, has been taken from consumable electrode in Baosteel plant.

Table I. Chemical composition of cold rolls steel MC5, mass%

Steel grade	Composition（wt%）						
	C	Si	Mn	Cr	Mo	V	Ni
MC5	0.51/0.55	0.45/0.60	0.50/0.65	4.85/5.35	0.52/0.58	0.15/0.23	0.42/0.48

Table II. Composition of ESR slag, mass%

Sample	CaF_2	CaO	Al_2O_3	SiO_2	MgO	Optical basicity
S1	70	0	30	0	0	0.665
S2	50	20	20	5	5	0.732
S3	40	25	25	7	3	0.756

Experimental Procedure and Sampling

Before all experiments, the bulk slags were prepared. Slag materials used in the experiment are analytical reagents and quantities were weighed using a high-precision electronic balance. Slags charged in an alumina crucible were calcined at 1123K for 4h after thoroughly mixing in an agate mortar. The alumina crucible was then stored in a desiccator before the experiment. Experiments were carried out with four slag mixtures, as shown in Table II.

The experimental apparatus is illustrated in Figure 1. The heating unit is made of molybdenum disilicide. At first, a sample of cold rolled steel about 800g in weight surrounding by a 0.3mm thick molybdenum film was placed into a magnesia crucible. Then the crucible was placed in a graphite crucible to prevent any leakage of molten steel. The power was switched after the crucible was placed in the chamber. The furnace was heated to the experimental temperature (1550°C) at a rate of 6K/min. Platinum rhodium thermocouple had been used for measure the furnace temperature.

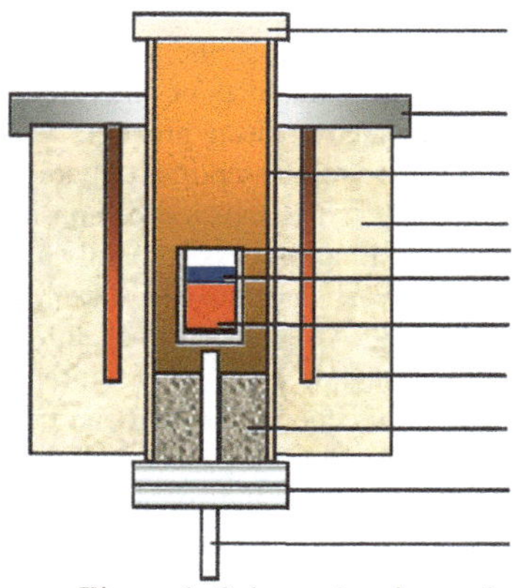

Figure 1. Schematic of experimental apparatus.

Argon has been used to protect the slag-metal reaction system from top and bottom of the furnace. The sample of No. 1 has been taken from the molten steel after the furnace temperature has up to 1550°C. A certain amount of slag was put onto the surface of molten steel. The samples from No. 2 to No. 5 have been taken from the molten steel at 5min, 10min, 15min, 20min and 30min after slag has melted respectively. The sample of No. 6 has been sampling before ending the experiment. And the sample No. 7 was sampling from the solid ingot. All the samples were water quenched and then preparation for analyzing.

Results and Discussion

All the samples have been analyzed by metallographic microscope and SEM-EDS. Quantitative metallography has been adopted for analyzing the amount and dimensions of non-metallic inclusions in samples. Figure 2 illustrates the amount of non-metallic inclusions for different time samples. The same original material has been adopted in the experiments; therefore, the amount of inclusions is the same. It can be seen from Figure 2 that the amount of inclusions decreases gradually during the experiments and arrives the lowest content at 15min, and then keep constant more or less. The value of final sample is the inclusions in solid ingot, which is higher than the situation of molten steel. The main reason should be the generation of new inclusions during steel solidification due to the reducing the solubility of oxygen in steel.

Size distribution of non-metallic inclusions has been shown in Figure 3. It can be seen that most inclusions in molten steel is less than 2 μm in dimension. There are less large size inclusions in steel especially for slag S2 and S3. In addition, the slag S1 is conventional slag used at the early stage of electroslag remelting birth. Results of Figure 2 and Figure 3 indicate that slag S2 and S3 is favor of controlling amount and dimensions in steel than S1.

Compositions of ten non-metallic inclusions have been analyzed for every sample. Results of SEM-EDS indicate that a lot of inclusions of $CaO-MgO-Al_2O_3$ type have been found in the molten steel without addition of slag on it. The inclusions are converted to mainly Al_2O_3 type for slag S2 after the slag addition has melted. And then these inclusions change gradually towards the direction from ① to ⑦ as shown in Figure 4. The value of ① to ⑦ represents the average value of ten non-metallic inclusions. That is to say that the composition of CaO in inclusions decreases gradually and its structure closes to the system of $MgO-Al_2O_3$. It is similar to S2 for S3 slag. But the original composition of non-metallic inclusions is different from the situation of slag S2. More CaO content exists in inclusion for slag S3 and then CaO in inclusions decreased gradually in the following time. Ultimately, the inclusion turned into $MgO-Al_2O_3$ just like the situation of slag S2.

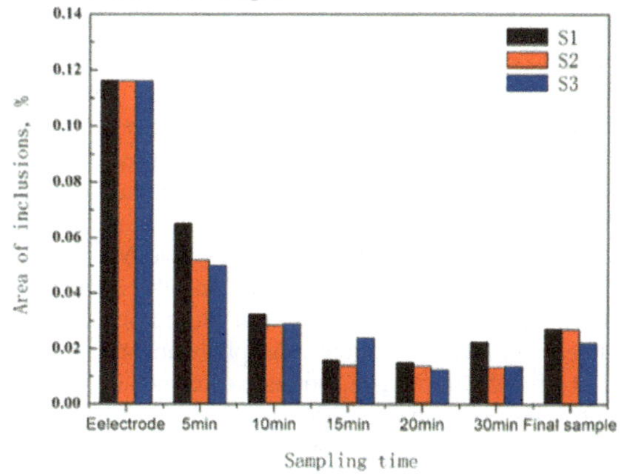

Figure 2. Amount of inclusions in steel.

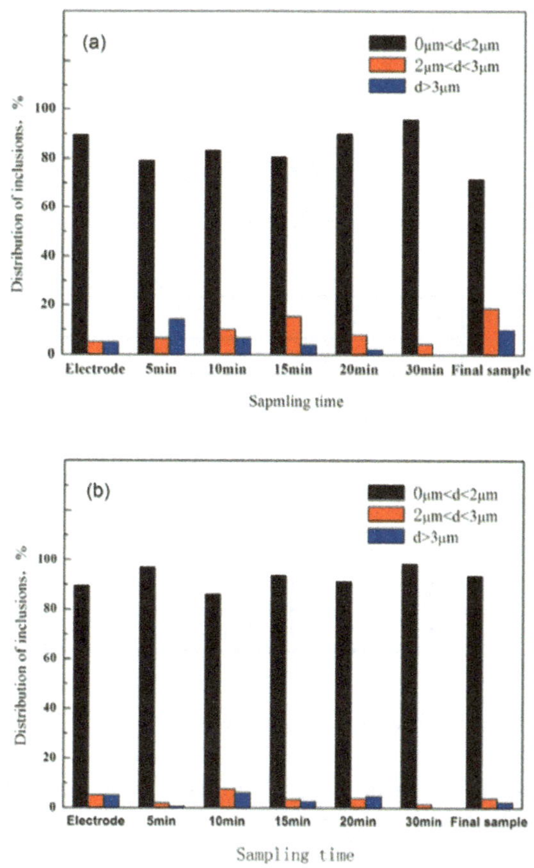

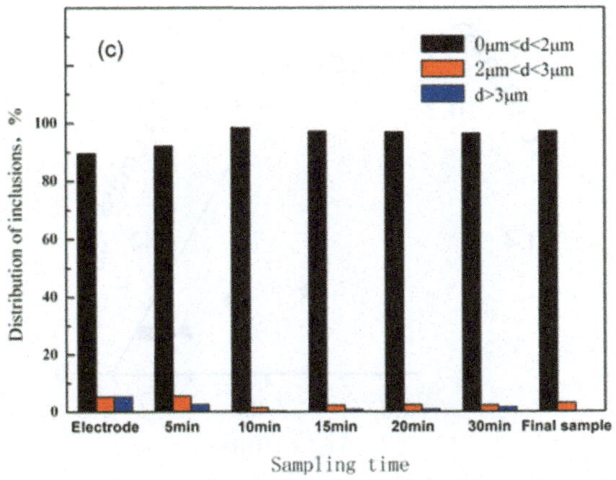

Figure 3. Distribution of inclusions in steel (a) S1, (b) S2, (c) S3.

In general, Al is a strong deoxidizer in steel, which can react with MgO and CaO in liquid slag under steelmaking temperature as shown in Eq. (1) and Eq. (2).

$$3(MgO)+2[Al]=3[Mg]+(Al_2O_3) \quad (1)$$

$$3(CaO)+2[Al]=3[Ca]+(Al_2O_3) \quad (2)$$

In addition, MgO is less stable than CaO and it is easier to be reduced by Al. Moreover the reduction reaction is very quick. It can be seen from Figure 4 that Most CaO and MgO in original inclusions in electrode have been reduced by Al. Inclusions changed into inclusions consisting mainly of Al_2O_3. Results indicate that these non-metallic inclusions will continue to react and vary with time. The possible reaction can be expressed as follows [6]:

$$4(Al_2O_3)+3[Mg]=3(MgO \cdot Al_2O_3)+2[Al] \quad (3)$$

Mg in this equation is from the reaction product of Eq. (1). But Ca obtained in Eq. (2) does not induce the increase of CaO in inclusions according to the results in Figure 4. The reason may be the influence of CaF_2 in slag. In addition, without CaO for slag S1, it may be the reason why the inclusions are mainly close to $MgO–Al_2O_3$. The results mentioned above are different from conventional refining process using fluoride free or less fluoride slag. Deng [7] has studied the progress of inclusions during conventional refining process. They found that inclusions vary with the route as Al_2O_3 inclusions →$MgO–Al_2O_3$ system inclusions →$CaO–MgO–Al_2O_3$ system inclusions.

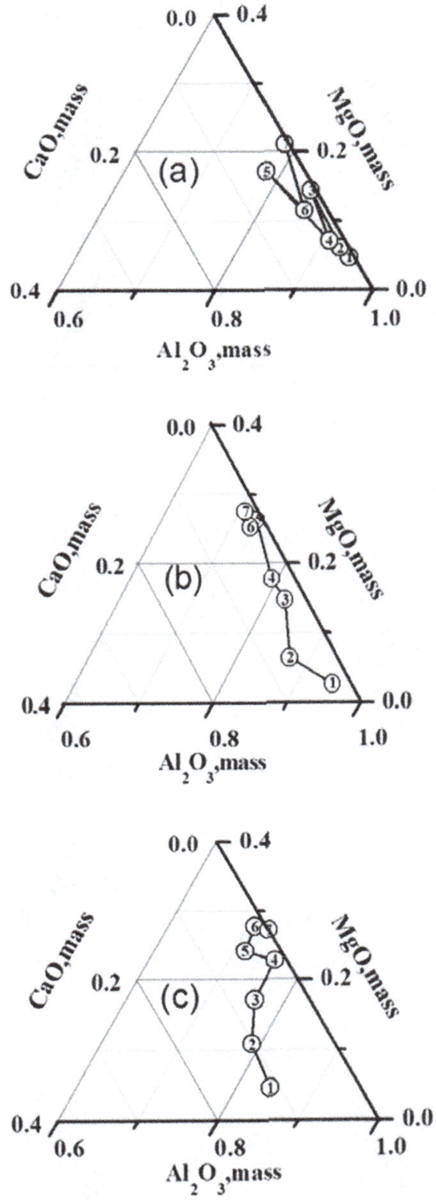

Figure 4. Compositions vary of inclusions with time for slag (a) S1, (b) S2, (c) S3.

The inclusions of the same time (10min) have been chosen in order to investigate the situation of different slag as shown Figure 5. In which solid "★" is the average value of non-metallic inclusions in original steel and hollow one stands for the measure value at 10min. As such, solid "▲", "●" and "■" represent the average value of inclusions in steel for slag S1, S2 and S3 at 10min and hollow ones are the measure value respectively.

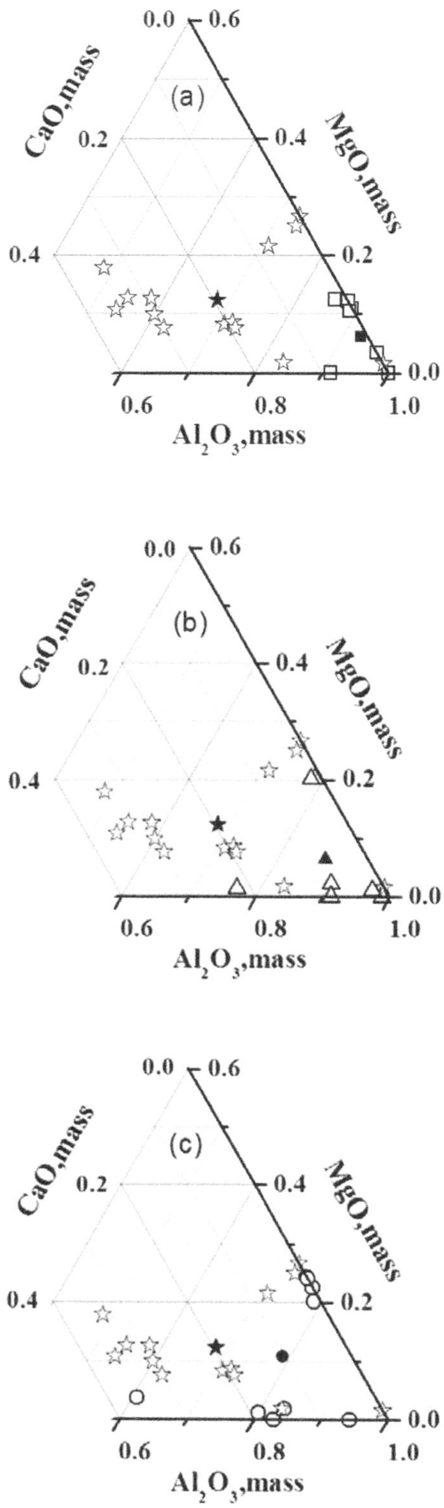

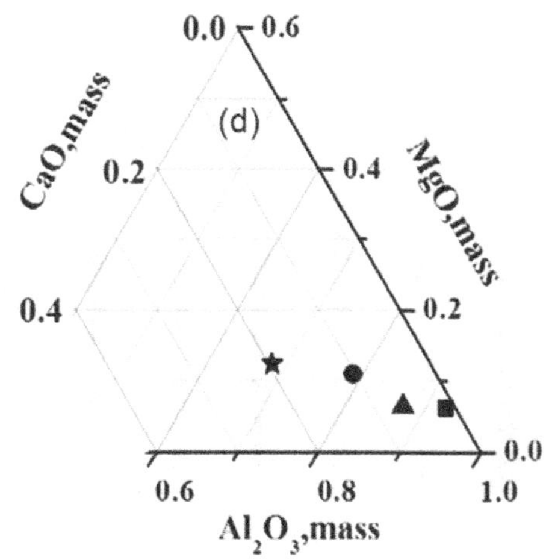

Figure 5. Compositions vary of inclusions at 10min for slag (a) S1, (b) S2, (c) S3 (d) total.

It can be seen from this figure that the compositions of non-metallic inclusions have obviously changed at 10min opposite to original one in steel. The composition is different for different slags at 10 min. Results indicate that the higher the optical basicity is the higher the CaO and MgO content is. Moreover, it is obvious for CaO content in inclusions, which may be due to the higher CaO content in slag.

The slag S1 and S2 have been employed in the 500kg ESR furnace. Results indicate that the amount of inclusions in electroslag steel by slag S1 is higher than slag S2, which is consistent with the findings mentioned above. The most inclusions in steel is Al_2O_3 for the electroslag steel by slag S1, however, it is MgO- Al_2O_3 for slag S2 as shown in Figure 6. In which "■" and "□" represent the average value and measure value of slag-metal reaction experiment, and "▲" and "△" stand for the average value and measure value of ESR process.

In general, all the non-metallic inclusions were thought deleterious [8]. But Al_2O_3 inclusion is more detrimental comparing to MgO-Al_2O_3 inclusions. Because interfacial energy between MgO- Al_2O_3 and steel is less than that of Al_2O_3 and steel. MgO- Al_2O_3 is easily dispersed in the steel without aggregation [9, 10]. However, Al_2O_3 is prone to aggregate and form large inclusions, which is detrimental to product performance. From this view point the slag S2 is superior to S1 for electroslag remelting process.

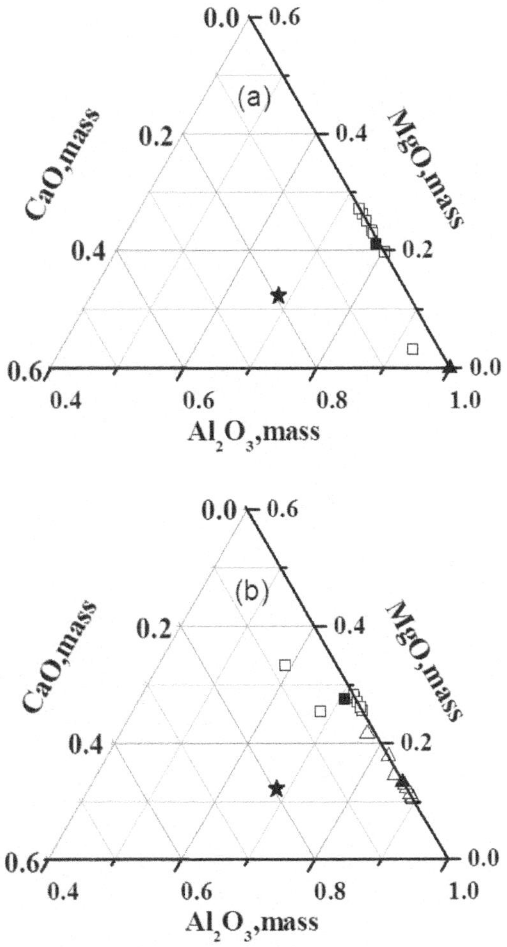

Figure 6. Results comparison of slag-metal reaction and ESR (a) S1 and (b) S2.

Conclusions

Slag-metal reaction has been done in this paper. Development regulation of non-metallic inclusions has been studied. Results indicate that inclusions are prone to MgO-Al_2O_3 inclusions for fluoride containing slag. Different slag has different capacity of adsorption and solution non-metallic inclusions. Results of ESR process also indicate that most non-metallic inclusions in electroslag steel is MgO-Al_2O_3 inclusions for multi-component slag, but it is Al_2O_3 inclusions when remelting process using conventional 70wt%CaF_2-30wt% Al_2O_3 slag.

Acknowledgements

This project supported by the National Nature Science Foundation of China No. 50904015 and No. 51274266. Also, this project supported by Fundamental Research Funds for Central University of China with the grant No. 120502002. In addition, the work is supported by Program for Liaoning Innovative Research Team in University with the grant No. LT20120008.

References

1. A. Sabih, P. Wanjara, J. Nemes, "Characterization of Internal Voids and Cracks in Cold Heading of Dual Phase Steel," *ISIJ International*, 45(8) (2005), 1179-1186.
2. Jiang Zhouhua, *Mathematical simulation and experiment analysis of thermal transfer for ESR*, Shenyang: Northeastern University Press, 1986), 1-20.
3. Li Zhengbang, *Electroslag casing*, Beijing: National Defense Industry Press, 1979), 17-24.
4. B. H. Yoon, K. H. Heo, J. S. Kim, et al., "Improvement of steel cleanliness by controlling slag composition," *Ironmak. and Steelmak.* 29(3) (2002), 214-217.
5. Dong Y W, Jiang Z H, xiao Z X, et al., "Influence of ESR process parameters on solidification quality of remelting ingots," *Journal of Northeastern University*, 30(11) (2009), 1598-1601.
6. H. Todoroki and K. Mizuno, "Variation of Inclusion Composition in 304 Stainless Steel Deoxidized with Aluminum," *Iron Steelmaker*, 30(2003), 60.
7. Deng Z Y, Zhu M Y, "Evolution mechanism of non-metallic inclusions in Al-killed alloyed steel during secondary refining process," *ISIJ International*, 53(3) (2013), 450-458.
8. Itoh H, Hino M, Barrya S, "Thermodynamics on the formation of spinel nonmetallic in liquid steel," *Metallurgical and materials transactions B*, 28(5) (1997), 953-956.
9. Mapelli C, Nicodemi W, Vedani M, et al., "Control of inclusions in a resulphurised steel," *Steel Research International*, 71(5) (2000), 161.
10. Ohta H, Suito H, "Calcium and magnesium deoxidation in Fe-Ni and Fe-Cr alloys equilibrated with CaO-Al_2O_3 and CaO-Al_2O_3-MgO slags", *ISIJ International*, 43(9) (2003), 1293-1300.

Proceedings of the 2013 International Symposium on Liquid Metal Processing & Casting
Edited by: Matthew J. M. Krane, Alain Jardy, Rodney L. Williamson, and Joseph J. Beaman
TMS (The Minerals, Metals & Materials Society), 2013

PRODUCTION OF HIGH QUALITY DIE STEELS FROM LARGE ESR SLAB INGOTS

Xin Geng, Zhou-hua Jiang, Hua-bing Li, Fu-bin Liu, Xing Li
School of Materials and Metallurgy, Northeastern University, Wenhua Rd., Heping district, Shenyang, Liaoning, 110004, PR China

Keywords: ESR, Die steels, Large slab ingot, Ultra-heavy plate.

Abstract

With the rapid development of manufacture industry in China, die steels are in great need of large slab ingot of high quality and large tonnage, such as P20, WSM718R and so on. Solidification structure and size of large slab ingots produced with conventional methods are not satisfied. However, large slab ingots manufactured by ESR process have a good solidification structure and enough section size. In the present research, the new slab ESR process was used to produce the die steels large slab ingots with the maximum size of 980×2000×3200mm. The compact and sound ingot can be manufactured by the slab ESR process. The ultra-heavy plates with the maximum thickness of 410 mm can be obtained after rolling the 49 tons ingots. Due to reducing the cogging and forging process, the ESR for large slab ingots process can increase greatly the yield and production efficiency, and evidently cut off product costs.

Introduction

The high hardness, large scale, high purity, high isotropic, long lifetime and other properties is very necessary for mold materials [1-3]. In order to achieve high quality die steels slab ingots, the furnace was designed to be able to produce ingots of several sizes as shown in Figure 1. The ESR for large slab ingot technology developed in our work mainly consists of low frequency power supply, bifilar mode, secondary cooling technology and packing mould, etc. And by successful application of the load cell for electrode weighing, the various stages of the main melting period is in the true sense of to achieve the control of constant remelting rate.
The secondary cooling technology developed improves greatly the quality of solidification, and it can reduce string-like segregation of slab ingots [4,5]. Reduction of hydrogen content in slab ingots has been made possible by the control of atmosphere and the use of liquid slag starting [4-6]. And in order to minimize the conduction losses, a low frequency in the range 2-5 Hz is used [7].

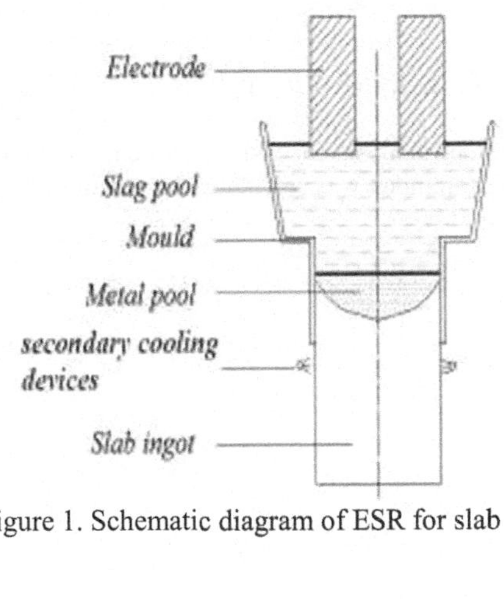

Figure 1. Schematic diagram of ESR for slab ingots

During the ESR process for large slab ingots, the two continuous casting slabs were connected as the electrodes with bifilar configuration. In order to improve the quality of solidification, short collar mould together with withdrawing system, secondary cooling technology and load cell for electrode weighing were applied in the large slab ESR process.

Experiment

The die steels ESR slab ingts with the maximum thickness of the world in size of 980 mm thickness, 2000 mm wideness and 2800 mm length have been produced successfully as shown in Figure 2. The ultra-heavy plates with 410 mm in thickness can be obtained after rolling the 49 tons ESR slab ingots. The experimental results of macro segregation show that the large ESR slab ingots exhibits excellent uniformity in different position. And the macrostructure of ultra-heavy plates produced by ESR slab ingots are superior to that of conventional materials.

Results and Discussion

Surface Quality

The good surface quality of large ESR slab ingts can be obtained as shown in Figure 3. There are no surface cracks, wrinkles, residue slag and other defects. As the ingot and mold for relative movement, it is easy to produce slag pool temperature fluctuations in the frequent changes. For reasons of selections of slag systems, melting rates and taper of the mold are much limited for large slab ingots comparing to small ones, adjusting taper of the mold and using CaF_2-CaO-Al_2O_3-SiO_2-MgO slag which can decrease the high temperature viscosity and increase the stability of slag viscosity are necessary for improving surface quality of the ESR slab [8].

Composition Control

The changes of composition between the electrode and the ingot were shown in Table I. From it we can obtain a conclusion that there were a little changes of composition between the electrode and the ingot which fully able to meet the requirements of the steel grades.

(a)

(b)

(c)

Figure 2. The ESR process for large slab ingot
(a) Liquid starting, (b) Remelting process,
(c) Appearance of slab ingot.

Figure 3. Surface quality of remelted slab ingot.

<u>Macro segregation</u>

In order to investigate the uniformity of ingot composition, segregation of the ingot was analyzed. The slab ingot were rolled to ultra-heavy plate, then the strip specimens were taken at the head and tail in the plate. The specific sampling locations are shown in Figure 4, a full-thickness specimen was taken at each location. Each sample with full-thickness at eight locations was taken, and then shavings were taken by planer for chemical composition analysis.

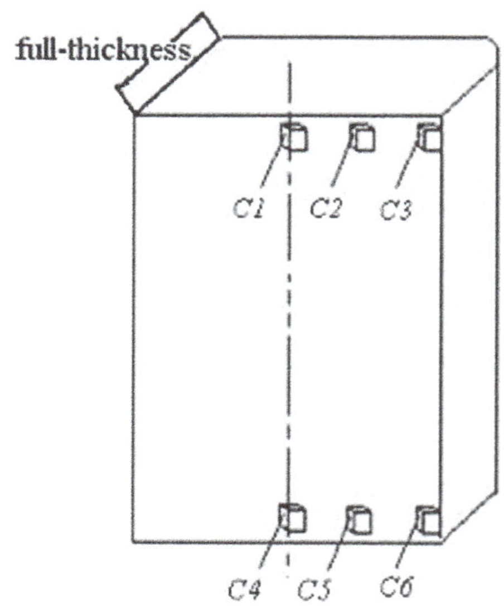

Figure 4. Sketch map of plates for sampling.

Macro-component analysis of each specimen was done, and the macro-segregation index was calculated by equation (1).

$$\delta = \frac{[\%M]}{[\%\overline{M}]} \tag{1}$$

Table I. Change of components of WSM718R steel, %.

	C	Si	Mn	P	S	Ni	Cr	Al	Mo
Electrode	0.36	0.38	1.18	0.013	0.003	0.91	1.83	0.040	0.45
Ingot	0.35	0.42	1.12	0.013	0.003	0.90	1.81	0.025	0.44
Changes	↓0.01	↑0.04	↓0.06	0	0	↓0.01	↓0.02	↓0.015	↓0.01

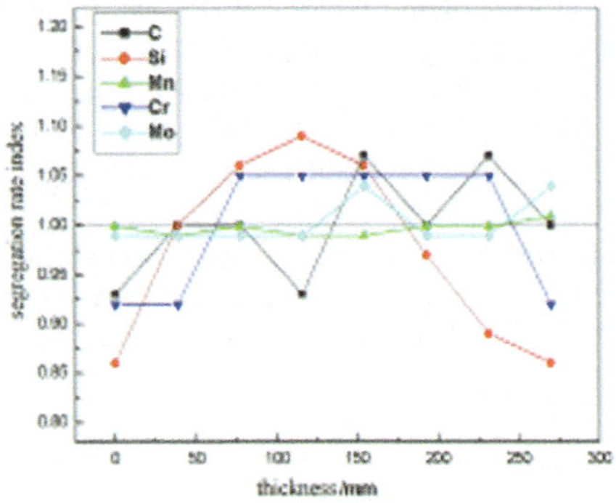

(C1)

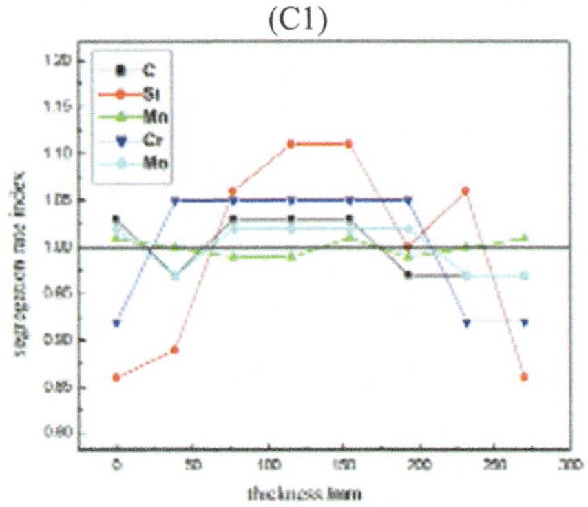

(C2)

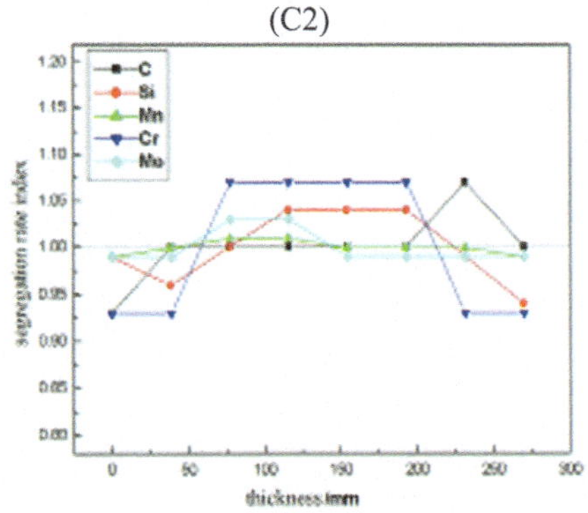

(C3)

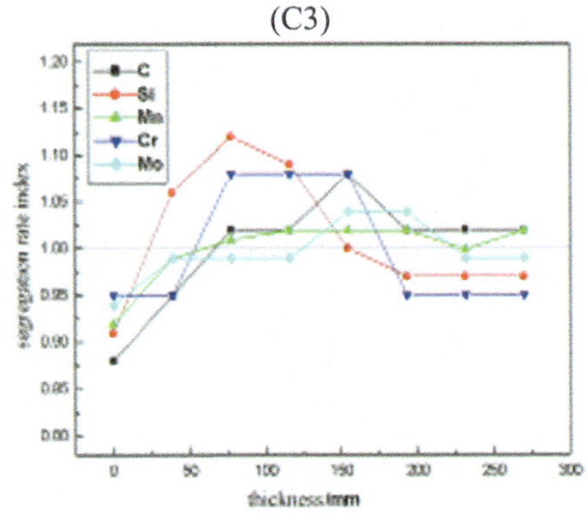

(C4)

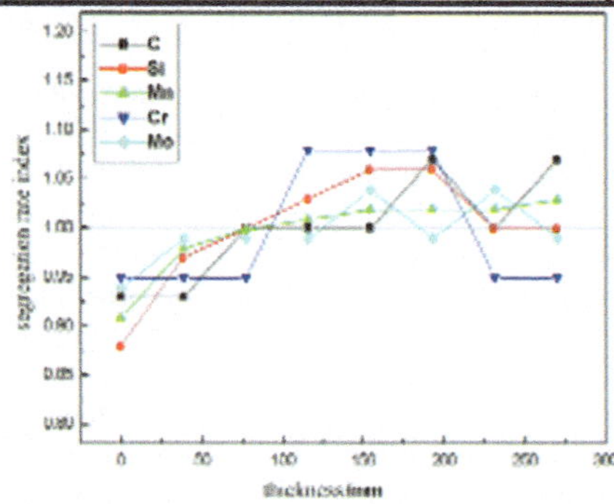

(C5)

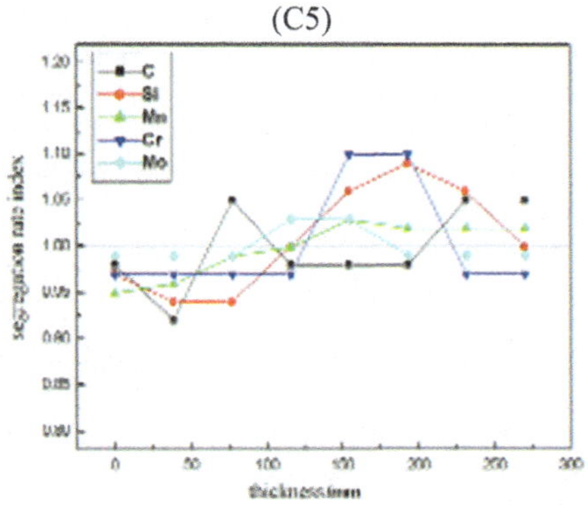

(C6)

Figure 5. Macrosegregation of each specimen.

The macro-segregation rate of the each sample is shown in Figure 5. On the whole, segregation rate of C1 and C2 was relatively large, the segregation rate of C3, C4, C5 and C6 sample was relatively small. The specimens of C1 and C2were at the top of the ingot where hot topping was carried out. Besides some segregation rates of individual elements of certain positions is 0.86~1.12, the segregation rates of the most elements are able to control in 0.95~1.05.

Macrostructure

Figure 6. Macrostructure of WSM718R steel ultra-heavy plates.

Figure 6 shows the macrostructure of the die steels ultra-heavy plates produced by ESR slab ingots tested according to GB/226-1991 standard test, respectively. The general porosity, center porosity and scattered segregation are all less than 0.5 levels according to GB/226-1991. The typical macrostructure defects such as white spots, slag entrapment and bubbles can not be found.

Conclusions

The die steels ESR slabs with the maximum thickness of the world in size of 980 mm thickness, 2000 mm wideness and 2800 mm length have been produced successfully. The ultra-heavy plates with the maximum thickness 410 mm can be obtained after rolling the 49 tons ESR slabs. which The good surface quality of large ESR slab ingots can be obtained by adjusting taper of the mold and using CaF_2-CaO-Al_2O_3-SiO_2-MgO slag. The results of macro segregation experiments show that the ultra-heavy plates produced by ESR slab ingots exhibits excellent uniformity in different positions. And the mechanical properties of ultra-heavy plates are superior to that of conventional materials. Due to reducing the cogging and forging process, the ESR for large slab ingots process can increase greatly the yield and production efficiency, and evidently cut off product costs. So the ESR is a satisfied method to produce the high quality die steel large slab ingots. The ESR process for large slab ingots will provide a great economic value and application prospect.

Acknowledgement

The authors would like to thank the National Natural Science Foundation of China Youth Project (No. 51104038), the Central University of basic scientific research expenses of nurture seeds for national research project (No. N120402008), the National High Technology Research and Development Program of China (863 Program) (No. 2012AA03A502). In addition, the work is supported by Program for Liaoning Innovative Research Team in University with the grant (No. LT20120008).

References

1. Zaizhi Chen and Dengshen Ma, "Advance in Die Steel Products at Home and Abroad", *Journal of SPECIAL STEEL*, 27 (5) (2006), 37-39.
2. Hongkui Zhang and Minghua Xu, "Superficial View on Development Room of Die Steel in China", *Journal of HEAT TREATMENT*, 23 (5) (2008), 7-15.
3. Zaizhi Chen and Dengshen Ma, "Analysis of Development Strategy for National Die Steel Industry", *Journal of Iron and Steel*, 41 (4) (2006), 5-9.
4. Hiroyuki Kajioka, Ko Yamaguchi, Nobuo Sato, et al., "Effects of Various Melting Parameters on the Qualities of Electro-Slag Remelted Ingots. Proc.", (paper presented at the 5th International Symposium on ESR, Pittsburgh, 1974), 102-114.
5. M. Nishiwaki, T. Yamaguchi, M. Koba, et al., "Operation of large bifilar ESR furnace for slab production and quality of slabs and heavy plates produced.", (paper presented at THE FIFTH INTERNATIONAL CONFERENCE ON VACUUM METALLURGY AND ELECTROSLAG REMELTING PROCESSES, MUNICH, 1976), 197-202.
6. Kozuo Okohira, Takaharu Shimizu, Nobuo Sato, et al., "Refining and Quality of Slabs and Heavy Plates Produced by 40t Electro-Slag Remelting Process.", *Journal of Tetsu to Hagane*, 63(13) (1977), 2208-2223.
7. L. Rohde and D. Lohr., "Operational results of the 50 ton ESR plant of Thyssen Henrichshütte AG.", (paper presented at the PROCEEDINGS OF THE FIFTH INTERNATIONAL CONFERENCE ON VACUUM METALLURGY AND ELECTROSLAG REMELTING PROCESSES, MUNICH, 1976), 177-180.
8. Renbo Yu, Zuxian Zhang, Yuwen Mao, "Sdudy on viscosity of CaF2-CaO-MgO-Al_2O_3-SiO_2 ESR slags.", *Journal of Iron and Steel Research*, 1 (2) (1989), 9-14.

Proceedings of the 2013 International Symposium on Liquid Metal Processing & Casting
Edited by: Matthew J. M. Krane, Alain Jardy, Rodney L. Williamson, and Joseph J. Beaman
TMS (The Minerals, Metals & Materials Society), 2013

A STUDY OF SLAG AND STEEL LEAKAGE INFLUENCE FACTORS DURING ELECTROSLAG REMELTING WITHDRAWING PROCESS

Ximin Zang[1], Zhouhua Jiang[2], Hua Song[1], Fubin Liu[2], Xin Deng[2], Xu Chen[2], Chong Han[1]
[1]University of Science and Technology Liaoning, Anshan, Liaoning, 114051, PR China
[2]Northeastern University, Shenyang, Liaoning, 110004, PR China

Keywords: Electroslag remelting withdrawing technology, Slag leakage and steel leakage, Slag physicochemical properties, Solidification.

Abstract

The electroslag remelting withdrawing (ESRW) process overcomes the shortcomings of the traditional electroslag remelting（ESR）process such as the low production efficiency and high cost, and so on. The main features of ESRW compared to traditional ESR are a continuous relative movement between ingot and mold, and electrode exchange with several times and ingot withdrawing. The key technology for ESRW is how to avoid the leakage of liquid slag and steel from the mold during ingot withdrawing process. It is found that the slag properties including shrinkage, high temperature strength and plasticity, lubrication and friction of slag skin on slag-metal interface, ingot solidification shrinkage behavior, process coordination control are the key factors which influence the leakage of liquid slag and steel in ESRW process. In addition, a multi-taper mold present in this paper is beneficial to obtain good surface quality of ingot based on mathematical modeling of ESRW process.

Introduction

E.O. Paton Electric Welding Institute developed ESRW technology in the 80's of last century [1]. In 2002, we also developed ESRW technology which improved production efficiency and reduce cost [2]. Compared with traditional electroslag remelting （ESR）process, the main difference of ESRW technology is the continuous relative movement between ingot and mold, which could result in low-surface quality of ingot and slag and steel leakage problems etc. in some cases. Figure 1 shows 300×340×6000mm ingot surface quality obtained by using different slag systems 60%CaF_2+20%Al_2O_3+20%CaO,40%CaF_2+30%Al_2O_3+30%CaO,60%CaF_2+30%Al_2O_3+10%CaO with the same process condition. There were wrinkles with varying degree on ingot surface formed by liquid metal from inerratic broken slag crust.

Figure 1. Surface quality of 300×340×6000mm ingot.

How to control the slag and steel leakage is the key technology in the ESRW process. The melting temperature and viscosity of slag system and the taper of mold are the most important factors of the slag and steel leakage. Figure 2 shows the physical model of ESRW system. In This paper the study of slag system and taper of mold for ESRW process are presented.

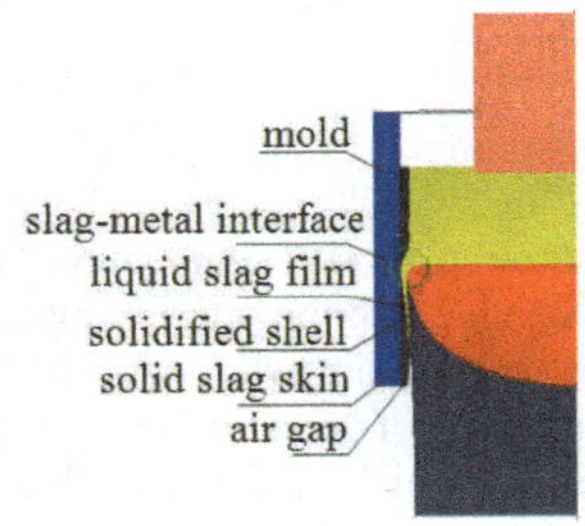

Figure 2. Physical model of ESRW system.

Research and Development of ESRW Slag System

Physicochemical Properties of Withdrawing Slag System

Figure 3. The fundamental of slag broken due to the friction.

It is crucial to study the physicochemical prosperities of slag system. Lower melting temperature and lower viscosity are beneficial to improve the surface quality of ingot in ESR. The liquid-solid transition behaviour of slag at the slag-metal interface plays a role of lubrication. The majority of slag has a very large shrinkage during cooling process. After moving certain distance, the slag and ingot detach from the mold, which will reduce the movement resistance. If the shrinkage is not enough to detach the slag from the mold during cooling process, the slag will be stuck by mold which will increase the tensile stress significantly. Then there will be metal wing surface defects after slag detaches and the worst case is slag and steel leakage accident. Figure 3 shows the fundamental of the process. High temperature strength, certain plasticity and ideal shrinkage ratio of slag are required for the withdrawing process. Liang Lianke pointed out that the friction resistance will decrease with SiO_2 increase while the friction resistance will increase with Al_2O_3 increase in the slag and also long slag is preferred in terms of ESRW process, because of the high temperature fluctuation caused by slag-metal interface fluctuation [3-5].

Following requirements should be considered for selection of the withdrawing slag system.

(1) Lower melting temperature and lower viscosity of slag is better to the uniform and thin crust formation on the ingot surface, which will improve the ingot surface quality. It is easy to generate temperature fluctuations in slag pool during withdrawing process, which required the viscosity of slag is insensitive to temperature.

(2) In order to save power consumption, slag system of higher resistance is developed. A large number of domestic and foreign researches show that the conductivity decrease with CaF_2 decrease while the resistance increase with the oxide content increase[6-7].

(3) The high temperature strength, plasticity and shrinkage of slag skin suits for ESRW process. The results studied by Zhakhovsky show that slag monoclinic ß -Ca_2SiO_4 will spontaneously change into the orthorhombic γ-Ca_2SiO_4 under 723K or lower temperature when ratio of CaO and SiO_2 is determined. This kind of crystal transition is with significant volume expansions which lead to the slag skin broken and this is the real cause of friction increasing between ingot and mold. A suitable of the SiO_2 content will prevent orthosilicate calcium creation process and improve the plasticity of slag significantly [8-10].

Development of Withdrawing Slag System

The detailed components of four slag systems designed in present work are shown in Table I. ANF-6 slag system is for comparison.

Table I. Composition of the slag, %.

slag	CaF_2	Al_2O_3	CaO	SiO_2
L1	50~70	20~30	10~20	≤10
L2	50~70	20~30	10~20	—
L3	60~70	20~30	≤10	≤5
L4	50~60	10~30	10~20	≤10
ANF-6	70	30	—	—

Calculation and Measurement of Physicochemical Properties of Slag

The melting temperature of four slag systems was measured with hemisphere method by using LZ- II slag melting test device developed in our lab. Density, conductivity, optical basicity of slag was calculated by using the empirical formula. The detailed data is shown in Table II, Table III, Table IV and Table V.

Table II. Experimental results of melting temperature for different slags, ℃.

L1	L2	L3	L4	ANF-6
1331	1434	1376	1315	1403

Table III. Calculated results of density for different slags of 1350 ℃ to 1400℃, g · cm^{-3}.

L1	L2	L3	L4	ANF-6
2.587	2.657	2.605	2.616	2.569

Table IV. Calculated results of optical basicity for different slags.

L1	L2	L3	L4	ANF-6
0.680	0.697	0.684	0.679	0.665

Table V. Calculated results of electric conductivity for different slags, $\Omega^{-1}\cdot cm^{-1}$.

	1550℃	1600℃	1650℃
L1	2.50	2.69	2.89
L2	2.54	2.74	2.93
L3	2.69	2.89	3.08
L4	1.94	2.14	2.33
ANF-6	2.81	3.01	3.20

With rotating cylinder method, four kinds of slag viscosity were tested by using DN- II slag viscosity test device developed in our lab as shown in Table VI. The test results of regression relation curves of viscosity and temperature are shown in Figure 4.

Table VI. Experimental results of viscosity for different slags, Pa·s.

	1500℃	1470℃	1440℃	1410℃	1380℃	1350℃	1341℃	1300℃
L1	0.0537	0.0564	0.0671	0.0697	0.0722	0.2483	0.8536	6.1634
L2	0.0456	0.0537	0.0819	0.0685	0.8671	4.3219	-	-
L3	0.0846	0.0792	0.0631	0.0577	0.0591	2.0804	5.3138	-
L4	0.1396	0.1342	0.1772	0.0846	0.0738	0.6429	1.7838	5.3930

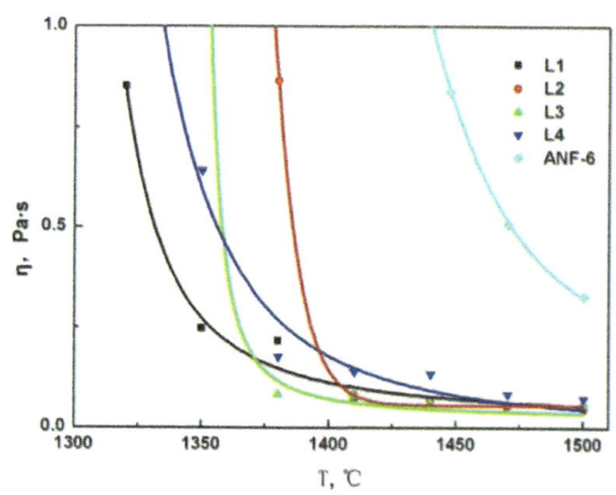

Figure 4. Curve of viscosity change with different temperature for different slag.

From the chart we can see that the four kinds of slag viscosity are lower than the ANF-6 slag viscosity. From 1350℃ to 1500℃, viscosity variation range is gentler than ANF-6 slag system, so it can improves the quality of ingot surface. L1 slag and L4 slag contains a certain amount of SiO_2. Because of large oxygen anion polymerization degree and poor crystallization ability, even when cooled to below liquids' temperature it can still remain super cooled liquid state. In terms of L1 slag system, L4 slag system, when the temperature decreases, viscosity increase slowly, which are the long slag systems. This characteristic is good to the ESR ingot surface quality improvement. Combined with the density, melting temperature, conductivity, L4 slag system was chosen to be the experimental slag[11].

Study on Withdrawing Mold Taper

As shown in Figure 2, ingot in the mold exit is basically completely solidification. Temperature gradient in ingot is very large, after the liquid-solid phase change shrinkage, solid-state cooling process of shrinking. Ingot solidification shrinkage to air gap formation between mold and ingot, which is one of the most important causes of the slag and steel leakage using without taper mold in ESRW process.

We used the ANSYS software to simulate the temperature field of ingot in order to study ingot solidification process from the slag-metal interface to the mold exit. Taking 300 ×340mm rectangular ingot as an example, the simulation results is shown in Figure 5, which includes the wide surface of ingot distance corresponding center shrinkage curves of different position[12].

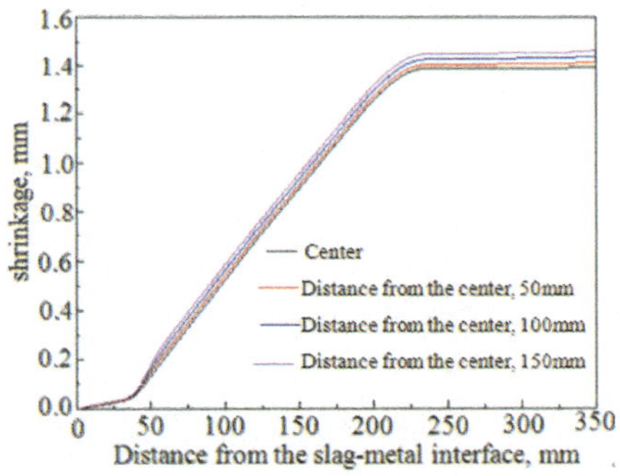

Figure 5. Shrinkage curves of different position of broad side surface.

As shown in Figure 5, ingot shrinkage is divided into three phases. At the slag-metal interface to the slag-metal interface 40mm region, ingot is most in the liquid region, this contraction is due to liquid phase temperature decreasing and the density of liquid phase increasing with little the density change, so the shrinkage in this area is slow. At a distance from slag-metal interface of 40mm to 225mm, the ingot is in the two-phase region of liquid and solid, ingot solidification with separating out austenitic phase, austenitic phase density is far greater than the liquid steel under the same temperature, so the shrinkage is larger. At a distance of slag-metal interface 225mm to the mold exit, the solidification of liquid transition solid phase is basically end and ingot is in the solid cooling stage, so shrinkage rate also decline. So without taper or single taper mold, it can not meet the requirements of ESRW process and multi-taper mold must be adopted to decrease the air gap between the ingot and mold.

Experimental Results

Rectangular Ingot Experiment

The steel grade of GCr15 (chemical composition is shown in Table VII), L4 slag, electrode size of 600×340×3000mm were adopted, and voltage of 88V, current of 16.8kA, melting rate of 850kg per hour, multi-taper mold were used to produce 300×340×6000mm ingot. Ingot surface was smooth, and its internal macroscopically quality was good, as shown in Figure 6[13].

Table VII. Composition of the consuming electrode, %.

C	Si	Mn	P	S	Cr	Ni
0.99	0.25	0.30	0.007	0.005	1.45	0.1

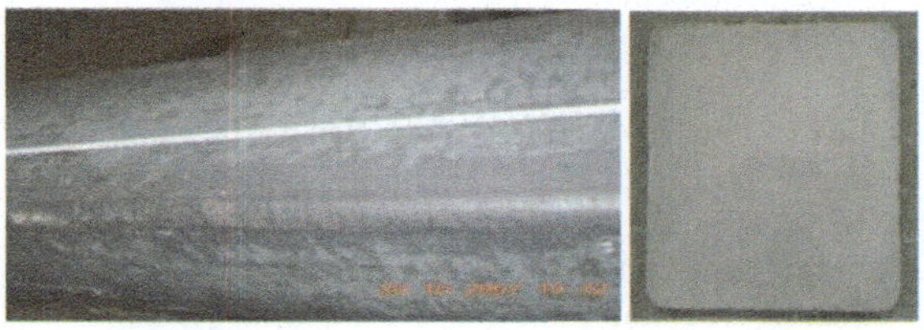

Figure 6. Surface quality and inner macroscopically quality of 300×340×6000mm ingot.

Round Ingot Experiment

The steel grade of 12Cr1MoVG (chemical composition is shown in Table VIII), L4 slag, electrode size of Φ 480 × 3000mm were adopted, and voltage of 86V, current of 13.6kA, melting rate of 750kg per hour, multi-taper mold with exchanging electrodes were applied to produce Φ 600 × 6000mm ingot. Ingot surface was smooth and its internal macroscopically quality was good, as shown in Figure 7.

Table VIII. Composition of the consuming electrode, %.

C	Si	Mn	P	S	Cr	Ni	Cu	Mo	V
0.12	0.29	0.55	0.009	0.005	1	0.03	0.09	0.3	0.2

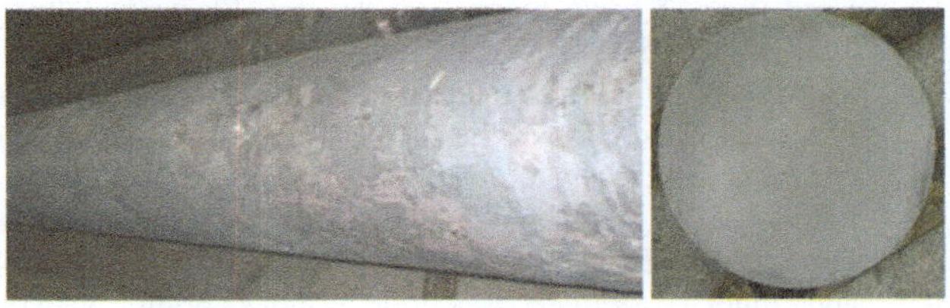

Figure 7. Surface quality and inner macroscopically quality of Φ600×6000mm ingot.

Conclusions

(1) The physicochemical and high-temperature dynamics characteristic of slag system and the taper of mold are the most important influence factors of ingot surface quality and slag and steel leakage in the ESRW process.
(2) Low melting temperature and low viscosity slag is suitable for ESRW technology. Moreover, appropriate high-temperature strength and plasticity dynamic characteristic are also required.

(3) Adding a certain amount of SiO_2 can improve the high-temperature strength and plasticity dynamic characteristic of slag system, but which should be further studied with detailed quantitative relationship.

(4) Without taper and single taper mold, it cannot meet the requirements of ESRW technology and multi-taper mold must be adopted to decrease the air gap between the ingot and mold.

Acknowledgement

This project supported by the National Nature Science Foundation of China (No. 51204041) and the National High Technology Research and Development Program of China (863 Program) (No. 2012AA03A502). In addition, the work is supported by Program for Liaoning Innovative Research Team in University with the grant No. LT20120008.

References

1. Jie Fu. The Development of the Second Generation of Large Ingot by Electroslag Metallurgy Technology [J]. *China Metallurgy*, 2010, 5, 20(5): 1-4.
2. Ximin Zang et al., Development of the Electroslag Continue Casting Technology [J], *China Metallurgy*, 2006, 5, 26(3): 10-13.
3. Lianke Liang et al., Metallurgical Thermodynamics and Dynamics [M]. *Shenyang, Publishing Company of Northeastern University*, 1990, 181.
4. Proceedings of Third International Symposium on Electroslag and Other Special Melting Technologies[C]. *Pittsburgh: Carnegie-Mellon Institute,* 1971.
5. Yinbo Yu et al., A Study of Higher Temperature Mechanical Properties of Slag System, *Journal of Iron and Steel Research*, 1991, 3(1): 17-23.
6. Lianke Liang, Huai Yang, Zhongwen Guo et al. Physical properties of slag effect on energy consumption of electroslag remelting process , Journal of Northeastern University, 1993,14(2):171
7. Lianke Liang, Zhongwen Guo, Fangzhi Wang et al. Study on the conductivity of the slag by AC four probe method , Journal of Northeastern University,1985, 6(3): 71
8. Martyn V.M. et al., Mechanical Properties of Certain Fluxes Used for EST. *Problems of Special Electrometallurgy*, 1978,iss.8, pp52-54
9. Zhakhovsky V.S. Investigation of Strength and Ductility of Industrial Fluxes for EST at High Temperatures. *Problems of Special Electrometallurgy*, 1980, iss.12, pp30-33
10. Paton B.E. et al. About the Mechanism of Ingot Surface Formation in ESR at the Relative Displacement of the Mold and Ingot. *"Refining remelting"*, Ed.2, Kiev, Naukova Dumka.
11. Fubin Liu. Mathematical Modeling of Electroslag Continuous Casting Process and Control of Ingot Quality [D]. Shenyang, Northeastern University, 2009
12. Fulei Zhu. Mold Taper Design of Electoslag Remelting Withdrawing Process. [D]. Shenyang, Northeastern University, 2011
13. Ximin Zang. Development of Electroslag Continuous Casting Technology and Its Process Research [D]. Shenyang, Northeastern University, 2008

Proceedings of the 2013 International Symposium on Liquid Metal Processing & Casting
Edited by: Matthew J. M. Krane, Alain Jardy, Rodney L. Williamson, and Joseph J. Beaman
TMS (The Minerals, Metals & Materials Society), 2013

FACTORS AFFECTING SURFACE QUALITY OF INGOT PRODUCED BY ELECTROSLAG CONTINUOUS CASTING WITH LIQUID METAL

Xin Deng[1], Zhouhua Jiang[1], Ximin Zang[2]
[1] Northeastern University; Heping district Wenhua Rd.; Shenyang, Liaoning, 110819, PR China
[2] University of Science and Technology Liaoning; Anshan, Liaoning, 114051, PR China

Keywords: electroslag continuous casting with liquid metal (ESCCLM); surface quality; premelted slag.

Abstract

The surface quality of ø1000mm ingot of steel 45# produced by electroslag continuous casting with liquid metal (ESCCLM) were investigated in this work. The results showed that the important effect factors for getting the better surface quality of ingot were appropriate casting temperature of molten steel (1606℃), higher voltage (80V) and steady current (18kA), the special premelted slag, uniformily and appropriate secondary cooling control and steady withdrawing speed (8~10mm/min).

Introduction

The electroslag continuous casting with liquid metal (ESCCLM) technique has been successfully developed by Northeastern University for the first time in China. The new equipment mainly consists of tundish heating, current conductive mold[1-3], low frequency power, control system for small flow of liquid steel casting , the level control at interface of slag and metal, etc. The characteristics of ESCCLM are different from the traditional ESR which needs the preparation of consumable electrode, and much faster casting speed which was 5~10 times of the traditional ESR melting raten, much lower power consumption that can overcome the traditional ESR process with " long production process, low production efficiency and high operation cost ". Its principle was shown in Figure 1.

Strict control of the surface quality of ingot produced by the ESCCLM, elimination of the ingot surface wrinkles and skin crack phenomenon can reduce the grinding processing of next working procedure[4].

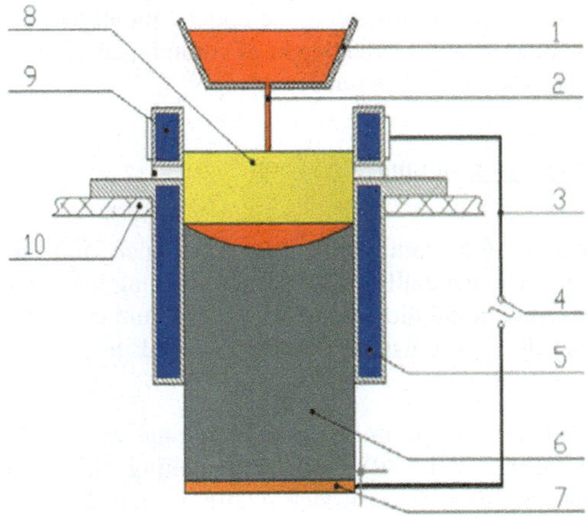

1-tundish; 2-molten steel; 3-power circuit; 4-transformer; 5-down mold; 6-remelting ingot; 7-watertank; 8-slag bath; 9-current conductive mold; 10- sopporting platform.
Figure 1. Principle of ESCCLM

Effect of Casting Temperature

The experimental steel grade was 45# steel, the chemical composition of which is shown in Table I. The ESCCLM ingot diameter was 1000mm.

Table I. Chemical composition of test steel, mass%

Steel grade	Composition（wt%）					
	C	Si	Mn	Cr	Ni	Cu
45#	0.47	0.25	0.52	0.21	0.27	0.2

The experiments of ESCCLM were performed in the same conditions of slag amount, slag system and power supply, but with the different casting temperature of 1576°C and 1606°C. The surface quality of remelting ingot was shown in Figure 2. It is not difficult to find that the shape of remelting ingot (a) is not full and the phenomenon of surface wrinkle is more serious, when the casting temperature of (a) was 1576°C. But when the casting temperature was increased to 1606°C, the surface of ingot (b) is very smooth and clean. It follows that the higher casting temperature improves the fluidity of molten steel, and then it is conducive to smooth the ingot surface quality.

(a)1576℃ (b)1606℃
Figure 2. Surface quality of ingot with different casting temperature

Effect of Power Supply System

The Function of Current Conductive Mold

The remelting current circuit consists of transformer, current conductive mold, slag bath, metal pool[5] and solid ingot. The current conductive mold is one of the important part of the circuit, which solves the problem of uneven distribution of heat in the slag bath and increases the temperature of slag bath edge. This device is beneficial to producing the upper column part of metallic molten pool and controls the height of column part in 10mm[6], in order to ensure the surface quality of ingot.

The current and Voltage

According to the traditional ESR experience and qualitative conclusion, it is generally believed that the slag temperature will be increased with the increase of the voltage and current[7], at the same time the slag crust will be thinned and the shape of ingot will be more full.

In the industrial experiment, the remelting voltage has been changed from 73V to 80V and the remelting current has been maintained at 18kA. The surface quality of remelting ingot was shown in Figure 3. The depth of slag ditch gradually becomes shallow with increasing of the remelting voltage so it is necessary to increase the remelting voltage for improving the ingot surface quality.

Figure 3. Surface quality of ingot with different voltage

Effect of the Special Premelted Slag

The special premelted slag was used in this experiment, which has both refining and lubricating functions[8]. It has the higher strength, lower fusion temperature and lower conductivity. Its viscosity with temperature changes very little, so that ingot surface quality has been obviously improveed.

Effect of Secondary Cooling Control

The secondary cooling control is one of the key factor for control of the ESCCLM ingot quality. In the secondary cooling area, the cooling capacity is too strong or weak will have an impact on the surface quality of ingot. The weak secondary cooling will cause that the ingot surface temperature is high, to speed up the formation of iron oxide scale and lead to the residual elements (Cu, Sn) assembled in the grain boundary, and then the surface crack will be found. The strong secondary cooling will cause that the ingot surface temperature decrease greatly and the temperature gradient of the solidified shell increase acutely. The trace elements (Al, Nb, B...) will deposit along the grain boundary and increase the surface crack sensitivity[9].

Therefore, the capacity of secondary cooling control should maintain a uniform, stable and moderate state. In the experiment, because the secondary cooling area was very uneven, the surface crack of ingot happened, as shown in Figure 4.

Figure 4. The surface crack of ingot

Effect of Withdrawing Speed

If the amplitude of variation of withdrawing speed was great, the ingot surface quality will be difficult to control. When the withdrawing speed was too fast, it will have the leakage risk of slag or molten steel. On the other hand, the slowly withdrawing speed will produce the phenomenon of thickness slag skin and ingot surface wrinkle. From the experimental results, in the premise of stable control at interface of slag and metal, the withdrawing speed that was in the range of 8 to 10mm/min can control the surface quality of ingot produced by ESCCLM more effectively.

Conclusions

The higher casting temperature can improve the fluidity of molten steel. The ingot surface produced with the casting temperature of 1606℃ was very smooth.

High voltage and stability current can be conducive to improving the ingot surface quality.

In the process of ESCCLM, it is necessary to use the special premelted slag, which has the higher strength, lower fusion temperature, lower conductivity and its viscosity with temperature changes very little, for improving the ingot surface quality.

The uniform, stable and moderate state of the secondary cooling area can reduce the occurrence of ingot surface crack.

The withdrawing speed in the range of 8 to 10mm/min can control the surface quality of ingot produced by ESCCLM more effectively.

Acknowledgements

This project is supported by Program for Liaoning Innovative Research Team in University with the grant No. LT20120008.

References

1. Holzgruber H, Holzgruber W, "Medovar Memorial Symposium" (Kiev: E O Paton Electric Welding Institute, 2001), 41-48.
2. Holzgruber W, Holzgruber H, "Production of High Quality Billets with the New Electroslag Rapid Remelting Process," *MPT International*, 19(5)1996, 48-50.
3. Medovar L B et al., "New Electroslag Technologies, Medovar Memorial Symposium" (Kiev: E O Paton Electric Welding Institute, 2001), 49-60.
4. Weiguo Xu et al., "Application of Centrifugal Electroslag Induction Casting Technology," *Special Steel*, 16(5)1995, 42-44.
5. Xchun Chen et al., "Recent Development of Electroslag Metallurgy," *Journal of Iron and Steel Research*, 15(2)2003, 63-67.
6. Zhouhua Jiang, *The Physical Chemistry and Metallurgical Appearance of Electroslag Metallurgy* (Shenyang: NEU press, 2000), 67.
7. Ximin Zang et al., "Effect Factors on Surface Quality of Electroslag Concasting Billet," *Special Steel,* 27(5)2006, 49-50.
8. Zhengbang Li, *Theory and Practice of electroslag metallurgy* (Beijing: Metallurgical Industry Press, 2010), 367.
9. Wei Xing et al., "Study on Secondary Cooling and Quality of Continuous Cast Billet," *Foundry Technology*, 33(1) 2012, 66-68.

Proceedings of the 2013 International Symposium on Liquid Metal Processing & Casting
Edited by: Matthew J. M. Krane, Alain Jardy, Rodney L. Williamson, and Joseph J. Beaman
TMS (The Minerals, Metals & Materials Society), 2013

MATHEMATICAL MODEL OF SOLIDIFICATION DURING ELECTROSLAG CASTING OF PILGER ROLL

Fubin Liu[1], Huabing Li[1], Zhouhua Jiang[1], Yanwu Dong[1], Xu Chen[1], Xin Geng[1], Ximin Zang[2]
[1]School of Materials and Metallurgy, Northeastern University, Wenhua Rd., Heping district, Shenyang, Liaoning, 110004, PR China;
[2]University of Science and Technology Liaoning, Anshan, Liaoning, 114051, PR China

Keywords: Electroslag casting, Pilger roll, Solidification, Mathematical modeling, Local solidification time.

Abstract

A mathematical model for describing the interaction of multiple physical fields in slag bath and solidification process in ingot during pilger roll casting with variable cross-section which is produced by the electroslag casting (ESC) process was developed. The commercial software ANSYS was applied to calculate the electromagnetic field, magnetic driven fluid flow, buoyancy-driven flow and heat transfer. The transportation phenomenon in slag bath and solidification characteristic of ingots are analyzed for variable cross-section with variable input power under the conditions of 9Cr3NiMo steel and 70%CaF_2 - 30%Al_2O_3 slag system. The calculated results show that characteristic of current density distribution, velocity patterns and temperature profiles in the slag bath and metal pool profiles in ingot have distinct difference at variable cross-sections due to difference of input power and cooling condition. The pool shape and the local solidification time (LST) during Pilger roll ESC process are analyzed.

Introduction

In recent years, with the rapid development of nuclear power, hydropower and other energy fields, seamless steel pipe of high quality, especially in the large diameter (OD 400mm~1000mm, wall thickness 50mm~110mm), is required. Rolling on the pilger mills is considered as one of the most effective manufacturing process in the case of large diameter tube production. Pilger roll is an important component of the pilger rolling mill where it will be subjected to complex thermo-mechanical loading. Thus better wearing resistance and higher other mechanical properties are required for pilger roll with a variable cross-section.

It has been universally recognized that the application of the electroslag casting (ESC) is rational for the production of large ingot. The ESC process has been developed based on electroslag remelting (ESR) process of ingots. During the ESC process, the steel melt is refined and shaped into a casting. The defects of shrinkage, porosity, inclusion and cracking are eliminated as soon as possible. At the same time it has high density, uniform chemical composition and higher mechanical properties.

Due to expensive costs and difficulties of physical modeling, mathematical modeling is a valuable tool for enhancing fundamental understanding of process, in which several physical phenomena (electromagnetic effects, fluid flow, heat transfer, phase transformation, etc.) occur [1-16]. However modeling and simulation of ESC process, especially of the casting formed in a mould with variable cross-section and variable input power are reported rarely.

In this paper, a mathematical model for describing the interaction of multiple physical fields in slag bath and solidification process in ingot during pilger roll casting with variable cross-section which is produced by the ESC process was developed. The commercial software ANSYS was applied to calculate the electromagnetic field, magnetic driven fluid flow, buoyancy-driven flow and heat transfer. The transportation phenomenon in slag bath and solidification characteristic of ingots are analyzed.

Numerical Model

A macroscopic physical model for simulation of the pilger roll ESC process has been derived and implemented as a new specific module into ANSYS, as shown in Figure 1. Three-dimensional Cartesian coordinate system is performed for simulating the ESC process. The main results obtained by this model are presented for the case a 9Cr3NiMo steel remelted ingot, with a slag composition of 70%CaF_2-30%Al_2O_3.

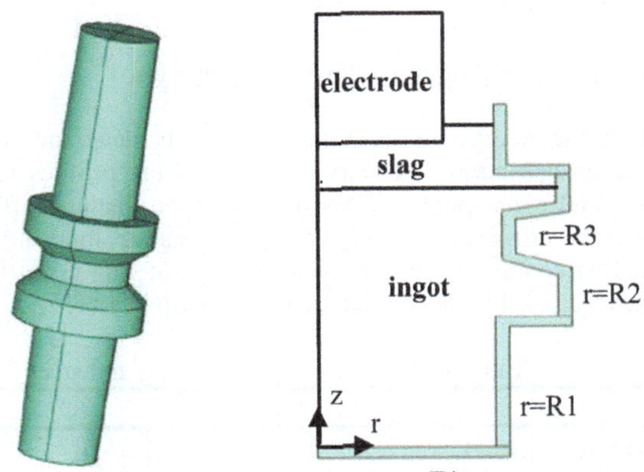

(A) (B)

Figure 1: Schematic sketch of pilger roll (A) and computational domain (B).

Numerical Procedures

To calculate the coupling fields in the system, the mathematical solver uses a standard FEM (Finite Element Method) to find the solution of the coupling of the quasi steady-state Maxwell's equations with the conservation equation of momentum and enthalpy on a level which enables a macroscopic process description. Maxwell equations are first solved to determine the electromagnetic force and Joule heating. Next, coupled fluid flow and heat transfer equations are written for the slag bath. The k-ε method is used to represent turbulence. The system of coupled partial differential equations is then solved, using a FEM (finite element method). Pressure-velocity coupling algorithm is SIMPLE (Semi-Implicit Method for Pressure Linked Equations) algorithm and the governing equations were solved by TDMA (tridiagonal matrix algorithm).

Boundary Conditions

The electromagnetic field is solved by using the electric field φ and the magnetic potential vector. A constant and uniform electric current flow is applied at the bottom boundary of the liquid domain, and a constant electric potential (φ=0) is applied at the top of the solid electrode. A constant radial current (J_r=0) is applied at the electrode top and the ingot bottom. At the slag free surface, the axial current is assumed to be zero (J_z=0). At the slag free surface and the outer surface of the ingot, the value of the magnetic flux density is specified according to Ampere's law in the following manner. The mould is assumed to be electrically insulated by the solidified slag layer, no current is allowed leave the domain through the lateral wall (mould).

A no-slip condition is applied at all the liquid/solid interfaces. At the slag free surface, a zero shear stress condition is imposed. Concerning the turbulence model, wall functions are applied to compute the turbulence kinetic energy and dissipation rate.

Heat transfer between the slag and the falling droplets is considered to be negligible. The tip of the electrode is the liquidus temperature (1745K). The heat transfer boundary conditions are most often Dirichlet conditions because of phase changes, which occur on the most boundaries. Heat flux at all the external surfaces and at the slag-metal interface is continuing. The slag free surface is supposed to exchange heat with the surrounding air through radiation.

Operating Conditions and Physical Properties

With the height of ingot increasing, characteristic of current density distribution, velocity and temperature profiles in the slag bath and metal pool profiles in ingot have distinct difference at variable cross-sections due to difference of input power and cooling condition. The physical property of slag taken from [17] and steel used in this model are given in Table 1.

Table 1: Physical property of slag and steel

Slag	
Density/ (kg·m^{-3})	2622
viscosity /(kg·m^{-1}·s^{-1}), 1673K	0.25
viscosity /(kg·m^{-1}·s^{-1}), 1773K	0.0326
Specific heat, liquid / (J·kg^{-1}·K^{-1})	1200
Thermal expans. coefficient / K^{-1}	2.5×10^{-4}
Thermal conductivity /(W m^{-1}·K^{-1}), 1873K	41.6
Thermal conductivity /(W m^{-1}·K^{-1}), 1973K	41.7
Electric conductivity/ (Ω·m)$^{-1}$	$\ln\sigma_s$=-6769/T+8.818
Steel	
Density/ (kg·m^{-3})	7800
Specific heat, liquid / (J·kg^{-1}·K^{-1})	750
Thermal conductivity /(W m^{-1}·K^{-1})	33
Electric conductivity/ (Ω·m)$^{-1}$	7×10^5

Variable operating process parameters with variable cross-sections are shown in Figure 2. In this paper, three quasi-steady state models computed and analyzed are slag bath in stage 1, 2 and 3, respectively. Geometrical and process parameters are listed in Table 2. With variable cross-sections, nominal slag height is change by adding and subtracting slag quality.

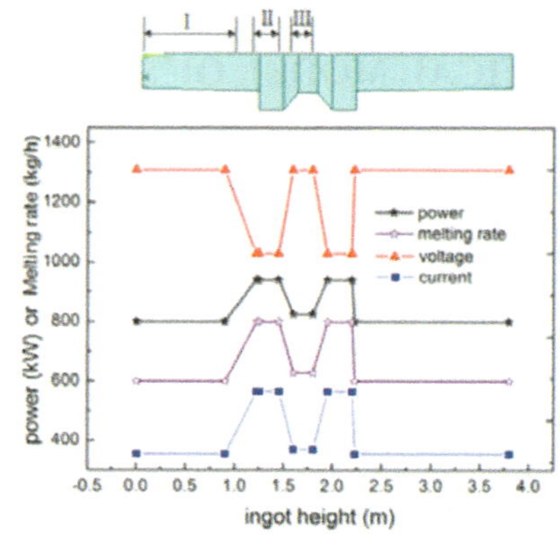

Figure 2: Variable operating process parameters with variable cross-sections.

Table 2: Geometrical and process parameters

	I	II	III
Nominal slag height / m	0.3	0. 2	0.29
Ingot radius(R1) / m	0.35		
Ingot radius(R2) / m		0.575	
Ingot radius(R3) / m			0.4
Electrode radius / m	0.275	0.275	0.275
Remelting current / A	16000	23500	20000
Remelting rate / (kg·h^{-1})	600	800	750
Frequency / Hz	50	50	50

Calculated Results and Discussion

Stage 1

Figure 3(a) shows the results of calculated voltage distribution in the slag bath of ESC process. There is a maximum at the electrode-slag interface due to magnitude electric resistance of slag is three time than that of steel. Distribution of current density is shown in the Figure 3(b). Maximum values occur close to the region between corner electrodes, in the other region is more uniform. As a result, the region of intensive heat generation is observed in the near-electrode zone, close to the slag-steel interface the distribution is more homogenous (Figure 3(d)).

Figure 3(c) shows the calculated Lorentz force vectors. As shown in the figure, the direction of Lorentz force vectors in slag bath is inwards and downwards. Lorentz force, current density and magnetic induction intensity are consistent with Fleming's left hand rule. Maximum Lorentz force values occur to corner electrodes. Lorentz force at the "slag-metal" interface is smaller due to less current density. The distribution of Lorentz force results in the pressure gradient at the axial direction of slag bath. As a result, the flow vortex turns in the anti-clockwise direction by Lorentz force.

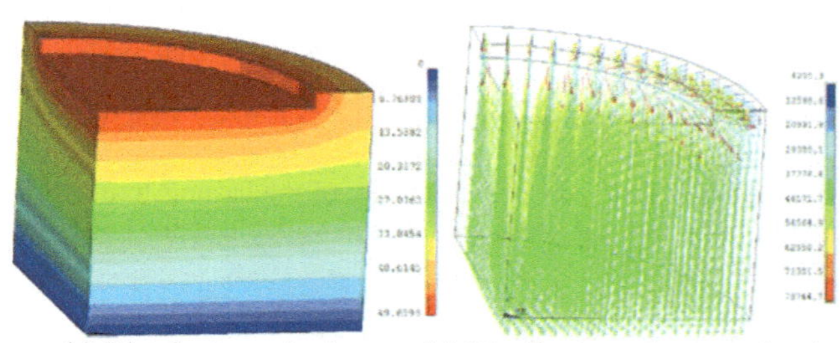

(a) Distribution of voltage (b) Distribution of current density

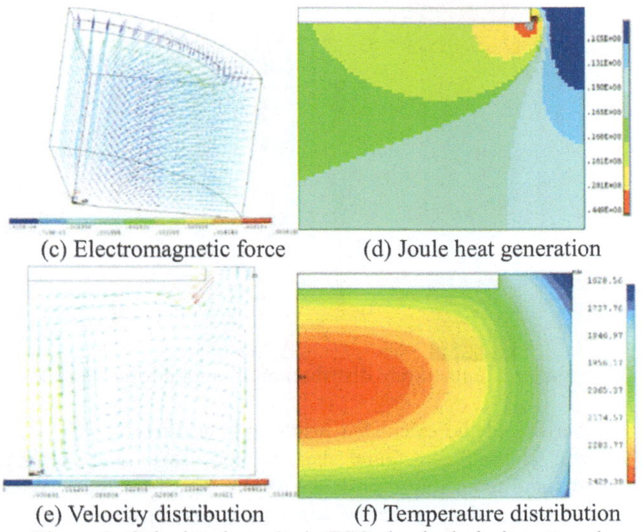

(c) Electromagnetic force (d) Joule heat generation

(e) Velocity distribution (f) Temperature distribution

Figure 3: Calculated results in ESC slag bath during stage 1.

The turbulent flow field in the slag bath is created by the Lorentz and buoyancy forces. Velocity distribution in slag bath is shown in Figure 3(e). There are two vortexes, one rotating clockwise at the vicinity of the mold and the slag/metal interface (in the region of Re≤r≤Rm), and the other rotating anti-clockwise under the electrode. The calculated velocities are in the range of 0-0.05 $m \cdot s^{-1}$ with maximum values occurring close to electrode corner. The temperature distribution is shown in Figure 3(f). Maximum temperature are located still under the electrode, providing thermodynamic and kinetic conditions for the removal of non-metallic inclusions.

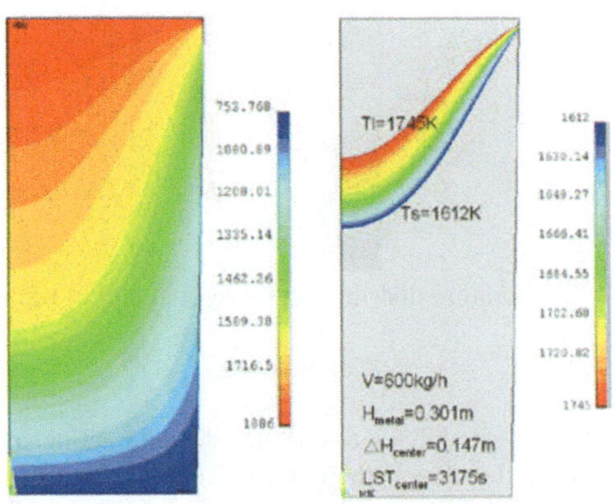

Figure 4: Temperature distribution in ingot during stage 1.

The temperature distribution results of the ingot shown in Figure 4 clearly indicate larger solidification range for 9Cr3NiMo steel. With the condition of 600kg/h remelting rate, the height of metal pool, maximum height of mushy zone and maximum local solidification time (LST) are 0.301m, 0.147m and 3175s, respectively.

Stage 2

When the ingot rises, the slag bath floats high to arriving variable cross-sections of mould. Additional slag is poured into due to lower of nominal slag height resulted from the change of mould radius (R1=0.35m, R2=0.575m). Calculated results in ESC slag bath during stage 2 are shown in Fig. 5 clearly indicate obvious distinctness comparison of calculated results of stage 1, due to the change of fill ration (from 0.6 to 0.23 (Re=0.275m)) and difference of input power and water cooling condition.

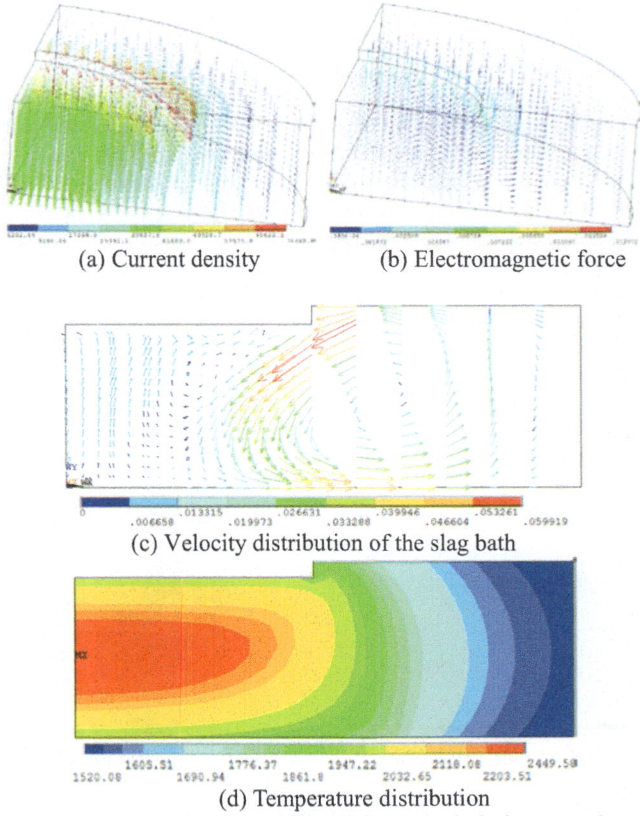

(a) Current density (b) Electromagnetic force

(c) Velocity distribution of the slag bath

(d) Temperature distribution

Figure 5: Calculated results in ESC slag bath during stage 2.

Figure 5 (a) shows the calculated current density results. As shown in the figure, the current density in slag bath is more inhomogeneous due to small fill ratio. Maximum values occur close to the region of electrodes corner. The current density in the region between corner electrodes and mould is far less than the current density of in the region between electrodes and slag-metal interface. With increasing of remelting current, Lorentz force is increasing shown Figure 5 (b).

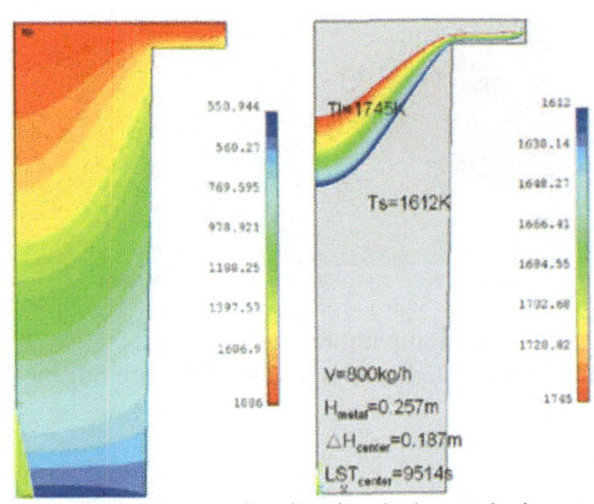

Figure 6: Temperature distribution in ingot during stage 2.

In Figure 5(c), the calculated velocity distribution in the slag bath of ESC process is visible. Also two vortexes are present, one rotating anti-clockwise in the region of $R_e \leq r \leq R_m$, and a second rotating clockwise direction located under the electrode. The calculated velocities are in the range of $0{\sim}0.06 m\cdot s^{-1}$ with maximum values occurring close to electrode corner. Maximum velocity values is more than the maximum velocity values of stage I ESC process ($0.05 m\cdot s^{-1}$). In Figure 5(d), the temperature near to mould wall is less than that of stage 1 ESC process due to difference of input power and water cooling condition.

The temperature distribution results of the ingot shown in Figure 6. With the condition of 800kg/h remelting rate, the height of metal pool, maximum height of mushy zone and maximum LST are 0.257m, 0.187m and 9514s, respectively.

Stage 3

When the ingot rises to arriving first slope between R2 and R3 of ingot, calculated results of ESC slag bath during stage 3 are shown in Figure 7. The temperature distribution results of the ingot shown in Figure 8. With the condition of 750kg/h remelting rate, the height of metal pool, maximum height of mushy zone and maximum LST are 0.214m, 0.141m and 8460s, respectively.

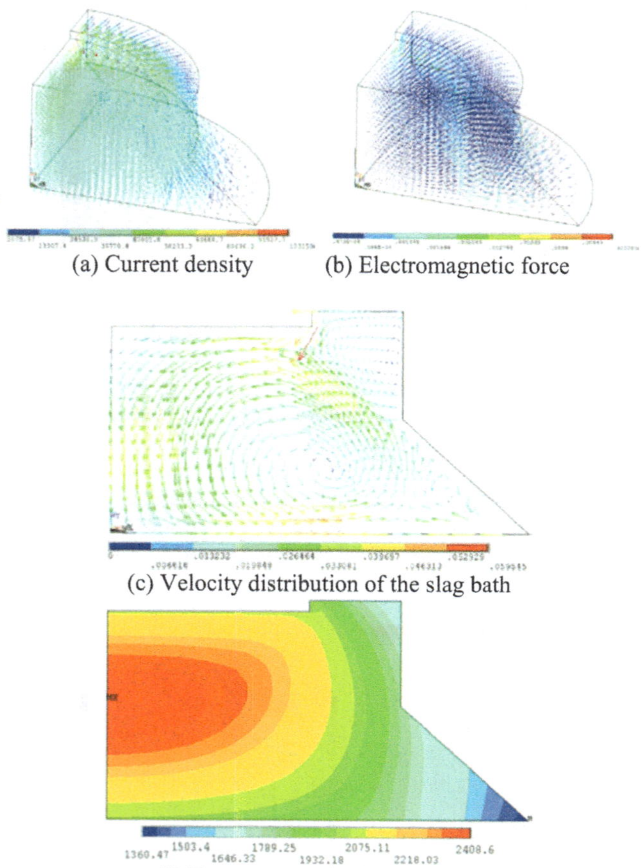

(a) Current density (b) Electromagnetic force

(c) Velocity distribution of the slag bath

(d) Temperature distribution

Figure 7: Calculated results in ESC slag bath during stage 3.

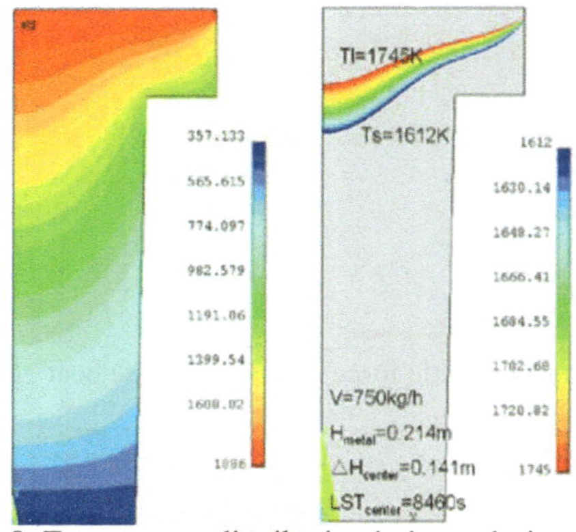

Figure 8: Temperature distribution in ingot during stage 3.

Comparison of the calculated results during different ESC stage process, it has been found that current distribution, velocity patterns and temperature profiles are obvious distinctness. When the ingot rises, the slag bath floats high to arriving variable cross-sections of mould. The interface of slag-mould wall and ingot-mould wall change lead to heating exchange condition change. Combine the different input power and fill ratio, the characteristic of current density distribution, velocity patterns and temperature profiles in the slag bath and metal pool profiles in ingot have distinct difference at variable cross-sections. The temperature distribution of ingot at end of remelting is shown in Figure 9.

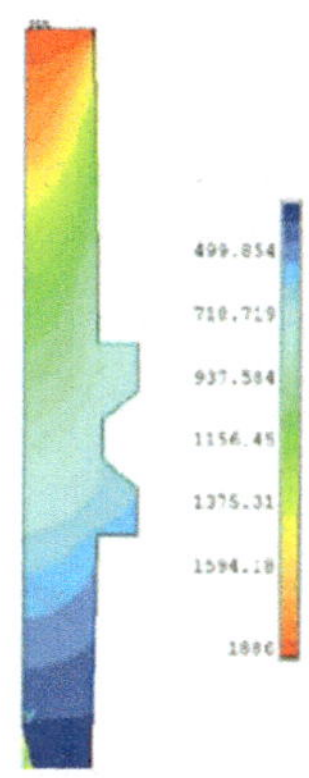

Figure 9: Temperature distribution of ingot at end of remelting.

Conclusions

A mathematical model for describing the interaction of multiple physical fields in slag bath and solidification process in ingot during pilger roll ESC process is developed. The model can be used to understand the ESC process with variable cross-section mould.

The characteristic of current density distribution, velocity patterns and temperature profiles in the slag bath and metal pool profiles in ingot have distinct difference at variable cross-sections due to difference of input power and water cooling condition.

The pool shape and the local solidification time are changing during pilger roll ESC process with variable cross-section mould. It is important to control ingot quality by adjusting operating parameters and melting rate due to the ingot structure controlled by the local solidification time.

Acknowledgement

This project supported by the National Nature Science Foundation of China (No. 51204041) and the National High Technology Research and Development Program of China (863 Program) (No. 2012AA03A502). In addition, the work is supported by Program for Liaoning Innovative Research Team in University with the grant No. LT20120008.

References

1. A.D. Patel, "Electrode immersion depth effects in the ESR process", (Paper presented at Proceedings of the 2011 International Symposium on Liquid Metal Processing and Casting, Nancy, France, 25 September 2011), 49.
2. M. Choudly and J. Szekely, "Modelling of fluid flow and heat transfer in industrial-scale ESR system", *Ironmak. Steelmak.*, 5(1981), 225-230.
3. K.M. Kelkar, J. Mok, S.V. Patankar, and A. Mitchell, "Computational modeling of electroslag remelting processes", *J. Phys. IV,* 120(2004), 421-425.
4. K.M. Kelkar, S.V. Patankar, and A. Mitchell, "Computational modeling of the electroslag remelting (ESR) process used for the production of ingots of high-performance alloys", (Paper presented at Proceedings of the 2005 International Symposium on Liquid Metal Processing and Casting, Materials Park, OH, 18 September 2005), 137.
5. A.H. Dilawari and J. Szekely, "Heat transfer and fluid flow phenomena in electroslag refining", *Metall. Trans. B,* 9(1978), 77-83.
6. A.H Dilawari and J. Szekely, "A mathematical model of slag and metal flow in the ESR process", *Metall. Trans. B,* 8(1977), 227-232.
7. J.H. Wei and Y.L. Ren, "Mathematical simulation of magnetic field in ESR system", *Acta Metall. Sin.*, 31(1995), 51-55.
8. J.H. Wei and Y.L. Ren, "Mathematical modeling of slag flow field in ESR system", *Acta Metall. Sin.*, 30(1994), 481-485.
9. A. Jardy, D. Ablitzer, and J.F. Wadier, "Magneto-hydrodynamic and thermal behavior of electroslag remlting slags", *Metall. Trans. B*, 22(1991), 111-117.
10. M. Choudly, J. Szekely, B.I. Medovar, and Yu G. Emelyanenko, The velocity filed in the molten slag region of ESR systems, *Metall. Trans. B*, 13(1982), p.35.
11. B.E. Launder and D.B. Spading, The numerical computation of turbulent flows, *Comput. Methods Appl. Mech. Eng.*, 3(1974), p.269.
12. A. Kharicha et al., "Multiphase modelling of the slag region in ESR Process", (Paper presented at Proceedings of the 2007 International Symposium on Liquid Metal Processing and Casting, Nancy, France, 2 September 2007), 107.
13. A. Kharicha, A. Mackenbrock, and A. Ludwig, Selected numerical investigations on ESR process, (Paper presented at Proceedings of the 2007 International Symposium on Liquid Metal Processing and Casting, Nancy, France, 2 September 2007), 113.
14. B.E. Paton et al., "ESR for titanium: yesterday, today, Tomorrow", (Paper presented at Proceedings of the 9th World Conference on Titanium, St. Petersburg, 1999,) 1385.
15. A.D. Patel, "An analysis of electromagnetic fields in ESR", (Paper presented at Proceedings of the 2005 International Symposium on Liquid Metal Processing and Casting, Materials Park, OH, 18 September 2005), 173.
16. V. Weber et al., "A comprehensive model of the electroslag remelting process: description and validation", *Metall. Trans. B,* 40(2009), 271-277.
17. Z. H. Jiang, *Mathematical Simulation and Experiment Analysis of Thermal Transfer for ESR.* (Shenyang: Northeastern University Press, 1986), 168.

INTRODUCING CARBON NANOPARTICLES IN TITANIUM DURING CHAMBER ELECTROSLAG REMELTING (CHESR)

A. Ryabtsev[1], B. Friedrich [2], F. Leokha[1], S. Ratiev[1], O. Snizhko[1], P. Spiess [2], S. Radwitz [2]
[1]Donetsk National Technical University, 58 Artema, Donetsk, 83001, Ukraine
[2]IME Process Metallurgy and Metal Recycling, RWTH Aachen University, 3 Intzestraße, 52056, Aachen, Germany

Keywords: chamber electroslag remelting, titanium, alloying, carbon nanoparticles

Abstract

The possibility of introducing carbon nanoparticles in titanium during ChESR is currently under investigation at DonNTU. Theoretical evaluation of phase and structure formation during crystallization was fulfilled by using the binary phase diagram of titanium and carbon. Within the investigations, consumable electrodes with previously introduced dispersed carbon particles were prepared by pressing of titanium sponge and nanoparticles of different types. The subsequent remelting was performed under a protective gas atmosphere in a 724 kW ChESR furnace. The average carbon content after remelting was in the range of 0.03 – 0.34 wt.-%. An increased carbon content simultaneously resulted in an increase of hardness from 140 to 220 HB. Metallographic investigations of precipitations in metal have shown phases mainly consisting of titanium-carbide particles with nonstoichiometric composition. As a result it could be stated, that structure as well as hardness show that ChESR provides good chemical and structural homogeneity of titanium ingots.

Introduction

Among structural materials, titanium and its alloys occupy a special position. Due to its properties, titanium is considered as a basic structural material for many industries, including the aero or medical sector. In the latter case, beside high specific strength and resistance to impact loading, the most important requirements for medical titanium alloys are corrosion resistance, biocompatibility and the absence of toxic elements. The most common titanium alloy for medical applications is alloy type Ti-6Al-4V (Grade 5). However, under certain conditions the presence of vanadium in this alloy can lead to the formation of toxic compounds in the human body [1-6]. To eliminate this drawback, vanadium can be replaced by a safer alloying component, oxygen [7-13] or carbon in particular [14]. Carbon is classified as a α-stabilizer - element that increases the temperature of the polymorphic transformation of titanium. Titanium reacts with carbon, forming a narrow field of β- and α-solutions and a chemically stable compound: titanium carbide (Figure 1).

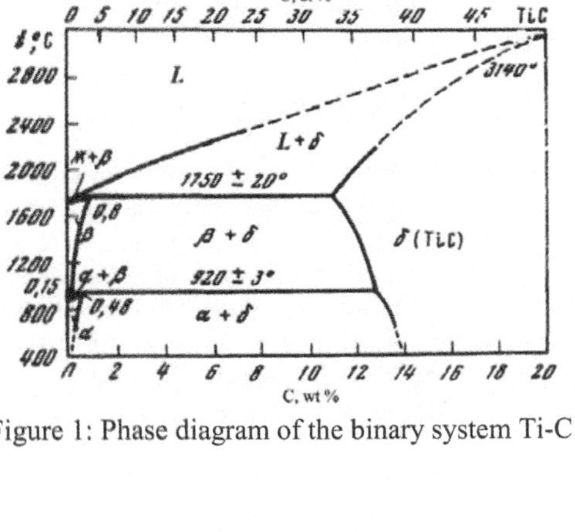

Figure 1: Phase diagram of the binary system Ti-C

The solubility of carbon in β-Ti at the peritectic temperature (1750 °C) is ~ 5 at.-% and almost constant at low temperatures. The maximum solubility of carbon in α-Ti at 920 °C is ~ 2 at.-% (0.48 wt.-%) and decreases as the temperature decreases from 0.48 wt.-% at 920 °C to 0.05 wt.-% 20 °C [15]. Therefore, when the carbon content is higher than 0.1 wt.-% the precipitation of carbides in the structure of titanium appears [16]. Carbon, like oxygen, is a good alloy strengthener. Its strengthening coefficient is 7-8 MPa per 0.01 wt.-% (Figure 2) [14-16].

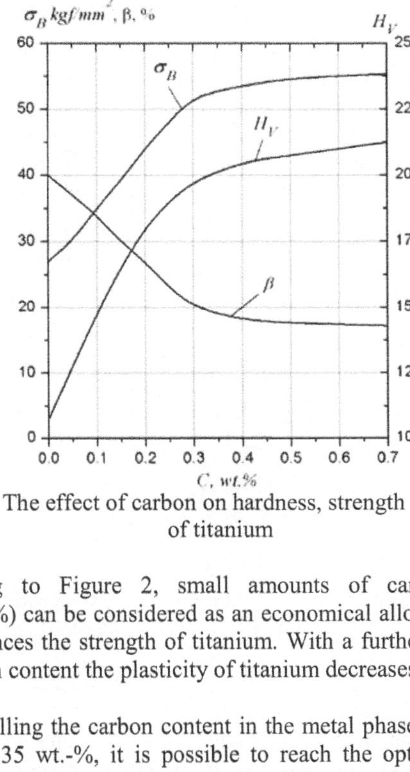

Figure 2: The effect of carbon on hardness, strength and ductility of titanium

According to Figure 2, small amounts of carbon (up to 0.35 wt.-%) can be considered as an economical alloying element that enhances the strength of titanium. With a further increase of the carbon content the plasticity of titanium decreases.

By controlling the carbon content in the metal phase in the range of 0.15-0.35 wt.-%, it is possible to reach the optimal ratio of plasticity and strength characteristics of the material. In this case it is very important to ensure a uniform distribution of carbon in the metal and the desirable form of its existence in it. This can be achieved by applying appropriate technologies for melting and the appropriate ligatures.

As carbon "ligatures" in this work micropowder of carbon and carbon nanotubes were used. The introduction of dispersed particles during crystallization of the material, or the formation of these particles in the material as a result of phase transformations during cooling is effective in means of enhancing of the structure. Even if the main structural elements (grains) are not granulated up to nano-scale, a nano-substructure can be formed in the material. This allows an increasing contribution of substructural strengthening in material properties. The introduction of nanosize particle which leads to the subsequent formation of a nano-substructure can also allow to realize the mechanisms of strengthening due to inhibition of dislocation motion, which allows to reach the increasing of strength and ductility of the alloy.

This mechanism can be realized by the introduction of carbon nanotubes in titanium. If the carbon in titanium is present in non-equilibrium concentrations, the formation of dispersed precipitates of titanium carbide during cooling phase is possible. The number and size of these particles can be controlled by changing the carbon content in the material or the heat treatment of the obtained alloy. For the practical implementation of this approach the chamber electroslag remelting (ChESR) process can be used. In addition to refining in a controlled atmosphere, this process offers the possibility of additional alloying during remelting. ChESR provides, as other remelting processes, a good structural and chemical homogeneity of the ingots with a typical as-cast structure.

Experimental

The semi products for setting up the consumable electrodes were made by pressing of titanium sponge grade TG-110 in blocks with a diameter of 41 mm and a length of 150 - 200 mm. In the obtained blanks axially holes with the diameter of 4.0 and 6.5 mm were drilled (Figure 3), in which carbon in the form of powder (~15 μm) and nanotubes (CNT ~15 nm) were pressed. Afterwards, the blocks were welded by argon-arc welding to consumable electrodes with the already mentioned diameter of 41 mm and a length of 550-650 mm. The content of the main impurities in the initial electrode (sponge TG-110 grade) is: C = 0.03 wt.-%; O = 0.04 wt.-%; N = 0.02 wt.-%.

Figure 3: Pressed block of titanium sponge with carbon powder filled into the drilled axial hole

The electrodes were remelted in a chamber electroslag furnace with a maximum power generator output of 724 kW. The ChESR is based on an A-550 unit (Figure 4) with a water cooled copper crucible. The crucible possesses a diameter of 70 mm.

The refining parameters and the results of the chemical analysis are shown in Table I. It can be seen that all the considered variants of melting parameters result in an accordance of the calculated and the achieved carbon contents in the range from 0.022 to 0.34 wt.-%. This indicates the good assimilation of carbon in the ChESR process. In this case, an increase of the oxygen content and the reduction of nitrogen is about 1.5 times compared with the initial contents in the titanium sponge.

Figure 4: The chamber electroslag furnace A-550 (724 kW)

The remelting was conducted under a flux of pure CaF_2 (TU 6-09-5335-88). The flux was melted directly in the crucible, using the technology of "solid" start. The starting mixture was prepared of titanium chips and the operating flux. The electrical parameters of the refining maintained at U = 36.0 V and I = 2.0-2.5 kA, providing a good surface quality of the remelted ingots (Figure 5).

U-0 U-2 U-4 U-6

Figure 5: Remelted titanium ingots of the various trials: U-0-without introduction of the carbon, U-2, U-4 and U-6 with the introduction of carbon

Metallographic studies

Samples were cut from the ingot (Figure 5) in order to conduct chemical analysis and metallographic studies. Samples for metallographic investigation of microstructure were sectioned at the ½ of ingot height level at distance ½ of radius Metallographic investigations at magnifications from ×50 to ×5000 were carried out by using optical microscopes (Axiovert 40 MAT Carl Zeiss and Neophot 2), an electron microscope (JEOL JSM-6490LV JEOL, Japan) equipped with an energy dispersive spectrometer

(INCA Penta FETx3, Oxford Instruments, England), a wave spectrometer (INCA Wave, Oxford Instruments, England) and a backscattered electron diffraction detector (HKL, Oxford Instruments, England).

Table I: Melt parameters and chemical composition of the metal

No	Electrode material	Estimated carbon content in the ingot in wt.-%	Resulting concentration of selected elements in the ingot in wt.-%		
			C	O	N
U-0	Titanium sponge TG-100	-	0.022	0.067	0.033
U-2	Titanium sponge TG-100+CNT	0.14	0.135	0.093	0.013
U-3	Titanium sponge TG-100+CNT	0.14	0.13	0.1	0.021
U-4	Titanium sponge TG-100+CNT	0.35	0.34	0.14	0.021
U-6	Titanium sponge TG-100+carbon	0.35	0.30	0.11	0.012

The chemical composition of the metal was determined with an optical emission spectrometer (SPECTROMAX, SPECTRO, Germany). The gas content was determined in the laboratories of the E.O. Paton Institute of Electric Welding of the National Academy of Science of Ukraine, Zaporozhye titanium and magnesium plant and RWTH Aachen University, Germany on different analyzers (TN-114, RO-316, RH-2, RH-3, Ströhlein O/N-MAT 8500). Mechanical tests and measurements of hardness were carried out by standard methods.

The analysis of the macrostructures (Figure 6) show that the introduction of carbon in form of nanoparticles has no significant impact on the ingots macrostructure (U-2-4) and the input of carbon in form of microparticles (U-6) leads to the formation of equiaxial crystallites with reduced size.

As can be seen (Figure 7, melting U-0), the structure of the metal melted without addition of carbon is typical for titanium of commercial purity and has crystallites with a large size. The carbon input of 0.135 wt.-% of carbon nanotubes leads to a dramatic structure refinement, and it changes to basket-weave morphology (Figure 7, melting U-2). With the increase of the carbon concentration to 0.340 wt.-%, the dispersion of the structure remains constant. However, the order of packets, which are typical for basket-weaving, is disrupted (Figure 7, melting U-4). When the carbon powder of micronized is used as ligature and the carbon content in the metal is 0.30 wt.-%, a structure of not oriented, approximately equiaxed, crystallites is forming (Figure 7, melting U-6). The basket-weave morphology indications are not observed.

In the present work additional investigations on the microstructure of titanium were carried out. The most typical microstructures are shown in Figure 7.

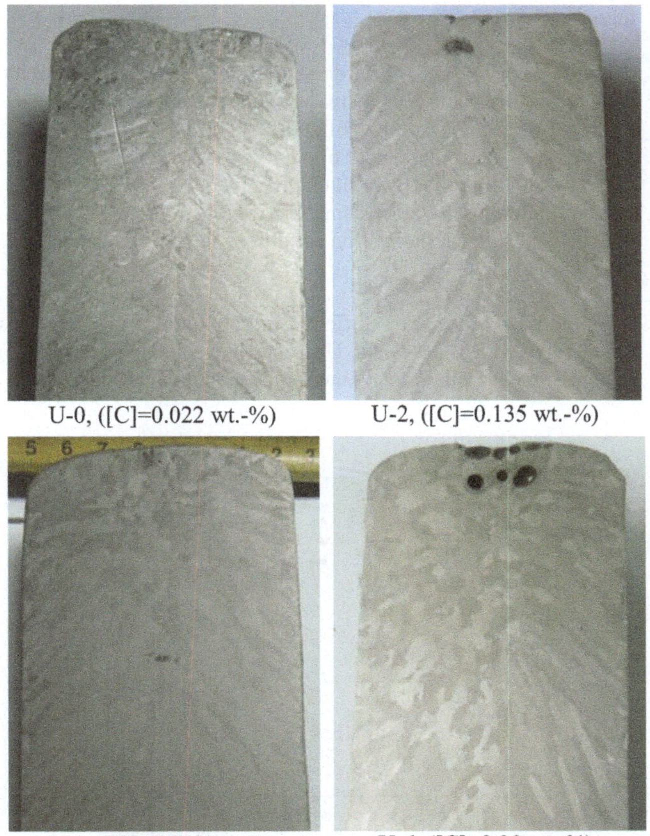

U-0, ([C]=0.022 wt.-%) U-2, ([C]=0.135 wt.-%)

U-4, ([C]=0.340 wt.-%) U-6, ([C]=0.30 wt.-%)

Figure 6: Macrostructure of the titanium ingots after ChESR

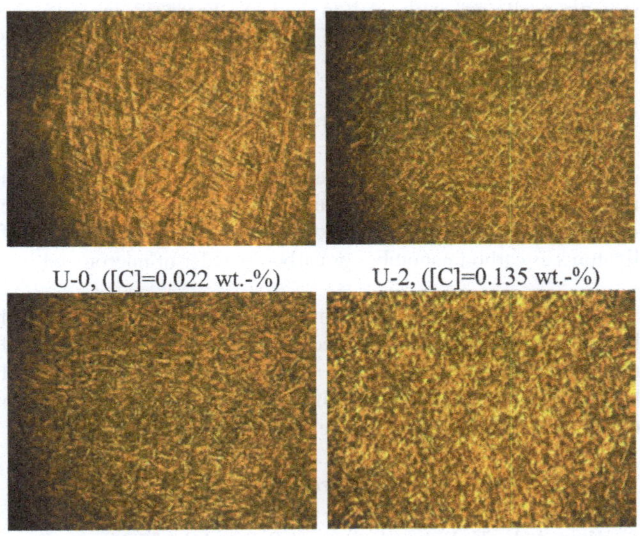

U-0, ([C]=0.022 wt.-%) U-2, ([C]=0.135 wt.-%)

U-4, ([C]=0.340 wt.-%) U-6, ([C]=0.30 wt.-%)

Figure 7: The titanium microstructure of ChESR ingots, × 100

An indirect indicator of the content and distribution of impurities in titanium is the hardness of the metal. For hardness measurement ingots were cut along the vertical axis for equal parts. Measurement was fulfilled along the central longitudinal axis of sample from bottom to top. Figure 8 shows the values of hardness, which have been measured at a height of experimental ingots.

As shown in Figure 8, the hardness of titanium correlates with its content of carbon and thus increases with its content in the metal. Thus, the maximum hardness of 200-220 HB is typical for samples with a carbon content of 0.340 wt.-% (melting U-4), and the lowest – 140-160 HB for titanium containing 0.022 wt.-% of carbon (melting U-0). Decreasing of hardness in the top of ingots may be deal with crystallization conditions in period of shrinkage elimination.

It should be noted that the different morphology of the structures of the metal melts U-4 and U-6 correlate with a difference in hardness, with approximately the same amount of carbon.

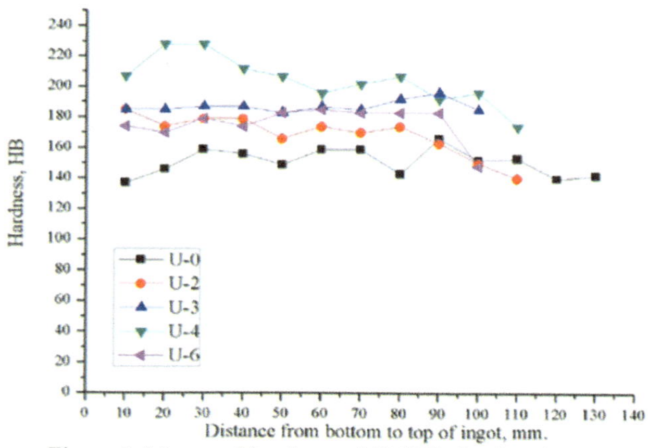

Figure 8: Measured hardness of ChESR titanium ingots

The obtained preliminary results of investigation of titanium structures alloyed with carbon and its hardness indicate the influence of carbon nanotubes on the processes of structure forming.

Conclusions

The chamber electroslag remelting as metallurgical process illustrates an efficient way of alloying titanium with carbon in the investigated range of 0.022 to 0.34 wt.-% at the application of ligatures as carbon nanotubes or carbon powder of micronized.

The preliminary results of the investigation of the structure and hardness measurements showed that ChESR provides good chemical and structural homogeneity of titanium ingots, alloyed by carbon. In this case, there an increase of strength characteristics and changes in the structure of titanium is found.

References

1. Ilyin A.A., Kolachev B.A., and Polkin I.S., *Titanium alloys: The composition, structure and properties* (Moscow: VILS-MATI, 2009).
2. Leyens Christoph, *Titanium and Titanium Alloys. Fundamentals and Applications* (Weinheim: Manfred Peters and Wiley-VCH, 2003).
3. Disegi J. A. "Titanium alloys for fracture fixation implants," *Injury*, 31 (2000), S-D14-17.
4. Hanawa Takao. "Recent Development of New Alloys for Biomedical Use," *Materials Science Forum*, 512 (2006), 243-248.
5. Kolobov Yu. R. "Nanotechnologies for the Formation of Medical Implants Based on Titanium Alloys with Bioactive Coatings," *Nanotechnologies in Russia*, 4 (2009), 758–775.
6. Oshida Yoshiki. *Bioscience and bioengineering of titanium materials* (Amsterdam, Boston, London: Elsevier, 2007).
7. Ryabtsev A. D., Troyanskyy A. A. "Electroslag remelting of metals and alloys under fluxes with active additions in furnaces of chamber type (ChESR)," *International Symposium on Liquid Metal Processing and Casting «LMPC 2005»* (Santa Fe, USA, 18-21 September, 2005).
8. Ryabtsev A. D., Troyanskyy A. A. "The electroslag remelting of metals and alloys under fluxes with active additive in the furnaces of chamber type," *Electrometallurgy*, 4 (2005), 25-32.
9. Ryabtsev A. D., Troyanskyy A. A., Davidov S. I. "The refining of titanium from oxygen and nitrogen during electroslag remelting" *Modern electrometallurgy*, 4 (2009), 1-3.
10. Troyanskyy A. A., Ryabtsev A. D. "About works of Donetsk national technical university on electroslag melting and refining of titanium," *Titanium*, 1 (2007), 28-31.
11. Benz M. G. et al., "ESR as a Fast Technique to Dissolve Nitrogen-rich Inclusions in Titanium," *Materials Research Innovations*, 6 (1999), 364-368.
12. Reitz J., Friedrich B., Stoephasius J. C. "Fundamentals of deoxidation behaviour of Ti-alloys by chamber ESR with Ca-reactive slags," *Proceedings of the European Metallurgical Conference EMC 2007* (Düsseldorf, Germany, 11-14 June, 2007).
13. Friedrich B., Reitz J., Stoephasius J. C. "PESR processing of TiAl-electrodes made by aluminothermic reduction," *International Symposium on Liquid Metal Processing and Casting «LMPC 2009»* (Santa Fe, USA, 20-23 September, 2009), 295-301.
14. Panotsky D. A., Boreslavsky A. L. "Adjustment of carbon alloys for titanium alloy carbon" *Intern. Conf. "Ti-2005 in CIS"* (Kiev, Ukraine, 22-25 May, 2005).
15. Kornilov I. I., *Titanium: Sources, properties, and application chemistry of metals* (Moscow: Nauka, 1975).
16. Kolachev B.A., Elagin V. I., Livanov V. A., *Metallurgy and heat treatment of non-ferrous metals and alloys* (Moscow: MISIS, 1999).

Proceedings of the 2013 International Symposium on Liquid Metal Processing & Casting
Edited by: Matthew J. M. Krane, Alain Jardy, Rodney L. Williamson, and Joseph J. Beaman
TMS (The Minerals, Metals & Materials Society), 2013

EVOLUTION OF ESR TECHNOLOGY AND EQUIPMENT FOR LONG HOLLOW INGOTS MANUFACTURE

Lev Medovar[1], Ganna Stovpchenko[1], Grigory Dudka[2], Alexander Kozminskiy[2], Borys Fedorovskii[3], Vitalii Lebid[1], Iaroslav Gusiev[1]

[1] E.O. Paton Electric Welding Institute of National academy of Science of Ukraine, 11 Bozhenko str.,Kyiv, 03150, Ukraine
[2] JSC Energomash Belgorod - BZEM, 111 Hmelnitskogo ave, Belgorod, 308002, Russia
[3] Elmet-Roll; P.O. box 259, Kyiv-150, 03150, Ukraine

Keywords: ESR, Hollow ingot, Tubular billet, Pipe, Continuous casting

Abstract

In this paper development of both ESR technology and equipment for hollow ingot manufacture review and analysis are presented. The real complications of hollow ingot manufacture and some tendentious issues which restrict process dissemination are discussed. An actual data of modern manufacture of as-cast pipes for heat and power engineering by traditional ESR with consumable electrode are given. Results of microstructure and nonmetal inclusion investigations have shown the high quality of as-cast ESR pipes. On the basis of these results the possibility to produce huge ESR hollows (up 5000 mm in dia) with final goal drastically to reduce setting ratio on forged shells and rings or even replace it by ESR hollows as-cast is grounded. Two new ESR technologies - consumable electrodes change and liquid metal usage - have passed pilot tests for heavy hollow production and shown very prospective results to be presented.

Introduction

On LMPC-2011 and some other recent conferences we discussed the latest improvements of the ESR technology for hollow ingots [1-4]. For more than 40 years of its application a big variety of hollow ingots (round and square billets with hole of a various form, thick-walled pipes, cylinders and various shells with thickness of walls from 20 mm, with a diameter up to 1600 mm and length of up to 5-6 m and weight up to 20 t) from various steel and alloys [5,6] were mastered.

Advantages of ESR hollow ingot/ billets are quite proved. The list of the advantages follows:

- the number of technological operations at forging and rolling (sadden, tag, upset and punch are excluded) decrease and so do the expenses for step operational heating;
- high metallurgical quality and metal density, stable chemical structure (the minimum liquation) are guaranteed due to the minimum volume and depth of liquid metal bath in comparison with a solid ESR ingot of the same outer diameter and even in comparison with the hollow ingot cast in a mold;
- the high yield of metal is provided;
- the formation of fast crystallization structures (thanks to two fronts of crystal growth in the opposite directions - from internal and external mold) guarantees excellent properties of metal as-cast.

However, despite obvious benefits the hollow ingots processing still has not become the main process for manufacture of products with holes. What is more, the solid billet has not been replaced by hollow one for pipe production yet.

There is a common-sense question. What are the reasons? The opponents of hollow ingots application repeatedly pointed out the following circumstances:

- the technology of hollow ingots casting is more complicated than technology of solid ingots manufacture. It requires more exact observance of melting parameters and allows only very insignificant variations;
- while solid ingot forging the worse quality central part of an ingot is removed, and while hollow ingot forging the macrosegregation remains in a body of a product;
- the quality of hollow products forged from hollow ingots is worse than the same made of solid ingots because the deformation ration is low;
- the yield increase at hollow ingots manufacture doesn't compensate the additional costs on their manufacture.

More detailed consideration of ESR of hollow ingots will allow us to disprove these objections.

Evolution of Traditional ESR for Hollow Ingots: Background, Opportunities and Restrictions

Pipes manufacture, especially heavy shells, cylinders and vessels, is one of the most technologically difficult processes in metallurgy and machinery. In long-term disputes over expediency of solid ingot replacement by hollow one, as at the brisk intersection, the very inconsistent opinions encounter. They are based both on unbiased and subjective estimations of complicity, deliveries of technology and achievable level of quality. The classical example is old disagreement between supporters of the cast metal and defenders of the deformed metal. Certainly, in many cases rolled and forged metal still surpasses the cast one in quality. At the same time, the near-net-shape casting is more expedient and prospective than casting of a big section ingot and forging it to final dimensions. The continuous casting process of a strip of 1-5 mm thick is a very bright example of the successful solution of this dilemma. Actually the slab casting, hot-rolled strip itself and the equipment for their production will consign to the past while cold rolled sheets will be made from a cast strip already. In our opinion, the technology development for many hollow products will go according to the similar scenario: solid ingots will be more and more displaced by hollow ones, which approximate to final size of ready product.
Let's consider now one by one the restriction that constrains dissemination of ESR hollow ingots manufacture.

Standard Technologies of ESR Hollow Ingots

Nowadays there are two types of ESR process: remelting of consumable electrodes and direct pouring of liquid metal. Both types can be realized in two options of inner (forming a hole) mold location relative to outer (forming an ingot) mold. In one case, the internal mold fastens on some kind of long mast, and in the other – on upper flange of an outer mold.

It is obvious that in order to receive long ingots the inner mold needs to be hung on an outer mold from above and should not prevent withdrawing of ingot down (theoretically infinitely).
The option with inner mold fastened from below on long mast worked in the industry for ten years, however, allowed small length ingots receiving. Today (according to our information) this method is not applied.

The multi electrode scheme (remelting of several consumable electrodes in a short collar mold and an ingot withdrawing) with the top arrangement of inner mold is being used in the industry. At the present day only in Russia and in Ukraine not less than 10 ESR furnaces make thousands of tons of hollow ingots annually. In this regard the question of complexity of technology, in our opinion, is a purely rhetorical one. Certainly, there are some difficulties in operation of the furnaces applying traditional process to melting of hollow ingots with application of consumable electrodes. However, they can be solved.

Let us take as an example production of "JSC Energomash (Belgorod) — BZEM" which is the traditional supplier of pipelines for thermal and nuclear stations of Russia and abroad and where the manufacture of cast ESR pipes is mastered instead of the forged ones. The multi electrode bifilar diagram with a hinged inner mold is used. Outer diameters of produced ESR pipes are 273 - 920 mm, wall thicknesses is from 22 to 100 mm.

And now we will focus not on complexity of ESR itself but on the difficulties directly connected with cast metal quality. In case of incorrectly chosen melting modes there can be superficial or internal defects with necessity of additional machining or deformation, respectively. Therefore, with an insufficient heat input the ripples arise on a surface; and, with an increased productivity of melting and high speed of withdrawing, the hot topping of solidified metal is interrupted which at thin section leads to porosity formation (it is the second argument against hollow ingots application and it will be discussed further).

Features of Solidification Process and Structure of ESR Hollow Ingots

Formation of structure of any ingot of ESR is defined by relationship of two opposite, in fact, processes, which go at remelting in parallel and at the same time. First is an inflow of liquid metal (with a certain speed and temperature) in a mold, and the second one is heat removal by its water cooled copper walls. The basics for understanding structure formation in ESR ingot is the fact that there are three main periods of process. The initial stage is one when the speed of solidification is higher than the speed of metal inflow. The liquid metal pool is minimum, all cooling surface (mold(s) walls and bottom plate) work effectively. At such situation dendrites grow mainly perpendicular to the bottom plate surface as it is promoted by gravitation.

If distance from the bottom plate increases its cooling action weakens and the vertical wall of a mold becomes the main surface of cooling. At the initial moment the speed of solidification is higher than the melting speed. That causes existence of slag films in near-bottom part of an ingot. After formation of some layer of solid metal on the bottom plate the speed of melting is increased and at a stationary stage of ESR the melting speed is equal to solidification speed with some correction for the size of liquid metal pool. However, if in solid ingots such equality of speeds can be provided in quite wide interval of melting speed values, the increase in withdrawing speed over optimum while hollow ingot melting will lead to drastically deepening of a liquid pool (as it is observed at continuous casting) and, the central porosity of a hollow ingot is inevitably formed.

It is known that the phenomenon of metal volume reduction at solidification in ingots and billets leads to shrinkage formation which can appear in the form of the concentrated cavity or the distributed central porosity. And, the higher temperature of the metal, the more the shrinkage as liquid metal reduces volume not only at transition to solid, but also while cooling in a liquid. In spite the fact that metal temperature at ESR is very high formation of the concentrated shrinkage cavity is almost excluded thanks to effective hot topping at the end of melting. Porosity can be observed in ESR when a big liquid metal pool and wide mushy zone are formed because of the increased melting productivity and high withdrawing speed. Thus, the solidification of the last portions of melt takes place in interdendritic space, when there is no inflow of new portions of melt to a zone of crystallization fronts connection.

For ESR hollows the strong development of the long axis columnar dendrites oriented at an angle to both ingot walls in the direction to heat inflow (a contour of a liquid metal pool) is inherent. These dendrites meet near ingot axis or on it in case of thin sections and big heat input.

Such limit case of transcrystallization can be observed at an excessive both heat inflow in a liquid metal pool and speed of withdrawing of thin-walled hollow ingots (Figure 1).

Figure 1. Transcrystallization in structure of longitudinal section of hollow ingot wall (thickness 50 mm) of HSLA steel. Etched in boiling HCl.

Behind the zone of frozen crystals near an ingot side the thick skeleton arms of dendrites (first order arms) start to grow transversely to liquid pool profile. The angle of dendrites to horizon from left side of sample (from main mold) is 3-5 degrees less than from the inner one; the line of dendrites connection displaces to the right. This part has more etched parts (usually caused by dendrite liquation). This testifies some less rate of solidification from inner mold than from the outer one. Therefore, that structure is escorted by marked dendrite liquation and porosity.

Minor segregation and porosity can be partially removed by a deformation. But in our case, when pipes are used as-cast, the melting process is operated by strict regimes, which provide small deep of liquid metal pool, and, consequently, the absence of transcrystallization into ingot wall.

Microsegregation in as-cast pipes can to big extent be eliminated by heat treatment. Heat treatment modes for as-cast ESR metal should differ from required for forged ones made of the same steel grade because ESR ingot has bigger ancestral austenite grain. As diffusion processes accelerate with temperature growth, the heat treatment of ESR metal is usually made at higher temperature in order to provide rational duration of treatment from practical and efficiency point of view. However, it should be noted that too high temperature can lead to grain growth. Yet, an insufficient one results in incomplete recrystallization as shown on the Figure 2.

Figure 2 – Cross section of ESR pipe which demonstrates effect of improper homogenization on macrostructure.

Cast pipes (hollow ingots) melted in optimal mode have high quality. Metal macrostructure is dense and dendrite with narrow zone of equiaxed crystals in central part (Figure3).

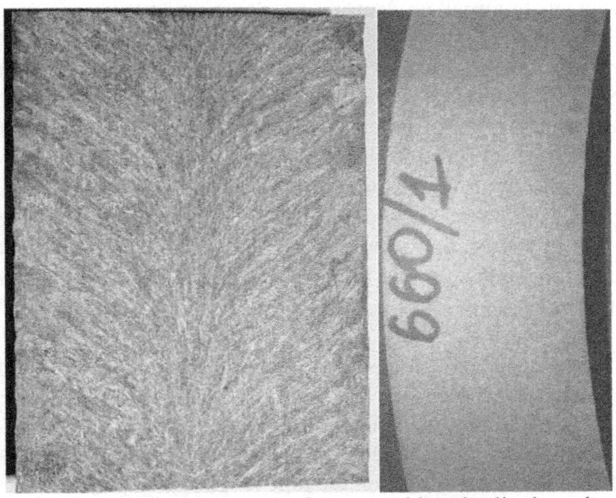

Figure 3. Macrostructure of cross and longitudinal sections of good quality ESR hollows.

The zone of crystallization connection is usually displaced on 0.1-0.15 of wall thickness to the hole side. Segregation defects are absent and nonmetal inclusions are dispersed and have uniform distribution.

The ESR ingot chemistry is practically on the same level as in consumable electrodes: both manganese and silicon losses are insufficient: sulphur content has been reduced. Macro- and micro segregation are not revealed.

The metal of ESR cast pipes is quite pure by nonmetal inclusion. There is a small number of round shape single small sulfides (up to 5 μm) and oxides, as well as complex oxysulfides (figure 4). The irregular shape inclusions are complex and consist mainly of several phases. Central part of such inclusion consists of high melting points, and the outer one has low melting point matters (MnS or glasses). The nature of observed in nonmetal inclusion testifies their formation in a liquid steel.

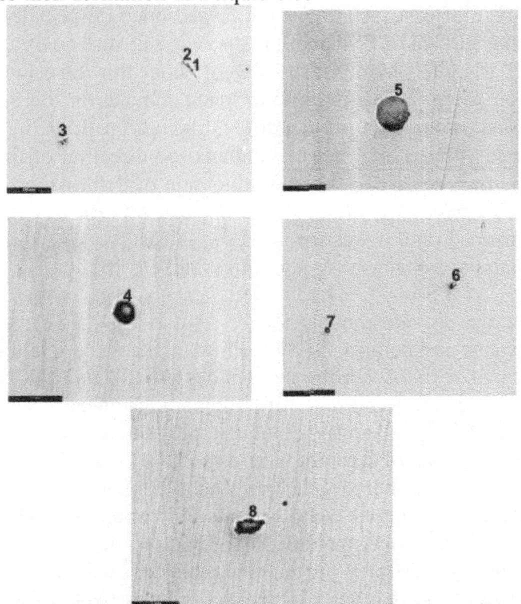

#	Content of elements, % weight						Max size,
	Al	Si	Mn	S	Ti	Ca	μm
1	87.6	6.7	-	1.3	0.6	3.756	29
2	53.4	3.5	9.1	4.3	5.3	24.5	
3	95.8	-	-	4.2	-	-	11
4	52.8	1.3	6.5	0.6	-	38.7	5
5	59.2	-	0.3	1.6	1.6	37.4	20
6	46.2	0.2	5.1	4.8	0.7	42.9	7
7	87.2	1.6	8.1	1.7	1.5	-	5
8	87.8	8.3	3.1	-	-	0.7	8

Figure 4. Typical nonmetal inclusion in cast ESR pipe of 650/554mm of 15 CrMoV steel.

Manganese sulfides are small as the sulfur content in metal is insignificant. Oxide and aluminum-calcium-silicon oxysulfides containing inclusions are the products of diffusion deoxidation of steel in slag. Their size is small, and their form and morphology are favorable. The composition of oxide nonmetal inclusions depends on activity ratio of aluminum and silicon (both present in metal and in slag) and calcium (present in a slag mainly in the form of oxide and fluoride). Existence of inclusions of such size in structure of steel has become admissible due to the fact that the probability of their action as fracture concentrators is almost insufficient.

The amount of oxide and oxysulfide nonmetallic inclusions increases with the increase of aluminum content, which is responsible for oxide part of nonmetal inclusions. There are almost pure alumina oxides which have the irregular shape and seldom quite big size (50 microns and more) that does them dangerous concentrators of fracture. Considering that alumina nonmetal inclusions are inclined to agglomeration deoxidation of steel is made by modes preventing their formation.

Application and Properties of As-Cast Metal of ESR Hollow Ingots

The most impressive result that was received more than 40 years ago by several independent researchers is that the as-cast ESR metal has mechanical properties up to the mark and even better than deformed metal does.

Advantage of cast metal over the forged one is especially bright when the surface of ESR ingot is the working surface of a tool or part. It is quite obvious that very often in this case there is a situation when the main working loads coincide in the direction with first order axis of dendrite structure in ESR metal. In deformed (forged and rolled) products the direction of the main loads is transversely to the main direction of deformation texture of metal. Under such circumstances the cast tool works better than the deformed one. It was recorded for as-cast working rolls of Z-mill, stamps, containers for pipes pressing, etc. [6].

Further we provide more evidence. Authors [7] give results of both as-cast and deformed ESR hollow ingots metal quality at the end of two years of various steel grades production (ball bearing, HSS and HSLA). Researches of hollow ingots metal from various types of HSS steels showed similar results – thinning of a carbide net and reduction of carbides sizes as well as further improvement of structure after homogenization and deformation. It is shown that the hollow ingots metal considerably surpasses solid ingot metal of both usual melting and ESR by degree of carbide heterogeneity, purity by nonmetallic inclusions and service life in ready tools. The durability of the hob cutters made of a hollow ingot without deformation was not worse than forged.

The paper of authors [8] is devoted to research of quality of electroslag hollow ingots of ball bearing (Russian grade ShH15) received by multielectrode scheme. It was proved that as-cast ESR metal of a hollow ingot has higher density in comparison with the deformed metal melted in ESF.

The comparison of mechanical properties of the ESR as cast metal (die steel) with properties of metal of open melting (OM) with deformation ratio 5.3 and 27. 5 (square billet of 80 mm and round billet of 30 mm in dia) was made by authors [9]. The performance of hot pressing matrixes made from ESR hollows with wall thickness of 60-80 mm surpass the same for matrix made from rolled billet (round shape of 180-200 mm in dia) more than twice. Hardness of ESR metal after annealing and OM metal for rounds of 30 mm made 46 and 48 HRC, and for square of 180 mm - 44 and 46 HRC, consequently. It was found that by strength characteristics as-cast steel 3Cr3Mo3V and H13 were on the same level, and steel 4Cr5MoVSi was a little bit lower than strength of OM metal in rounds of 30 mm. Mechanical properties of ESR steel at high temperatures were also higher. That allowed authors to conclude that as-cast ESR die steel in sections up to 150 mm has higher properties than OM with deformation ratio 5.3 [9].
The manufacturing technology and quality of hollow ingots of 508/254 mm from heat resistant alloys (Haynes 718) and the low-alloyed AISI 4330 steel were investigated. It is shown that the ingot of Haynes 718 alloy is free from a liquation through whole length, and the content of both titan and aluminum is controllable. The same is achieved for ESR hollow ingots from AISI 4330Mod steel. Excellent properties of hollow ingots as-cast were received. As an example: impact strength KCV was 51.5 Nm at yield strength of 28.5 Nm. Such properties make it possible to use as-cast hollows instead forged materials, which are used now [10].
The research of structure and properties of hollows for cast-welded vessels from the austenitic Cr-Mo steel 12Cr18Ni10 for cryogenic temperatures and high pressures [11, 12] showed that metal structure is dispersed and consists mainly of austenite. The quantity of ferrite phase is insignificant (from 2.5 to 7%). There is no sulfur segregation. The very small quantity (permissible by standard specification) of carbonitride inclusions is revealed. Properties of electroslag hollow ingots metal of steel tempering were similar to properties of forged metal of the same grade.

The hollow billets of the super low carbon steel grade 03Cr20Ni16Mn6 strengthened by nitrogen were applied to production of cast-welded cryogenic vessels. Weights of hollow ingots were 0.15 t (stationary mold 320x140mm) and 6 t (T-shaped short collar mold 700x345 mm). It is shown that the steel chemistry after remelting practically didn't differ from consumable electrodes. The oxygen content in metal was reduced. Degree of anisotropy of mechanical properties of ESR hollow ingots was 1.073…1.081 against 1.4-1.8 for forged metal of open melting. It is shown that mechanical properties of steel 03Cr20Ni16Mn6 are as good as the same of the deformed metal by the main aspects defining its workability.

Authors [13] investigated a pilot batch of hollow electroslag billets (length 1600 mm, outer diameter 348 mm, wall thickness 50 mm) from steel 38Cr2Ni2Mo subjected to heat treatment (quenching and annealing). Structural and chemical heterogeneity, mechanical properties and purity of electroslag metal by nonmetallic inclusions were under study. It is shown that ESR metal differs by uniform quality of a chemical composition as well as by high density through the whole length. Nonmetallic inclusions are generally presented by oxides (1.5…2.5 marks) and nitrides (0.5 marks). Quality of as-cast electroslag metal met requirements to forged products.

The results received earlier were taken as a reference point for mastering technology of as-cast pipes ESR production used today. As shown above, the structure of pipes received by regular melting regimes is dense, without the revealed segregation.
On the basis of an extensive complex of studies of structure, quality and mechanical properties of as-cast electroslag pipes the Technical Specification was developed and industrial ESR method production of pipes for pipelines of thermal and nuclear stations has begun.

Pipes produced nowadays completely meet quite rigid requirements of geometry:

- maximum deviations on outer diameter (+3.5 mm; – 1.3 mm), maximum deviations on wall thickness – (+10 mm; - 2mm), maximum deviations on the pipe length (±20 mm);
- the most admissible curvature of pipes on any part of 1 m long: with a wall of 20 mm – 2.0 mm, more than 30 mm – 4 mm: the same on the whole length – 15.0 mm.

ESR pipes are supplied after heat treatment and surface machining. Each pipe is exposed to ultrasonic control before final heat treatment or after fair machining.

Mechanical properties have to be not less than specified in table 1. However, in practice indicators of impact strength are significantly (by 2-3 times) higher. This fact confirms high quality of ESR metal once again.

Table 1. Technical specification requirement to mechanical properties of cast ESR pipes (after heat treatment)

Steel grade	Tensile strength, MPA	Yield strength, MPA	Elongation, %	Reduction of area, %	KCU, J/cm2	KCV, J/cm2
20	410-550	215	22	40		
15 MnSi	490-610	295	16	40		
16 Mn Si	490-635	275	18	40	59	54
15CrMoV	490-635	315	18	50		
10Cr9MoVNb	>600	400	17	50		

Durability tests of cast pipes showed high rates of both yield strength and strength rupture at increased temperatures (up to the 500°C for steels 20, 15MnSi and 16MnSi and up to 600°C for steels 15NiMoV and 10Ni9MoVB). The mechanical properties defining working capacity of ESR pipes hold out for the requirements to forged pipes from the same steel grade. Thus, that allows their application in pipelines.

Accumulated experience of ESR of hollow ingot production for cast pipes for heat and nuclear power plants shows that technology today is steady and well proven, allowing production of competitive products of high quality. And today, on the basis of the analysis of long-term successful experience we can claim that ESR provides the best quality of hollow ingots.

Emergence of new modifications of ESR process, which will be discussed below, allows us to increase flexibility and technological effectiveness of process, to lower expenses and, even to set the goal of development of continuous casting of hollow billets.

New ESR Technologies Potentiality

In our opinion, using the standard ESR with consumable electrodes without change within melts is good for relatively small ingots up to 20-30 tons weight only. In that case perfect quality can be achieved at simple standard equipment. On the other hand, for long (tall) and heavy hollows usage of standard ESR technology leads to the dead end due to the unacceptable increase of the furnace height, complicity of it design, high cost of the equipment and day by day operation. Moreover, the cost of the consumable electrodes has become very high too. List of the problems could be continued. Although, consumable electrodes are the leading problem, or even barrier on the way of steady operation.

From our point of view, further development of standard hollow ESR technology could be achieved due to consumable electrode change. A separate article about this method will be presented at our conference [14]. So we just will mention its main characteristics. In general this method looks like standard ESR for hollow ingots. The real difference is that so known double circuit ESR, ESR TC is realized for hollows instead of standard ESR [15]. Principles of that method are given in Figure 5.

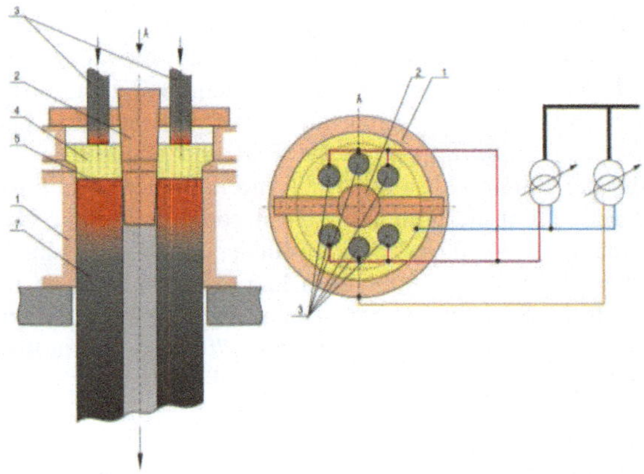

Figure.5. Concept diagram of hollow ingots ESC TC: 1- Current Supplying Mold (CSM®), 2 - Inner Mold, 3- Consumable electrode, 4-Molten slag, 5-Molten Metal, 7-Ingot.

Only one of the various possible arrangements of high current loop is shown in the fig.5. This diagram is similar to the ESR TC of the solid ingot. The main point is that such diagram makes the power input and speed of ingot withdrawing almost independent. It is very important for practical use as it is possible to maintain constant power input, to stop or reduce drastically the speed of the relative movement of the hollow ingot along the inner mold, and to avoid the inner mold bite by solidifying ingot.

Several heats show that it is enough to keep such relative movement between inner mold and ingot on the level of just 1mm per min. It means that enough time could be provided for electrode change. For example, 10 min for electrode change with such speed of movement will lead to the slag and liquid metal level change for just 10 mm and such figure is not dangerous for hollow ingot formation.

In addition the described problem of the wall center segregation, porosity, etc will disappear as it is easy to combine necessary level of the power inflow and ingot growth speed is suitable for sound structure formation.

This variant of technology has been realized recently in new 25 t ESR furnace for huge hollow ingot production with length about 6000mm*. Photo of such ingot presented in the fig.6. In the area of electrode change that takes just 2 minutes it was no ripples or other marks on both the inner and outer surfaces. Current supplying molds (CSM®) for 650 mm outer diameter hollows are used. Such T-shape mold has 200mm stairs between mold work diameter and mold upper part. Such design opens the possibility to replace rolled electrode by casted ones.

Figure 6. ESC TC hollow ingot produced with electrode change

In our opinion, the best perspective of huge hollows ESR technology is the one with direct processing of liquid metal, so called ESR LM. On one hand, such method is a good alternative to the ESR with consumable electrodes. On the other hand, it combines positive futures of the ESR and continuous casting. Method of ESR LM has practically no limitation in hollow wall thickness. From the minimum side it could be 25mm, from the maximum side – 500mm or even bigger. Important point is that small wall thickness could be achieved for big radius hollow ingot. It is the way for cast shells manufacturing. Maximum thickness could be bigger, for example, 700 mm. But in that case central segregation could not be prevented.

After several laboratory tests ESR LM for hollow ingots was tested in the industrial conditions [1, 2, 4]. Industrial trials were performed using furnace for surfacing of rolling mill roll by ESS LM (electroslag surfacing by liquid metal) owned by Ukrainian company NKMZ. ESR LM process sketch and appearances of the hollow ingot diam. 770mm x 70mm are presented in the Fig.7

* This project was realized together with Prof . Z. Jiang and his team, North Eastern University, Shenyang, China, see ref.[14]

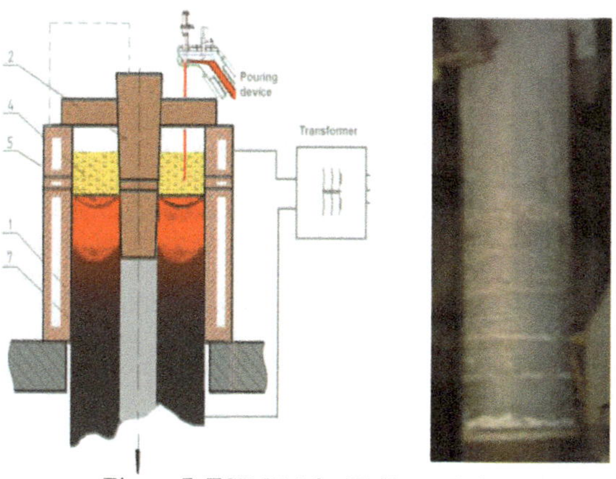

Figure 7. ESR LM for Hollows (left) and ESR LM Hollow ingot (right)

ESR LM process for hollows guarantees at least two main advantages:

- low overheat over liquidus temperature allows to receive very dense metal of an ingot with the minimum liquation on height and section thanks to the accelerated crystallization (between two copper water-cooled molds) as well as because of both permanent and slow supply of liquid metal (from the pouring-holding device) that provides excellent properties of metal even in as-cast condition;
- an increase of economical effectiveness ESR due to elimination of the consumable electrode cost starting from their casting till consumable electrode clamping on the ESR furnace

In addition, as it was mentioned, ESR LM of the hollows is the only way to realize old idea to combine advantages of ESR and continuous casting and realize real semi concast and, finally, concast of the hollow tubular billet [16]

Finally, it's necessary to answer the last question about economic efficiency. Let us use information from two companies that use hollow ingots, including ESR hollows. Well know company AREVA-Creosout Forge using hollows instead of solid ingots within last 30 years for manufacturing forged vessels for various applications [17]. It is known that in the heavy-, power- and petrochemical machine building industry about 50% of heavy forging has hollow shape. It is clear that higher weight of forging leads to the higher coefficient of ingot metal usage in case of hollow ingots. According to the various estimations it could be even 5 times higher in comparison with such figure for forgings made from solid ingot [1-4].

The second example connected with the mentioned above plant of the Russian company JSC Energomash Belgorod - BZEM Manufacturing of the ESR thick wall cast tubes replaced the forging tubes. Even with usage of purchased rolled electrodes in the given company the as-cast ESR tubes cost is 15% less.

Conclusion

Hollow ingot solidification process leads to less intensive segregation in comparison with the solid ingot of the similar weight and height due to the minimum volume of the liquid metal bath. Favorable structure and high level of the mechanical properties give the possibility to use as-cast thick wall pipes for nuclear and fossil power station and to replace forged tubes. Utilizing of the CSM® and ESR TC offers the hollow ingot casting using electrode change and makes possible to use short electrodes and to reduce overall furnace height.

Hollow ingots manufacturing makes it possible to reach three goals:

increasing ESR cast metal quality and more wide usage of the cast ESR metal instead of forged or rolled ones;

reducing the metal scrapping and decreasing of the complicity of the hollow production;

creation of the semi-concast and concast technology of the hollow tubular billet production.

References

1. L. Medovar et al. "ESR of the Heavy Hollow Ingots" (Paper presented at the 1st International Conference on Casting, Rolling and Forging, Aachen, German, June 2012), CD.
2. B. Fedorovskii et al. "ESR of Hollow Ingots. New Approaches to a Traditional Problem"(Paper presented at the LMPC, Nancy, France, September 2011), CD.
3. G. Stovpchenko e.a. "Features of slag-metal interaction at electroslag remelting of heavy ingot" (Paper presented at the International Conference on Molten Slags, Fluxes and Salts (MOLTEN 12), Beijing, May, 2012), CD.
4. L. Medovar et al. "ESR of heavy hollow ingots: New approaches to a traditional problem" (Paper presented at the 18 International Forgemasters Meeting, Pittsburgh, USA, 2001, USA, 183-188.
5. B.I. Medovar et al. *Electroslag Metal*, Naukova dumka, Kiev, 1981, 677.
6. B. Paton, B. Medovar and G. Boiko, *Electroslag Casting,* Kiev, Naukova dumka, 1982, 282.
7. A.Petrov et al., "Quality of the ESR Hollow Ingot Metal Investigations as Cast and After Deformation", Special electrometallurgy,1973, 20,18-24.
8. A. Petrov et al., "Quality of the Ball Bearing Steel ESR hollow ingot", Ibid, 1979, 40, 40-42.
9. A. Koval et al., "ESR Die Steel Properties and Workability", Problem of the Special Electrometallurgy, 16, 1982, 23-26.
10. G. Klein et al. "Electroslag Casting of Hollow" " (Paper presented at the ICVM-6, 1977).

11.B. Paton et al. "Electroslag melting of hollow ingots from Cr-No-Mo-V steel", *Special Electrometallurgy (Kiev: Naukova dumka, 1987), 55,3-10.*
12.A.Chepurnoi at al. "Research of quality of cast electroslag billets from steel 12Cr18Ni10Ti for high pressure vessels", *Special Electrometallurgy (Kiev: Naukova dumka, 1980),12,65-70).*
13. L. Gluskin et al., "High Pressure Vessels Manufacturing from Two Phase Steel", Problem of the Special Electrometallurgy, 3, 1988, 15-18.
14. Xu Chen et al., "Hot Test and Simulation of ESR Hollow Ingots Formation in Current Supplying Mold with Electrodes Change", " (Paper presented at the LMPC, Austin, USA, September 2013)
15. L. Medovar, et al. ESR with two power sources and process control" (Paper presented at the LMPC,Santa Fe, USA,2005), CD.
15. L.B. Medovar, Z.Jiang, A.P. Stovpchenko, "Continuous Casting of Hollow Tubular Billets," " (Paper presented at the ICS, Gifu, Japan, October 2008), 117-120.
17. G. Girardin, Hollow Ingots: Thirty Years of Use to Control Segregation and Quality for Nuclear and Petrochemical Large Shells" " (Paper presented at the IFM-18, Pittsburgh, PA, USA, September 2011), 170-174.

Proceedings of the 2013 International Symposium on Liquid Metal Processing & Casting
Edited by: Matthew J. M. Krane, Alain Jardy, Rodney L. Williamson, and Joseph J. Beaman
TMS (The Minerals, Metals & Materials Society), 2013

VARIATION OF THE RESISTANCE DURING THE ELECTRODE MOVEMENT IN THE ELECTROSLAG REMELTING PROCESS

A. Kharicha [1,2], M. Wu [1,2], A. Ludwig [2]
[1] Christian Doppler Laboratory for Advanced Process Simulation of Solidification and Melting, Univ. of Leoben, Austria
[2] Chair of Simulation and Modeling of Metallurgical Processes, Univ. of Leoben, Austria

Keywords: Electroslag Remelting (ESR), Magnetohydrodynamics (MHD), Control, VOF, Slag Skin

Abstract

In the present paper a transient 2D Magnetohydrodynamic model is used to explore the influence of the interface movements on the resistance swing in the static mould version of the Electro-Slag-Remelting (ESR) process. The model couple efficiently the electric current distribution with the movement of the electrode, the slag/gas, and the slag/metal pool interfaces. The solid slag skin is computed with a help of a 1D model that accounts for the mould current. Two electrode configurations are explored, a shallow and a 50 mm immersion depths.

Introduction

The Electro-Slag-Remelting (ESR) is an advanced technology for the production of components of e.g. high quality steels. To produce a high quality homogeneous ingot with good surface quality, the deviations in the process, such as melting rate or the immersion depth of the electrode need to be minimized. Optimisation of the process efficiency and surface quality are directly linked with the electrode immersion depth. The best results are believed to be obtained with constant immersion depth and as shallow as possible. Nevertheless, shallower is the immersion depth, the more difficult is to control the electric parameters (power, voltage). The extreme variations in voltage observed during such shallow immersion are attributed to the formation of air gap under the electrode, that can possibly lead to arcing and to deleterious oxidizing reactions. In opposite a too deep immersion depth is known to create poor surface and metallurgical quality in the final ingot.

No system currently exists to measure the depth directly, so it must be inferred from measured parameters of the process. The variation of the resistance, known as resistance swing, is the mostly used method for the control of the electrode position [1]. However, the increase in resistance swing can be reliably, but not quantitatively, related with the immersion depth. In addition, it is known that the control currently used in industry has experienced unexpected and unexplained difficulties, resulting in imperfections in the ingot being produced. This is why some efforts must be applied to the identification of process state, solely through analysis of electric process parameters.

The aim of the present work is to compute numerically the time evolution of the resistance. To achieve this goal it is important to identify the phenomena that can generate these electric fluctuations. Assuming that most of the resistance is generated within the slag cap, our analysis will focus on the electric properties of this region. The slag region experiences strong flow turbulence that can induce locally strong temperature fluctuations. The electric conductivity of a typical slag is not constant, instead it increases with temperature. If the temperature at a point located within the slag fluctuates, the Joule released at this point fluctuates as well. Recently, it was shown that these fluctuations have a considerable effect on the power generated in the slag [2]. A correction factor must be added to the Joule heating source, especially in area with strong thermal turbulence, i.e. under the electrode and at the vicinity of the mould [2]. Nevertheless, the estimated standard deviation of the total resistance was found to not exceed 3 %. By modifying the chemical composition of the slag, the electrochemical reactions may modify the average electric conductivity.

Large and sharp fluctuations of the resistance can only be generated by modifying the shape of the slag cap. The shape of the electrode tip being melted represents the first boundary. The time scale associated with a shape modification is in the order of minutes (5-20 min), assuming a constant immersion depth it cannot induce fast resistance fluctuations. The solid slag that develops at the mould (referred as slag skin) is a boundary that was considered for a long time as an electric insulator [3,4]. But recent experimental and numerical investigations on static mould ESR have shown that typically 20% (but up to 90 %) of the total current can cross it to enter directly into the mould [2,5]. The ratio mould current over the vertical current depends on the ratio between the electrode-mould radial distance and the slag cap height. A second factor is the ratio between the liquid slag and the slag skin electric conductivities. It is clear that a time fluctuation of the mould current intensity can induce a variation of the global resistance. The slag/ pool interface and the exposed slag surface are boundaries that are susceptible to move. Physically, the development of the heat and mass transfer at these interfaces is important for the final ingot quality, composition and cleanliness. Visual observations of the slag/air surface show a surface strongly affected by the slag eddies. Due to the presence of high temperatures, opacity of the materials, and the presence of the mould it is not possible to directly observe the behaviour the slag/pool interface. Although usually assumed flat, a previous work [6] using a Volume of fluid (VOF) model has shown that the interface between a layer of slag and steel layer in a cylindrical cavity is highly coupled with the distribution of the electric current. A full scale simulation of the ESR process using a VOF model has shown that the shape of the pool interface is likely to be non flat. Depending on whether a “flat” or “free” interface is assumed, an appreciable difference was found in the prediction of the pool shape and depth [7, 8]. This difference was due to a different magnitude and distribution of the Joule heat generated.

The current work presents the results given by a 2D Magnetohydrodynamic model coupled with the phases (steel, slag, gas) distribution. The model allows the movement of the electrode within the slag. The movements of the liquid interfaces are resolved in time and in space. The solidified slag skin thickness is considered with a 1D model which includes the influence of the electric current and the heat fluxes at the contact with the mould. Two different states are explored, one with shallow immersion depth, and a second with a deep penetration depth. In the present study, the influences of the melting, the solidification of the metallic pool and of the falling droplets on the electric current distribution are not taken into account.

Numerical Model

The axisymmetric calculation domain is presented in figure. 1. A rigid electrode is put in contact with a cylindrical container filled with a 10cm height layer of liquid slag and an equal quantity of liquid steel. Most of properties of steel, slag and gas (N_2) are assumed to be constant. The electrode supplies a total 5Hz AC current of I_0=5000 Amperes. The operating conditions as well as the material properties are presented in Table 1.

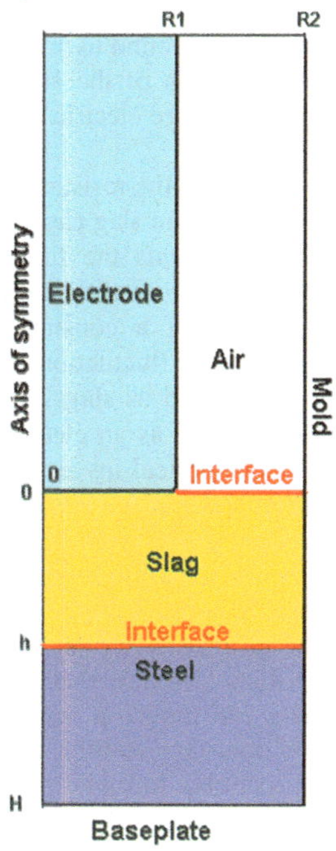

Figure 1. Calculation domain

Interface tracking

The Volume of fluid (VOF) method provides the possibility of tracking immiscible interfaces over a fixed Eulerian mesh. It is designed for two or more immiscible fluids where the position of the interface between the fluids is of interest. In the VOF method the motion of the interface between immiscible liquids of different properties is governed by a phase indicator, the so-called volume fraction f, and an interface tracking method. The volume fraction f_k is equal to 0 outside of liquid k, and equal to 1 inside. The evolution of the interface is calculated using the geometrical reconstruction scheme.

The local values of a physical property Θ (such as density, viscosity, electric conductivity) are interpolated by the following formula:

$$\Theta = \Theta_1 f_1 + \Theta_2 f_2 + \Theta_3 f_3 + \Theta_4 f_4 , \qquad (1)$$

where the subscripts 1-4 indicate the corresponding phase, 1 for slag, 2 for steel, 3 for solid steel (electrode), and 4 for air. An explicit version of the VOF technique was used for the present calculations. The value of the surface tension is fixed to 1 N/m at both air/slag and slag/melt interfaces. Depending on the dynamic of the interfaces, the typical calculation time step lies in the range of 10^{-2}-10^{-4} second.

Fluid flow

The motion of the slag, gas, and liquid steel is computed with the continuity and the Navier-Stokes equations. The no-slip condition is applied at all the walls. The electrode and the top air surface are modeled as velocity inlets with fixed velocities. To conserve the volume, whenever the electrode is pushed at a certain speed downwards or upward ($\vec{u} = \vec{u}_{ele}$), the equivalent speed is applied in the opposite direction at the top air surface $\vec{u} = -\frac{R_1^2}{R_2^2 - R_1^2}\vec{u}_{ele}$. The effect of the turbulence is estimated through an effective viscosity μ_t with the help of the Realizable k-ε. The no-slip condition is applied at all the walls (electrode, mould and baseplate boundary).

Electromagnetics

To solve the electromagnetic field, two techniques can be used, one based on the induced magnetic field $\vec{H}_\theta$, and another based on the electric potential ϕ and magnetic potential $\vec{A}$. Due to its simplicity, the method based induced magnetic field is the most widely used technique [2-8]. With no mould current, a single (or double in AC) equation must be solved with very simple boundary conditions. When the possibility of having currents crossing the solid slag skin directly into the mould, the induction equation needs to be solved with the thin slag layer with enough grid point to correctly resolve the strong decrease in electric conductivity [2,5].

In the present work, the potential method A-ϕ has been used the solid slag layer is modeled and not directly solved [see next part]. The potential at the mould and at the base plate is fixed to 0. The method based on the electric potential allows us to write the mould current in the simple form:

$$j_m = \sigma_W \frac{\phi}{\delta}, \qquad (2)$$

Where σ_w is the average electric conductivity within the slag skin, assumed to be 100 times smaller than the conductivity of the slag at liquidus temperature. At the level of the metal pool no current is allowed to cross the slag skin $(j_m = 0)$.

Since the time resolution of the interface needs the use of very small time step, the choice is made here to resolve the oscillation of the imposed electric current. The boundary condition at the mould consists in equalizing the electric flux to the mould current when the adjacent cell is filled with slag, or to 0 in the other cases (air or metal). Usually the Lorentz force and the Joule heat source are introduced in the corresponding equations only after time averaging [2-8]. In the present approach these sources are oscillating in time around their main values, allowing a full coupling between the hydrodynamic, the electromagnetic and the thermal phenomena.

Heat Equation

The heat balance is of importance here in order to estimate the thickness of the solid slag skin. The melting and solidification of the metal is not considered. The energy equation is solved in the fluid domain.

$$\frac{D(\rho C_p T)}{Dt} = \nabla(k\nabla T) + \frac{j^2}{\sigma} \tag{3}$$

Where C_p , k, σ, are the mixture heat capacity, the heat conductivity, and the electric conductivity. The temperature at the electrode/slag contact surface is fixed at the alloy liquidus temperature. The P1 radiation model is used to compute the radiation in the gas media. This radiation model is necessary in order to correctly estimate the heat radiation fluxes for any shape of the exposed slag surface. At the gas level, the radiative emissivity of the electrode and of the mould is taken equal to 0.8. At the lateral, and at the bottom boundaries, the temperature is fixed at the slag liquidus temperature (1650 K).

Modeling of the solid slag thickness

The solid slag skin is not directly resolved but rather implicitly modelled. The present approach is inspired from concept of wall functions used in turbulence modelling when the mesh is not fine enough to resolve the turbulent and the viscous boundary layers. From the heat balance through the solid slag layer it is possible to extract a simple 1 D equation for the evolution of its thickness $\delta(z,t)$:

$$\underset{(a)}{\rho(C_p\overline{T} - L)\left(\frac{\partial\delta}{\partial t} + u_{melt}\frac{\partial\delta}{\partial z}\right)} = \underset{(b)}{-\rho C_p\delta\left(\frac{\partial\overline{T}}{\partial t} + u_{melt}\frac{\partial\overline{T}}{\partial z}\right)} \underset{(c)}{- Q_{s->\delta}} \underset{(d)}{+ Q_{\delta->m}} \underset{(e)}{+ \int\frac{j_m^2(r,z)}{\sigma(T)}dr}$$

(4)

The terms from left to the right are the contributions of the (a) latent heat released or absorbed, (b) the evolution of the average temperature of the solidified slag layer $\overline{T}$, (c) heat lost from the liquid slag to the solidified slag layer, (d) the heat lost from the solid slag to the cooper mould, and (e) the Joule heat dissipated within the solid slag layer.

A simple linear temperature profile is assumed through the solid slag layer, the volume average temperature within the layer is:

$$\overline{T} = \frac{T_{Liquidus} + T_{mould}}{2} \tag{5}$$

The effect of the elevation of the slag level during the melting is taken into account through the velocity u_{melt}, related to a fixed melt rate (2kg/min). This approach allows also the slag skin thickness to vary with the melt rate.

In the present calculation the Joule heating source is simplified by using the average slag skin electric conductivity:

$$\int_{R_2}^{R_2+\delta} \frac{j_m^2}{\sigma(T)} 2\pi r dr = 2\pi R_2 \frac{j_m^2}{\sigma_w}\delta \tag{6}$$

The heat lost by the liquid slag is:

$$Q_{s->\delta} = -k\frac{\partial T}{\partial r}. \tag{7}$$

an excellent heat contact is assumed between the slag skin and the mould, the temperature at the interface slag skin/mould is assumed constant:

$$Q_{\delta->m} = k_{ss}\frac{T_{Liquidus} - T_{mould}}{\delta} \tag{8}$$

The variation of the slag skin is considered at the level of the slag. When the slag skin reaches the pool region it is simply shifted towards the bottom with a speed related to the melt rate.

In the next iteration the mould current is recomputed from the given slag thickness. Without the presence of mould current, the solid slag thickness is only controlled by the melting rate and the balance of heat fluxes at the mould. By adding a heat source within the slag skin, the presence of mould currents decreases this thickness, which in turn increases the quantity of mould current (Eq.2). This cycle of increases is stopped when the electric current has found the pattern that minimizes the global resistance. This simple approach can model the creation of the slag skin at the level of the exposed slag surface where time oscillations of the interface are expected. After having computed the electric currents pattern, the total electric resistance include the resistance of the liquid slag (volume integral) and of the slag skin (integrated over the slag height):

$$Res(t) = \frac{1}{I_0^2}\left(\int\frac{j^2(r,z,t)}{\sigma(T,t)}dv + \int\frac{j_m^2(z,t)}{\sigma_w}dz\right). \tag{9}$$

The time average resistance is :

$$\overline{Res(t)} = \frac{1}{2T_0}\int_{-T_0}^{T_0} Res(t)dt, \tag{10}$$

Table 1. Parameters used in the simulations.

Steel	
Density ($kg \cdot m^{-3}$)	6800
Viscosity ($Kg \cdot m^{-1} \cdot s^{-1}$)	0.006
Specific heat, liquid ($J \cdot Kg^{-1} \cdot K^{-1}$)	800
Thermal Conductivity, liquid($W \cdot m^{-1} \cdot K^{-1}$)	40
Electric Conductivity, liquid($ohm^{-1} \cdot m^{-1}$)	880000
Slag	
Density, liquid ($kg \cdot m^{-3}$) Solid	2700 3000
Viscosity ($Kg \cdot m^{-1} \cdot s^{-1}$)	0.0025
Specific heat, liquid ($J \cdot Kg^{-1} \cdot K^{-1}$) solid	1200 1000
Thermal Conductivity, liquid($W \cdot m^{-1} \cdot K^{-1}$) solid	10 0.5
Electric Conductivity, liquid($ohm^{-1} \cdot m^{-1}$) , solid	120 1.2
Thermal expansion coefficient (K^{-1})	2.5×10^{-4}
Liquidus temperature (K)	1650
Latent heat of fusion(W/kg)	4.5×10^5
Geometry Mould Thickness (mm) Slag and metal height(mm) Electrode diameter, R_1 (mm) Mould diameter, R_2 (mm)	 30 100 135 200
Process parameter	
T_{mould}	480 K
I_0, f(A, 5Hz)	5000

Application of the model

Shallow immersion depth

In the case when the electrode is just put in contact with the slag surface, the simulation shows a strongly fluctuating slag/gas interface (Figure 2). During the calculations, the electric current intensity was progressively increased. Far before reaching the desired intensity, at about 1400 Amps, a large gas bubble was created between the electrode and the slag media (Figure 2). During the insertion of the gas, the electric current lines are shifted towards the electrode centre, this phenomena is at the origin of the strong increase of the resistance (Figure 3 point B). The strong decrease of resistance at the stage C is due to occurrence of a small surface contact between the slag and the electrode. Due to some flow turbulence, the electric contact is very unstable. Combined with hazardous electric contacts, the forward movement of the gas gap led to a new increase in resistance (Figure 3 stage D).

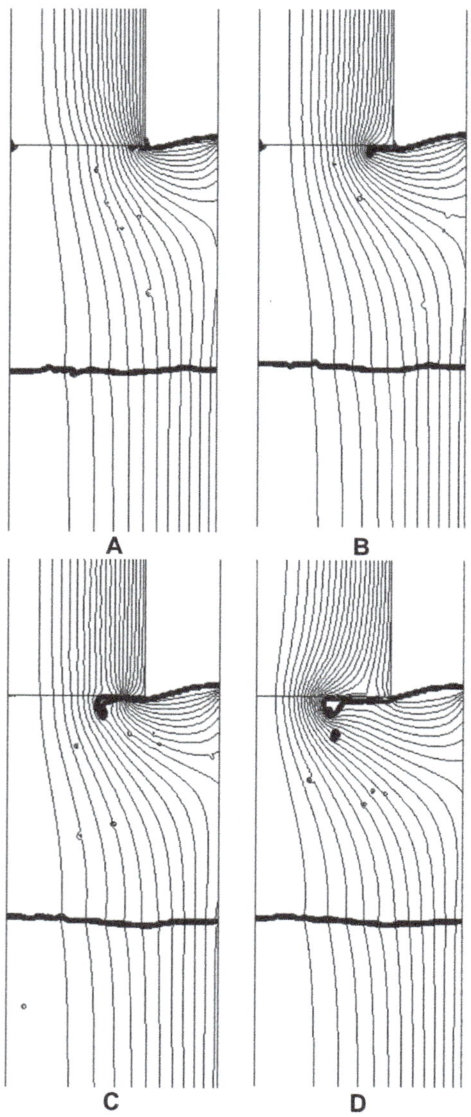

Figure 2: Electric current lines during the formation of an air gap between the slag and the electrode (see also figure 3)

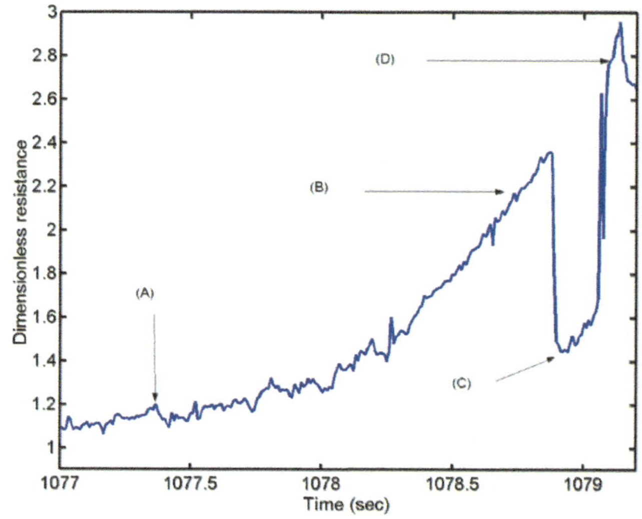

Figure 3: Evolution of the resistance during the formation of an air gap under the electrode

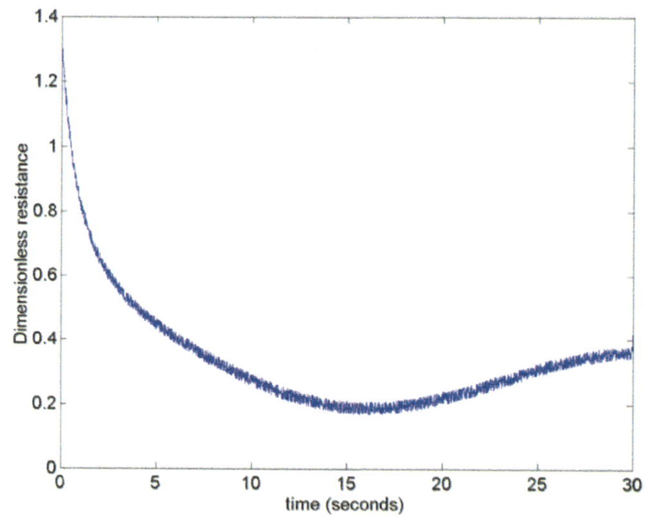

Figure 4: Average resistance during the electrode penetration into the slag

Immersion depth of 50 mm

The electrode is moved downward at a speed of 1.8 mm/s a shift of about 30 mm, corresponding to 50 mm electrode immersion depth. This speed, is relatively small compared to the maximum velocity calculated (~20 mm/s). During the electrode drive, the main resistance decreases dramatically (Figure 4). In opposite to the slag/pool interface, the slag/gas interface does not seem to be affected by the electrode movement (Figure 6-7). During the electrode movement, the elevation of the slag surface increases the surface through which the electric current enters the mould. Although the electrode is now closer to the liquid pool, the average proportion of mould current increased from 30 to 50 %. After 20 seconds, the calculations show an increase of the average resistance with time, it is due to the slow formation of the solid slag thickness and to the redirection of the electric current lines towards the liquid pool (Figure 6). The maximum amplitude of the interface slag/pool movement is about 2 cm, with main frequencies in the 1-30 Hz range. The computed solid slag thickness (Figure 7) is larger than in the previous case, except at the level of the electrode. This behaviour can be explained by a lower bulk temperature and lower amount of mould current usually associated with deep electrode penetration depth. The thinner layer observed at the level of the electrode is due to the presence of mould current and to flow convection.

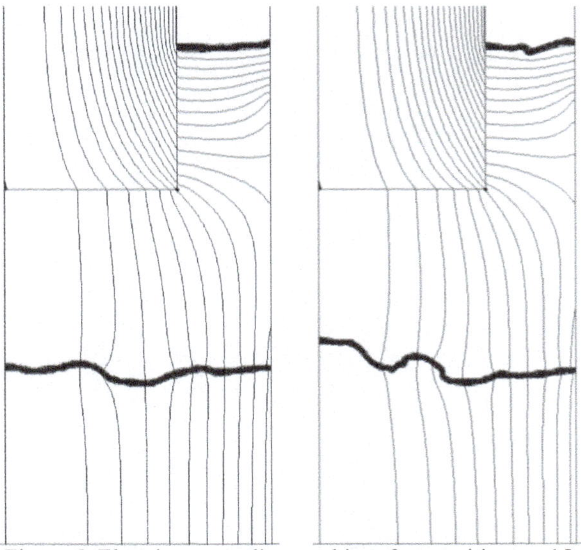

Figure 5: Electric current lines and interface positions at 15 seconds after end of electrode movement

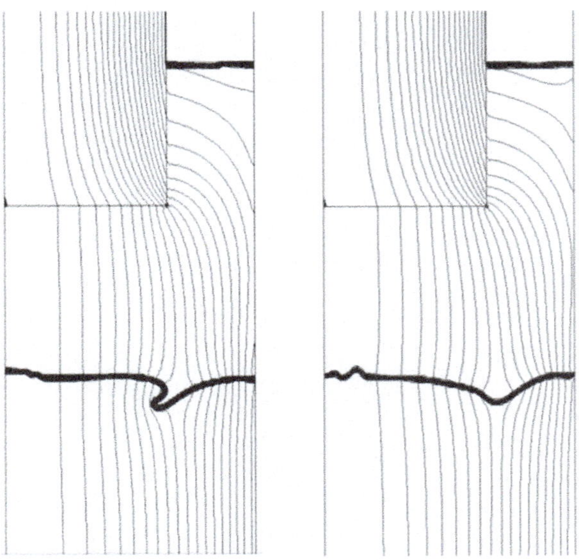

Figure 6: Three successive MHD configurations during steady state reached about 10 minutes after the electrode movement.

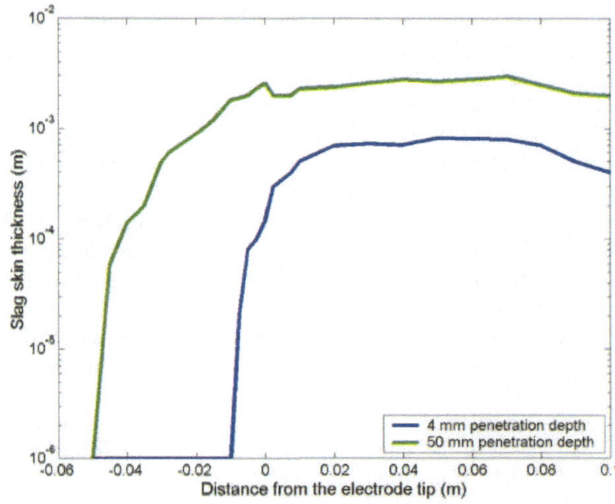

Figure 7: Evolution of the solid slag thickness along the mould wall during steady state.

Conclusion

ESR process is inherently a dynamic process. Analysis of time data of the resistance has the potential to expose a number of abnormal or undesired process attributes (arcing, too deep electrode penetration depth). In order to improve process diagnostics, the identification of these attributes remained an area of interest for a number of years. The increasing capability of computational power has opened the possibility to explore with CFD techniques complex coupled phenomena. A numerical model was built with good ability to handle free surfaces, and good flexibility to handle the dynamic electromagnetic fields. The model was applied to a small ESR process to explore the interaction between the interfaces and the quantity of mould current. The time evolution of the resistance was recorder for each simulation run. The strong coupling between the slag/pool, slag/air interfaces, and mould current gives a new insight into the nature of the MHD oscillation present in the ESR process. Considerable work needs to be performed to achieve a full understanding of the interface movement under the action of MHD forces. However in the real process, these fluctuations are likely to be three dimensional. Although numerical requirements for 3D MHD flows with free interfaces are considerable, the building of a 3D model of the ESR process is a necessity.

Acknowledgements

The authors acknowledge the financial support by the Austrian Federal Ministry of Economy, Family and Youth and the National Foundation for Research, Technology and Development.

References

1. D.K. Melgaard, R.L. Williamson, and J. J Beaman. "Controlling Remelting Processes for Superalloys and Aerospace Ti Alloys" *JOM,* 1998 March, p13-17
2. A. Kharicha , W. Schützenhöfer, A. Ludwig , G. Reiter, "Re formulation of the Joule heating in presence of turbulent fluctuation," *Int. J. Cast Metals Research*, 22 (2009), 155-9.

3. B. Hernandez-Morales and A. Mitchell, "Review of mathematical models of fluid flow, heat transfer, and mass transfer in electroslag remelting process", *Ironmaking & Steelmaking,* Vol. 26 No. 6 (1999), p.423-438
4. V. Weber, A. Jardy, B. Dussoubs, D. Ablitzer, S. Ryberon, V. Schmitt, S. Hans, H. Poisson, "A comprehensive model of the electroslag remelting process: description and validation," *Metall. Mater. Trans. B.* 14 january
5. A. *Kharicha* , W. *Schützenhöfer*, A. Ludwig , G. Reiter, M. Wu, "On the importance of electric currents flowing directly into the mould during an ESR process," *Steel Research Int.*, 79 (2008) 632-36.
6. Kharicha A., Ludwig A., Wu M. "Shape and Stability of the Slag/Melt Interface in a Small DC ESR Process" *Mater. Sci. Eng. A,* 413-4(2005)129-34.
7. Kharicha A., Schützenhöfer W., Ludwig A., Reiter G. "Influence of the slag/pool interface on the solidification in an electroslag remelting process" *Mater.Sci.Forum,* 649 (2010) 229-36.
8. A. Kharicha,_A. Mackenbrock, A. Ludwig, W. Schützenhöfer, V. Maronnier, M. Wu, O. Köser, R. Tanzer: "Selected Numerical Investigations on ESR Process" *Int. Symp. Liquid Metal Processing and Casting (LMPC-07),* Sept. 2-5, 2007, Nancy, France. Eds. P.D. Lee, A. Mitchell, J.P. Bellot, A. Jardy, p 113-119

Proceedings of the 2013 International Symposium on Liquid Metal Processing & Casting
Edited by: Matthew J. M. Krane, Alain Jardy, Rodney L. Williamson, and Joseph J. Beaman
TMS (The Minerals, Metals & Materials Society), 2013

MODELING MACROSEGREGATION DURING ELECTROSLAG REMELTING OF ALLOY 625

Kyle Fezi[1], Jeff Yanke[1,2], and Matthew J.M. Krane[1]
[1]Purdue University School of Materials Engineering; 701 West Stadium Avenue; West Lafayette, IN, 47906, USA
[2]currently Carpenter Technology Corporation, Reading, Pennsylvania USA

Keywords: ESR, Macrosegregation, Alloy 625

Abstract

A numerical model of ESR is used to examine the effects of ingot diameter, process current levels, and initial composition on macrosegregation in alloy 625. Composition variations are made within the standard ranges for alloy 625 to evaluate their effect on buoyancy driven flows. Overall steady state macrosegregation, along with composition distributions are compared, to identify processing conditions that best ameliorate this defect. The composition distributions are analyzed to illustrate how macrosegregation affects the ability to meet the compositional specification. As expected, increasing the ingot size from the standard processing condition of 50 cm (20 inch) to 76.2 cm (30 inch) increases the segregation level overall and at the centerline. The segregation tends to decrease with decreasing mushy zone velocity and sump depth. Processing ingots with a low current and an initial composition in the low end of the specification range is the best choice to reduce macrosegregation.

Introduction

ESR is used to enhance the metallurgical structure and refine the chemistry of an ingot. However, there are limitations to ESR due to increasing segregation levels with ingot diameter. In order to understand issues associated with production of ingots larger than current practice and to explain possible ways to minimize the segregation levels, several constant current ESR runs are simulated.

In this study, a continuum mixture model consisting of the conservation equations for mass, energy, species, momentum, and electromagnetics for both slag and metal is used. The model is coupled with a multicomponent primary solidification model to simulate the production of alloy 625 ingots. The metal-slag interface is tracked using a volume of fluid method. Using constant process currents, predictions of compositional fields and macrosegregation are compared for two ingot diameters. The effect of the initial composition and process current are also studied.

Numerical Model

The conservation equations are solved on a staggered-grid using the finite volume method and the SIMPLER algorithm for the flow field [1]. The continuum conservation equations in axisymmetric coordinates for mass and momentum are shown in equations (1-3).

$$\frac{\partial}{\partial t}(\rho)+\nabla\bullet\left(\rho\overline{V}\right)=0 \tag{1}$$

$$\begin{aligned}\frac{\partial}{\partial t}(\rho u)+\nabla\bullet\left(\rho\overline{V}u\right)&=\nabla\bullet(\mu\nabla u)-\frac{\partial P}{\partial z}-\frac{\mu}{K}u-\mu_0 J_z H_\theta\\&+g\rho\left[\beta_T(T-T_0)+\sum\beta_C^i\left(C^i-C_0^i\right)\right]\end{aligned} \tag{2}$$

$$\frac{\partial}{\partial t}(\rho v)+\nabla\bullet\left(\rho\overline{V}v\right)=\nabla\bullet(\mu\nabla v)-\frac{\partial P}{\partial r}-\frac{\mu}{K}v-\mu_0 J_r H_\theta-\mu\frac{v}{r^2} \tag{3}$$

Thermal and solutal buoyancy drive the flow in the slag and metal. The solid dendrites slow the flow in the metal mushy zone due to drag, which is taken in to account through the Blake-Kozeny equation and is related to the primary dendrite arm spacing.

$$K=\frac{\lambda_1^2}{180}\frac{(1-f_s)^3}{f_s^2} \tag{4}$$

The enthalpy and species conservation equations are given by:

$$\begin{aligned}\frac{\partial}{\partial t}(\rho h)+\nabla\bullet\left(\rho\overline{V}h\right)&=\nabla\bullet\left(\frac{k}{c}\nabla h\right)+\nabla\bullet\left(\frac{k}{c}\nabla(h_s-h)\right)\\&-\nabla\bullet\left(\rho f_s\overline{V}(h_s-h)\right)+\dot{q}_{Joule}\end{aligned} \tag{5}$$

$$\begin{aligned}\frac{\partial}{\partial t}\left(\rho C^i\right)+\nabla\bullet\left(\rho f_l\overline{V}C^i\right)&=\nabla\bullet\left(\rho f_l D\nabla C^i\right)+\nabla\bullet\left(\rho D\nabla\left(C_l^i-C^i\right)\right)\\&-\nabla\bullet\left(\rho f_s\overline{V}\left(C_l^i-C^i\right)\right)\end{aligned} \tag{6}$$

Alloy 625 is comprised of many elements but only 4 composition equations (6) are solved. Cr, Fe, Mo, and Nb are chosen to represent this alloy, based on their contributions to solutal buoyancy and segregation characteristics. The liquidus temperature (in K) used in the thermodynamic model is taken from Yang et al. [2]:

$$\begin{aligned}T_{liquidus}&=1728-14.8365+75C_l^{Cr}-16C_l^{Fe}\\&-524C_l^{Mo}-1230C_l^{Nb}\end{aligned} \tag{7}$$

The partition coefficients for the alloying elements are obtained from the thermodynamic database program PANDAT. Those coefficients are 1.04 for Cr, 1.31 for Fe, 0.83 for Mo, and 0.54 for Nb.

The electrode melt rate is calculated based on an input current through a slag-electrode heat transfer model. The electrode is assumed to have a 1-D temperature profile and the melting temperature is calculated based on the initial alloy composition. The liquid metal passing through the slag layer is not modeled and instead the metal is added uniformly just below the slag layer at the initial composition and the temperature of the slag nearest the metal-slag interface. To account for the growing domain, a moving mesh is used, with the top row of cells fixed to the top of the slag and the bottom row fixed to the bottom surface of the metal. A row of cells in the solid just below the bottom of the sump expands proportionally to the filling rate of the ingot. The growing row of control volumes spawns a new row of control volumes once they are 1.5 times the height of the other rows.

Using this model several cases are run in which the ingot diameter, initial composition, permeability of the mushy zone, and process current are varied. The cases are outlined in Table II.

Table I. Thermophysical properties of metal and slag

Material Property	Alloy 625	Slag
Density [kg/m^3] [3,4]	8440	2490
Specific Heat [J/kg K] [3,4]	670	1260
Thermal Conductivity [W/m K] [3,4]	25.2	6
Electrical Conductivity [1/Ω m] [3,4]	7.1x10^5	313
Thermal Expansion [1/K] [5]	1.28x10^{-4}	-
Solutal Expansion [1/wt. fraction]	Cr: 0.201 Fe: 0.108 Mo: -0.184 Nb: -0.255	-
Diffusivity [m^2/s]	1x10^{-9}	-
Latent Heat [J/Kg] [6]	2.9x10^5	-
Eutectic Temperature [K] [2,3]	1430	1390
Permeability Constant	20 in ingot: 5.5x10^{-11} 30 in ingot: 1.4x10^{-9}	-
Partition Coefficients	Cr: 1.04 Fe: 1.31 Mo: 0.83 Nb: 0.54	-

Table II. Parametric study cases

Case	Ingot Diameter [m]	Process Current [kA]	Cr [wt. %]	Fe [wt. %]	Mo [wt. %]	Nb [wt. %]	Permeability Constant [m^2]
1	0.508	14.5	21.5	2.5	9.0	3.65	5.5x10^{-11}
2	0.762	25.5	21.5	2.5	9.0	3.65	5.5x10^{-11}
3	0.762	25.5	21.5	2.5	9.0	3.65	1.4x10^{-9}
4	0.762	25.5	20.5	0.833	8.3	3.3	1.4x10^{-9}
5	0.762	25.5	22.5	4.2	9.7	4.0	1.4x10^{-9}
6	0.762	24.0	21.5	2.5	9.0	3.65	1.4x10^{-9}
7	0.762	27.5	21.5	2.5	9.0	3.65	1.4x10^{-9}
8	0.762	31.0	21.5	2.5	9.0	3.65	1.4x10^{-9}

Results and Discussion

A standard case is simulated based on industrial practice for a 50 cm diameter ingot for alloy 625 in order to provide a basis for comparison. The flow in the slag had two dominant flow cells as shown in Figure 1. The flow in the slag is more vigorous throughout than the flow in the metal pool. Both flow cells are thermally driven due to the heat extraction out of the mold wall and through the electrode. The two flow cells in the slag layer oscillate as to which cell is more dominant. The clockwise flow cell does not extend very far radially near the slag-electrode interface but does have the tendency to extend near the midradius and even near the centerline at the slag-metal interface. The counter clockwise cell is dominant at the slag-electrode interface in the radial direction, but the cell does not have much interaction with the flow in the metal pool. The clockwise flow cell interacts with the stronger flow in the molten metal pool at the slag-metal interface [7].

The fluid flow pattern in the metal pool is determined by thermal and solutal density differences. As the sump grows vertically during start-up, so does the shape of the flow cell. The fluid moves down the mold wall, in the metal head, toward the bottom of the ingot and then flows briefly across the solidification front towards the centerline. The fluid then flows up towards the slag-metal interface and then toward the mold wall thus completing the flow cell. Across the solidification front, from the mold wall toward the centerline, the strength of the flow field decreases and is practically negligible near the centerline as the sump is thermally and solutally stratified.

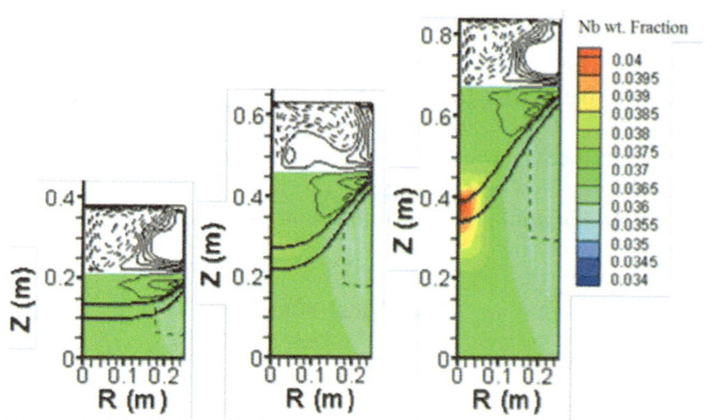

Figure 1. Evolution of Nb segregation, sump shape, and fluid flow during the start-up regime. The slag is shown in white and the dark lines show the extent of the mushy zone. Dotted lines represent counter clockwise flows, while solid thin lines represent clockwise flows in which -0.13 kg/s ≥ $\rho\psi \leq$ 0.1 kg/s with $\rho\Delta\psi$ =0.02 kg/s.

The melt rate of the electrode was converted to the filling velocity of the mold for comparison to other cases and mold diameters' using equation (8).

$$V_f = \frac{\dot{m}}{\rho A_C} \tag{8}$$

The steady state average filling velocity for Case 1 is 7.9x10^{-5}m/s. With the constant process current, the steady state filling velocity is established early on in the process, while the steady state sump depth takes longer to develop. Figure 2 shows the sump depth at two radial locations as well as the height of the metal head. The steady state depths are 0.040 m for the metal head, 0.16 m at the mid-radius, and an average of 0.35 m for the centerline. The centerline sump depth variation can be attributed partly to the segregation characteristics and the small temperature gradient at the centerline. Thus a small change in local composition can occur in front of the solidification front due to the fluid motion. The compositional change causes a decrease in the liquidus temperature, leading to solid forming further up in the metal pool. When the centerline sump depth reaches steady state, the ingot height is approximately 0.4 m tall which is slightly less than the ingot diameter of 0.5 m.

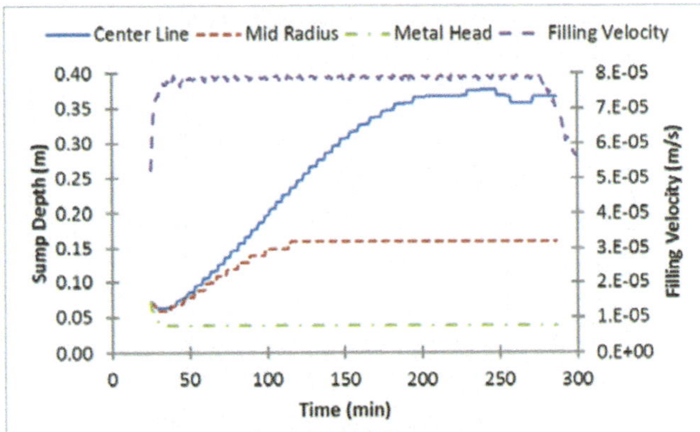

Figure 2. Plot of the sump depth at the centerline and mid-radius, and the metal head height, as well as the filling velocity. The process time taken to reach steady state is based on the fs=0.2 contour.

During steady state melting, the shape of the sump and the metal flow pattern remains relatively constant except for a few small oscillations. Due to the partitioning of the alloying elements, the liquid composition in a control volume becomes enriched in niobium and molybdenum and depleted in chromium and iron. The flow down the solidification front causes the interdendritic liquid to be replaced by liquid with a composition lower in Cr and Fe and higher in Nb and Mo. Therefore, the outer radius of the ingot is rich in chromium and iron and is depleted near the mid-radius and centerline, and the opposite is true for molybdenum and niobium, as shown in Figure 3. Figure 4 shows the steady state Nb composition field, sump shape, and the fluid flow in the slag and metal pool for 3 different axial positions in the ingot. To quantify the amount of macrosegregation in the ingot for comparison to other cases, the steady state macrosegregation number is calculated from the radial profiles in Figure 3. In this case the steady state macrosegregation number is 0.0239.

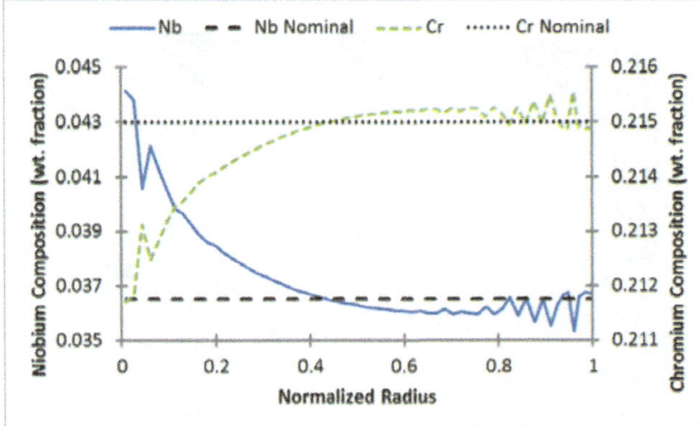

Figure 3. Plot of the Nb and Cr compositions across the normalized radius.

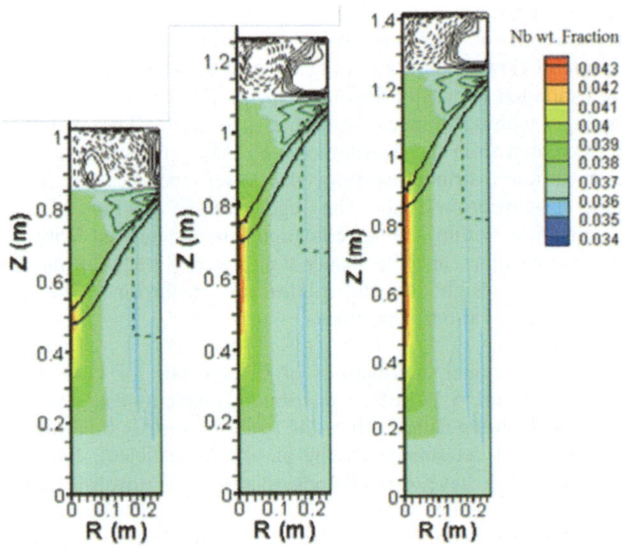

Figure 4. Contour plots showing steady state Nb composition field, sump shape (0.05 and 0.95 fraction solid contours), and streamlines in the slag and metal pool. Snapshots are taken at (a) t = 192 min, (b) t = 242 min, and (c) t = 276 min. Streamlines are plotted for -0.13 kg/s $\geq \rho\psi \leq$ 0.1 kg/s with $\rho\Delta\psi$ =0.02 kg/s.

The sump takes on a V shape during the steady state process, as in Figures 4 and 5. As described above, the interdendritic liquid moves towards the centerline due to buoyancy, but it opposed by the drag from the solidifying surface. Across the radius, the interdendritic liquid flows a short distance before becoming entrained in the rising solidification front. The overall effect causes a shift in the interdendritic liquid toward the centerline. Depending on the partitioning of the alloying elements, the interdendritic liquid will either be enriched (k <1), causing a lower ingot surface composition, or will be depleted (k > 1), causing the surface to be enriched. This phenomenon can be seen in Figures 4 and 5 in which the niobium composition is lower than the nominal at the outer radius and increases to well above the nominal at the centerline and crosses the nominal composition line near the mid-radius. Although the positive segregation appears to deviate from the nominal much more than the negative segregation, the volume near the centerline is much less than the outer radius.

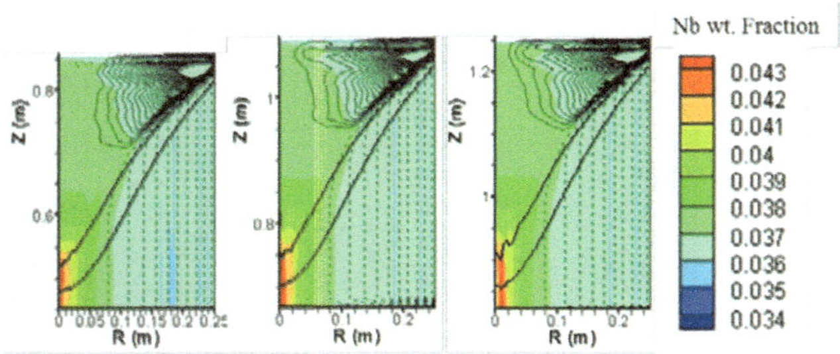

Figure 5. Contour plots showing steady state Nb composition field, sump shape (0.05 and 0.95 fraction solid contours), and streamlines. Snapshots are taken at (a) t = 192 min, (b) t = 242 min, and (c) t = 276 min. Streamlines are plotted for -0.02 kg/s $\geq \rho\psi \leq 0.04$ kg/s with $\rho\Delta\psi = 0.002$ kg/s.

The initial composition for Case 1 is chosen to be in the middle of the specification range for each alloying element and the distributions of the final solid compositions for all control volumes below the bottom of the sump (not including the hot top) are examined for each alloying element. The volume fraction of the ingot within the specification limits for each alloying element is calculated from the distributions, by taking the integral of the composition distributions from the lower specification to the upper specification limit. The ingot volume fraction of each composition within the specification limits, coupled with the distribution plots, and the steady state macrosegregation number will provide insight into the solidification behavior of the ESR process under various conditions.

The compositional distributions of Cr and Nb for Case 1 are shown in Figure 6. All of the distributions have a peak at or near the average composition. From the peak the distribution tails off in a direction determined by the partition coefficient. Elements with a $k < 1$ have distributions that tail off towards higher compositions, while elements with $k > 1$ have distributions that tail off to lower compositions. Nb is the only alloying element in which the entire volume of the ingot is not between the specification limits, although only 0.18% of the ingot volume has a composition above the upper specification limit. Fe and Cr did not segregate very far from the initial composition and occupy only 6.7 and 13.4% of the specification range respectively, while Mo has a larger distribution range (38.1% of the range). Because Nb has a partition coefficient that has the largest deviation from unity and is the only element to fall anywhere outside the specification limits, Nb is the alloying element plotted to analyze the macrosegregation behavior in the ingot.

Two 76.2 cm diameter ingot simulations are compared to the 50 cm diameter base case (Case 1). The first (Case 2) has the same permeability constant as Case 1, and only changes the ingot diameter and input current level. The current is increased in order to prevent the slag-metal interface from solidifying. The other 76.2 cm case (Case 3) has a larger permeability constant which takes into account the larger average dendrite arm spacing typically seen in larger diameter ingots.

The strongest liquid metal flows for the larger diameter ingot cases are 126% stronger than in the 50 cm diameter ingot case, which indicates the flow in the mushy zone is also stronger. Cases 2 and 3 have increased macrosegregation because the interdendritic fluid is able to travel farther down the solidification front before becoming entrained as can be seen in Figure 7. Case 2 has a larger volume near the centerline that is enriched in Nb, to the same degree as Case 1, leading to a slightly higher level of macrosegregation. The centerline segregation is not more severe because the two cases have the same permeability constant. Increasing the permeability constant drastically increases the macrosegregation level compared to Case 1 and 2. The increased macrosegregation level from Case 1 to Case 3 is due to a combination of the increased ingot size as well as the increase in the dendrite arm spacing in the mushy zone. Along with the macrosegregation level, the sump depth and filling velocity also increase when the diameter is increased.

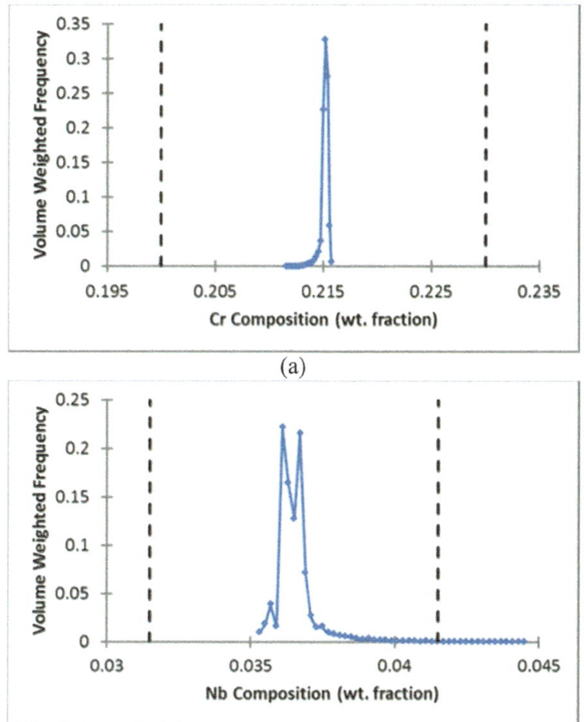

(a)

(b)

Figure 6. Case 1 composition distributions for (a) Cr and (b) Nb, weighted by volume fraction and showing the upper and lower specification limits.

For Cases 2 and 3, the sump shapes and depths are very similar, with a metal head of 0.05 m for both, mid-radius sump depths of 0.26 and 0.30 m, and average centerline sump depths of 0.68 and 0.73 m, respectively. Increasing the ingot diameter by about 50% causes the centerline sump depth to roughly double, and the liquidus interface becomes steeper. Consequently the solid ingot height when steady state is reached nearly doubled from 0.34 m to 0.60 m. The average filling velocity is also higher in Cases 2 and 3 than in Case 1, increasing from 7.9×10^{-5} to 1.7×10^{-4}. The larger process current for Cases 2 and 3 increases the Joule heating in the slag and therefore increases the filling velocity. The deeper sump and higher filling velocities in Cases 2 and 3 contribute to the higher macrosegregation compared to Case 1.

Increasing the ingot diameter from Case 1 to Case 2 causes an increase in the global Nb macrosegregation number of 14.6% from 0.0239 to 0.0280. Increasing the permeability constant from Case 2 to 3 causes an increase in the Nb macrosegregation number of 80.6% because the flow in the mushy zone is not as restricted by the solid dendrites. As expected from the macrosegregation comparison, the volume of ingot outside of the specification limits increases for both Cases 2 and 3, as seen in Figure 8. For Case 2, all of the alloying elements remain within the specifications except niobium, for which 0.35% of the volume is above the upper limit near the center of the ingot. The Nb compositions outside of the specification limit occupy a small volume fraction of the total ingot, but, according to Figure 7, the segregated region occupies a long axial length, ~0.60 m, at the centerline of the steady state region of the ingot, which would be harder to remove by machining than if the high composition was at the outer radius or near the ends of the ingot. By increasing the ingot diameter, a larger volume fraction of the ingot is outside of the specification. For Case 3, Fe is the only alloying element that remained fully between the specification limits, as 7.72% of the volume of the ingot is outside of the Nb specification and, 2.10% of the volume of the ingot is outside of the Mo specification.

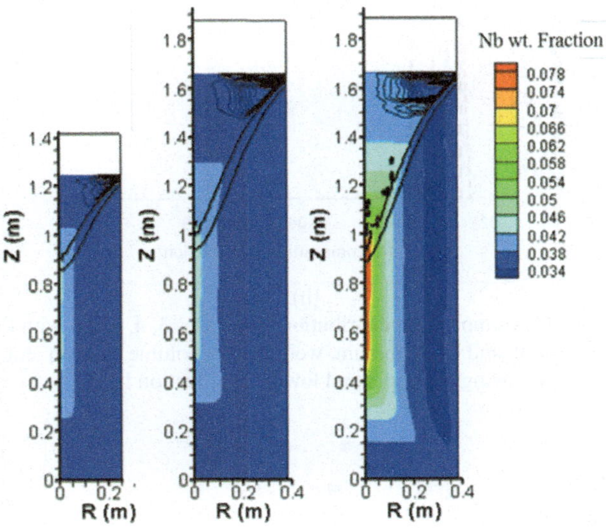

(a) (b) (c)

Figure 7. Nb composition field, sump shape (0.05 and 0.95 fraction solid contours), and streamlines in the metal pool for (a) Case 1, (b) Case 2, and (c) Case 3 Streamlines are plotted for (a) 0.005 kg/s ≥ ρψ ≤ 0.075 kg/s with ρΔψ =0.005 kg/s, (b) and (c) 0.0075 kg/s ≥ ρψ ≤ 0.1725 kg/s with ρΔψ =0.0075 kg/s.

As a possible method of decreasing macrosegregation, the composition of the alloy was varied to change the solutal buoyancy and to shift the composition distributions to be more inside the specification limits. The initial composition of the alloy is shifted towards the limits of the specification in Cases 4 and 5 in order to increase and decrease the solutal contribution to the buoyancy driven flow in the molten pool. Figure 9 shows how shifting the initial composition toward the high or low specification limit affects the solutal buoyancy and the comparison to the thermal buoyancy for Case 3. Increasing the initial composition of each alloying element increases the solutal buoyancy while decreasing the initial composition decreases the solutal buoyancy. In this study of alloy 625, each of the alloying elements that are tracked has the same sign for their solutal buoyancy contribution. In other words, by decreasing the composition, the solutal buoyancy also decreases and vice versa. Also, by changing the initial composition the highly segregated low volume part of the ingot does shift so that more of the ingot is within the specification range.

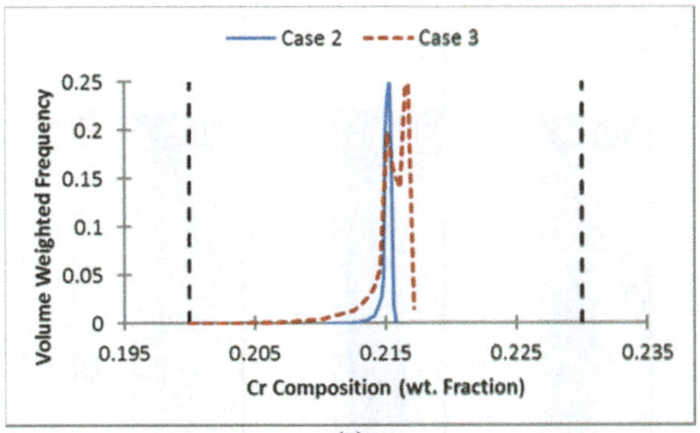

(a)

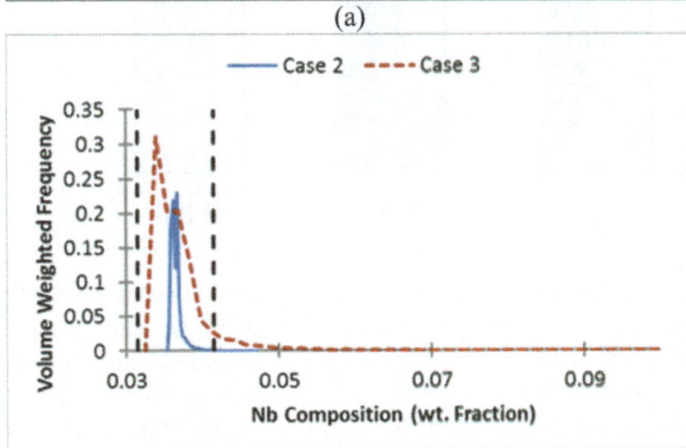

(b)

Figure 8. Composition distributions for Cases 2 and 3 with (a) Cr and (b) Nb, weighted by volume fraction and showing the upper and lower specification limits.

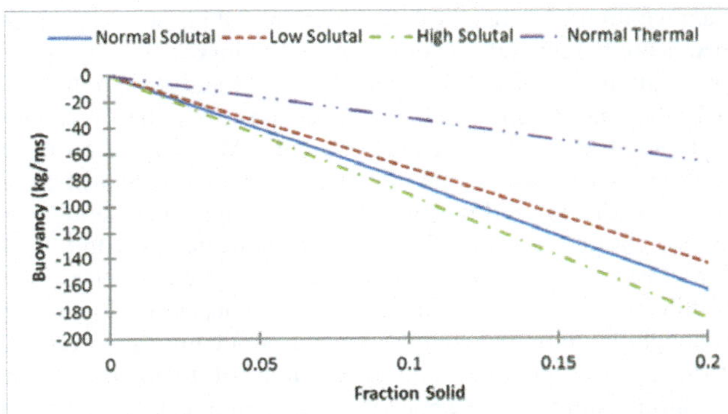

Figure 9. Plot of the solutal and thermal contributions to the buoyancy driven flow in the molten metal pool, showing the effect of changing the initial composition and fraction solid.

The shape of the sump and flow cell in the metal pool is very similar for Cases 2 and 3. Figure 10 shows the Nb composition fields, the sump shapes, and flow patterns in the metal pools from Cases 3-5. The macrosegregation number of the steady state region increases from Case 4 (low composition, M^{Nb} = 0.137) to Case 3 (average composition, M^{Nb} = 0.144), and again to Case 5 (high composition, M^{Nb} = 0.155). Changing the initial composition in this way to alter the solutal buoyancy while keeping it within the specification changed the macrosegregation number +/- 5%. Obviously, the solutal buoyancy can only be reduced by a small amount before the initial composition is outside the specification limits.

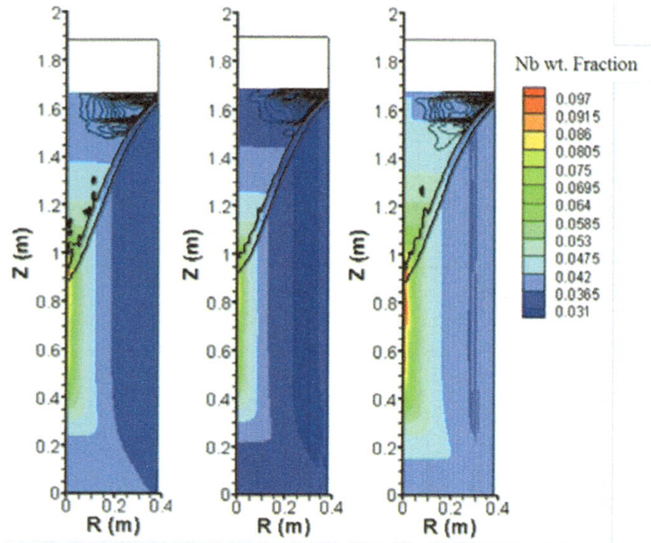

(a) (b) (c)

Figure 10. Nb composition field, sump shape (0.05 and 0.95 fraction solid contours), and streamlines in the metal pool for (a) Case 3, (b) Case 4, and (c) Case 5 Streamlines are plotted for (a) $0.0075\ kg/s \geq \rho\psi \leq 0.18\ kg/s$ with $\rho\Delta\psi = 0.0075\ kg/s$, (b) $0.0075\ kg/s \geq \rho\psi \leq 0.1725\ kg/s$ with $\rho\Delta\psi = 0.0075\ kg/s$, and (c) $0.0075\ kg/s \geq \rho\psi \leq 0.1577\ kg/s$ with $\rho\Delta\psi = 0.0075\ kg/s$.

Based on the steady state global Nb macrosegregation number, shifting the initial compositions to the low end of the specification range the ingot quality seems to improve ingot quality, but examination of the ingot volume fraction within the composition specification limit tells a different story. Figure 11 shows the composition distribution for Cases 3-5 for Cr and Nb. For Case 4, each alloying element, with the exception of Fe, has a volume fraction outside of the specification limits. Also, Cr has a similar distribution as Case 3 but because the initial composition is shifted to the lower limit, and the maximum composition, the volume fraction outside of the specification increases from less than 0.01% to 1.78%. Also, all of the volume outside the specification is at the center of ingot as indicated by Figure 10. Although the initial Fe composition was shifted to the lower end of the specification range, all of the volume of the ingot is within the specification. The degree of Fe macrosegregation is larger than for Cr (Figure 12), but the specification range for Cr is narrower. This result highlights the importance of comparing the distribution of each alloying element to its specification range to characterize this defect, in addition to using the overall macrosegregation number. The Mo volume fraction of the ingot outside of the specification limit decreased to 0.27% from 2.1% by lowering the initial composition from Case 3 to 4. The volume fraction of Nb above the upper specification limit decreases from Case 3 to 4, but the overall volume fraction outside of the specification limits increases to 48%. The large increase is due to left side of the curve in the depleted region of the ingot moving outside the lower specification limit. If the initial composition of Nb was shifted slightly higher, away from the lower limit, the overall volume outside the specification would decrease. Even though more of the volume is outside the specification for Nb, the location of the defect has shifted from the center to the outer radius of the ingot.

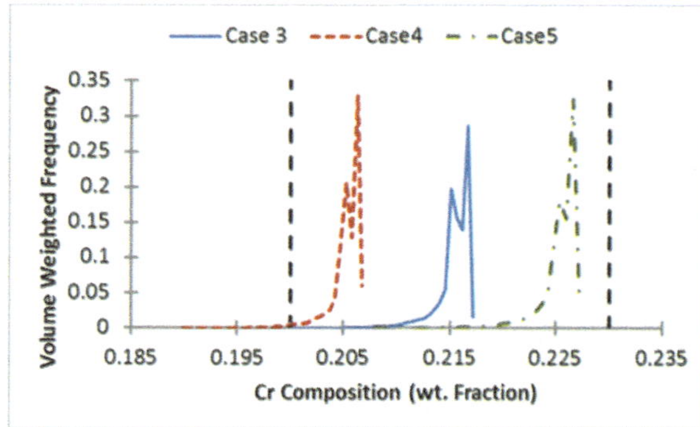

(a)

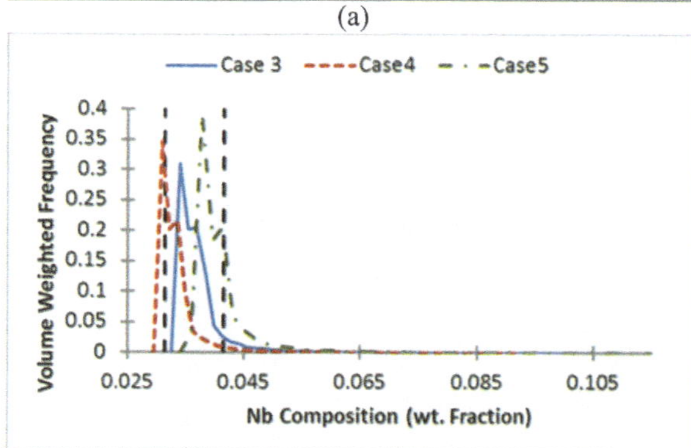

(b)

Figure 11. Composition distributions for Cases 3, 4, and 5 with (a) chromium and (b) niobioum, weighted by volume fraction and showing the upper and lower specification limits.

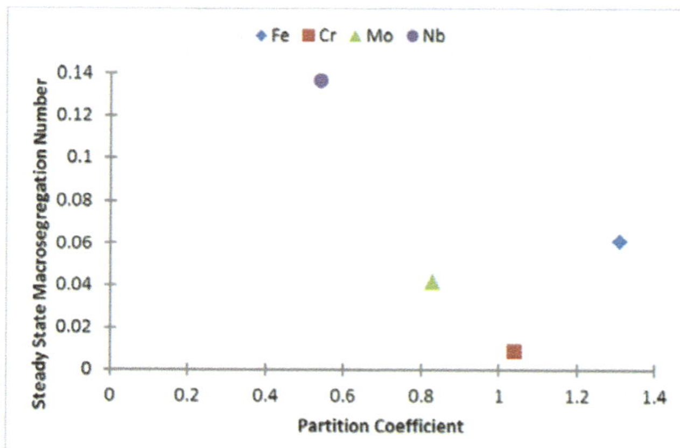

Figure 12. Steady state macrosegregation of each alloying element for case 4.

The steady state current is varied in Cases 6-8 to examine its effect on the filling velocity and the steady state macrosegregation number, as well as the composition distributions, sump shape, and fluid flow. Cases 6-8 are compared to Case 3, which is treated as a baseline for the processes with a 76.2 cm diameter. From Case 3, in which the current was 25.5 kA, the current is decreased to 24 kA for Case 6, increased to 27.5 kA for Case 7, and increased again to 31 kA for Case 8. If the current is reduced much below 24 kA (Case 6), then the slag-metal interface begins to freeze, causing the fluid mechanics model to break down and ultimately stop the simulation from running. The initial composition was left in the middle of the specification range these cases.

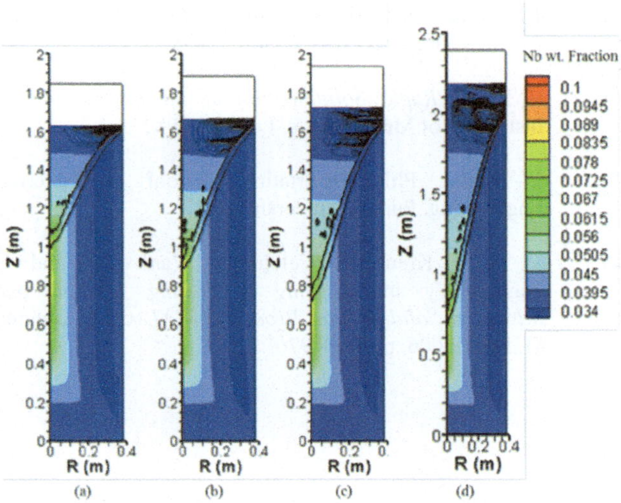

Figure 13. Nb composition field, sump shape (0.05 and 0.95 fraction solid contours), and streamlines in the metal pool for (a) Case 6, (b) Case 3, (c) Case 7, and (d) Case 8. Streamlines are plotted for (a) 0.0075 kg/s $\geq \rho\psi \leq$ 0.18 kg/s with $\rho\Delta\psi$ =0.0075 kg/s, (b) 0.0075 kg/s $\geq \rho\psi \leq$ 0.195 kg/s with $\rho\Delta\psi$ =0.0075 kg/s, (c) 0.0075 kg/s $\geq \rho\psi \leq$ 0.12 kg/s with $\rho\Delta\psi$ =0.0075 kg/s, and (d) 0.0075 kg/s $\geq \rho\psi \leq$ 0.135 kg/s with $\rho\Delta\psi$ =0.0075 kg/s.

Table III. Average centerline sump depth and filling velocity for various current levels.

Case	Current [kA]	Average Centerline Sump Depth [m]	Filling Velocity [m/s]
1	14.5	0.35	7.86×10^{-5}
2	25.5	0.68	1.71×10^{-4}
3	25.5	0.73	1.71×10^{-4}
4	25.5	0.73	1.72×10^{-4}
5	25.5	0.73	1.71×10^{-4}
6	24	0.58	1.44×10^{-4}
7	27.5	0.94	2.16×10^{-4}
8	31	1.4	3.01×10^{-4}

The sump depth and filling velocity increase and decrease with the process current, while the stream function in the metal pool fluctuates slightly between the cases. Figure 13 shows the Nb composition field along with the steady state sump shape and fluid flow field in the metal pool for Cases 6-8. Decreasing the current level from Case 3 to Case 6 causes the average centerline sump depth, as determined by the 0.2 fraction solid contour, to decrease from 0.73 m to 0.58 m, a difference of 21%. Increasing the current causes the average centerline sump depth to increase by 22% to 0.935 m for Case 7 and by 49% to 1.4 m for Case 8. Although the centerline sump depth varies with the current, the height of the bottom of the sump in which the process reaches steady state, determined by the change in centerline sump depth, does not vary. Each case reaches steady state once the solid ingot height below the sump reaches approximately 0.6 meters. The average filling velocity for each case varies in the same way as the centerline sump depth. For Case 6, the filling velocity decreases from 1.71×10^{-4} to 1.44×10^{-4}. For Cases 7 and 8 the filling velocity increases to 2.16×10^{-4} and 3.05×10^{-4} respectively. The average centerline sump depth and filling velocity is summarized in Table III.

The composition distributions have only small variations between the cases. The volume fraction outside of the specification limits for each alloying element and case is summarized in Table IV. Iron remains within the specification for all 4 cases and chromium has negligible amounts of the ingot outside the specification. Increasing the process current from Case 6 (24 kA) to 3 (25.5 kA) increases the volume fraction of the ingot outside the specification limits, and further increasing the current to Case 7 (27.5 kA) decreases the volume fraction outside the specification. Increasing the current to Case 8 (31 kA) slightly decreases the volume fraction from Case 7. Relating the filling velocity to the current and the composition distribution range to the steady state macrosegregation number displays a similar trend.

Table IV. Volume fraction of the ingot outside the specification limits for each alloying element for cases in which the current level was varied.

Case	Cr	Fe	Mo	Nb
3	<0.01%	0%	2.1%	7.72%
6	0%	0%	2.06%	7.66%
7	0%	0%	1.83%	7.56%
8	<0.01%	0%	1.91%	7.58%

Figure 14 shows the relationship between the steady state Nb macrosegregation number (M^{Nb}) and the filling velocity. Case 3 is a local maximum of (M^{Nb} = 0.144) and changing the current (and consequently the filling velocity) in either direction from this local maximum causes the M^{Nb} to decrease. Decreasing the current causes a steeper drop in the macrosegregation than increasing the current. A similar relationship was discussed by Krane et al [8]. A filling velocity approaching the limit of zero causes no sump to form and the flow to decrease, leading to no segregation. As the filling velocity increases, the flow in the mushy zone increases with a steeper sump until the maximum M^{Nb} is reached. With further increases in the filling velocity, the flow in the mushy zone does not increase fast enough to move the enriched interdendritic fluid very far before becoming immobilized by the ever increasing solidification front. With less time to move in the mushy zone, there is less macroscale redistribution of solute from the surface to the centerline.

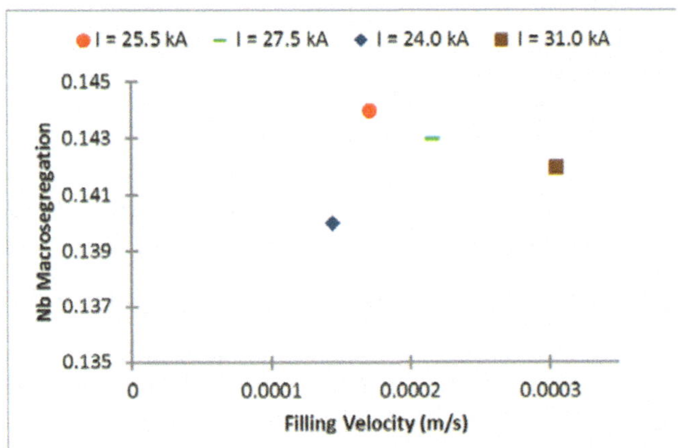

Figure 14. Plot of the Nb macrosegregation number and filling velocity showing the trend with increasing filling velocity and the local maximum of Case 3.

Reducing the current to a lower level than in Case 6 would theoretically cause the filling velocity to decrease further; leading to an even lower M^{Nb}. However the surface quality of the ingot would most likely be sacrificed. The model was unable to predict the process behavior due to freezing over of the slag-metal interface, which might signal a process in which the surface texture has large variations. (The problem may also arise from uncertainty in the properties of the slag and metal.) The trade-off of M^{Nb} and surface quality may be one to consider in order to reduce the macrosegregation in the ingot. Another way to reduce the macrosegregation even further would be to combine the lower steady state current of Case 6 with the lower solutal buoyancy force of Case 4.

Summary

A 50 cm (20 inch) diameter ingot case is used as the baseline to simulate the behavior of industrial practice. The 50 cm (20 inch) diameter ingot is also compared to two 76.2 cm (30 inch) diameter cases in which two different permeability constants are used. Increasing the diameter alone increases the macrosegregation in the ingot as well as the volume fraction of the ingot outside of the specification limits. Decreasing the permeability constant causes the mushy zone to become more permeable and increase the overall macrosegregation, as well as the volume fraction of the ingot outside of the specification limits. Shifting the initial composition causes changes in the solutal buoyancy. Decreasing the solutal buoyancy decreases the interdendritic flow velocities and the macrosegregation, while increasing it had the opposite effect. The composition speicification for this highly segregation-prone alloy is very narrow. Shifting the initial composition to decrease the solutal buoyancy and composition distribution tail so that it is within the specification limits is the best option.

Changing the process current changes the macrosegregation number, but not monotonically. The segregation increases with filling velocity up to a maximum, beyond which the segregation slowly decreases. This phenomenon is due to the competition between the speed of the flow of the segregated liquid down through the mushy zone to the rate with which the solidification front freezes this liquid. Coupling a low process current with a low solutal buoyancy force will further decrease the macrosegregation present in the larger ingots.

Acknowledgements

Funding for this work was provided by PCC.

References

[1] Suhas V. Patankar: *Numerical Heat Transfer and Fluid Flow*, Mcgraw-Hill, New York, 1980.

[2] Wanhong Yang, Wei Chen, Keh-minn Chang, Sarwan Mannan, and John DeBarbadillo: in *Superalloys 2000*, T.M. Pollock, R.D. Kissinger, R.R. Bowman, K.A. Green, M. McLean, S. Olson, and J.J. Schirra, eds., TMS, 2000, pp. 75–84.

[3] G Hoyle: *Electroslag Processes*, Elsevier Science Publishing Co., New York, 1983.

[4] Special Metals Corporation Technical Bulletin: *INCONEL Alloy 625*, 2006, pp. 1–20.

[5] T Iida and R Guthrie: *The Physical Properties of Liquid Metals*, Clarendon Press, Oxford, 1988.

[6] H Jones: *Rapid Solidification of Metals and Alloys*, Institution of Metallurgists, London, 1982.

[7] J. Yanke: PhD dissertation, School of Materials Engineering, Purdue University, 2013.

[8] M J M Krane, S Cefalu, K VanEvery, and D Zagrebelnyy: in *Modeling of Casting, Welding and Advanced Solidification Processes - XI*, C.-A. Gandin, M. Bellet2006, pp. 969–976.

Proceedings of the 2013 International Symposium on Liquid Metal Processing & Casting
Edited by: Matthew J. M. Krane, Alain Jardy, Rodney L. Williamson, and Joseph J. Beaman
TMS (The Minerals, Metals & Materials Society), 2013

COMPARISON OF ARC SLAG REMELTING VS. P-ESR MELTING FOR HIGH NITROGEN STEELS

Dr. Roman Ritzenhoff[1], Prof. Lev Medovar[2], Volodymyr Petrenko[2], Ganna Stovpchenko[2],
[1]Head of Quality and Technology, Energietechnik Essen GmbH, Westendstraße 15, 45143 Essen
[2] E.O. Paton Electric Welding Institute, Academy of Sciences of Ukraine, 11 Bozhenko str., Ukraine 03150, Kyiv.

Keywords: High Nitrogen Steels, Electroslag remelting, P-ESR, ASR, nitrogen, ingot.

Abstract

High Nitrogen Steels with nitrogen contents above atmospheric solubility are usually melted using an advanced P-ESR process under elevated pressure. The nitrogen pick up can be controlled by Si_3N_4 or nitrogen gas. This paper provide a comparison of alloying by nitrogen gas by ASR vs P-ESR process in terms of processing and nitrogen distribution. In two series of ASR and P-ESR separate trials the nitrogen content was increased at remelting in nitrogen atmosphere practically twice in spite of big difference in nitrogen pressure. Comparison of P-ESR and ASR processes abilities in steel alloying by nitrogen from gas phase shown quite close results that gives us reasons for experiments continuation.

Introduction

Arc Slag Remelting (ASR) is a variant of standard ESR that accumulated advantages of ESR and VAR. At the beginning of the seventies ASR was developed at the E.O. Paton Electric Welding Institute of the National Academy of Sciences of Ukraine [1]. A lot of the credit must go to the Prof. Boris Medovar and his research team in the creation of the ASR and substantiation of principal priority trends of its development and industrial implementation. Summary of their work is given in the book entitled "ASR of steel and alloys"[2]. P-ESR process also is a variant of standard ESR. In fact P-ESR is main industrial technology of high nitrogen steel manufacture [3,4].

This work is an attempt of specialists of Energietechnik Essen GmbH and E.O. Paton Electric Welding Institute to resume experimental researches of ESR of High Nitrogen Steels and to check an idea on expediency of ASR application at elevated pressure of nitrogen in the course of HNS receiving by a direct alloying of nitrogen from furnace atmosphere.

It should be noted that process of alloying by nitrogen at ESR for steels with quite low nitrogen content, i.e. at microalloying, was studied in detail. There is also an extensive literature data on the matter. With the appearance of the new type of nitrogen bearing steels, called HNS, investigations of ASR efficiency for these grades [5,6] were started. Results of P-ESR researches [4,7], which outline main advantages and shortcomings of known methods of alloying by nitrogen were used as basis (Table I).
Thus, both expediency and efficiency of realization of one kind of combination of two known versions of ESR in the new process – slag remelting (P-ARS) is the purpose of comparative researches
Taking into consideration that main principles and distinctive features of P-ESR were presented in numerous publications and known for specialists much better than ASR we will consider last process in more details.

Table I. Advantages and disadvantages of different nidriding mediums [7].

Alloying method	Advantages	Disadvantages
Si_3N_4	-Nontoxic -Ease of operation and storage -Ease of dissociation	-Very abrasive (joints and gaskets, valves) -Kinetic of dissociation of N3- -ion must be regarded. -Non-continuous allowance on slags -Silicium increase
N_2 -Gas	Continuous allowance possible -Simple regulation over the pressure -High equal distribution in ingot -Appropriate for Si-critical steel	-Slag composition very important -Sievertrsches law at high pressure not ideally achieved. - Diffusions conditions in system slag-metal must be known

Figure 1 shows the general appearance of P-ESR furnace of Energietechnik Essen GmbH were this process is well developed [7]

Figure 1. View of the PESR furnace at Energietechnik Essen GmbH for ingot weight up to 20 t and diam.1030 mm.

ASR Main Principles

Let's remind that modern ESR processes began from discovery of phenomena of current passing through the melted slag. It is well known this phenomena was revealed at usual arc welding of metal of big thicknesses. Welders surprisingly found out that the arch went out but both the current and welding process continue to go. The process of stable burning of an arch under a layer of slag is used not just at submerge arc welding (SAW), but also in metallurgy, for example, when melting of ferroalloys in Submerge Arc Furnaces. Moreover, so-called transition processes are well familiar to metallurgists too, especially at initial stage of melting in EAF. Thus, ASR was created as a way of use of the effects occurring at frontier regimes in arc and electroslag processes.The various options of ASR implementation are rather diverse and given in literature.

In Figure 2 two main options of ASR implementation in a stationary mould under controlled atmosphere are shown. On the left diagram, the arch burns between electrode edge and slag surface when slag layer is thick. On the right figure – the amount of slag is reduced and arch pressure is big enough to remove slag to mould walls.

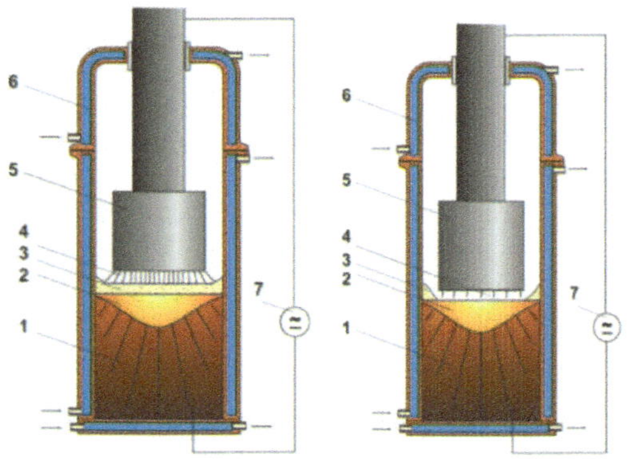

Figure 2. Diagrams of arc-slag remelting of the solid consumable electrode: 1 - ingot; 2 - metal pool; 3 - slag pool; 4 - electric arc; 5 - consumable electrode; 6 – water-cooled mould; 7 - power source.

The pressure force of arc (F_0) on slag is related to the current value by the relationship [6]:

$$F_0 = \frac{\mu_0 I^2}{4\pi} \qquad (1)$$

where μ_0 is the absolute magnetic permeability, H/m; I is the current, A.

Besides the ASR in stationary mould this process can be realized in short collar mould with an ingot withdrawal. For alloying by nitrogen from a gas phase naturally ASR in the nitrogen atmosphere should be organized. Researches of ASR opportunities showed that this process potentially has a number of interesting features. In particular, ARS requires smaller amount of slag and specific power consumption, than at classical ESR [2,6].Table II presents data of both ESR and ASR processes at melting of ingots of the same size on ESR industrial furnace from cast electrodes 375 x375 mm in a mould 565 x565 mm.

Table II. Main parameters of standard ESR and ASR at 565 x565 mm ingot melting

#	Performances	ESR	ASR
1	Voltage between electrode clamping and bottom plate, v	70	120
2	Working current, кA	15	5
3	Voltage drop on electrode, V	14,77	5, 45
4	Slag pool resistance, Om	$3 \cdot 10^{-3}$	$1{,}08 \cdot 10^{-3}$
5	Voltage drop on slag pool, V	45	5
6	Power delivered from power source, kVA	1050	600
7	cosφ	0,77	0,973
8	Specific power consumption, kWh/t	1500	1000

Alloying by Nitrogen

Starting from the very first days of ASR existence the researches were directed as for technological parameters optimizations in order to reduce of power consumption as for usage of potential of alloying from gas phase.

It was found that even small admixtures of oxygen in used nitrogen atmosphere retards nitrogen absorption by metal. The calculated coefficients of mass transfer of nitrogen for 18Cr-18Mn steel have the following values. At oxygen concentration in the nitrogen media of 0.001%, mass transfer coefficient β equals to $1.85 \cdot 10^{-2}$cm/s, and at 0.11% it decreases and makes $1.2 \cdot 10^{-2}$cm/s. On the basis of these investigations results it may be concluded that the effect of oxygen in the gas phase in any melting method decreases the rate of absorption of nitrogen. This takes place due to the fact that the oxygen, adsorbed on the liquid metal surface, occupied part of vacancies and thus decreases the number of adsorbed nitrogen atoms per unit time. Detail investigation of this effect was described in [5].

The vast investigations of nitrogen behavior in various slag systems were made. For example, important data for nitrogen solubility in typical ESR slag were presented in [6] see Table III.

Effects of slag basicity on nitrogen solubility kinetics are given of figure 3 [6].

Results of this researches allow us to make very important conclusion that metal nitriding at ASR can be made without carbon bearing slag, which means that harmful admixtures like CN^- (cyanide or cyanamide) will not appear.

In parallel the researches of alloying by nitrogen were made at ASR of the same steel when remelting under various ESR slag for steel 18Cr-18Mn. The figure 4 generalizes kinetics of nitrogen solubility at remelting under slag of five different compounds and without slag. Figure 4. shows the kinetics of Nitrogen absorption by 18Cr-18Mn steel. The oxygen content of in nitrogen atmosphere was identical in all cases and didn't exceed 0,001%.

Table III. Solubility of nitrogen in ESR slags of different composition at 1723 K

#	Content, %							Basicity of slags $\frac{nCaO+0,45nMgO}{nSiO_2+0,16nAl_2O_3}$
	CaF_2	Al_2O_3	CaO	SiO_2	MgO	C	(N)	
1	84.80	8.10	2.40	2.00	1.00	0.010	0.285/0.586	2.9
2	66.90	8.90	22.16	0.90	0.70	0.025	0.231/0.534	2.1
3	49.50	11.10	20.00	7.20	10.50	0.010	0.133/0.318	1.5
4	38.00	35.90	1.30	20.10	2.80	0.005	0.050/0.902	1.8
5	39.00	36.80	2.00	19.20	1.80	0.015	0.110/0.939	1.9
6	31.00	37.00	13.00	18.00	0.90	0.009	0.210/0.483	1.8

Note. Numerator - initial nitrogen content; denominator -limiting one.

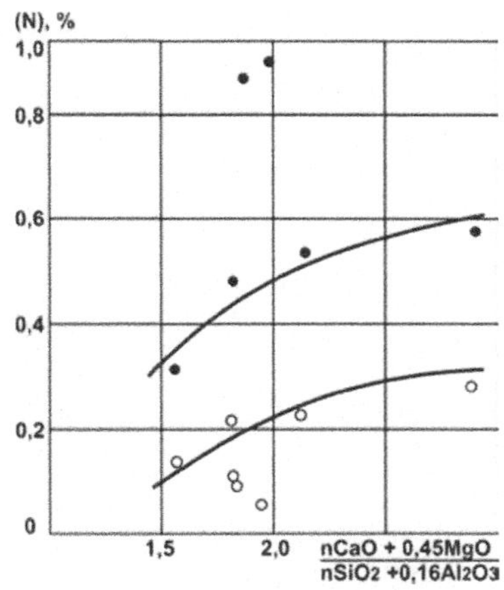

Figure 3. Dependence of nitrogen content in ESR fluxes on their basicity

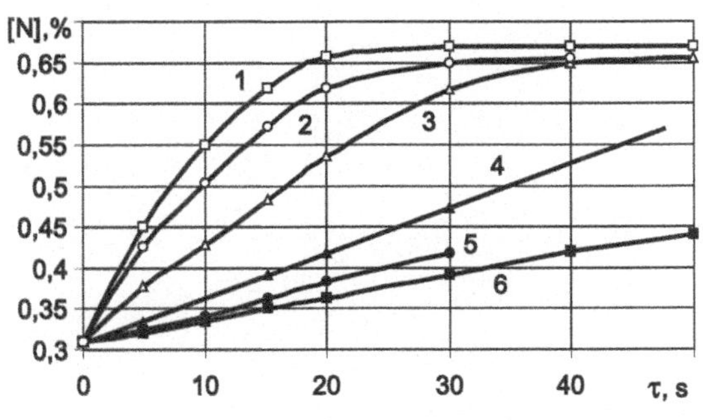

Figure 4. ASR conditions 1) without slag; 2) under slag with composition 100%CaF_2; 3) 70%CaF_2 + 15%CaO + 15%Al_2O_3; 4) 40%CaF_2 + 32%CaO + 23%Al_2O_3 + 5%SiO_2; 5) 15%CaF_2 + 29%Ca0 + 52%Al_2O_3 + 2.5%SiO_2; 6) 50%Ca0 + 50% Al_2O_3[5].

Experimental Data

In order to have bench mark for two process comparison with ETE P-ESR furnace use some experiments on steel alloying from nitrogen atmosphere in P-ESR were made. On the first stage Ø630mm Cronidur30 ingot (0,3%C, 15%Cr, 0,95%Mo) was received. The nitrogen content of the electrode was 0,13%. After remelting with N2 pressure at 30 bar a nitrogen content of 0,38% was achieved in the ingot.

At the next stage the ingot of Ø830mm of the same grade with the same slag was received. It was Wacker slag S2027 (CaO+MgO 17%, CaF2 67%, Al2O3 15%) + 3%SiO2. The pressure was set to appr. 40 bar. The nitrogen content of the electrode was 0,18%. Nitrogen content in ingot was analyzed through its height and cross section (on the surface, in the center and on half radius). Nitrogen content deviation was less than ±0.01% and content was on the same level as in first ingot (but twice increased comparing with electrode).

The natural desire of our team was to make direct comparative experiments on the same P-ESR furnace and to realize ASR trials with the same steel grade and the same slag. At this stage such opportunity wasn't owing to the organizational reasons. Therefore for first approach available data of previous ASR trials which have been carried out on the ESR standard furnaces were used. That trials ASR run under controlled atmosphere was made by means of flux gate [2]. The furnace used an industrial frequency of (50Hz). Consumable electrodes of Ø 600 mm from 18Cr-18Mn steel were melted into an ingot of 860 mm in dia with a partial pressure of nitrogen equal to 1 bar. The content of oxygen in nitrogen didn't exceed 20 ppm. The nitrogen content in the electrode was 0,255%, and within the ingot - 0,42% (1.6 times more).In both cases nitrogen content was increased practically in two times.

Conclusion

Comparison of P-ESR and ASR processes abilities in steel alloying by nitrogen from gas phase shown quite close results that gives us reasons for experiments continuation. Next stage will be direct comparison of processes results and performances on same furnace, with same steel and slag. Then combination of both processes P-ASR can be tested.

References

[1] B.E. Paton et al., Method of consumable electrode remelting, Author's certificate 520784 USSR, Filed 09.08.1974, Publ. in Bull. of Inventions, 1982, 20-22.

[2] Lev Medovar et al., *Arc-slag remelting of steels and alloys* (Cambridge International Science Publishing, 1996), 160.

[3] Gerald Stein and Joachim Lueg, "High Nitrogen Steels - Applications. Present and Future"(Paper presented at the High Nitrogen Steels Conference, Kiev, Ukraine, September 1993), 31-44.

[4] A. Carosi et al., "Mastering P-ESR technology for high nitrogen steel grades for high value applications"(Final report for

European Commission, Contract No RFSR-CT-2005-00004, 2010).

[5] Lev Medovar et al., "Influence of the flux composition and state of the gas atmosphere over the kinetics of nitrogen absorption by steels and alloys at the arc slag remelting" (Paper presented at High Nitrogen Steels Conference, Espoo, Finland, Stockholm, Sweden, May 1998), 53-60.

[6] Yu.M. Pomarin et al., " Importance of slag in producing ingots of high nitrogen steels by the method of Arc Slag Remelting", " (Paper presented at High Nitrogen Steels Conference, Ostend, Belgium, September 2004), 311-316

[7] Roman Ritzenhoff and Andre Hahn, "Corrosion Resistance of High Nitrogen Steels", *Corrosion Resistance*, Ed.Dr Shih, 2012.

Defects

Liquid Metal Processing & Casting 2013

Proceedings of the 2013 International Symposium on Liquid Metal Processing & Casting
Edited by: Matthew J. M. Krane, Alain Jardy, Rodney L. Williamson, and Joseph J. Beaman
TMS (The Minerals, Metals & Materials Society), 2013

NUMERICAL SIMULATION OF MACROSEGREGATION IN 570-TON LOW-ALLOYED STEEL INGOT

Tomoki Sawada, Koji Kajikawa
The Japan Steel Works, Ltd; 4 Chatsu-Machi, Muroran, Hokkaido, 051-8505, JAPAN

Keywords: Macrosegregation, Ingot Casting, Numerical Analysis

Abstract

We developed a numerical analysis code for macrosegregation in large ingots of Fe-C based multi-component alloys. A numerical simulation of macrosegregation in a 570-ton low-alloyed steel ingot was performed using the developed code in 2-dimensional axisymmetric coordinates. The calculated segregation pattern was compared to the macrostructure of the actual ingot to verify the practical effectiveness of the developed code. A solidification-microsegregation model that takes into account both the equilibrium redistribution of carbon and the non-equilibrium redistribution of other elements was formulated and used in this simulation. In the simulation, numerous inverted V segregations formed as was observed in the longitudinal macrostructure of the actual ingot. The computed carbon profile along the centerline of the ingot presents a similar tendency to that of the actual measurement value. The developed code can be applied to the prediction of macrosegregation in large steel ingots.

Introduction

Large ingots take a long time to solidify, for example, the solidification of a 600 t ingot takes over three days. During this time, the concentrated liquid that is redistributed at the solid/liquid interface is transported throughout the ingot by thermo-solutal convection. As a result, macrosegregation ineluctably occurs. The positive segregation that is caused by gravity segregation and the negative segregation that is caused by the settlement of crystals are examples of macrosegregation. An ingot that contains such positive or negative segregation region can be regarded as defective when the concentration in that region is not acceptable in a product. Thus, it is important to remove the segregation region from a product region of an ingot by optimizing the ingot shape, amount of riser, and so forth. This requires a deep understanding of the macrosegregation, including the concentration distribution in various ingot shapes and the cooling conditions under which these optimizations are performed. However, it is not practical to produce ingots that are more than 600 t experimentally in order to investigate the distribution. Also, it is very difficult to extrapolate the distribution of large ingots from laboratory-scale experiments, since it is difficult to reproduce unique phenomena in large ingots, such as size effect, settling crystals, and channel type segregation, and fit all dimensionless numbers regarding the transportation phenomena. Therefore, it is important to employ numerical analysis of macrosegregation on the optimizations.

In the 1990's, many researchers studied numerical analysis for macrosegregation.[1-3] At present, basic methods of numerical analysis for macrosegregation to solve a fluid-solidification coupled problem are described by the momentum equation of fluid with considering Darcy law, continuity equation, energy conservation law, and species conservation law. Although the systems applied in early studies are applied to small ingots, now, analyses of commercial scale ingots can be performed[4,5] because of the rapid advances in computer performance. In recent studies, some models take account of the movement of equiaxed crystals and settling crystals[5], then, transportation phenomena in ingots are clarified.

While it important to develop models of such macroscopic kinetics and transportation, modeling a unique microsegregation behavior is important in macrosegregation in steel ingots. In the solid phase, the diffusion coefficient of carbon elements contained in steel alloy is much higher than the coefficient of other added elements. The microsegregation behaviors of most added elements are assumed as non-equilibrium redistribution, generally. On the other hand, the behavior of carbon is often assumed as equilibrium redistribution. CALPHAD[6] is an effective calculation method that can compute the time evolution of solid fractions, solid and liquid concentrations in such complex redistributions. Thus, the fluid-CALPHAD coupled method[7] is considered one of the most accurate methods. However coupling CALPHAD may increase the calculation time. In addition, an explicit evolution equation of a solid fraction cannot be obtained, when CALPHAD is employed in numerical analysis. Therefore it is difficult to understand solidification phenomena by analytical manipulation of the governing equation, such as integration, derivation, and stability analysis. In this study, a microsegregation and solidification model was formulated that considers both equilibrium and non-equilibrium redistribution of microsegregation. Numerical analysis of macrosegregation in a 570-ton low-alloyed steel ingot was performed according to the formulated model. This numerical result was compared with the concentration profile and macrostructure of an actual ingot.

Numerical Method

Governing Equations

It was assumed that the melt was an incompressible fluid and flow in the mushy zone can be expressed by Darcy's law. The thermo-solutal convection of the melt was taken into account with a density variation by temperature and liquid concentration; however, it was assumed that divergence free, Eq. (2) is satisfied. Solutes in the liquid were moved by the fluid flow and diffusion and the redistribution of the solute at the interface between the liquid and the solid was taken into account. The diffusion of the solute was neglected, because it was smaller than the numerical diffusion. The governing equations were as follows:

$$\frac{\partial \boldsymbol{u}}{\partial t} + \boldsymbol{u} \cdot \nabla \boldsymbol{u} = -\frac{\nabla p}{\rho} - \frac{\nu}{K}\boldsymbol{u} + \nu \nabla^2 \boldsymbol{u} + g \tag{1}$$

$$\nabla \cdot \boldsymbol{u}_E = 0 \tag{2}$$

$$\frac{\partial T}{\partial t} + \nabla \cdot T\boldsymbol{u}_E = \frac{\nabla \cdot \boldsymbol{q}}{\rho c} + \frac{H}{c}\frac{\partial f_S}{\partial t} \tag{3}$$

$$\frac{\partial C_{L,i}}{\partial t} + \nabla \cdot C_{L,i} \boldsymbol{u}_E = D_{L,i} \nabla^2 C_{L,i} + \frac{\partial C_{L,i}}{\partial f_S} \frac{\partial f_S}{\partial t} \tag{4}$$

Equations (1), (2), (3) and (4) are the momentum equation, the equation of continuity, the energy equation, and the solute conservation equation in liquid, respectively, where, t is the time, $\boldsymbol{u}$ and $\boldsymbol{u}_E$ is the flow velocity and empty tower velocity, ρ is the density, p is the pressure, ν is the kinetic viscosity, K is the permeability, $\boldsymbol{g}$ is the gravity vector, T is the temperature, c is the specific heat, $\boldsymbol{q}$ is the vector of the heat flux, H is the latent heat in fusion, fs is the solid fraction, $C_{L,i}$ is the concentration of a i-th solute element in liquid. The second term on the right hand side in the solute conservation equation, *i.e.* the derivative of liquid concentration with respect to the solid fraction, shows micro segregation behavior. This derivative is described in detail in the following section.

Description of Microsegregation

Researchers have proposed various microsegregation models. These models describe the relationship between the liquid concentration and the solid fraction in a closed system and are expressed as an integration form of mass conservation equation. However, in numerical analysis of macrosegregation, each control volume cannot be regarded as a closed system, since there are various mass transfers among all control volumes. Thus, in this model, microsegregation behavior and other phenomena are described with a differential form of microsegregation. Equations (5) and (6) are differential forms of redistribution of the equilibrium solidification and Scheil-Gulliver model, respectively.

$$\frac{\partial C_L}{\partial f_S} = \frac{(1-k)C_L}{1+(k-1)f_S} \tag{5}$$

$$\frac{\partial C_L}{\partial f_S} = \frac{(1-k)C_L}{1-f_S} \tag{6}$$

Using a differential form, it is unnecessary to assume distribution coefficient k as constant. Other microsegregation models that consider more complicated back diffusion can also be applied to this model. Most back diffusion models[8-10] are expressed by Eq. (7). The differential form of this equation is Eq. (8).

$$C_L = C_0 [1-(1-\beta k) f_S]^{\frac{k-1}{1-\beta k}} \tag{7}$$

$$\frac{\partial C_L}{\partial f_S} = \frac{(1-k)C_L}{1-(1-\beta k)f_S} \tag{8}$$

Where, β is the back diffusion parameter. Although such back diffusion model can be employed in the proposed model, numerical analysis was performed with the simple equilibrium model and the Scheil-Gulliver model because of difficulty in deciding the parameter β.

However, there is a peritectic transformation point in temperature during the solidification of most steel alloys. Therefore, the two solid phases, δ and γ, precipitate during solidification and each have different partition coefficients. In this simulation, as the partition coefficient of an element that is assumed to be equilibrium solidification, the average value of the δ phase and γ phase was used. As the partition coefficient of the Scheil type element, one of the coefficients of δ or γ was used.

Evolution Equation of Solid Fraction

The temperature recovery method was used in this simulation. However, the evolution equation of the solid fraction should be formulated by an assumption in this method. In this study, we assumed that the following condition is satisfied in the temperature field during solidification.

$$T = T_L(C_{L,1}, C_{L,2}, C_{L,3}, \ldots, C_{L,N}) \tag{9}$$

Eq. (9) means that the temperature is consistently equal to the liquidus temperature in each control volume. However, the evolution equation of the solid fraction cannot be derived from Eq. (9) directly. Thus, the temporal differential form of Eq. (9), which is expressed as Eq. (10), is used in this model, assuming the liquidus temperature depends only on the liquid concentration.

$$\frac{\partial T}{\partial t} = \sum_{i=1}^{N} \frac{\partial T_L}{\partial C_{L,i}} \frac{\partial C_{L,i}}{\partial t} \tag{10}$$

By substituting Eq. (3) and (4) into Eq. (10), an equation about the time variation of the solid fraction is obtained.

$$F_T + \frac{H}{c}\frac{\partial f_S}{\partial t} = \sum_{i=1}^{N} \frac{\partial T_L}{\partial C_{L,i}} \left(F_{C_{L,i}} + \frac{\partial C_{L,i}}{\partial f_S}\frac{\partial f_S}{\partial t} \right) \tag{11}$$

$$F_T = -\nabla \cdot T\boldsymbol{u}_E + \frac{\nabla \cdot \boldsymbol{q}}{\rho c} \tag{12}$$

$$F_{C_{L,i}} = -\nabla \cdot C_{L,i}\boldsymbol{u}_E + D_{L,i}\nabla^2 C_{L,i} \tag{13}$$

Then, the evolution equation of the solid fraction can be expressed as Eq. (14), rewriting Eq. (11).

$$\frac{\partial f_S}{\partial t} = \left(\sum_{i=1}^{N} \frac{\partial T_L}{\partial C_{L,i}} \frac{\partial C_{L,i}}{\partial f_S} - \frac{H}{c} \right)^{-1} \left(F_T - \sum_{i=1}^{N} \frac{\partial T_L}{\partial C_{L,i}} F_{C_{L,i}} \right) \tag{14}$$

However, eq. (10) is implicit expression that describes temperature is equal to liquidus temperature. When difference formula that Eq. (14) is directly discretized is used in numerical analysis, the difference between the temperature and the liquidus temperature might be caused by numerical error. Therefore, instead of Eq. (10), the following equation, which is a first-order temporal integration, was used to stabilize the numerical analysis.

$$T + \Delta t \frac{\partial T}{\partial t} = T_L + \Delta t \sum_{i=1}^{N} \frac{\partial T_L}{\partial C_{L,i}} \frac{\partial C_{L,i}}{\partial t} \tag{15}$$

A semi-discretized evolution equation is derived from Eq. (15), (3) and (4).

$$\frac{\partial f_S}{\partial t} = \left(\sum_{i=1}^{N} \frac{\partial T_L}{\partial C_{L,i}} \frac{\partial C_{L,i}}{\partial f_S} - \frac{H}{c} \right)^{-1} \left(\frac{T - T_L}{\Delta t} + F_T - \sum_{i=1}^{N} F_{C_{L,i}} \frac{\partial T_L}{\partial C_{L,i}} \right) \tag{16}$$

As is clear from Eq. (16), when the temperature is equal to the liquidus temperature, Eq. (14) and (16) are equivalent.

Flow Analysis

The fractional step method[11] was used as the basic algorithm in the flow analysis and modified to allow analysis of the flow in the mushy zone stably. The Kozeny-Carman model[12] was used to express permeability in Darcy term. An overview of the algorithm is shown below. Eq. (1) is discretized as Eq. (17).

$$\frac{\boldsymbol{u}^{n+1}-\boldsymbol{u}^{n}}{\Delta t}+\boldsymbol{u}^{n}\cdot\nabla\boldsymbol{u}^{n}=-\frac{\nabla p^{n+1}}{\rho^{n}}-\frac{\nu^{n}}{K^{n}}\boldsymbol{u}^{n+1}+\nu\nabla^{2}\boldsymbol{u}^{n}+\boldsymbol{g} \tag{17}$$

Here, the liquid density, ρ^n, was computed for a given temperature and liquid concentration. An advection term, the second term on the left-hand side, was discretized using a Constraint Interpolated Profile (CIP) method[13] that has time-space 3rd-order accuracy. By introducing an intermediate velocity field, $\boldsymbol{u}^*$, Eq. (5) was rewritten as Eq. (18) and (19).

$$\frac{\boldsymbol{u}^{*}-\boldsymbol{u}^{n}}{\Delta t}+\boldsymbol{u}^{n}\cdot\nabla\boldsymbol{u}^{n}=\nu\nabla^{2}\boldsymbol{u}^{n}+\boldsymbol{g} \tag{18}$$

$$\frac{\boldsymbol{u}^{n+1}-\boldsymbol{u}^{*}}{\Delta t}=-\frac{\nabla p^{n+1}}{\rho^{n}}-\frac{\nu^{n}}{K^{n}}\boldsymbol{u}^{n+1} \tag{19}$$

The pressure Poisson equation (Eq. (21)) was obtained by applying a divergence operator to Eq. (19). Here, the dimensionless number E, defined in Eq. (20), is introduced.

$$E^{n}=\frac{K^{n}}{K^{n}+\Delta t\nu^{n}} \tag{20}$$

$$\nabla\cdot\left(E^{n}\frac{\nabla p^{n+1}}{\rho^{n}}\right)=\nabla\cdot\left(\frac{E^{n}\boldsymbol{u}_{E}^{*}}{\Delta t}\right) \tag{21}$$

Empty tower velocity $\boldsymbol{u}_E$ is estimated by Eq. (22).

$$\boldsymbol{u}_{E}=(1-f_{S})\boldsymbol{u} \tag{22}$$

Use of E instead of K makes the numerical calculation stable because E may change from 0 (solid phase) to 1 (liquid phase), while K may change from 0 (solid phase) to infinity (liquid phase).

Discretization of Energy and Solute Conservation

In temporal discretization of the energy and solute conservation law, intermediate temperature and species fields (T^* and C_{Li}^*) are also introduced. Intermediate fields are explicitly calculated by Eq. (23) and (24).

$$\frac{T^{*}-T^{n}}{\Delta t}=-\nabla\cdot\left(T^{n}\boldsymbol{u}_{E}{}^{n}+\frac{\boldsymbol{q}^{n}}{\rho c}\right) \tag{23}$$

$$\frac{C_{L,i}^{*}-C_{L,i}^{n}}{\Delta t}=-\nabla\cdot\left(C_{L,i}^{n}\boldsymbol{u}_{E}-D_{L,i}\nabla\cdot C_{L,i}^{n}\right) \tag{24}$$

In this simulation, when advection terms of Eq. (23) and (24) were solved, second-order TVD upwind scheme and second order Runge-Kutta time integration was used. Then, the solid fraction is computed by discretized Eq. (16) shown as Eq. (25)

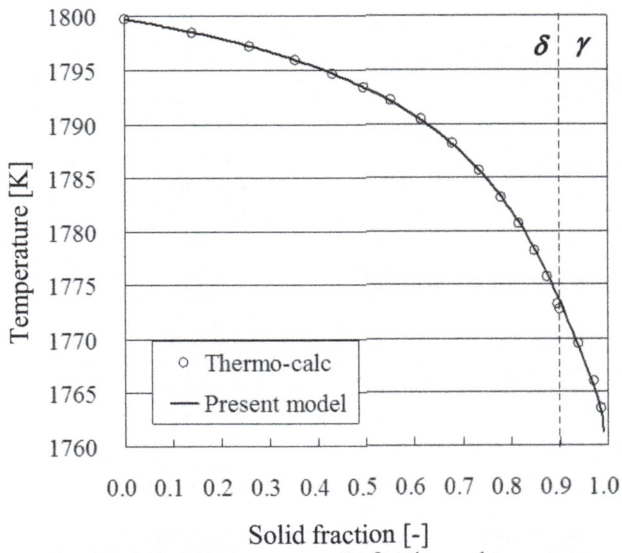

Fig. 1 Relationship between solid fraction and temperature obtained with present model (solid line) compared with Thermo-calc (hollowed circle).

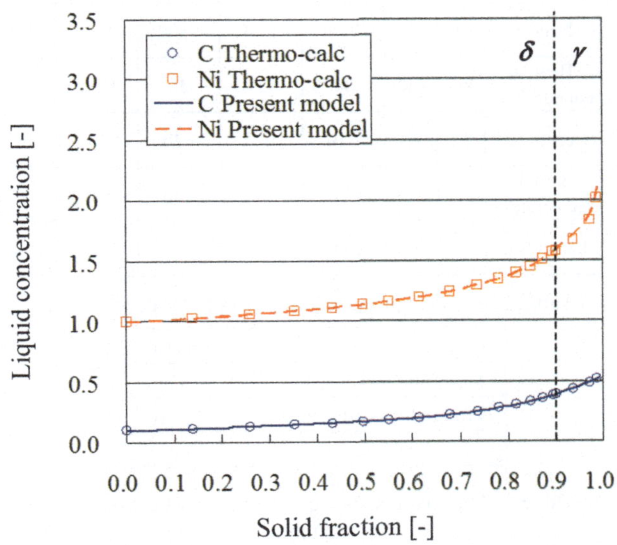

Fig. 2 Relationship between liquid concentration and solid fraction obtained with present model (solid line) and Thermo-calc (hollowed circle).

$$\frac{f_{S}{}^{n+1}-f_{S}{}^{n}}{\Delta t}=\frac{\dfrac{T^{n}-T_{L}{}^{n}}{\Delta t}+\left(T^{*}-T^{n}\right)-\sum_{i=1}^{N}\left(C_{L,i}{}^{*}-C_{L,i}{}^{n}\right)\dfrac{\partial T_{L}}{\partial C_{L,i}}}{\sum_{i=1}^{N}\dfrac{\partial T_{L}}{\partial C_{L,i}}\dfrac{\partial C_{L,i}}{\partial f_{S}}-\dfrac{H}{c}} \tag{25}$$

Finally, the solid fraction term in Eq. (3) and Eq. (4) was computed to update the temperature and liquid concentration fields.

$$\frac{T^{n+1}-T^{*}}{\Delta t}=\frac{H}{c}\frac{f_{S}{}^{n+1}-f_{S}{}^{n}}{\Delta t} \tag{26}$$

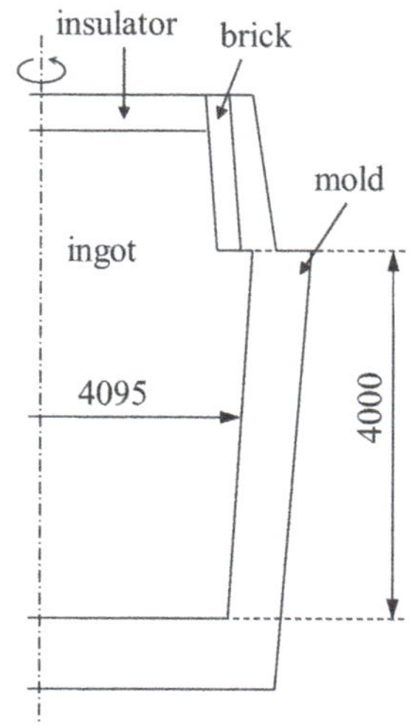

Fig. 3 Schematic view of computational domain of ingot

Table 1: Physical properties of material used in simulation

		Ingot	Air	Mold	Brick	Insulator
Density	kg/m^3	7000	1.33	7800	2300	1500
Thermal conductivity	J/(s·m·K)	35	1	25	0.42	3.256
Specific heat	J/(kg·K)	680	10^{+300}	670	1130	1000
Initial temperature	K	1802	300	300	300	300
Latent heat	kJ/kg	260	-	-	-	-

Table 2: Physical properties and parameter used in flow

Kinetic viscosity	m^2/s	2.0×10^{-6}
Max. Courant number	-	0.35
Permeability coefficient	m^2	1.25×10^{-8}
Grid size	m	0.02

Table 3: Thermal resistances of boundary between each material

		Ingot	Air	Mold	Brick	Insulator
Ingot	s·m^2·K/J	-	10^{+300}	0.012	0.028	10^{+300}
Air		-	-	10^{+300}	10^{+300}	10^{+300}
Mold		-	-	-	0.02	10^{+300}
Brick		-	-	-	-	0.02
Insulator		-	-	-	-	-

$$\frac{C_{L,i}^{\,n+1}-C_{L,i}^{\,*}}{\Delta t}=\frac{\partial C_{L,i}}{\partial f_S}\frac{f_S^{\,n+1}-f_S^{\,n}}{\Delta t} \qquad (27)$$

Results

Validation of Microsegregation-Solidification Model

The proposed method is expected to be capable of tracing the microsegregation of CALPHAD or the solidification pass of an actual alloy when the distribution coefficient and liquidus temperature are known as functions of the liquid concentration

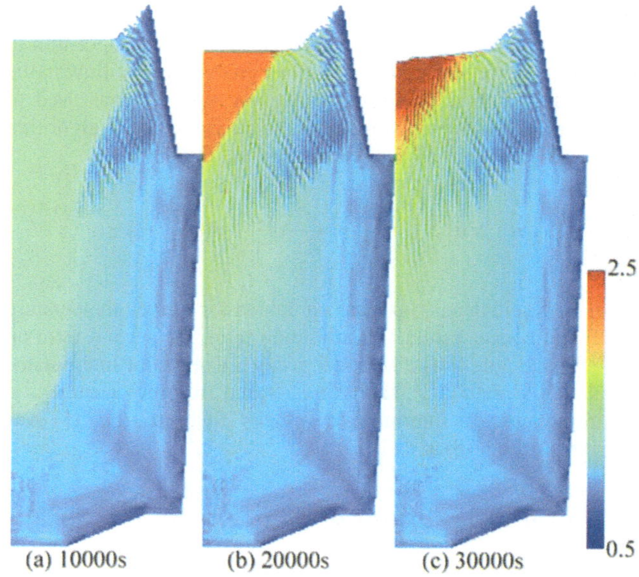

Fig. 4 Transition of carbon segregation ratio distribution obtained with numerical simulation.

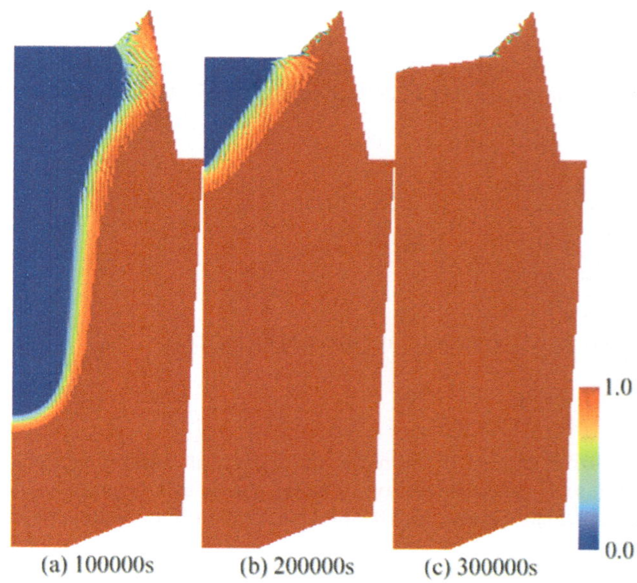

Fig. 5 Transition of computed solid fraction distribution

and temperature. To confirm this, the microsegregation behavior was compared between the computed results of Thermo-calc[14] and the proposed model.

In this comparison, the chemical composition of the model alloy is Fe-0.1mass%C-1.0mass%Ni. In computation of the Thermo-calc, it is assumed that the redistribution of carbon can be regarded as equilibrium and nickel obeys the Scheil model. In calculation by the model, the liquidus temperature was described as a polynomial function of liquid concentrations, and partition coefficients were also expressed as similar functions of temperature and liquid concentrations. The functions are constructed from the result of Thermo-calc by least square method and consist of second and first order terms of all variables and cross terms between all variables.

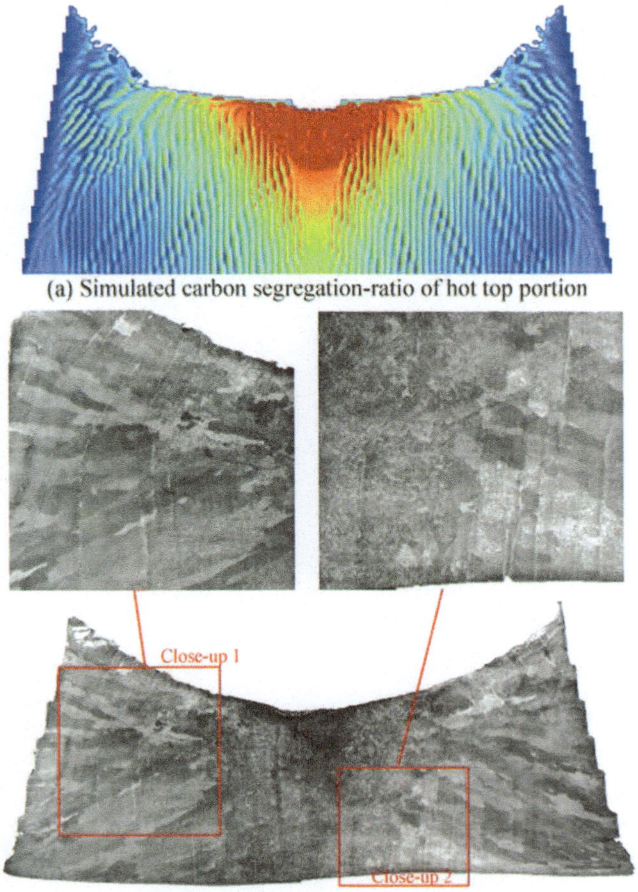

(a) Simulated carbon segregation-ratio of hot top portion

(b) Macrostructure of hot top portion

Fig. 6 Computed C segregation ratio of hot top and macrostructure of actual ingot

The results are shown in Figs. 1 and 2. Figure 1 shows the solid fraction-temperature curve that is computed by Thermo-calc and the present model. Also, Fig. 2 shows C (equilibrium) and Ni (Scheil) microsegregation behavior obtained with both calculations. The vertical dashed line located at solid fraction 0.90 in both figures indicates the peritectic transformation point of this model alloy.

In each case, the result of the present method is in good agreement with the result of Thermo-calc. In addition, the computation rate of the present model is four hundred times faster than Thermo-calc. Thus, the present method can replicate accurate analysis of microsegregation rapidly, when appropriate liquidus temperature and partition coefficients are obtained.

Macrosegregation of 570 t ingot

Numerical analysis of macrosegregation in a 570-ton low-alloyed steel ingot was performed according to the present method discretized in a two-dimensional axisymmetric coordinate. The average diameter and height of the ingot were 4095 mm and 4000 mm, respectively. As can be seen in the schematic view of the computational system in Fig. 3, the system consists of an ingot, mold, brick, and insulator. Their physical properties, other parameters and properties, and thermal resistance of the boundary between each material used in the simulation are shown in Tables 1, 2 and 3. The Carman-Kozeny model was used in this simulation, as the isotropic permeability model. The Carman-Kozeny model is expressed by Eq. (28).

$$K = \frac{d^2}{180}\frac{f_S^{\,2}}{(1-f_S)^3} \tag{28}$$

Here, d denotes the characteristic scale in a porous medium. Primary dendrite-arm-spacing (PDAS) was adopted as the characteristic scale in this simulation. PDAS in a 570-ton ingot was assumed to be 1.50 mm, since the result of average PDAS measurement was about 1.50 mm in the macrostructure of a 600-ton scale ingot.

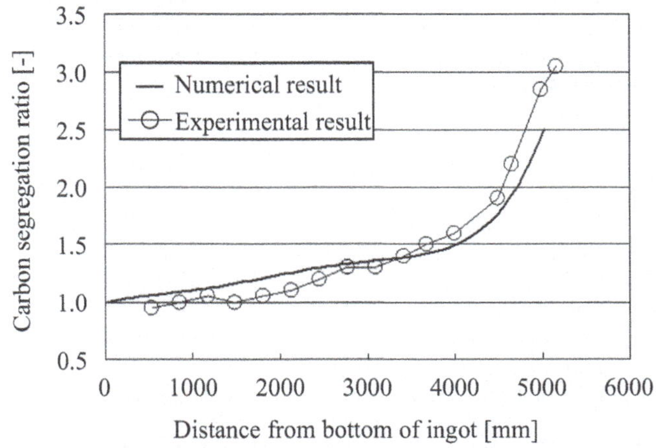

Fig. 7 Carbon segregation ratio profile along ingot axis obtained with present method and experimental result

The segregation-ratios of carbon and solid fraction distributions obtained with the present model are shown in Figs. 4 and 5. In each figure, three distributions denote 100000 s, 200000 s and 300000 s from the start of solidification. Numerous high-concentrated streaks appear in the segregation-ratio distribution, and a low solid fraction channel can be observed at the same position in the solid fraction distribution. According to this characteristic, these streaks can be thought of as inverted V segregation. For comparison of inverted V segregation between the numerical result and the actual macro structure, the computed C segregation-ratio in the hot-top portion was examined in detail, as shown in Fig. 6, together with the macrostructure of the hot-top portion in the actual ingot. In the macrostructure of the hot-top, most segregations were V-shaped (Close-up 2) and the streak directions of some segregations were curved (Close-up 1). Such streak in the downside is toward the mold side and in the upside is toward the ingot center. In the numerical result, also, the shapes of most channels show a similar inverted V-shape, and bending streaks are present similar to the ones observed in the macrostructure. Thus the proposed method is able to replicate the tendency of inverted V segregation in an actual 570-ton ingot.

Concentration profiles along the ingot axis are shown in Fig. 6 for comparison between the numerical result and the experimental result of the actual ingot. The computed values are slightly higher than the actual measurement value at a distance of less than 2500 mm from the bottom. On the other hand, at a distance of more 2500 mm from the bottom, the computed values are lower than the actual measurement values. However, it is confirmed that the result computed by the present method replicates a solution that is close to the actual segregation profile.

Conclusion

A solidification-microsegregation model that is applicable to Fe-C alloy was formulated. This model can deal with the redistribution behaviors of both equilibrium and non-equilibrium and is capable of reproducing accurate microsegregation behavior carried out by Thermo-calc rapidly.

The proposed solidification-microsegregation model was used to perform numerical analysis of macrosegregation in a 570-ton low-alloyed steel ingot. The inverted V segregation that appears in the numerical result was compared with the segregation in the macrostructure of the hot-top portion. The computed result has a similar feature of actual inverted V segregation, in terms of the direction and shape of the streak. The computed carbon profile along the centerline of the ingot presents a similar tendency to the actual measurement value.

References

1. M. C. Schneider and C. Beckermann: *Metall. Mater. Trans.* A, 1995, Vol. 26A, pp. 2373-2388.
2. D. G. Neilson and F. P. Incropera: *Int. J. Heat Tran.*, 1991, Vol. 34, No. 7, pp. 1717-1732.
3. S. D. Felicelli, D. R. Poirier, and J. C. Heinrich: *Metall. Mater. Trans.* B, 1998, Vol. 29B, pp. 847-855.
4. H. Combeau, M. Založnik, S. Hans, Richy, and E. Pierre: *Metall. Mater. Trans.* B, 2009, Vol. 40B, pp. 289-304.
5. Y. Ebisu: *Metall. Mater. Trans.* B, 2011, Vol. 42B, pp. 341-369.
6. T. G. Chart, J. F. Counsell, W. Slough, and P. J. Spencer: *Int. Met. Rev.,* 1975, Vol. 20, p. 57-82.
7. M. C. Schneider, J. P. Gu, C. Beckermann, W. J. Boettinger, and U. R. Kattner: *Metall. Mater. Trans.* A, 1997, Vol. 28A, pp. 1517-1531.
8. H. D. Brody and M. C. Flemings: *Trans.* AIME, 1966, Vol. 236, pp. 615-624.
9. T. W. Clyne and W. Kurz: *Metall. Trans.* A, 1981, Vol. 12A, pp. 965-972.
10. I. Ohnaka: *Trans.* ISIJ, 1986, Vol. 26, pp. 1045-1051.
11. J. Kim and P. Moin, *J. Comp. Phys.*, 1985, Vol. 59, pp. 308-323.
12. C. Pequet, M. Gremaud, and M. Rappaz, *Met. Mater. Trans.* A, 2002, Vol. 33A, pp. 2095-2106.
13. F. Xiao, T. Yabe, G. Nizam and T. Ito, *Comput. Phys. Commun.*, 1996, Vol. 93, 103-118.
14. B. Sundman, B. Jansson, and J. O. Anderson, *CALPHAD* 9, 1985, Vol. 2, 153-190.

Proceedings of the 2013 International Symposium on Liquid Metal Processing & Casting
Edited by: Matthew J. M. Krane, Alain Jardy, Rodney L. Williamson, and Joseph J. Beaman
TMS (The Minerals, Metals & Materials Society), 2013

USING A THREE-PHASE MIXED COLUMNAR-EQUIAXED SOLIDIFICATION MODEL TO STUDY MACROSEGREGATION IN INGOT CASTINGS: PERSPECTIVES AND LIMITATIONS

M. Wu[1,2], J. Li[1,2], A. Kharicha[1,2], A. Ludwig[2]
[1] Christian Doppler Laboratory for Advanced Process Simulation of Solidification and Melting, Univ. of Leoben, Austria
[2] Chair of Simulation and Modelling of Metallurgical Processes, Univ. of Leoben, Austria

Keywords: Steel, Ingot Casting, Macrosegregation, Multiphase Simulation

Abstract

A three-phase model for mixed columnar-equiaxed solidification was recently developed. The most critical features, necessary for modelling the macrosegregation, were considered: the progressive growth of the columnar dendrite trunks from the ingot surface, the nucleation and growth of the equiaxed crystals including the motion of the equiaxed crystals, the thermal and solutal buoyancy flow and its interactions with the growing crystals (equiaxed and columnar), the transport of solute due to melt convection and equiaxed sedimentation, and the columnar-to-equiaxed transition (CET). Application of the aforementioned model is mainly limited by two factors: one is the extreme computational expense; one is the lack of reliable parameters required by the model. In order to perform a calculation of industry ingot (up to hundreds of tons) on the basis of the current computer resources, a compromise is often made between the model capability and the computational feasibility, i.e. some necessary model simplifications have to be made. In this article the ongoing efforts to scale-up the current model for industry applications are reported on.

Introduction

Most valuable experimental research on the macrosegregation in large steel ingots was performed approximately one century ago [1-2]. A series of steel ingots, scaled from a few hundred kilograms up to 172 tons, were poured and cut for segregation analysis. Primary knowledge was obtained, and a typical segregation map of the large steel ingots was drawn [3-5], as shown in Figure 1. By now, most segregation phenomena can be physically explained. Multiphase flow such as thermo-solutal convection, happening in the interdendritic and bulk regions, and crystal sedimentation during solidification are the key mechanism for the formation of segregation. The thermodynamics, solidification kinetics and thermal mechanics are also coupled with the flow phenomena, and contribute to the final segregation results.

Today due to extremely high costs these kinds of experimental trials are only carried out occasionally and with caution [6-9]. Instead, the mathematical (both analytical and numerical) modelling approach becomes a most efficient tool for this purpose. Some progress has been reviewed by other authors [9-12]. Understanding of the segregation mechanism was significantly improved by the mathematical models. Unfortunately, the patterns schematically shown in Figure 1 are still not quantitatively reproducible with sufficient details by current numerical models. The great challenge arises from the multiphase nature of the solidification phenomenon. The solution of the segregation problem demands a precise description of the multiphase flow, which occurs and interacts with the solidifying microstructure (dendritic morphology) at different length scales. From the flow dynamic point of view, at least three (hydrodynamic) phases are involved in a typical ingot casting during solidification: two moving phases (liquid and equiaxed crystals) and one stationary phase (columnar dendrite trunks). In other words, a model able to reproduce the patterns of Figure 1 needs at least to consider these three phases. However, the limitation of early computational hardware resources has prevented people from considering so many phases. A compromise has to be made between the model capability and the computational feasibility. For example, Gu and Beckermann used a mixture liquid-columnar solidification model [13] and Combeau and co-workers used a two-phase equiaxed solidification model [8] to simulate the segregation in steel ingots, and some successes were achieved.

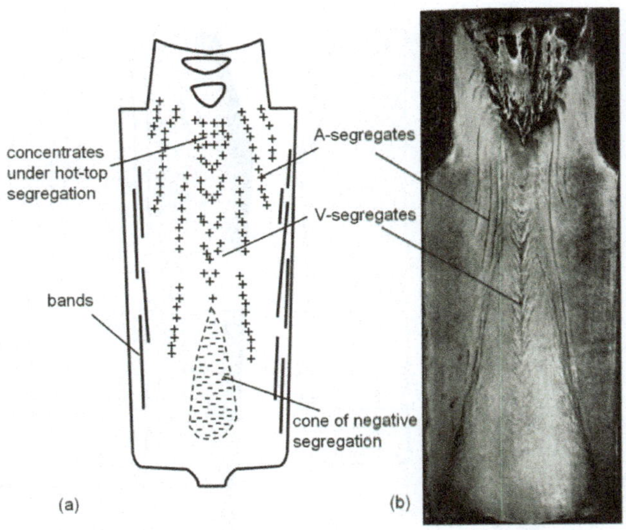

Figure 1: Typical segregation map in steel ingots (the figure is reproduced from literature [3]). (a) schematic representation with '+' for positive and '-' for negative segregation; (b) sulphur print of a 10-ton ingot.

This article does not give a comprehensive review of the topic of macrosegregation models, but focuses on the relevant activities being performed by the current authors at the University of Leoben. On the basis of previous work by Beckermann [11, 14-17], a series of multiphase solidification and macrosegregation models were proposed. These include a two-phase globular equiaxed solidification model [18-20], a two-phase monotectic solidification model [21-22], a three-phase mixed columnar-equiaxed solidification model [23-24], an equiaxed solidification model with dendritic morphology [25-26] and a five-phase mixed columnar-equiaxed solidification model with dendritic morphology [27-28]. As the computational expense increases relative to the increasing number of phases, the modelling activities take two directions. One is to further develop comprehensive models by including as many phases and physical phenomena as necessary in order to on one hand solve as much as possible segregation features. Application of this kind of model may still rely on future enhancement of the hardware resource. On the other hand, we can base the calculations on the available hardware resource by using as simple as possible a model to solve the principal segregation phenomena of the industry ingots. This article reports on some modelling examples by using a three-phase mixed columnar-equiaxed model [23-24]. The applicability of this model to the industry ingots is investigated, and some perspectives and limitations are discussed.

Characterisation of the Three-phase Model

To characterise the three-phase mixed columnar-equiaxed solidification model, a benchmark (ϕ 66 mm x 170 mm) of a steel ingot was simulated. Macrosegregation formation due to the combined thermosolutal convection, grain sedimentation, and sedimentation-induced convection was modelled. Details about the settings for this benchmark stem from previous publications [23, 24, 29]. Model assumptions are summarised as follows:

- Solidification starts with an initial concentration Fe-0.34 wt.%C and an initial temperature of 1785 K; mould filling is ignored;
- The three phases considered are: the melt, globular equiaxed crystals and columnar dendrite trunks;
- Morphologies are approximated by step-wise growing cylinders for columnar dendrite trunks and spheres for globular equiaxed crystals;
- Columnar trunks grow from the side and bottom walls, and the columnar tip front is explicitly tracked;
- A three-parameter heterogeneous nucleation law is implemented for the nucleation of the equiaxed crystals [30]. The 3 parameters are maximum grain number density n_{max} [m^{-3}], Gaussian distribution width of nucleation ΔT_σ [K] and undercooling at the maximum grain production rate ΔT_N [K].
- Solidification shrinkage is ignored. The buoyancy force for the thermosolutal convection and crystal sedimentation is accounted for by a Boussinesq approximation;
- The equiaxed crystals ahead of the columnar tip front can move freely, but they can be captured by the columnar trunks as the local columnar volume fraction is beyond 0.2;
- Hunt's blocking mechanism [31] is applied for predicting CET (columnar-to-equiaxed transition);
- Constant heat transfer coefficients and constant ambient temperatures are assumed [23].

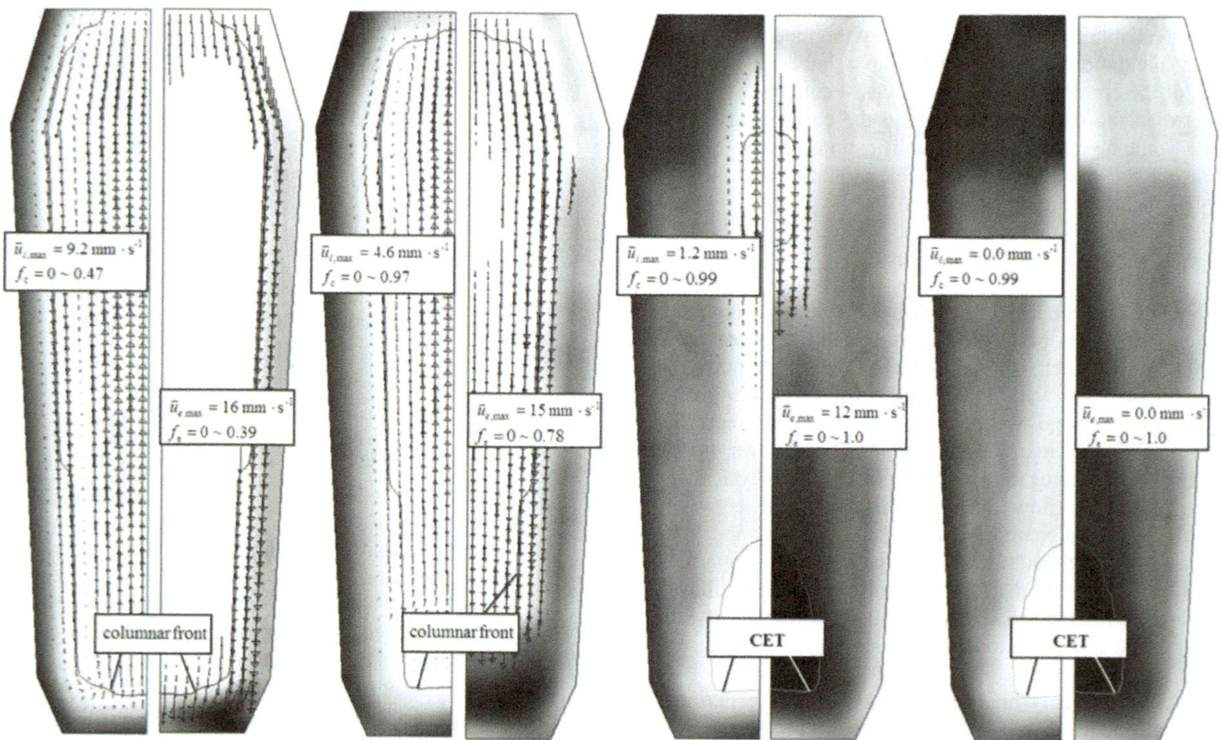

(a) 5 s (b) 20 s (c) 60 s (d) 90 s

Figure 2: Simulated solidification sequence of a benchmark steel ingot. The left half of each figure shows the volume fraction of the columnar phase f_c in greyscale from minimum (bright) to maximum (black), together with the liquid velocity vectors $\vec{u}_\ell$. The right half of each figure shows the volume fraction of the equiaxed phase f_e in greyscale, together with the velocity vectors of equiaxed crystals $\vec{u}_e$. The columnar tip front and CET are indicated with a solid line. Here an arbitrary set of nucleation parameters is used to characterise the formation and sedimentation of equiaxed crystals: $n_{max} = 5\times10^9\ m^{-3}$, $\Delta T_\sigma = 2\ K$, $\Delta T_N = 5\ K$.

The solidification sequence including sedimentation of the equiaxed crystals, the sedimentation-induced and thermosolutal buoyancy-induced melt convection are shown in Figure 2. The simulated solidification sequence agrees with the explanation of steel ingot solidification, as summarised by Campbell [32]. The columnar dendrites grow from the mould wall and the columnar tip front moves inwards. The equiaxed grains nucleate near the mould walls and in the bulk melt. The columnar dendrites are stationary, whereas the equiaxed grains sink and settle in the base region of the ingot. The accumulation of such grains at the base of the ingot has a characteristic cone-shape. The sedimentation of grains and the melt convection influence the macroscopic solidification sequence and, thus, the final phase distribution. More equiaxed grains will be found at the bottom and in the base region, while columnar structure will be predominant in the upper part of the ingot

As the columnar tip front is explicitly tracked, the simulation shows that the columnar tip fronts from both sides tend to meet in the casting centre. However, in the lower part of the casting the accumulation of equiaxed grains stops the propagation of the columnar tip front. Its final position indicates the CET position. The CET separates the areas where only equiaxed grains appear from the areas where both columnar dendrites and equiaxed grains coexist.

The final macrosegregation pattern is predicted, as shown in Figure 3(a). From the simulation results it appears evident that the main mechanism for the cone-shaped negative segregation in the base region is the grain sedimentation. The settling grains are poor in solute elements, thus their pile-up results in negative segregation at the bottom of the ingot. A further contributing factor to the strength of negative segregation arises from the flow divergence of the residual liquid through this zone at a late solidification stage. The positive segregation at the top region of the ingot is caused by the flow of the enriched melt in the bulk region. This kind of positive segregation coincides with classical experimental results [32]. It should be noted that channel segregations, which are frequently found in large steel ingots, are not predicted in this kind of a benchmark ingot with reduced dimensions.

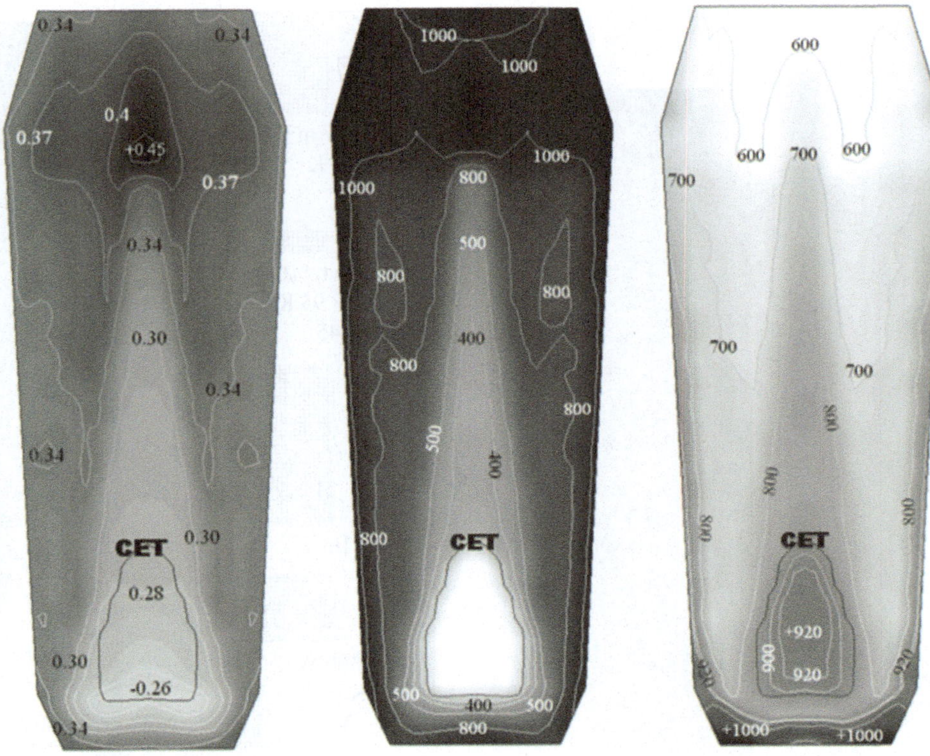

(a) c_{mix} (0.23 – 0.46) (b) d_c in μm (0~1000) (c) d_e in μm (394~1262)

Figure 3: Result in the finally solidified ingot. (a) Macrosegregation pattern, c_{mix} (local average concentration of columnar and equiaxed phases); (b) distribution of the columnar trunk diameter, d_c, and (c) distribution of the diameter of the equiaxed grains, d_e. The quantities are shown with grey scales from minimum (bright) to maximum (dark), together with isolines. The CET is shown by a black line.

With the given nucleation parameters $n_{max} = 5.\ x\ 10^9\ m^{-3}$, ΔT_N = 5 K, ΔT_σ = 2 K, and a constant primary dendrite spacing of λ_1 = 1000 μm, the average equiaxed grain size d_e, and the average dendrite trunk diameter d_c, are predicted as shown in Figure 3(b)-(c). The absolute value of d_e and d_c depend on the aforementioned modelling parameters, but the predicted size distribution pattern reflects the special characteristic of the mixed columnar-equiaxed solidification. In the upper part of the ingot, the average dendrite trunk diameter has reached the maximal possible value, namely λ_1. This is due to the fact that (i) equiaxed crystals are very scarce ($f_e < 1\%$), and (ii) no interdendritic eutectic solidification was modelled in the present simulation. Contrary to the upper region, a totally equiaxed zone without any columnar dendrites exists in the area enclosed by the CET line. The cone shape distribution pattern for both d_e and d_c is similar to the pattern of the phase volume fractions f_c and f_e, as shown in Figure 2(d).

2.34-ton Ingot

The experimentally measured macrosegregation of a 2.45 ton big-end-up ingot (Fe-0.45 wt.%C) was reported [1]. The ingot had a section of square and was cast in a chilled mould. As a reference, the segregation pattern in this ingot is numerically simulated and compared with the experiment, as shown in Figure 4. Due to the lack of a precise process description, some process parameters and boundary conditions have to be derived on the basis of assumptions. The sulphur print of this ingot is shown in Figure 4(a). The measured (nominal) mixture concentration $((c_{mix}-c_0)/c_0)$ map is shown in Figure 4(b). Configuration of this ingot, together with necessary boundary and initial conditions used for the calculation, is described in Figure 4(c). More details about the simulation configurations are presented elsewhere [33], and the same three-phase model (Section 2.0) is used. 2D axis symmetrical simulations are performed to approximate the solidification behaviour in the square section ingot. The predicted solidification sequence is shown in Figure 5 and the segregation map is shown in Figure 4(d).

The global solidification sequence in this 2.45 ton ingot (Figure 5) is actually similar to what was characterised by the previous small benchmark ingot (Figure 2-3). The sinking of the equiaxed crystals in front of the columnar dendrite tips leads to an accumulation of equiaxed phase in the base region of the ingot. The accumulation of the equiaxed phase in the base region will block the growth of the columnar dendrite tips, i.e. CET occurs there, hence finally causing a characteristic cone-shape distribution of the equiaxed zone, being enveloped in the CET line. Relatively strong negative segregation is predicted in the low-bottom equiaxed zone. With the sedimentation of a large amount of equiaxed crystals downwards, the solute-enriched melt is pushed upwards in the casting centre, hence causing a positive segregation in the upper region. Despite the above similarity between the 2.45 ton ingot (Figure 4) and the small benchmark ingot (Figure 2), significant differences are identified, which are described below.

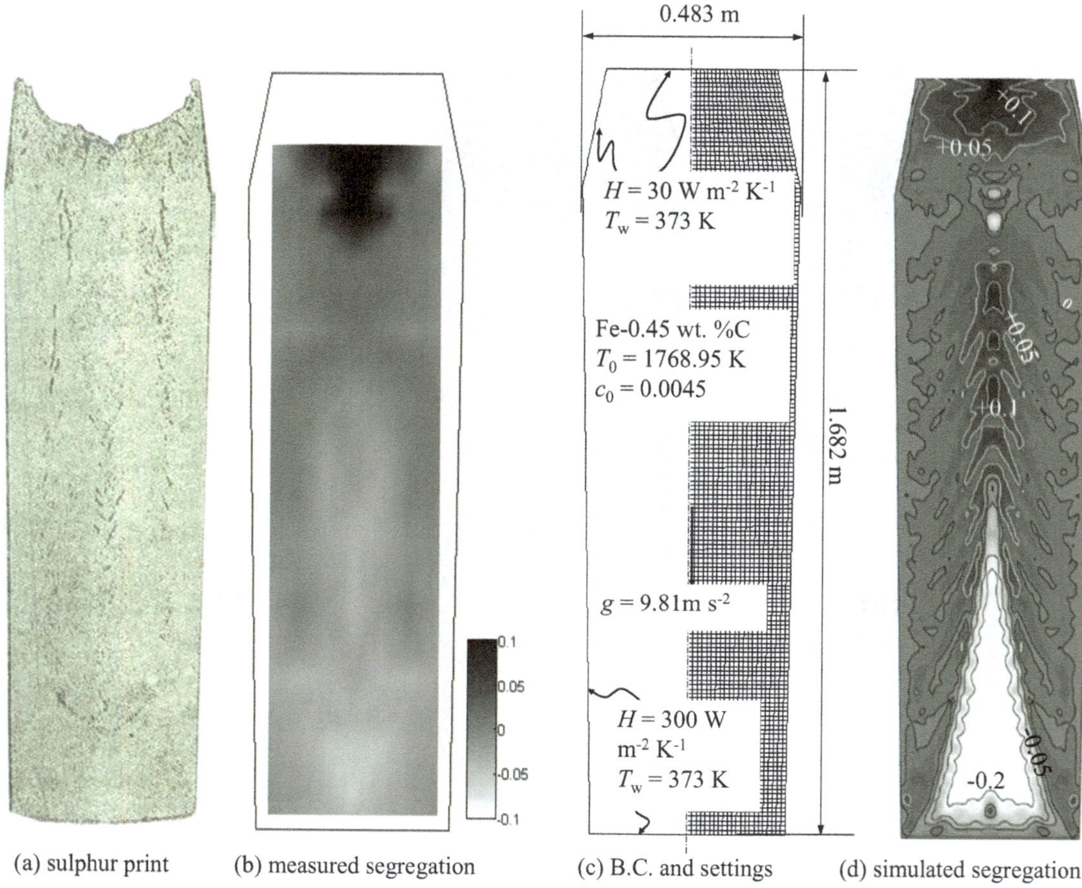

Figure 4: Configuration of a 2.45-ton industry-scale steel ingot. (a)-(b) experiment [1], (c) simulation configuration and (d) simulated macrosegregation in greyscale (black for the positive segregation and light for the negative segregation), overlapped with iso-lines. The macrosegregation, both experimental (b) and simulated (d), is shown for the nominal mixture concentration $((c_{mix}-c_0)/c_0)$. Nucleation parameters: $n_{max} = 5\times10^9$ m^{-3}, $\Delta T_\sigma = 2$ K, $\Delta T_N = 5$ K.

Firstly, the flow is much more instable (Figure 5). The melt flow in the bulk region ahead of the columnar dendrite tip front is driven by three mechanisms: the solutal buoyancy driving upwards; the thermal buoyancy driving downwards; and the equiaxed sedimentation which drags the surrounding melt downwards. Generally the two downward driving forces dominate, and the melt flows downwards along the columnar dendrite tip front. This downward flow along the columnar tips will force the melt to rise in the ingot centre. This rising melt will interact with the falling equiaxed crystals and with the downward flow near the columnar tip front, to form many local convection cells. The pattern of melt convection and crystal sedimentation becomes chaotic. These local convection cells are developed or suppressed dynamically, and the flow direction in the cells changes with time as well. The flow instability and the flow chaotic behaviour are dependent on the ingot size (ingot height). Therefore, to explain the influence of the ingot size on the macrosegregation, knowledge about the influence of the ingot size on the flow pattern is required.

Secondly, a streak-like segregation pattern (Figure 4(d)) in the mixed columnar-equiaxed region is predicted, which does not occur in the small ingot (Figure 3(5)). For a concrete explanation of this segregation pattern, a more detailed analysis of the flow and sedimentation and their interaction with the solidification is necessary; nevertheless a tentative hypothesis is proposed, as follows. As the equiaxed crystal can be captured (crystal entrapment) by the growing columnar trunks, the entrapment of the equiaxed crystals will lead to a heterogeneous, i.e. streak-like, phase distribution between the columnar and equiaxed crystals immediately behind the columnar tip front, as seen in Figure 5(b)-(d). The resistance to the interdendritic flow by the columnar trunks and the entrapped equiaxed crystals is different; therefore the flow direction of the melt in this region is slightly diverted by the heterogeneous phase distribution. This diverted-flow can only be visible in the carefully zoomed view. As the macrosegregation is extremely sensitive to the interdendritic flow, it is not surprising that the induced macrosegregation (Figure 4(d)) takes the similar streak-like pattern of the phase distribution (Figure 5(d)).

One may notice that this streak-like segregation has a similar contour to the classical A-segregation, but it is still not clear if the classical A-segregation is the same as streak-like segregation or originates from this kind of streak-like segregation. According to the most widely accepted empirical explanation, A-segregation belongs to a kind of channel segregation in large steel ingots, which originates and develops in the stationary dendritic mushy zone. A recent study by the authors [34-35] in a Sn-Pb laboratory casting has found that the channel segregation can originate and develop in a pure columnar solidification, where no equiaxed crystal exists. Therefore, we name the streak-like segregation here a quasi-A-segregation. To form this quasi-A-segregation, the sedimentation of equiaxed crystals and its interaction with the columnar tip front and melt flow seem to play an important role. Details about the formation mechanism for this kind of quasi-A-segregation are still to be verified.

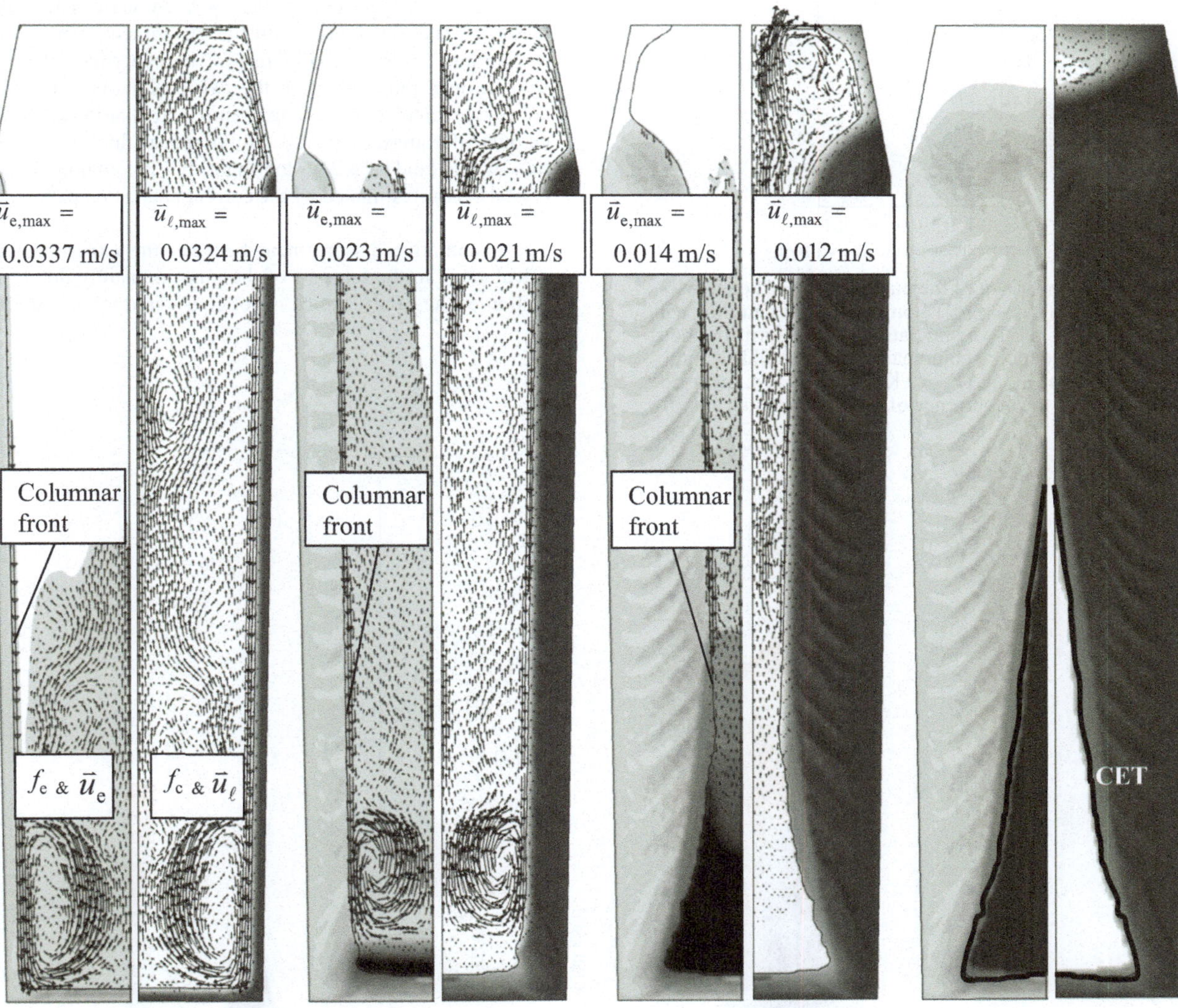

(a) 100 s (b) 500 s (c) 1500 s (d) 4000 s

Figure 5: Solidification sequence of the 2.45-ton ingot. The volume fraction of each phase (f_e or f_c) is shown in greyscale from minimum (bright) to maximum (dark). The left half of each figure shows the evolution of the equiaxed volume fraction (f_e) together with the equiaxed sedimentation velocity ($\vec{u}_e$) in black arrows. The right half of each figure shows the evolution of the columnar volume fraction (f_c) together with the melt velocity ($\vec{u}_\ell$) in black arrows. The columnar dendrite tip position is also marked with a black solid line.

Thirdly, the simulation of the 2.45-ton ingot shows an isolated hot spot in the upper part (Figure 5(d)), which takes a long time to solidify. As the middle part of the ingot is already blocked by the columnar trunks, the solidification of the hot spot behaves like a mini-ingot. Sedimentation of the equiaxed crystals in the mini-ingot will cause a small region of negative segregation, as shown in Figure 4(d). This kind of phenomenon happens very often in long (small section) ingot casting or in the continuously-cast round billet casting, and is called 'bridging and mini-ingotism' [36]. The experimental result of Figure 4(b) seems to show that no such 'bridging and mini-ingotism' occurs, as no such negative segregation zone is identified. It implies that the heat transfer boundary conditions applied in the current simulation might not be coincident with the reality.

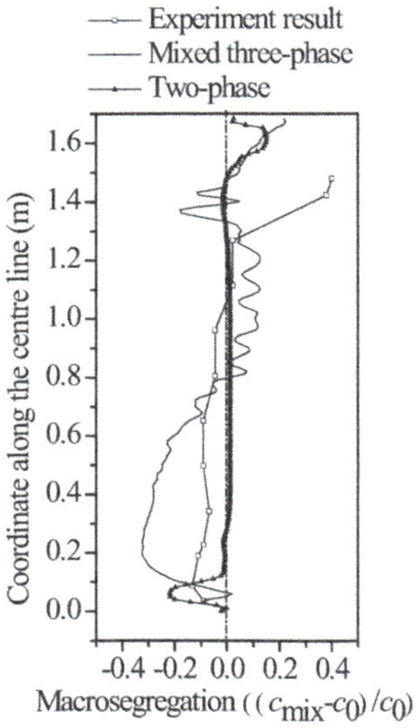

Figure 6: Comparison of the numerically predicted macrosegregation ($(c_{mix}$-$c_o)/c_o$) along the ingot centreline with the experiment [1]. Two simulations are performed: one is to consider the mixed columnar-equiaxed solidification ($n_{max} = 5\times10^9$ m^{-3}, $\Delta T_\sigma = 2$ K, $\Delta T_N = 5$ K); one is to ignore the occurrence of equiaxed crystal.

The segregation along the ingot centreline is analysed and compared with the experiment, as shown in Figure 6. The experiment shows the negative segregation in the lower part and positive segregation in the upper part. The model also shows the same tendency. They agree with each other qualitatively. However, the negative segregation in the lower part is predicted more severely than the experimental result. The overestimation of the negative segregation in the lower part by the model may result from two aspects. One is the assumption of globular equiaxed morphology, which can cause significant overestimation of the sedimentation-induced negative segregation. The other aspect is the error assumption of the equiaxed nucleation parameters.

In order to demonstrate the role of the equiaxed phase in the formation of segregation, an additional calculation is performed by ignoring the occurrence of the equiaxed phase. This case seems to show better agreement with the experiment, especially in the middle and lower part (Figure 6). The experimental result falls actually in a range between the two calculations. Based on the above two simulations, one may anticipate that in reality a certain amount of equiaxed crystals would appear during the solidification of this kind of a 2.45-ton ingot, but the amount of equiaxed crystals is overestimated by the current nucleation parameters.

The quantitative disagreement between the experiment and the calculations in the top part of the ingot is mainly due to the formation of the cavity, which is not considered by the current model.

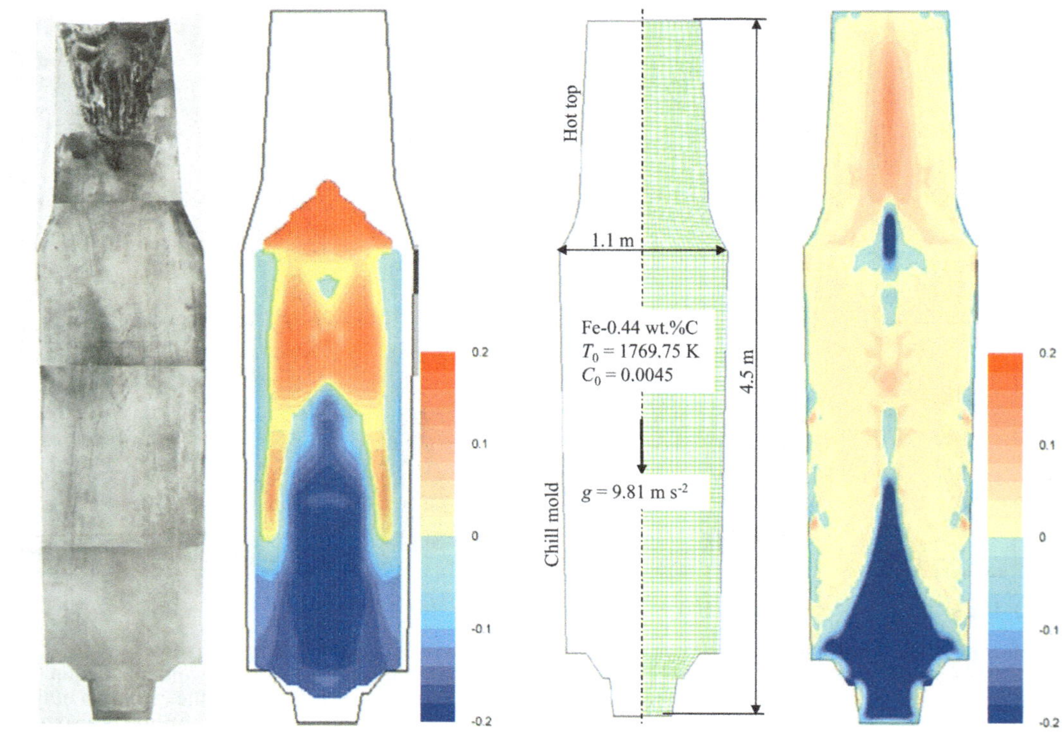

Figure 7: Configuration of a 25-ton industry-scale steel ingot. (a)-(b) experiment [1], (c) simulation configuration and (d) simulated macrosegregation. The macrosegregation, both experimental (b) and simulated (d), is shown for the nominal mixture concentration ($(c_{mix}$-$c_o)/c_o$). Nucleation parameters: $n_{max} = 5\times10^9$ m^{-3}, $\Delta T_\sigma = 2$ K, $\Delta T_N = 5$ K.

25-ton Ingot

As a further step to validate the current three-phase solidification model, a 25 ton ingot is also simulated (Figure 7). The shape of the ingot is octagonal, and the real industry alloy is multicomponent, but here only a 2D axis symmetrical calculation for a simplified binary alloy (Fe-0.44 wt.%C) is performed. Process parameters and material data have to be assumed based on the information provided by the original report [1]. Mould filling is ignored, and the nucleation parameters are assumed as: $n_{max} = 5\times10^9\ m^{-3}$, $\Delta T_\sigma = 2\,K$, $\Delta T_N = 5\,K$. The same three-phase model as presented in Section 2 is used.

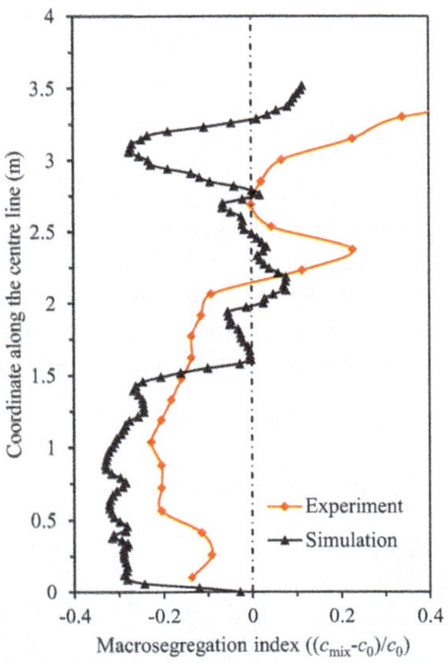

Figure 8: Comparison of the numerically predicted macrosegregation $((c_{mix}-c_o)/c_o)$ along the ingot centreline with the experiment [1].

The predicted macrosegregation are compared with the experimentally reported results, as shown in Figure 7 and Figure 8. Qualitatively, the predicted macrosegregation pattern follows the experimentally measured one. The bottom equiaxed zone, accompanying relatively strong negative segregation (cone shape), is predicted. This cone-shaped negative segregation is mainly due to the equiaxed crystal sedimentation. Above the tail of the cone-shaped negative segregation zone, there is a positive segregation zone. The mechanism to form such positive segregation zone is due to the transport of the solute-enriched melt in the bulk. In the upper region of the ingot, just below the hot top, when the columnar dendrite tips, growing from side walls, meet together in the casting centre, some solute-enriched melt is 'frozen' to form this positive segregation. The late solidification of the hot top behaves like a mini-ingot. Sedimentation of the equiaxed crystals continues in the mini-ingot, but the crystals can only settle in the base region of the mini-ingot, causing a secondary negative segregation zone in the mini-ingot. Finally, a large positive segregation occurs in the hot top. Segregates along the centerline (Figure 8) of both experiment and simulation show the same tendency: the negative segregation in the lower part and positive segregation in the upper part, and negative segregation region between ½ and ¾ height (just below the hot top) of the steel ingot.

Despite the above agreement, the quantitative discrepancy between numerically predicted and experimentally reported results is still significant. A cone-shaped equiaxed zone with a narrow tail extending along the centreline of the ingot is numerically predicted (Figure 7 (d)). However, the experimentally found equiaxed zone shows that the equiaxed zone is broader and it extends to a higher position, as shown in Figure 7 (a)-(b). The negative segregation in the cone-shape sedimentation zone is predicted more severely than the experimental result. The main reasons for the overestimation of the bottom negative segregation have previously been discussed: one is the deficiency of the crystal dendritic morphology, and one is the assumption of the crystal nucleation parameters. Additionally, the secondary negative segregation zone due to the mini-ingotism below the hot top is numerically predicted at a position higher than the experiment shows, and the negative segregation is predicted more intense. The reasons for this discrepancy could be the uncertainty of the thermal boundary conditions and ignorance of the formation of the shrinkage cavity in the top.

For the calculation of this 25-ton ingot, a very coarse grid (between 15 and 32 mm) was used. The number of total volume elements is 2640. The calculation time step used is between 0.01 and 0.02 s. The solidification time takes about 6.7 hours. The calculation takes about 20 days in Intel Nehalem Cluster (8 cores in parallel, 2.GHz/core). With such a coarse grid, the quasi-A-segregation cannot be calculated.

Discussions

This article related the on-going activities of the group of authors to scale-up a mixed columnar-equiaxed solidification model for industry applications. Three examples were analysed in order to evaluate the potentials and limitations of the model.

Model Potentials

Firstly, the simulated solidification sequence, the sedimentation of the equiaxed crystals, the growth of the columnar tip front and the formation of the final macroscopic phase distribution fit with the widely accepted explanations of experimental findings, as summarised by Campbell [32]: "The fragments (equiaxed grains) fall at a rate somewhere between that of a stone and snow. They are likely to grow as they fall if they travel through the undercooled liquid just ahead of the growing columnar front, possibly by rolling or tumbling down this front. The heap of such grains at the base of the ingot has a characteristic cone shape." These kinds of multiphase flow dynamics and interactions among the melt, equiaxed crystals and growing columnar trunks are the very important phenomena for modelling the segregation pattern of Figure 1. They are considered by the current model.

Secondly, it is also verified by the modelling examples above that the most typical segregates, the concentrated positive segregation under hot-top and the cone of negative segregation at the base of the ingot, can be simulated by the model. A widely accepted explanation for the formation of the cone-shaped negative segregation is verified, again in Campbell's words: "The heap of equiaxed grains at the base of the ingot has a characteristic cone shape. Because it is composed of dendritic fragments, its average composition is that of rather pure iron, having less solute than the average for the ingot." The simulated negative segregation formation process by equiaxed crystal sedimentation (Figure 2 and 5) seems to have reproduced the experimental phenomenon. Mechanisms for positive segregations under the hot-top in steel ingots are diverse. It is generally agreed that they are caused by the melt convection in the bulk region or the partially solidified and/or remelted mushy zone. For example, the upper positive segregation is explained by the melt convection in the bulk region, because the light solute-rich melt rises. In actual fact, according to the recent modelling results, with the sedimentation of a large amount of equiaxed crystals downwards, the relatively positive segregated melt is pushed upwards, instead of 'rising' by itself, in the casting centre, hence causing a positive segregation zone in the upper region.

Thirdly, the possibility to calculate the distribution of the columnar and equiaxed structure has been demonstrated. The upper region of the ingot mainly consists of columnar dendrites, whereas a larger amount of equiaxed grains are predicted in the base region. Within the CET enclosed region, only the equiaxed phase exists, while outside of the CET region both columnar and equiaxed phases coexist. The macrostructure strongly depends on certain modelling and process parameters, i.e. the equiaxed nucleation parameters ΔT_N, $n_{\max}$, ΔT_σ, the primary columnar space λ_1, and boundary conditions.

Finally, the capability of the current model for the interdendritic-flow-induced channel segregation was also demonstrated [34-35], but it was not clearly shown in the above examples. The modelling result for the channel segregation is extremely sensitive to the grid resolution. Grid size less than 0.1 mm is often required, and this is unrealistic for the large industry ingots on the basis of the current computer resources. One interesting finding by the current three-phase solidification model, worth mentioning here, is the streak-like (quasi-A) segregation pattern, which occurs in such large ingots and is strengthened by the columnar-equiaxed interaction at the columnar tip front. The streak-like segregation pattern has some similarity to the classical A-segregation, but it is not clear if the classical A-segregation is the same as or whether it originates from the streak-like segregation. This is still to be verified.

Limitation of the Model

The importance of the applied process conditions, e.g. the pouring temperature, pouring method, mould materials and interfacial heat transfer between the ingot and the mould, etc., for the quantitative accuracy of the simulated solidification process and, hence, for the accuracy of the macrosegregation is evident. It is not discussed here. Following discussions focus on the aspect of numerical model.

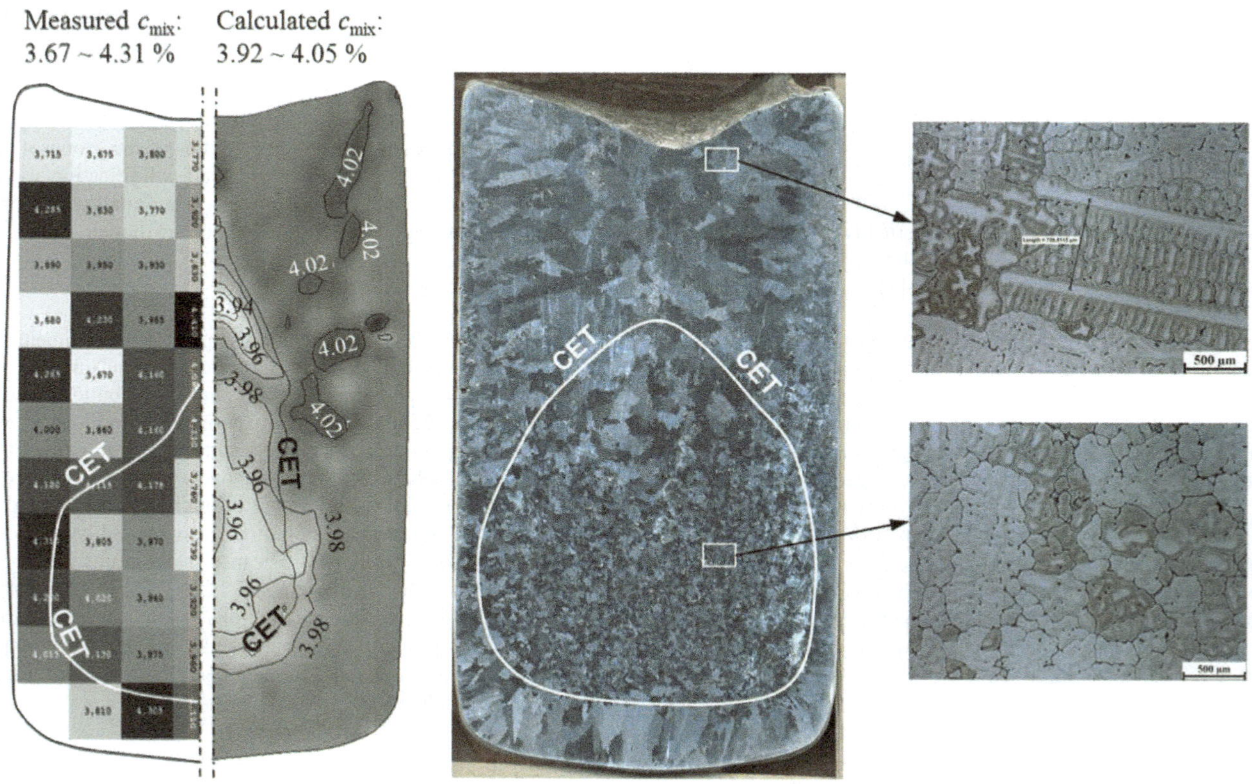

a) macrosegregation b) macrograph c) micrograph

Figure 9: An example of the modelling result of an Al-4wt.%Cu ingot with a five-phase mixed columnar-equiaxed solidification model with dendritic morphology. (a) Comparison of the measured (spark analysis) macrosegregation (left half) with the calculated one (right half). The casting is poured at 800 °C. The mixture concentration c_{mix} is shown in greyscale (dark for the highest and light for the lowest value). CET positions are plotted. This numerical simulation result shows satisfactory agreement with the as-cast macrostructure (b) macrograph, (c) micrograph.

Firstly, the influence of the nucleation event on macrosegregation was addressed in the example of the 2.45-ton ingot. The origin of the equiaxed grains may be due to different mechanisms, e.g. heterogeneous nucleation, and/or fragmentation and detachment of dendrites by re-melting, and/or nucleation formed during pouring by contact with the initial chilling of the mould. The recent model condenses all of these phenomena into a single effective nucleation description. Here, a three-parameter heterogeneous nucleation law [30] is applied for the origin of equiaxed crystals. It is only possible to obtain the reliable nucleation parameters experimentally.

Secondly, no shrinkage cavity and porosity are considered. This deficiency will influence the accuracy of the calculation, especially in the hot-top region. As shrinkage contributes to or influences the interdendritic flow, it will also influence the final distribution of the channel segregation. However, the global segregation pattern, e.g., the concentrated positive segregation in the upper region and the cone of negative segregation at the base of the ingot, will not be significantly influenced by the shrinkage.

Thirdly, no thermal mechanics is considered. The thermal mechanical shrinkage of the solidified outer shell of the ingot will influence the internal flow, but this may not be particularly significant. What is most important is the deformation of the growing crystals due to the thermal shrinkage or the solid phase transition, which would have great impact on the flow near the end of solidification at the centreline. The 'V' segregation is very much related to this kind of deformation. This 'V' segregation is not modelled by the current model.

Finally, the current three-phase model does not include dendritic morphology. This deficiency has overestimated the cone of the negative segregation at the base, like what we see in Figure 6 and 8. In order to consider the dendritic morphology, more phases, i.e. the interdendritic melt, must be separately considered. A five-phase model was developed by the group of authors to consider the mixed columnar-equiaxed solidification with dendritic morphology [27-28]. The calculation expense is so high as to prevent application in industry ingots. The validity of this model for such a purpose has been verified, but in a laboratory Al-4.0 wt.%Cu ingot casting, as shown in Figure 9. A cylindrical casting (ϕ 75 mm x 133 mm) was poured and was analysed for both macrostructure and macrosegregation. The experimental results were used to validate the numerical simulations. Satisfactory agreement between them was obtained, as reported elsewhere [37-39].

Outlook

The future modelling activities for macrosegregation in large steel ingots will progress in two directions. One is to further enhance the model capability by including more physical phenomena such as crystal morphology, solidification shrinkage, thermal mechanics, dendrite fragmentation, etc. Another direction is to further validate and improve the existing multiphase model, and to apply it for the purposes of solving engineering problems and enhancing fundamental understanding of different segregation phenomena.

1. Thanks to the work of the Iron Steel Institute [1], many steel ingots scaled from 600 kg to 172 tons were poured and sectioned for segregation analysis. This work provides most valuable information for the validation of the numerical models. Although many process parameters for those ingots are unknown and have to be assumed, the capability to reproduce segregation patterns of all (most) those ingots numerically is an important step for the development of macrosegregation model.
2. The existing model can be applied for studying the process parameters. Despite the difficulty of quantitatively reproducing the segregation pattern of reality, the influence of the process parameters, such as casting geometry, mould materials, pouring temperature, and other engineering measures on segregation can be well described by the model. By performing this kind of a parameter study, metallurgists would acquire ideas for process optimisation.
3. Any segregation mechanism, as proposed from experimental observation, can (should) be verified quantitatively by the mathematical (numerical or analytical) model. The three-phase model can help to explain many well-known segregation phenomena in detail. It may also help to explore the new segregation phenomena, which are caused by the multiphase flow. For example, the question of streak-like segregation, here referred to as quasi-A-segregation, is raised by the authors for the first time on the basis of the current modelling result. The equiaxed-columnar interaction at the columnar dendrite tip front and its influence on the melt flow seems to induce or enhance this kind of a streak-like macrosegregation

Acknowledgements

The authors acknowledge the financial support by the Austrian Federal Ministry of Economy, Family and Youth and the National Foundation for Research, Technology and Development.

References

1. Iron Steel Institute, Report on the heterogeneity of steel ingots, *J. Iron Steel Institute*, 103 (1926), 39-151.
2. E. Marburg, "Accelerated Solidification in Ingots: Its Influence on Ingot Soundness," *J. Met.*, 5 (1953), 157-172.
3. J.J. Moore and N.A. Shah, "Mechanisms of Formation of A- and V-Segregation in Cast Steel," *Int. Metals Rev.*, 28 (1983), 338-356.
4. J.R. Blank and F.B. Pickering, "Effect of Solidification in Large Ingots on Segregation of Non-Metallic Inclusions, in: The Solidification of Metals," *The solidification of metals* (London: The Iron & Steel inst., 1968), 370-376.
5. H. Yamada, T. Sakurai, T. Takemonchi and K. Suzuki, "The Critical Conditions for the Formation of "A" Segregation in Forging Ingots," (Proc. Annual Meeting of AIME, Dallas, A82-39, 1982), 1-6.
6. K. Kajikawa, S. Suzuki, F. Takahashi, S. Yamamoto, T. Suzuki, S. Ueda, T. Shibata, H. Yoshida, "Development of 650-ton-class Ingot for Turbine Rotor Shaft Forging Application," (Paper presented at the 1st Int. Conf. Ingot Casting, Rolling and Forging, Session – Ingot Casting Technology, 5 June 2012, Aachen, Germany), 1-6.
7. J. Wang, P. Fu, H. Liu, D. Li, Y. Li, "Shrinkage Porosity Criteria and Optimized Design of a 100-ton $30Cr_2Ni_4MoV$ Forging Ingot," *Mater. Design*, 35 (2012), 446-456.
8. H. Combeau, M. Zaloznik, S. Hans and P.E. Richy, "Prediction of Macrosegregation in Steel Ingot: Influence of the Motion and the Morphology of Equiaxed Grains," *Metall. Mater. Trans.*, 40B (2009), 289-304.
9. G. Lesoult, "Macrosegregation in Steel Strands and Ingots: Characterization, Formation and Consequences," *Mater. Sci. Eng.* A, 413-414 (2005), 19-30.
10. M.C. Flemings, "Our Understanding of Macrosegregation: Past and Present,", *ISIJ Int.*, 40 (2000), 833-841.

11. C. Beckermann, "Modelling of Macrosegregation: Applications and Future Needs," *Int. Mater. Rev.*, 47 (2002), 243-261.
12. I. Ohnaka, "Modeling of Microsegregation and Macrosegregation", *ASM Handbook*, 15-*Casting*, ASM Int., USA, 136.
13. J. P. Gu, C. Beckermann, "Simulation of Convection and Macrosegregation in a Large Steel Ingot," *Metall. Mater. Trans.*, 30A(1999), 1357-1366.
14. J. Ni, C. Beckermann, "A Volume-Averaged Two-Phase Model for Solidification Transport Phenomena," *Metall. Mater. Trans.*, 22B (1991), 349-361.
15. C.Y. Wang, C. Beckermann, "A multiphase solute diffusion model for dendritic alloy solidification," *Metall. Trans.*, 24A (1993), 2787-2802.
16. C. Beckermann, R. Viskanta, "Mathematical Modeling of Transport Phenomena during Alloy Solidification," *Appl. Mech. Rev.*, 46 (1993), 1-27.
17. C.Y. Wang, C. Beckermann, "Equiaxed Dendritic Solidification with Convection: Part I. Multi-Scale/-Phase Modeling," *Metall. Mater. Trans.*, 27A (1996), 2754-2764.
18. M. Wu, A. Ludwig, A. Bührig-Polaczek and P. Sahm, "Influence of Convection and Grain Movement on Globular Equiaxed Solidification," *Int. J. Heat Mass Transfer*, 46 (2003), 2819.
19. A. Ludwig, M. Wu, "Modeling of Globular Equiaxed Solidification with a Two-Phase Approach," *Metall. Mater. Trans.*, 33A (2002), 3673-3683.
20. M. Wu, A. Ludwig, J. Luo, "Numerical Study of the Thermal-Solutal Convection and Grain Sedimentation during Globular Equiaxed Solidification," *Mater. Sci. Forum*, 475-479 (2005), 2725-2730.
21. M. Wu, A. Ludwig, L. Ratke, "Modeling of Marangoni Induced Droplet Motion and Convection during Solidification of Hypermonotectic Alloys," *Metall. Mater. Trans.*, 34A (2003), 3009-19.
22. M. Wu, A. Ludwig, L. Ratke, "Modelling the solidification of hypermonotectic alloys," *Modell. Simu. Mat. Sci. Eng.*, 11 (2003), 755-769.
23. M. Wu, A. Ludwig, "A 3-Phase Model for Mixed Columnar-Equiaxed Solidification," *Metall. Mater. Trans.*, 37A (2006), 1613-1631.
24. M. Wu, A. Ludwig, "Using a Three-Phase Deterministic Model for the Columnar-to-Equiaxed Transition (CET)," *Metall. Mater. Trans.*, 38A (2007), 1465-1475.
25. M. Wu, A. Ludwig, "Modeling Equiaxed Dendritic Solidification with Melt Convection and Grain Sedimentation, Part I: The Model," *Acta Mater.*, 57 (2009), 5621-5631.
26. M. Wu, A. Ludwig, "Modeling Equiaxed Dendritic Solidification with Melt Convection and Grain Sedimentation, Part II: The Results and Verifications," *Acta Mater.*, 57 (2009), 5632-5644.
27. M. Wu, A. Fjeld, A. Ludwig, "Modeling Mixed Columnar-Equiaxed Solidification with Melt Convection and Grain Sedimentation-Part I: Model Description," *Comp. Mater. Sci.*, 50 (2010), 32-42.
28. M. Wu, A. Fjeld, A. Ludwig, "Modeling Mixed Columnar-Equiaxed Solidification with Melt Convection and Grain Sedimentation-Part II: Illustrative Modeling Results and Parameter Studies," *Comp. Mater. Sci.*, 50 (2010), 43-58.
29. M. Wu, L. Könözsy, A. Ludwig, W. Schützenhöfer, R. Tanzer, "On the Formation of Macrosegregations in Steel Ingot Castings," *Steel Res. Int.*, 79 (2008), 637-644.
30. M. Rappaz, "Modeling of Microsegregation Formation in Solidification Processes," *Int. Mater. Rev.*, 34 (1989), 93-122.
31. J.D. Hunt, "Steady State Columnar and Equiaxed Growth of Dendrites and Eutectic," *Mater. Sci. Eng.*, 65 (1984), 75-83.
32. J. Campbell, *Castings* (Butterworth Heinemann Ltd, Oxford, 1991).
33. J. Li, M. Wu, A. Ludwig, A. Kharicha, "Modelling Macrosegregation in a 2.45 Ton Steel Ingot," *IOP Conf. Series: Mater. Sci. Eng.*, 27 (2012), doi: 10.1088/1757-899X/33/1/012091.
34. J. Li, M. Wu, J. Hao, A. Ludwig, "Simulation of Channel Segregation Using a Two-Phase Columnar Solidification Model, Part I: Model Description and Verification," *Comp. Mater. Sci.*, 55 (2012), 407-418.
35. J. Li, M. Wu, J. Hao, A. Kharicha, A. Ludwig, "Simulation of Channel Segregation Using a Two-Phase Columnar Solidification Model, Part II: Mechanism and Parameter Study," *Comp. Mater. Sci.*, 55 (2012), 419-429.
36. J.J. Moore, "Review of Axial Segregation in Continuously Cast Steel," *Iron and Steel Soc.*, 3 (1984), 11-20.
37. M. Wu, G. Nunner, A. Ludwig, J.H. Li, P. Schumacher, "Evaluation of a Mixed Columnar-Equiaxed Solidification Model with Laboratory Castings," *IOP Conf. Series: Mater. Sci. Eng.*, 27 (2011), doi: 1088/1757-899X/27/1/012018.
38. M. Wu, M. Ahmadein, A. Ludwig, A. Kharicha, J.H. Li, P. Schumacher, "Simulation of the As-Cast Structure of Al-4.0wt.%Cu Ingots with a 5-Phase Mixed Columnar-Equiaxed Solidification Model," *IOP Conf. Series: Mater. Sci. Eng.*, 27 (2012), doi: 10.1088/1757-899X/33/1/012075.
39. M. Ahmadein, M. Wu, A. Ludwig, A. Kharicha, J.H. Li, P. Schumacher, "Prediction of the As-Cast Structure of Al-4.0 wt.% Cu Ingots," *Metall. Mater. Trans.*, 44 (2013), 2895-2903.

Proceedings of the 2013 International Symposium on Liquid Metal Processing & Casting
Edited by: Matthew J. M. Krane, Alain Jardy, Rodney L. Williamson, and Joseph J. Beaman
TMS (The Minerals, Metals & Materials Society), 2013

EFFECT OF SOLIDIFICATION FRONT ANGLE ON FRECKLE FORMATION IN ALLOY 625

Koji Kajikawa, Masaru Tanaka, Tomoki Sawada, and Shigeru Suzuki
The Japan Steel Works, Ltd.; 4 Chatsu-machi, Muroran, Hokkaido, 051-0006, JAPAN

Keywords: Freckle, Channel segregation, solidification, solidification front angle, Ni-base alloy

Abstract

Freckle formation depends on solidification front angle in ingots. Freckles can be divided into two types; one is a solute-enriched liquid trail growing upward and the other is a solute enriched liquid trail growing downward. The effect of solidification front angle on upward type freckling alloy was reported by Sakurai *et al*. However there are few studies on the effect on downward type freckling alloys. In this study, the effect of solidification front angle on freckle formation was investigated using downward type freckling Alloy 625. Although freckle becomes difficult with decrease in the solidification front angle as expected, the effect on its solidification front angle was larger than that on upward type alloy A286. It is suggested that the difference is due to the freckle formation mechanism

Introduction

"A" segregation (inverted "V" segregation) is one form of macrosegregation that appears in conventional cast ingots. This form of segregation is often termed "channel type segregation" or "freckle" (the term "freckle" is commonly used in Ni-based superalloy field). It is widely accepted that freckle (channel type segregation) is induced by gravity-driven flow. Freckle appears in some specific solid fraction.[1)]

Experimental literatures reported the effect of solidification front angle on freckle formation. Sakurai *et al.*[2)] studied the effect of the solidification front angle on freckle formation in A286 alloy using a horizontal directional solidification furnace and assessed the effect quantitatively. The result revealed that lowering Si and Ti contents and reducing the liquid pool angle, which corresponded to shallow liquid pool in an actual ESR ingot, were favorable for reducing "A" segregation (channel type segregation).

Wan-hong Yang *et al.*[3)] conducted similar experiments, however, they did not distinguish between upward type and downward type segregation. Also, they did not show the macrostructure of their tilted solidification experiment ingots and it is difficult to know the details of the experimental results. Auburtin *et al.*[4)] showed the effect of the solidification front angle on freckle formation. However, their freckle is not channel type segregation but a freckle in dendrite arms.

In terms of liquid density difference, researchers have identified two types of freckle based on the liquid density difference between bulk liquid and solute enriched liquid. Grugel and Brush[5)] used Sn-Pn binary alloy and two alloy compositions, Pb-45mass%Sn and Pb-75mass%Sn, were selected for investigation. The 45mass% tin alloy is a less-dense solute alloy. In contrast, the 75mass% tin alloy is a dense solute alloy. Although the 45mass% tin alloy showed freckles in the ingots, there are few freckles in the 75mass% alloy ingots.

Weinberg's group performed several vertical directional experiments. Kaempffer and Weinberg[6)] used a Cu-8mass%Ag alloy whose interdendritic liquid was denser than the bulk liquid. They found that there seems to be no segregation plume and only the displacement of interdendritic liquid occurred at the bottom of the ingot. Also, Streat and Weinberg[7)] used Sn-Pn binary alloy and, by changing its composition ratio, two types of segregation tendencies were studied. In upward type segregation alloy Pb-20mass%Sn, macro segregation plumes appeared. These originated in the interior of the ingot as an interdendritic channel, which subsequently widened and moved towards the mold wall. However, Sn-4mass%Pb, which was selected as an example of a composition where the interdendritic liquid would be denseer than the bulk liquid above, did not show any segregation plumes. These results suggest that the freckle tendency of the downward type alloy is different from that of the upward type alloy.

Our previous study[8)] showed that freckling tendency can be marshaled by using liquid density difference and freckle formation index proposed by Suzuki and Miyamoto[8)]. The study indicated the importance of distinguishing upward type segregation from downward type segregation. There are few studies on the effect of the solidification front angle on freckle formation in downward segregation alloys. Also few articles discuss formation of the two freckle types.

In this study, we investigated the effect of the solidification front angle on freckle formation in the case of the downward type segregation alloy.

Experimental procedure

Alloy 625 was selected for this experiment, because the value of the freckle formation index[2)] of the alloy is close to that of A286 that was reported by Sakurai *et al.* If the value is too small, it may be impossible to form freckle in the tilted solidification front condition. Also Alloy 625 is a typical downward type freckling alloy[8)].

The horizontal directional solidification furnace used for this study is schematically illustrated in Figure 1. The furnace is a rectangular-shaped resistant heat furnace using a Kanthal Super heater. An Al_2O_3 crucible was used for the experiment. The inner size of the crucible was 55Wx190Lx230Hmm. The chiller was cooled by compressed air. In order to simulate the titled solidification front condition, specially designed chillers were used for this experiment. The designs of the chillers are shown in Figure 2.

The chillers are composed of silicate sand filled in stainless steel casing. In this series of experiments, the chiller angle of the original straight shape was 90 degrees. 60, 45, and 30 degree chillers were made in addition to 90 degrees. In order to work the slope portion as the major cooling interface, silicate sand on the straight portion of the top side of the chiller was thicker than that on the slope portion. This thicker sand portion acted as a type of insulation.

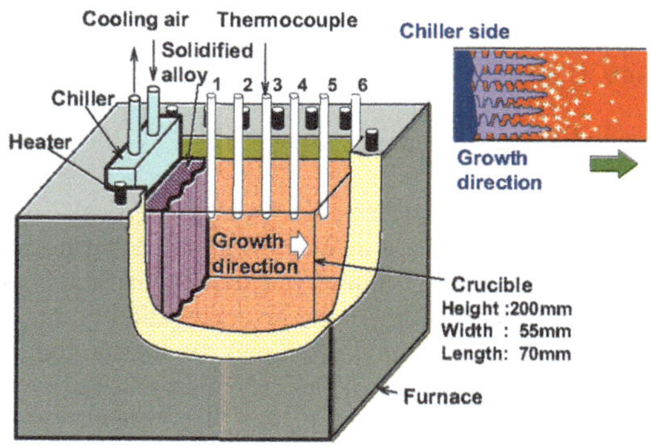

Figure 1. Schematic of the horizontal directional solidification furnace

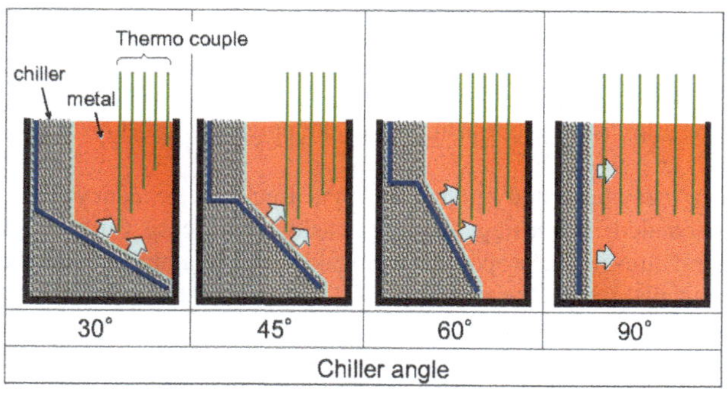

Figure 2. The chillers used for the experiment

The alloy was melted in an induction furnace and poured into the crucible using a small ladle. Just after pouring, cooling air was blown into the chiller. The cooling rates were controlled by the flow rate of the compressed air. Thermocouples were inserted into the crucible in order to acquire the temperature profile during solidification. The ingots were cut perpendicular to the cooling interface. The cut plane was polished and macro etched to confirm the existence or non-existence of freckles in the ingots. If freckles were found in the ingot, the ingot was cut on the vertical cross section of the freckle spot that is close to the chiller face. The cooling rate ε(K/min) and growth rate R(mm/min) are obtained from the thermal history of the ingot. By using those two values, freckle formation index $\alpha=\varepsilon \cdot R^{1.1}$ was calculated. Chemical analysis of the alloy was conducted by an X-ray fluorescence method.

Results

Examples of the thermal histories of the ingots are shown in Figures 3 and 4. In the 30 degree chiller experiment, the sequence of 4CH and 5CH was reversed, i.e., 5CH was solidified faster than 4CH. This may be because the location of 5CH is close to the face and the liquid mass was small. However the freckle initiation point was far from 5CH; therefore, data was considered usable for freckle formation condition analysis.

The vertical cross section of each ingot is shown in Figures 5-6. Hereafter, the straight chiller is described as a 90 degree chiller.

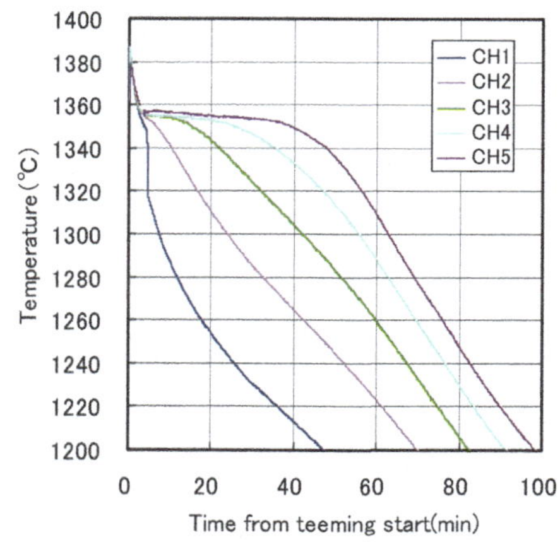

Figure 3. Thermal history of 60 degree chiller ingot

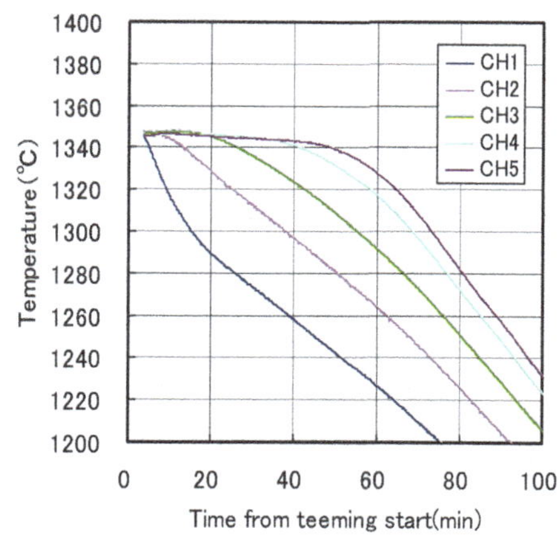

Figure 4. Thermal history of 30 degree chiller ingot

In each photograph, left hand side arrays are the chiller side. The directional solidified structure was obtained in each condition although the primary crystal structures were not perpendicular to the chiller face. Yellow dotted lines in both figures indicate freckles observed in the ingots

Chiller face angle, chemical composition, liquid density difference, and freckle formation index in each experiment are shown in Table 1. Liquid density data[10-13] and Thermo-calc[14] with Ni-data[15] were used for the liquid density calculation. There are some deviations in the chemical composition.

Accordingly, liquid densities vary slightly. However, the deviation is not too large to analyze the condition of freckle formation. The relationship between freckle formation indexes and chiller angles is shown in Figure 7. In this figure, the freckle formation criterion proposed by Suzuki and Miyamoto[8] was used for the vertical axis. The criterion is an empirical value come from their experimental results. The equation of the criterion is expressed as follows:

$$\alpha = \varepsilon \ R^{1.1} \tag{1}$$

where α is a critical freckle formation index, ε is the cooling rate, and R is the growth rate. This α means that if the actual solidification condition exceeds the value, freckle forms. There seems to be a relationship between the chiller face angle and freckle formation index, i.e., if the chiller angle is smaller, the freckle formation index becomes smaller. This tendency is almost the same as the result of upward segregation alloy A286[2]. However, it is noteworthy that the result of 45 and 30 degree chiller. In this angle, freckling tendency is steeply decreases compared to that of A286. This figure suggests that the effect of the solidification front angle on freckle formation in downward type segregation alloy is larger than that of the upward type segregation alloy.

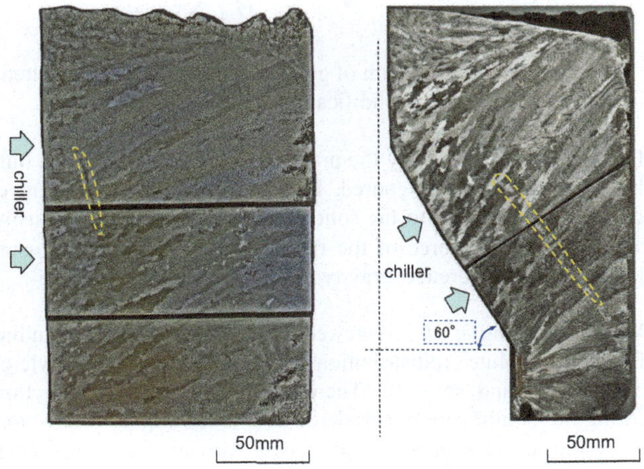

Figure 5. Vertical cross section of 90 and 60 degree chiller ingots

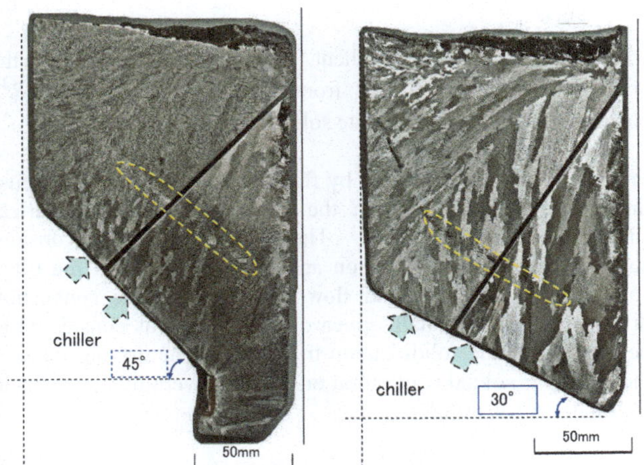

Figure 6. Vertical cross section of 45 and 30 degree chiller ingot

Table 1: Chemical composition of test alloys (mass %)

Angle	Chemical composition								$\Delta\rho$
deg	C	Ni	Cr	Mo	Al	Ti	Nb	Fe	g/cm3
90	0.055	64.06	20.35	9.00	0.12	0.23	3.56	2.34	0.0115
60	0.046	63.30	21.18	8.86	0.24	0.19	3.60	2.28	0.0134
45	0.049	63.04	21.23	8.96	0.21	0.19	3.59	2.62	0.0118
30	0.025	63.10	21.36	8.93	0.16	0.17	3.57	2.57	0.0132

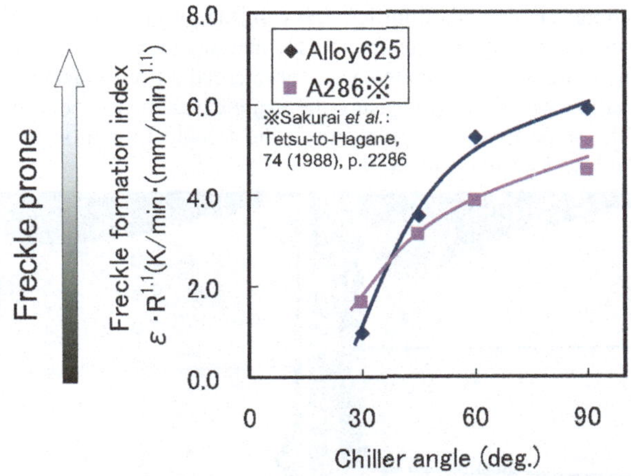

Figure 7. The relationship between freckle formation indexes and chiller angles

Solidification front angles are slightly different from the chiller angles. The solidification front angles were estimated from primary dendrites morphology in the ingots. The angles in A286 ingots were estimated from the photographs of each macro etched structure. The relationship between freckle formation indexes and solidification front angles is shown in Figure 8. Although the tendency of the curvature is the same as the diagram using the chiller angle, the convex shape shifts toward the lower angle side. This figure suggests that the effect of the solidification front angle on freckle formation may be marshaled factor that utilize the angle.

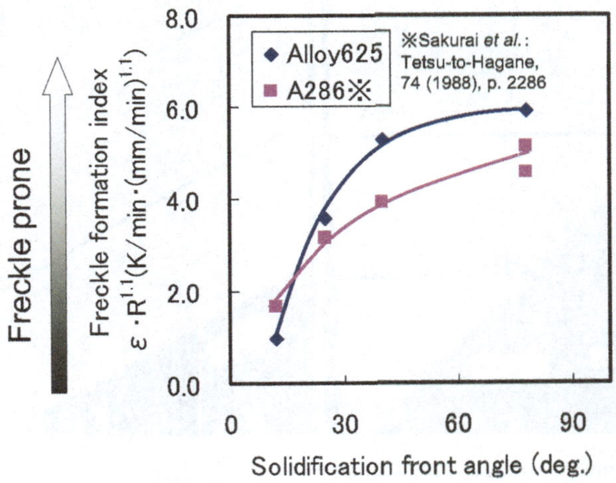

Figure 8. The relationship between freckle formation indexes and solidification front angles

Discussion

As described above, the chiller face angle is not equal to the solidification front angle. The solidification front angle in the horizontal directional solidification experiment is around 70 – 80 degrees in the case of the straight shape chiller that gives a 90 degree angle. The freckle formation index proposed by Suzuki and Miyamoto[9] is considered to be the index for the solidification front angle of 70 – 80 degrees. The freckle formation index indicates good prediction for freckling location in conventional ingots [16,17], because the solidification front angle in the side mold is in the same range as the solidification front angle in the horizontal directional solidified ingot. In the previous study[8], the 90 degree chiller and the freckle formation index were used to evaluate the freckling tendency of commercial alloys. Our unified evaluation was successful using the same experimental condition and the same index for each alloy. The freckle formation index worked well for the evaluation.

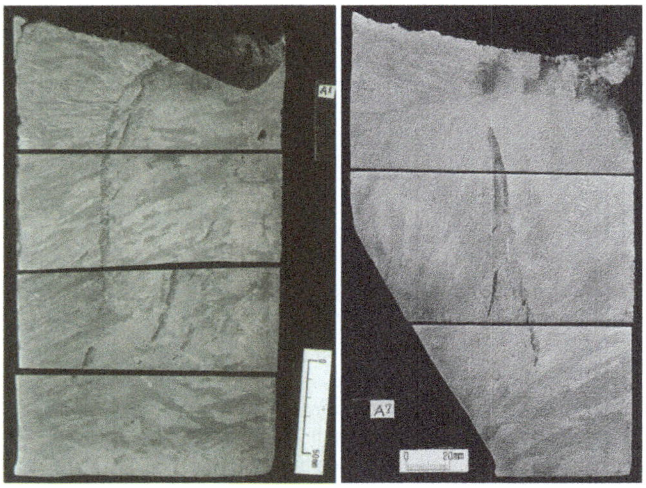

Figure 9. Vertical cross section of 90 and 60 degree chiller A286 ingot

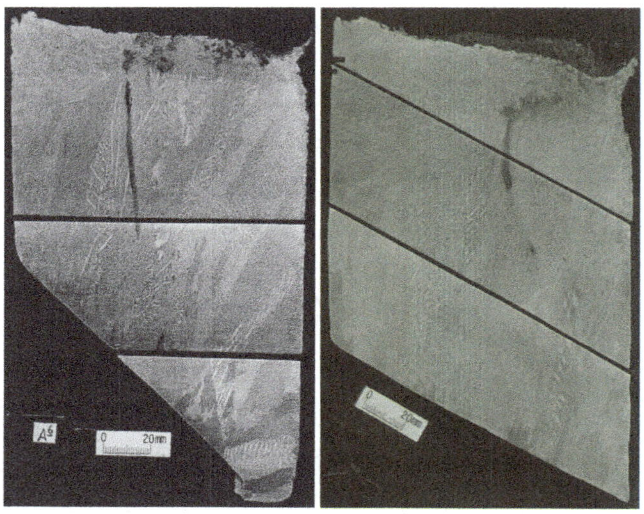

Figure10. Vertical cross section of 45 and 30 degree chiller A286 ingot

The macro structure of A286 alloy horizontal directional solidified ingots using tilted chillers is shown in Figures 9, 10. All freckle streaks shown here grow upward and the directions are almost the same regardless of those solidification front angles. Meanwhile, in contrast to conventional cast ingots, solidification proceeds with a more angled solidification front in the secondary melt process such as ESR and VAR. The experiment aimed to simulate the tilted solidification front. The effect of the solidification front angle is considered using a vector diagram. A vector diagram of gravity induced force at the tilted solidification front is shown in Figure 9.

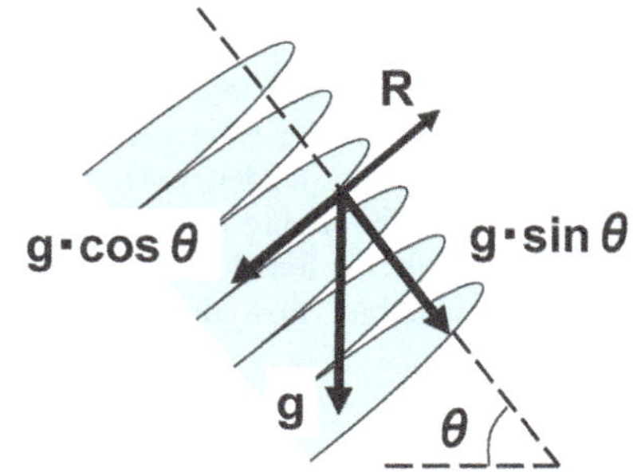

Figure 9. A vector diagram of gravity induced force at the tilted solidification front

Here, in order to simplify the problem, other effects such as bulk liquid convection are ignored. In this figure, the component of gravity force parallel to the solidification front is extracted using sinθ. The driving force of the flow direction is expressed using liquid density difference, gravity and the angle.

Wan-hong Yang *et al.* [3] reviewed all freckling criteria, including Fleming's solute redistribution equation[19], several Rayleigh criteria[20-26], and so forth. There should be interdendritic flow owing to solidification shrinkage and bulk liquid convection. They ultimately proposed that freckle formation criterion G·R divided by sinθ is the freckle formation index for the tilted solidification front angle. They proposed the following equation for freckling criterion.

$$\alpha = \frac{G \cdot R^{2.1}}{\sin\theta} \tag{2}$$

where G is temperature gradient, R is growth rate and θ is the angle of solidification front from the horizontal line. $G \cdot R^{2.1}$ equals $\varepsilon \cdot R^{1.1}$ in the steady state solidification.

Freckle formation is affected by fluid flow [18]. In order to discuss the effect of fluid flow on the freckle formation, numerical simulation was performed. The model did not consider solidification for simplification and only the effect of the tilted solidification front on fluid flow is discussed. The convection driven by the intentionally given driving force was considered to be affected by the solidification front angle. The driving force in the system is solutal convection only. The governing equations in the model are as follows:

$$\frac{\partial \mathbf{u}}{\partial t} + \mathbf{u} \cdot \nabla \mathbf{u} = -\frac{\nabla p}{\rho} + \nu \nabla^2 \mathbf{u} + \left(1 + \frac{c_L}{\rho_0}\frac{\partial \rho}{\partial c_L}\right)\mathbf{g} \tag{3}$$

$$\frac{\partial c_L}{\partial t} + \mathbf{u} \cdot \nabla c_L = D \nabla^2 c_L \tag{4}$$

where **u** is flow rate, D is Diffusion coefficient, p is pressure, ν is dynamic viscosity, g is gravity force, ρ is liquid density, and C_L is chemical composition of the liquid. In the calculation, the order of the ratio of dynamic viscosity and diffusion coefficient (Pr number in solute) was accorded with liquid metal and viscosity was increased to prevent turbulent flow.

Maximum flow rates were compared at the steady state using the model. It may be necessary to consider the solute distribution in the liquid. The flow rate equation by solutal effect is expressed as the following formula:

$$U^2 = \frac{D}{\nu} g' \frac{1}{\rho} \frac{\partial \rho}{\partial c_L} \Delta c_L L \quad (5)$$

Where g' is the component of gravity force and L is characteristic length. The value g' along with the solidification front is g·sinθ. Accordingly, the flow rate would relate to the solidification front angle θ. Figure 10 shows the fluid flow simulation result.

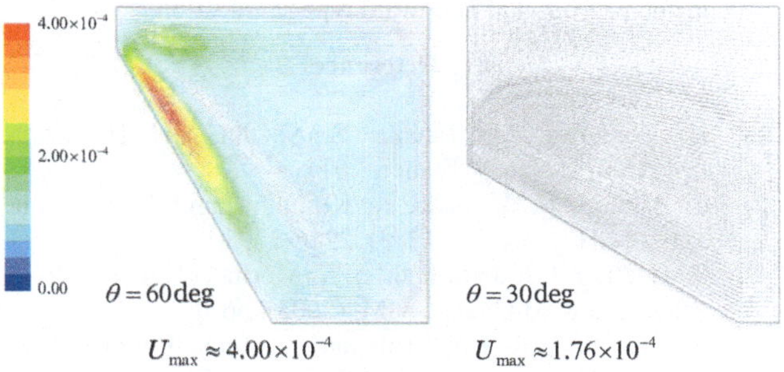

Figure 10. Fluid flow simulation result in tilted side wall

The result indicates that the flow rate at the tilted wall depends on the angle. If the gradient becomes shelved, the flow rate is small. It is important to note that the flow rate is not exactly proportional to sinθ. ΔC_L in equation (5) also depends on the side wall angle because solute distribution in the vertical direction varies in the calculation. The result shows that if the solidification front angle decreases, flow rate at the solidification front decreases.

There should be other effects on the freckle formation in the tilted solidification front. Modification of freckle formation criterion may be required. Figure 11 shows the value derived from multiplying the freckle formation index at the 90 degree chiller by sinθ

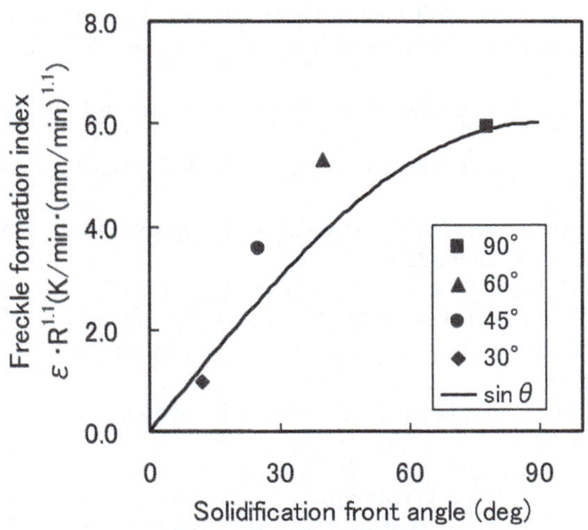

Figure 11. The effect of sinθ for the freckle formation indexes

The values of the indexes at 60 degrees and 45 degrees are larger than that of multiplying the index at 90 degrees by sinθ. Since the deviation of 60 degree and 45 degree values is large, simple modification using sinθ may not be able to evaluate the effect of the solidification front angle on freckle formation.

Thermosolutal convection in the mushy zone affects freckle formation in an ingot, which may account for the deviation. Although there is some deviation in freckle formation indexes derived from the horizontal directional solidification experiment, this method showed the tendency of freckling. Simple criterion such as $\varepsilon \cdot R^{1.1}$ or $G \cdot R^{2.1}$ is desirable from an industrial perspective. Criterion using ε, R, G may need further modification to make more precise estimation for freckle formation at tilted solidification front.

The effect of the solidification front angle on freckle formation varies depending on the type of freckle, such as upward type and downward type. There are two types of freckle suppression mechanism. Figure 12 is a schematic of freckle formation in the tilted solidification front for upward type and downward type segregation.

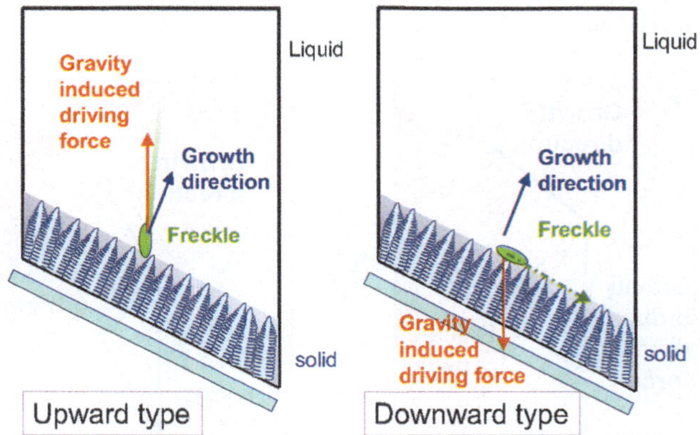

Figure 12. The effect of tilted solidification front for freckle formation

One is in the case of upward type segregation formation alloy. Freckle streak formed in solidification grow upward owing to its density difference from bulk liquid. If the solidification front is tilted, the direction of freckle floating up is close to the solidification direction. The formed freckle streak proceeds toward low solid fraction direction. Morikawa *et.al.* made a series of experiment that cast ingots were inclined in some solidification period. Carbon steels that are typical upward type segregation alloy were used for their study. Figure 13 is schematic drawing of freckle observation in the inclined ingot [9] based on the result of Morikawa *et.al.* experiment[27].

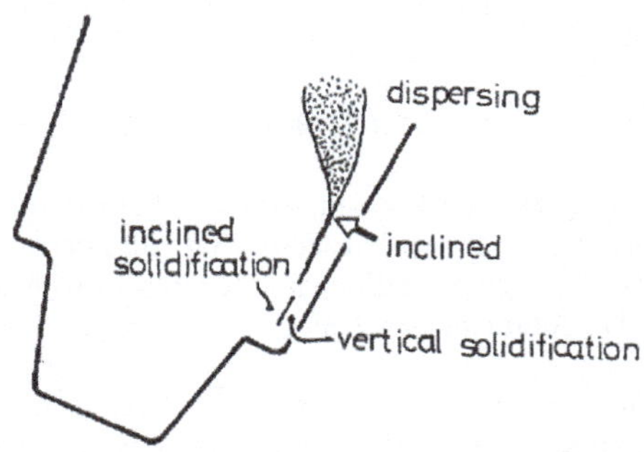

Figure 13. Schematic model of inclined ingot[26]

They reported that if a carbon steel ingot is inclined during solidification, the diameter of the freckle streak becomes large and it disperses toward the upper direction. The result suggests that a solute-enriched liquid would be easy to diffuse at the tilted solidification front in upward type segregation alloys. This may be the effect of freckle suppression by the tilted solidification front for upward type segregation.

In contrast, solute-enriched liquid is dense in downward type freckling alloys. In the case of downward type segregation, gravity driven force owing to liquid density difference is in the direction of the reverse of upward type segregation. Therefore, solute-enriched liquid tends to move toward the bottom of the ingot. If the solidification front is tilted much (freckle growth direction arrow in the figure is stuck in the mushy zone), it is impossible to penetrate into the high solid fraction dendrite forest. Therefore freckle streak grows in low solid fraction regions, i.e., freckle streak grows along the solidification plane. Since downward type segregation may be difficult to grow upward, the growing direction toward high solid fraction suppresses freckle formation.

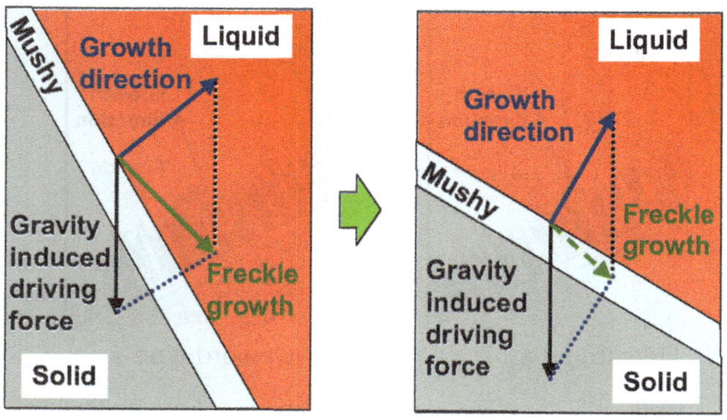

Figure 14 Schematic of assumed freckle suppression mechanism by solidification front tilting in downward type freckle

Figure 14 shows the schematic of the assumed freckle suppression mechanism by solidification tilting in the case of downward type segregation. Solute-enriched liquid tends to be trapped by the solidified layer before freckle formation in the right side of the solidification condition in Figure 14. This may be the suppression mechanism of downward type segregation due to the effect of solidification front tilting. Streat and Weinberg's experimental result[7)] could support this idea. The result showed no freckle was found in their vertical directional solidification experiment using dense solute alloy. If the angle of the solidification front becomes much smaller, freckle would not form in Alloy625 because of this mechanism.

Conclusions

How the solidification front angle affects freckle formation in the downward type segregation alloy was investigated using the horizontal directional solidification experiment with tilted surface chillers. The following conclusions can be drawn from this study.

1. If the chiller angle decreases, the freckle formation index becomes smaller. The tilting solidification front suppressed freckle formation in downward type segregation Alloy625.
2. The effect of the solidification front angle on freckle formation in the downward type segregation alloy is larger than that of the upward type segregation alloy.
3. Simple modification using $\sin\theta$ may not be able to evaluate the effect of the solidification front angle on freckle formation.
4. The effect of the solidification front angle on freckle formation varies depending on the type of freckle, such as upward type and downward type.
5. There are two types of freckle suppression mechanism. The suppression mechanism in downward type segregation is different from that in upward type segregation.

References

1) K. Kajikawa, M. Tanaka, S. Suzuki, and H. Yamada: LMPC2011, Nancy, France, 2011.
2) T. Sakurai, H. Yamada, K. Kitamura, and T. Takenouchi: Tetsu-to-Hagane, 74, (1988), 2286-2293
3) W-H Yang, J. J. deBarbadillo, K. Morita, T. Suzuki, W. Chen, and K-M Chang: JOM, (2004), 56-61
4) P. Auburtin, S. L. Cockcroft, and A. Mitchell: Superalloys 1996, TMS, Warrendale PA, (1996) 443-450
5) R. N. Grugel and L. N. Brush: JOM, (1997), 26-30
6) F. Kaempffer and F. Weinberg: Metall. Trans., **2** (1971) 2477-2483
7) N. Streat and F. Weinberg: Metall. Trans., **5** (1974) 2539-2548
8) K. Kajikawa, T. Sato, and H. Yamada: LMPC2009, Santa Fe, NM, 2009.
9) K. Suzuki, T. Miyamoto: Tetsu-to-Hagane, 63, (1977), 45-52
10) K. Mukai, Z. Liu, and K. C. Mills: Metall. Mater. Trans. 36B, (2005), 255-262
11) D. J. Steinberg: Metall. Trans., 5, (1974), 1341-1343
12) Z. Li, and K. C. Mills: Metall. Mater. Trans. B, 37B, (2006), 781-790
13) K. Mukai, Z. Li, and L. Fang: Mater. Trans. JIM, 45, (2004), 2987-2993
14) B. Sundman, B. Jansson and J.-O. Andersson: CALPHAD, **9** (1985) 153-190.
15) N. Saunders: Ni-database. Thermo Tech Ltd., Guilford, 1995.
16) H. Yamada, T. Sakurai, T. Takenouchi: Tetsu-to-Hagane, 75, (1989), 105-112
17) H. Yamada, T. Sakurai, T. Takenouchi: Tetsu-to-Hagane, 75, (1989), 97-104
18) T. Sawada and K. Kajikawa: LMPC2011, Nancy, France, 2011
19) R. Mehrabian, M. Keane and M.C. Flemings, Metall. Trans., 1 (1970), 1209-1220.
20) M.G. Worster: *Ann. Rev. Fluid Mech.*, 29 (1997), 91-122.
21) H.E. Huppert: J. Fluid Mech., 212 (1990), 209-40.
22) P.W. Emms and A.C. Fowler: J. Fluid Mech., 262 (1994), 111-39.
23) F. Chen, J.W. Lu, and T.L. Yang: J. Fluid Mech., 276 (1994), 163-87
24) M.H. McCay and T.D. McCay: J. Thermophys. Heat Transfer, 2 (1988), 197-202.
25) J.R. Sarazin, A. Hellawell, and R.S. Steube: Phil. Trans. R. Soc. London, 345 (1993), 507-44.
26) A.K. Sample and A. Hellawell: Metall. Trans. A, 15A (1984), 2163-73.
27) H. Morikawa, S. Onodera, and Y. Arakida: Tetsu-to-Hagane, 44 (1958) 1259-1265

Proceedings of the 2013 International Symposium on Liquid Metal Processing & Casting
Edited by: Matthew J. M. Krane, Alain Jardy, Rodney L. Williamson, and Joseph J. Beaman
TMS (The Minerals, Metals & Materials Society), 2013

THE BEHAVIOUR OF ENTRAINMENT DEFECTS IN ALUMINIUM ALLOY CASTINGS

[1]W. D. Griffiths, [1]A. J. Caden and [2]M. A. El-Sayed
[1]School of Metallurgy and Materials Science, College of Engineering and Physical Sciences, University of Birmingham, Edgbaston, Birmingham, United Kingdom. B15 2TT.
[2]Department of Industrial and Management Engineering, Arab Academy for Science, Technology and Maritime Transport Abu Qir, P.O. Box: 1029, Alexandria, Egypt

Keywords: Aluminium alloys, oxide films, entrainment, mechanical properties.

Abstract

The results of experiments are reported in which air bubbles were held for varying lengths of time in melts of three different Al alloys, (commercially pure Al, Al-7Si-0.3Mg, (2L99), and Al-5Mg alloy), with the air bubbles acting as analogs for the behavior of the interior atmosphere of double oxide film defects. After solidification a Pore Gas Analyser was used to measure the contents of the bubble, to determine how their oxygen, nitrogen and hydrogen content varied over time in the different alloys. The results showed that, firstly, oxygen should be consumed by reaction with the surrounding melt to form an oxide, followed by nitrogen, forming a nitride, probably AlN. Simultaneously, hydrogen dissolved in the melt passed in to the air bubble. At the end of the 40 minute holding period the air bubbles, and by analogy the interior atmosphere of a double oxide film defect, largely consisted of a nitrogen/hydrogen atmosphere. The reaction rates obtained from these experiments were used to estimate the time taken for the interior atmosphere of a typical double oxide film defect to be consumed, which was found to vary with alloy type, and to be about 25 minutes in the case of a defect in the Al-7Si-0.3 Mg alloy.

Introduction

The transfer of liquid Al alloys, and the action of pouring into a mould, is often accompanied by entrainment of the oxidized surface skin of the alloy. The very high reactivity of liquid Al with oxygen in the air means that the surface of the liquid metal can be considered to be permanently coated with an oxide film. Therefore, as the oxide film is ruptured by movement of the liquid metal underneath, the rupture must be almost instantaneously sealed by reaction between the Al and oxygen in the air. Pouring or transfer of the liquid metal results in considerable splashing of the liquid metal, (which has a viscosity similar to that of water), and as the splashes coalesce and merge with the bulk liquid, a doubled-over oxide film can be carried into the liquid metal. The sides of the double oxide film defect are unbonded, and the defect would capture a portion of the local atmosphere to be carried into the melt, from where it may be swept into the mould cavity, (see Figure 1) [1].

Double oxide film defects therefore consist of a crevice in the solid metal, formed by the unbonded oxide films, and containing a layer of gas. They form a defect in the final solidified casting of variable size, and with variable orientation with respect to any applied load experienced by the casting in service. It is therefore thought that double oxide films play a major role in influencing the variability of mechanical properties in Al alloys, [2], Mg alloys, [3], and possibly other alloys as well [4].

Since the defects consist of unbonded oxide surfaces containing a layer of gas, this gives rise to the speculation that the interior atmosphere of the defect can continue to react with the surrounding bulk liquid metal once it has been entrained. If the interior atmosphere becomes consumed, it may be possible for the oxide surfaces of the defect to be brought together where they may bond in some way, reducing the deleterious effect of the oxide film defect.

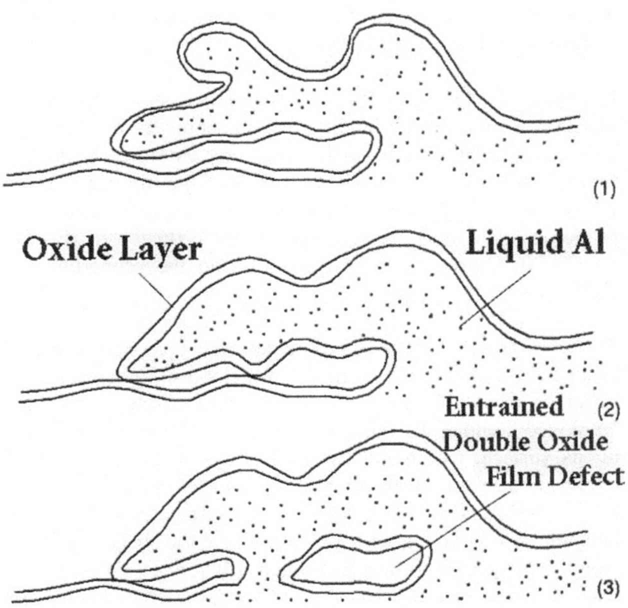

Figure 1. Sketch of the formation of an entrained double oxide film defect.

Nyahumwa et al. [5] speculated that if the interior gas in the defect was air, then, initially, the oxygen would react with the surrounding liquid metal to form alumina, and once all of the oxygen was consumed, the remaining nitrogen would react with Al to form AlN. Once all the internal atmosphere was consumed, Ar would remain, (air consists of 1 vol.% Ar), which would be inert and insoluble. The entrapment of an atmosphere consisting of air would be reasonable if a die casting is considered. In the case of a sand casting, the breakdown of the binder would produce a local atmosphere containing gases such as water vapour, hydrogen and hydrocarbons, of varying amounts, depending on whether a greensand or chemically-bonded sand was used.

Raiszadeh and Griffiths performed an experiment in which a bubble of air, acting as an analogue for a double oxide film defect, was held in a melt of commercial purity Al for several hours [6].

The volume of the air bubble was monitored using a real-time X-ray instrument, which showed a reduction in the bubble volume corresponding to the reaction of the air bubble with the surrounding melt. It was inferred that the air bubble contracted initially due to the reaction of Al with oxygen to form Al_2O_3, followed by the reaction of Al with nitrogen to form AlN. Examination of the surfaces of the solidified sample of the air bubble surface confirmed the occurrence of Al_2O_3 and AlN. It was suggested that these reactions occurred as the skin of the air bubble was ruptured and resealed, rather than by diffusion of the interior gases through the oxide skin of the bubble.

In melts containing relatively low hydrogen contents the bubble volume initially contracted, and then steadily shrank. In melts containing relatively high hydrogen contents, the bubble volume initially expanded, and then contracted, and from this it was inferred that hydrogen was passing through the oxide film into the air bubble, showing how double oxide film defects could be initiators of hydrogen-related porosity. Reaction rates obtained from observation of the rate of reduction of the bubble volume were used to infer likely lifetimes of the interior atmosphere of double oxide film defects of a couple of minutes, about the typical solidification time of an Al casting.

El-Sayed et al. suggested the occurrence of two competing mechanisms during the holding of Al castings in the liquid state before solidification that could affect any entrained bifilms. The first mechanism was the consumption of the air inside bifilms, which reduced their size and in turn improved mechanical properties. The other mechanism was the diffusion of H into the defects, increasing their size and making the properties worse [7,8].

The work presented here is a more detailed study of the behaviour of air bubbles held in Al melts, as a method of understanding the behavior of double oxide film defects. Previous work mostly employed commercial purity Al alloy, which would form Al_2O_3 on its surface. Commercially used alloys would form different oxides, and in this work other alloys were also used, namely, an Al-7Si-0.3Mg alloy, (2L99), where the surface oxide film would be spinel, (Mg_2AlO_4), and Al-5Mg, where the surface oxide film would be expected to be MgO. Such different oxide films would have different characteristics, and this may affect double oxide film defect behavior. In the previous work, reaction rates were inferred by changes in bubble volume, whereas in this work the air bubbles held in the melt were preserved by solidification of the surrounding liquid metal, and the contents of the air bubbles subsequently analysed using mass spectroscopy.

The results obtained allow a more detailed understanding of double oxide film defects in commercial alloys, and may point towards methods of eliminating the defects once they are formed.

Experimental Procedure

Air bubbles, to represent double oxide film defects, were trapped inside Al alloy melts by immersing a steel strip of length 50 mm, containing two blind holes of height 5 mm and diameter 6 mm. The steel strip was immersed in a melt to a depth of 20 mm, and then rotated at 540 rpm, giving an angular velocity at the position of the trapped bubbles of 1.4 ms^{-1}, (see Figure 2). This procedure was intended to represent the shearing action that might occur as a double oxide film defect was carried through the melt during mould filling.

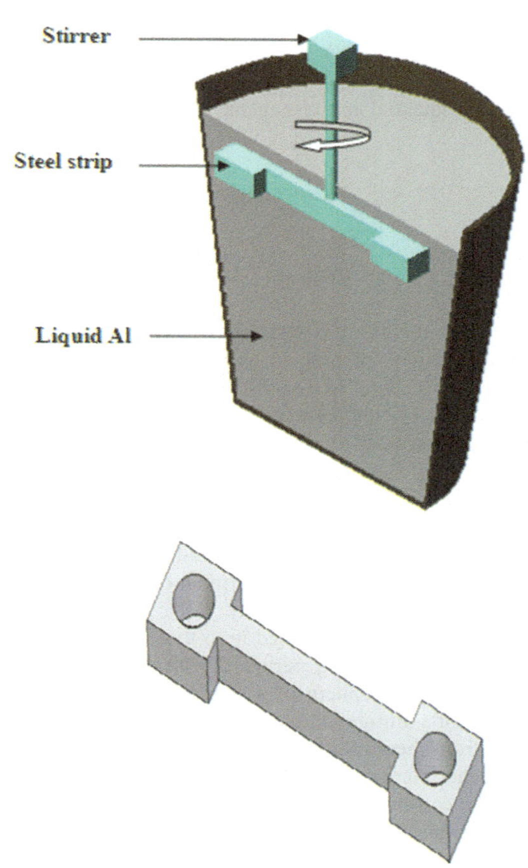

Figure 2. (a). Schematic of the experimental technique. (b). Illustration of the steel strip immersed in the melt.

The steel strip was rotated in the melt for periods of time ranging from 2 to 40 minutes, during which time the contents of the air bubble reacted with the surrounding Al alloy melt. Following the desired holding period the rotation was halted, and the alloy melt quickly solidified by the insertion of cold alloy blocks. The trapped air bubble was then sectioned out of the solidified melt and placed in a Pore Gas Analyser, (built by Hiden Analytical Ltd., Warrington, UK).

The Pore Gas Analyser was evacuated to a pressure of less than about 10^{-8} bar, and then the air bubble was pierced, and its contents passed through a mass spectrometer for analysis. In order to test the accuracy and reproducibility of the Pore Gas Analyser, air was trapped inside a piece of Cu tube of similar dimensions, which was soldered shut at both ends, and then its contents analysed in the same way. Table 1 shows the measurement of the composition of the air trapped inside three Cu tubes, with the different samples agreeing largely with the expected analysis of air, although the oxygen content, about 18%, was slightly lower than expected. However, the reproducibility and relative accuracy of the measurements gave confidence in the method used. The determination of the composition of the bubbles held in the melt for varying time periods was obtained from the Pore Gas Analyser by comparison with the results obtained from the analysis of air.

Experiments were carried out using bubbles of air trapped in three alloys having variable Mg contents, namely, commercial purity Al alloy, 2L99 alloy, (Al-7Si-0.3Mg), and Al-5Mg alloy. The varying Mg contents meant that each alloy should form a different surface oxide film. These reactants, formed on the surface of the alloy melt exposed to the air bubble, were analysed using scanning electron microscopy and EDX spectroscopic analysis.

Table 1. Measured compositions of samples containing air, as measured using the Pore Gas Analyser.

Gas	Vol.% of gases inside bubble			% Variation
	Sample 1	Sample 2	Sample 3	
Ar	0.98	0.96	0.96	2.0
H	0.71	0.73	0.73	2.0
N	79.47	78.68	78.68	1.0
O	18.24	18.42	18.42	1.0
H_2O	0.59	0.59	0.59	1.0

The change in composition of the air bubbles was used to estimate the rates of reaction of the bubble contents with the surrounding melt, (i.e., the change in oxygen and nitrogen content), and the rate of diffusion of hydrogen into the bubble. From these estimated rates of reaction, estimates were made of the time taken for consumption of the interior atmosphere of a double oxide film defect.

Results

Figure 3 shows a typical output from the Pore Gas Analyser, showing the variation in pressure obtained from the mass spectrometer as it analysed a sample of an air bubble in an Al-7Si-0.3Mg alloy melt, after 2 minutes holding. The largest peak was associated with nitrogen, as was to be expected, followed by oxygen, and hydrogen and argon, with a trace amount of water vapour.

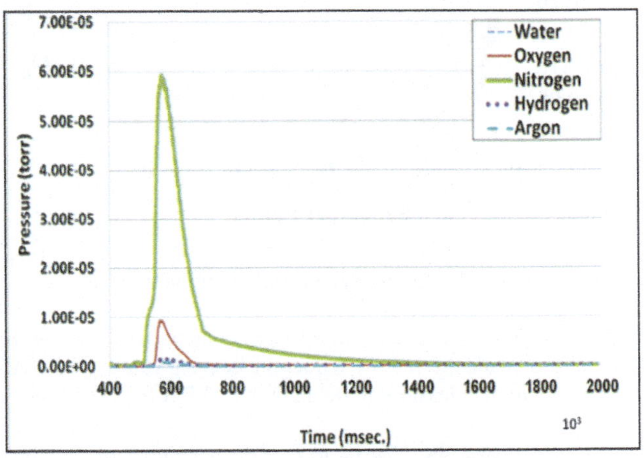

Figure 3. An example of the output from the Pore Gas Analyser. (The largest peak is due to nitrogen, the smaller peak is due to oxygen. There was little hydrogen in this sample).

Figures 4(a) to (c) show the change in volume for air bubbles trapped in the three different alloys. In the case of the commercially purity Al alloy, and the Al-5Mg alloy, the bubble volume decreased to about 40% of its original volume, with 40 minutes holding time, (see Figures 4(a) and 4(c)). In contrast, in the Al-7Si-0.3Mg alloy (Figure 4(b)), the reduction in bubble volume was much less, reaching about 70% of its initial volume.

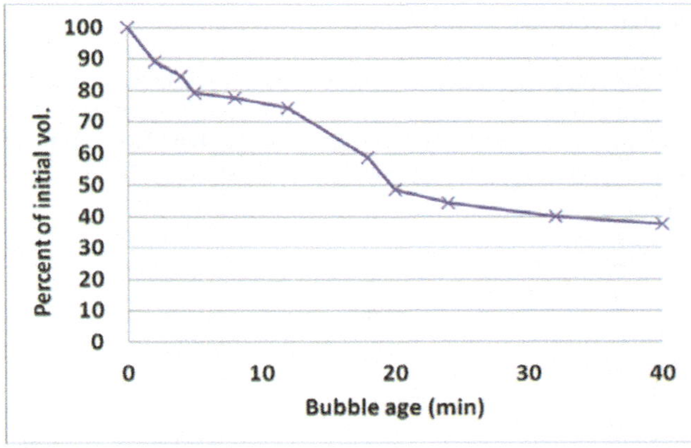

Figure 4 (a). The reduction in air bubble volume with time for the commercial purity Al alloy melt.

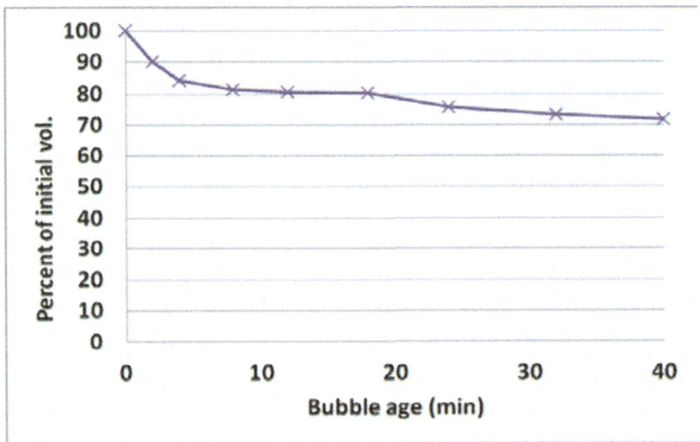

Figure 4(b). The reduction in air bubble volume with time for the Al-7Si-0.3Mg alloy melt.

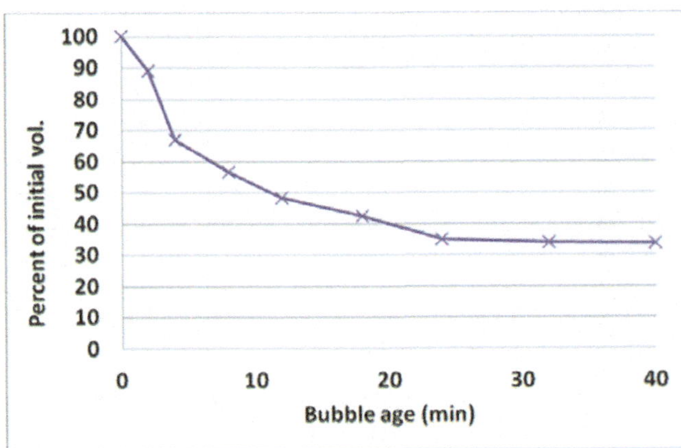

Figure 4(c). The reduction in air bubble volume with time for the Al-5Mg alloy melt.

Figure 5 shows the change in composition of the air bubbles trapped in the different alloy melts. Figure 5(a) shows that for commercial purity Al alloy, there was an initial reduction in oxygen content, from about 20% to nearly 0%. As the oxygen content declined there was a corresponding increase in nitrogen content. Once most, (but note, not all), of the oxygen had been consumed, the nitrogen content began to decrease, presumably due to the formation of AlN.

As first the oxygen content was decreased and then the nitrogen content decreased, the hydrogen content steadily increased. At the end of the 40 minute holding period, the bubble consisted of approximately 50% nitrogen and 50% hydrogen.

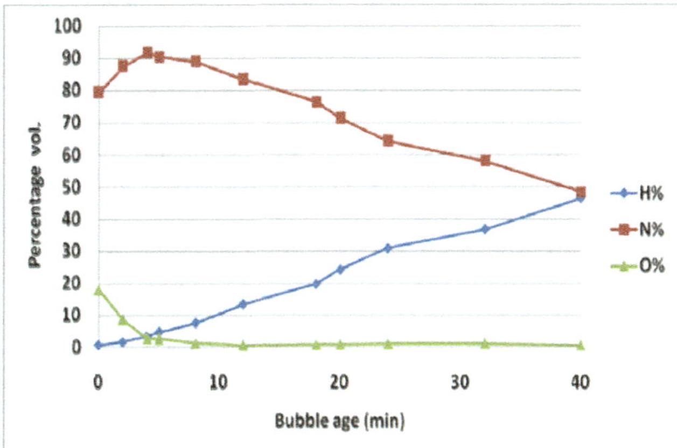

Figure 5(a). The change in composition for an air bubble held in a commercial purity Al alloy melt for up to 40 minutes.

Figure 5(b) shows the change in composition with the air bubble being held in a melt of Al-7Si-0.3Mg alloy. As Figure 4(b) showed, the rates of reaction were much less, but with the same trends. There was an initial reduction in oxygen, and a corresponding increase in nitrogen content until most of the oxygen was consumed, followed by a reduction in nitrogen content. During this time there was also an increase in hydrogen content. At the end of the 40 minute holding period the air bubble had a composition that was about 80% nitrogen, and 20% hydrogen.

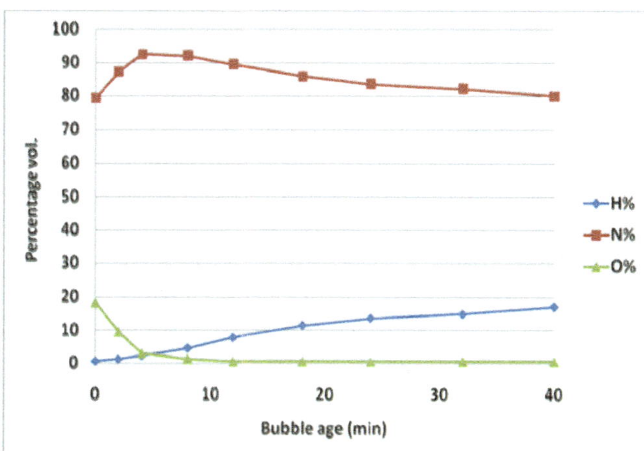

Figure 5(b). The change in composition for an air bubble held in an Al-7Si-0.3Mg alloy melt for up to 40 minutes.

Figure 5(c) shows the change in composition of an air bubble held in an Al-5Mg alloy melt. This also showed a reduction in oxygen content to almost 0% over the first five minutes, and a corresponding increase in nitrogen content, followed by a decrease. The hydrogen content also increased during this period, but slowed after about 25 minutes. The final composition of the bubble after 40 minutes holding time was about 70% hydrogen and about 25 to 30% nitrogen.

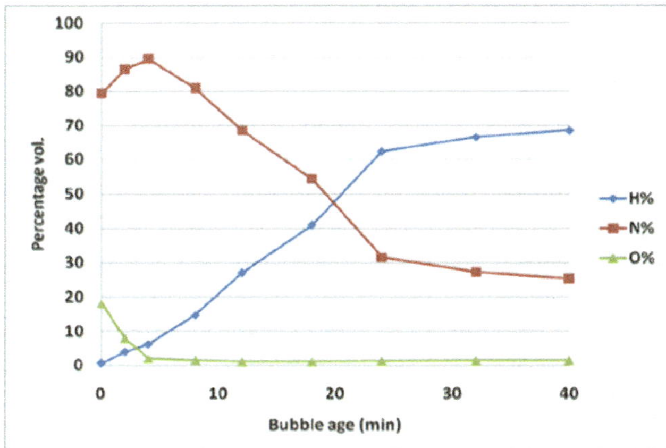

Figure 5(c). The change in composition for an air bubble held in an Al-5Mg alloy melt for up to 40 minutes.

Figure 6 shows the changes in composition for each element in the case of the different alloys. Figure 6(a) shows the change in oxygen content over time, which shows that for each alloy the reduction in oxygen occurred over about 5 minutes, with the Al-5Mg alloy, expected to have a porous MgO interface, having the quickest rate of loss.

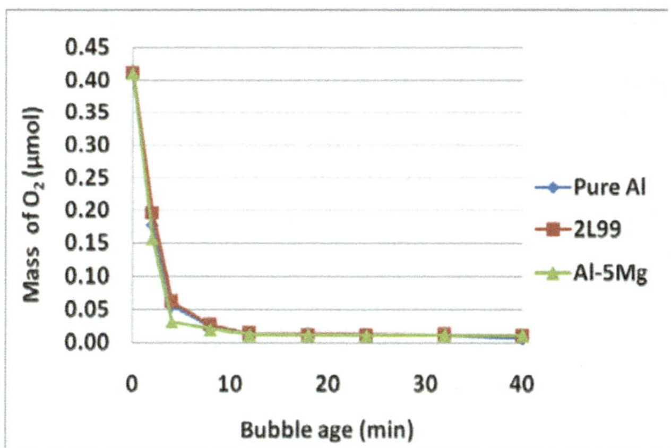

Figure 6(a). Change in oxygen content in the three alloys.

Figure 6(b) shows the change in nitrogen content over time for the three alloys, with the rate of reduction being greatest in the Al-5Mg alloy, and the least rapid being in the Al-7Si-0.3 Mg alloy.

Figure 6(c) shows the change in hydrogen content of the three alloys, with hydrogen passing into the air bubble most quickly in the case of the Al-5Mg alloy, almost reaching equilibrium, and least quickly in the case of the Al-7Si-0.3Mg alloy.

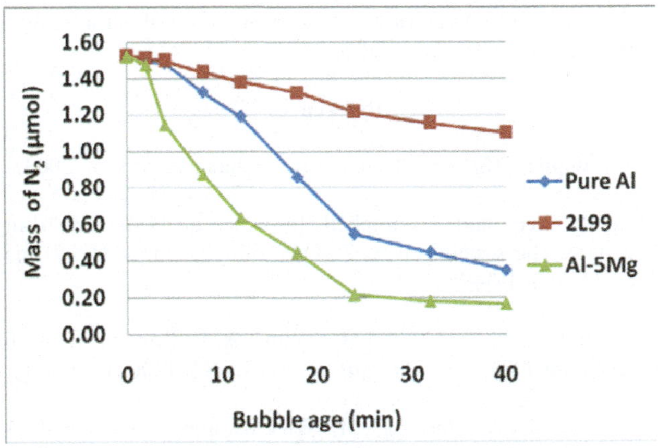

Figure 6(b). Change in nitrogen content in the three alloys.

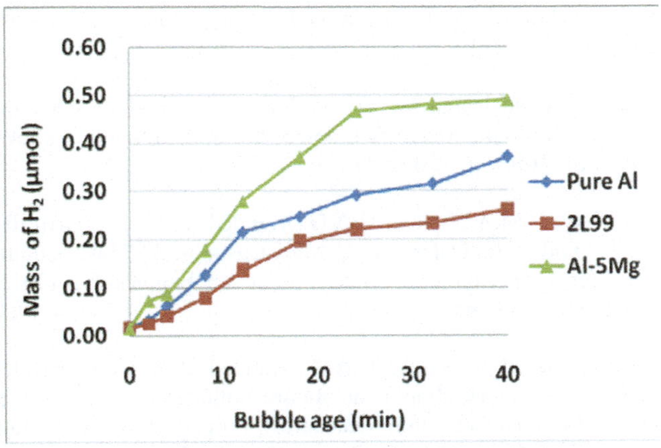

Figure 6(c). Change in hydrogen content in the three alloys.

Following analysis of the air bubble contents the interface of the alloy with the air bubble was examined using scanning electron microscopy and EDX to infer the compounds formed by reaction between the bubble and the melt. The oxides formed in the case of the three alloys appeared to be as expected, namely, alumina in the case of the commercial purity Al alloy, spinel ($MgAl_2O_4$) in the case of the Al-7Si-0.3Mg alloy, and MgO in the case of the Al-5Mg alloy. A nitride, probably AlN of Mg_3N_2, was also detected in all of the three cases.

Discussion

The reaction rates of oxygen and nitrogen obtained from the Pore Gas Analysis results were used to estimate the rate of consumption of the interior atmosphere of a typical double oxide film defect. This required an assumption of a typical double oxide film defect size, and specifically dimensions for the surface area of the defect, through which the reactions must occur, and the interior volume of the defect, and hence the amount of gas available for reaction. Previous work, using commercial purity Al, [9] had assumed a defect dimension of 10 x 5 mm with a thickness of 5 µm, based on a suggestion by Campbell [1]. This led to an estimate for the duration of the defect atmosphere of up to around two minutes.

More recently microfocus X-ray studies of porosity in cast Al alloys have been carried out, [10] which suggested dimensions of oxide film defects of around 2.2 x 2.2 x 0.1 mm. The duration of the atmosphere was estimated assuming these dimensions, and found to vary considerably with alloy composition, (see Figure 7). This was presumably due to the different nature of the oxides formed in each case, leading to variations in the reaction rates. Figure 7 shows that the atmosphere in a double oxide film defect in Al-5Mg alloy may have a duration of about 6 minutes, in commercial purity Al about 8 minutes, and in Al-7Si-0.3Mg alloy about 25 minutes. That the shortest predicted duration was associated with the Al-5Mg alloy is suggested to be due to the surface oxide film expected to form on the liquid alloy being (porous) MgO, which would allow a faster reaction rate. Note that the predicted time period for the presence of oxygen in a double oxide film defect was similar, and it was the varying rates of reaction with nitrogen which led to the predicted variation in the time taken for the interior atmosphere to be consumed.

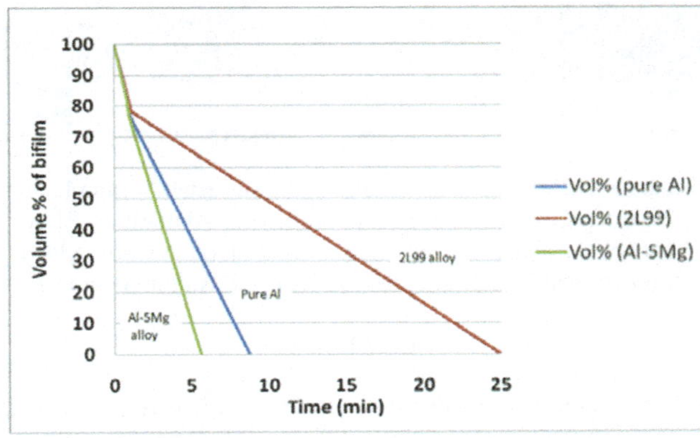

Figure 7. Estimates of the duration of the atmosphere in a double oxide film defect of assumed dimensions, for the three different alloys.

These predicted time periods for the survival of the interior atmosphere of double oxide film defects were longer than the expected solidification times of most thin section Al castings, showing how double oxide film defects formed during pouring would persist into the solidified casting. Furthermore, these estimates of duration of the atmosphere would be greatly extended by the passage of dissolved hydrogen from the melt into the defect, as shown in Figure 5.

Assuming that the melt hydrogen content was approximately equal in the tests, the greatest hydrogen pick-up by the bubble atmosphere occurred in the case of the Al-5Mg alloy, (0.49 µmol, in 40 minutes), while the commercial purity Al alloy was associated with a slightly less hydrogen pick-up of 0.37 µmol, and the Al-7Si-0.3Mg alloy was associated with the smallest hydrogen pick up of 0.26 µmol. Again, the variations in hydrogen pick-up were probably related to the variations in the nature of the oxide film associated with each alloy, with the greatest hydrogen pick-up being associated with the porous MgO film expected to form on the Al-5Mg alloy melt.

Finally, Figure 8 shows an example of an SEM picture taken from a sample of commercially pure Al alloy held for 40 minutes. EDX analyses at points X1, X2 and X3 revealed alumina, probably AlN and pure Al alloy, respectively, one layer over another. At this time, all or most of the oxygen in the air bubble had been consumed, (see Figure 5(a)), and the bare Al alloy surface (labeled X3), would have been in contact with an atmosphere that was largely nitrogen and hydrogen. That the bare Al alloy surface can occur in the presence of nitrogen, suggests that the formation of AlN may require some sort of incubation period, (at least in the case of the commercial purity Al alloy).

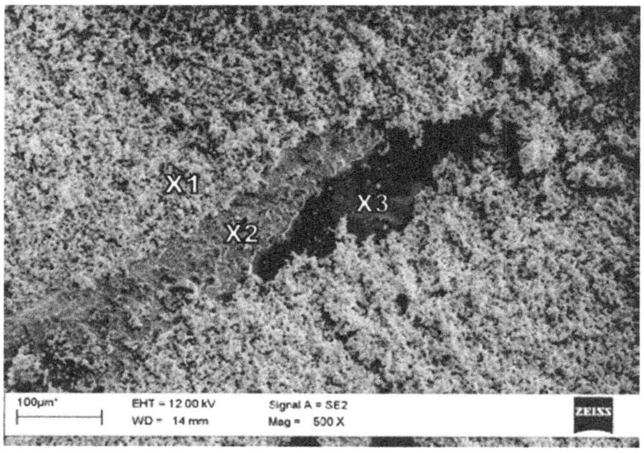

Figure 8. SEM micrograph showing the surface of a sample after holding in commercial purity Al alloy for 40 minutes. EDX analysis revealed the presence of alumina (at position X1), a nitride, probably AlN, (at position X2), and Al, (at position X3).

Conclusions

1. Pore Gas Analysis experiments were carried out on air bubbles held in three different Al alloys for periods of up to 40 minutes, to act as analogues for double oxide film behaviour in liquid Al. The measured changes in bubble composition were used to infer reaction rates to predict the duration of the atmosphere in typical double oxide film defects, leading to estimates of between 6 to 25 minutes, depending upon alloy composition.

2. The Pore Gas Analysis results also showed an increase in hydrogen in the bubble atmosphere, as it diffused in from the melt. The rate of diffusion also varied with alloy composition, and was again most rapid in the case of the Al-5Mg alloy. The estimated durations of the oxide film defect atmospheres would be greatly extended by the presence of hydrogen.

3. Examination of the compounds formed during reaction between the melt and the air bubbles indicated the formation of nitrides. In the case of the commercial purity Al alloy, circumstantial evidence suggests an incubation period may be necessary before the formation of AlN can take place.

Acknowledgements

The authors would like to gratefully acknowledge the technical assistance of Mr. Adrian Caden and Mr. Robert Goodwin of the University of Birmingham, and the financial assistance of Prof. A. Y. Kandeil of the Arab Academy for Science, Technology and Maritime Transport, Alexandria, Egypt, and the technical assistance of Prof. H. Salem of the American University in Cairo, Egypt. In addition, partial support from the EPSRC Centre for Innovative Manufacturing in Liquid Metal Engineering (EP/H026177/1) is acknowledged.

References

1. J. Campbell, *Castings*, Butterworth-Heineman, Oxford, 2003.

2. N. R. Green and J. Campbell, Influence of oxide film filling defects on the strength of Al-7Si-Mg alloy castings, *AFS Trans.*, 102 (1994) 341-347.

3. W. D. Griffiths and N-W Lai, Double oxide film defects in cast magnesium alloy, *Met. and Mat. Trans. A*, 38A, (2007) 190-196.

4. J. Mi, R. A. Harding, M. Wickins and J. Campbell, J., Entrained oxide films in TiAl castings, *Intermetallics*, 11 (2003) 377-385.

5. C. Nyahumwa, N. R. Green, and J. Campbell, J., *AFS Trans.*, 58 (1998) 215-223.

6. R. Raeiszadeh and W. D. Griffiths, A method to study the history of a double oxide film defect in liquid aluminium alloys, *Met. Mat. Trans. B*, 37B (2006) 865-871.

7. El-Sayed, M.A., Salem, H.A.G., Kandeil, A.Y., and Griffiths, W.D., Oxide Film Defects in Al Alloy Castings, 22nd International Conference on Computer-Aided Production Engineering, (2011) Edinburgh, Scotland.

8. El-Sayed, M.A., Salem, H.A.G., Kandeil, A.Y., and Griffiths, W.D., Effect of Holding Time Before Solidification on Double-Oxide Film Defects and Mechanical Properties of Aluminum Alloys, *Met. Mat. Trans. B* 42B (2011) 1104-1109.

9. R. Raeiszadeh and W. D. Griffiths, A semi-empirical mathematical model to estimate the duration of the atmosphere within a double oxide film defect in pure Al alloy, 39B, *Met. Mat. Trans. B,* 39B (2008) 298-303.

10. J. M. Park, Behaviour of bifilms in A356 alloy during solidification: developing observation techniques with 3-D micro x-ray tomography, *MRes thesis*, University of Birmingham, 2010.

Proceedings of the 2013 International Symposium on Liquid Metal Processing & Casting
Edited by: Matthew J. M. Krane, Alain Jardy, Rodney L. Williamson, and Joseph J. Beaman
TMS (The Minerals, Metals & Materials Society), 2013

INCLUSION BEHAVIOR IN STEEL AND ALUMINUM MAKING REACTORS

J.P. Bellot, O. Mirgaux, A. Jardy
Institut Jean Lamour, UMR 7198, CNRS ('LabEx DAMAS')/ Université de Lorraine, F-54011 Nancy, France
Key words: Inclusion, steel, aluminum, ladle, simulation, population balance

Abstract

The control of inclusion population remains an important issue for the steelmaking and aluminum making industries where the removal of particles is mainly operated by flotation and gravity separation. Argon gas is injected through porous plugs in the ladle of liquid steel whereas a rotating impeller and a mixture of argon and chlorine stir up the aluminum liquid bath. Nowadays the modeling of such complex three-phase-reactors is possible by combining Population Balance with convective transport of the inclusions so as to calculate the time evolution of the size distribution of the inclusion population. An operator splitting technique is employed to solve the coupled Population Balance Equation and the transport equation. Results are provided for either pilot or industrial scales and allow us to compare the respective roles of mechanisms (flotation, entrapment on the liquid surface, gravity separation, agglomeration) on the particle size distribution and on the inclusion removal rate.

Introduction

The control of metal cleanliness has always been an issue of great concern for the metallurgists since the inclusions influence directly the mechanical properties of the alloys mainly the workability, surface quality and fatigue strength. In most of the metallurgical routes a refining treatment of the molten alloy has been introduced in particular with the aim of improving the metal cleanliness which means a better control of the amount, the size and morphology and finally the composition of inclusions. After solidification of the metal these properties can be quantified by either destructive or non-destructive assessment techniques but the cleanliness can be hardly modified by thermal treatments. That is one very good reason why the inclusion treatment occurs before casting. Since the beginning of the XXIst century this issue is a new challenge for the metallurgists for two main reasons: on one hand the reduction in the weight of high performance materials together with the improvement of mechanical properties requires an improvement of the metal cleanliness. On the other hand the increase of the recycling of used metal reinforces the role of molten metal refining. Gas-stirring ladle treatment of liquid metal has been pointed out for a long time as the processing stage mainly responsible for the inclusion population of aluminum and specialty steels. A particular attention has been paid on the optimization of this complex three-phase-reactor, where strongly dispersed inclusions are transported by the turbulent liquid metal / bubbles flow.

The numerical simulation of such complex metallurgical reactor has already been the purpose of many studies. The mathematical modeling became in the early '70s an intrinsic part of materials engineering and the numerical simulation of refining reactors has been initially driven by Szekely. Hence as an example, Nakamishi and Szekely [1] initiate the use of k-ε turbulent model for the prediction of agglomeration kernels in order to estimate the Al deoxidation rate of liquid steel reactor. Other authors followed this approach among them Guthrie [2,3] who improved step by step the numerical simulation of gas stirred ladle systems. Mazundar and Guthrie published in 1995 a relatively exhaustive review addressing physical and mathematical models of ladles. Since then, the simulations of liquid metal processing were developed with an increasing level of sophistication. This type of work, performed under the leadership of KTH in Stockholm [4] and the University of Urbana-Champaign [5], can be considered as a benchmark in this field. KTH and Mefos in Sweden have carried out a number of projects within Jernkontoret'research especially focused on non-metallic inclusions in ladle and tundish. Jonsson [6] pointed out that the understanding of particle behavior nearby the liquid surface remains very poor and no model exists for predicting the deposition rate. The authors gave a first approach by Lagrangian calculation of particles near the slag-liquid interface and concluded on the role of particle inertia on the deposition rate. A similar approach has been followed by Gardin [7] revealing a lack of a detailed description of the non-isotropic and damped turbulence near the interface.

An integrated model coupling CFD and a nucleation-growth-removal model has been recently developed in Postech by Kwon et al. [8]. Even if the authors do not provide all the details of the numerical methods employed the numerical platform is probably one of the most integrated since in addition to the bubble-liquid flow calculated by an Eulerian-Lagrangian method the model simulates nucleation, growth and agglomeration of inclusions. Recently, a very innovative work was performed by Claudotte et al. [9] by coupling a population balance equation (PBE) with the computational fluid dynamics (CFD) to model the convective transport of particles and the probability of collisions due to the turbulent molten steel flow and the changes of particle properties due to breakup and aggregation. A variant of the quadrature method of moments (QMOM) has been adapted by the authors for the prediction of inclusion population in terms of chemical composition and size taken into consideration nucleation, diffusion growth and aggregation. A comparison of two numerical methods for solving the PBE is given by Daoud et al. [10] and reveals that the QMOM and the CM (class method) lead to similar results. However, most models used by these authors should be regarded as general models of processes that do not accurately describe the behavior and the capture of particles at interfaces (refractory wall, surface and bubbles).

In this paper we present the development, using the CFD code as a basis, of two- or three-dimensional simulation models applied to aluminum and steel stirred reactors.

The metallurgical reactors

Liquid metal processing shows similarities between aluminum and steel since in both cases the argon injection is operated. Flotation is one of the aims of gas injection which is a process widely used in industry to separate particles from a continuous liquid phase. During their ascension through the bulk, the bubbles collect the dispersed particles and release them at the surface, where they accumulate in the dross layer for Al and in the slag for steel. Figures 1a and 1b give a schematic diagram of the ladles used in the two industries. In the Al flotation tank, the gas (often a mixture of chlorine and argon) is injected into the melt through a rotating impeller. This impeller generates a turbulent fluid flow in the reactor, enhancing the probability of collisions between bubbles and unwanted inclusions. In the steel ladle , an injection of argon through one or more porous plugs at the bottom of the reactor provides both mixing of the liquid metal to achieve thermal and chemical homogeneity and the entrapment of the inclusions by the flotation mechanism.

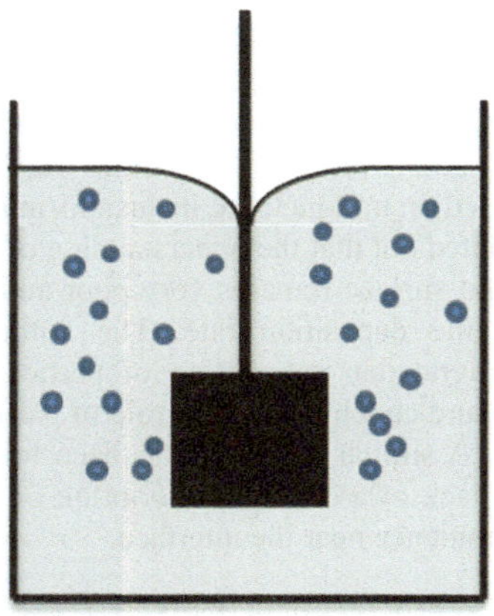

Figure 1a: Al flotation tank

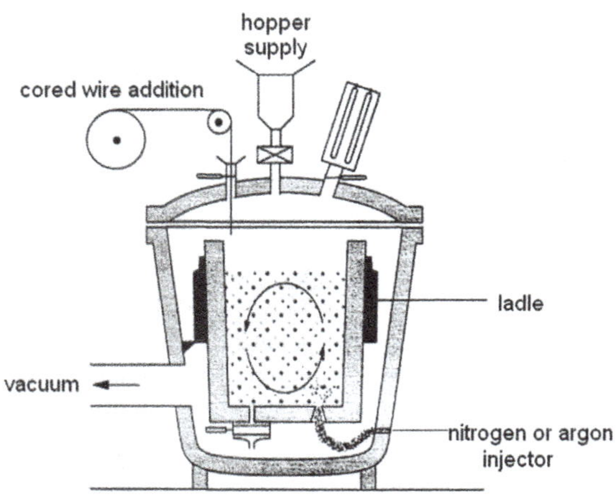

Figure 1b: Steel gas stirred ladle

Inclusions most frequently found in molten aluminum are oxide films, refractory particles and aluminum carbide (originating from refractory degradation or refractory metal reactions). Inclusion population in molten steel is mainly composed of non-metallic oxide inclusions such as calcium aluminate inclusions. Size of these inclusions may vary from one micrometer to a few hundreds micrometers, for the coarsest ones.

Hence, the technology applied to remove the inclusions remains quite similar. The mechanisms involved are:

- the collisions between inclusions, which can lead to aggregation and even to agglomeration if reconstruction of the aggregate occurs,
- the collisions with bubbles, which lead to the mechanism of flotation,
- the entrapment at the interface between the liquid bath and slag coverage,
- the separation induced by gravity.
- the entrapment at the ladle walls.

Mathematical modeling

The approach adopted for reactor modeling is divided into two parts. The bubbles plume play an important role in metal bath mixing while, owing to their small weight fraction (<0.01%), the inclusions do not affect the flow. First, the two-phase flow turbulent bubble-liquid metal is simulated for the 2D or 3D geometry of the ladle and a strong coupling is achieved between the liquid metal and the bubbles. This resolution provides the velocity fields as well as the maps of local flow turbulence and retention rate (gas volume fraction in the mixture); these data define the conditions of inclusion interaction (aggregation, flotation, entrapment). Details of the hydrodynamic modeling will be found in ref. [11]. In the second step, the behavior of the inclusion population is caused to both the transport at the macroscopic scale of the ladle and the interaction mechanisms at the mesoscopic scale of the particle. The main features of this mathematical modeling are presented below.

As presented by Kumar and Ramkrishna [12], population balance is a powerful way of synthesizing the behavior of a population of discrete particles from the behavior of single particles in their local environment. The behavior of the inclusion population, defined by a distribution function of particle size (N_i is the number of inclusions of class "i" –volume of the particle is between v_i and v_{i+1} -per m^3 of liquid metal), is described by the population balance equation (PBE). Equation (4) represents the macroscopic transport phenomena of inclusions (left member), and mesoscopic phenomena such as bubble-inclusion (flotation Z_{bi}) and inclusion-inclusion (aggregation B_i-D_i) interactions:

$$\frac{\partial \alpha_l N_i}{\partial t} + div(\alpha_l \mathbf{u}_l N_i) = \alpha_l \left(B_i - D_i\right) - \alpha_l Z_{bi} - S_i \quad (1)$$

In equation (4), S_i is the inclusion gravity separation term.

The transient solution to this equation can be obtained by separating the transport and collision operators [13]. In the first step, the transport equation of the quantity N_i is solved using the Finite Volume Method:

$$\frac{\partial \alpha_l N_i}{\partial t} + div(\alpha_l \boldsymbol{u}_l N_i) = 0 \quad (2)$$

In the second step, the population balance (equation 6) is solved in each control volume applying the cell average technique [14] which is a variant of the fixed pivot method of Kumar and Ramkrishna [12] leading to a significant reduction in numerical diffusion:

$$\frac{\partial \alpha_l N_i}{\partial t} = \alpha_l \left(B_i - D_i\right) - \alpha_l Z_{bi} - S_i \quad (3)$$

The aggregation and flotation kernels are an issue of great development and the reader will find details of the physical phenomena and of the applied models in [10,11]. Concerning the separation induced by gravity, following the decomposition of particle velocity into local fluid velocity and Stokes velocity, the source term S_i for the transport equation is:

$$S_i = div(\alpha_l \boldsymbol{u}_s N_i) \quad (4)$$

where u_s is the vertical Stokes velocity in the case of small inclusions whose particle Reynolds number is lower than 1.
A particular attention has been paid on inclusion entrapment at the liquid metal / slag interface. It is modeled following the approach based on a deposition law developed initially by Wood [15] and adapted to a free liquid surface condition by Xayasenh [16]. The entrapment of inclusions at the ladle walls is not considered.

Concerning the aluminum stirred reactor, the axisymmetric feature allows us to develop a 2D homemade code to solve the equation 2. On the contrary, since the flow in the steel ladle is 3D in nature (due to the location of the porous plugs), the equation 2 is solved within the frame of the CFD code FLUENT.

Numerical simulation of the Al flotation tank

A cylindrical laboratory scale apparatus with an inner diameter of 33 cm and containing 70 kg of molten aluminum at 1000 K was modeled (figure 2). At the tip of each blade, a gas injector blows a mixture of argon and chlorine into the melt. Bubble size was calculated with correlations established from water experiments and transposed to liquid aluminum by Waz [17].

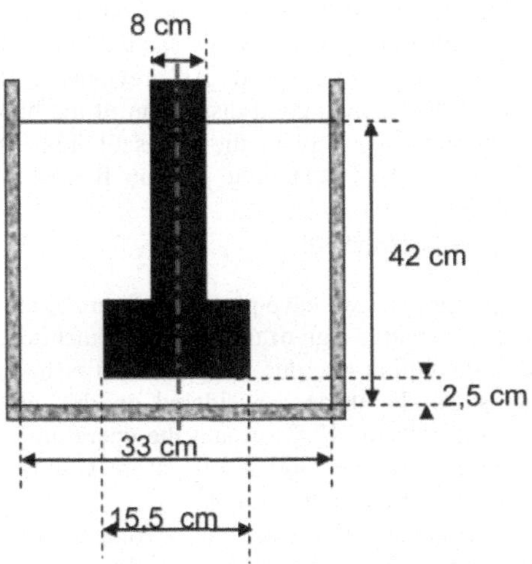

Figure 2: Schematic representation of the laboratory scale apparatus

In an industrial cast house, tuning of the flotation process is achieved with three parameters that are gas flow rate, rotor speed and location of the rotor in the tank. Dimensions of the pilot tank being relatively small, the rotor is positioned closed to the bottom in order to allow significant residence time of the bubbles into the melt. Influence of the gas flow rate and rotor speed on the process efficiency is studied through three different cases, referenced in table I.

Table I: Operating conditions

	N (RPM)	Q_g (Nm^3/h)
A	250	0.5
B	250	1.5
C	500	0.5

In the following calculations, we consider alumina inclusions with a 3900 $kg.m^{-3}$ density. Initial size distribution of the inclusions was established using the mean of several measurements performed with a Liquid Metal Cleanliness Analyzer (LIMCA™). Twenty-three classes of inclusions were considered with representative diameters spanning from 2.5 µm to 205 µm. As the resolution of LIMCA™ does not go below 20 µm, the measured distribution was extrapolated (the four smallest classes) to obtain a more realistic distribution. The complete distribution is reported in figure 3 and is used as the initial PSD (Particle Size Distribution).

At time t=0, the inclusions spatial distribution is supposed to be homogeneous in the melt.

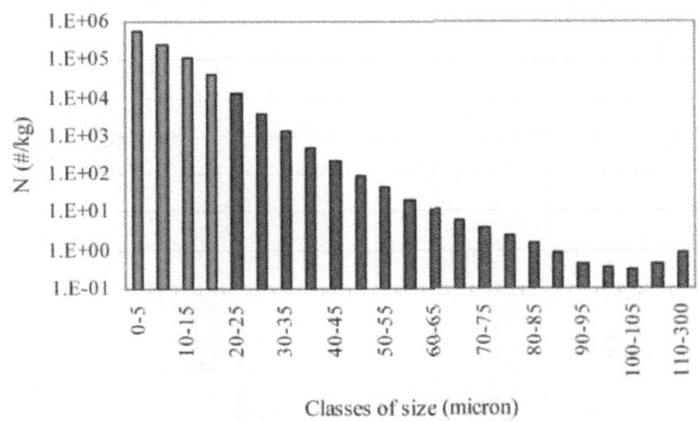

Figure 3: Initial PSD. In gray, the extrapolated part of the distribution

Fluid flow and bubble repartition

CFD calculations show a relatively weak influence of the dispersed gas phase on the liquid metal flow pattern in the bulk of the reactor. Figure 4 gives the calculated time-averaged streamlines in the steady state for case B. A weak swirl is noticed near the shaft, close to the surface of the bath. It becomes more pronounced at higher rotation speed. Turbulence properties (ε and k) reach their maximum value around the blades of the rotor where the shear is strong.
As shown in figure 5, increasing rotor speed with a constant gas flow rate improves bubbles dispersion in the reactor, leading to an upward trend in mean gas holdup. No significant difference between bubbles dispersion in cases A and B (same rotor speed and higher flow rate) is noticed.
On the other hand, the highest mean gas holdup is predicted for case B, which has a gas flow rate 3 times higher than cases A and C. Average residence time of the bubble into the melt ranges from 0.6 to 0.86 depending on the case considered.

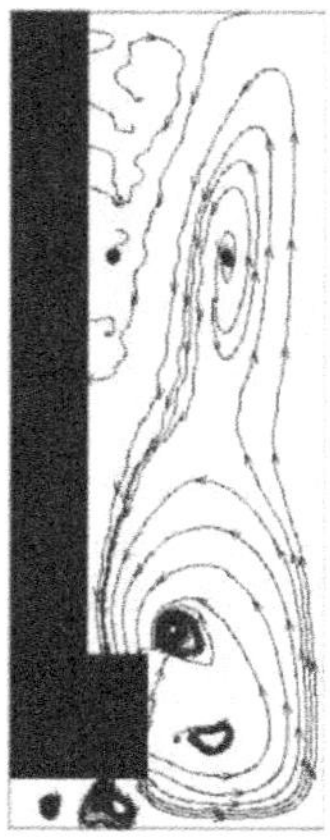

Figure 4: Computed streamlines for case B

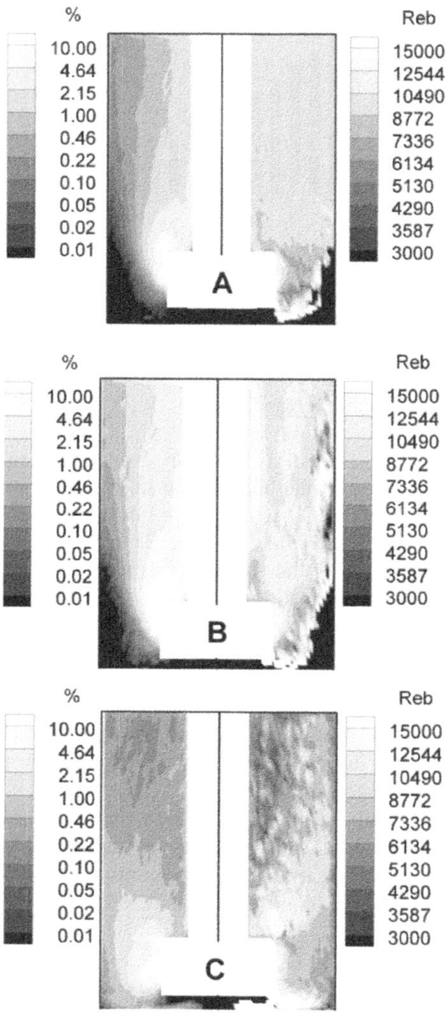

Figure 5: Computed local gas holdup (volume fraction in %) and bubbles Reynolds number for cases A, B and C

Time evolution of the inclusion population

A ten minute treatment was simulated for the three cases studied. Time evolution of the total number of inclusions into the melt is reported in figure 6. After a strong decrease in the total number of inclusions at the beginning of the treatment, the rate of removal slowly softens. This change is due only to the depletion of the number of inclusions in the melt (driving force of agglomeration and flotation phenomena).

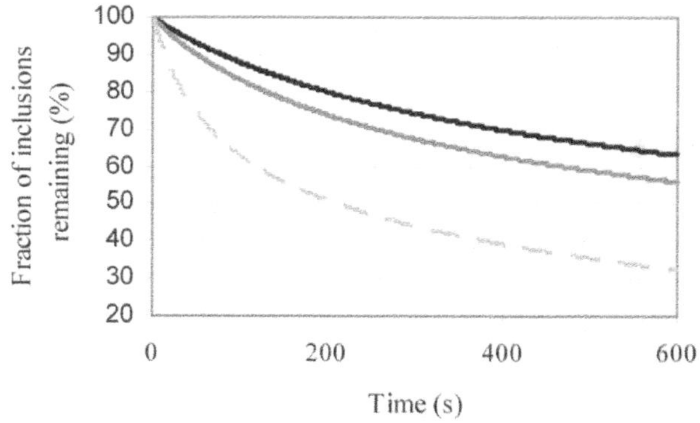

Figure 6: Computed evolution of the total number of inclusions into the melt for case A (black), B (gray) and C (doted gray).

Cases A and B have the same rotor speed, but the gas flow rate of case B is three times higher than that of case A. It appears that cases A and B are relatively close to each other, case B being slightly more efficient. Case C has a low gas flow rate (equivalent to case A) but a high rotor speed. As previously seen, those operating parameters allow a good dispersion of the bubbles into the liquid bath and especially in the zones of high turbulence intensity, which is the best situation to promote flotation.

Agglomeration versus flotation

For comparison purposes we have plotted in figure 7, the flotation removal rate (the negative sign of the rate is not mentioned on the figure) and the evolution rate due to agglomeration for the same class of inclusions. Inclusions considered on this plot have a representative diameter of 47.5 μm and the operating conditions correspond to case A. The plotted values are computed at the beginning of the treatment (initial time).

Under those conditions the computed evolution rate due to agglomeration is positive in the whole tank, which means that the birth rate is greater than the death rate.

Comparison between the left and right parts of figure 7 reveals that, in a large part of the tank, removal by flotation is in the same order of magnitude as birth by agglomeration. This means that equilibrium between agglomeration and flotation may be reached locally for certain classes of inclusions, which results in very low evolution rates.

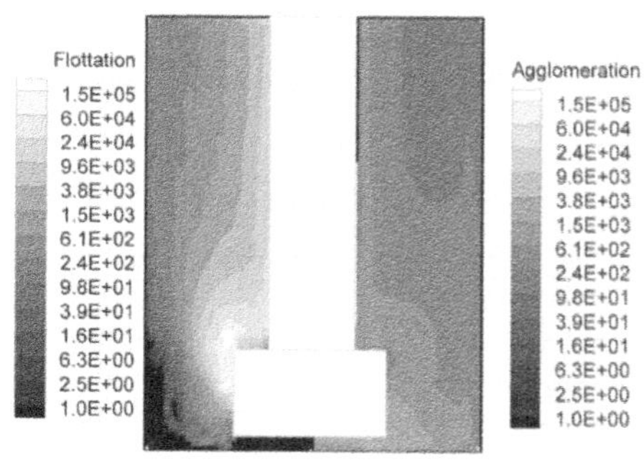

Figure 7: Computed flotation removal rate (left) and agglomeration feed rate (right) ($m^{-3}.s^{-1}$)

PSD after 5 and 10 minutes of treatment are shown for case B in figure 8. No significant difference between those two PSD can be observed, except for the smallest classes that keep on loosing inclusions.

At the beginning of the treatment coarser inclusions ($d_p > 70$ µm) are quickly removed from the melt since they are more likely to collide with bubbles, and because of their high settling rate. Evolution rates (agglomeration, flotation and settling) of the smallest classes remain quasi constant with respect to time since the probability of collision and flotation are very low. Intermediate classes ($40 < d_p < 70$ µm) reach a balance between removal by flotation and feed by agglomeration, as previously explained.

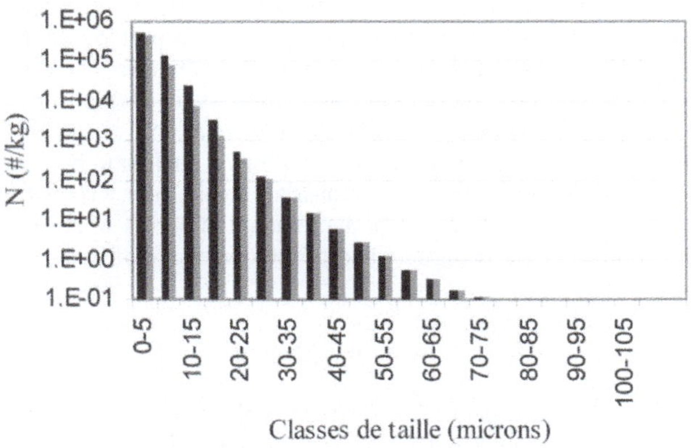

Figure 8: PSD after 300 sec. (black) and 600 sec (gray), for case B

Numerical simulation of the steel gas stirred ladle

The numerical simulation is applied to a full scale 60 t steel ladle. Argon is injected through two porous plugs, located at the base of the ladle, into a bath of molten steel contained in a slightly tapered cylindrical vessel. Argon flow rate used for modeling is small (less than 30 Nl/min for each plug), a relatively low flow rate for such a quantity of liquid steel. The model assumes a uniform liquid temperature equal to 1600 °C.

Fluid flow and bubble repartition

Two-phases (liquid/bubbles) flow is calculated by the CFD commercial code Fluent and the liquid steel velocity and argon plume region on a plane passing through the porous plugs of the ladle are shown in Figure 9. One can observe the shapes of the two bubble plumes rising from the two porous plugs (the isosurface of gas volume fraction equal to 1% is drawn). These regions are characterized by weak bath aeration as the distance from the porous plug increases. The gas volume fraction was found to be below 5% in each plume. The liquid metal flow is associated with two recirculation zones in each half of the plane of symmetry. Turbulence is strongly non-uniform and the dissipation rate of the turbulence kinetic energy ranges between 10^{-6} and 10^{-1} $m^2 s^{-3}$.

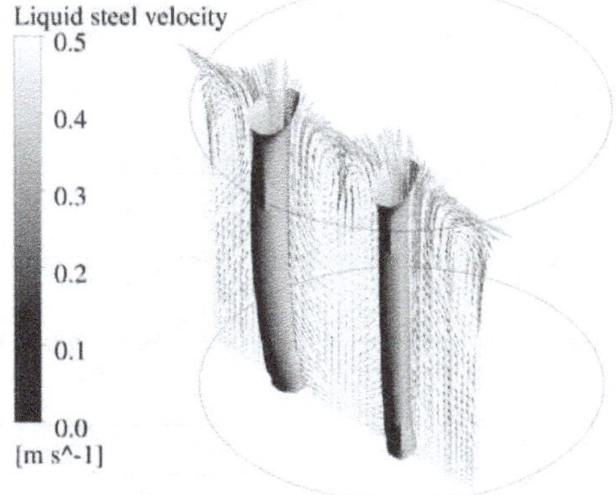

Figure 9: Predicted velocity of the liquid steel along with the argon plumes (isosurface of the 1% gas volume fraction) in a vertical plane passing through the porous plugs.

Time evolution of the inclusion population

With the aim of highlighting the capability of such numerical model to predict the time evolution of an inclusion population, a typical initial distribution of inclusions is considered in the 60 t stirred ladle for the hydrodynamics condition simulated above. For confidential reasons, the initial distribution does not correspond to real industrial conditions but is sufficiently close so as to the conclusions might be significant.

The inclusion population is discretized into twenty different size classes in the range [1-200 µm]. Figure 10 compares the PSD after 100 s of gas stirring (blue) with the one obtained at the end of the 600 s of treatment (red). The computed distribution continuously evolves with time, as larger size particles appear due to the agglomeration of small size particles leading to an increase of the Sauter diameter d_{32} (diameter of a sphere that has the same volume/surface area ratio as the inclusion of interest) over time as shown (see figure 11). Since the flotation and sedimentation mechanisms are strongly dependant of the particle size, the large size aggregates are preferentially removed. As a consequence the total mass of inclusions in the ladle diminishes and the associated rate of removal is not constant but slowly increases.

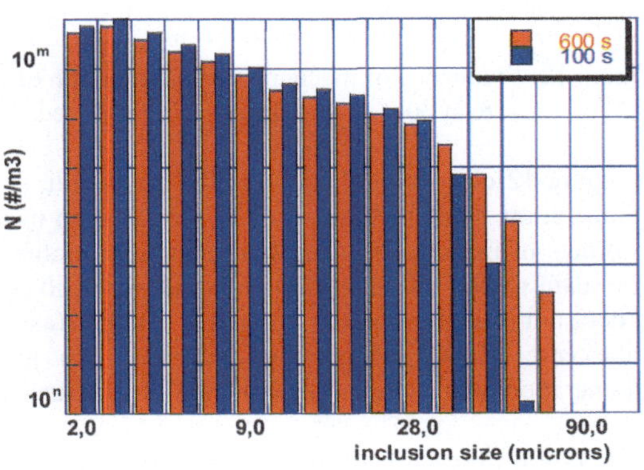

Figure 10: PSD after 100 and 600 seconds (arbitrary Y-coordinate).

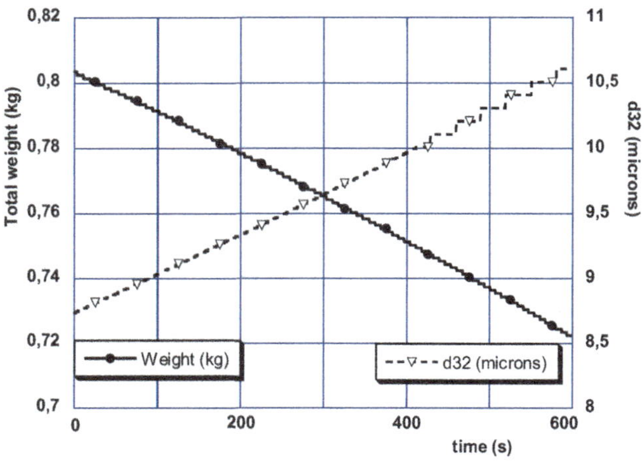

Figure 11: Total mass of the inclusion population and Sauter diameter of the PSD versus time

The numerical simulation allows us to compare the relative importance of the different removal mechanisms on the inclusion population. The frequencies of the aggregation, flotation, settling and capture on the free surface mechanisms have been reported in figure 12 (at the beginning of the process i.e. at time equal to 30 s). The sign (-) or (+) indicates that the numerical density of a given class decreases or increases following either the aggregation, flotation, separation and entrapment processes. As an example, the aggregation frequency F_A is calculated as:

$$F_A = \int_V (B_i - D_i)dV \tag{5}$$

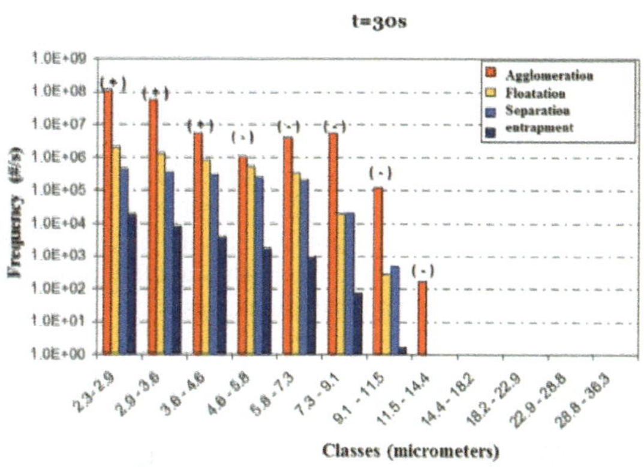

Figure 12: Frequency of mechanisms as a function of inclusion size at 30 s – (-): reduced (+): produced

The figure 12 clearly highlights that the major role is played by agglomeration, since the frequency of the flotation mechanism is two orders of magnitude lower than the agglomeration one. This is the main reason why the PSD continuously evolves with time and does not reach the equilibrium in this example (as was noticed in the case of Al). Large size inclusions are produced by agglomeration with a rate larger than the flotation removal rate.

This strong difference between Al and steel reactors stems from the blowing rate which can be much stronger in Al making. The efficiency of the flotation mechanism is then directly influenced by the gas holdup and by the dispersion of the bubbles into the bath. The turbine blades allow a large dispersion of the bubbles within the Al reactor whereas the bubble swarms are weakly dispersed in the gas stirred ladle.

Conclusion

Nowadays the numerical modelling affords the possibility to simulate complex three-phase metallurgical reactor by combining CFD and Population Balance resolution. Hence a mathematical model of inclusion behaviour in a stirred ladle has been built up step by step. First the turbulent flow is simulated using the Euler-Euler method, taking into account all the interaction forces between the two phases (gas bubbles and liquid steel). Second the numerical model handles both agglomeration and removal mechanisms (flotation, gravity separation, free surface entrapment) together with the convective transport of inclusions into the melt. The coupling of convective transport equation and the PBE is achieved using a time-splitting technique. The discrete class method with the cell average feature was adopted to solve the PBE and was implemented into CFD codes.
As an illustration, the numerical modeling is applied to both an Al laboratory reactor and an industrial steel ladle. The results emphasize the relative role of agglomeration versus flotation as a function of the blowing rate and of the dispersion of bubbles into the bath.

Acknowledgement

This study was a part of the CIREM project (inclusion behaviour in the metallurgical reactors) supported by the French National Research Agency (ANR06 MATPRO 0005).

References

[1]Nakanishi K., Szekely J., "Deoxidation kinetics in a turbulent-flow field", *Transaction of the Iron and steel institute of Japan*, Vol. 15, 10 (1975), 522-530

[2] Sahai Y., Guthrie R.I.L., "Hydrodynamics of gas stirred melts 1. Gas liquid coupling", *Met. Trans B*, Vol. 13 , 2, (1982), 193-202

[3] Sahai Y., Guthrie R.I.L., "Hydrodynamics of gas stirred melts 2. Axisymmetrric flows", *Met. Trans B*, Vol. 13 , 2, (1982), 203-211

[4] Hallberg M., Jönsson P.G., Jonsson T.L.I., Erikson R., *Scandinavian Journal of Metallurgy*, (2005), Vol. 34, 41-56

[5] Aoki J., Thomas B. G., Peter J., Peaslee K. D., Proc. of Iron and Steel Technology Conference AISTech, 2004, 1045

[6] Söder M., Jönsson O., Jonsson L., Nzotta M., *Steel Research Int.*, (2005), 481

[7] Gardin P., Gauthier S., Simonnet M., "Multiphase Model for Predicting the Elimination of Inclusions inside a Liquid-Steel Ladle", *Advanced Engineering Materials*, (2011), Vol.13, 538-542

[8] Kwon Y.-J., Zhang J., Lee H.-G., "A CFD-based Nucleation-growth-removal Model for Inclusion", *ISIJ International*, (2008), 48, 7-14.

[9] Claudotte L., Rimbert N., Gardin P., Simonnet M., Lehmann J., Oesterle B., "A Multi-QMOM Framework to Describe Multi-Component Agglomerates in Liquid Steel", *AICHE Journal*, (2010), Vol. 56, 2347-2355

[10] Daoud I. L. A. , Rimbert N., Jardy A., Oesterle B., Hans S., Bellot J.P., "3D modelling of the aggregation of oxides inclusions in a liquid steel ladle: two numerical approaches", *Advanced Engineering Materials*, (2011), Vol.13, 543-549

[11] Mirgaux O., Ablitzer D., Waz E., Bellot J.P., "Mathematical modelling and computer simulation of molten aluminium purification by flotation in stirred reactor", *Met Trans B*, Vol. 40B, (2009), 363-375.

[12] Kumar S., Ramkrishna D., *Chem. Eng. Sci.*, (1996), Vol.51, 1311-1332.

[13] Toro E.F., *Riemann solver and numerical methods for fluid dynamics: a pratical introduction*, (Springer, Berlin, 1999).

[14] Kumar J., Peglow M., Warnecke G., Heinrich S., Mörl L., Chem. Eng. Sci., 2006, vol.61, 3327-3342

[15] Wood N.B., *Journal of Aerosol Science*, (1981), Vol.12, 275

[16] Xayasenh A., Duval H., Joly L., "Analyse qualitative du transport turbulent d'inclusions à l'interface métal liquide/laitier" (Proc. of the Symposium Matériaux, Nantes, France, edited by the Materials French Society on CD-ROM, 2010)

[17] Waz E., Carré J., Le Brun P., Jardy A., Xuereb C. and Ablitzer D., "Physical modeling of the aluminum degassing process: Experimental and mathematical approaches", *Light Metals. TMS*, (2003), 901-907

Proceedings of the 2013 International Symposium on Liquid Metal Processing & Casting
Edited by: Matthew J. M. Krane, Alain Jardy, Rodney L. Williamson, and Joseph J. Beaman
TMS (The Minerals, Metals & Materials Society), 2013

MODELING THE TITANIUM NITRIDE (TiN) GERMINATION AND GROWTH DURING THE SOLIDIFICATION OF A MARAGING STEEL

Vincent Descotes[1,2], Jean-Pierre Bellot[2], Sylvain Witzke[1], Alain Jardy[2]
[1]APERAM Alloys Imphy; BP1 Centre de recherches P. Chevenard; F-58160 Imphy, France
[2]Institut Jean Lamour - UMR CNRS 7198 - Université de Lorraine; Parc de Saurupt - CS 50840; 54011 Nancy, France. (Laboratoire d'Excellence DAMAS).

Keywords: Nucleation, TiN, Segregation, Inclusions, Steel

Abstract

A maraging steel containing nitrogen and titanium is considered. As solidification proceeds, the segregation accounts for an increase of Ti and N mass fraction in the liquid phase. This eventually leads to the formation of TiN if the supersaturation is high enough. A model has been developed to calculate the creation and evolution of the TiN distributions in both phases. Microsegregation is modeled using the lever rule, while the kinetics of precipitation is mainly driven by the supersaturation of the liquid bath. The model enables one to investigate the competition between segregation and precipitation regarding the solute concentrations and the shape of the particle size distributions in the liquid and solid phases. A parametric study on the solidification time reveals the existence of a maximal inclusion size. It also confirms the influence of the initial composition on the final size of TiN particles.

Introduction

Aperam Alloys Imphy (France) steel plant produces a maraging steel devoted to automotive applications. This Fe-based alloy contains 18%Ni, 9%Co, 5%Mo, and 0.45%Ti. Its casting process involves a passage through the liquid state and a solidification step. Nitrogen is present in traces amount in the initial material, and it is supposed to be completely soluble at the liquidus temperature. However, as solidification proceeds, micro-segregation may lead to the formation of titanium nitride (TiN) inclusions. The largest TiN inclusions are recognized as a major defect for this alloy regarding its fatigue properties, because they increase the vulnerability of the steel to fatigue cracks propagation [1]. A deeper understanding of the kinetics of nucleation and growth of those inclusions during the steel casting would help steel producers to find ways to reduce both number and size of TiN inclusions.

A model has therefore been proposed to study the coupling of segregation and precipitation of TiN inclusions during the solidification of steel. This model, that is rather simple, is intended to be later included in a more complete simulation of the actual remelting process to study the effect of operational parameters on the inclusion population. In its present state, it enables one to quantify the effect of various parameters with fewer side effects than in a larger, more complex calculation. The equations of the model are presented as well as some results that give some insight on the inclusion behavior in this particular alloy.

Equations of the Model

A liquid steel bath at liquidus temperature in an arbitrary volume is considered. This bath contains oxides that are supposed to be inert and to follow a precisely known distribution. The nucleation is supposed to be only heterogeneous and to occur during the solidification on the preexisting oxide distribution in the liquid phase. The TiN particles which have been engulfed in the solid phase become inert until the end of the solidification. This assumption is supported by the much lower rate of the transport of solutes in the solid phase compared with the transport in the liquid phase. A sketch of the system is provided in figure 1.

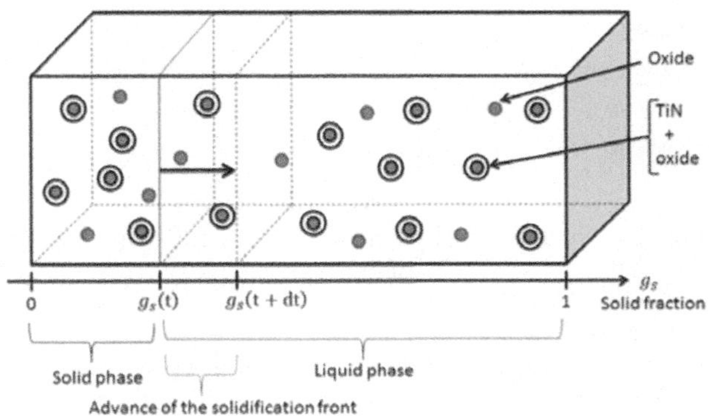

Figure 1: Drawing of the system modeled

No heat equation is modeled and the evolution of the solidification is simply calculated by applying an arbitrary evolution of the temperature of the bath with time. The relationship between the temperature and solid fraction is also supposed to be known, so the solidification state of the bath is completely defined at any moment. The attention is turned on the evolution of dissolved Ti, N and solid TiN concentrations. A competition takes place between the segregation which brings solutes in the liquid phase and the precipitation which consumes solutes.

All equations are written in terms of the molar concentrations of element X: C_X (mol.m^{-3}). This choice considerably simplifies the equations. It is possible to derive the mass fraction from the molar concentration by the relation:

$$\omega_X = C_X \frac{M_X}{\rho_{steel}} \tag{1}$$

where ω_X is the mass fraction of element X, C_X its molar concentration, M_X its molar mass (kg.mol^{-1}), and ρ_{steel} the density of the alloy (kg.m^{-3}).

Germination of the TiN Inclusions

An arbitrary distribution of spherical oxide particles in the bath, following a log-normal distribution, is considered. The density function and the cumulative distribution function are given below.

$$f(r) = \frac{1}{r\,\sigma\sqrt{2\pi}} \cdot e^{-\frac{1}{2}\left(\frac{\ln r-\mu}{\sigma}\right)^2} \tag{2}$$

$$F(r) = \frac{1}{2} + \frac{1}{2}\operatorname{erf}\left(\frac{\ln r - \mu}{\sigma\sqrt{2}}\right) \tag{3}$$

The normalized numeral density obtained with arbitrary (μ, σ) parameters is plotted on figure 2. This arbitrary oxide distribution is kept in all our further calculations.

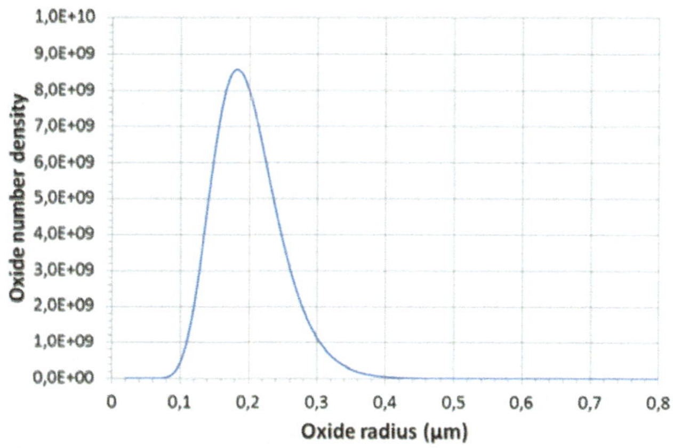

Figure 2: Numeral density function of the log-normal distribution of the oxides

The kinetic reaction Ti + N ⇆ $TiN_{(s)}$ is characterized by a dissolution product corresponding to the inverse of the thermodynamic constant associated with the precipitation reaction. The dissolution product K_{TiN} is dependent on temperature and is defined by:

$$K_{\mathrm{TiN}} = C_{\mathrm{Ti}}^{eq} \cdot C_{\mathrm{N}}^{eq} \tag{4}$$

$$\ln K_{\mathrm{TiN}} = \left(-\frac{\alpha}{T(t)} + \beta\right) \tag{5}$$

where α and β are two constant that need to be determined experimentally, t is the time and T is the temperature of the liquid bath (K).

The supersaturation is then introduced as:

$$S = \frac{\gamma_{\mathrm{N}}\gamma_{\mathrm{Ti}}C_{\mathrm{Ti}}C_{\mathrm{N}}}{K_{\mathrm{TiN}}} \tag{6}$$

where γ_{N} and γ_{Ti} are the activity coefficients of N and Ti in the liquid steel, independent of the temperature.

The nucleation model is derived from the classical one reported in [2] and developed in a similar way for athermal nucleation of solids in [3]. It is based on the computation of the Gibbs energy variation of an elemental volume of supersaturated liquid. The Gibbs energy variation is composed of a chemical enthalpy term (negative when $S \geq 1$) and a positive term coming from the surface tension forces that act against nucleation. In order to create precipitates, the supersaturation in the liquid should be sufficiently higher than 1 to compensate the creation of new surfaces. It can be shown (see [3]) that the curvature radius of a stable TiN nucleus is depending on the supersaturation through the formula:

$$r_{\mathrm{nucl}}^{\mathrm{TiN}} = \frac{2\gamma_{\mathrm{TiN/\,steel}}\,M_{\mathrm{TiN}}}{RT\ln(S)\,\rho_{\mathrm{TiN}}} \tag{7}$$

where $r_{\mathrm{nucl}}^{\mathrm{TiN}}$ represents the radius of a TiN nucleus (m), $\gamma_{\mathrm{TiN/\,steel}}$ the interfacial energy of TiN in the steel ($J.m^{-2}$), M_{TiN} the molar mass of TiN ($mol.kg^{-1}$), R the ideal gas constant ($J.mol^{-1}.K^{-1}$), T the temperature (K) of the liquid steel, S the supersaturation of the bulk, and ρ_{TiN} the TiN density ($kg.m^{-3}$).

It is considered that nucleation always takes place upon spherical oxides of radii ϕ. The curvature radius of the TiN nucleus should be high enough to enable the connection on the underlying oxide in accordance with the wetting angle of the TiN on the oxide, because geometrically:

$$\sin\theta \approx \frac{\phi}{r_{\mathrm{nucl}}^{\mathrm{TiN}}} \tag{8}$$

where θ represents the theoretical wetting angle of a TiN on an oxide in steel and ϕ is the oxide radius (m).

The geometric condition $r_{\mathrm{nucl}}^{\mathrm{TiN}}\sin\theta \leq \phi$ combined with equation 7 induces the existence of a minimum supersaturation written S_g, needed to activate an oxide as a nucleation site for a TiN inclusion. We end up therefore with the following relationship giving the minimum supersaturation of nucleation S_g as a function of the oxide radius:

$$S_g(\phi; T) = \exp\left(\frac{2\gamma_{\mathrm{TiN/\,steel}}\frac{M_{\mathrm{TiN}}}{\rho_{\mathrm{TiN}}}\sin\theta}{RT\phi}\right) \tag{9}$$

The nucleation is easier on large oxides than on small ones. It is considered that once $S \geq S_g(\phi; T)$, all oxides of the same size are immediately activated, because the nucleation frequencies are usually extremely high. In this sense, nucleation is said to be athermal [3].

The increase in TiN particle mass will be taken into account during the growth of the inclusion. However, the size of the initial oxide will be recalled in memory in order to have a proper solute balance.

Growth of the TiN Inclusions

We consider the following kinetic law:

$$\frac{dr}{dt} = \dot{r} = G(S) = k_c(S-1) \tag{10}$$

where r is the radius of an oxide-TiN assembly (m), G is the growing speed of the inclusion, S the supersaturation, and k_c a kinetic factor which gathers the chemical phenomena at the interface and the Ti and N mass transport from the supersaturated bulk liquid to the TiN particle surface. This law was inspired by a similar model from reference [4]. It describes the fact that the more the bath is supersaturated, the faster an inclusion will grow. Ti and N solutes play an equal role in the mechanism of inclusion growth.

From the speed of growth, a transport equation in the internal coordinate space (inclusion size) for the particle distribution can be derived:

$$\frac{\partial n^l_{\mathrm{TiN}}(r,t)}{\partial t} = -G\,\frac{\partial n^l_{\mathrm{TiN}}(r,t)}{\partial r} \tag{11}$$

where $n^l_{TiN}(r,t)$ is the distribution of TiN in the liquid phase (number of inclusions of radius r per unit volume of liquid metal and per unit length of particle radius).

This type of differential equation is similar to the advection equation encountered in fluid flow modeling. Its resolution is performed using a high resolution TVD scheme.

The TiN distribution in the liquid phase is related to the TiN concentration in the liquid through the equation:

$$C^l_{\mathrm{TiN}}(t) = \frac{\rho_{\mathrm{TiN}}}{M_{\mathrm{TiN}}}\left[\int_0^{+\infty} n^l_{\mathrm{TiN}}(r,t)\frac{4}{3}\pi r^3 \mathrm{d}r - \int_0^{+\infty} n^l_{\mathrm{OxAct}}(r,t)\frac{4}{3}\pi r^3 \mathrm{d}r\right] \tag{12}$$

where $n^l_{\mathrm{TiN}}(r,t)$ represents the distribution of TiN in the liquid phase (number of inclusions of radius r per m^4) and $n^l_{\mathrm{OxAct}}(r,t)$ the distribution of oxides which were activated (number of oxides of radius r per m^4).

It is then considered that the increase of volume of the inclusions is the result of the chemical advance of reaction: Ti + N $\leftrightharpoons$ $TiN_{(s)}$. The differential increment of the volume chemical advance ($mol.m^{-3}$) of the reaction in the liquid phase is therefore obtained by equation:

$$\mathrm{d}\xi = \mathrm{d}C^l_{\mathrm{TiN}} \tag{13}$$

Solidification Model

A lever rule model has been used to describe the solidification of the alloy, thus assuming infinite diffusion of the solutes within the two phases. The theory can be found in reference [5]. Without any precipitation, the mass conservation leads to:

$$(1 - g_s + k_X g_s)\,\mathrm{d}C^l_X = C^l_X(1 - k_X)\,\mathrm{d}g_s \tag{14}$$

$$\Leftrightarrow \mathrm{d}C^l_X = \frac{(1-k_X)C^l_X \mathrm{d}g_s}{1-(1-k_X)g_s} \tag{15}$$

where g_s is the solid fraction of the system, k_X is the partition coefficient, and C^l_X the concentration of a solute (Ti or N) in the liquid phase.

This equation expresses the transfer of solutes from the liquid phase volume occupied by the corresponding $\mathrm{d}g_s$ mass fraction of solid solidifying, to the solid and liquid phases.

When precipitation occurs, some of the solute is redistributed into the TiN phase. The balance equation is then modified as below:

$$\mathrm{d}C^l_X = \frac{(1-k_X)C^l_X \mathrm{d}g_s}{1-(1-k_X)g_s} - \frac{(1-g_s)\mathrm{d}C^l_{\mathrm{TiN}}}{1-(1-k_X)g_s} \tag{16}$$

It is noted here that $\mathrm{d}C^l_{\mathrm{TiN}} = \mathrm{d}\xi$, the chemical advance, so that the segregation is indeed strongly coupled to the precipitation of the TiN in the liquid phase.

The Ti and N concentrations in the solid phases are then calculated by the lever rule:

$$C^s_X = k_X C^l_X \tag{17}$$

where X stands for Ti or N and k_X is the partition coefficient of element X.

Temperature and Solidification Path

The temperature is arbitrary given by a linear variation between the liquidus and solidus temperatures, so that the whole solidification is controlled by the solidification time t_f. The solidification path is also arbitrary set as linear between 0 and 1:

$$\begin{cases} T = T_{\mathrm{liq}} - \left(T_{\mathrm{liq}} - T_{\mathrm{sol}}\right)\dfrac{t}{t_f} \\ g_s = \dfrac{T - T_{\mathrm{sol}}}{T_{\mathrm{liq}} - T_{\mathrm{sol}}} \end{cases} \tag{18}$$

$$\Leftrightarrow \begin{cases} g_s = \dfrac{t}{t_f} \\ T = T_{\mathrm{liq}} - \left(T_{\mathrm{liq}} - T_{\mathrm{sol}}\right)g_s \end{cases} \tag{19}$$

where T is the temperature (K), g_s the solid fraction, T_{liq} and T_{sol} the liquidus and solidus temperatures (K).

It is noted that these two assumptions can be far away from the real alloy behavior and solidification process. However, it enables one to focus on the competition between segregation and precipitation, without introducing more phenomena in the problem. As already stated, more sophisticated models can be derived from this one by implementing it into a process simulation software.

Engulfment of TiN Inclusions

When solidification proceeds, the solidification front engulfs some TiN inclusions from the liquid into the solid. Those inclusions will no more be available in the liquid to grow further, so they become inert: as stated previously, the model assumes no diffusion of solutes in the solid phase at the macroscopic scale.

Considering a uniform spatial distribution of the TiN inclusions in the liquid at each time, the solidification of a mass fraction $\mathrm{d}g_s$ of the bath during the interval $\mathrm{d}t$ transfers the density $\alpha(r)\, n^l_{\mathrm{TiN}}(r,t)\, \mathrm{d}g_s$ of TiN of radius r from the liquid to the solid phase.

The following equation is therefore obtained:

$$d\left(g_s n_{\text{TiN}}^s(r,t)\right) = +\alpha\,(r)\; n_{\text{TiN}}^l(r,t)\;\mathrm{d}g_s \qquad (20)$$

$$d\left((1-g_s)n_{\text{TiN}}^l(r,t)\right) = -\alpha(r) n_{\text{TiN}}^l(r,t)\mathrm{d}g_s \qquad (21)$$

The factor $\alpha\,(r) \in [0;1]$ takes into account a possible rejection of TiN at the solidification front. In the rest of the article, it will be considered that $\forall r \in \mathbb{R}^+, \alpha(r) = 1$. This expresses that no obstacle is set on the engulfment of an inclusion into the solid phase. The equations therefore become:

$$d\left(g_s n_{\text{TiN}}^s(r,t)\right) = +\; n_{\text{TiN}}^l(r,t)\;\mathrm{d}g_s \qquad (22)$$

$$d\left((1-g_s)n_{\text{TiN}}^l(r,t)\right) = -n_{\text{TiN}}^l(r,t)\mathrm{d}g_s \qquad (23)$$

$$\Rightarrow dn_{\text{TiN}}^l(r,t) = 0 \qquad (24)$$

As it can be seen from equation 24, the TiN distribution in the liquid will not be affected by the engulfment of particles whereas the one in the solid phase will be progressively filled in by engulfing inclusions from the liquid phase (eq. 22).

Numerical Aspects

A FORTRAN code has been written to simulate those coupled equations: germination (eq. 9), calculation of the speed of growth (eq. 10), transport of the TiN particles in the liquid (eq. 11), calculation of the chemical advance (eq. 12 and 13), calculation of the segregation and consumption of Ti and N in the liquid phase (eq. 16), calculation of the concentrations in the solid phase (eq. 17), evolution of the temperature and solid fraction (eq. 18), and engulfment of TiN particles in the solid phase (eq. 22 and 24). All equations were discretized in a conservative manner. In addition, the transport equation of the distribution in the liquid (eq. 11) is solved using a total variation diminishing scheme (TVD), described in reference [6]. The discretization involves a time interval Δt and a radius interval Δr.

Results and Analysis

Our model enables one to calculate the evolution of the TiN distributions in the liquid and solid phase. It is especially instructive to look at the distribution of inclusions in the solid phase after complete solidification, and in the very last liquid droplet. As the final size of the largest TiN inclusions is the most interesting information for industrial concerns, a distribution will be characterized by its $r_{97\%}$ radius, which corresponds to the radius which is above 97% of the radii of the TiN distribution. The variation of this indicator with some input parameters such as the initial nitrogen mass fraction, the total solidification time and the density of oxides will be investigated. It must be noted that a convergence study has been previously carried out on the time step to ensure the convergence of the numerical model, and that the total Ti and N balances are systematically verified.

Nominal Case

A nominal case has been defined and is reported in table 1. Most of the parameters had to be estimated. The parameters whose effect will be investigated on the model are separated from the ones which are set to a constant value in all simulations.

Table I: Nominal parameters of the model

(i) Variable parameters

Parameter	Value
ω_N^0	15 ppm $= 1.5 \cdot 10^{-4}\%$
t_f	200s
n_{Ox}	$1 \cdot 10^{11}$oxides. m^{-3}

(ii) Constant parameters

Parameter	Value
Δr	$1 \cdot 10^{-8}$m
Δt	$1 \cdot 10^{-4}$s
γ_{Ti}	0.4
γ_{N}	1.09
ω_{Ti}^0	0.47%
k_{Ti}	0.74
k_{N}	0.125
ρ_{TiN}	5100 kg.m^{-3}
ρ_{steel}	7450 kg.m^{-3}
k_c	$8.23 \cdot 10^{-9}$m.s^{-1}
$\gamma_{\text{TiN}/steel}$	1 J.m^{-2}
θ	$\frac{\pi}{2}$rad
T_{liq}	1470°C
T_{sol}	1440°C

In the nominal case, precipitation occurs from the very beginning of the solidification because the initial melted bath is largely supersaturated. Figure 3 shows the variation of the r_{97}^l and r_{97}^s radii at each time. Both continuously rise as segregation always provides more solutes available for the precipitation. A departure between the two curves is observed, because when solidification proceeds, the apparent growth of the TiN particles in the solid is only caused by the engulfment of inclusions from the liquid. The TiN particles already present in the solid keep the size that they had previously reached in the liquid phase.

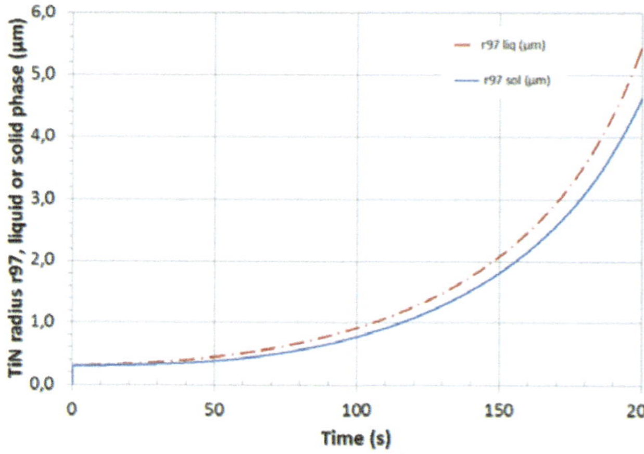

Figure 3: Evolution of the radii r_{97}^l and $r_{97}^s$of the TiN distributions in the liquid and solid phases during solidification

Figure 4a shows the evolution of the mass fractions in the liquid phase. In this particular case, segregation is sufficiently high to be little affected by the consumption of solutes from the precipitation. The mass fractions in the solid phase (fig. 4b) show a similar evolution as they are proportional to the ones in the liquid.

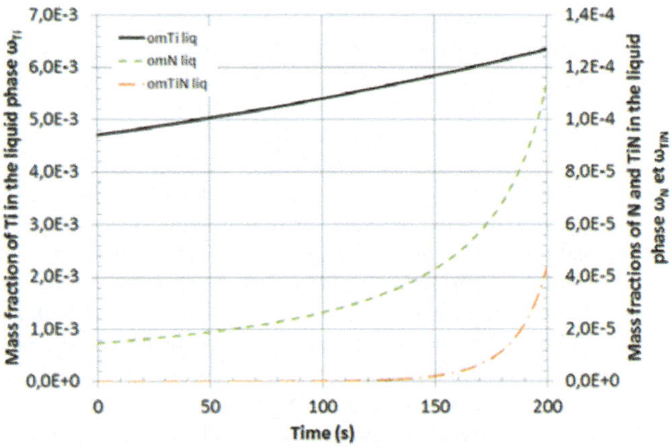

(a) Liquid phase

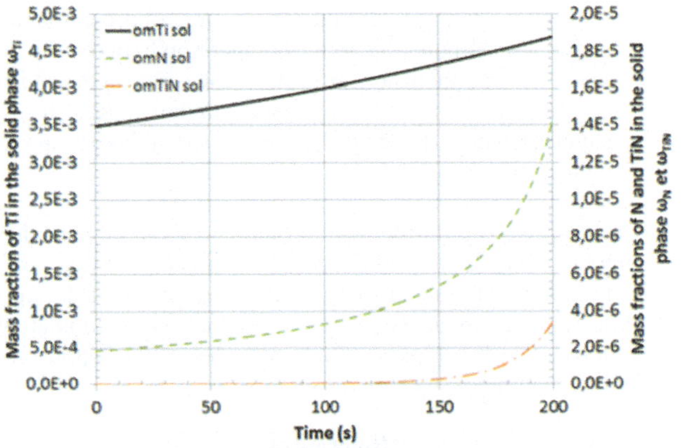

(b) Solid phase

Figure 4: Evolution of the mass fractions of Ti, N and TiN in the two phases

Finally, figure 5 shows the final distribution of TiN inclusions in the liquid and solid phases. The only distribution that would be experimentally observable is obviously the one in the solid phase. The initial oxide distribution of the liquid is also drawn. It appears that all oxides were used as a base for precipitation, because the supersaturation was quite high.

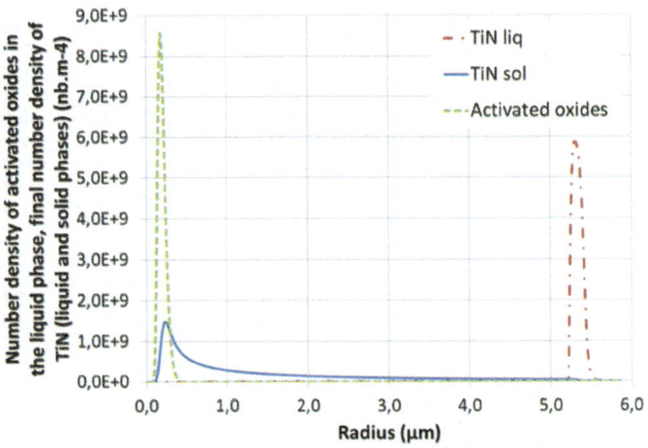

Figure 5: Initial oxide distribution, and final TiN distributions in the liquid and solid phases

It is observed that the numerical treatment of the transport equation is not perfect, because the TiN distribution in the liquid should exhibit exactly the same shape as the one of the oxide distribution. It however exhibits a rather close shape, which seems acceptable.

The TiN distribution in the solid phase shows a very different shape from the distribution in the liquid. This comes from the progressive engulfment of the TiN inclusions in the solid where they become inert. Figure 6 plots this curve with more appropriate scales. It is noticed that the distribution is much flattened, as compared to the one of the initial oxides.

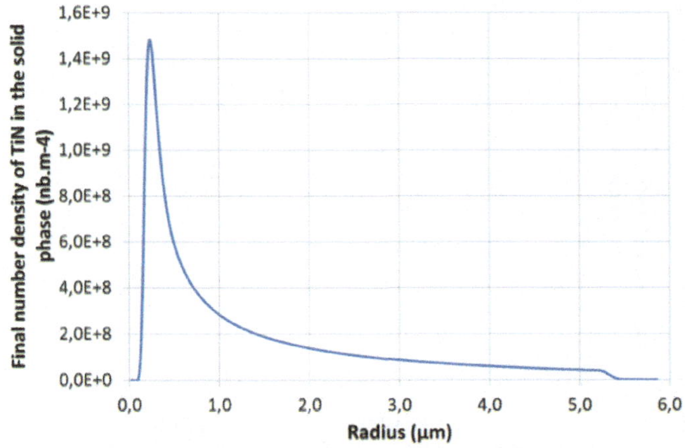

Figure 6: Final TiN distribution in the solid phase

Effect of Varying the Initial Numeral Oxide Density

The effect of a variation on the initial numeral oxide density is plotted on figure 7 in semi-logarithmic scale. For relatively high densities, an increase of the oxide density leads to a reduction of the r_{97}^{s} radius reached by the final TiN distribution in the solid phase. This comes from the fact that the more TiN are present in the bath, the more solute they will consume in order to grow. As the total amount of solutes remains the same, the final size of the TiN is lowered.

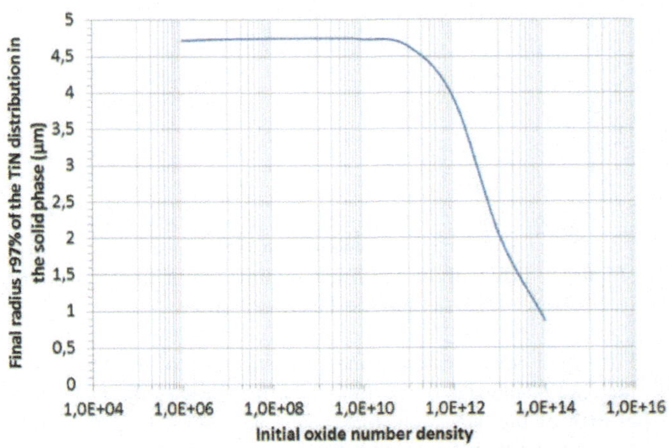

Figure 7: Influence of the oxide density on the TiN distribution. Semi-logarithmic scale

Effect of Varying the Initial Mass Fraction of N

Increasing the initial mass fraction of nitrogen in the bath leads to an increase in the r_{97}^{s} radius of the final TiN distribution in the solid phase as shown on figure 8. However, for relatively high solidification times, an inflection on the curve is observed. This indicates that the consumption of solutes by TiN particles becomes much higher than the segregation contribution, which leads to a decrease of the growing speed of the inclusions. This is mostly observed when the solidification time is high, so that the TiN growth can develop and become significant.

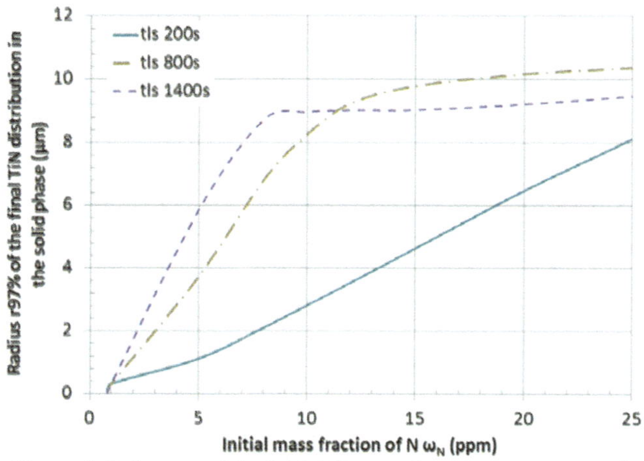

Figure 8: Influence of the initial N mass fraction on the final r_{97}^{s} of the TiN distribution in the solid phase

Effect of Varying the Total Solidification Time

Figure 9 shows the effect of varying the solidification time for a given initial nitrogen mass fraction (15 ppm).

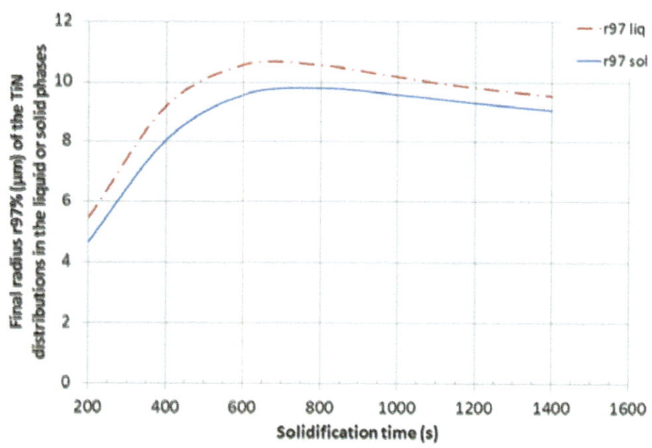

Figure 9: Effect of the solidification time on the final r_{97}^{l} and r_{97}^{s} radii of the TiN distributions

For small solidification times, an increase of t_f leads to an increase of the size of the final TiN. It expresses the fact that the growth of the TiN inclusions takes place on a longer period of time, so that their final size is larger.

However, a maximum value is reached, followed by a decrease in inclusion size with the solidification time. A detailed analysis has shown that in those cases, the particles develop so much that the consumption of solutes is larger than the contribution of segregation, which leads to a decrease in the supersaturation and hence in the growing speed of the TiN inclusions. If the total solidification time is high enough, this effect is very probably dominant.

Conclusion

A nucleation and growth model has been developed to describe the formation of TiN particles in a maraging steel during its solidification. This work enables one to understand the competition which takes place between segregation and precipitation. The effect of the initial composition of the bath and total solidification time on the final size of the inclusions has been investigated. It reveals some unattended trends:

- the oxide density influences the final inclusion size;
- the final inclusion size increases with the initial nitrogen content, but this effect is reduced when the solidification time is high enough;
- the value of r_{97}^{s} as a function of the solidification time shows a maximum, due to the competition between the time given for the inclusion to grow, which increases with the solidification time, and the diminishing of the supersaturation (hence the growing speed) as solutes are consumed.

As stated previously, further use of this model can be considered by implementing it in a more complete process simulation, in order to simulate the inclusion cleanliness of an ingot during its solidification. A similar model could also be used to simulate the dissolution of a distribution of TiN particles in the liquid phase during the melting of steel, *mutatis mutandis*.

References

[1] Y. Murakami, *Metal fatigue: Effects of Small Defects and Nonmetallic Inclusions* (Elsevier, 2002).

[2] M. Soustelle, *Cinétique hétérogène* (Hermès Science Publication, 2007).

[3] J. Dantzig, M. Rappaz, *Solidification* (EPFL Press, 2009).

[4] J. Lehman, P. Rocabois and H. Gaye, "Kinetic model of non-metallic inclusions' precipitation during steel solidification", *Journal of Non-Crystalline Solids* 282 (2001), 61-71.

[5] W. Kurz, D. Fischer, *Fundamentals of Solidification* (Trans Tech publication LTD, 1998).

[6] R. Leveque, *Numerical Methods for Conservation Laws* (Basel, Switzerland: Birkhäuser Verlag, 2008).

Proceedings of the 2013 International Symposium on Liquid Metal Processing & Casting
Edited by: Matthew J. M. Krane, Alain Jardy, Rodney L. Williamson, and Joseph J. Beaman
TMS (The Minerals, Metals & Materials Society), 2013

THERMODYNAMICS FOR THE INFLUENCE OF SLAG COMPOSITION ON THE INCLUSION CONTROL IN SEMI-KILLED LIQUID STEELS

Joo Hyun Park[1,]*, Jun Seok Park[1]
[1]Department of Materials Engineering, Hanyang University, Ansan 426-791, Korea
* contact email ; basicity@hanyang.ac.kr

Keywords: $CaO-Al_2O_3-SiO_2-CaF_2-MgO(-MnO)$ slag, Si-Mn-killed steel, Inclusion, Slag basicity, Slag-Steel-Inclusion multiphase equilibria, Alumina ladle

Abstract

Thermodynamic equilibria between $CaO-Al_2O_3-SiO_2-CaF_2-MgO$ (-MnO) slag and Fe-1.5%Mn-0.5%Si-0.5%Cr steel was investigated at 1873 K to understand the effect of slag composition on the concentration of Al_2O_3 in the inclusions in Si-Mn-killed steels. The composition of the inclusions were mainly equal to $(MnO)/(SiO_2)=0.8(\pm0.06)$ (mole ratio) with various Al_2O_3 content which increased by increasing the basicity of slag. The concentration ratio of the inclusion components, $(MnO\times Al_2O_3/SiO_2)$ and the activity ratio of the steel components showed a good linear relationship in a logarithmic scale, indicating that the activity coefficient ratio of the inclusion components was not significantly changed. From the slag-metal-inclusion multiphase equilibria, the concentration of Al_2O_3 in the inclusions was expressed as a linear function of the activity ratio of the slag components in a logarithmic scale. Consequently, the compositional window of the slag for obtaining the inclusions with low melting point in the Si-Mn-killed steel treated in an alumina refractory was recommended.

Introduction

Non-metallic inclusions in steel cause deterioration of mechanical properties as well as severe problems during steel processing.[1-3] The cleanliness of the steel is important in spring steel and tire cord steel since nonmetallic inclusions act as crack initiation sites when subjected to cyclic stress. However, no steel can be totally free from inclusions. For this reason, Si and Mn are generally used as deoxidizers in spring steel and tire cord steel in order to avoid harmful Al_2O_3-rich inclusions.[4,5] A number of studies have been carried out to identify techniques that achieve inclusions that are as harmless as possible.[4-18] The application of optimized slag to suppress the formation of harmful inclusions such as alumina and spinel is one such method used for various steel grades including Si-Mn-killed steel.

For these reasons, the $MnO-Al_2O_3-SiO_2$ ternary system has gained attention for several decades. Many studies have been performed to fundamentally understand this oxide system in terms of thermodynamics.[6-10] Thanks to previous work, if we determine the concentration range of Mn, Si, Al, and O in molten steel, we can predict the corresponding composition window for MnO, SiO_2 and Al_2O_3 in the above inclusion system. Based on this background, several researchers suggested deoxidation for liquid inclusion in the $MnO-Al_2O_3-SiO_2$ system.

Kang and Lee reported that inclusions having a MnO/SiO_2 mass% ratio near unity and Al_2O_3 content in the range of 10~20 mass% have low liquidus temperatures (~1473 K) and suggested that a Mn/Si ratio of 2~5 meets these conditions when Mn+Si=1.0% in the steel.[9] Bertrand et al.[12] found that slag composition during secondary metallurgy must have a low basicity (CaO/SiO_2), and Al content in the steel should be maintained under 15 ppm to avoid the formation of Al_2O_3. Chen et al.[5] studied the effect of top slag on the inclusion composition in tire cord steel and found that inclusion plasticization can be achieved by controlling the binary basicity of top slag (CaO/SiO_2) around 1.0 and maintaining the Al_2O_3 content in top slag below 10%. However, an experimental study to identify conditions for obtaining liquid inclusion when the slag has relatively high basicity (CaO/SiO_2) coupled with Al_2O_3 saturation, which corresponds to the alumina refractory lining, has not yet been conducted.

Therefore, in the present study, the equilibria between $CaO-Al_2O_3-SiO_2-CaF_2-MgO(-MnO)$ slag and Fe-1.5%Mn-0.5%Si-0.5%Cr melt was investigated at 1873 K in order to understand the effect of slag composition on the concentration of Al_2O_3 in the inclusions.

Experimental

A Fe-1.5%Mn-0.5%Si-0.5%Cr alloy (235 g), Fe-0.23(±0.02)%O alloy (65 g) and $CaO-Al_2O_3-SiO_2-CaF_2-MgO(-MnO)$ slag (CaO/SiO_2=0.5~2.0) were equilibrated for an hour at 1873 K in a fused alumina crucible (60 mm × 52 mm × 120 mm) with a graphite heater under an Ar-3%H_2 gas atmosphere using an induction furnace. The Fe-0.23(±0.02)%O alloy was prepared in an MgO crucible by blowing oxygen gas on Fe melt for 20 minutes using an induction furnace. The slags were pre-melted in a graphite crucible for two hours in an electric resistance furnace. Some slags, disintegrated into fine powders due to high basicity, were pelletized to enable accurate addition to the melt surface.

Temperature was controlled by a B-type thermocouple. After the temperature reached 1903 K, Fe-O alloy was added to the Fe-Si-Mn-Cr melt under an inert atmosphere to control the initial oxygen content. The slag was then added to the melt when the temperature was recovered to 1873 K. After a period of 60 minutes, which was preliminarily determined to achieve equilibrium, sampling was performed using a quartz tube and then the samples were directly quenched by dipping them in brine.

The chemical composition of each steel sample was determined by ICP-AES, and the nitrogen and oxygen content was analyzed using a combustion analyzer. The characteristics of inclusions including 3-dimensional morphology and chemical composition were analyzed using the potentiostatic electrolytic extraction method and SEM-EDS. Details regarding characterization of inclusions are given in previously published articles.[19,20] The composition of the inclusion in each sample was taken as the average of about ten inclusions.

Results and Discussion

The following reactions occurred at the slag-metal interface:[21,22]

$$2[\mathrm{Al}] + 3[\mathrm{O}] = (\mathrm{Al_2O_3})^{\mathrm{slag}} \quad [1]$$

$$\log K_{[1]} = \log\left[\frac{a^{s}_{\mathrm{Al_2O_3}}}{a^{2}_{\mathrm{Al}} \cdot a^{3}_{\mathrm{O}}}\right] = 13.60 \quad (\text{at } 1873\,\mathrm{K}) \quad [2]$$

$$[\mathrm{Si}] + 2[\mathrm{O}] = (\mathrm{SiO_2})^{\mathrm{slag}} \quad [3]$$

$$\log K_{[3]} = \log\left[\frac{a^{s}_{\mathrm{SiO_2}}}{a_{\mathrm{Si}} \cdot a^{2}_{\mathrm{O}}}\right] = 4.67 \quad (\text{at } 1873\,\mathrm{K}) \quad [4]$$

where $K_{[n]}$ and a_i are the equilibrium constant of Eq. [n] and the activity of element i in molten steel. The superscript s represents the 'slag' phase. The activity of oxygen determined by the equilibrium of $[\mathrm{Al}]/(\mathrm{Al_2O_3})$, $a^{\mathrm{Al}}_{\mathrm{O}}$ and by $[\mathrm{Si}]/(\mathrm{SiO_2})$, $a^{\mathrm{Si}}_{\mathrm{O}}$ can be estimated from Eqs. [5] and [6], respectively.

$$\log a^{\mathrm{Al}}_{\mathrm{O}} = \frac{1}{3}(\log a^{s}_{\mathrm{Al_2O_3}} - 2\log a_{\mathrm{Al}} - \log K_{[1]}) \quad [5]$$

$$\log a^{\mathrm{Si}}_{\mathrm{O}} = \frac{1}{2}(\log a^{s}_{\mathrm{SiO_2}} - \log a_{\mathrm{Si}} - \log K_{[3]}) \quad [6]$$

In addition, the activity coefficient of solute element (M) can be calculated by classical Wagner formalism using the first- and second-order interaction parameters, which are listed in Table I.[21-28]

$$\log f_{\mathrm{M}} = \sum_{i=\mathrm{Cr,\,Si,\,Mn,\,Al,\,O}} (e^{i}_{\mathrm{M}} \times [\%i] + r^{i}_{\mathrm{M}} \times [\%i]^2) \quad [7]$$

where f_{M}, e^{i}_{M} and r^{i}_{M} are the Henrian activity coefficient of element M, and the first- and second-order interaction parameters between M and i, respectively.

Table I. Interaction Parameters Used in the Present Study.[21]

e^{j}_{i} (r^{j}_{i})	Cr	Si	Mn	Al	O
Al	0.03[27]	0.056	0.035[28]	0.043	-1.98 (39.8)[26]
Si	-0.021[23]	0.1	-0.007[17]	0.058	-0.12
O	-0.032[24]	-0.066	-0.037[25]	-1.17 (-0.01)[26]	-0.17

The activities of oxygen calculated from the $[\mathrm{Al}]/(\mathrm{Al_2O_3})$ and $[\mathrm{Si}]/(\mathrm{SiO_2})$ equilibria are plotted against the activity of oxygen determined from classical Wagner formalism as shown in Figure 1. The activity of $\mathrm{Al_2O_3}$ and $\mathrm{SiO_2}$ in the slag at 1873 K was also calculated using thermodynamic software, FactSage™6.3, assuming that the $\mathrm{CaF_2}$ (~10%) does not seriously affect the activity of slag components. This software has been successfully applied to compute the phase equilibria of multicomponent steel and oxide systems.[15-20,29-47]

The activities of oxygen calculated from the $[\mathrm{Si}]/(\mathrm{SiO_2})$ and $[\mathrm{Al}]/(\mathrm{Al_2O_3})$ equilibria were in good agreement with that from classical Wagner formalism, while several points for the activity of oxygen calculated from the $[\mathrm{Al}]/(\mathrm{Al_2O_3})$ equilibrium were slightly higher than those determined using Wagner formalism. This scatter is probably due to the fact that the concentration of total Al, instead of dissolved Al, was used when the activity of Al in molten steel was calculated.[48] Nevertheless, the slag-metal reaction is believed to have been in equilibrium in the present experiments.

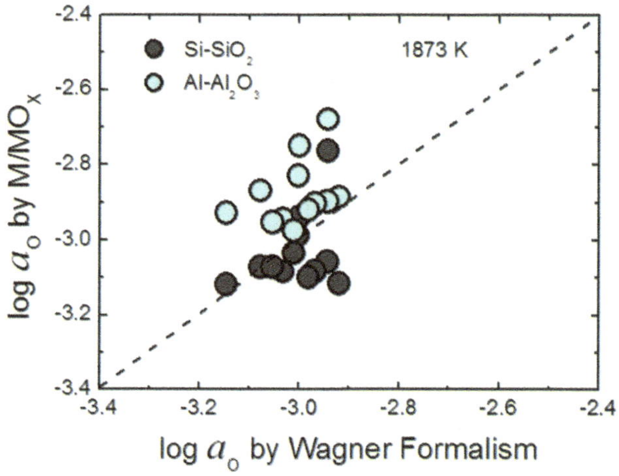

Fig. 1. Relationship between activities of oxygen calculated from classical Wagner formalism and deoxidation equilibria by aluminum and silicon.

The composition of the inclusions is plotted on the MnO-$\mathrm{Al_2O_3}$-$\mathrm{SiO_2}$ ternary phase diagram as shown in Figure 2, which was generated by FactSage™6.3 with the FToxid database at 1873 K and $p(\mathrm{O_2})=10^{-10}$ atm. Five solid phases, $\mathrm{SiO_2}$, MnO, $\mathrm{MnAl_2O_4}$, $\mathrm{Al_2O_3}$ and $\mathrm{Al_6Si_2O_{13}}$ (mullite), are shown to be in equilibrium with the liquid phase. All of the inclusions were located in the liquid phase and their compositions were nearly equal to $(\mathrm{MnO})/(\mathrm{SiO_2})=0.8(\pm 0.06)$ in mole ratio. The only exception was sample #1, the composition of which corresponded to high silica area because of the relatively low aluminum content in molten steel in conjunction with the high activity of $\mathrm{SiO_2}$ in slag.

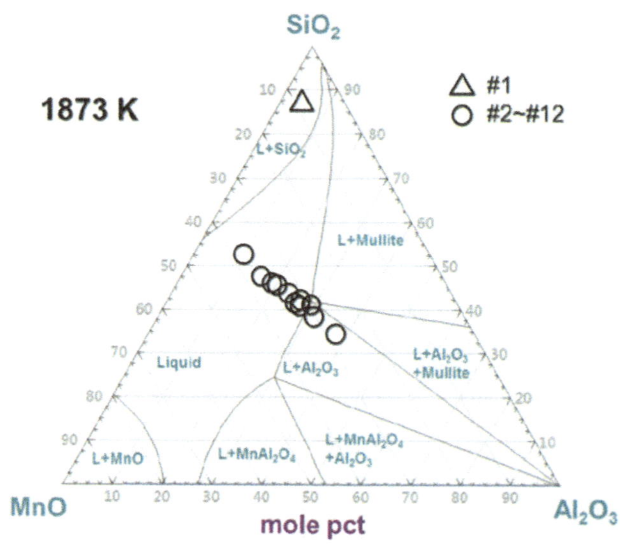

Fig. 2. Computed phase diagram of the MnO-$\mathrm{Al_2O_3}$-$\mathrm{SiO_2}$ system at 1873 K and $p(\mathrm{O_2})=10^{-10}$ atm. The triangles and circles are the average composition of ten inclusions per sample.

The effect of slag basicity (=CaO/SiO_2 ratio) on the concentration of alumina in the inclusions at 1873 K is shown in Figure 3. The alumina content in the inclusions increased with increasing slag basicity up to about C/S=1.5, followed by some variation in the Al_2O_3 content around 40(±10)% when the C/S ratio was greater than 1.5. When the basicity of the slag was greater than 1.5 (#4~#12), the slag was saturated by Al_2O_3-rich phase such as CaO•2Al_2O_3 or CaO•6Al_2O_3, whereas the alumina content was dependent on the slag basicity when the basicity was lower than 1.5 (#1~#3). Hence, the content of alumina in the inclusions is strongly affected by the activity of alumina in the slag in equilibrium with molten steel.

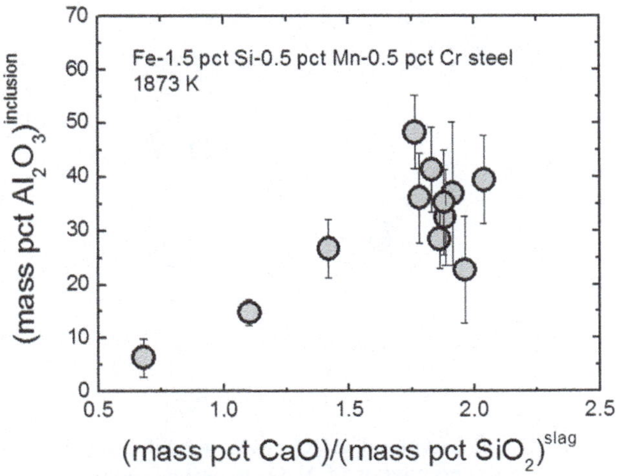

Fig. 3. Effect of the ratio of CaO/SiO_2 in the slag on the concentration of Al_2O_3 in the inclusions.

The effect of alumina content on the melting point and the primary crystalline phase of the inclusion is shown in Figure 4. The liquidus surface of the (MnO)/(SiO_2)=1.0 join in the MnO-SiO_2-Al_2O_3 inclusion system at 1873 K was calculated using the FactSage^{TM}6.3 program. Increasing the concentration of alumina in the inclusions changed the primary phase in the order of $MnSiO_3$ (rhodonite), $Mn_3Al_2Si_3O_{12}$ (spessartite), $MnAl_2O_4$ (galaxite) and Al_2O_3. The melting point of the inclusion decreased with increasing alumina concentration up to about 8%, followed by a rebound after this composition.

When the slag basicity was relatively low, i.e. C/S < 1.5, the melting point of the inclusion was lower than 1623 K and the primary crystalline phases were spessartite and galaxite. In contrast, the melting point of inclusions ranged from about 1523 to 2023 K and the crystalline phase was mainly alumina when the basicity of the slag was greater than 1.5. Therefore, controlling the slag basicity to be less than 1.5 is recommended in order to avoid Al_2O_3-rich inclusions during both the refining stage and the solidification process.

The oxide-forming elements including Al, Si and Mn in molten steel are in equilibrium with not only the slag phase as shown in Eqs. [1] and [3], but also the inclusion phase as indicated in Eqs. [8], [10] and [12].

$$2[\mathrm{Al}] + 3[\mathrm{O}] = (\mathrm{Al_2O_3})^{\mathrm{inclusion}} \quad [8]$$

$$\log K_{[8]} = \log X^i_{\mathrm{Al_2O_3}} + \log \gamma^i_{\mathrm{Al_2O_3}} - 2\log a_{\mathrm{Al}} - 3\log a_{\mathrm{O}} \quad [9]$$

$$[\mathrm{Si}] + 2[\mathrm{O}] = (\mathrm{SiO_2})^{\mathrm{inclusion}} \quad [10]$$

$$\log K_{[10]} = \log X^i_{\mathrm{SiO_2}} + \log \gamma^i_{\mathrm{SiO_2}} - \log a_{\mathrm{Si}} - 2\log a_{\mathrm{O}} \quad [11]$$

$$[\mathrm{Mn}] + [\mathrm{O}] = (\mathrm{MnO})^{\mathrm{inclusion}} \quad [12]$$

$$\log K_{[12]} = \log X^i_{\mathrm{MnO}} + \log \gamma^i_{\mathrm{MnO}} - \log a_{\mathrm{Mn}} - \log a_{\mathrm{O}} \quad [13]$$

where the superscript i represents the 'inclusion' phase. By combining Eqs. [9], [11] and [13], the following equation can be deduced:

$$\log\left(\frac{X^i_{\mathrm{Al_2O_3}} \cdot X^i_{\mathrm{MnO}}}{X^i_{\mathrm{SiO_2}}}\right) = \log\left[\frac{a^2_{\mathrm{Al}} \cdot a_{\mathrm{Mn}} \cdot a^2_{\mathrm{O}}}{a_{\mathrm{Si}}}\right] + \log\left(\frac{\gamma^i_{\mathrm{SiO_2}}}{\gamma^i_{\mathrm{Al_2O_3}} \cdot \gamma^i_{\mathrm{MnO}}}\right) + C \quad [14]$$

where C is constant at a given temperature. From Eq. [14], the molar composition ratio of the inclusions, viz. MnO×Al_2O_3 to SiO_2 will be directly proportional to the activity ratio of steel composition (Al^2×Mn×O^2 to Si) on a logarithmic scale if the activity coefficient ratio of the inclusions would not be seriously changed. In order to calculate the activity of steel components, the activity coefficient of each element was calculated using Eq. [7], and the interaction parameters listed in Table I.

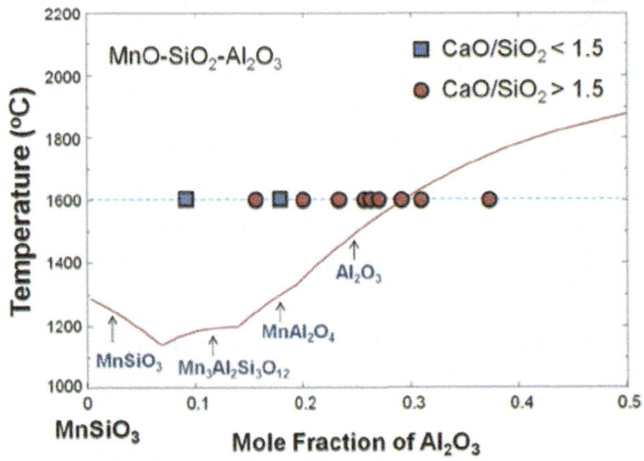

Fig. 4. Calculated liquidus temperatures in the MnO-SiO_2-Al_2O_3 systems as a function of the concentration of Al_2O_3 when the ratio of MnO to SiO_2 (mol%) is unity.

The relationship between $\log(X_{\mathrm{Al_2O_3}} \cdot X_{\mathrm{MnO}} / X_{\mathrm{SiO_2}})$ and $\log[a^2_{\mathrm{Al}} \cdot a_{\mathrm{Mn}} \cdot a^2_{\mathrm{O}} / a_{\mathrm{Si}}]$ is shown in Figure 5. A linear correlation between them is evident with a slope of 1.2 within experimental scatters, which is relatively close to unity within the composition range investigated in the present study. Therefore, it is concluded that the components in molten steel and each oxide in the inclusions are in equilibrium and that the activity coefficient ratio of oxide components in the inclusions (second term on the right-hand side in Eq. [14]) would be constant under the present experimental conditions.

Alternatively, the activities of the components in molten steel are also affected by those of their oxide components in the slag phase as indicated by Eqs. [1], [3] and [15].[21]

$$[\mathrm{Mn}] + [\mathrm{O}] = (\mathrm{MnO})^{\mathrm{slag}} \qquad [15]$$

$$\log K_{[15]} = \log\left[\frac{a^{s}_{\mathrm{MnO}}}{a_{\mathrm{Mn}} \cdot a_{\mathrm{O}}}\right] = 1.27 \quad \text{(at 1873 K)} \qquad [16]$$

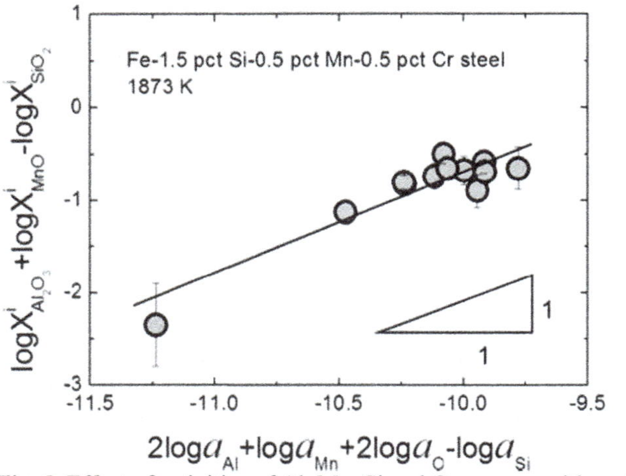

Fig. 5. Effect of activities of Al, Mn, Si and O on composition of inclusions in steel at 1873 K.

By combining Eqs. [2], [4], [14] and [16], the following equation can be deduced:

$$\log X^{i}_{\mathrm{Al_2O_3}} + \log\left(\frac{X^{i}_{\mathrm{MnO}}}{X^{i}_{\mathrm{SiO_2}}}\right) = \log\left(\frac{a^{s}_{\mathrm{Al_2O_3}} \cdot a^{s}_{\mathrm{MnO}}}{a^{s}_{\mathrm{SiO_2}}}\right) + \log\left(\frac{\gamma^{i}_{\mathrm{SiO_2}}}{\gamma^{i}_{\mathrm{Al_2O_3}} \cdot \gamma^{i}_{\mathrm{MnO}}}\right) + C \qquad [17]$$

This equation implies that the activity ratio of slag components directly affects the concentration of the inclusions since the activity coefficient ratio of the inclusion components can be assumed to be constant from Figure 5 and Eq. [14]. Furthermore, Figure 2 (excluding sample #1) indicates that the molar concentration ratio of MnO to SiO_2 in the inclusions is constant, i.e., 0.8(±0.06). Consequently, Eq. [17] can be simplified to Eq. [18].

$$\log X^{i}_{\mathrm{Al_2O_3}} = \log\left(\frac{a^{s}_{\mathrm{Al_2O_3}} \cdot a^{s}_{\mathrm{MnO}}}{a^{s}_{\mathrm{SiO_2}}}\right) + C' \qquad [18]$$

Thus, it is simply expected that there is a linear relationship between $\log X^{i}_{\mathrm{Al_2O_3}}$ and $\log(a^{s}_{\mathrm{Al_2O_3}} \cdot a^{s}_{\mathrm{MnO}} / a^{s}_{\mathrm{SiO_2}})$ with a slope of unity, which is confirmed by the result shown in Figure 6. The slope of the line was found to be 1.1 from linear regression analysis. Even though there were some experimental scatters, this result qualitatively implies that the activity ratio of slag components directly affects the concentration of alumina in the inclusions.

Actually, from the FactSage™6.3 computation, the activity of Al_2O_3 and MnO in the slags investigated in the present study was found to range from about 0.5 to 0.9 and from 0.015 to 0.03 (approximately two-fold increase), respectively. However, the activity of SiO_2 varied from about 0.01 to 0.1, i.e. a ten-fold increase within the present composition range. The iso-activity contours of Al_2O_3, MnO, and SiO_2 in the CaO-SiO_2-Al_2O_3-5% MnO slag system at 1873 K are shown in Figure 7.

The results shown in Figure 4 indicate that the Al_2O_3 content in the inclusions should be kept lower than about 18% in order to maintain the melting point of the inclusions below 1573 K. Hence, from Figure 6, the activity ratio of the slag components, $(a^{s}_{\mathrm{Al_2O_3}} \cdot a^{s}_{\mathrm{MnO}} / a^{s}_{\mathrm{SiO_2}})$, should be lower than unity.

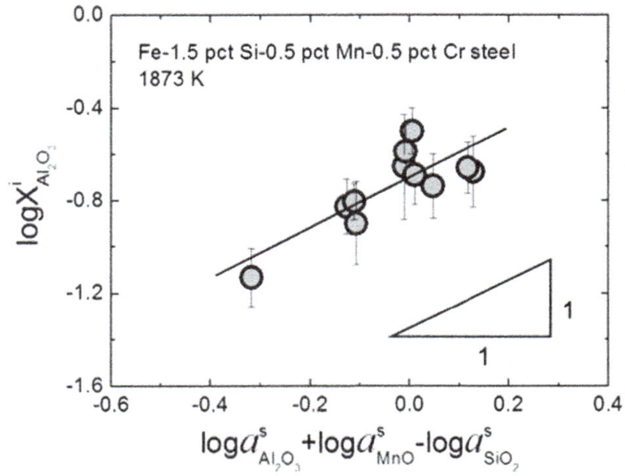

Fig. 6. Effect of activities of Al_2O_3, MnO and SiO_2 in the slag on the concentration of Al_2O_3 in the inclusion.

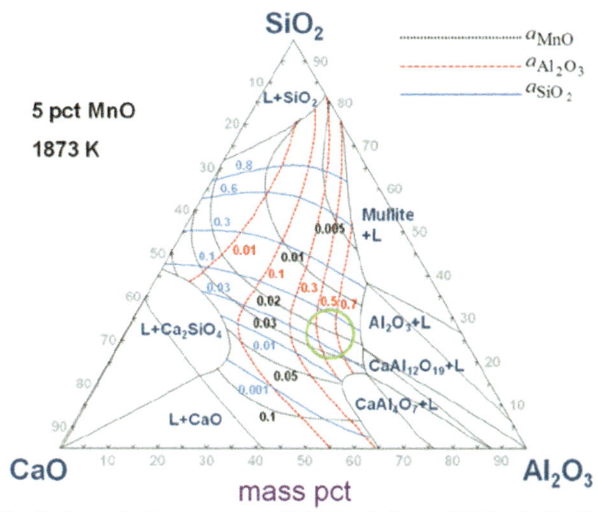

Fig. 7. Iso-activity contours of MnO, Al_2O_3 and SiO_2 in the CaO-SiO_2-Al_2O_3-5%MnO slag at 1873 K. The circle is the recommended slag composition for obtaining liquid inclusion of which melting point is lower than 1573 K in Si-Mn-killed steel refined in the alumina ladle.

Because the activity of Al_2O_3 in the slags ranged from about 0.5 to 0.9, the activity of MnO should be lower than that of SiO_2 to satisfy the above condition. The compositional window corresponding to this condition is also shown as a circle in Fig. 7.

In summary, this experimental study of the slag-steel-inclusion multiphase equilibria for Si-Mn-killed steels found that the

MnO/SiO_2 ratio of the inclusions was relatively close to unity with varied Al_2O_3 content. From thermodynamic analyses of the steel-inclusion and the steel-slag systems, quantitative thermodynamic information for the slag-inclusion system was obtained. Finally, we propose the operating window of ladle slag in which the activity of MnO should be lower than that of SiO_2 for obtaining inclusions with a low melting point (< 1573 K) in Si-Mn-killed steel, which is treated in an alumina-lined ladle.

Conclusions

The thermodynamic equilibria between $CaO\text{-}Al_2O_3\text{-}SiO_2\text{-}CaF_2\text{-}MgO(\text{-}MnO)$ slag and Fe-1.5%Mn-0.5%Si-0.5%Cr melt was investigated at 1873 K in order to understand the effect of slag composition on the concentration of Al_2O_3 in the inclusions. Our major findings are as follows:

1. The activities of oxygen calculated from the [Si]/(SiO_2) and [Al]/(Al_2O_3) equilibria were in good agreement with those determined from classical Wagner formalism. The slag-metal reaction was believed to be in equilibrium in the present experiments.

2. The composition of the inclusions were mainly equal to (MnO)/(SiO_2)=0.8(±0.06) (mol%) with Al_2O_3 content that was increased from about 10 to 40 mol% by increasing the basicity of the slag from about 0.7 to 2.1.

3. The $\log(X_{Al_2O_3} \cdot X_{MnO} / X_{SiO_2})$ of the inclusion was expressed as a linear function of the $\log[a_{Al}^2 \cdot a_{Mn} \cdot a_O^2 / a_{Si}]$ of the steel, indicating that the activity coefficient ratio of the inclusion components, $\gamma^i_{SiO_2} / (\gamma^i_{Al_2O_3} \cdot \gamma^i_{MnO})$ was not significantly changed under the present experimental conditions.

4. The concentration of Al_2O_3 in the inclusions linearly increased with an increase in the activity ratio of slag components, $(a^s_{Al_2O_3} \cdot a^s_{MnO} / a^s_{SiO_2})$, in a logarithmic scale. This linear relationship enabled determination of the appropriate compositional window of slag for obtaining inclusions with a low melting point in Si-Mn-killed steel refined in an alumina ladle.

References

1. R. Kiessling and N. Lange: *Nonmetallic Inclusions in Steels*, The Metals Society, London, 1978, Part. I.
2. A. Ghosh: *Secondary Steelmaking – Principles and Applications*, CRC press, Boca Raton, FL, 2001, pp. 255-69.
3. M.E. Fine: *Metall. Trans. A*, 1980, vol. 11A, pp. 365-79.
4. K. Kirihara: *Kobelco Technology Review*, 2011, no. 30, pp. 62-65.
5. S. Chen, M. Jiang, X. He and X. Wang: *Int. J. Min. Metall. Mater.*, 2012, vol. 19, pp. 490-98.
6. T. Fujisawa and H. Sakao: *Tetsu-to-Hagané*, 1977, vol. 63, pp. 1494-503.
7. T. Fujisawa and H. Sakao: *Tetsu-to-Hagané*, 1977, vol. 63, pp. 1504-11.
8. H. Suito and R. Inoue: *ISIJ Int.*, 1996, vol. 36, pp. 528-36.
9. Y.B. Kang and H.G. Lee: *ISIJ Int.*, 2004, vol. 44, pp.1006-15.
10. I.H. Jung, Y.B. Kang, S.A. Decterov and A.D. Pelton: *Metall. Mater. Trans. B*, 2004, vol. 35B, pp. 259-68.
11. C. Garlick, M. Griffiths, P. Whitehouse and C. Gore: *Ironmaking & Steelmaking*, 2002, vol. 29, pp. 140-46.
12. C. Bertrand, J. Molinero, S. Landa, R. Elvira, M. Wild, G. Barthold, P. Valentin and H. Schifferl: *Ironmaking & Steelmaking*, 2003, vol. 30, pp. 165-69.
13. S.K. Choudhary: *ISIJ Int.*, 2011, vol. 51, pp. 557-65.
14. G. Okuyama, K. Yamaguchi, S. Takeuchi and K. Sorimachi: *ISIJ Int.*, 2000, vol. 40, pp. 121-28.
15. J.H. Park, D.S. Kim and S.B. Lee: *Metall. Mater. Trans. B*, 2005, vol. 36B, pp. 67-73.
16. J.H. Park and D.S. Kim: *Metall. Mater. Trans. B*, 2005, vol. 36B, pp. 495-502.
17. J.H. Park, S.B. Lee and H.R. Gaye: *Metall. Mater. Trans. B*, 2008, vol. 39B, pp. 853-61.
18. J.H. Park and H. Todoroki: *ISIJ Int.*, 2010, vol. 50, pp. 1333-46.
19. J.H. Park, D.J. Kim, and D.J. Min: *Metall. Mater. Trans. A*, 2012, vol. 43A, pp. 2316-24.
20. J.S. Park, C. Lee, and J.H. Park: *Metall. Mater. Trans. B*, 2012, vol. 43B, pp. 1550-64.
21. *Thermodynamic Data for Steelmaking*, ed. by M. Hino and K. Ito, The Japan Society for the Promotion of Science, The 19th Committee on Steelmaking, Tohoku University Press, Sendai, 2010.
22. F. Ishii and S. Ban-ya: *ISIJ Int.*, 1992, vol. 32, pp. 1091-96.
23. K. Suzuki, S. Ban-ya and M. Hino: *ISIJ Int.*, 2001, vol. 41, pp. 813-17.
24. T. Itoh, T. Nagasaka and M. Hino: *ISIJ Int.*, 2000, vol. 40, pp. 1051-58.
25. K. Takahashi and M. Hino: *High Temp. Mater. Proc.*, 2000, vol. 19, pp. 1-10.
26. H. Itoh, M. Hino and S. Ban-ya: *Tetsu-to-Hagané*, 1997, vol. 83, pp. 773-78.
27. H.R. Gaye: *in The Making, Shaping and Treating of Steel*, 11th ed., Casting Volume, The AISE Steel Foundation, Pittsburgh, PA, 2003.
28. Z. Hong, X. Wu and C. Kun: *Steel Res.*, 1995, vol. 66, pp. 72-76.
29. www.factsage.com.
30. C.W. Bale, E. Belisle, P. Chartrand, S.A. Decterov, G. Eriksson, K. Hack, I.H. Jung, Y.B. Kang, J. Melancon, A.D. Pelton, C. Robelin and S. Petersen: *Calphad*, 2009, vol. 33, pp. 295-311.
31. M.O. Suk and J.H. Park: *J. Am. Ceram. Soc.*, 2009, vol. 92, pp. 717-23.
32. J.H. Park, I.H. Jung and S.B. Lee: *Met. Mater. Int.*, 2009, vol. 15, pp. 677-81.
33. J.H. Park: *Met. Mater. Int.*, 2010, vol. 16, pp. 987-92.
34. J.H. Park, J.G. Park, D.J. Min, Y.E. Lee and Y.B. Kang: *J. Eur. Ceram. Soc.*, 2010, vol. 30, pp. 3181-86.
35. J.H. Park, M.O. Suk, I.H. Jung, M. Guo and B. Blanpain: *Steel Res. Int.*, 2010, vol. 81, pp. 860-68.
36. J.H. Park, G.H. Park and Y.E. Lee: *ISIJ Int.*, 2010, vol. 50, pp. 1078-83.
37. G.H. Park, Y.B. Kang and J.H. Park: *ISIJ Int.*, 2011, vol. 51, pp. 1375-82.
38. Y.B. Kang and J.H. Park: *Metall. Mater. Trans. B*, 2011, vol. 42B, pp. 1211-17.
39. K.Y. Ko and J.H. Park: *Metall. Mater. Trans. B*, 2011, vol. 42B, pp. 1224-30.
40. J.H. Park and G.H. Park: *ISIJ Int.*, 2012, vol. 52, pp. 764-69.
41. K.Y. Ko and J.H. Park: *Metall. Mater. Trans. B*, 2012, vol. 43B, pp. 440-42.
42. D.J. Kim and J.H. Park: *Metall. Mater. Trans. B*, 2012, vol. 43B, pp. 875-86.
43. J.H. Park: *ISIJ Int.*, 2012, vol. 52, pp. 1627-36.
44. J.H. Heo, S.S. Park and J.H. Park: *Metall. Mater. Trans. B*, 2012, vol. 43B, pp. 1098-105.
45. J.H. Park: *ISIJ Int.*, 2012, vol. 52, pp. 2303-04.
46. J.H. Park: *Met. Mater. Int.*, 2013, vol. 19, pp. 577-84.
47. J.H. Park: *Steel Res. Int.*, 2013, vol. 84, accepted, in print. DOI: 10.1002/srin.201200210.
48. K. Tomioka, K. Ogawa and H. Matsumoto: *ISIJ Int.*, 1996, vol. 36, pp. S101-04.

Proceedings of the 2013 International Symposium on Liquid Metal Processing & Casting
Edited by: Matthew J. M. Krane, Alain Jardy, Rodney L. Williamson, and Joseph J. Beaman
TMS (The Minerals, Metals & Materials Society), 2013

THE INFLUENCE OF DIFFERENT MELTING AND REMELTING ROUTES ON THE CLEANLINESS OF HIGH ALLOYED STEELS

Gerhard Reiter, Wolfgang Schuetzenhoefer, Angelika Tazreiter, Carlos Martinez, Paul Wuerzinger, Christian Loecker
Bohler Edelstahl GmbH&CoKG, P.O.-box 96, A-8605 Kapfenberg, Austria

Keywords: cleanliness, NMI, nonmetallic inclusions, high alloyed steel, EAF, remelting, ESR, PESR, VAR, ASTM E45

Abstract

The cleanliness of steel has an essential impact on many properties like corrosion resistance or toughness. Beside the deoxidation treatment, remelting or even multiple remelting strongly influences the cleanliness and can result in huge improvements. After a rough overview on measurement methods and the melting and remelting facility at Böhler Edelstahl, this work first shows typical cleanliness results according to the ASTM E45 measurement of two different steel types taken from standard production. The investigated steels range from hot work tool steel to austenitic stainless steel covering also different kinds of deoxidation procedures. Furthermore the differences of various melting and remelting routes are described and compared to each other, including the process routes EAF, EAF+ESR, EAF+PESR and EAF+VAR. Additionally, the morphology and composition of the different inclusion types found in the investigated material are described and compared.

Introduction

One of the efforts of today's steel melting is very simple: keep the material clean. The reason is clear; cleanliness influences strongly required properties like corrosion resistance, toughness and fatigue. Beside a state of the art metal treatment during primary melting, performing a subsequent remelting step either in open, protective ESR or VAR is the most effective method in order to guarantee a low content of inclusions and, hence, a high cleanliness [1, 2]. The requirements on materials for aviation industry, power generation or for human implants can only be fulfilled by remelted grades. Cleanliness is not the only, but a very important factor for this.

The potential for improvements due to remelting has already been reported [1, 2, 3, 5], but due to the progress in melting, it is worth having a look on the current situation.

Cleanliness Measurement By Means Of Optical Microscopy

Among many different ways to judge the cleanliness of steel, the standard methods use optical microscopy to classify the cleanliness based on

- the amount,
- the size and
- the optical appearance

of inclusions in steel samples taken from converted hot formed material. Many standards were developed that can be basically distinguished between worst field measurements, that give only an information about the field with the highest inclusion content, and frequency measurements, that also give information about how frequent inclusions appear. **Figure 1** gives an overview on several of the available standards [6, 7, 8, 9, 10].

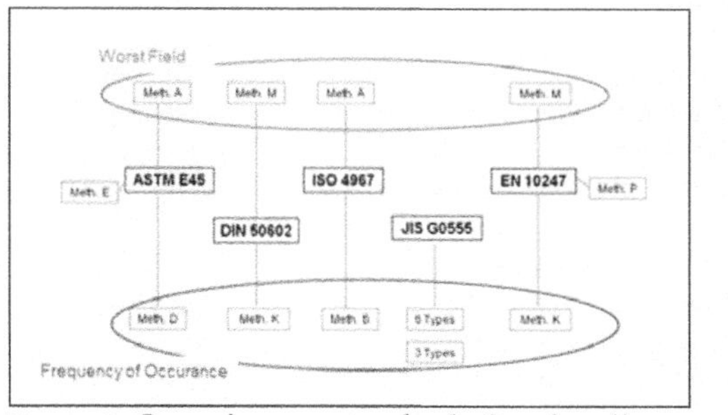

Figure 1: Overview on standards for cleanliness measurements [6]

In order to evaluate rather clean material, worst field methods don't give a good picture and, hence, they are not a proper way for comparing and judge remelted material. Frequency measurements allow a better description of the occurring inclusions and are a better method for comparison. Among the frequency measurements, ASTM E45 method D is a widely used standard method capable to handle the content of small inclusions. Therefore, the focus of this investigation is on ASTM E45 method D.

Melting And Remelting Of High Alloyed Steels

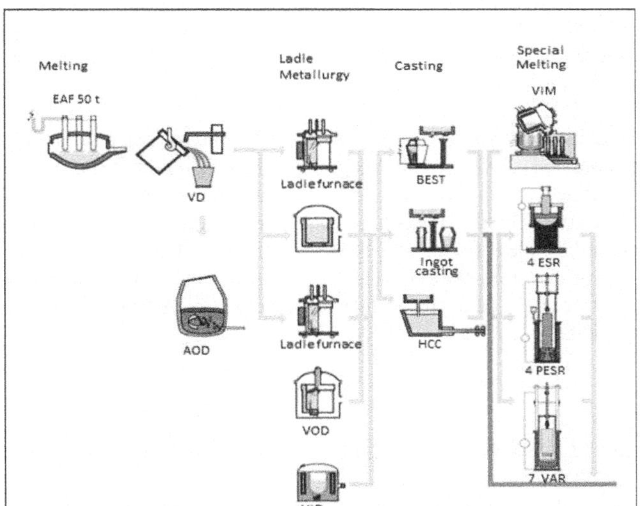

Figure 2: Melting and remelting facilities at Böhler Edelstahl in Kapfenberg [4]

The samples for investigation are taken from standard production at Böhler Edelstahl GmbH & CoKG in Kapfenberg (Austria).

Figure 2 shows the melting and remelting facilities at the production site in Kapfenberg. Conventionally melted material is melted in EAF, followed by ladle metallurgy, ingot casting and conversion. Remelted material is also melted in EAF, followed by secondary metallurgy. For casting either ingot or continuous casting are possible. A subsequent remelting step is carried out either in open, protective ESR or in VAR facilities. For primary melting, VID and VIM processes are also available.

Note that the focus of the work is on the comparison of final material. Hence, the results for conventionally melted material are taken from final products and not from electrodes (prior to remelting). Due to different deoxidation treatments, the cleanliness of electrodes and final products can differ.

The investigation for this work is carried out on three different steel grades, two comparable hot work tool steel grades and a Cr-Ni steel grade. Conventionally melted material is compared to remelted material, including ESR, protective ESR and VAR processes.
Table 1 gives an overview on the investigated steel grades and melting and remelting routes. E45 states that an ASTM measurement was carried out; SEM indicates that additional SEM measurements were performed.

Table 1: Overview on investigated grades and melting routes

Grade	EAF	ESR	PESR	VAR
Tool steels	E45/SEM	E45	E45/SEM	E45/SEM
Cr-Ni steel	E45		E45/SEM	E45/SEM

The investigated material was produced in production facilities at the plant in Kapfenberg. For remelting in open and protective ESR a similar standard Wacker slag was used.

Methods of Investigation

The cleanliness measurements were carried out according to ASTM E45 method D on a fully automated Olympus light microscope using the software package *analySIS auto Inclusion Inspector5.2*.

Additionally to the light microscopy measurements, a fully automated JEOL 6490 scanning electron microscope (SEM) equipped with an Oxford Instruments energy dispersive spectrometer (EDS) was used to scan the surface of polished steel samples, identify the existing inclusions and characterize their chemical composition. The acquisition and analysis of the measurement data was performed with the INCA Feature software package.

Results

Cleanliness Ratings According To ASTM E45 Method D

Each result represents the mean value of several measurements carried out on different ingots. Hence, the results represent an average for standard production. As indicated in
Table 1 measurements were carried out on all steel grades and all possible routes, except open ESR for the Cr-Ni steel grade.

Hot Work Tool Steels

First, two hot work tool steel grades were investigated; the chemical composition is shown in

Grade	C	Si	Mn	Cr	Mo	V
X38CrMoV 5 1	0.38%	1.10%	0.40%	5%	1.3%	0.4%
X40CrMoV 5 1	0.39%	1.10%	0.40%	5.20%	1.4%	0.95%

.

Table 2: Composition of the investigated tool steels [4]

Grade	C	Si	Mn	Cr	Mo	V
X38CrMoV 5 1	0.38%	1.10%	0.40%	5%	1.3%	0.4%
X40CrMoV 5 1	0.39%	1.10%	0.40%	5.20%	1.4%	0.95%

Generally, a 3-dimensional graph is used to illustrate the ASTM results that usually come in tables. The x axis of the graph shows the inclusion type, starting from A thin on the left hand side to D thick on the far right hand side. The y axis shows the severity from 0.5 to 1.5. The frequency of inclusions is plotted on the z axis.

Figure 3 displays the cleanliness results of material melted in EAF without a subsequent remelting step. It shows that there are mainly D thin type inclusions (small globular oxides) up to a severity of 1.5 and some D heavy types of severity 0.5. Very rarely B thin type inclusions can be found. A type and C type inclusions are not apparent in the investigated material.

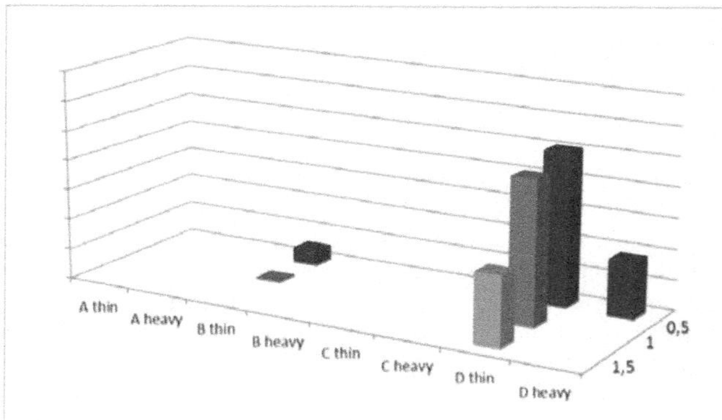

Figure 3: ASTM E45 results of conventionally melted material (EAF/AOD); (X38CrMoV51)

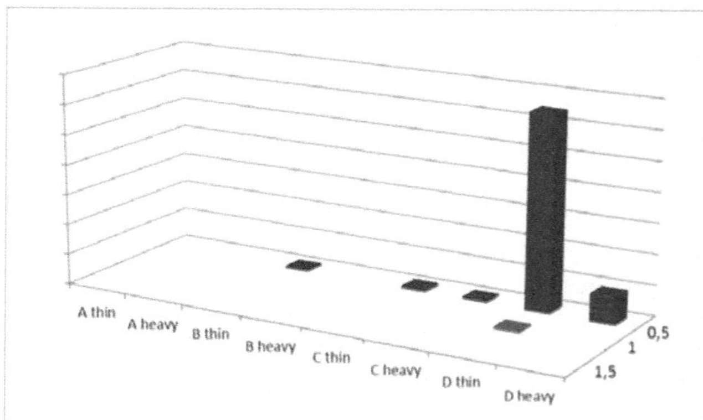

Figure 4: ASTM E45 results of material remelted in open ESR (X38CrMoV51)

Using open ESR as a subsequent remelting step changes the cleanliness as shown in **Figure 4**. D thin type inclusions (globular oxides, severity 0.5) are predominant. All other inclusions appear rather rarely and only up to severity 0.5.
B thin type (elongated oxides) and D heavy (massive globular oxides) are reduced in number and size compared to conventionally melted material. Their overall occurrence is very rare (lower than 1 field per sample).

Remelting in protective ESR instead of open ESR changes the cleanliness in such that only both thin and heavy D 0.5 type inclusions occur, as shown in **Figure 5**. A, B and C type inclusion cannot be found. The amount of fields showing D thin 0.5 type inclusions is reduced compared to open ESR.

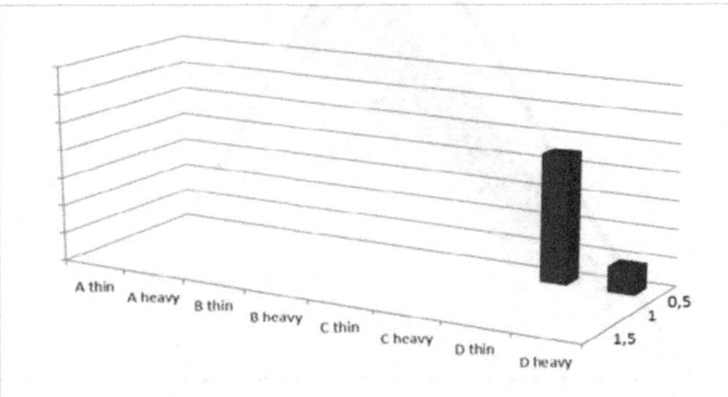

Figure 5: ASTM E45 results of material remelted in protective ESR (X40CrMoV51)

A similar result shows **Figure 6** for material remelted in vacuum. D thin 0.5 are predominant, whereas D heavy type inclusions are almost completely removed compared to protective ESR material. A, B and C type inclusions cannot be found.

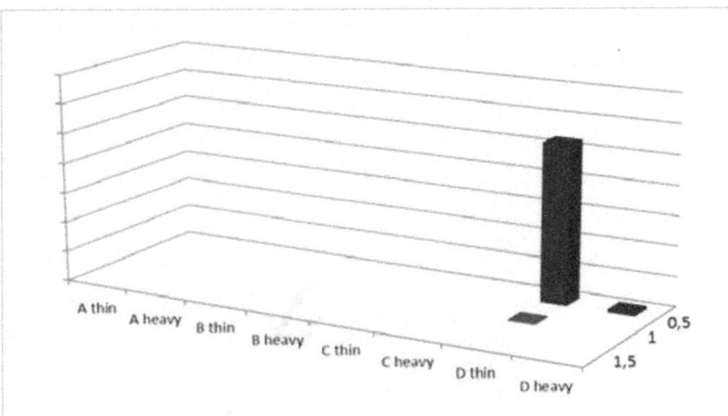

Figure 6: ASTM E45 results of material remelted in VAR (~X38CrMoV51)

Martensitic Cr-Ni Steel

Compared to tool steels in Cr-Ni steels usually a lower Al and Si content are requested which results in a lower deoxidation potential. The chemical composition is shown in

Grade	C	Si	Cr	Ni	Cu	Nb
X5CrNiCuNb 17 4 4	0.04%	0.25%	15.3%	4.5%	3.25%	0.30%

.

Table 3: Chemical composition of grade X5CrNiCuNb 17 4 [4]

Grade	C	Si	Cr	Ni	Cu	Nb
X5CrNiCuNb 17 4 4	0.04%	0.25%	15.3%	4.5%	3.25%	0.30%

Figure 7 shows the cleanliness rating of conventionally melted material. Similar to hot work tool steel mainly D type inclusions (globular oxides) and B thin 0.5 type occur. B and C type inclusions, except B thin 0.5; appear only in very low frequency of less than one field per sample. A type inclusions (sulfides) cannot be found.

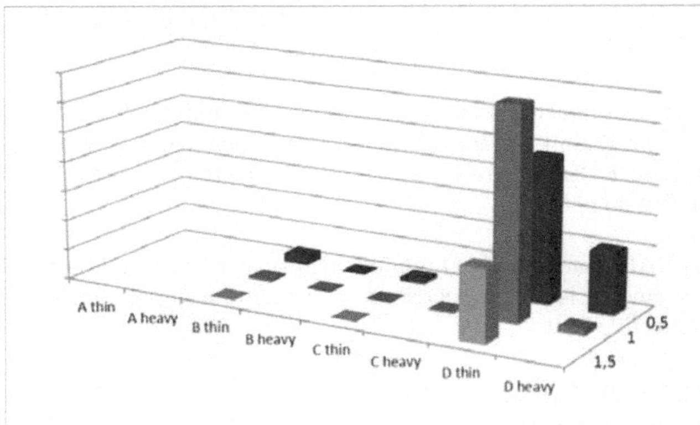

Figure 7: ASTM E45 rating of conventionally melted material (X5CrNiCuNb 17 4)

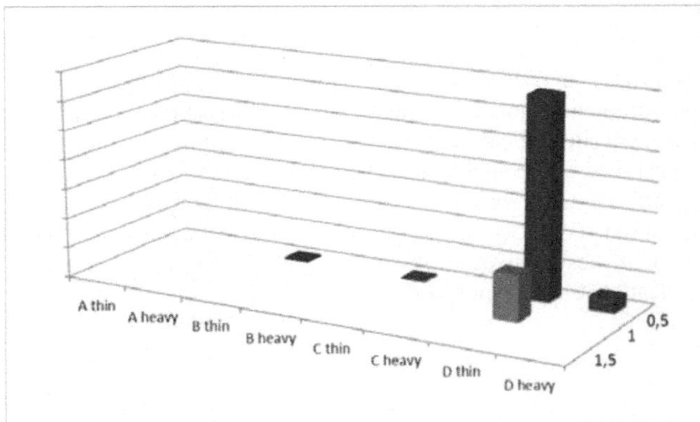

Figure 8: ASTM E45 results of material remelted in protective ESR (X5CrNiCuNb 17 4)

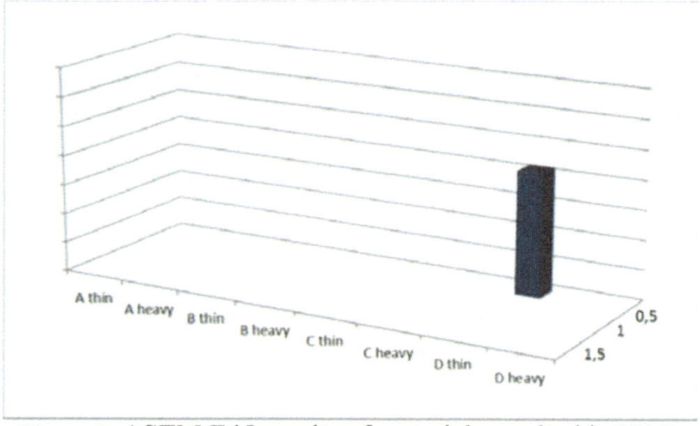

Figure 9: ASTM E45 results of material remelted in VAR (X5CrNiCuNb 17 4)

The steel remelted in protective ESR shows only globular oxides of type D thin 0.5 and 1.0 as well as D heavy 0.5 as shown in **Figure 8**. Except for type D thin, inclusions of severity 1.0 or higher do not occur.

In VAR material only D thin 0.5 inclusions occur (see **Figure 9**), the amount is further decreased compared to material remelted in protective ESR.

Composition Of Inclusions

As indicated in **Table 1,** SEM measurements were carried out on samples taken from conventionally melted as well as from material PESR and VAR remelted tool steel.

Generally, the composition of inclusions according to SEM measurement cannot be compared to ASTM E45 ratings, since the judgment in optical microscopy is only based on the optical appearance of the inclusions. All results are shown in the ternary system Al_2O_3 – CaO – MgO. Each dot corresponds to the measured composition of a single inclusion, and its color distinguishes between oxides, oxisulfides (typically sulfides nucleated at oxidic inclusions) and sulfides.

Hot Work Tool Steel

Figure 10 shows the distribution of inclusions in the mentioned ternary system for conventionally melted hot work tool steel. A rather complex inclusion landscape can be seen, mainly oxides (MA spinel and different CA types) and oxisulfides (containing CaS).

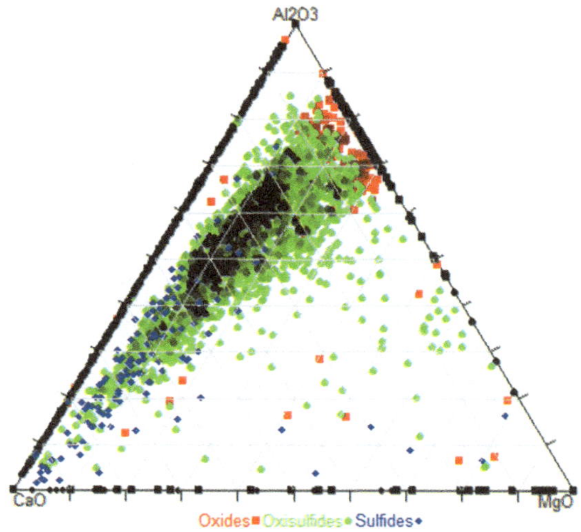

Figure 10: Distribution of inclusions in EAF melted Cr-Ni-steel (X38CrMoV51)

Figure 11 shows the composition of inclusion in hot work tool steel remelted in protective ESR. The inclusions are mainly oxides and can be found in the Al_2O_3-rich corner of the ternary system. Most of the particles contain MA spinel or have a very similar composition. Sulfides (CaS) are not predominant and can only be found very rarely.

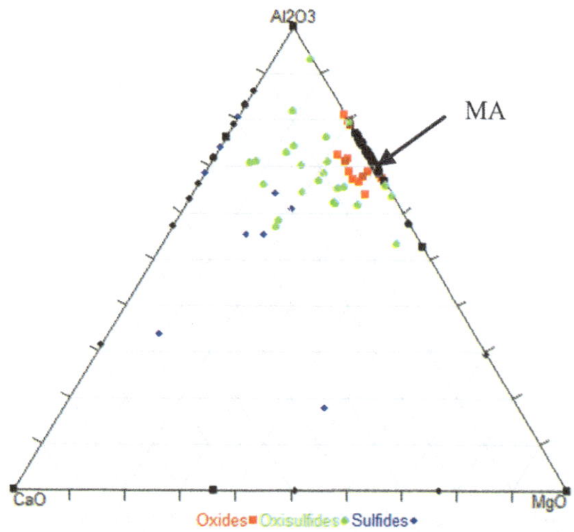

Figure 11: Distribution of inclusions in hot work tool steel (X38CrMoV51) remelted in protective ESR

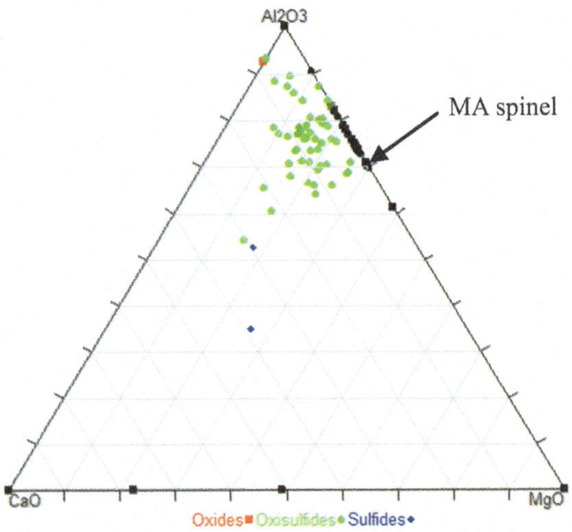

Figure 12: Distribution of inclusions in hot work tool steel (X38CrMoV51) remelted in VAR

In hot work tool steel remelted in vacuum MA spinel is the predominant type of inclusions as can be seen in **Figure 12**. Similar to all other samples sulfides are very rare.

Martensitic Cr-Ni steel

Since the basic results are similar, here only the differences to the results reported for tool steel are pointed out.

Compared to tool steel the inclusion landscape of Cr-Ni steel is different, showing slightly higher Al_2O_3 and CaO contents. Hence, the amount of CA type inclusions is increased and the composition of MA type inclusions is shifted to a higher Al_2O_3 content, as shown in **Figure 13**.

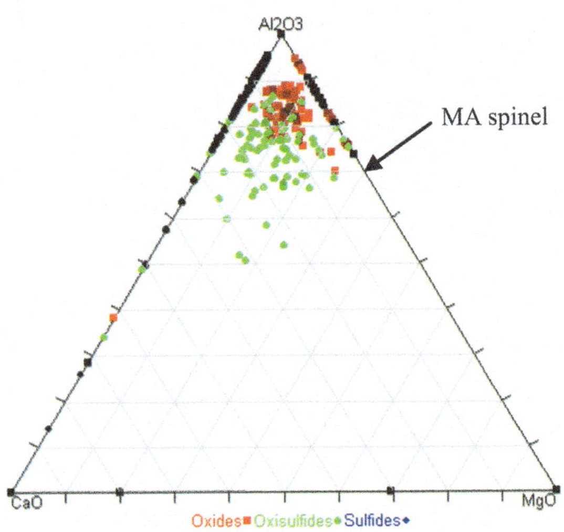

Figure 13: Distribution of inclusions in Cr-Ni-steel (X5CrNiCuNb 17 4) remelted in PESR

A similar effect is apparent in VAR material. As in the tool steel MA type inclusions are predominant, but a slight content of calcium aluminates can also be seen in **Figure 14**.

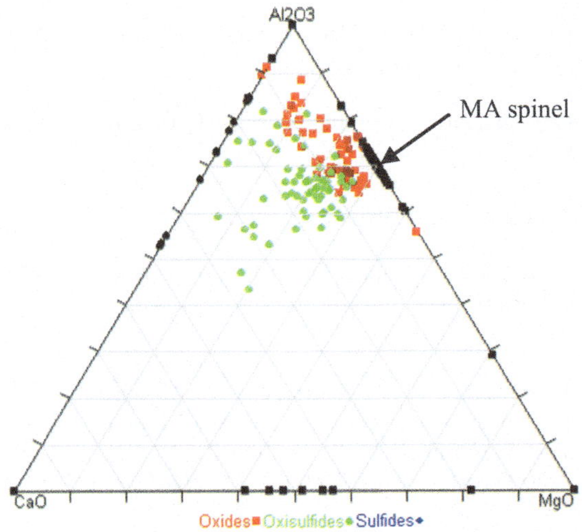

Figure 14: Distribution of inclusions in material remelted in VAR in the Al_2O_3-CaO-MgO system (X5CrNiCuNb 17 4)

Discussion

The ASTM results quite clearly show the improvement in cleanliness from conventionally melted material to the alloys remelted in ESR, protective ESR and VAR. The most effective cleaning effect is shown – not surprisingly – by the protective ESR process and even more by the VAR process. Globular oxides are predominant, silicates and elongated oxides only occur in conventionally melted steels and – to a very low extent – in material that is remelted in open ESR facilities. PESR or VAR remelted material does not show any of these inclusion types.

According to the ASTM measurements it seems that remelting simply decreases the amount and the size of inclusions, which corresponds to a lower frequency of occurrence and a lower severity, but the SEM measurements show a change in the inclusion composition, too. The rather complex inclusion landscape of EAF melted material is modified after remelting to inclusions with a very high Al_2O_3 content, although the cleaning mechanism in ESR and VAR are different.

It is also shown, that the small difference between protective ESR and VAR material according to the ASTM results, which is "just" a lower content of small globular oxides in VAR material, corresponds as well to slightly different inclusions, with a small portion of CA type inclusions in PESR remelted steel, whereas in VAR mainly MA type inclusions can be found.

Finally, different steel grades with different deoxidation treatment and different solubility for trace elements also have an influence on the type of inclusions, as can be seen from the comparison between the SEM results of tool steel and Cr-Ni steel.

References

[1] Hoyle, G: Electroslag processes; Applied Science Publishers LTD (1983)

[2] Duckworth, W.E., Hoyle, G.: Electro-slag refining; Chapman and Hall LTD (1969)

[3] Gmelin Durrer, Metallurgy of Iron, Volume 9; Springer Verlag (1988)

[4] www.bohler-edelstahl.at

[5] Mayerhofer, Johann: Internal report, Böhler Edelstahl GmbH&CoKG

[6] Tazreiter, A: Statistische Absicherung von Reinheitsgradbestimmungen bei Stählen unterschiedlicher Verfahrensrouten mittels automatisierter, lichtmikroskopischer Methode, Bachelorarbeit MU Leoben, 2012

[7] American Society for Testing and Materials (Ed.): ASTM E45, Standard Test Methods for Determining the Inclusion Content of Steel, Beuth Verlag GmbH (2011).

[8] Deutsches Institut für Normung (Ed.): DIN EN 10247, Metallographische Prüfung des Gehaltes nichtmetallischer Einschlüsse in Stählen mit Bildreihen, Beuth Verlag GmbH, Berlin (2007).

[9] International Organization of Standardization (Ed.): ISO 4967, Steel – Determination of content of nonmetallic inclusions – Micrographic method using standard diagrams, (1998).

[10] JIS G0555, Microscopic testing method of the non-metallic inclusion in steel, Beuth Verlag GmbH, Berlin (2003).

Proceedings of the 2013 International Symposium on Liquid Metal Processing & Casting
Edited by: Matthew J. M. Krane, Alain Jardy, Rodney L. Williamson, and Joseph J. Beaman
TMS (The Minerals, Metals & Materials Society), 2013

REFINEMENT OF PRIMARY AND EUTECTIC SILICON PARTICLES IN HYPEREUTECTIC Al-Si ALLOYS USING AN APPLIED ELECTRIC POTENTIAL

A. Plotkowski[1] and P. N. Anyalebechi[2]

[1]Purdue Center for Metal Casting Research, School of Materials Engineering,
Purdue University, West Lafayette, Indiana USA 47907
Formerly of Padnos College of Engineering and Computing, Grand Valley State University

[2]Padnos College of Engineering and Computing, Grand Valley State University
Grand Rapids, Michigan USA 49504

Keywords: Cast microstructure, solidification, electric current, Al-Si alloys

Abstract

The effects of applied electric current on the cast microstructure of hypoeutectic Al-13 wt.% Si and Al-20 wt.% Si alloys have been investigated. This involved, with a constant voltage power supply, an application of an electric current density of about 500 mA/cm^2 of melt surface area to laboratory-size ingots during solidification in a metal mold. In general, the applied electric current refined the cast microstructure of the hypereutectic Al-Si alloys. Specifically, the electric current did not modify the eutectic silicon particle, but changed the size distribution of the primary silicon particles by increasing the population of comparatively smaller size particles. The extent of the observed cast microstructure refinement was less than the reported effects of applied electric current in the technical literature and is significantly less than the effects of the traditional refinement obtained by addition of strontium and phosphorus to the molten metal prior to casting.

Introduction

Hypereutectic Al-Si alloys are widely used in the automobile and aerospace industries because of their excellent wear resistance, high strength-to-weight ratio, low coefficient of thermal expansion, good corrosion resistance, excellent fluidity, and good castability [1–3]. They are used in various applications such as liner-less engine blocks [4], automotive pistons [5], compressor bodies, and pumps [6]. Hypereutectic Al-Si alloys are used to produce engine blocks without cylinder liners, automotive pistons, and a number of other products primarily because of their high wear resistance which results from a large volume fraction of the hard primary silicon phase.

However, all of the aforementioned desirable properties of hypereutectic Al-Si alloys depend on the characteristics of their cast microstructures, namely secondary dendrite cell size or arm spacing, and the size, morphology, and distribution of eutectic and primary silicon particles. The morphology of primary silicon particles can be rather complex, such as plate-like [7], star-shaped [7,8], polygonal, blocky, or feathery [9], varying with solidification conditions, chemical composition, and alloying elements. In these forms, primary silicon particles compromise the machinability, wear resistance, and mechanical properties of the alloy castings. Refinement and control of the eutectic and primary silicon particles is an effective way of improving the properties of the hypereutectic Al-Si alloys. For example, hypereutectic Al-Si alloys with a uniform distribution of fine primary silicon particles have higher strength and better wear resistance.

Several different techniques have been proposed for the refinement of eutectic and primary silicon particles in hypereutectic Al-Si alloys. They include: (i) chemical treatments by addition of elements such as Na, P, Sr, La, etc. [3,10,11], (ii) mechanical stirring [12], (iii) electromagnetic stirring [9,13,14] and vibration [15], (iv) ultrasonic treatment [16], and (v) application of an electric current during solidification [5,17].

Interest in using an applied electric current to modify cast microstructures began with several early studies that involved investigation of segregation, solute redistribution, and planar front stability [18,19]. The studies in the mid-1980s examined the potential use of applied electric current to refine cast microstructures and improve mechanical properties. These studies were performed on a variety of materials, but can be divided into two groups based on processing conditions: (i) high density pulsed current [5,17,20–28], and (ii) low density steady current [29–33]. Several studies that used pulsed current sources attributed the observed refinement to shearing of dendrite arms caused by the pinch force [17,20,21,24,25]. However, this mechanism does not explain refinement in the case of low current density, steady current application. Other proposed mechanisms include a change in the free energy barrier to nucleation, resulting in an increased undercooling and subsequent increase in nucleation rate [22,23], and increased interfacial stability caused by joule and Peltier heating [30,33].

The purpose of this study is to investigate the effect of applied electric current on the microstructure of hypereutectic Al-Si alloys. The goal is to develop a process that can be used to effectively refine the primary and eutectic silicon particles without the associated technological, practical, and environmental drawbacks associated with traditional methods [11,34]. While results of previous studies show that the application of electric current during solidification can produce significant refinement of the cast microstructure of different metallic alloys [22,26,32,33], few studies have been published on the application of electric current to Al-Si alloys [5,28,27].

Experimental Procedure

Two different hypereutectic Al-Si alloy compositions were used for this experimental study, namely, a near-eutectic Al-13 wt.% Si and hypereutectic Al-20 wt.% Si alloys, to investigate the effects of electric current on the eutectic silicon particles and primary silicon particles, respectively. A natural gas furnace was used to melt the charge material, in a graphite crucible, to a superheat of about 150°C. Each casting consisted of approximately 1.5 kg of charge material. None of the castings was modified, so no strontium, sodium, or phosphorus was added to any of the alloy melts. The molten metal was then manually poured into a steel mold with dimensions 177.8 mm x 101.6 mm x 25.4 mm with thermocouples located at 38.1, 88.9, and 139.7 mm from the top of the mold. Temperature data were recorded at 2 Hz with k-type thermocouples. The temperature data were used to derive the average solidification rates at the known distances from the top of the castings.

A steady electric current density of approximately 500 mA/cm^2 was applied using a constant voltage power supply and graphite electrodes. A power resistor was added to the circuit to avoid shorting the power supply. Figure 1 shows a schematic of the experimental setup including the circuit design, electrode placement, and thermocouple locations. The experimental conditions are given in Table I.

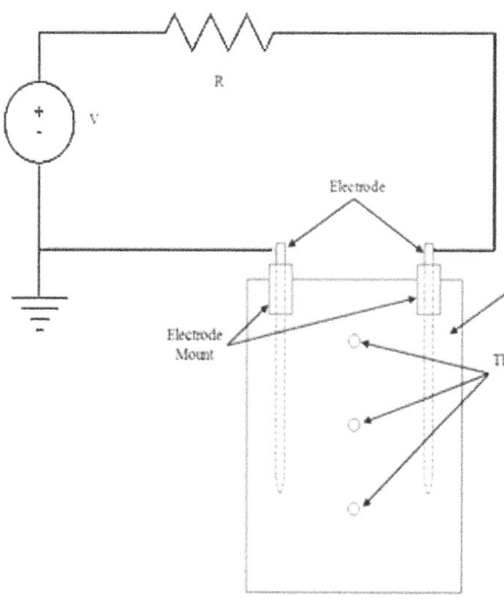

Figure 1: Schematic of experimental setup and circuit design for application of electric current to castings during solidification.

Four castings of each alloy were made with and without applied electric current. The applied electric current was calculated with Ohm's law using the resistance of the circuit, measured after solidification, and the voltage set on the power supply. The current density was calculated using the cross sectional area normal to the direction of current flow. Samples of the cast ingots were taken near each thermocouple location, metallographically prepared, and examined with a metallurgical microscope. The secondary dendrite cell size and the sizes of the primary silicon particles were determined with the Leco IA32 image analyzer. The limited quantitative measurement of primary silicon particle size was conducted on only the Al-20 wt.% Si alloy castings.

Results

The average solidification rates derived from the measured cooling curve data are given in Table II. It is apparent that application of the electric current during solidification did not affect the rate of solidification of any of the castings.

As shown in Figures 2 and 3, the cast microstructure of the unmodified Al-13 wt.% Si alloy consists of primary Si particles, fairly coarse eutectic dendrite-like aluminum and coarse eutectic Si particles. As apparent in Figure 2 (i), in the near-eutectic Al-13 wt.% Si alloy, the eutectic structure is dominant; amount of primary phase (the angular/blocky primary silicon particles or **α**-Al dendrites) is relatively small. The cast microstructure of the unmodified Al-20 wt.% Si alloy consisted of primary silicon particles in a matrix of aluminum and eutectic silicon (Figures 4 and 5). In the cast microstructures of both alloys, there are primary aluminum solid solution dendrites and unmodified primary and eutectic silicon particles.

As shown in the measured secondary dendrite cell size data in Table III, application of electric current during solidification reduced the average dendrite cell size. It also reduced the maximum dendrite cell size. The effect appeared to be independent of solidification rate but slightly dependent on location in the casting. The observed reduction in average dendrite cell size appeared to be greatest at mid-length and towards the bottom of the castings. In the Al-13 wt.% Si alloy castings, applied electric current reduced dendrite cell size by 7.80% near the top, but by 25.7% and 26.5%, at mid-length and near the bottom of the casting, respectively (Table IV). In the Al-20 wt.% Si alloy castings, the reduction in dendrite cell size is significantly less, varying from 3.40% near the top, to 10.4% at mid-length, and no change near the bottom of the casting.

Table I: The experimental casting conditions.

Experimental conditions	Al-13 wt.% Si				Al-20 wt.% Si			
	A	B	C	D	E	F	G	H
Molten metal pouring temperature , °C	825	826	825	826	852	850	856	851
Resistor, Ω	-	-	4.5	4.5	-	-	3.0	4.5
Measured resistance of circuit, Ω	-	-	5.17	4.85	-	-	4.05	4.7
Applied voltage, V	-	-	110	110	-	-	85	110
Calculated electric current, A	-	-	21.3	22.7	-	-	20.98	23.4
Electric current density, mA/cm^2	-	-	472	503	-	-	464	518

Table II: Average solidification rate at known distances from the top of the castings of hypereutectic Al-Si alloys with and without application of electric current

Distance from the top of the mold, mm	Average solidification rate, °C/s							
	Al-13 wt.% Si				Al-20 wt.% Si			
	No current A	No current B	472 mA/cm^2 C	503 mA/cm^2 D	No current E	No current F	464 mA/cm^2 G	518 mA/cm^2 H
38.1	0.08	0.08	0.08	0.08	1.09	0.93	0.98	1.11
88.9	0.09	0.09	0.12	0.10	1.15	1.20	1.06	1.20
139.7	0.12	0.12	0.12	0.13	1.52	1.61	1.55	1.52

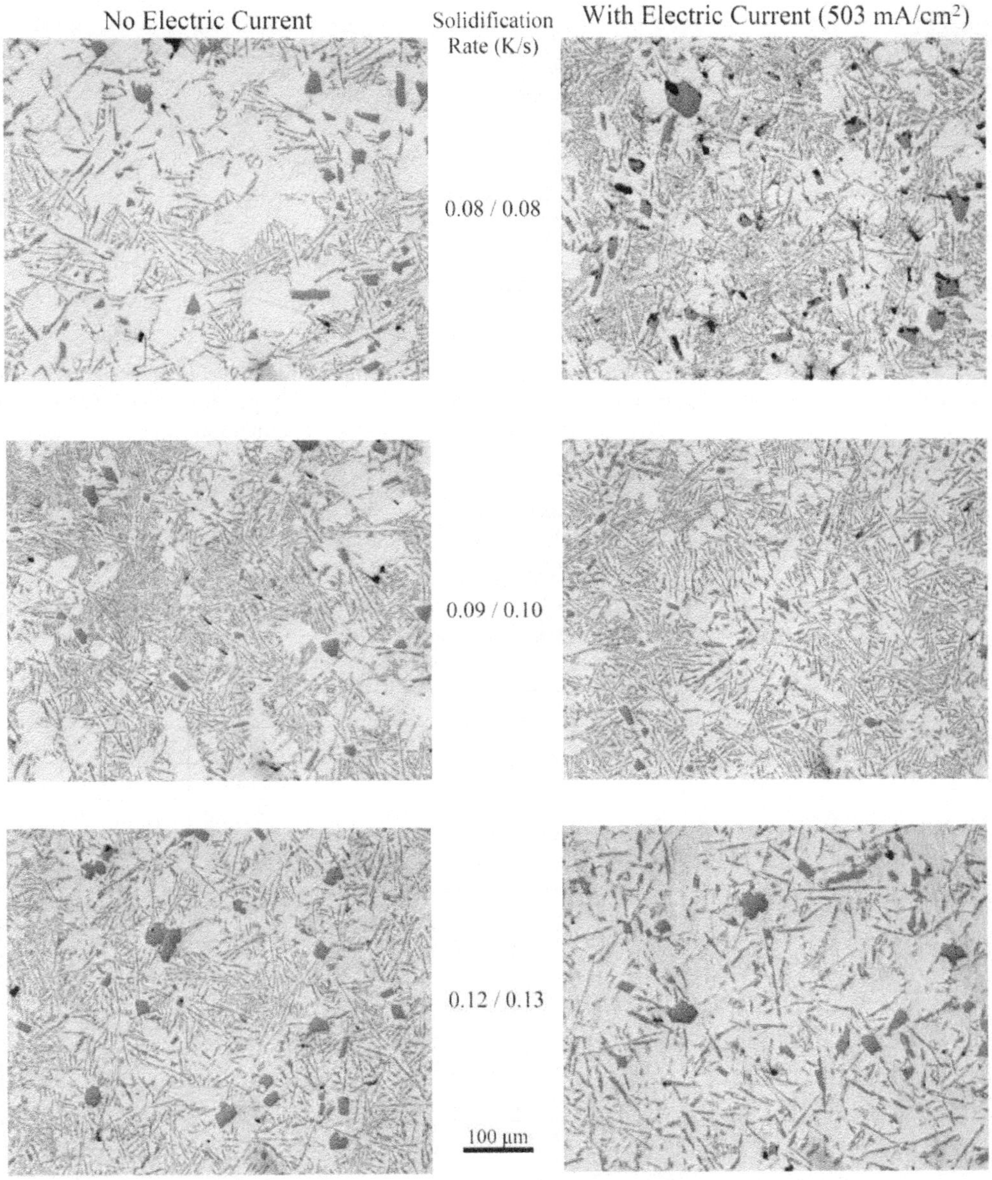

Figure 2: Optical photomicrographs showing the effects of application of electric current during solidification on the cast microstructure of hypereutectic Al-13 wt.% Si alloy.

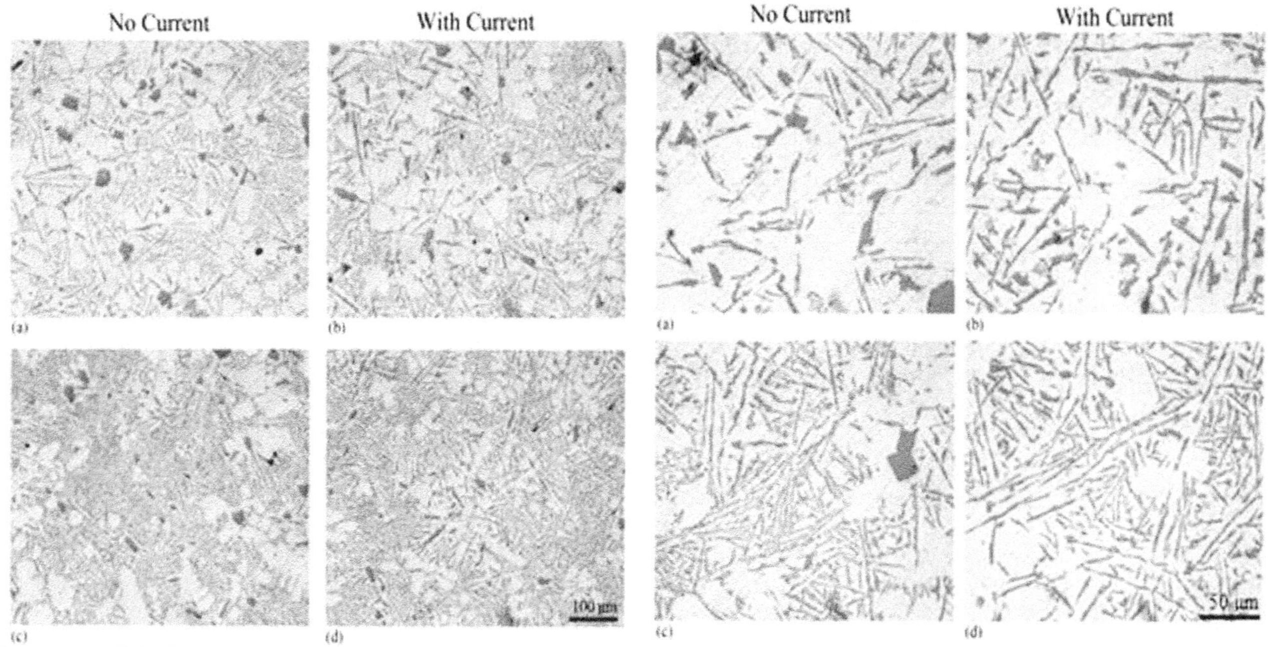

(i) (ii)

Figure 3: Optical photomicrographs showing the effects of application of electric current (471-502 mA/cm^2) during solidification on the cast microstructure of hypereutectic Al-13wt.% Si alloy solidified at 0.08°C/s. Note the lack of effect of electric current on the eutectic Si particles in (ii).

Table III: Effects of application of electric current during solidification on the dendrite cell size in castings of hypereutectic Al-Si alloys (based on the measurement of the sizes of 60 dendrites per condition for Al-13 wt.% Si castings, and 120 for Al-20 wt.% Si castings).

Alloy	Condition	Distance from top of casting, mm	Average Solidification Rae, °C/s	Dendrite cell size, µm			
				Average	Std. Dev.	Max.	Min.
Al-13 wt.%Si	No Current	38.1	0.08	37.8	12.2	89.9	16.6
	With current*		0.08	34.9	10.4	66.4	14.5
	No Current	88.9	0.09	34.7	12.4	70.4	16.2
	With current		0.09	25.8	12	51.7	12.2
	No Current	139.7	0.12	31.4	10.1	58.5	7.66
	With current		0.12	23.1	8.7	45.5	9.06
Al-20 wt.% Si	No Current	38.1	0.93-1.09	38	11.74	72	18
	With current#		0.98-1.11	36.7	12.98	81.7	14.4
	No Current	88.9	1.15-1.20	36.7	9.33	60.3	18
	With current		1.06-1.20	32.9	13.01	75.3	8
	No Current	139.7	1.52-1.61	26.1	9.79	51.8	9.66
	With current		1.52-1.55	26.2	10.23	58.1	7.96

Two-tailed t-tests showed that the observed reductions in dendrite cell size by application of electric current during solidification of the castings of both alloys were statistically significant. However, the observed reductions in dendrite cell size in Al-13 wt.% Si alloy castings were more statistically significant than those observed in the Al-20 wt.% Si alloy castings.

The optical photomicrographs of the cast microstructures of the castings of both alloys in Figures 2 and 3 suggest that application of electric current during solidification reduced the average size of the primary silicon particles. Specifically, the electric current appeared to have increased the population of the very small size primary silicon particles. Similar to the effect on dendrite cell size, image analysis results show that the electric current reduced both the average and maximum primary silicon particle sizes in

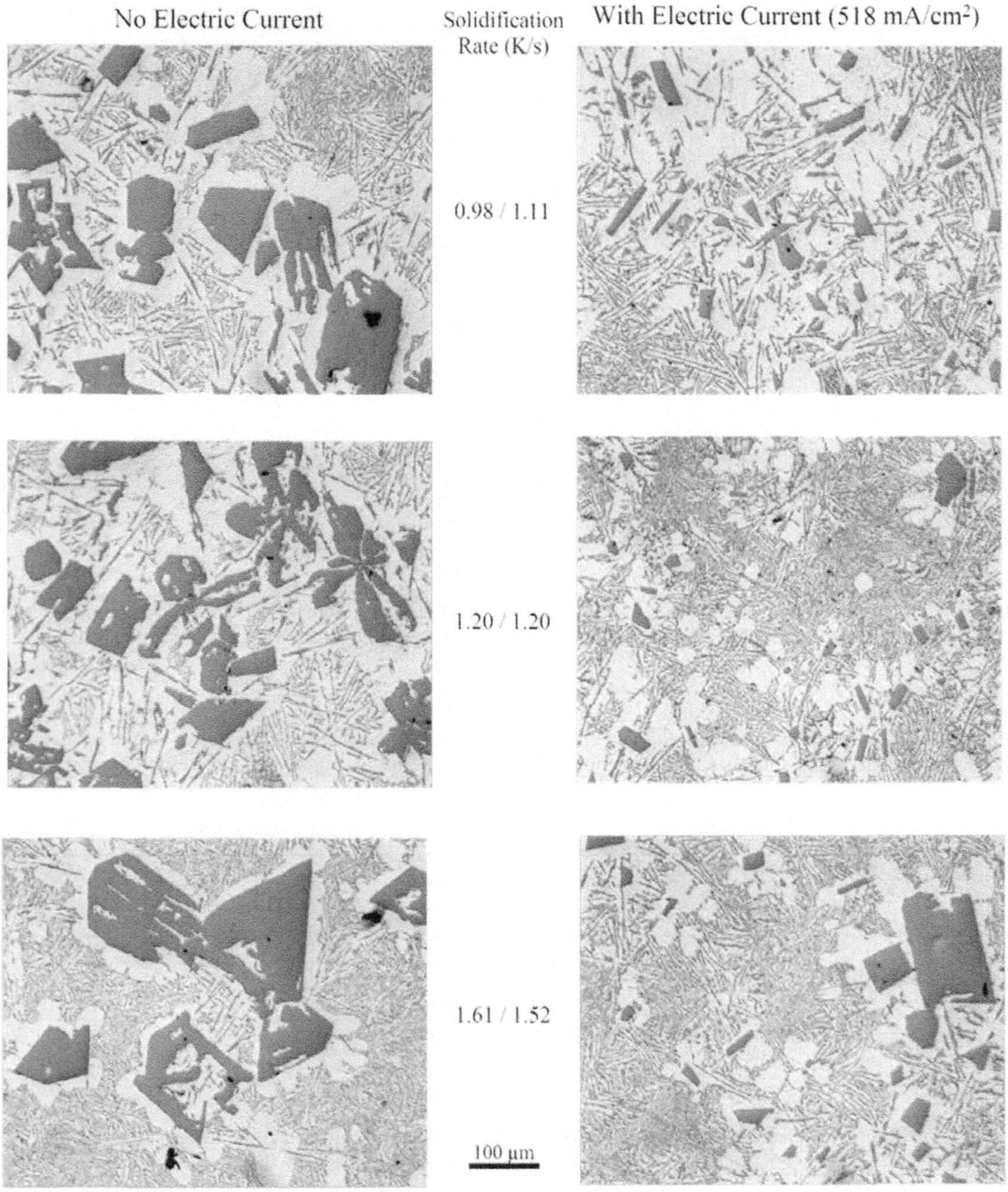

Figure 4: Optical photomicrographs showing the effects of application of electric current during solidification on the cast microstructure of hypereutectic Al-20 wt.% Si.

the Al-20 wt.% Si castings (Table V). In addition, the lower standard deviation of sizes of the primary silicon particles in the castings solidified with application of electric current suggest that the electric current improved their size distribution. The observed percent reduction in primary silicon particle size appeared to be significant at all of the locations of the casting examined, varying from 58.0% near the top, to 36% and 64%, at mid-length and near bottom, respectively (Table VI). Two-tailed t-tests showed that the observed reductions in primary silicon particle size in the Al-20 wt,% Si castings by application of electric current during solidification were statistically significant.

However, unlike the effects on dendrite cell and primary silicon particle sizes, application of electric current during solidification did not appear to have significant effects on the size and morphology of eutectic silicon particles in both alloys (Figures 3 and 5).

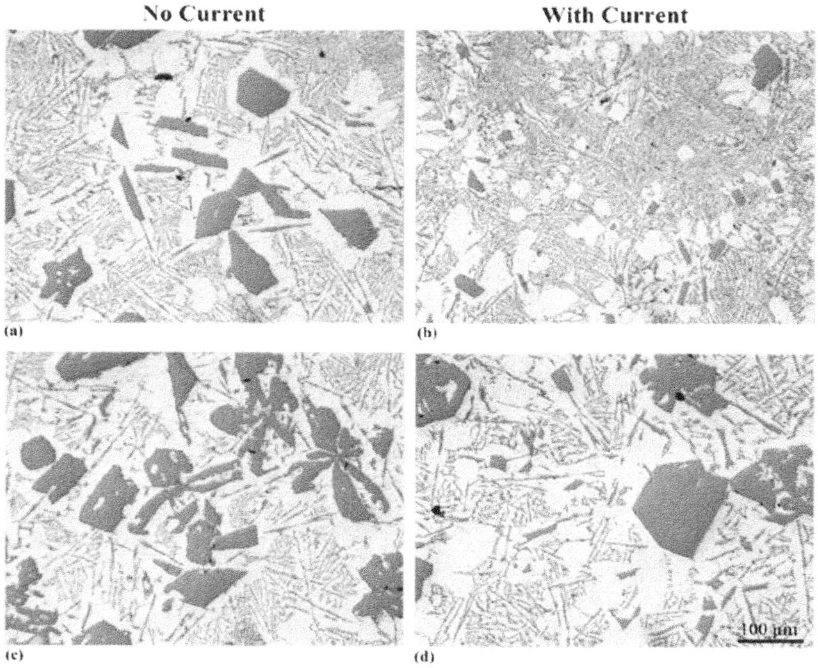

Figure 5: Optical photomicrographs showing the lack of effect application of electric current (465-518 mA/cm^2) during solidification on the sizes and morphology of the unmodified eutectic silicon particles in hypereutectic Al-20 wt.% Si alloy solidified at 0.93-0.98°C/s.

Table IV: Percent reductions in average dendrite cell size in castings of hypereutectic Al-Si alloys caused by application of electric current during solidification and p-values resulting from t-tests comparing the corresponding data sets.

Alloy	Distance from top of casting, mm	Average solidification rate, °C/s	Reduction in average dendrite cell size, %	p-value
Al-13 wt.% Si *	38.2	0.08	7.80	0.079
	88.9	0.09	25.7	0.055
	139.7	0.12	26.5	0.078
Al-20 wt.% Si #	38.2	0.98-1.11	3.40	0.213
	88.9	1.06-1.20	10.40	0.005
	139.7	1.52-1.55	−0.30	0.473

*Current density – 472-503 mA/cm^2, #Current density – 465-518 mA/cm^2

Table V: Effects of application of electric current during solidification on primary silicon particle size in castings of hypereutectic Al-20 wt.% Si alloy.

Condition	Distance from top of casting, mm	Average solidification rate, °C/s	Primary Si particle size, μm^2				No. of particles measured
			Average	Std. dev.	Max.	Min.	
No Current	38.1	0.93-1.09	5,744	7,118	37,708	10	98
With current#		0.98-1.11	2,394	3,850	22,430	10	137
No Current	88.9	1.15-1.20	4,863	6,018	33,717	26	108
With current		1.06-1.20	3,110	8,415	52,295	21	164
No Current	139.7	1.52-1.61	5,352	6,602	35,235	40	95
With current		1.52-1.55	1,918	4,711	35,610	27	156

Table VI: Percent reduction in mean primary silicon particle size in castings of hypereutectic Al-20 wt.% Si alloys caused by application of electric current during solidification and p-values resulting from t-tests comparing the corresponding data sets.

Distance from top of casting, mm	Average solidification rate, °C/s	Reduction in mean primary Si particle area, %	p-value
38.2	0.98-1.11	58.0	0.000022
88.9	1.06-1.20	36.0	0.019
139.7	1.52-1.55	64.0	0.000009

Discussion

Dendrite Cell Size

The observed reduction in dendrite cell size in the Al-13 wt.% Si castings in this study (Table III and IV) is consistent with published results. Anyalebechi and Tomaswick [33] reported a decrease of 29% in dendrite cell size in aluminum alloy 7050 castings solidified with an applied electric current density ranging from 465 to 930 mA/cm^2. They found that the level of refinement was independent of both the current density and the application mode (steady or pulsed). The results of this study also compare favorably with those found by Vaschenko et al. [29], who reported 25% to 40% reduction in the size of graphite flakes in gray cast iron.

Anyalebechi and Tomaswick [33] speculated that localized convection currents produced by Joule and Peltier heating at the solid-liquid interface caused shear stress induced fragmentation of dendrite arms, which would then float into the melt and serve as additional nuclei. This mechanism may explain the reduction in dendrite cell size found in this study since similar electrical conditions were used. Other studies [20,21] proposed similar theories, but speculated that the shearing mechanism was the pinch force caused by high density electric pulses. This theory does not explain the observations of this study since a low density steady current was used.

Anyalebechi and Tomaswick [33] also suggested that a secondary mechanism in which the applied electric current dramatically changed the solid-liquid interfacial energy, leading to an increase in nucleation rate. Qin and Zhou [23] theoretically studied this refinement mechanism and calculated the level of refinement as a function of electric current density for a variety of pure metals. Their calculations did not predict significant refinement at electric current densities below 10^3 A/cm^2. However, they made no predictions for comparatively more complex multi-component systems. In the case of dendrite arms in hypereutectic Al-Si alloys, the interfacial energy of concern is between solid primary silicon particles and eutectic composition liquid transforming into eutectic aluminum dendrites. It is possible that this interfacial energy is affected to a greater extent by the application of electric current in pure metals. Alternatively, this mechanism may make a small contribution in combination with other mechanisms to result in the observed refinement of dendrite cell size.

For the Al-20 wt.% Si alloys, the change in the dendrite size was not statistically significant in the samples from the top and bottom of the castings. Good statistical significance ($p = 0.005$) was found at the center of the casting, but the reduction was modest in comparison to the Al-13 wt.% Si alloy at 10%. This result suggests that the presence of a large volume fraction of primary silicon particles in the melt at the time when the eutectic aluminum dendrites formed changed the interaction between the electric current and the solidifying eutectic aluminum dendrite cells. This may have been a result of increased Joule and Peltier heating caused by the large volume fraction of primary silicon particles. At the eutectic temperature for the Al-Si binary system, pure silicon has an electrical resistivity approximately four orders of magnitude greater than that of aluminum [35-37]. Joule heating is directly proportional to resistance, and as such, for a given current density, local heating caused by the Joule effect may be increased by a larger volume fraction of primary silicon particles. Peltier heating is a function of the interface between two dissimilar materials. As such, this heating would increase with the increased volume fraction of primary silicon particles.

Primary Silicon Particles

Only two previous studies examined the effects of an applied electric current on solidification of primary silicon particles in Al-Si alloys. Ban et al. [17] observed an obvious refinement of the primary silicon particles when using high density current pulses. They attributed this refinement to the pinch force, which, when caused by each current pulse, produced a vibration in the melt which broke apart larger primary silicon particles. This mechanism does not explain the refinement observed in this study since a steady current source was used, with a current density several orders of magnitude smaller.

Hongsheng et al. [5] observed that high density current pulses reduced the aspect ratio of primary silicon particles. They determined that the presence of an electric current changed the free energy barrier for nucleation in the melt, resulting in the promotion of Al nucleation, but suppression of Si nuclei. This would cause an increase in the undercooling at which primary silicon nucleation begins, and a subsequent increase in nucleation rate, resulting in smaller particles with a more uniform size distribution. This mechanism might explain the refinement observed in this study, although it is unclear if this mechanism alone fully describes this phenomenon.

An additional explanation for the primary silicon refinement may be localized joule heating. Since silicon is much more electrically resistive than aluminum, the electric current may cause localized high temperature regions, which, while not large enough to change the solidification condition of the entire melt, might locally impede the growth of the primary silicon particles.

Eutectic Silicon Particles

The lack of an observed effects of the applied electric current during solidification on the size and morphology of the eutectic silicon particles in this study is not consistent with the reported effects of pulsed electric current in the technical literature [5,28,27]. For example, Hongsheng et al. reported that the application of a high density pulsed electric current during solidification reduced the length of the eutectic silicon particles [5]. The electric current density used for this study was not reported, but the voltage used was 2000V, discharged via a capacitor bank. As such, it may be assumed that the electric current density was much larger than that used for the present study. Xu et al. [28] experimented with both direct and alternating electric currents on hypoeutectic Al-Si alloys. They reported similar results as Hongsheng et al. [5], in that the length of the eutectic particles was decreased, and noted that the result was more dramatic with the alternating current. Again, the electric current density was not reported, but electric currents of 220A were used, so assuming that the cross sectional area of the casting was on the same order of magnitude as this study, the current density was much greater.

The electric current density used in this study was several orders of magnitude less than that used by Hongsheng et al. [5] and Xu et al. [28]. Additionally, electric current was administered in a steady rather than pulsed [5] or alternating [28] manner. Considering that no change in the eutectic silicon particles was observed in this study, it may be concluded that the effect of the application of electric current during solidification to Al-Si alloys on the eutectic particles is either a function of electric current density, or the manner of application. Hongsheng et al. [5] suggested that the mechanism of refinement was suppression of the nucleation of silicon. This effect is likely a function of current density, and thus may not have been significant in this study with a current density of 500 mA/cm^2. Xu et al. [28] attributed refinement of the eutectic silicon particles to vibration induced by the alternating current, similar to refinement via electromagnetic vibration. The current in this study was applied in a direct manner, so this effect was not applicable.

Zhang et al. [27] applied pulsed electric current densities ranging from 800 A/cm^2 to 2400 A/cm^2 to Al-Si alloys of eutectic composition. They reported that this procedure produced areas of fine lamellar structures of eutectic silicon, distributed among the usual, randomly oriented needle or plate-like eutectic structure. Zhang et al. [27] concluded that the fine lamellar eutectic silicon structure was formed because the Lorentz force generated by the pulsed electric current constantly fed solute to the solidifying region. However, the Lorentz force is only generated under a time dependent magnetic field [37]. Steady applied electric current, such as used in this study, is not time dependent, and therefore, does not generate the Lorentz force. Therefore, the lack of such fine lamellar structures in the samples from this study is not surprising.

Conclusions

1. The application of approximately 500 mA/cm^2 of steady electric current to Al-20 wt.% Si alloy during solidification reduced the mean primary silicon particle size by 36% - 64% with statistical significance ($8.98 \times 10^{-6} < p < 0.019$).

2. The application of 465-518 mA/cm^2 of steady electric current to Al-13 wt.% Si alloy during solidification reduced the average dendrite cell size through the casting by 7.8% - 26.5% with statistical significance ($p < 0.1$). The application of electric current during solidification to the Al-20 wt.% Si alloy did not significantly change the dendrite cell size at the top or bottom of the casting, but it reduced the average dendrite cell size at the center of the casting by 10.4% ($p = 0.005$).

3. Qualitatively, the application of a steady electric current during solidification did not appear to have any effect on the size, size distribution, or morphology of the eutectic silicon particles in both the Al-13 wt.% Si and Al-20 wt.% Si alloys.

References

1. B. K. Prasad, K. Venkateswarlu, O. P. Modi, A. K. Jha, S. Das, and R. Dasgupta, "Sliding Wear Behavior of Some Al-Si Alloys : Role of Shape and Size of Si Particles and Test Conditions," *Metallurgical and Materials Transactions A*, 29 (1998), 2747–2752.

2. D. Dwivedi, "Wear behaviour of cast hypereutectic aluminium silicon alloys," *Materials & Design*, 27 (7) (2006), 610–616.

3. S. Hegde and K. N. Prabhu, "Modification of eutectic silicon in Al–Si alloys," *Journal of Materials Science*, 43 (9) (2008), 3009–3027.

4. J. Jorstad, "The hypereutectic aluminum-silicon alloy used to cast the Vega engine block," *Modern Casting,* 60 (4) (1971), 59–64.

5. D. Hongsheng, Z. Yong, J. Sanyong, C. Ruirun, and Z. Zhilong, "Influences of pulse electric current treatment on solidification microstructures and mechanical properties of Al-Si piston alloys," *China Foundry*, 6 (1) 2008, 24–31.

6. J. Yeh, S. Yuan, and C. Peng, "A reciprocating extrusion process for producing hypereutectic Al – 20wt .% Si wrought alloys," *Materials Science and Engineering A*, 252 (1998), 212–221.

7. J. Y. Chang, G. H. Kim, I. G. Moon, and C. S. Choi, "Rare Earth Concentration in the Primary Si Crystal in Rare Earth Added Al-21wt.%Si Alloy," *Acta Metallurgica*, vol. 39 (3) (1998), 307–314.

8. Q. C. Jiang, C. L. Xu, M. Lu, and H. Y. Wang, "Effect of new Al–P–Ti–TiC–Y modifier on primary silicon in hypereutectic Al–Si alloys," *Materials Letters*, 59 (6) (2005), 624–628.

9. F. C. Hernández and J. H. Sokolowski, "Comparison among chemical and electromagnetic stirring and vibration melt treatments for Al–Si hypereutectic alloys," *Journal of Alloys and Compounds*, 426 (1-2) 2006, 205–212.

10. H. S. Dai and X. F. Liu, "Optimal holding temperatures and phosphorus additions for primary silicon refinement in Al–high Si alloys," *Materials Science and Technology*, 25 (10) (2009), 1183–1188.

11. D. P. Kanicki, "Processing molten aluminum - part 1 : understanding silicon modification," *Modern Casting*, (1990), 24.

12. K. Sukumaran, B. C. Pai, and M. Chakraborty, "The effect of isothermal mechanical stirring on an Al–Si alloy in the semisolid condition," *Materials Science and Engineering: A*, 369 (1–2) (2004) 275–283.

13. D. Lu, Y. Jiang, G. Guan, R. Zhou, and Z. Li, "Refinement of primary Si in hypereutectic Al–Si alloy by electromagnetic stirring," *Journal of Materials Processing Technology*, 189 (1-3) (2007), 13–18.

14. B. I. Jung, C. H. Jung, T. K. Han, and Y. H. Kim, "Electromagnetic stirring and Sr modification in A356 alloy," *Journal of Materials Processing Technology*, 111 (2001), 69–73.

15. J. Yu, Z. Ren, and K. Deng, "Refinement and migrating behaviors in Al-Si hypereutectic alloys solidified under electromagnetic vibration," *Acta Metallurgica Sinica*, 24 (4) (2011), 301–308.

16. X. Jian, T. T. Meek, and Q. Han, "Refinement of eutectic silicon phase of aluminum A356 alloy using high-intensity ultrasonic vibration," *Scripta Materialia*, 54 (5) (2006), 893–896.

17. C. Ban, Y. Han, Q. Ba, and J. Cui, "Influence of Pulse Electric Current on Solidification Structures of Al-Si Alloys," *Materials Science Forum*, 546-549 (2007), 723–728.

18. J. Angus, D. Ragone, and E. Hucke, "The Effect of an Electric Field on the Segregation of Solute Atoms at a Freezing Interface," in *Metallurgical Society Conference*, (1961), 833–843.

19. J. Verhoeven, "The Effect of an Electric Field upon the Solidification of Bismuth-Tin Alloys," *Transactions of the Metallurgical Society of AIME*, 233 (1965), 1156–1163.

20. M. Nakada, Y. Shiohara, and M. Flemings, "Modification of Solidification Structures by Pulse Electric Discharging," *ISIJ International*, 30 (1) (1990), 27–33.

21. L. Jianming, L. Shengli, and L. Hantong, "Modification of Solidification Structure by Pulse Electric Discharging," *Scripta metallurgica et materialia*, 31 (12) (1994), 1691–1694.

22. J. P. Barnak and A. F. Sprecher, "Colony (grain) size reduction in eutectic Pb-Sn castings by electroplusing," *Scripta metallurgica et materialia*, 32 (6) (1995), 879–884.

23. R. Qin and B. Zhou, "Effect of Electric Current Pulses on Grain Size in Castings," *International Journal of non-Equilibrium Processing*, 11 (1998), 77–86.

24. H. Shu-xian, W. Jun, S. Bao-de, and Z. Yao-he, "Effect of high density pulse electric current on solidification structure of low temperature melt of A356 alloy," *Transactions of Non-ferrous Metals Society of China*, 12 (3) (2002), 414–418.

25]. H. Shu-xian, W. Jun, J. Wan, S. Bao-de, and Z. Yao-he, "Effect of melt pulse electric current and thermal treatment on A356 alloy," *Transactions of Nonferrous Metals Society of China*, 13 (1) (2003), 126–130.

26. X. Liao, Q. Zhai, J. Luo, W. Chen, and Y. Gong, "Refining mechanism of the electric current pulse on the solidification structure of pure aluminum," *Acta Materialia*, 55 (9) 2007, 3103–3109.

27. Y. Zhang, C. Song, L. Zhu, H. Zheng, H. Zhong, Q. Han, and Q. Zhai, "Influence of Electric-Current Pulse Treatment on the Formation of Regular Eutectic Morphology in an Al-Si Eutectic Alloy," *Metallurgical and Materials Transactions B*, 42 (3) (2011), 604–611.

28. G. Xu, J. Zheng, Y. Liu, and J. Cui, "Effect of electric current on the cast micro- structure of Al-Si alloy," *China Foundry*, 2 (3) (2005), 171–175.

29. K. Vashchenko, D. Chernega, S. Vorob'ev, N. Lysenko, and Y. Yakovchuk, "Effect of Electric Current on the Solidification of Cast Iron," *Metallovedenie i Termicheskaya Obrabotka Metallov*, 3 (1974) 62–65.

30. A. Misra, "Effect of Electric Potentials on Solidification of Near Eutectic Pb-Sb-Sn Alloy," *Materials Letters*, 4 (3) (1986), 176–177.

31. A. Misra, "A Novel Solidification Technique of Metals and Alloys: Under the Influence of Applied Potential," *Metallurgical Transactions A*, 24 (1985) 1354–1355.

32. A. Misra, "Misra Technique Applied to Solidification of Cast Iron," *Metallurgical Transaccctions A*, 17 (2) (1986), 358–360.

33. P. N. Anyalebechi and K. M. Tomaswick, "Effect of Application of Electric Current During Solidification on the Cast Microstructure of Aluminum Alloy 7050," in *EPD Congress 2009*, (2009), 541–554.

34. N. Abu-Dheir, M. Khraisheh, K. Saito, and A. Male, "Silicon morphology modification in the eutectic Al–Si alloy using mechanical mold vibration," *Materials Science and Engineering: A*, 393 (1–2) (2005), 109–117.

35. S. G. Teodorescu, R. A. Overfelt, and S. I. Bakhtiyarov, "An Inductive Technique for Electrical Conductivity Measurements on Molten Metals," *International Journal of Thermophysics*, 22 (5) (2001), 1521–1535.

36. G. Pearson and J. Bardeen, "Electrical Properties of Pure Silicon and Silicon Alloys Containing Boron and Phosphorus," *Physical Review*, 75 (5) (1949), 865–883.

37. H. K. Moffatt, "Electromagnetic Stirring," *Physics of Fluids A*, 3 (5) (1991), 1336–1343.

Proceedings of the 2013 International Symposium on Liquid Metal Processing & Casting
Edited by: Matthew J. M. Krane, Alain Jardy, Rodney L. Williamson, and Joseph J. Beaman
TMS (The Minerals, Metals & Materials Society), 2013

CHARACTERIZATION OF BIFILMS AND OXIDE INCLUSIONS IN INVESTMENT CAST IN100

Max A. Kaplan[1], G. E. Fuchs[1]
[1]University of Florida, Materials Science and Engineering Dept., P.O. Box 116400, 116 Rhines Hall, Gainesville, FL 32611-6400, USA

Keywords: Superalloys; Bifilms; Sulfur; Auger

Abstract

Oxide bifilms are a proposed casting inclusion reported to have been observed in vacuum investment cast polycrystalline Ni-base superalloys. Ongoing research seeks to determine if current superalloy casting practices can result in the formation of oxide bifilms, and subsequently if it is possible to observe and characterize this phenomenon. The effect of casting atmosphere, turbulence, filtering, hot isostatic pressing (HIP), and heat treatment have been investigated to identify the critical parameters that can result in bifilm formation in Ni-based superalloy IN100. Tensile and fatigue testing is used to identify the effects of each casting and processing parameter on bifilm formation and the resultant effects on mechanical properties. Characterization of mechanical test specimens seeks to identify the role of bifilms in the fracture mechanics of the conditions utilized, as well as the prevalence of bifilm formation and the chemical composition(s) of oxide bifilms. The characterization methods used were scanning electron microscopy (SEM), energy dispersive x-ray spectroscopy (EDS), and Auger electron spectroscopy (AES). Across the range of process parameters investigated, bifilms were not indicated by mechanical testing or identified by the characterization methods employed. A nano-layer of sulfur has been identified on the surface of dendritic casting porosity and characterized by in-situ fractographic Auger analysis.

Introduction

Nickel-base superalloy components are often produced using the investment casting process. Many types of casting anomalies are possible, and it is necessary that each be minimized to optimize performance. A newly proposed casting anomaly known as oxide bifilms has been recently reported. Oxide bifilms are theorized to form if a liquid metal capable of growing a solid oxide is subjected to turbulence during casting. It has been reported that molten metal reacts with its casting atmosphere, forming a nanoscale thickness oxide skin [1-4]. It has also been reported that the formation of this oxide skin is possible with vacuum cast components [1, 2, 5-7]. This oxide skin folds due to casting induced turbulence. When folded, the oxide forms a bilayer structure and can be entrained into the melt [1-4]. The entrained bifilm behaves as a preexisting material discontinuity. For this reason, bifilms have been reported to decrease the mechanical properties of nickel-base superalloys and other alloy systems. This degradation of mechanical properties includes a significant reduction in tensile strength and fatigue life [2, 5, 6, 8-11]. In addition, bifilms are theorized to play a major role in the formation of casting porosity [1-3, 5, 6, 8, 9, 12-16]. This type of casting porosity is created during solidification after the bifilm is entrained into the melt. As dendrites form and grow, the oxide bifilm is wedged between the growing dendrites. As the alloy cools, the bifilm theoretically unfolds and fills the space between dendrite arms, thus creating casting porosity [1, 15]. This process can be exacerbated if gas content from the casting environment has been trapped between the layers of the bifilm during the entrainment process [1, 15].

Bifilms are often distinguished categorically as being either "new" or "old" bifilms [1, 5, 7-9, 11, 17]. New bifilms are described as the alumina films that are present immediately following turbulent casting [1, 17]. Old bifilms are described as being an oxide film which has thickened [1, 5, 7, 17]. New bifilms have been characterized as exhibiting a thickness of only about 20 nm [1], whereas thickened bifilms have been characterized as being as much as three orders of magnitude greater in thickness [5, 7]. The old films are stated to originate from the growth of "new" alumina films [1, 5, 7, 17]. Such growth can occur from the thickening of new bifilms that have been entrained at any stage during liquid metal processing, and such accumulation of new bifilms is considered to be additive with each stage of molten processing [1, 7]. Oxide film growth is usually attributed to exposure of the alloy to elevated temperatures following the casting of ingots [7]. Such thickening is also theorized to coincide with a conversion of the oxide structure from alumina to spinel [1, 17]. The new alumina bifilms can therefore function as a type of precursor for the formation of "old" thicker spinel bifilms. New bifilms are difficult to characterize, therefore when bifilms are observed in the literature it is most often thick spinel bifilms that are being characterized [5, 7]. Characterization of new bifilms is rarely reported, with only a few exceptions [1, 16].

Several publications report degradation of numerous material properties due to the formation of bifilms [1-6, 8, 9, 11, 14, 17-24]. This includes a reduction of tensile strength, bend strength, impact strength, creep strength, ductility, and fatigue life. This also includes an increase in casting porosity, hot tearing, scatter in mechanical test data, inclusions, secondary phase embrittlement, and intergranular embrittlement. In addition, there have been reported predictions of astounding increases in mechanical properties if bifilms can be successfully avoided [1, 2, 5, 23]. For this reason, there has been significant interest in methods of bifilm prevention and/or bifilm remediation. One suggested method of bifilm prevention, particularly in Ni-base superalloy components, is to bottom fill investment castings [6]. Reports suggest that a properly designed bottom-gated casting would eliminate turbulence during casting, therefore preventing bifilm creation. A suggested method for reducing the impact bifilms have on mechanical properties is hot isostatic pressing (HIP) [1, 8, 9, 11]. Reports suggest that the use of HIP can cause perforations in the oxide structure, allowing for metal bonding across the oxide film, and can reduce the effect of bifilms on properties. This process is sometimes referred to as bifilm "healing".

Since bifilms have a strongly negative impact on tensile and fatigue properties, bifilms are often located by fractographic analysis of mechanical test specimens [1, 5, 7-9, 12, 17]. The existence of the thin alumina films is frequently determined by inference. For example, planar surfaces seen during fractographic analysis can in many cases be considered the result of cracking along the bilayer of oxide, effectively splitting the oxide bifilm [1, 5, 7-9, 17]. This is a common method seen in the literature for identifying the presence of new bifilms during fractographic analysis. A wide variety of cracking observed during microstructural analysis is also frequently attributed to a similar mechanism of bifilm splitting [1, 2, 5, 7-9, 11, 12]. However, sufficient evidence to support these conclusions is currently limited [12, 25]. Furthermore, quantitative characterization beyond the observation of old bifilms identified through fractographic analysis is also limited [5, 7]. Other locations expected to exhibit new bifilms such as porosity, cleavage planes, grain boundaries, dendrite arms, etc. are observed and also often assumed to be due to bifilm formation [1, 8, 9, 12, 13, 15, 17]. Therefore, extensive and detailed characterization of such locations to indicate the presence of an oxide film is imperative to confirming the existence of oxide bifilms.

Experimental Procedure

Casting/Processing

The alloy selected for investigation was polycrystalline investment cast IN100.

Table I: Nominal composition of IN100 (wt%) [26]

Cr	Ni	Co	Mo	Ti	Al	Fe	C	B	Zr	V
10	Bal.	15	3.0	4.7	5.5	<0.6	0.15	0.015	0.06	1.0

A wide variety of processing and casting parameters were investigated to determine the effect of each on material properties and inclusion formation. A full sample matrix can be seen in Table III. Charges were melted by induction and investment cast into cluster molds. Each mold contained both cylindrical bar samples to be machined into mechanical test specimens, and generic power turbine blades to be used for further microstructural characterization. Vacuum casting was conducted using standard VIM and investment casting procedures. Melting and casting atmospheres were carefully controlled and maintained at less than 15 mTorr. Argon casting was conducted by backfilling the melt chamber with Ar to a pressure of 10 Torr. Air casting was achieved by casting in atmospheric air. For each casting environment, samples were either top filled or bottom filled. Top filled samples utilized a gating design which would allow the molten metal to turbulently fill the mold cavity from the top down. Bottom filled samples were gated to fill the mold cavity from the bottom up to minimize turbulence. In addition, some vacuum cast samples employed turbulent gating, where turbulators were installed in the gating of the mold setup to induce additional turbulence. An extruded ceramic filter was utilized for some air cast samples. Castings were provided by Alcoa Howmet Research Center in Whitehall, MI.

Some samples from each casting condition were hot isostatically pressed (HIP) and heat treated. Industry standard HIP and solution heat treatment cycles were used and can be seen in Table II. The remaining samples were examined in both the as cast and heat treated condition.

Table II: Sample processing parameters

Processing	Procedure
HIP	2200 °F, 15 KSI, 4hrs
Solution Heat Treatment	1975 °F in vacuum, hold 4hrs; gas fan cool to 1600 °F, hold 12hrs; air cool to room temperature

Mechanical Testing

Cylindrical bar castings where machined into threaded tensile samples and threaded high cycle fatigue (HCF) samples. Tensile samples were pulled to failure, and 0.2% yield stress, ultimate tensile stress, and elongation were measured. Fracture surfaces from each tensile bar pulled to failure were removed and mounted for fractographic analysis. The opposite half of each tensile bar was cross-sectioned along the gauge length, then mounted and polished for metallographic analysis, allowing for a transverse examination of the microstructure near the fracture surface. Samples prepared for metallographic analysis were swab etched with γ′ etchant Pratt & Whitney #17 (100 mL H_20 + 100 mL HCl + 100 mL HNO_3 + 3 g MoO_3). Fractographic and metallographic characterization was conducted using scanning electron microscopy (SEM) and energy dispersive x-ray spectroscopy (EDS).

All HCF samples were subjected to stress-controlled HCF testing at room temperature until failure, and the number of cycles to failure was recorded. HCF was conducted using a tension-tension stress condition with σ_{max}=690 MPa, σ_{min}=69 MPa, $\Delta\sigma$=621 MPa, and R=0.1. A triangular waveform was used with a frequency of 15 Hz. Fracture surfaces from each bar fatigued to failure were removed and mounted for fractographic analysis via SEM.

Table III: Complete Sample Matrix

Environment	Filtering	Fill Condition	Gating	Processing	Heat Treatment	Tensile Bars (#)	Fatigue Bars (#)	Blades (#)
Vacuum	No Filter	Top Fill	Conventional	HIP	Heat Treat	6	6	18
				No HIP	Heat Treat	2	6	6
			Turbulent	HIP	Heat Treat	2	4	4
				No HIP	Heat Treat	2	4	6
		Bottom Fill	Conventional	HIP	Heat Treat	4	4	10
				No HIP	Heat Treat	2	4	6
			Turbulent	HIP	Heat Treat	2	4	4
				No HIP	Heat Treat	2	3	6
10 Torr Argon	No Filter	Top Fill	Conventional	HIP	Heat Treat	4	4	14
				No HIP	Heat Treat	2	4	4
		Bottom Fill	Conventional	HIP	Heat Treat	4	4	10
				No HIP	Heat Treat	2	4	6
Air	No Filter	Top Fill	Conventional	HIP	Heat Treat	2	4	8
				No HIP	Heat Treat	0	4	0
					As Cast	2	6	3
		Bottom Fill	Conventional	HIP	Heat Treat	1	0	4
	Filter	Top Fill	Conventional	HIP	Heat Treat	4	4	0
				No HIP	Heat Treat	4	4	0
					As Cast	4	4	6

Auger Analysis

Cast blades were machined by electrical discharge machining (EDM) to remove several small Auger pin samples from each blade. The Auger pins were cylindrical notched samples 20.3 mm in length and 3.18 mm in diameter. A blunt notch was machined 11.4 mm from the top of the sample, and to a depth of 0.51 mm. Each Auger pin was loaded into a PHI 660 scanning Auger microscope and fractured in-situ with an actuating hammer. In-situ fracture occurred in ultra-high vacuum conditions of approximately 0.3-0.6 nTorr. Samples were immediately analyzed fractographically.

Results and Discussion

Tensile Testing

There was no discernible improvement in tensile properties from bottom filling investment castings (Figures 1-3). In either condition, fracture was dominated by the formation of microvoids in the presence of cracked M(Ti, Mo)C carbides. Carbide chemistry was confirmed by EDS analysis. Examples of carbide cracking observed during fractographic and metallographic analysis is shown in Figure 4. Casting atmosphere did exhibit an effect on tensile properties. In particular, air casting increased scatter and decreased ductility. This is likely due to the presence of nonmetallic inclusions (NMIs). SEM and EDS analysis indicates that the inclusions observed in air cast samples are heterogeneous in structure, consisting of stringy M(Ti,Mo)C carbide structures and highly discontinuous alumina. Examples of NMI structures observed during fractographic and metallographic analysis is shown in Figure 5. Turbulators in the gating of castings did not exhibit a significant effect on tensile properties. Microstructure and fracture mechanisms were unaffected by turbulators. Filtering also did not have a significant effect on tensile properties (Figures 6 and 7).

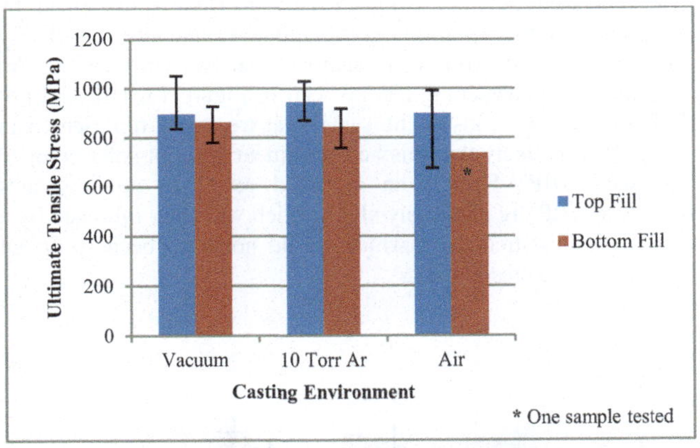

Figure 1. Mean UTS across fill conditions and casting atmospheres. Error bars representative of data set range.

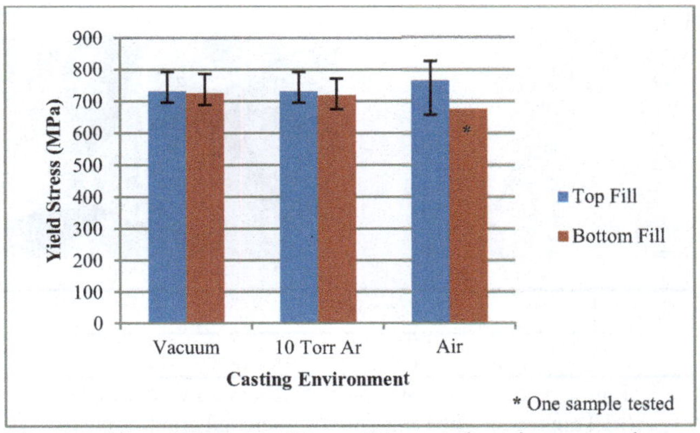

Figure 2. Mean YS across fill conditions and casting atmospheres. Error bars representative of data set range.

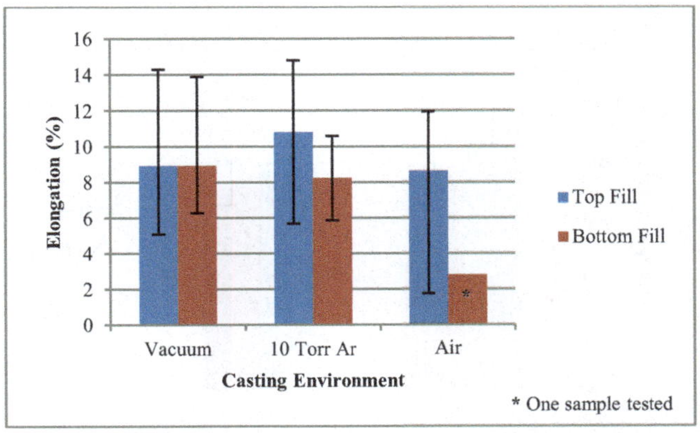

Figure 3. Mean elongation across fill conditions and casting atmospheres. Error bars representative of data set range.

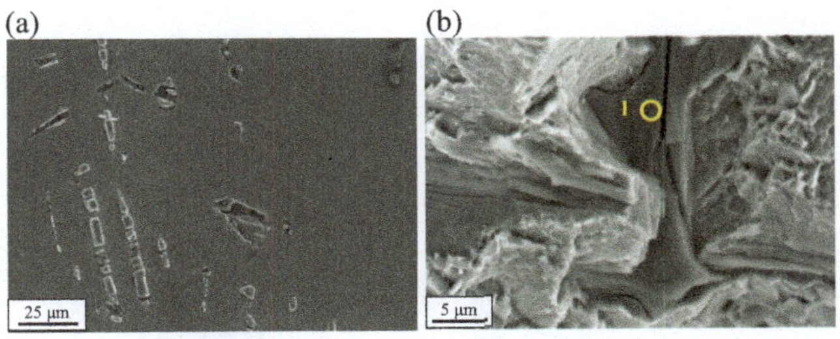

(c)

Spectrum (at%)	*C*	*Ti*	*Mo*	*Total*
1	44.18	46.54	9.29	100

Figure 4. Examples of typical carbide cracking. (a) Transverse SEM imaging of vacuum cast sample; (b) Fractographic SEM imaging of argon cast sample; (c) Quantitative EDS of spot indicated, matrix signal omitted

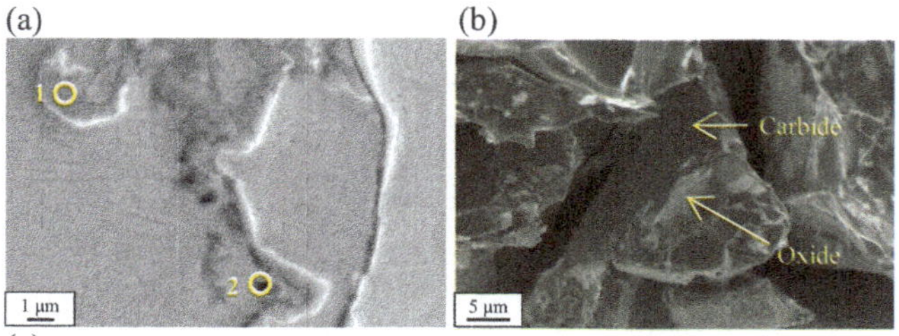

(c)

Spectrum (at%)	*C*	*O*	*Al*	*Mo*	*Ti*	*V*	*Cr*	*Co*	*Ni*
1	50.03	-----	4.81	2.52	23.31	1.98	2.71	3.88	10.76
2	7.29	56.18	13.69	1.41	17.03	1.36	-----	-----	3.04

Figure 5. Examples of typical NMIs in air cast sample. (a) Transverse SEM imaging of unetched condition; (b) Fractographic SEM imaging; (c) Quantitative EDS of spots indicated

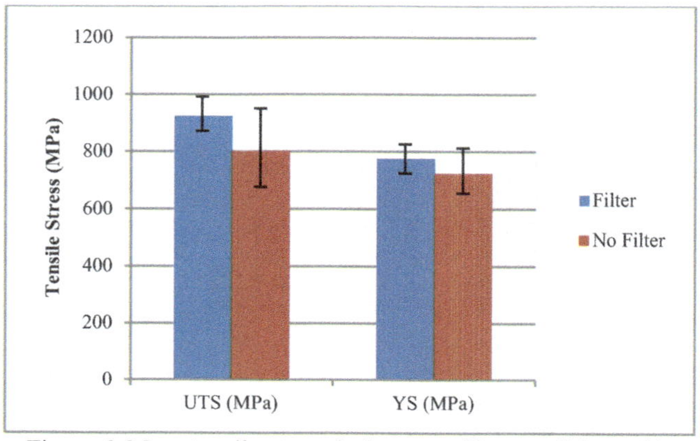

Figure 6. Mean tensile strengths between filter conditions. Air cast samples. Error bars representative of data set range.

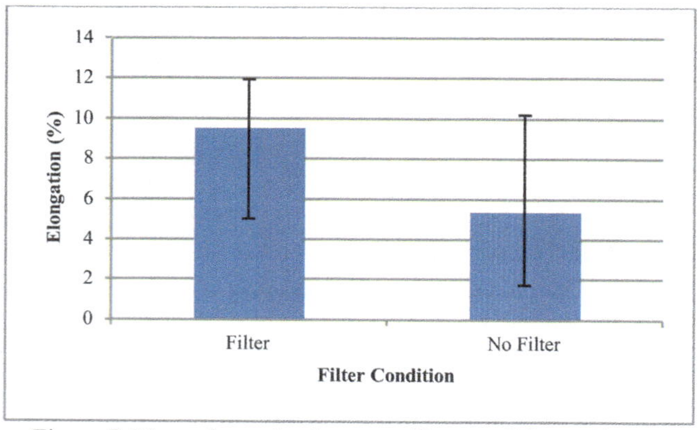

Figure 7. Mean elongation between filter conditions. Air cast samples. Error bars representative of data set range.

The results of this study do not indicate that tensile strength trends strongly with relative amounts of casting turbulence (Figures 8 and 9). Instead, it seems clear that of the casting conditions investigated, the most critical was oxygen content of the casting atmosphere. Excessive oxygen content led to the formation of deleterious non-metallic inclusions.

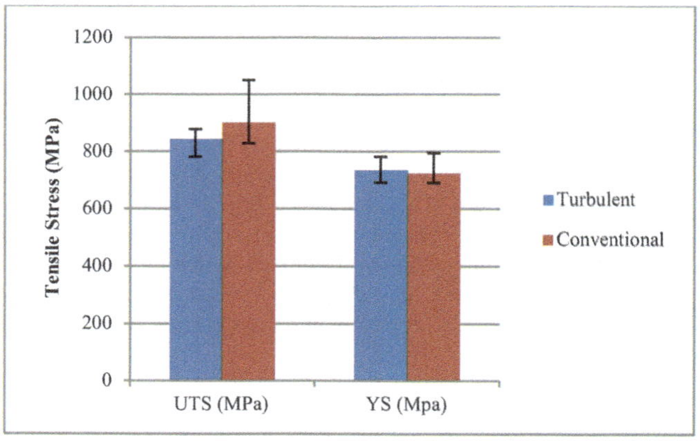

Figure 8. Mean tensile strengths between gating conditions. Vacuum cast samples. Error bars representative of data set range.

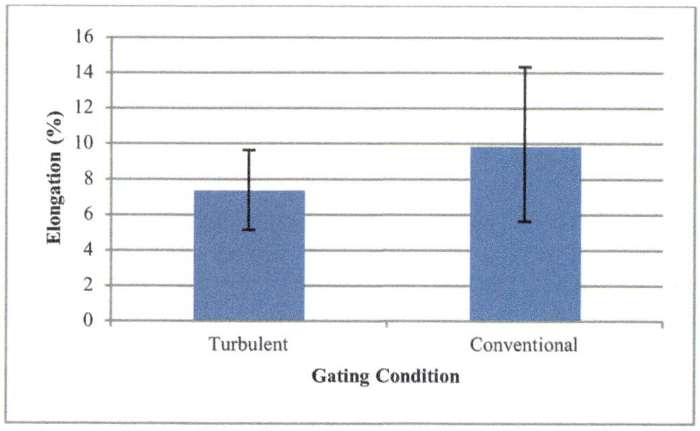

Figure 9. Mean elongation between gating conditions. Vacuum cast samples. Error bars representative of data set range.

Utilization of HIP had a discernible effect on tensile properties. HIP followed by heat treatment generated reduced yield strength and increased ductility (Figures 10-12). This is explained by softening due to γ' coarsening. Precipitate coarsening was clearly observed microstructurally in samples that had been HIPed. An example of γ' coarsening can be seen in Figure 13. HIP and no-HIP samples were given the same heat treatment represented in Table II. It is likely that this heat treatment was not solutioning γ' following HIP. This is an expected result as cooling rates following HIP are extremely slow, which was then followed by a subsolvus heat treatment, which would not have been sufficient temperatures to resolution γ'.

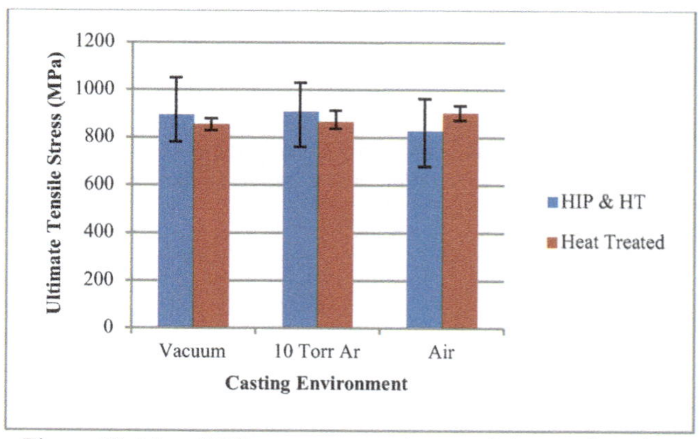

Figure 10. Mean UTS across processing conditions and casting atmospheres. Error bars representative of data set range.

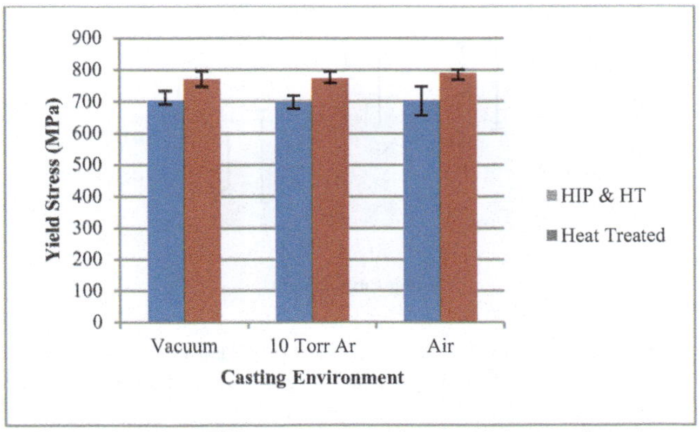

Figure 11. Mean YS across processing conditions and casting atmospheres. Error bars representative of data set range.

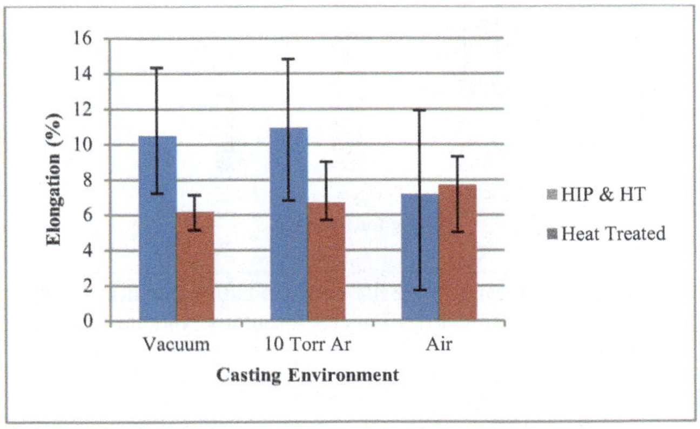

Figure 12. Mean elongation across processing conditions and casting atmospheres. Error bars representative of data set range.

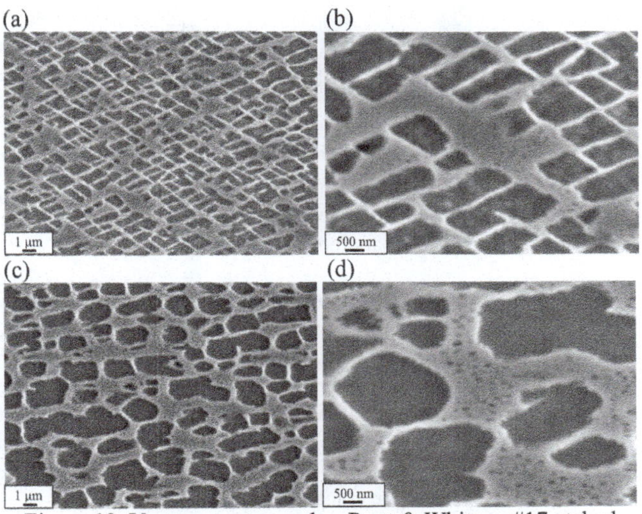

Figure 13. Vacuum cast samples, Pratt & Whitney #17 etched. (a, b) Typical heat treated microstructure; (c, d) Typical HIP and heat treated microstructure

The oxide inclusion structure was examined fractographically by SEM for air castings in the "HIP and heat treat" and "as cast" conditions. There was no discernible difference between the nonmetallic inclusions found in either condition. Both consisted of a heterogeneous M(Ti, Mo)C carbide and discontinuous alumina structure, as seen in Figures 14 and 15. Composition was confirmed by EDS. It has been theorized that such oxide inclusions possess a continuous morphology following casting, and are fractured during HIP [8, 9]. These observations indicate that the oxide structure is discontinuous regardless of HIP. The discontinuity is observable on the microscale, where the oxide appears patchy. In addition, nanoscale voids in the oxide structure are observable in Figures 14 and 15. The implication of such discontinuities is that the oxide would not be capable of "inflating" as had been previously theorized [1].

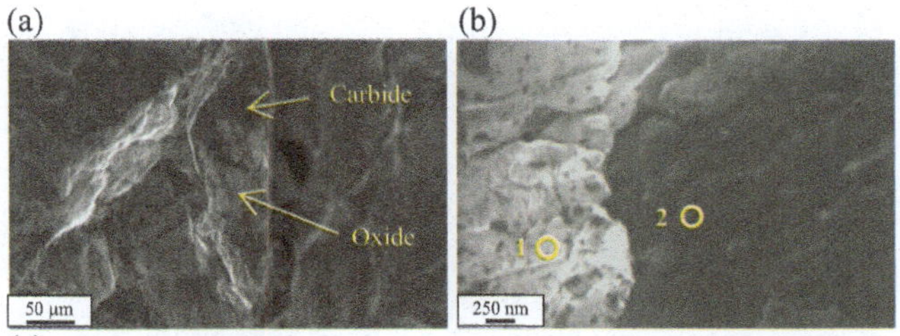

(c)

Spectrum (at%)	C	O	Al	Mo	Ti	V	Cr	Co	Ni
1	16.17	58.35	13.25	1.17	5.75	0.33	0.87	0.94	3.17
2	68.05	-----	4.20	2.02	11.00	0.59	2.02	1.85	10.28

Figure 14. NMI in HIP and heat treated air cast sample. (a) SEM fractography showing inclusion discontinuity; (b) EDS location showing micro and nanoscale gaps in oxide structure; (c) Quantitative EDS of spots indicated

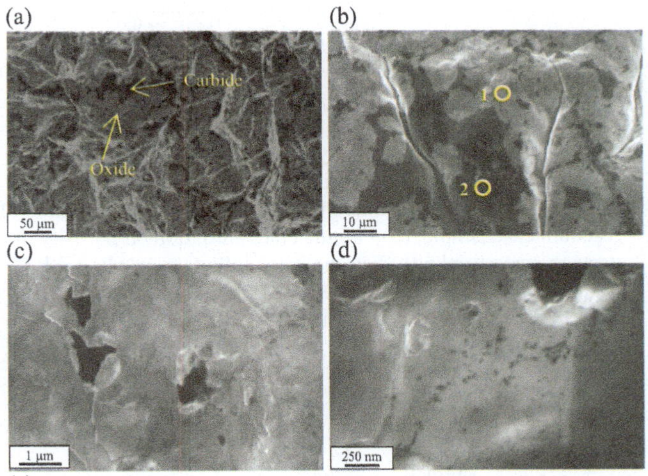

(e)

Spectrum (at%)	C	O	Al	Mo	Ti	V	Cr	Co	Ni
1	4.99	28.40	13.54	1.48	9.91	1.12	4.90	7.01	28.64
2	38.58	-----	0.75	5.95	34.81	2.20	2.45	2.86	12.41

Figure 15. NMI in as cast air cast sample. (a) SEM fractography showing inclusion discontinuity; (b) EDS location; (c) Micro-scale oxide discontinuities (d) Nano-scale oxide discontinuities (e) Quantitative EDS of spots indicated

Fatigue Testing

There was no discernible improvement in fatigue life from bottom filling investment castings. There was also no significant degradation of fatigue properties from the utilization of turbulators. For vacuum or argon cast samples in the no-HIP condition, fatigue crack initiation occurred by slip band formation in the presence of near-surface M(Ti, Mo)C carbides, near-surface dendritic casting porosity, or both. For vacuum or argon cast samples in the HIPed condition, fatigue cracking only initiated at near-surface M(Ti, Mo)C carbides. Carbide chemistry was confirmed by EDS analysis. An example of each fatigue crack initiation mechanism observed during fractographic analysis is shown in Figure 16. Casting atmosphere did have an effect on fatigue life (Figures 17-19). Air cast samples exhibited greater relative scatter and reduced fatigue life. Filtering air cast samples did not have a significant impact on fatigue life (Figure 20).

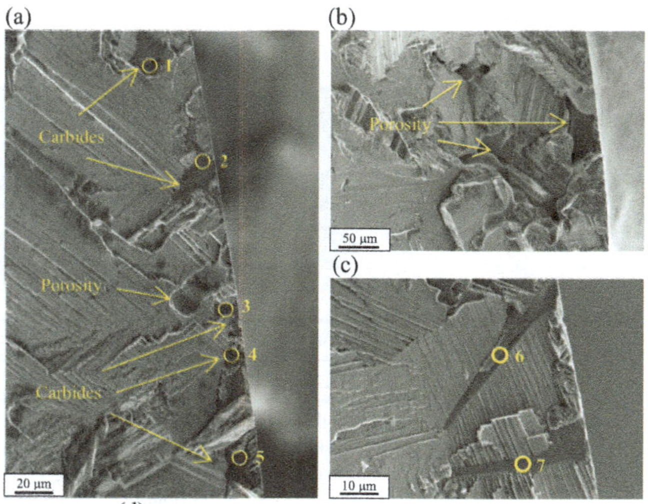

(d)

Spectrum (at%)	*C*	*Ti*	*V*	*Mo*
1	69.50	21.23	2.25	7.01
2	56.25	31.65	2.93	9.17
3	64.82	26.28	1.60	7.29
4	67.55	24.07	1.34	7.05
5	65.60	26.00	1.51	6.89
6	54.84	32.60	2.80	9.76
7	69.90	20.92	2.26	6.92

Figure 16. Fatigue crack initiation examples in no-HIP samples. (a) Combined M(Ti, Mo)C and porosity fatigue crack initiation of vacuum cast sample; (b) Fatigue crack initiation on porosity of vacuum cast sample; (c) Fatigue crack initiation on M(Ti, Mo)C carbides of argon cast sample; (d) Quantitative EDS of spots indicated

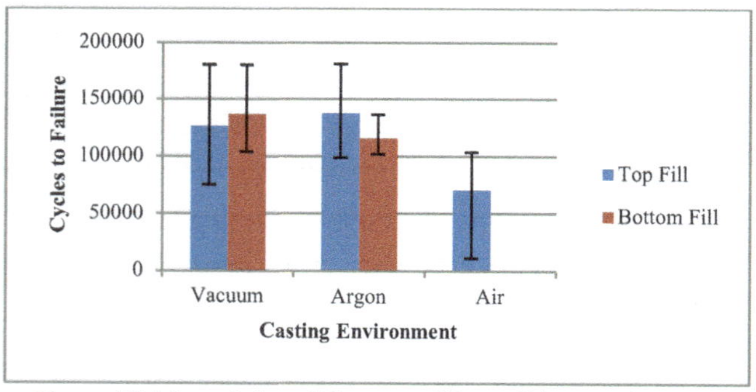

Figure 17. Mean fatigue life across fill conditions and casting atmospheres. Error bars representative of data set range.

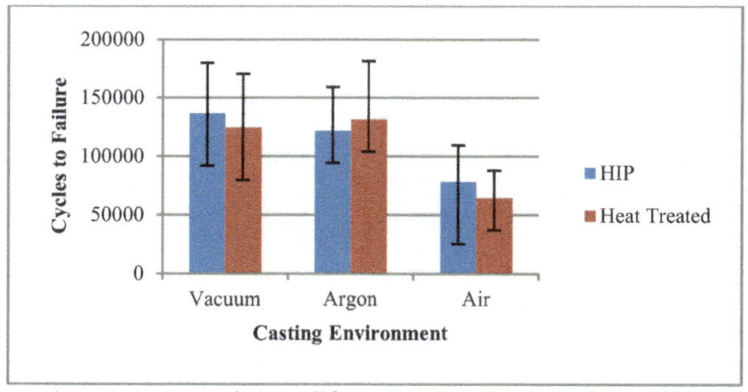

Figure 18. Mean fatigue life across processing conditions and casting atmospheres. Error bars representative of data set range.

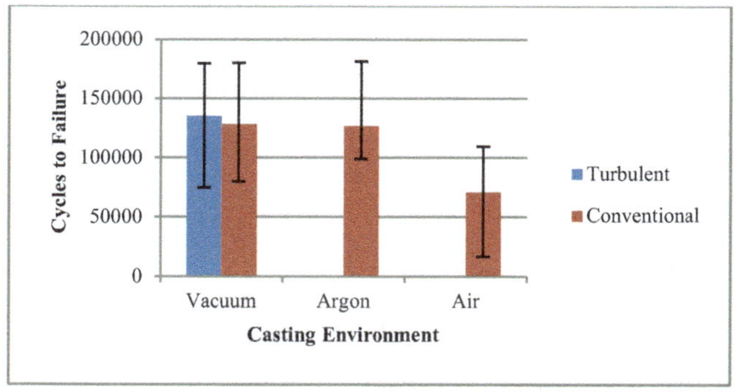

Figure 19. Mean fatigue life between gating conditions and casting atmospheres. Error bars representative of data set range.

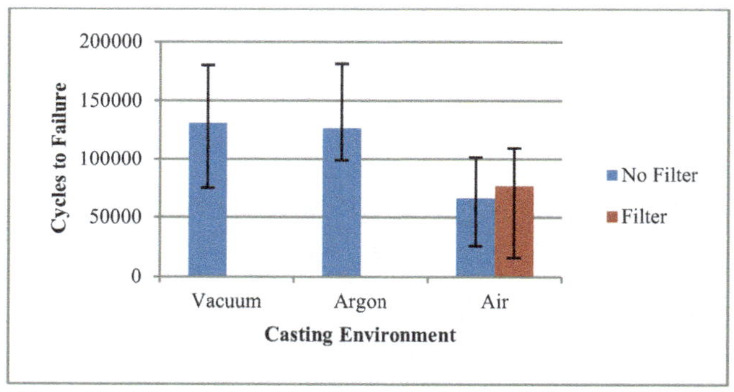

Figure 20. Mean fatigue life between filter conditions and casting atmospheres. Error bars representative of data set range.

The results of this study do not indicate that fatigue life is strongly related to casting turbulence. Instead, it seems clear that of the casting conditions investigated, the most critical was oxygen content of the casting atmosphere. Interestingly, HIP did not have a significant effect on fatigue life. This is likely related to the fact that for many no-HIP samples, the initiation site did not exhibit porosity. This suggests that near-surface porosity is not always the optimal location for fatigue crack initiation.

Auger Analysis

At least one blade from each blade sample condition (Table III) was used to machine Auger pins. Pins from each condition were used for in-situ fractographic Auger analysis. AES spectra were collected from fractographic regions of interest including: Cleavage planes, dendritic porosity, cracked carbides, fracture surface facets, and non-metallic inclusions (Figures 21-24). AES results did not indicate the presence of bifilms in any sample condition. Oxide was only detected on the surface of the alumina structure of NMIs found in pins machined from air cast samples. Sulfur was detected on the surface of dendritic casting porosity (Figures 24 and 25).

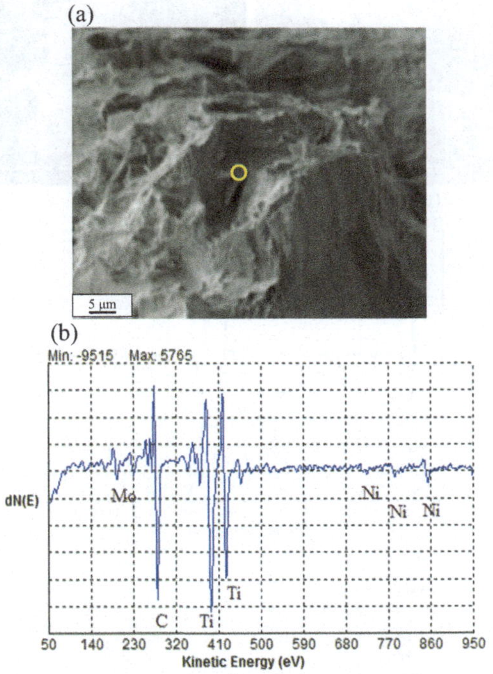

Figure 21. Fractured Auger pin from top filled air cast blade. (a) Cracked carbide in microvoid; (b) Corresponding AES spectrum

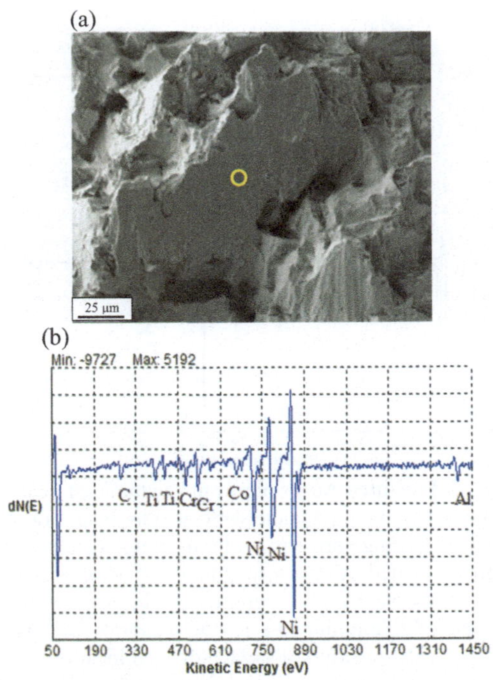

Figure 22. Fractured Auger pin from top filled air cast blade. (a) Cleavage plane; (b) Corresponding AES spectrum

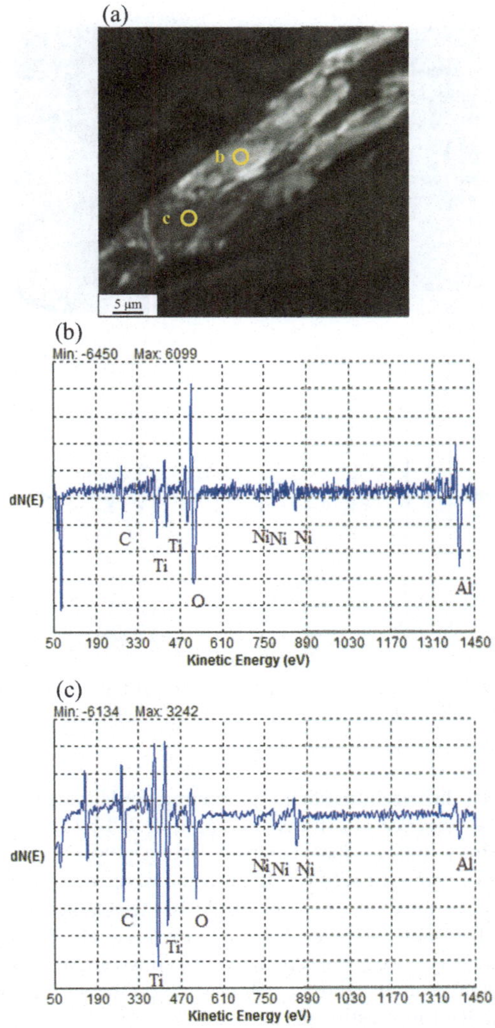

Figure 23. Fractured Auger pin from top fill air cast blade. (a) NMI exhibiting discontinuous oxide structure; (b) Corresponding AES spectrum taken from oxide region; (c) Corresponding AES spectrum taken from carbide region

It has been theorized that MC carbide cracking seen in Ni-based alloys is also due to the presence of bifilms [1, 5, 7]. The bifilm would behave as a heterogeneous nucleation site for the carbide, such that carbide cracking observed in fracture samples can be due to bifilm splitting. In this study, oxide films were not observed on the surface of cracked carbides, indicating that they would not have nucleated in this manner. Cleavage planes can, in some cases, be used as an indication that bifilm splitting has occurred [1, 5, 7-9, 17]. In this study, the cleavage planes observed did not exhibit oxide films. The oxide inclusions that were detected were similar to those seen in SEM fractography of tensile samples. Auger analysis indicates that this is a heterogeneous structure, consisting of discrete locations of oxide on M(Ti, Mo)C carbide.

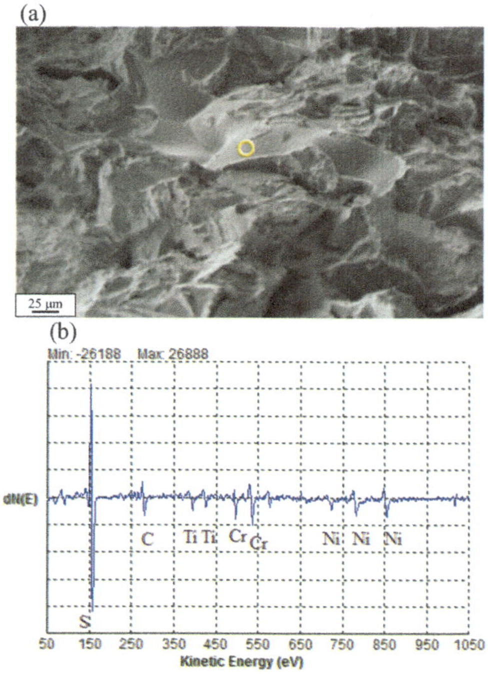

Figure 24. Fractured Auger pin from top filled air cast as cast blade. (a) Dendritic casting porosity; (b) Corresponding AES spectrum

It has been theorized that dendritic casting porosity can be formed due to the presence of bifilms [1-3, 5, 6, 8, 9, 12-16]. The results of this study indicate that dendritic casting porosity investigated did not form due to bifilms. If porosity were to form from the expansion of an oxide film, it should be detectable on the pore surface. This was not the case for the samples analyzed. Instead, a continuous layer of sulfur was observed. Interestingly, this sulfur layer was ubiquitous, and was observable on the surface of every analyzed dendritic casting pore.

It has been theorized that non-metallic inclusions observed in Ni-based superalloy castings originate from the thickening of a new bifilm [7]. This thickening is due to exposure of the material to elevated temperatures during alloy manufacture, with the bifilm acting as a precursor to the formation of more classical oxide inclusions. In order to properly state that oxide inclusions found in such samples are due to thickening of bifilms, the new bifilm precursor would need to be characterized. The absence of bifilms in the in-situ fractographic Auger samples indicates that the oxide inclusions characterized in this study likely did not form due to the thickening of bifilms.

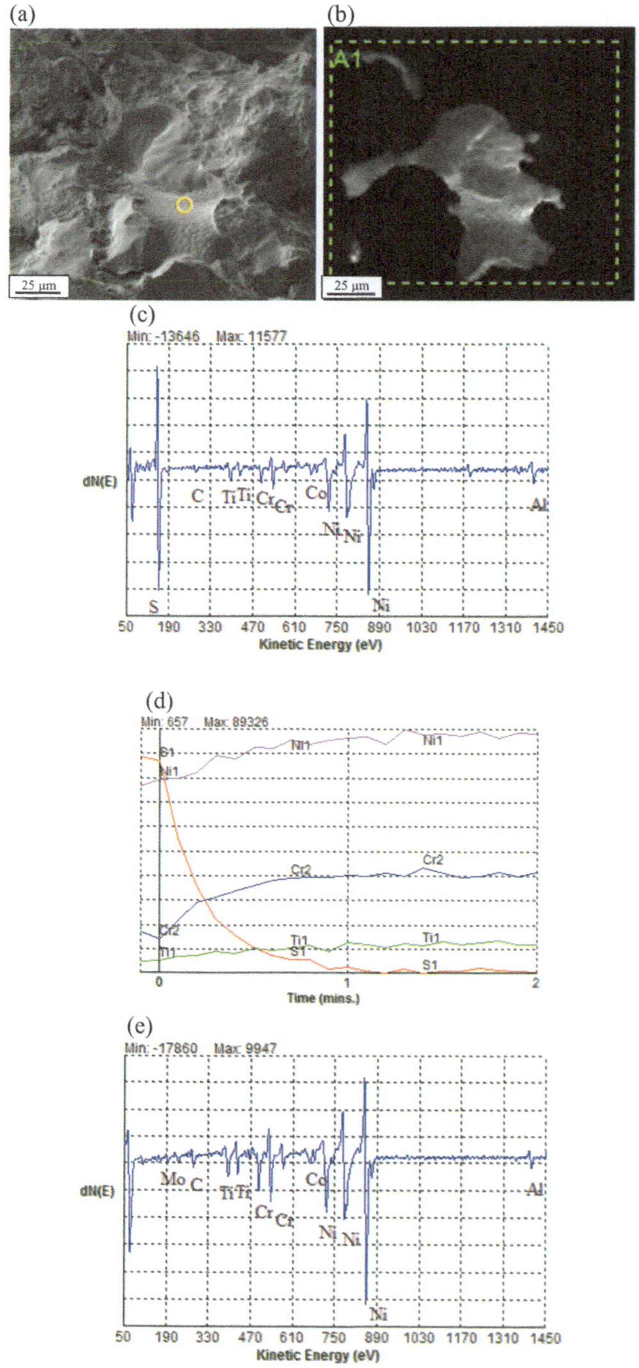

Figure 25. Fractured Auger pin from vacuum cast heat treated sample. Marked region is the same area in images (a) & (b). (a) Dendritic casting porosity; (b) Elemental Auger map of sulfur for region indicated; (c) Corresponding AES spectrum; (d) depth profile of spot indicated in (a). Sputter conditions used produced a sputter rate of 2.5nm/min on a Ta_2O_5 standard; (e) Post-sputter AES spectrum of same indicated location

Conclusions

Tensile properties of IN100 castings were not strongly influenced by casting turbulence. HIP has a softening effect, likely due to improper γ' resolutioning during subsolvus heat treatment following HIP. Air casting had a deleterious effect on tensile properties, increasing scatter and decreasing ductility. Tensile failure was due to microvoid formation on cracked M(Ti, Mo)C carbides. Non-metallic inclusions observed in tensile samples exhibited a heterogeneous morphology, consisting of highly discontinuous alumina on stringy M(Ti, Mo)C carbide. These inclusions likely did not form due to the thickening of a new bifilm precursor. Fatigue life was not strongly influenced by casting turbulence. Fatigue crack initiation occurred due to slip band formation on near-surface porosity, near surface M(Ti, Mo)C carbides, or both. Auger analysis indicates that in these samples carbide cracking, cleavage plane formation, and dendritic casting porosity was not due to the presence of bifilms. A sulfur nano-layer was observed on the surface of dendritic casting pores. The findings of this study conflict strongly with previous reports pertaining to bifilms. For this reason, substantial further investigation is necessary. It is suggested that characterization techniques with nanoscale spatial resolution or surface sensitivity be used for future efforts to characterize oxide bifilms.

Acknowledgements

The authors would like to thank our collaborators at the Alcoa Howmet Research Center, Whitehall, MI, and the University of Florida Major Analytical and Instrumentation Center.

References

[1] Campbell, J., 2006, "Entrainment defects," Materials Science and Technology, 22(2), pp. 127-145.

[2] Campbell, J., 2012, "Stop Pouring, Start Casting," International Journal of Metalcasting, 6(3), pp. 7-18.

[3] Dispinar, D., and Campbell, J., 2011, "Porosity, hydrogen and bifilm content in Al alloy castings," Materials Science and Engineering a-Structural Materials Properties Microstructure and Processing, 528(10-11), pp. 3860-3865.

[4] Dispinar, D., Akhtar, S., Nordmark, A., Di Sabatino, M., and Arnberg, L., 2010, "Degassing, hydrogen and porosity phenomena in A356," Materials Science and Engineering a-Structural Materials Properties Microstructure and Processing, 527(16-17), pp. 3719-3725.

[5] Campbell, J., and Tiryakioglu, M., 2012, "Bifilm Defects in Ni-Based Alloy Castings," Metallurgical and Materials Transactions B-Process Metallurgy and Materials Processing Science, 43(4), pp. 902-914.

[6] Li, D., Campbell, J., and Li, Y., 2004, "Filling system for investment cast Ni-base turbine blades," Journal of Materials Processing Technology, 148(3), pp. 310-316.

[7] Rashid, A., and Campbell, J., 2004, "Oxide defects in a vacuum investment-cast Ni-based turbine blade," Metallurgical and Materials Transactions a-Physical Metallurgy and Materials Science, 35A(7), pp. 2063-2071.

[8] Staley, J. T., Tiryakioglu, M., and Campbell, J., 2007, "The effect of hot isostatic pressing (HIP) on the fatigue life of A206-T71 aluminum castings," Materials Science and Engineering a-Structural Materials Properties Microstructure and Processing, 465(1-2), pp. 136-145.

[9] Staley, J. T., Tiryakioglu, M., and Campbell, J., 2007, "The effect of increased HIP temperatures on bifilms and tensile properties of A206-T71 aluminum castings," Materials Science and Engineering a-Structural Materials Properties Microstructure and Processing, 460, pp. 324-334.

[10] Tiryakioglu, M., 2010, "On fatigue life variability in cast Al-10%Si-Mg alloys," Materials Science and Engineering a-Structural Materials Properties Microstructure and Processing, 527(6), pp. 1560-1564.

[11] Tiryakioglu, M., Staley, J. T., and Campbell, J., 2008, "The effect of structural integrity on the tensile deformation characteristics of A206-T71 alloy castings," Materials Science and Engineering a-Structural Materials Properties Microstructure and Processing, 487(1-2), pp. 383-387.

[12] Campbell, J., 2008, "Evidence for entrainment defects," Materials Science and Technology, 24(7), pp. 875-879.

[13] Dispinar, D., and Campbell, J., 2006, "Effect of casting conditions on aluminium metal quality," Journal of Materials Processing Technology, 182(1-3), pp. 405-410.

[14] Dai, X., Yang, X., Campbell, J., and Wood, J., 2003, "Effects of runner system design on the mechanical strength of Al-7Si-Mg alloy castings," Materials Science and Engineering a-Structural Materials Properties Microstructure and Processing, 354(1-2), pp. 315-325.

[15] Fox, S., and Campbell, J., 2000, "Visualisation of oxide film defects during solidification of aluminium alloys," Scripta Materialia, 43(10), pp. 881-886.

[16] Felberbaum, M., and Rappaz, M., 2011, "Curvature of micropores in Al-Cu alloys: An X-ray tomography study," Acta Materialia, 59(18), pp. 6849-6860.

[17] Wang, Q., Davidson, C., Griffiths, J., and Crepeau, P., 2006, "Oxide films, pores and the fatigue lives of cast aluminum alloys," Metallurgical and Materials Transactions B-Process Metallurgy and Materials Processing Science, 37(6), pp. 887-895.

[18] Bozchaloei, G., Varahram, N., Davami, P., and Kim, S., 2012, "Effect of oxide bifilms on the mechanical properties of cast Al-7Si-0.3Mg alloy and the roll of runner height after filter on their formation," Materials Science and Engineering a-Structural Materials Properties Microstructure and Processing, 548, pp. 99-105.

[19] Eisaabadi, G., Davami, P., Kim, S., and Varahram, N., 2012, "Effects of hydrogen and oxides on tensile properties of Al-Si-Mg

cast alloys," Materials Science and Engineering a-Structural Materials Properties Microstructure and Processing, 552, pp. 36-47.

[20] Tiryakioglu, M., Campbell, J., and Nyahumwa, C., 2011, "Fracture Surface Facets and Fatigue Life Potential of Castings," Metallurgical and Materials Transactions B-Process Metallurgy and Materials Processing Science, 42(6), pp. 1098-1103.

[21] Rincon, E., Lopez, H., Cisneros, M., and Mancha, H., 2009, "Temperature effects on the tensile properties of cast and heat treated aluminum alloy A319," Materials Science and Engineering a-Structural Materials Properties Microstructure and Processing, 519(1-2), pp. 128-140.

[22] Tiryakioglu, M., Campbell, J., and Staley, J., 2003, "The influence of structural integrity on the tensile deformation of cast Al-7wt.%Si-0.6wt.%Mg alloys," Scripta Materialia, 49(9), pp. 873-878.

[23] Campbell, J., 2011, "The Origin of Griffith Cracks," Metallurgical and Materials Transactions B-Process Metallurgy and Materials Processing Science, 42(6), pp. 1091-1097.

[24] Lee, C. D., 2013, "Variability in the impact properties of A356 aluminum alloy on microporosity variation," Materials Science and Engineering: A, 565(0), pp. 187-195.

[25] Knott, J., Beeley, P. R., Griffiths, J. R., Green, N. R., Newton, C. J., and Campbell, J., 2006, "Commentaries on 'Entrainment defects' by J Campbell," Materials Science and Technology, 22(8), pp. 999-1008.

[26] Furrer, D. U., and Semiatin, S. L., 2005, "Forging of Nickel-Base Alloys," ASM Handbook, ASM International, pp. 324-330.

Vacuum Arc Remelting

LMPC 2013

Liquid Metal Processing & Casting 2013

Proceedings of the 2013 International Symposium on Liquid Metal Processing & Casting
Edited by: Matthew J. M. Krane, Alain Jardy, Rodney L. Williamson, and Joseph J. Beaman
TMS (The Minerals, Metals & Materials Society), 2013

APPLICATION OF A MODEL FOR SIMULATING THE VACUUM ARC REMELTING PROCESS IN TITANIUM ALLOYS

Ashish Patel, David W. Tripp, and Daniel Fiore
TIMET, 900 Hemlock Road, Morgantown Business Park, Morgantown PA 19543

Keywords: Modeling, VAR

Abstract

Mathematical modeling is routinely used in the process development and production of advanced aerospace alloys to gain greater insight into system dynamics and to predict the effect of process modifications or upsets on final properties. This article describes the application of a 2-D mathematical VAR model presented in previous LMPC meetings. The impact of process parameters on melt pool geometry, solidification behavior, fluid-flow and chemistry in Ti-6Al-4V ingots will be discussed. Model predictions were first validated against the measured characteristics of industrially produced ingots, and process inputs and model formulation were adjusted to match macro-etched pool shapes. The results are compared to published data in the literature. Finally, the model is used to examine ingot chemistry during successive VAR melts.

Introduction

The basic process of VAR is the continuous melting of a consumable electrode by means of a direct current arc in a vacuum on the order of 0.1 to 1 Pa. In titanium the consumable electrode is manufactured using compacted sponge and master alloys or is the output of a previous melt cycle (either VAR or Cold Hearth melted) /1/. The consumable electrode is suspended in a water cooled copper crucible. A DC electric arc is struck between the bottom of the VAR crucible (stool or baseplate) and the bottom of the electrode. The electrode melts creating an ingot within the crucible. For titanium and titanium alloys, currents can range as high as 35 kA with arc voltages typically between 24 and 45 volts.

Titanium VAR melting furnaces have an external magnetic coil installed around the water jacket as shown in Figure 1. This magnetic coil is used to control the motion of the arc and prevent its attachment to any particular location on the water cooled crucible wall. Any such prolonged attachment could breach the crucible and produce a catastrophic failure of the furnace by either steam or hydrogen explosion. /2/. Interacting with the current in the furnace, the induced magnetic field also stirs the liquid which can impact segregation during solidification.

The mathematical model, SOLAR, (SOLidification during Arc Remelting) is routinely used at TIMET for new process development and process improvement efforts. The model was developed at the École des Mines, Nancy, and over the past decade, several papers have been published (including LMPC proceedings) describing the model and highlighting its capabilities. In short, the ingot growth during VAR is simulated using a two-dimensional, axi-symmetric representation. The model computes the current, temperature, turbulent flow and species distribution in the ingot by iteratively solving the appropriate governing equations with prescribed boundary conditions using the finite volume method. /3-4/.

The primary focus of this paper is validation of this model using available data from published literature /5-6/. Specifically, the predicted pool profile is compared against previously reported measurements from 0.914 m (36 inch) and 0.864 m (34 inch) diameter Ti-6Al-4V ingots. Nominal thermo-physical properties were used in these simulations /7/.

Results

VAR ingot 1

As reported /5/, a 0.914 m dia. ingot was melted at a constant current of 33 kA and voltage of 44.7 V, which resulted in a constant melt rate of 0.45 kg/sec. A 0.002 T axial stirring field, reversing every 60 seconds, was applied during this melt. These parameters, along with the nominal thermo-physical properties were input into SOLAR. A uniform mesh, 15mm by 9 mm was used in this simulation, and using a time step of 0.015 sec, run times on machine with a 2.66 GHz processor was about 24 hrs. The output from the model is shown in Figure 2. The VAR conditions described produce a very deep liquid pool, reaching a depth of nearly 1.5 m at the center. Also, the pool walls are very steep in the top half of the ingot. Due to the narrow solidification range of the alloy, the mushy zone size is relatively shallow. The predicted pool depth and profile compare reasonably well with measurements reported in the literature /5/. The flow of liquid metal in the longitudinal plane (Figure 1A) consists of two circulating cells positioned one on top of the other. Metal moves down the central axis and returns along the ingot wall, indicating that the flow is driven primarily by electromagnetic forces (note that the direction of flow is dependent on the polarity of the stirring field). Maximum velocity is of the order of 0.02 m/sec.

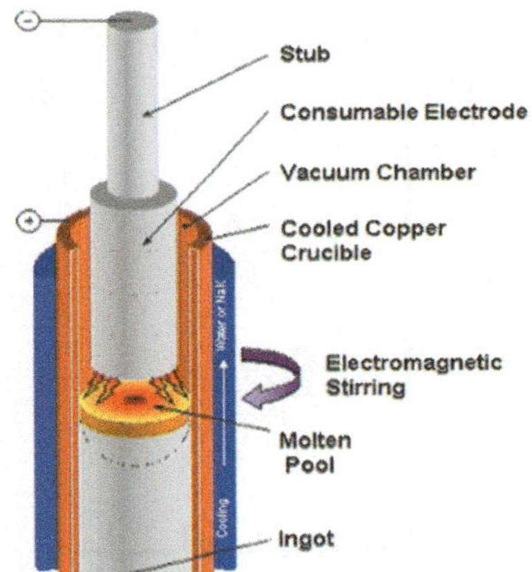

Figure 1: Sketch of the VAR Process used in Titanium alloys.

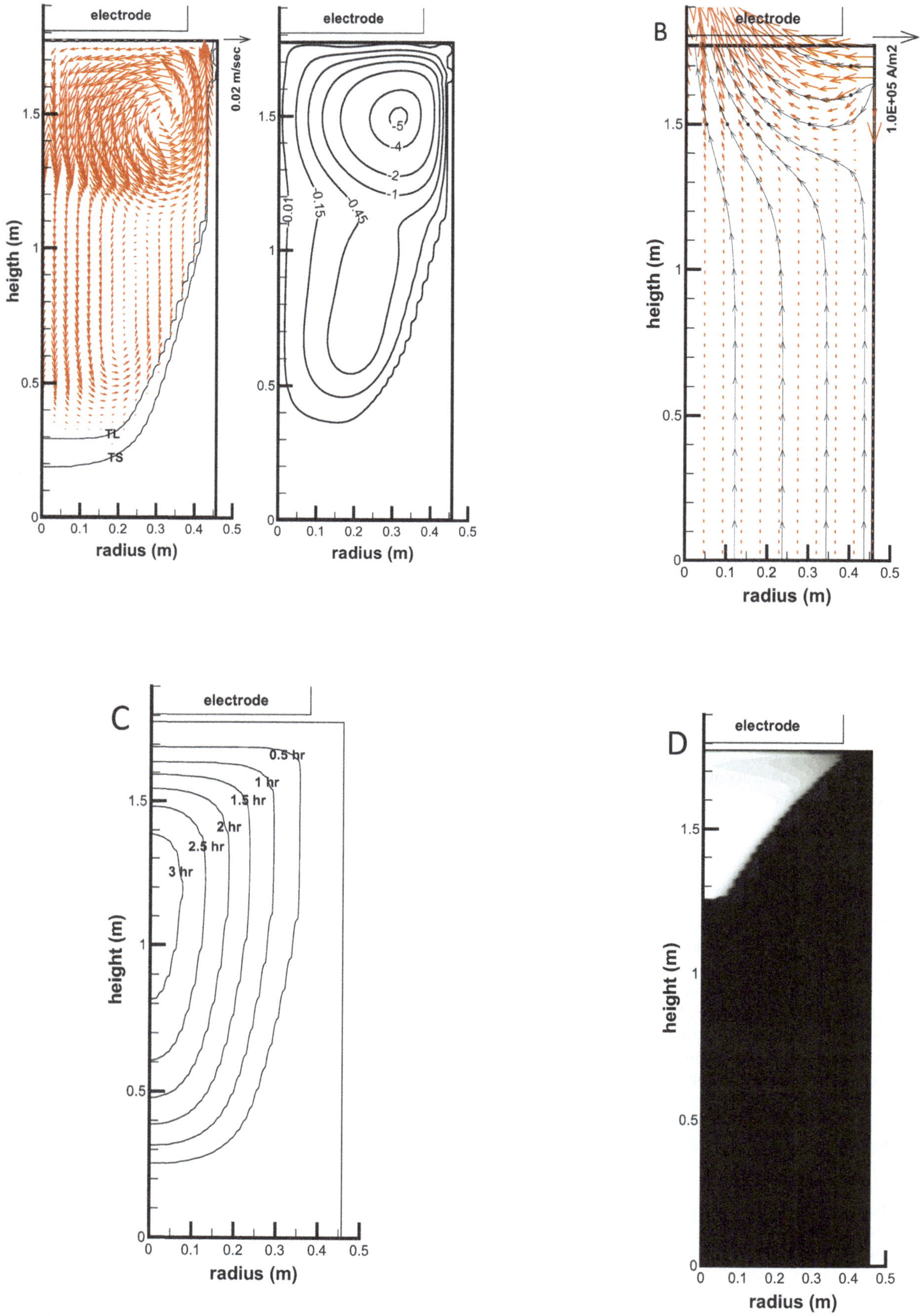

Figure 2: Model prediction for the 0.914 m (36 inch) diameter ingot, electrode diameter is 0.762 m (30 inch). A: Pool profile and flow in the longitudinal plane, B: current distribution, C: liquidus isotherm at different times after power off, D: likely location of macro-shrinkage porosity at the top of ingot, liquid feeding is cut-off at f_l=0.45. Stream lines are in Kg/sec.

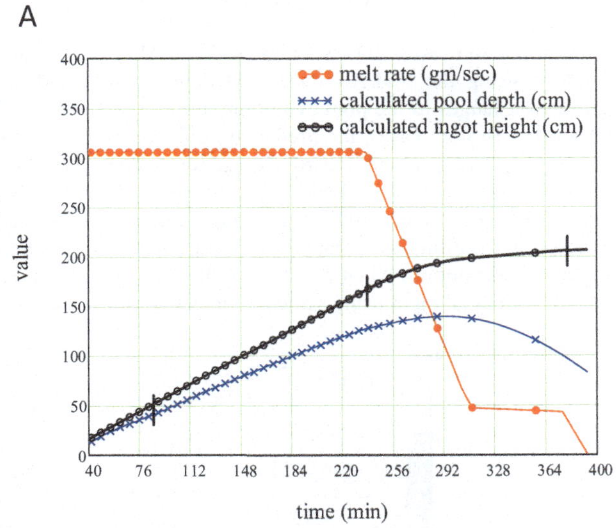

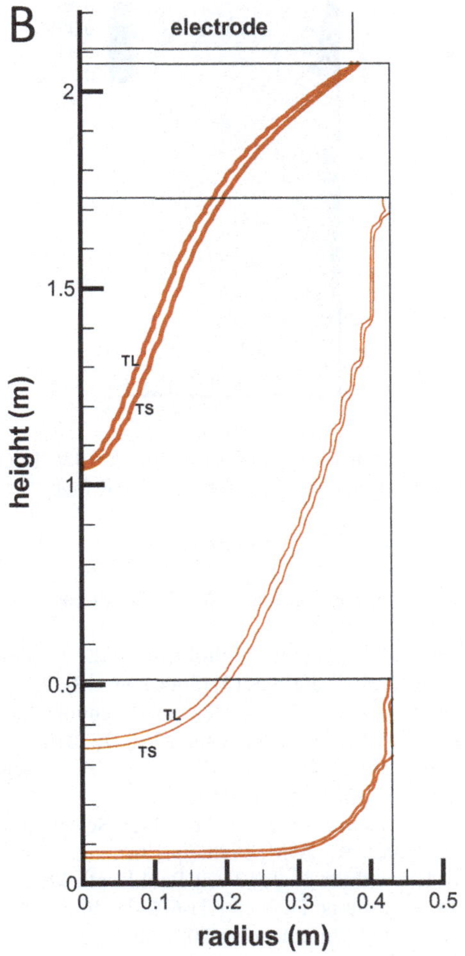

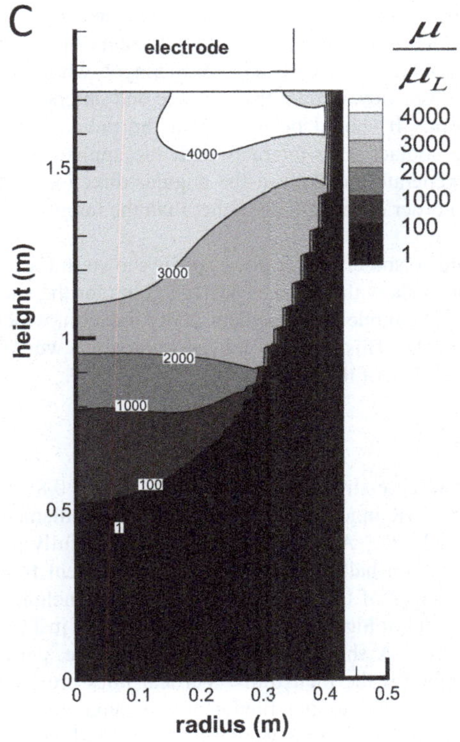

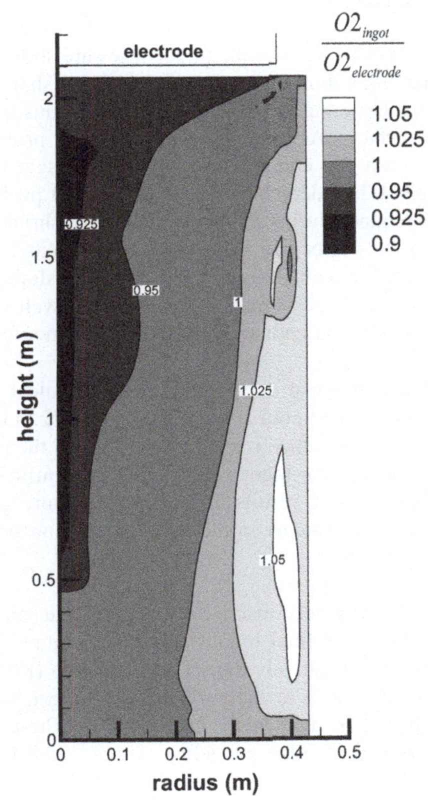

Figure 3: Computed results from the 0.86 m diameter VAR ingot. A: the melt rate profile, B: pool profiles during different stages of VAR, C: Turbulence intensity in the ingot, D: Predicted oxygen segregation in the ingot.

Figure 2-B shows the computed current distribution in the ingot. The magnitude is illustrated by the individual vector lines, and the current flow path is shown by the stream-lines (solid lines). As seen, a fraction of the applied current enters the ingot from the crucible contact zone on the side of the ingot, while the remainder of the current flows down the crucible and enters the ingot from the bottom. It should be noted that the radial current entering from the contact zone interacts with the applied axial magnetic field to generate the flow in the angular direction. This flow is nearly an order of magnitude higher than the longitudinal flow.

The liquidus isotherm after power off is shown in Fig. 2-C. These calculations show that it takes nearly 3.5 hrs for the entire ingot to solidify. The model also predicts cavity location, which is shown in Figure 1-D. This analysis indicates that voids would be present in the top 0.5 m of the ingot.

<u>VAR ingot 2</u>

The second case simulated in this study, was a 0.86 m (34 inch) diameter, VAR ingot from a 0.76 m (30 inch) diameter, Ti-6Al-4V electrode /6/. As reported, this was a carefully planned trial where tantalum balls were used at three different times to mark different stages of the melt. The three stages include: bottom of the ingot, end of high power, and end of hot top, just before power off. Figure 3-A shows the calculated melt rate, pool depth and ingot height for the melt trail. Vertical lines crossing the ingot height correspond to the time that the tantalum markers were added. As in the previous validation exercise, the processing data was entered into the model, and simulations were carried out with the same set of properties. Note, the power input was estimated from correlations provided in the available literature /8/, and a stirring field strength of 0.006 T, with 60 second reversal was used in this simulation.

The predicted pool shapes at the three stages are shown in Figure 2-B. At the ingot bottom, the pool is "dish" shaped with an essentially flat bottom and a steeply sloping liquidus in proximity to the ingot surface. As the ingot grows the pool profile changes significantly. At the end of the high-power stage, the pool is approximately 1.3 m deep and has a "U" shaped profile. When the power is reduced during the hot-top, the pool shrinks upwards, and just prior to power off, the simulation predicts a more "V" shaped pool profile with walls of nearly uniform slope. All three predicted pool shapes match reasonably well with the measurements of pool depth and profile reported in reference /6/.

The predicted turbulence intensity (ratio of turbulent to laminar viscosity) in the molted pool at ingot length of 1.72 m is shown in figure 3-C. The turbulence is not uniform within the pool, but is higher at the top of the ingot where there is significant angular velocity. This, in turn, results in lower temperature gradients at the top of the pool. Deeper in the pool, the stirring subsides and turbulence diminishes.

The model also predicts macro-segregation, and the predicted oxygen content in the final ingot is shown in figure 3-D. Since oxygen segregates negatively during solidification (K>1), there is enrichment of 2 to 5 % at the surface of the ingot, whereas the center is depleted in oxygen by nearly 10%. These trends are consistent with a similar modeling study published in the literature [9].

Concluding Remarks

Two production size ingots Ti-6Al-4V VAR ingots were simulated using the SOLAR code. Modeling results indicate, significant changes to the pool evolution during a complete VAR melt cycle. In both cases, the predicted pool shapes, location of the shrinkage cavity and macro-segregation pattern compared reasonably against data available in the literature /5, 6, 9/. This model is being routinely used to determine the effect of melting variables on metallurgical parameters. Figure 4 below shows the oxygen re-distribution, when the 0.86 m dia. electrode is melted into a 0.914 m dia. ingot using same melting conditions.

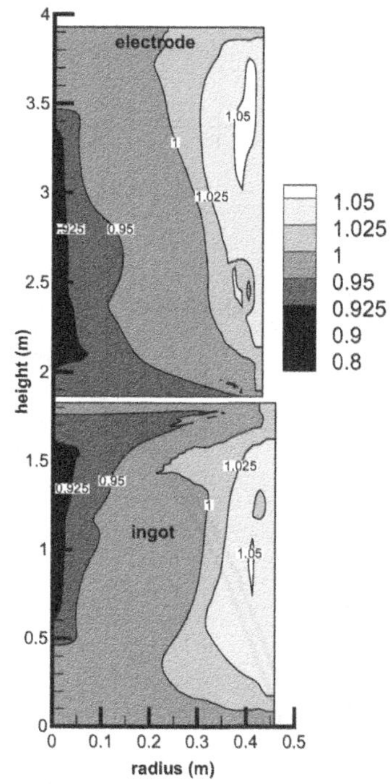

Figure 4: Computed oxygen distribution in a 0.914 m dia. VAR ingot, melting conditions are from VAR ingot 2, Fig. 3.

References

[1] "Vacuum Arc Remelting", ASM Handbook, Vol. 15, 2008, pp.132-138.
[2] E. Poulsen, "Safety-Related Problems in the Titanium Industry in the Last 50 Years", JOM 52(5) (2000), pp. 13-17.
[3] A. Jardy, et.al, "Segregation in Vacuum Arc Remelted Zirconium Alloy Ingots", ASTM Intl., Vol. 7, 2010, pp. 1-17.
[4] A. Jardy A, et.al, "", LMPC 1994, A. Mitchell and J. Fernihough Eds, AVS, pp. 143-154
[5] L. A. Bertram, et al., The MacroScale Simulation of Remelting Processes", JOM, 1998, pp 18-21
[6] R. S. Minisandram et.al ," VAR Pool Depth Measurement and Simulation for a Large Diameter Ti-6Al-4V Ingot", LMPC, P. D. Lee et al.,Ed., ASM International 2005 pp. 7-13.
[7] K. Mills, "Recommended values of thermophysical properties for selected commercial alloys", ASM 2002, pp. 205-217.
[8] J. J. Beaman et al, "A Nonlinear Reduced Order Model for Estimation and Control of VAR of Metal Alloys", ASME Exposition, 2005, pp. 1-9.
[9] K. M. Kelkar et al, "Computational Analysis of the Vacuum Arc Remelting and Electroslag Remelting (ESR) Processes", ASM Handbook, Vol. 22B, 2010, pp. 196-213.

Proceedings of the 2013 International Symposium on Liquid Metal Processing & Casting
Edited by: Matthew J. M. Krane, Alain Jardy, Rodney L. Williamson, and Joseph J. Beaman
TMS (The Minerals, Metals & Materials Society), 2013

Controlling liquid pool depth in VAR of a 21.6 cm diameter ingot of Alloy 718

Felipe Lopez[1], Joseph Beaman[1], Rodney Williamson[1], Eric Taleff[1], Trevor Watt[1]

[1]Advanced Manufacturing Center, Department of Mechanical Engineering, The University of Texas at Austin, Austin, TX 78712-0292, USA

Keywords: vacuum arc remelting, pool depth control

Abstract

It is believed that the final microstructure in vacuum arc remelted (VAR) ingots is strongly influenced by the molten metal pool profile. Thus, if the pool profile was properly controlled during the melt then defect-free microstructures would be obtained. The recent development of a reduced-order model of VAR solidification allowed the design of a pool depth controller to accomplish this task. The controller used a linear quadratic regulator and a Kalman filter to stabilize the melt pool solidification front under the effect of uncertain process dynamics and noisy measurements. Basic Axisymmetric Remelting (BAR), a high-fidelity VAR ingot model, was used in real time to provide pool depth measurements that were incorporated in the control loop. The controller was tested at Los Alamos National Laboratory in a 21.6 diameter Alloy 718 ingot. Details of the controller design will be presented, along with comparisons to experimentally-measured pool depths.

Introduction

Vacuum Arc Remelting (VAR) is a secondary melting process used for producing fully-dense homogenous ingots. These ingots are expected to be free from macrosegregation, porosity, shrinkage cavities, and other defects that result from uncontrolled solidification processes. The VAR process involves the loading of a cylindrical metal electrode into a water-cooled copper crucible. Current passes through the electrode resulting in a DC arc being struck between the electrode (cathode) and some start material on the bottom of the crucible (anode). The arc eventually melts the bottom of the electrode and the start material. Molten metal will fall into the bottom of the crucible and form the ingot. During the solidification process, a liquid metal pool exists on the top of the ingot; the metal will be in three phases: liquid, liquid-solid mixture (mushy zone), and solid. As the electrode melts, it is translated down toward the anode pool to keep the gap between the electrode and metal pool constant.

Industrial VAR practice usually involves production of ingots of 50 cm diameter. Even larger diameter forging stock is desirable but the increasing risk of forming solidification defects limits the size of VAR ingots. Inadequate control of the solidification front in large ingots may result in macrosegregation defects known as "freckles", which appear as thin channels extending towards the center of the ingot. It is believed that the collective effects of interdendritic fluid flow and concentration gradients within the mushy zone result in the formation of freckles. It was reported by Kou [11] that macrosegregation increases with increasing depth of mushy zone and decreasing solidification rate (higher local solidification time). The opposing case, a shallow pool depth and a higher solidification time, may result in a different kind of macrosegregation defects called solidification white spots. Therefore, it is clear that it is necessary to understand and control the factors affecting the geometry of the liquid pool profile - and, as a result, the local solidification time - in order to prevent macrosegregation defects from occurring in vacuum arc remelted ingots.

Although ingot solidification is the crucial part of the remelting process, which defines the microstructural properties of the obtained material, it is not directly controlled in usual industrial VAR practice. Years ago, it was common for remelting companies to control the DC current sent to the furnace and assume that, by doing so, the solidification front will be controlled. Modern day VAR practice usually calls for some form of melt rate control using load cell feedback, although it is possible to perform melt rate control without them [6]. Current control is still employed at the beginning and end of melting where melt rate control using load cell feedback is difficult. Extensive testing was used for companies to develop what is called a melt recipe, which defined how melt rate and current should be changed during the melt in order to obtain a defect-free microstructure. Such an approach lacks robustness and treats the whole process as a black box as there is no information of how the solidification front is being affected by the varying current.

In the past years, the Specialty Metals Processing Consortium (SMPC) has worked on the development of model-based controllers for the VAR process. These controllers combine uncertain measurements with knowledge of the physics of the process to provide improved estimates of process parameters used to stabilize the melt even under process upsets. The developed control algorithms have been implemented into the SMPC Advanced VAR Controller (AVARC). This controller has separate routines for the electrode gap and for the solidification, which is controlled differently depending on the operating mode. The electrode gap is controlled by changing the ram speed using estimates coming from voltage measurements or drip-short frequency, as proposed by Hysinger [9]. The current reference is defined by the process controller to stabilize the melt. The SMPC VAR controller has three operating modes: current control, melt rate control, and pool power control, which is defined as the total amount of power flowing into the liquid pool [13]. These controllers were seen as significant improvements but there was still no way to control solidification. The main impediment came from the lack of an appropriate ingot solidification model that could be used for estimation and control of VAR.

A reduced-order model of ingot solidification in VAR was developed exclusively to be used for pool depth control [5]. A first implementation of the pool depth controller, running in an open-loop mode, was reported in [14]. The author suggested that the performance of the controller could be improved with the inclusion of some means of measurement feedback for pool depth, and proposed that a high-fidelity ingot simulation could be used to provide the required pool depth measurements. Based on the idea described by Williamson, Basic Axisymmetric Remelting (BAR) was embedded in the pool depth controller to provide real-time measurements of the geometry of the liquid pool profile.

This paper describes the implementation and testing of a method of dynamic pool depth control for the VAR process, using feedback from a multiphysics ingot model, capable of stabilizing the solidification front at a desired reference. The experimental validation was performed at Los Alamos National Laboratory (Los Alamos, NM). A full treatment of the mathematical development underlying the controller has been reported elsewhere [4]. A concise description of the proposed process controller and estimator are given in the upcoming sections, followed by a depiction of the implementation of a high-fidelity computational model in the control loop. Experimental results confirming the accuracy of the proposed controller in the steady state part of the melt and conclusions are given at the end of the paper.

Reduced-order thermal model

Modeling of VAR has been studied extensively both in the industrial and academic worlds [16, 7, 10, 1] at the macroscale and microscale levels. VAR is a multiphysics process with involves not only heat transfer, but also fluid dynamics, mass transfer, phase transformations, and electromagnetics [15]. Macroscale models have been used for prediction and for comparative study of the interaction between all energy domains, while microscale models were used to study the mechanisms that produce defects and segregation problems.

The models found in the literature show good accuracy, but they are not suitable for real-time control applications due to the simplifying assumptions used in their development or due to the extensive number of state variables used to model the process. A reduced-order model developed specifically for estimation and control purposes was presented in [5]. This approximate thermal model seeks to predict liquid pool shape dynamically when input current and melt efficiency are given. The nonlinear system is solved using Galerkin spectral methods [8], linearized about nominal conditions and finally reduced when fast modes are removed. However, because this model accounts only for the solidification process it must be used with a different reduced-order model that accounts for the melting of the electrode [6]. The models have been linearized about the nominal operating point because in a real melt there is a short start-up period after which the majority of the melt occurs in quasi-steady conditions.

The dynamic equations of the reduced-order VAR model have the form

$$\delta\dot{x} = A\delta x + B_\mu \delta\mu + B_I \delta I \tag{1}$$

$$\delta y = C\delta x \tag{2}$$

where δx is the perturbation in the n state variables, $\delta\mu$ is the perturbation in melt efficiency which is modeled as an uncontrolled input, δI is the perturbation in melting current, and δy is the perturbation in the vector of pool depths at the radii where pool profile is measured.

Pool depth control

A diagram showing the overall controller architecture for the pool depth controller, using feedback from a computer simulation, is shown in Figure 1. The reduced-order model is used to design a pool depth controller and a pool depth (ingot) estimator. These new tools are designed to work with the electrode estimator, developed by Beaman [6].

Reference values are given to the controller as an electrode gap reference and a pool depth reference. In this first implementation, only one pool depth reference is sent to the controller; however, the pool depth controller is also capable of handling the more general case of having more than one reference pool depth at different radii. The accuracy of the control system in this case could be decreased as the controller would attempt to meet all the given references by minimizing deviations from desired values. The VAR pool depth controller is internally composed by two separate controllers: one for the electrode gap, and another for the liquid pool profile. The dynamics of the two systems are assumed independent.

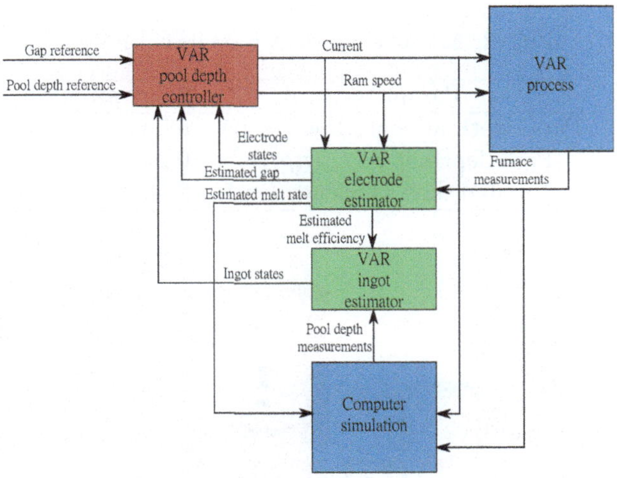

Figure 1: Pool depth control of VAR

Electrode gap is controlled by changing the ram speed in the furnace independently of the melting conditions. However, control of ingot solidification involves using the reduced-order model to invert the dynamic equations of the system and compute the required value of current that would drive the system to desired conditions. The pool depth controller computes the inputs (current and ram speed) to be sent to the furnace based on estimated electrode states, ingot states, and gap. Two separate estimators are used to reconstruct the states: one for the electrode and one for the ingot.

Furnace measurements (i.e., voltage, current, electrode weight, drip-short frequency) are sent to the electrode estimator along with the inputs fed to the furnace. The electrode estimator returns electrode states (electrode thermal boundary layer), estimated electrode gap, estimated instantaneous melt rate, and estimated melt efficiency. A high-fidelity computational model is run in parallel with the melt using the instantaneous melt rate obtained from the electrode estimator, and current and voltage coming from the furnace. More details on the implementation of this computational model are given in the upcoming sections. The high-fidelity model is used to provide measurements of the liquid pool profile, which could not be measured in real-time in any other way, that are sent to the ingot estimator. The ingot estimator returns estimates of the ingot states (i.e., indicators of the enthalpy distribution in the ingot) that are sent back to the pool depth controller.

The model used for control is only an approximation of the highly complex dynamics of VAR and all measurements in this process are known to be noisy. Therefore, uncertainty plays an important role in the design of the process controller. A Linear-Quadratic-Gaussian (LQG) strategy was used to design an optimal controller solving the linear quadratic regulator equations, and a Kalman Filter for optimal state estimation. The LQG controller returns the required optimal input even under model uncertainty and noisy measurements. A detailed description of the Linear Quadratic Gaussian controller can be found in the tutorial paper by Athans [3].

Current corrections are computed to make liquid pool depth coincide with the desired values at m radii r_i as shown

$$I = I_0 + \delta I \tag{3}$$

$$\delta I = -K_x \delta x - K_\mu \delta \mu - K_{ref} \delta y_{ref} \tag{4}$$

$$\delta y_{ref} = \begin{bmatrix} \delta S_{pool_{ref}}(r_1, t) \\ \cdots \\ \delta S_{pool_{ref}}(r_m, t) \end{bmatrix} \tag{5}$$

where the control gains matrices K_x, K_μ and K_{ref} are defined by the linear quadratic regulator. A detailed description of the linear quadratic controller is given in the full paper [4].

Pool depth estimation

The proposed structure for the controller would be able to drive the system to the desired pool depth if both the model and the measurements are perfect, i.e. process and measurement noises are zero. However, it is known that electrode gap measurements, based on voltage or drip-short frequency, are inherently noisy and uncertain, and there is no way to measure the liquid pool depth directly in real-time.

Kalman filters have been successfully used in the estimation and control of remelting processes [6, 2]. Kalman filters are used to construct optimal estimates based on uncertain measurements and a low-order approximation of the physical process.

In the Kalman filter, everything is developed under the assumption that all random variables are modeled as white noise processes. However, the change in melt efficiency is better described by a random walk process, so a way to express perturbation in melt efficiency as a white noise is required. Although we cannot model melt efficiency as a white noise sequence we can still do so with its increments. Thus, the state vector is augmented to include an extra variable: perturbation in melt efficiency, whose increments are modeled as white noise. In this case, it is assumed that process uncertainty comes only from inaccuracies in the values of I and μ.

The continuous-time dynamic system was transformed into a discrete-time process, because most measurements are available only at discrete times. The continuous-time equations are transformed into the discrete-time dynamic equations of the augmented system, given by equation (7) where Φ denotes the state transition matrix, $\mathbf{x}$ is the augmented state vector, $\mathbf{u}$ is the input vector, and $\mathbf{w}$ is

Table 1: PROCESS NOISE LEVELS.

Parameter	Noise strength
Current (w_I)	120 A
Melt efficiency rate (w_μ)	0.01 μ_0

Table 2: MEASUREMENT NOISE LEVELS.

Parameter	Noise strength
Pool depth at centerline (n_{PD})	6.0 cm
Pool depth at $0.7r_i$ (n_{PD})	0.2 cm
Estimated melt efficiency (n_μ)	2.33 $\times 10^{-5}$

the process uncertainty.

$$\begin{bmatrix} \delta x \\ \delta \mu \end{bmatrix}_{n+1} = \Phi \begin{bmatrix} \delta x \\ \delta \mu \end{bmatrix}_n + \Lambda \delta I_n + \Gamma \begin{bmatrix} w_I \Delta t \\ w_\mu \end{bmatrix}_n \quad (6)$$

$$\mathbf{x}_{n+1} = \Phi \mathbf{x}_n + \Lambda \mathbf{u}_n + \Gamma \mathbf{w}_n \quad (7)$$

The measurement equation is modified as well to include melt efficiency, which is now a state that can be "measured" from the electrode estimator, as shown in Figure 1. Uncertainty caused by noisy measurements, both for pool depth and melt efficiency, is included in the form of the vector $\mathbf{v}$ and is observed in the vector of augmented measurements $\mathbf{y}$ shown in equation (9).

$$\begin{bmatrix} \delta y \\ \delta \mu \end{bmatrix}_n = \begin{bmatrix} C & 0 \\ 0 & 1 \end{bmatrix} \begin{bmatrix} \delta x \\ \delta \mu \end{bmatrix}_n + \begin{bmatrix} v_{PD} \\ v_\mu \end{bmatrix}_n \quad (8)$$

$$\mathbf{y}_n = \mathbf{H}\mathbf{x}_n + \mathbf{v}_n \quad (9)$$

One more set of matrices are required to derive the Kalman gain matrix, namely, the covariance matrices corresponding to the process and measurement noise vectors $\mathbf{w}_n$ and $\mathbf{v}_n$. Noise terms are modeled as zero-mean white-noise random sequences. As a result they are defined completely by their covariance matrices. The process covariance, usually denoted $\mathbf{Q}$, is defined as a diagonal matrix to denote that the noise terms are not correlated. The measurement covariance, usually denoted $\mathbf{R}$, contains the uncertainties related to the pool depth measurements at m radii and the estimated melt efficiency, which is a measurement returned by the electrode estimator described in [6].

The levels of process noise and measurement noise used for the design of the Kalman filter are shown in Tables 1 and 2. The noise strengths are the standard deviation of the parameters. Some of these standard deviations were obtained from experiments, as in the case of the current. Melt efficiency noise strength was obtained from computer simulation, and the others were defined based on empirically realistic values.

The estimates obtained with the Kalman filter are used by the LQR to compute the optimal inputs as the true value of the state variables is never known exactly. A detailed description of the Kalman filter is given in [4].

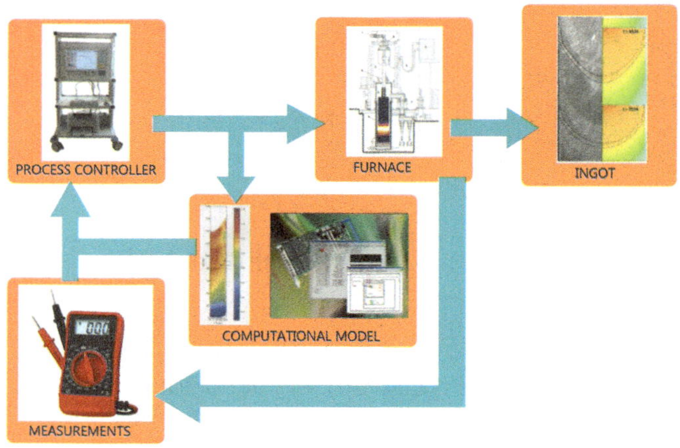

Figure 2: Proposed control strategy

LIQUID POOL DEPTH VIRTUAL MEASUREMENTS

An obvious inconvenience with the proposed controller and estimator is that no direct measurements are available for the liquid pool profile of the ingot in the furnace. If no measurements are available from the furnace then the system is unobservable and, therefore, accurate process control is not possible. Although today's technology does not provide with an accurate way of measuring the pool profile or the temperature distribution in the ingot, a high-fidelity computational model such as Basic Axisymmetric Remelting (BAR) could be used to improve the observability of the system and the overall performance of the controller. BAR, which is the closest approximation available to the actual melt, would be used as a non-invasive method to provide virtual measurements of the parameters that cannot be measured in any other way. Because liquid pool profile is known to be related to defect formation in VAR, it will be monitored and controlled throughout the melt. If other parameters are reported to be related to other solidification defects, they would be included in the high-fidelity model and the control system in a similar way.

BAR was not intended to be used in process control systems but to study the physical conditions that would lead to the formation of defects in remelting processes [7]. The structure of the model had to be modified so that it could be run in parallel to the actual melt. BAR is run with the same inputs that are sent to the actual furnace (i.e., current, voltage, melt rate, and helium pressure) to ensure that the model and the furnace are synchronized. The desktop computer was used for the real-time implementation of the computational model. BAR is used to provide a measurement of the temperature distribution in the solidifying ingot. These temperature measurements can be used to get an estimate of the liquid pool profile using a linear interpolation to find the location where the liquidus temperature occurs. The pool depth measurements can be obtained for as many radial locations as required.

In the case presented in this paper, only two virtual measurements were used to improve the observability of the control system. The locations where the measurements are taken are carefully chosen to get measurements that are representative to describe the liquid pool profile. The centerline pool depth and that of $r = 0.7r_i$ are used. The centerline pool depth provides information about the maximum liquid pool depth in the ingot, and $0.7r_i$ is a representative mean radius for which the internal cross-sectional area is the same as the external cross-sectional area. The two virtual measurements are not treated with the same uncertainty, as shown in Table 2. The physical properties used for the thermal model were calibrated to match BAR liquid pool depth predictions at $0.7r_i$ in steady-state, but the same properties result in an offset for the pool depth predictions at the centerline. In order to follow closely the prediction at the mean radius, a very small uncertainty is used for this virtual measurement compared to the one of the centerline.

IMPLEMENTATION OF THE POOL DEPTH CONTROLLER

The LQG pool depth controller was successfully implemented in SMPC AVARC. The new pool depth control tool was tested at a laboratory-scale furnace in Los Alamos National Laboratory using Alloy 718 in July 2011. A 15.2 cm electrode was melted into a 21.6 cm ingot. More information about the furnace and the material used for the test is given in Tables 3 and 4.

The goal of the experiment was to prove that the proposed estimation strategy, using a computational model in real-time to provide pool depth measurements, resulted in stable pool depth control around the reference pool depth; and that the pool depths predicted by the pool depth controller coincide with profiles measured experimentally.

As a first step in the development of this new generation of controllers, only one pool depth was monitored and controlled. The pool depth at $0.7r_i$ was chosen as the most representative pool depth because it is an average value for which the internal area equals the external one. Although only one pool depth will be monitored in the controller, two measurements will be used in the estimator.

Table 3: FURNACE PARAMETERS.

Parameter	Value
Electrode radius	7.62 cm
Ingot radius	10.8 cm
Nominal current	2800 A
Nominal voltage	23.5 V
Nominal melt rate	34 g/s
Nominal electrode gap	1.0 cm

Table 4: THERMOPHYSICAL PROPERTIES OF ALLOY 718.

Property	Value
Melt temperature	1623 K
Density	7.75 g/cm^3
Heat capacity	0.65 J/g K
Latent heat	210 J/g
Solidus enthalpy	762.4 J/g
Liquidus enthalpy	1070.0 J/g
Thermal diffusivity (solid phase)	0.0637 cm^2/s
Thermal diffusivity (liquid phase)	0.0676 cm^2/s

Virtual pool depth measurements are recorded from the beginning of the melt. However, they are included for estimation only when the measurements are close to the estimated pool shape, starting with the line denoted as BAR Feedback in Figure 3. It can be seen that the dynamics of the estimates and measurements are a lot closer after feedback is turned on. The reason for this is that the reduced-order model is based on a steady-state approximation so it is valid only close to steady-state. In the beginning of the melt the ingot is very small and just starting to solidify, so controlling the pool profile would not make much sense. The melt was started in current control mode and pool depth control was turned on approximately at 0.05 hours, as shown in Figure 3 with the first vertical dashed line. The pool depth estimator was turned on at 0.28 hours, as marked with the second vertical dashed line. After the pool depth estimator was turned, a series of step inputs were used to perturb the pool depth reference in order to test the behavior of the controller in a tracking problem. Experimental results show that once the virtual measurements have been included in the estimator, the measured pool depth for the mean radius and the estimated one match. A small amount of steady state error can be seen, but its value is negligible compared to the desired liquid pool depths.

Small oscillations are shown after step changes in pool depth reference but they disappear quickly.

In a real melt the desired liquid pool shape will, most likely, be kept a constant value in order to prevent macrosegregation. The case of a constant reference pool depth of $12.6cm$ was tested from 0.86 to 1.06 hours. In this part of the experiment the controller showed good performance and was able to maintain the predicted liquid pool depth close to the desired value. It should be noted that the current required to stabilize the system at the desired pool depth was $2300A$, which is below the nominal value of current because the pool profile in this experiment was a little shallower than the nominal one.

Pool profiles

A comparison between the pool profiles predicted by BAR and those measured experimentally is given in Figure 4. To measure the pool profiles, the ingot was sectioned lengthwise to expose the mid-plane, which was then ground and macro-etched to reveal the solidification microstructure. The etched surface was then digitally photographed at different locations across the surface using a SLR camera with a macro lens, stitched to form a mosaic, and digitally merged into a single image. Exposed pool outlines due to tree rings or freckle channels where then manually marked and exported into Matlab for plotting and comparison. Additional details on this procedure can be found in [12]. BAR predictions of liquid pool profiles were obtained from the temperature distributions returned by the computer simulation for five different instants of the melt: 0.5, 0.7, 0.9, 1.1, and 1.3 hours after the arc was struck. The correspondent profiles are shown as profiles A to E in the figure.

Agreement between predicted and measured pool depths is seen in the quasi steady part of the melt, as seen in profiles C and D which are close to the pool depths measured manually. A large discrepancy is observed in the early stage of the melt, shown in profiles A and B. Also, different behavior can be seen on each side of the axis close to B. The profiles observed on the right hand side show a steeper slope than the profiles on the left hand side. A similar asymmetric profile is observed close to the profile A. The VAR furnace at Los Alamos National Laboratory does not have co-axial power feeds and this may have resulted in asymmetric pools due to arc steering by stray magnetic fields. BAR, the high-fidelity model, is an axisymmetric model and will not be able to describe any asymmetric behavior. A three-dimensional model would be required for such a task.

The discrepancy between the computational model used for pool depth feedback and the pool depths measured experimentally can be very problematic for the process controller. Mainly because the controller attempts to

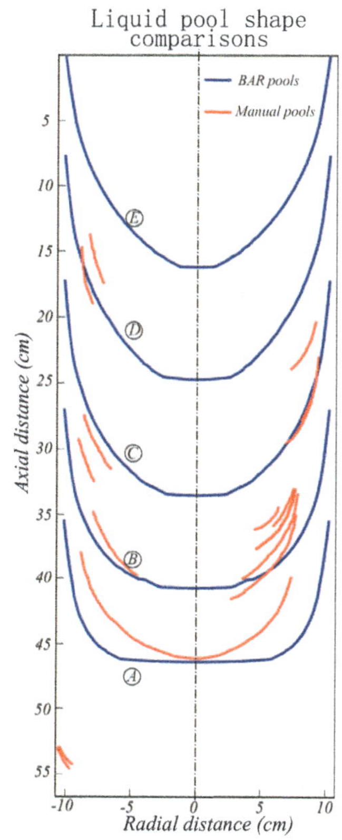

Figure 4: Comparison between predicted and measured liquid pool profiles

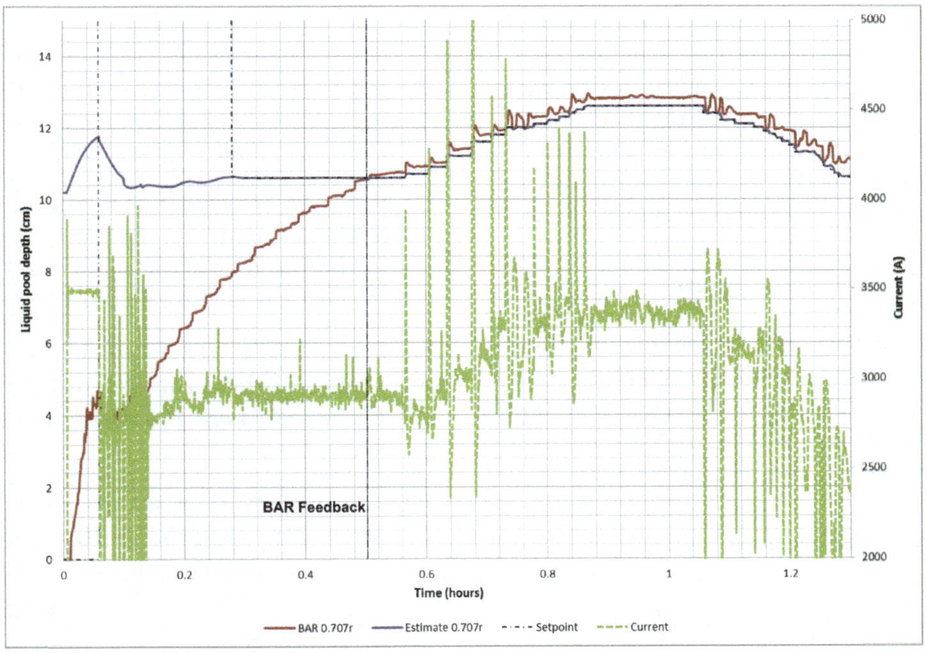

Figure 3: Experimental results

control pool depth based on the assumption that the pool depth measurements are accurate. Better agreement has been reported in past experiments for this same model [7]. A more accurate computational model may also result in a better pool depth controller that could be used for the entire melt. The current version of the pool depth controller is based on a linear approximation of highly nonlinear dynamics, and therefore, is valid only when close to the nominal operating conditions. The high-fidelity model is also accurate in this region proving that the pool depth controller using feedback from a high-fidelity model results in accurate quasi-steady control of the liquid pool depth in the solidifying ingot.

Conclusions

It is believed that if liquid pool shape, closely related to local solidification time (LST), could be controlled precisely to stabilize the solidification zone in the ingot, larger diameter defect-free ingot alloys could be produced. A new generation of VAR controllers was designed to control the liquid pool shape based on the stochastic Linear-Quadratic-Gaussian (LQG) problem. One obvious problem with this approach is that the liquid pool shape cannot be measured directly from the ingot. In order to solve this observability problem a high-fidelity model that describes the solidification of the ingot in a VAR process is incorporated in the control system. The furnace and the computational model are synchronized by using the same inputs in both systems (i.e., current, voltage, and melt rate) and using equal time steps.

The measurements coming from the high-fidelity model, just like the real ones, are subject to bias and noise. An estimator based in a steady-state Kalman filter was designed to provide accurate state estimates given a reduced-order solidification model and measurements coming from the high-fidelity model. The new controller was successfully implemented in a laboratory-scale furnace in Los Alamos National Laboratory in July 2011 showing a stable response and that the liquid pool depth predicted by the computational model could be driven to any desired reference. The pool depth controller showed a good performance under the assumption that the high-fidelity model accurately predicted the liquid pool profiles obtained in reality. Pool profiles were measured experimentally in the ingot to validate the hypothesis.

Agreement between predicted and measured pool depths is observed for the steady-state part of the melt. However, large discrepancies were observed between the pool profiles predicted by the high-fidelity model and the ones measured manually at the beginning of the melt. The accuracy of the pool depth controller is highly dependent on the accuracy of the high-fidelity model. Therefore, a more accurate model is required for more accurate pool depth control, especially for the highly-transient parts of the melt, such as start up.

The controller proposed in this paper is based on a linear model which is valid when close to steady-state melting conditions. In the experiment reported in this paper, the melt was just starting to reach steady-state when the melt ended. The performance of the controller, and the accuracy of the computational model used for feedback, are expected to be better in the case of a larger ingot where most of the melt occurs in steady-state. A dynamic controller would be required to improve accuracy in cases where the melt does not reach steady-state conditions, e.g. when melting titanium alloys.

Acknowledgements

The authors gratefully acknowledge funding for this work from ONR through contracts number N00014-11-1-0366 and N00014-12-1-0811. They would also like to thank Dr. Robert Aikin and staff, at Los Alamos National Laboratory, for graciously hosting the controller tests.

References

[1] C.B. Adasczik et al. Quantitative simulation of a superalloy VAR ingot at the macroscale. In *Proceesings of the AVS Vacuum Metallurgy Conference*, 1997.

[2] Seokyoung Ahn, Joseph J. Beaman, and Rodney L. Williamson. Model-based control of electroslag remelting process using unscented kalman filter. *ASME Journal of Dynamic Systems, Measurements and Control*, 132(1):011011, 2010.

[3] Michael Athans. The role and use of the stochastic linear-quadratic-gaussian problem in control system design. *IEEE Transactions on Automatic Control*, AC-16:529–552, 1971.

[4] Joseph J. Beaman, Luis F. Lopez, and Rodney L. Williamson. Model-based control of liquid pool profile in the vacuum arc remelting process. *ASME Journal of Dynamic Systems, Measurements and Control*, 2013. (In review).

[5] Joseph J. Beaman, Rodney L. Williamson, and Luis F. Lopez. A reduced-order thermal model for dynamic var pool depth control. In *Proceedings of the International Symposium on Liquid Metal Processing and Casting*, 2011.

[6] Joseph J. Beaman et al. A nonlinear reduced order model for estimation and control of Vacuum Arc Remelting of metal alloys. In *2005 ASME International Mechanical Engineering Congress and Exposition*, 2005.

[7] L.A. Bertram et al. The macroscale simulation of remelting processes. *Journal of Metals*, 50(3):18–21, 1998.

[8] John P. Boyd. *Chebyshev and Fourier Spectral Methods*. Dover, 2001.

[9] Christopher L. Hysinger, Joseph J. Beaman, Rodney L. Williamson, and D. K. Melgaard. Multiple input electrode gap control during vacuum arc remelting. In *International Symposium on Liquid Metal Processing and Casting*, 1999.

[10] Kanchan M. Kelkar et al. Computational modeling of the Vacuum Arc Remelting (VAR) process used for the production of ingots of Titanium alloys. In *Proceedings of the Ti-2007 Conference*, 2007.

[11] S. Kou, D. R. Poirier, and M. C. Flemings. Macrosegregation in electroslag remelted ingots. In *Electric Furnace Proceedings*, volume 35, page 221, 1977.

[12] Trevor J. Watt et al. Solidification mapping of a Nickel 718 laboratory VAR ingot. In *Proceedings of the International Symposium on Liquid Metal Processing and Casting*, 2013. In press.

[13] Rodney L. Williamson and Joseph J. Beaman. Modern control theory applied to remelting os superalloys. *Materials Science Forum*, 706-709:2484–2489, 2012.

[14] Rodney L. Williamson, Joseph J. Beaman, and Robert M. Aikin. VAR control using a reduced-order ingot pool depth model. In *Proceedings of the International Symposium on Liquid Metal Processing and Casting*, 2011.

[15] Kuang-O Yu, editor. *Modeling for casting and solidification processing*. CRC Press, 2002.

[16] Lang Yuan et al. Multiscale modeling of the Vacuum Arc Remelting process for prediction on microstructure formation. *International Journal of Modern Physics B*, 23:1584–1590, 2009.

Proceedings of the 2013 International Symposium on Liquid Metal Processing & Casting
Edited by: Matthew J. M. Krane, Alain Jardy, Rodney L. Williamson, and Joseph J. Beaman
TMS (The Minerals, Metals & Materials Society), 2013

SIMULATION OF RADIATION HEAT TRANSFER IN A VAR FURNACE USING AN ELECTRICAL RESISTANCE NETWORK

A. Stewart Ballantyne
MeltMet Technologies LLC; 3208 Mattia Ct; Charlotte, NC 28270, USA

Keywords: VAR, radiation, arc, model, electrical analogy, resistance network

Abstract

The use of electrical resistance networks to simulate heat transfer is a well known analytical technique that greatly simplifies the solution of radiation heat transfer problems. In a VAR furnace, radiative heat transfer occurs between the ingot, electrode, and crucible wall; and the arc when the latter is present during melting. To explore the relative heat exchange between these elements, a resistive network model was developed to simulate the heat exchange between the electrode, ingot, and crucible with and without the presence of an arc. This model was then combined with an ingot model to simulate the VAR process and permit a comparison between calculated and observed results during steady state melting. Results from simulations of a variety of alloys of different sizes have demonstrated the validity of the model. Subsequent simulations demonstrate the application of the model to the optimization of both steady state and hot top melt practices, and raises questions concerning heat flux assumptions at the ingot top surface.

Introduction

One of the important elements in the formulation of a comprehensive model for the VAR process is the simulation of the heat distribution in the vacuum arc. From a process modelling standpoint, a time-averaged description of heat transfer between the arc, electrode, ingot, and crucible wall would enable specification of the ingot top boundary condition in terms of furnace operating parameters which are readily available to a process engineer. At the completion of melting when no arc is present and the furnace is still under vacuum, heat transfer between the ingot, electrode, and crucible wall occurs essentially by means of radiative exchange between these surfaces. The solution of such heat transfer problems involving multiple reflections and re-reflections between "shiny" surfaces can be a very cumbersome, time-consuming exercise if ray tracing methods are used. However such radiation heat transfer problems can be solved much more readily through the use of an analogous electrical network in which current flow corresponds to heat flow, and voltage nodes correspond to radiation emitting surfaces. Once this formulation has been completed, a power source representing the arc may be introduced to permit calculation of the appropriate heat exchange between the relevant surfaces. While this approach ignores the detailed spatial energy distribution within the arc such as that due to anode and cathode falls, it does capture the overall energy balance in the system and, if successful, could lead to development of a macro process model to assist in the development of VAR practices.

Basic Network Elements

In the electrical analogy approach to solving radiation heat transfer problems, a typical surface element (Figure 1a) consists of a surface node (ω) having a "voltage" representing the Stefan-Boltzmann radiation ($\varepsilon\sigma T^4$), connected through a resistor to a radiosity node (J) representing the net radiation (both emitted and reflected) emanating from the surface. The value of the resistor connecting the surface node and the radiosity node can be shown[1] to be $(1-\varepsilon)/A\bullet\varepsilon$ where ε is the surface emissivity and A is the surface area. A simple network involving heat exchange between two surfaces (Figure 1b) consists of the two surface nodes connected to two radiosity nodes which are then joined to each other by a connecting resistor. This connecting resistor represents the amount of radiation emanating from one surface which reaches the second surface, and depends on the emitting surface area and the view factor or fraction of emitted radiation from surface 1 which reaches surface 2. The value of this connecting resistor is given by $1/A_1\bullet F_{12}$ where F_{12} is the view factor from surface 1 to surface 2; or by $1/A_2\bullet F_{21}$ by the reciprocity theorem[1]. In the resistive network, the potentials of the radiosity nodes can be determined by applying Kirchoff's law (∑Heat flows = 0) at each node and solving the resulting two simultaneous equations. This approach can then be expanded to construct a resistive network for a multi-surface problem as found in a VAR furnace.

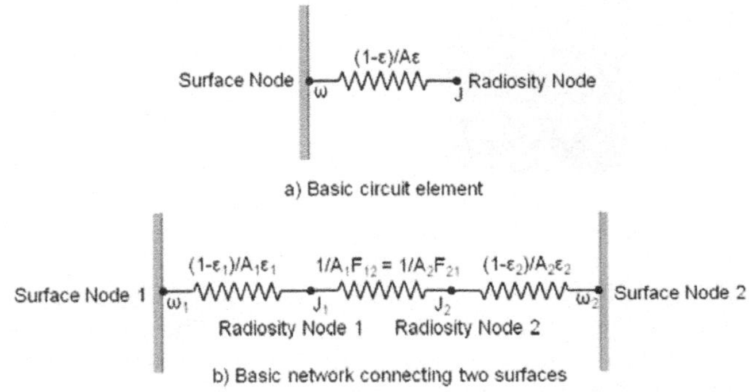

Figure 1. Network analysis elements

Resistive Network Without an Arc

Using a resistive network approach, the heat exchange between the ingot, electrode and crucible wall may be represented by the simple network shown in Figure 2. Nodes ω_I, ω_E, and ω_C representing the emissive power ($\varepsilon\sigma T^4$) of the ingot, electrode, and crucible wall surfaces are connected to nodes J1, J2, and J3 representing the radiosity of each surface. The resistors RI, RE, and RC relate the radiosity to the Stefan-Boltzmann radiation from each surface. The resistors RIE, RIC, and RCE which join pairs of radiosity nodes are evaluated from the respective surface area and view factor for each radiosity pair. Since some of the heat that reaches the electrode surface will be conducted into the electrode and lost to the crucible wall, an additional resistance representing this loss has been added to complete the circuit (Figure 3). The current flow (heat flow) in this leg is not calculated as part of the network solution, but is evaluated separately with a one-dimensional electrode model.

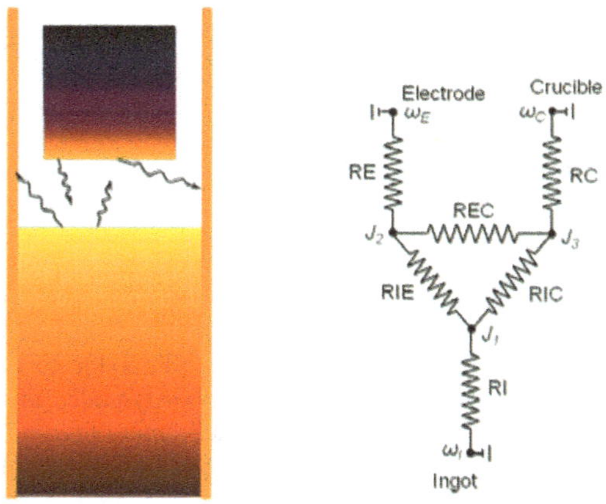

Figure 2. Network analysis for radiative exchange between electrode tip, ingot, and crucible wall

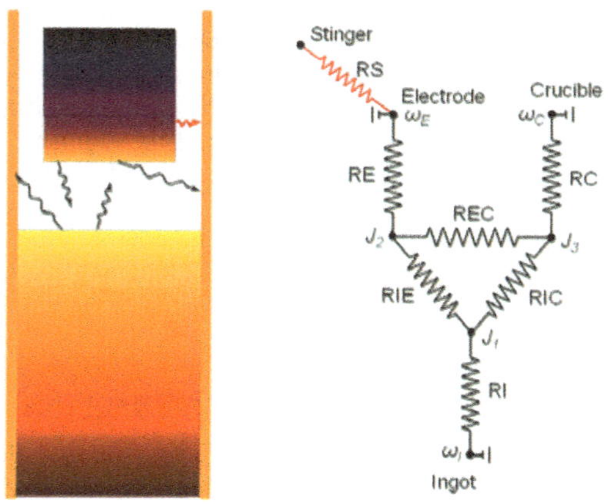

Figure 3. Network analysis incorporating conduction into stinger

The underlying assumptions made in this network (Figure 3) are that all surfaces are grey bodies and are isothermal, the latter assumption being reasonably valid for the electrode and the crucible wall, but decidedly invalid for the ingot surface. To accommodate this fact, the ingot surface was divided into eleven concentric rings each of which was assumed to be isothermal. As each ring was added to the network, the group of resistors within the encircled region was replicated with the resistor values being adjusted for the different ring areas and view factors (Figure 4). The result was a network comprised of eleven radiosity nodes for the ingot and one each for the electrode and crucible wall. View factors between the rings, the electrode and the crucible were calculated from the geometrical relationship between these elements. The application of Kirchoff's law at each radiosity node resulted in a system of thirteen simultaneous equations which were solved repetitively to calculate the heat transfer rates within the furnace in the absence of an arc. In this form, the electrical network in combination with an ingot model could be used to predict the post-melt solidification of an ingot once the arc has been extinguished. However, greater value could be derived from the model if it could be expanded to include a heat source representing the arc, and then coupled with an ingot model where it could provide the top boundary condition.

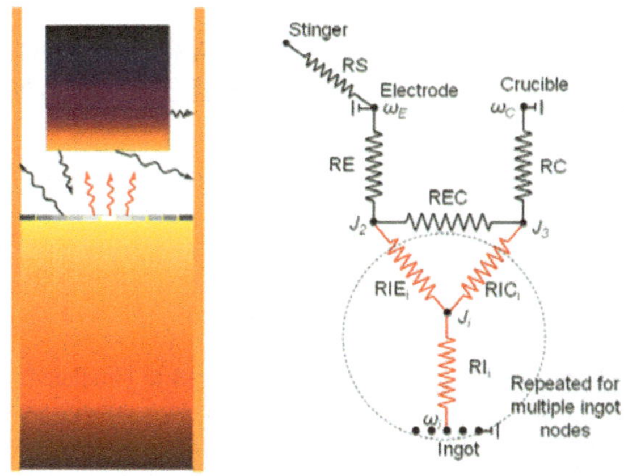

Figure 4. Modified network analysis to incorporate multiple ingot nodes

Resistive Network With an Arc

The introduction of a heat source into the resistive network first required a decision as to what physical form the source should take to represent a diffuse arc between the electrode and the ingot. The geometry of this pseudo-arc in relation to the electrode, ingot and crucible would determine the view factors needed to formulate the resistance network. Neither a point source nor a spherical source was felt to be physically compatible with the geometry of a gap between the ends of two cylinders.

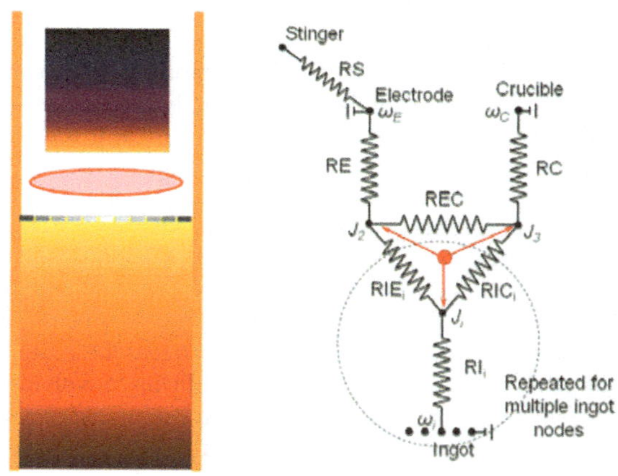

Figure 5. Introduction of planar heat source to simulate arc

As a result, the source was assumed to be represented by an opaque circular plane situated midway between the electrode and the ingot, and having a diameter equal to the average of the electrode and ingot diameters. This plane divided the arc gap horizontally (Figure 5) and was assumed to emit radiation uniformly and equally above and below the plane as would a uniform diffuse arc. This modification required the insertion of an arc node into the network at the location shown in Figure 5.

Because the arc was assumed to be opaque, the crucible wall was divided into two regions, above and below the plane of the arc since radiation from the upper surface of the arc plane would not fall on the lower region of the crucible wall and vice versa (Figure 6).

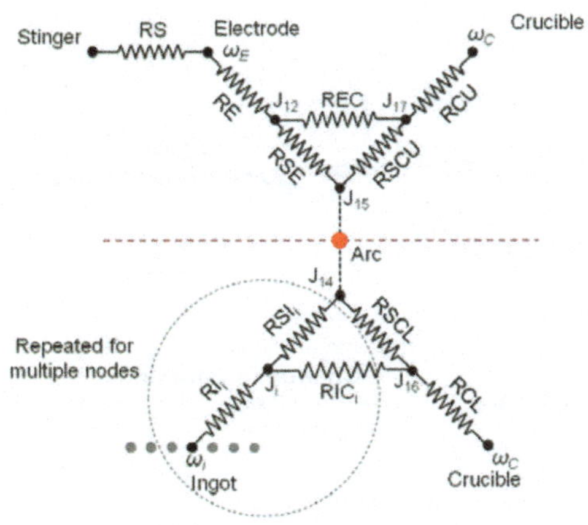

Figure 6. Network analysis for an opaque arc

Initial calculations with this model however yielded ingot top temperatures that were thought to be excessively high. Since the arc consists of a plasma and intermittent arc columns which continually move over the entire electrode surface, it was felt that a more realistic approach would be to assume that the arc plane was transparent. This meant that the electrode and ingot could see each other for a significant fraction of the time, and that the electrode and ingot surfaces could also see an expanded portion of the crucible wall through the arc. This assumption required the addition of a series of resistors (RIE_i) connecting the radiosity node of each annular ring on the ingot to the electrode radiosity node (Figure 7).

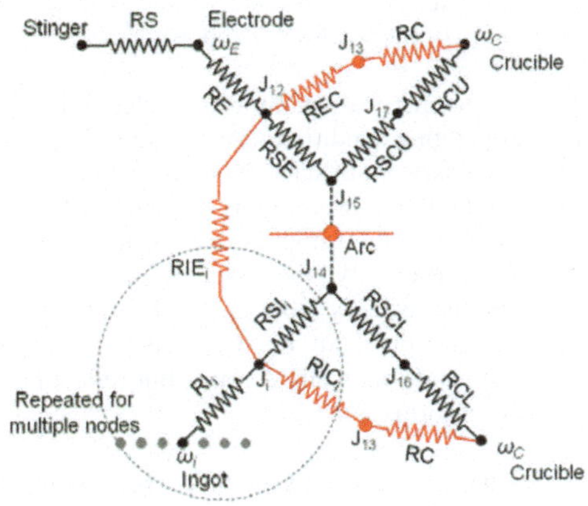

Figure 7. Modified network analysis for a transparent arc

Additionally, the resistances between the electrode and the crucible (REC + RC), and the ingot and the crucible (RIC_i + RC) had to be modified since each of these surfaces could now see an expanded crucible wall area through the transparent arc. Recognizing that the ω_C and J13 radiosity nodes above and below the arc plane are identical, the network was rearranged to that shown in Figure 8.

The assumption that liquid metal leaves the electrode at the liquidus temperature and arrives at the ingot at a superheated temperature above the liquidus requires the addition of a heat sink to withdraw this energy from the arc, depending on the specific heat of the alloy and the melt rate of the process.

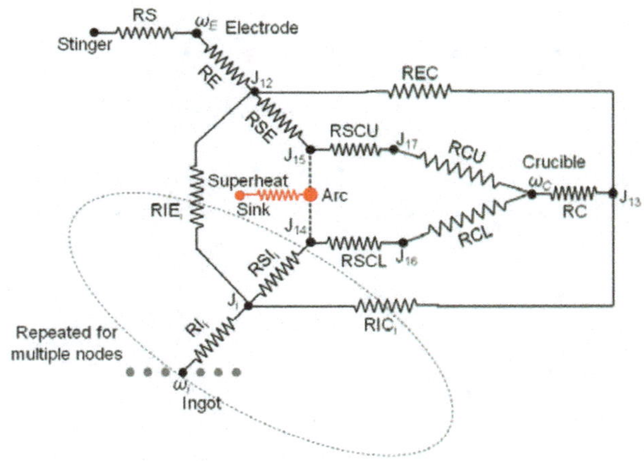

Figure 8. Modified network combining ω_C and J_{13} nodes, and adding a latent heat sink

The final modification to the resistance network was the inclusion of an additional power source representing the current flow from the electrode to the crucible wall above the arc gap. Axial voltage measurements along the crucible wall during melting confirm the presence of this side arc which has been estimated to account for 45% of the total current flow[4]. For the purpose of inclusion in the model, the physical representation of this arc was assumed to be that of a transparent concentric cylinder located midway between the electrode surface and the crucible wall (Figure 9).

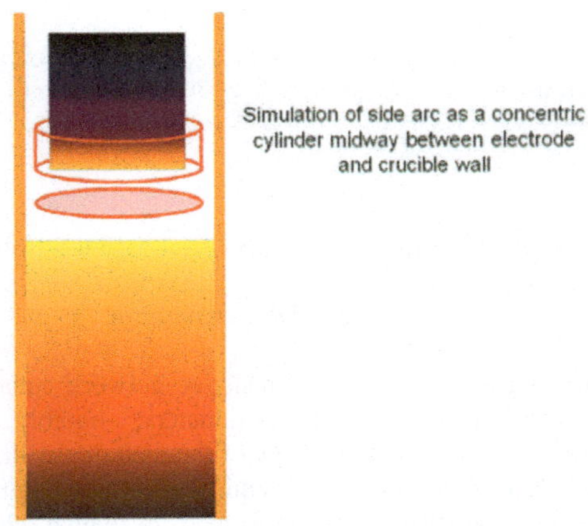

Figure 9. Physical representation of the side arc in VAR

The power generated in this side arc was defined as a percentage of the total furnace current, and was apportioned between the electrode and the crucible wall based on the view factors between these surfaces and the arc cylinder (Figure 10). Since the spatial distribution in the axial direction was of secondary concern with regard to the heat balance, the heat flows from the side arc to the electrode and crucible wall were concentrated in single nodes, J18 and J19. The end result of these changes is a resistive network (Figure 10) with nineteen radiosity nodes for which a solution must be obtained. Removal of the arc at the completion of melting leads to the network shown in Figure 11 which is equivalent to the original post-melt solidification network of Figure 2.

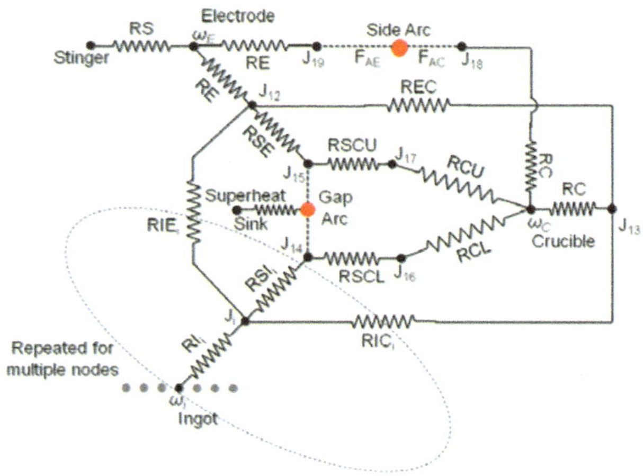

Figure 10. Final resistance network incorporating axial and side arcs in VAR

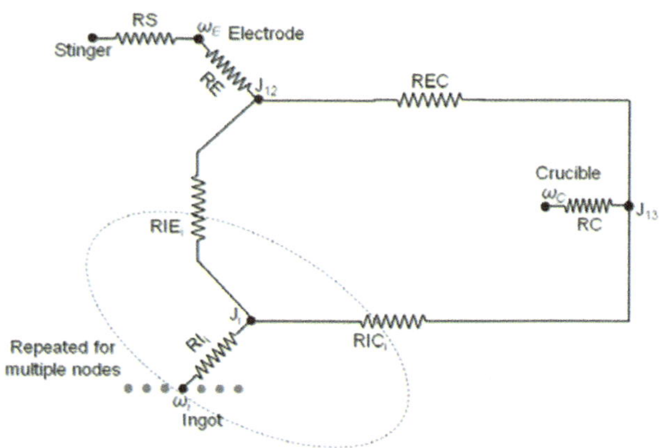

Figure 11. Post-melt network for final ingot solidification

Model Verification

Obtaining a solution to the resistive network equations requires specification of several parameters, notably the surface emissivities and the percentage of the melt power dissipated in the side arc. To determine which emissivities were the most critical, a sensitivity analysis was performed in which changes in the predicted melt rate were related to changes in the independent variable. The percentage change in melt rate due to a comparable percentage change in the selected emissivity (Figure 12) reveals that the electrode tip emissivity is the most critical parameter followed closely by that of the ingot pool. The emissivities of the crucible wall and electrode/stinger surface were of intermediate importance while that of the solidified ingot surface was minimal. The emissivities of the electrode tip and ingot pool were each assumed to be 0.2 since both are reflective molten metal surfaces. The crucible and stinger emissivities were assumed to be 0.4, being associated with duller, rougher surfaces.

An estimate of the power distribution between the axial arc and the side arc was obtained from the relative resistances of the arc gap and the annulus with the assumption that these resistances are

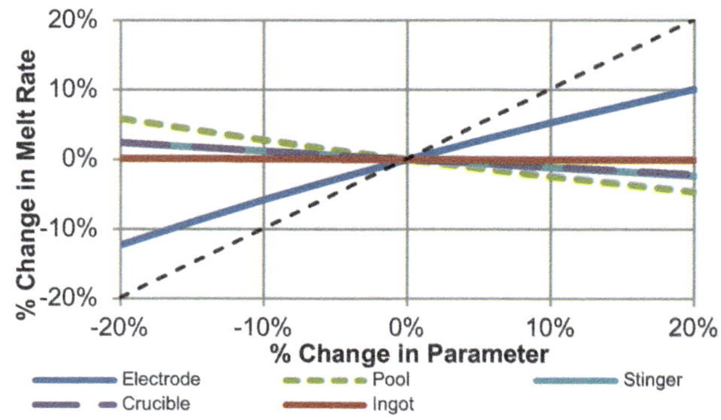

Figure 12. Effect of emissivity changes on melt rate

much greater than other resistances in the circuit[7]. The resistance of the electrode-ingot gap (ρL/A) is calculated to be $4\rho g/\pi D^2$ while that of the electrode-crucible gap is $\rho a/\pi Dh$ where g represents the arc gap; a, the annulus; h, the electrode height over which the side arc is distributed, and D, the axial and side arc diameters which are assumed to be equal. The ratio of gap resistance to annulus resistance is thus given as 4g•h/a•D. Voltage measurements[3] on the crucible wall during melting of a 500 mm diameter ingot of alloy 718 reveal that the voltage gradient drops to zero approximately 100 mm above the arc gap. For a typical ~430 mm diameter electrode, the height of the side arc would be ~0.25•D. The ratio equation thus becomes g/a, which for an annulus of 38 mm (432 mm electrode, 508 mm ingot) and an arc gap of 10 mm, leads to an estimate of the side arc power of approximately 20% of the total furnace power.

The remaining geometrical parameters (electrode and ingot diameters) and operating conditions (voltage, current, and arc gap) complete the necessary input data. Notably, voltage and current combine to specify the power input with no distinction being made between low voltage – high current or high voltage – low current conditions. Presently, the assumption is made that the power is partitioned equally above and below the arc plane recognizing that this may not be the case in practice due to differences in the anode and cathode falls, but reflecting a lack of more quantitative information.

Verification of the model results involved several phases. Heat balances over the electrode, arc, and ingot domains were calculated and compared to ensure internal consistency (Figure 13). The arc balance input is calculated from the melt power while the various losses are calculated from the solution to the resistive network. Agreement between input and output is very good with a closure of 0.07%. The electrode balance is superficial since the energy from which melt rate is calculated is not determined independently, but is simply the difference between the net energy into the electrode tip minus the stinger loss. The ingot balance, derived from the results of the ingot model, shows a heat balance difference of -7.8 kw compared to a calculated enthalpy difference of 1.8 kw out of a total energy flow of 100 kw. The heat accumulation within the ingot can appear as a source or a sink depending on whether the sensible heat of the ingot is decreasing or increasing respectively.

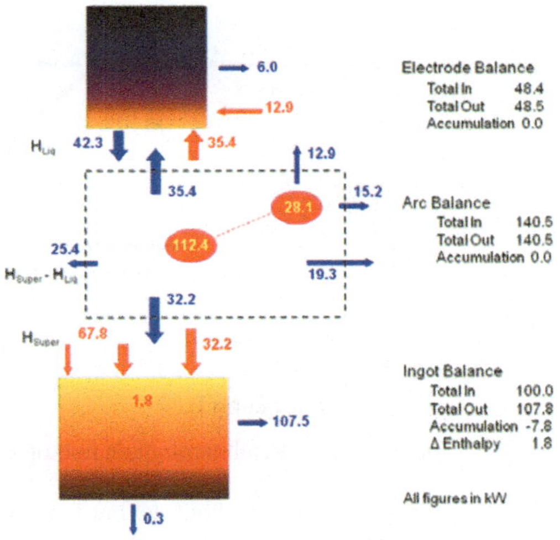

Figure 13. Heat balance - 508 mm 718 ingot at 6000 A

The next phase of verification involved comparisons between model predicted and observed melt rates. In reported examples where sufficient data (voltage, current and arc gap) were available, simulations of several different alloys of different diameter were performed. The alloys involved were 718, AISI 4340, M2 high speed steel, Waspaloy, and a marage steel; in sizes ranging from 300 mm to 1520 mm. Based on reported operating conditions[6,9,10], the calculated melt rate plotted against the observed melt rate (Figure 14) exhibited very good agreement. The specific power consumption over the examined range of sizes averaged 650 kwh/tonne (Figure 15) with the value for a 508 mm ingot of 718 (617 kwh/tonne) agreeing well with reported values[2].

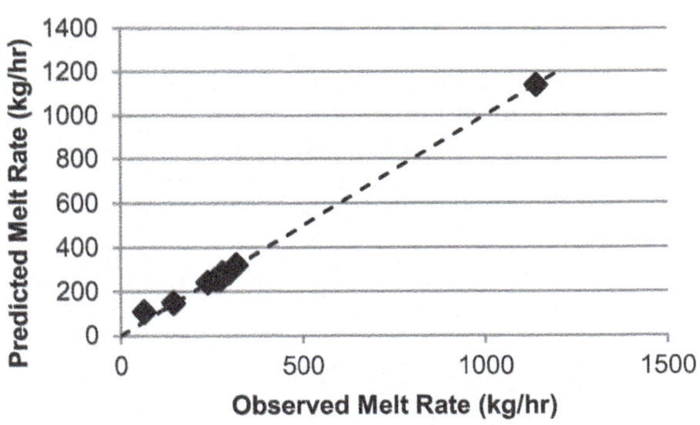

Figure 14. Predicted melt rate versus observed melt rate

Finally, for each of the aforementioned cases, the pool depth was calculated and compared to reported values[5,9,10]. In most cases, there was reasonable agreement between calculated and reported pool depths (Figure 16) with the discrepancies being attributed to uncertainty in the material thermophysical properties.

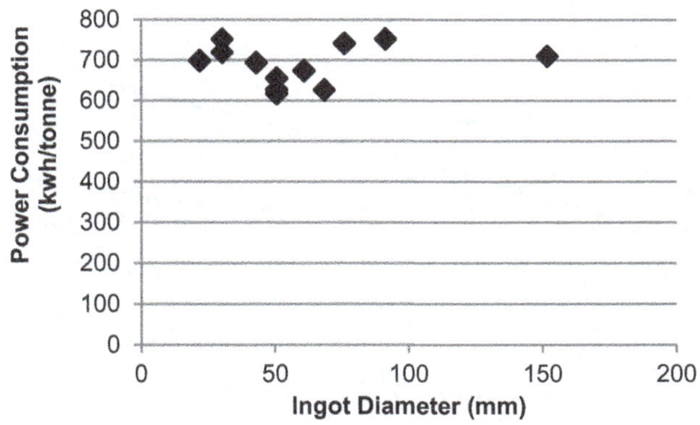

Figure 15. Specific power consumption versus ingot diameter

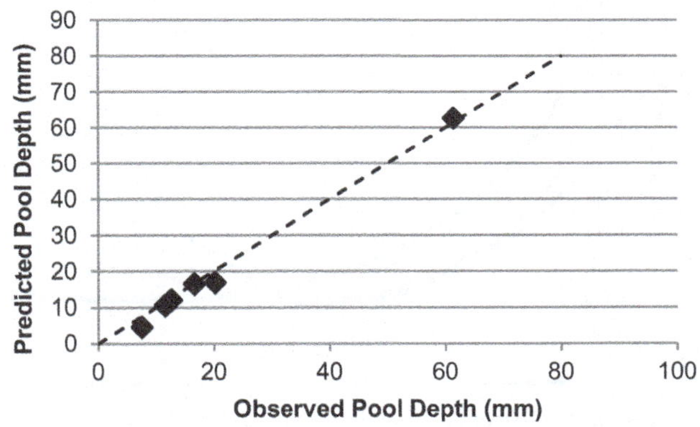

Figure 16. Predicted pool depth versus observed pool depth

Model Applications

Several examples of applications of the model illustrate how the model can be used to assist in the development a melt practice and contribute to a deeper understanding of the process. The first was that of using the model to examine the solidification conditions in an ingot during steady state melting.

It is well known that solidification conditions vary as a function of melt rate (Figure 17). In particular, the existence of a minimum in the local solidification time (LST) was predicted from modeling work several years ago[11], and selection of a melt rate to minimize the LST is known to convey benefits in terms of structure refinement. However, it is also interesting to examine some hypothetical cases to see how this effect may vary with different alloy systems. In the following example, alloy 718 is compared with two hypothetical alloys similar to 718 in all respects, but having narrower (1300-1350) and wider (1100-1350) liquidus-solidus ranges. The LST has been normalized to the minimum for each respective alloy and plotted as a function of melt rate (Figure 18), and the results reveal that as the liquidus-solidus range widens, the minimum LST moves to a higher melt rate. This unfortunately is not the trend we would like to see since a more highly alloyed system would tend to exhibit a wider liquidus-solidus gap, but would also likely exhibit a greater tendency toward freckle formation. Under such circumstances, use of a well substantiated model may assist in the definition of an acceptable melt practice window.

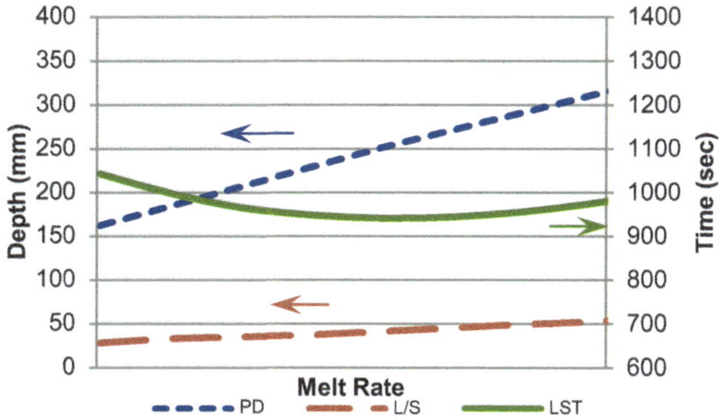

Figure 17. Variation in solidification conditions with melt rate

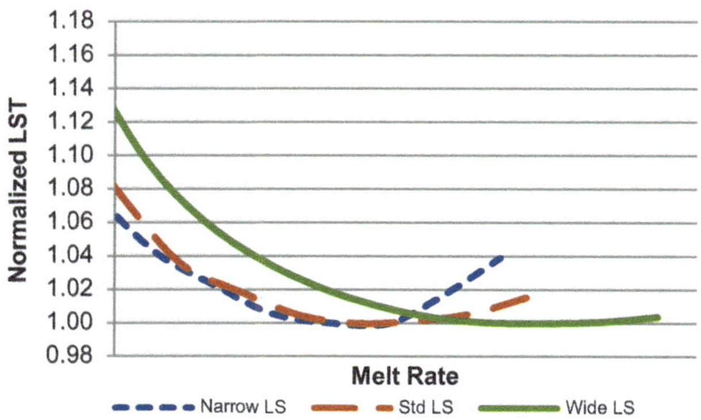

Figure 18. Effect of liquidus-solidus range on LST-melt rate minimum

The second application illustrates the use of the model to examine solidification conditions throughout a hot top cycle and the subsequent post-melt cooling cycle for a 508 mm diameter ingot of alloy 718. Two cases consisting of a full power abort from steady-state melting and a stepped reduction in melt power over the last 10% of the ingot were examined (Figure 19). As expected, the stepped power reduction led to a diminishing pool depth throughout the hot top cycle (Figure 20). Predicted post-melt solidification times varied from 46 minutes for the non-hot topped ingot to 26 minutes for the hot topped case, both times being within the realm of expected values. The final solidification void, assumed to be associated with the maximum final temperature, was calculated to be 60 mm below the ingot top in reasonable agreement with literature results[8].

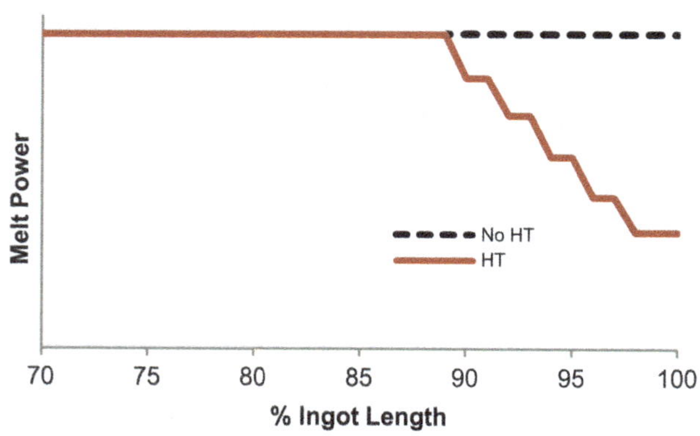

Figure 19. Simulated hot top profile

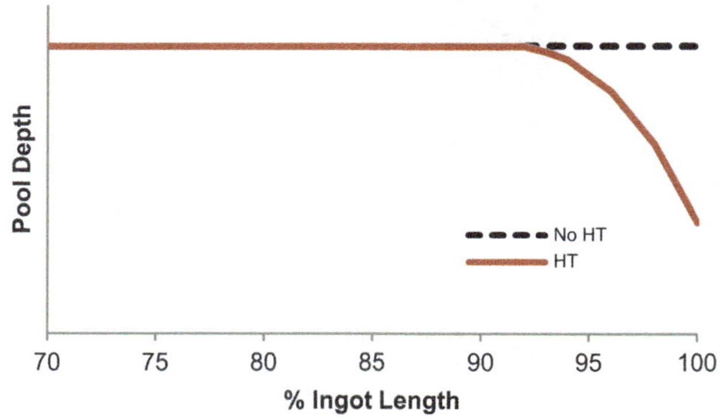

Figure 20. Variation in pool depth throughout hot top cycle

The predicted liquid-solid gap throughout the hot top cycle does not exhibit a large variation (Figure 21), however the LST shows significant (>50%) difference between hot topped and non-hot topped ingots (Figure 22). The final ingot quality and cropped yield will depend on the relationship between structure and LST, and the extent to which these final solidification conditions can be manipulated by operating parameters.

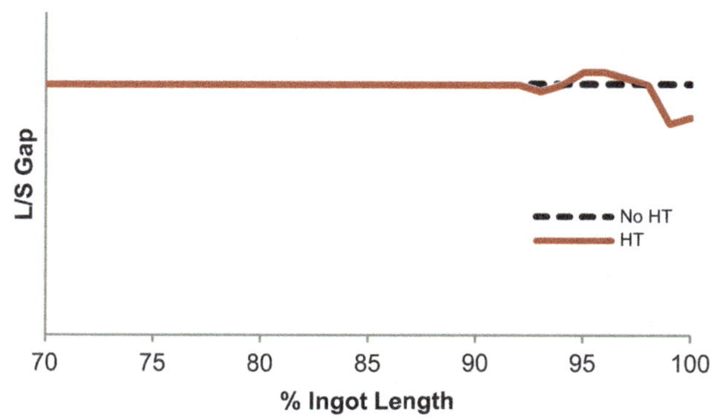

Figure 21. Variation in liquid-solid gap throughout hot top cycle

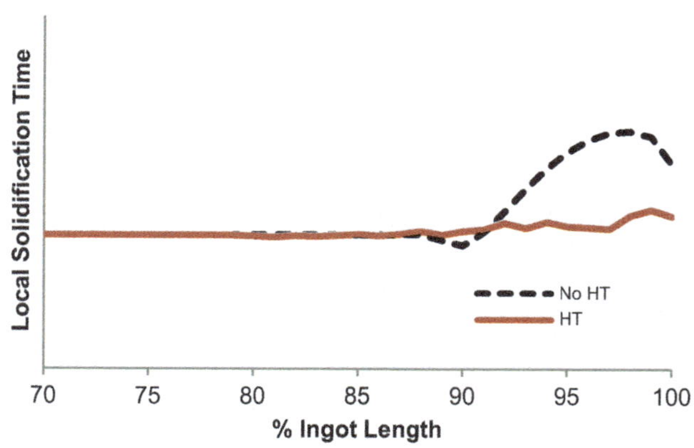

Figure 22. Variation in local solidification time throughout hot top cycle

Finally, it is interesting to examine the energy distribution over the top ingot surface. The assumption of a uniform incoming metal flow leads to a uniform sensible heat input to the top of ingot. However, the calculated net radiation flux based on the assumption of a uniform diffuse arc and incorporating the various internal reflections does not yield a uniform radiation flux to the ingot, but rather yields a flux distribution which increases by about 25% from the ingot center to the edge of the electrode shadow on the ingot and which then drops within the annulus region (Figure 23). This variation reflects the variation in the electrode-ingot view factor and the ingot-electrode temperature difference from ingot center to edge. While this distribution to the ingot is in part a consequence of the assumed uniform diffuse arc, the question arises as to how the usual assumption of a Gaussian distribution[5] for the radiation heat input is affected by reflections within the arc gap. Preliminary results from a model incorporating a multi-zone source (Figure 24) suggest that the internal reflections reduce the Gaussian distribution within the arc to a more uniform distribution at the ingot surface.

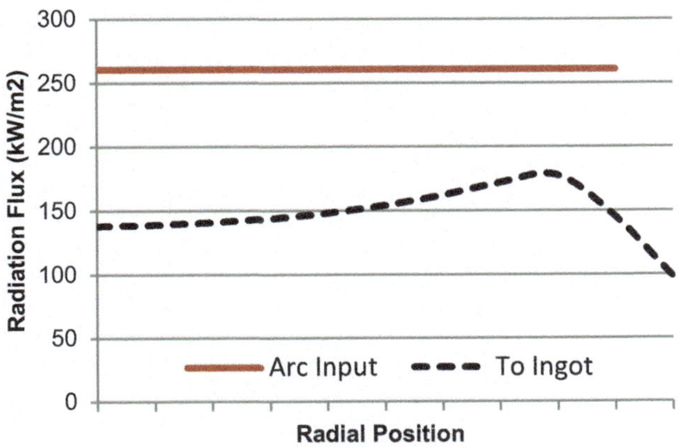

Figure 23. Heat fluxes for a uniform arc flux input.

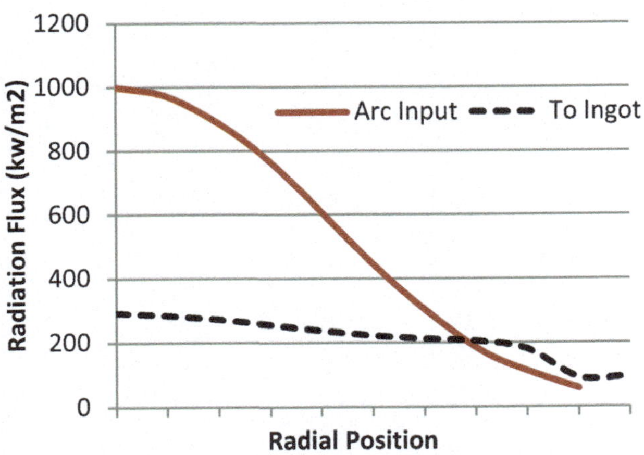

Figure 24. Heat fluxes for a Gaussian arc flux input.

Summary

The solution of the radiative heat transfer problem between the ingot, electrode and crucible wall within a VAR furnace under vacuum was accomplished through the use of an analogous electrical resistance network. This network, subsequently modified to include planar and cylindrical radiating heat sources representing the gap and side arcs, led to a comprehensive model capable of predicting the steady state melt rate of a variety of alloys of different diameters using only power input and arc gap as input variables. The combination of this arc model with an ingot model provides a useful macro engineering tool for simulation of the VAR process from steady state through the hot top cycle and post-melt solidification which should aid process engineers in the optimization of a melt practice. Preliminary results from a multi-zone arc model indicate that internal reflections between the electrode, ingot, and crucible alter the incoming arc energy distribution to yield a significantly different distribution at the ingot pool surface..

References

1. Sucec, J., "Heat Transfer", Simon & Schuster, 1975.
2. Zanner, F.W., Williamson, R.L., & Harrison, R., Proc. of 9th Intl. Conf. on Special Melting, AVS, 1988, p. 551-563.
3. Ward, R.M. & Jacobs, M.H., Proc. Int. Symp. On Liquid Metal and Castings, 2001, p. 244-257.
4. Shevchenko, D.M. & Ward, R.M., Met. Trans. B, 2009, June, p. 248-253.
5. Bertram, L.A., Adasczik, C.B., et al, Proc. Int. Symp. On Liquid Metal Processing and Castings, 1997, p. 110-132.
6. Risacher, A., Chapelle, P. et al, Proc. Int. Symp. On Liquid Metal and Castings, 2011, p. 1-8.
7. Williamson, R.L. & Shelmidine, G.J., Superalloys 718, 625, 706, and Various Derivatives, 2001, 91-102.
8. Reiter, G., Maronnier, C. et al, Proc. Int. Symp. On Liquid Metal and Castings, 2003, p. 77-86.
9. Busby, P.E., Trans. Vac. Met. Conf., Amer. Vac. Soc., 1964, p. 380-398.
10. Preston, J., Trans. Vac. Met. Conf., Amer. Vac. Soc., 1964, p. 366-379.
11. Ballantyne, A.S., Ph.D. Thesis, University of British Columbia, 1978.

Proceedings of the 2013 International Symposium on Liquid Metal Processing & Casting
Edited by: Matthew J. M. Krane, Alain Jardy, Rodney L. Williamson, and Joseph J. Beaman
TMS (The Minerals, Metals & Materials Society), 2013

SOLIDIFICATION MAPPING OF A NICKEL ALLOY 718 LABORATORY VAR INGOT

Trevor J. Watt, Eric M. Taleff, Felipe Lopez, Joe Beaman, Rodney Williamson

The University of Texas at Austin, Department of Mechanical Engineering

Keywords: vacuum arc remelting, solidification microstructure, image processing

Abstract

The solidification microstructure of a laboratory-scale Nickel alloy 718 vacuum arc remelted (VAR) ingot was analyzed. The cylindrical, 210-mm-diameter ingot was sectioned along a plane bisecting it length-wise, and this mid-plane surface was ground and etched using Canada's reagent to reveal segregation contrast. Over 350 photographs were taken of the etched mid-plane surface and stitched together to form a single mosaic image. Image data in the resulting mosaic were processed using a variety of algorithms to extract quantities such as primary dendrite orientation, primary dendrite arm spacing (PDAS), and secondary dendrite arm spacing (SDAS) as a function of location. These quantities were used to calculate pool shape and solidification rate during solidification using existing empirical relationships for Nickel Alloy 718. The details and outcomes of this approach, along with the resulting comparison to experimental processing conditions and computational models, are presented.

Introduction

Vacuum Arc Remelting (VAR) is a secondary melting process used to remove impurities from ingots cast using Vacuum Induction Melting (VIM) or similar primary techniques. During VAR, a continuous arc is struck between an electrode and a solidifying ingot, which results in material melting off the electrode and falling into a melt pool that sits atop the new ingot (Figure 1). By performing this procedure in a vacuum (typically 10^{-6} Torr), oxides and other volatile compounds are removed from the ingot material, which improves material properties [1]. Despite this advantage, the VAR process is susceptible to ingot defects produced by un-dissolved material fall-in (white spots), macro-segregation due to convection instabilities (freckle channels), and discontinuous solidification structures due to thermal perturbations in the melt pool (tree rings) [2].

To prevent such defects from forming, certain melt conditions should be maintained throughout the VAR production cycle. In order to prevent solute-lean white spots, deep melt pools need to be used so that any fall-in material has time to dissolve into the melt pool [3]. To prevent solute-rich freckle channels, the solidification front must be resistant to the formation and propagation of thermosolutal convection caused by local Rayleigh-Taylor instabilities. This can be accomplished by reducing the buoyancy driving force for convection or by using a shallow pool [4]. Preventing both white spots and freckles is therefore a balancing act that requires using a melt pool that is shallow enough to avoid freckle channels, while being deep enough to dissolve any solid material that falls in from the shelf, crown, or electrode regions.

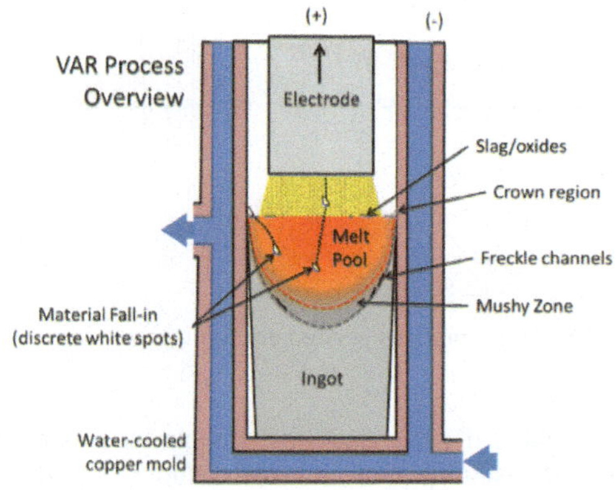

Figure 1. VAR process and solidification defects

Tree rings, the third of the aforementioned defects, are caused by thermal instabilities in the melt pool that produce local constitutional undercooling ahead of the advancing solidification front [5]. With sufficient time and undercooling, randomly-oriented "equiaxed" dendrites can nucleate and block the approaching solidification dendrites. These features show up as "tree rings" when a VAR ingot is sectioned by a plane perpendicular its length and this cross-section is examined with a macro-etchant. Tree rings are not as deleterious as white spots or freckles insofar as fatigue strength is concerned, but they are still considered undesirable [2]. Preventing tree rings relies primarily on controlling the current input into the melt pool such that temporal variations in heating do not occur. This is achieved through robust power supply design and careful control over the arc gap distance, but these alone are not 100% effective due to drip-shorts that occur randomly and result in unpredictable voltage and current fluctuations [6].

Controlling the pool shape during solidification is a critical aspect of being able to minimize defects in VAR ingots. To accomplish this task, the software package BAR (Basic Axisymmetric Remelting) has been used to exert greater control over the pool profile during the VAR process. BAR is a finite volume model developed by Sandia National Laboratories to simulate solidification in the vacuum arc remelting process. The model simultaneously solves conservation equations for energy, momentum, charge, and mass. It has been used extensively both in research environments as well as in industry, showing good agreement with experimental results [7]. During the summer of 2011 at Los Alamos National Laboratory (LANL), BAR was used in closed-looped control for the solidification of a Nickel Alloy 718 VAR ingot with different stages of melt pool control. This was the first time that the pool depth controller incorporated feedback from the VAR process measurements. The resulting laboratory-scale, cylindrical ingot was approximately 60 cm tall and 20 cm in diameter. Plots of the arc current and voltage *vs.* position from the base of the ingot are shown in Figure 2. The region from approximately 38 to 47 cm corresponds to a constant pool depth set-point for BAR.

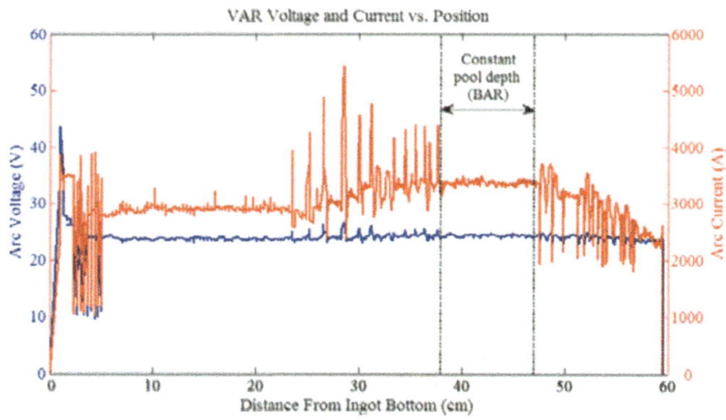

Figure 2. VAR ingot process current and voltage profiles

This ingot was sectioned on a plane that bisects its length for observation of its solidification microstructure. Additional sections were then cut simply to ease handling and preparation of the section surface for observations. Photographs of the ingot cross-section after grain-etching are shown in Figure 3. As shown in the figure, the ingot was separated into four pieces to facilitate preparation.

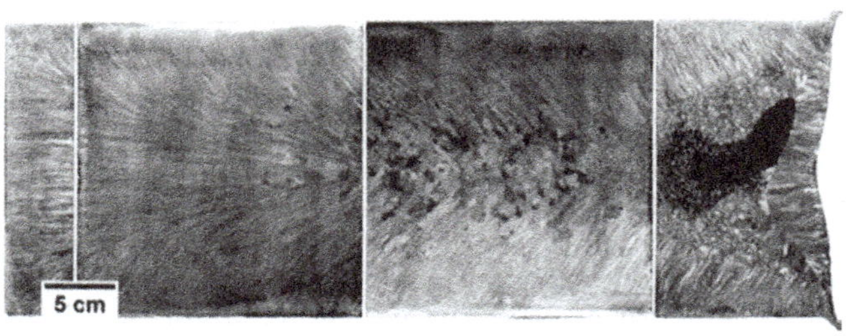

Figure 3. LANL VAR ingot cross-section photograph (grain etch) (Image is rotated 90° clockwise relative to other ingot images. The ingot diameter is oriented vertically in this image.)

The goal of this study is to analyze the VAR ingot and compare the results against the predictions made by BAR. This is accomplished using optical photography, hand measurements, and various image-processing techniques. Pool shapes are determined using two methods: First, by recording tree rings or freckle channels that serve as pool markers. Second, by examining the local primary dendrite orientation, which is an indicator of the local thermal gradient, and hence a local normal to the melt pool surface [8]. In addition to pool shape, the local solidification rate can be estimated by examining the primary dendrite arm spacing (PDAS), as well as the secondary dendrite arm spacing (SDAS). Both of these features are correlated with the local solidification rate, which can further be decomposed into a product of thermal gradient and solidification front velocity [9-11]. PDAS and SDAS measurements are historically made by hand from select regions of interest (ROIs). While there has been some work using semi-automated measuring techniques, to the authors' knowledge a fully-automated PDAS/SDAS measurement technique has not been attempted to date. Such a technique will be employed here, largely out of necessity given the large number of measurements needed to construct a reasonably dense solidification map.

Image Processing

The LANL ingot was first grain-etched and photographed, then ground and etched with a segregation etch (Canada etch). The photograph shown in Figure 3 was taken with each slab of the ingot photographed separately as a single image. The resulting resolution is sufficient for an overall assessment of ingot microstructure and for extracting some pool profiles, but it is insufficient for assessing any fine-scale features. To obtain higher resolution data, the Canada-etched ingot slabs were photographed as a mosaic using an 8 megapixel SLR camera (Canon EOS Digital Rebel XT) and a 90 mm macro lens (Sigma f/2.8), mounted on a vertical stand. Various lighting techniques were used, but it was ultimately determined that ambient room (overhead fluorescent) lighting was best for producing a consistent exposure with sufficient contrast. An example image is shown in Figure 4, which also provides a magnified view showing primary and secondary dendrites. The spacing between secondary dendrites was generally around 4 to 6 pixels, which is small but sufficient for calculating dendrite spacing.

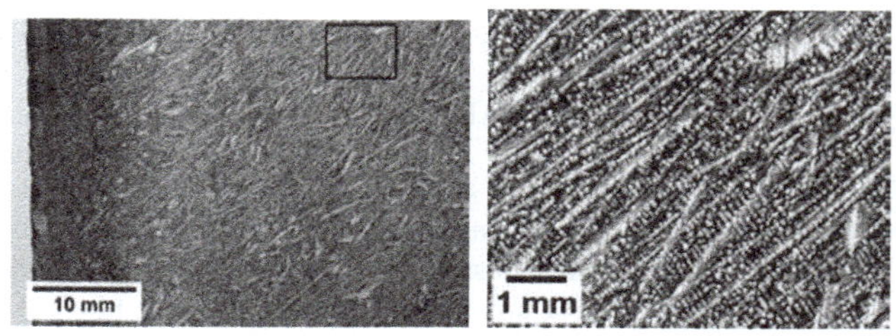

Figure 4. Example photograph of the Canada-etched LANL ingot surface (center-left of 3rd slab)

A single 8MP photograph of the ingot corresponds to a region approximately 4.3×2.9 cm in size (at 12.7 μm per pixel). The four slabs shown in Figure 3 (in order from left to right) required 26, 126, 124, and 75 individual photographs, respectively. A linear overlap of 20 to 50% was used for each photograph so that stitching could be performed to reconstruct a single, continuous image. FIJI (FIJI is Just ImageJ) was used for the majority of the image stitching and processing [12]. FIJI is a custom-build of ImageJ that incorporates numerous plugins designed for image processing, segmentation, and registration used in the life sciences, and it can take advantage of the increased memory capacity of 64-bit computers needed for working with large data sets. To perform the image stitching, the TrakEm2 plugin was used, as it best facilitated initial placement of the large image array prior to optimization and employs the SIFT (Scale Invariant Feature Transform) algorithm for robust automatic feature detection [13-16]. The stitching algorithm calculates a transformation for each individual image such that the SIFT-selected points within an image overlay with the same points in the surrounding images. The results of the initial stitching for the second slab are shown in Figure 5 (left).

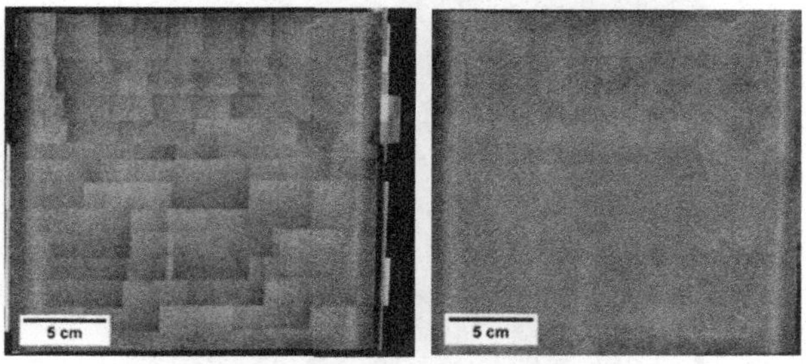

Figure 5. VAR ingot composite photographs (before and after affine-trans/blending)

The stitched photographs for the second slab are shown after blending in Figure 5 (right). The blending procedure creates a mask for each photograph that results in a minimum change in brightness between the overlapping pixels. Once this is done, the mosaic was "flattened" and exported as a single TIF image. Each of the four resulting mosaics was then combined into a single mosaic for the entire ingot, shown in Figure 6. The four image mosaics were all processed using the automatic SIFT algorithm, as described above, but were manually set in place since there were no overlapping pixels between slabs. Statistics for the composite images are provided in Table I.

Figure 6. VAR ingot mosaic (full size: 782 megapixels) (Image is rotated 90° clockwise relative to other ingot images)

Table I. Image Processing Statistics

	Mosaic Size *pixels (MP)*	Particle ROIs (FIJI)
Slab 1	18,215 × 5,116 (93)	13,709
Slab 2	20,624 × 16,337 (337)	73,400
Slab 3	19,636 × 16,434 (323)	184,892
Slab 4	18,244 × 11,051 (202)	84,520
Mosaic	17,216 × 45,408 (782)	356,521

ROI = region of interest

After stitching, the primary dendrite arms were analyzed in FIJI. The image processing and analysis used the following steps (see Figure 7): The mosaic was first processed using contrast-limited adaptive histogram equalization (CLAHE), which establishes a uniform histogram across the image area [17]. A bin size of 127×127 pixels was used for the CLAHE. A uniform histogram is required for proper thresholding, which was the next step in the process, since variations in image illumination can result in entire image regions being incorrectly thresholded. Thresholding converts the grayscale image into a binary image, which can then be processed using various "morphological" techniques. The mosaics were thresholded at a value of 127, or one-half of maximum 8-bit image intensity of 255. The following morphological processes involved erosion (removal of feature edges), followed by dilation (growth of feature edges). The erosion algorithm was applied twice, followed by dilation. This resulted in the effective detachment of secondary dendrite arms from the primary dendrite arms, which serves to isolate most of the primary dendrite arms so they could be subjected to FIJI's particle analysis algorithm. This last step was performed using a minimum particle area of 100 pixels, and a maximum circularity of 0.5 to help filter out globular features. The particle analysis routine in FIJI searches for all regions of connected pixels, and calculates various feature properties such as center of mass, bounding box coordinates, and elliptical fitting. The results of the analysis can then be exported to a text file for further processing.

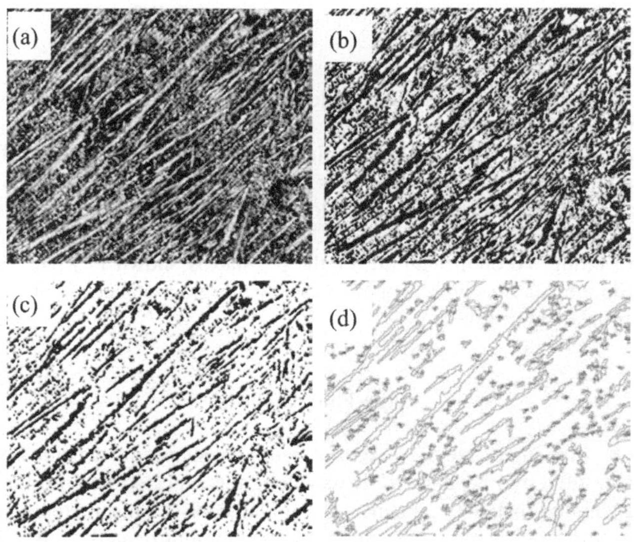

Figure 7. VAR ingot primary dendrite images (Slab 2): (a) After CLAHE filter, (b) thresholding, (c) erosion ×2 + dilation, (d) particle analysis ROIs

After processing in FIJI, the particle region of interest (ROI) data was loaded into Matlab for additional filtering. The following conditions were used for filtering the initial ROI list: (1) Aspect ratios for a fitted ellipse of less than 5 were rejected to avoid clusters of dendrites and nearly equiaxed features. (2) A minimum object length of 20 pixels (254 μm) was used to filter out inadvertent secondary dendrite arms and other small-scale noise. (3) A maximum length of 1000 pixels (12.7 mm) was used to remove excessively long dendrites or image boundary artifacts. (4) ROIs with a dendrite pixel intensity variation of greater than 50% were rejected to filter out non-lineal shapes. This step was performed by calculating the standard deviation of the pixel intensity across the appropriate diagonal of the ROI bounding box and dividing it by the corresponding mean pixel density. ROIs with a resulting error of greater than 50% were rejected. The result of these four filtering steps was an almost tenfold reduction in ROIs. A line plot of the resulting primary dendrites for the 3rd slab is shown in Figure 8.

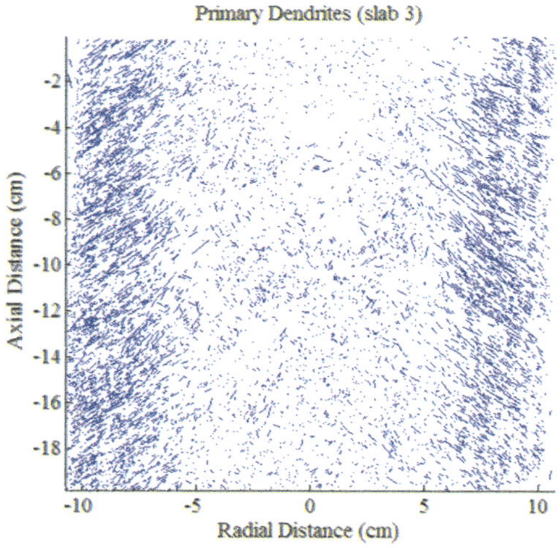

Figure 8. VAR Ingot primary dendrites (3rd slab)

An easier to interpret view of dendrite orientation is obtained by using a Voronoi diagram, which calculates the cells surrounding a collection of irregular data points [18]. Each cell contains all points in space whose location is closer to the contained data point than to any surrounding data location. The size of the cells is therefore a direct indication of the local spatial resolution. The resulting Voronoi diagram for the 3rd slab is shown in Figure 9 (bottom), along with a photograph of the grain-etched ingot for comparison (top).

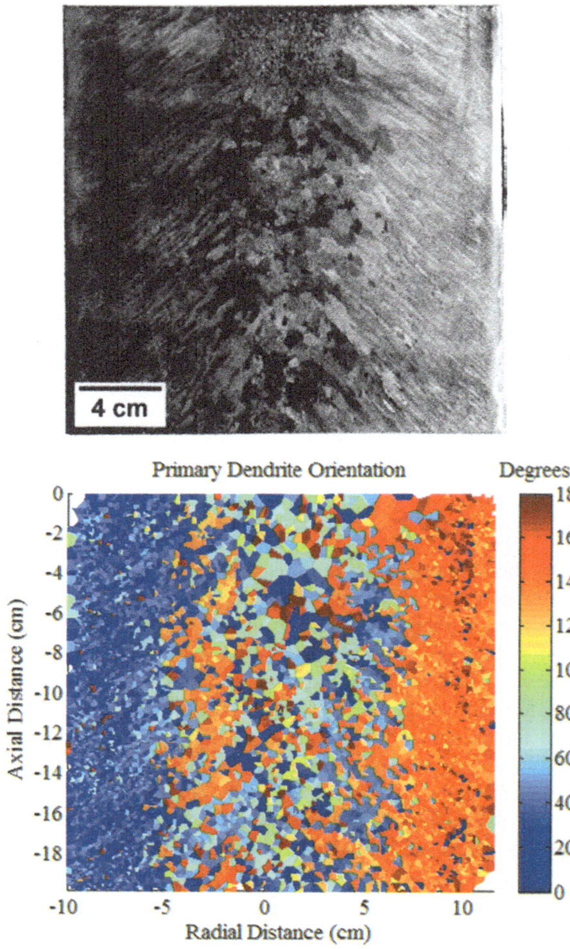

Figure 9. Grain etch (top), and primary dendrite orientation map (bottom) for 3rd slab. The scale bar indicates degrees counter-clock-wise from the positive radial direction, as shown on the horizontal axis.

Melt Pool Analysis

Once the image mosaics were formed, pool shapes were manually measured by tracking dark bands along the expected pool profile, attributed to freckle channels or similar dark-etching regions, or by tracking regions with clusters of randomly-oriented dendrites akin to tree rings [4,5,19,20]. This technique is a common practice, though some intrinsic subjectivity exists in the resulting measurements. The pools were marked using the multi-point selection tool in FIJI, exported to a text file, and imported into Matlab for additional processing. Several manually-tracked pool shapes for the third slab are shown in Figure 10.

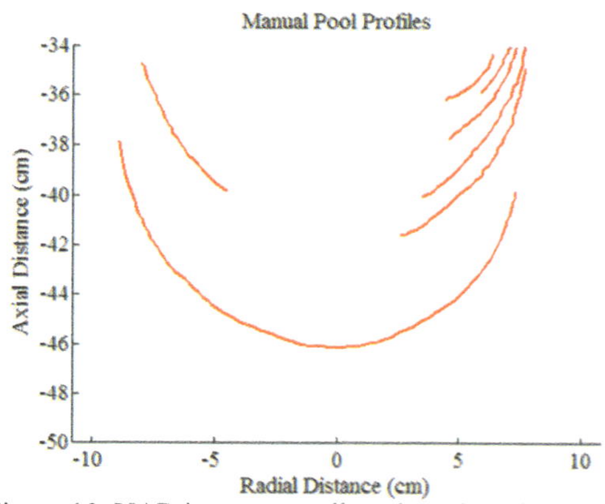

Figure 10. VAR ingot manually-selected pool profiles

For pool shape estimates, the angle of the primary dendrite arms can be used as an indicator of the pool shape normal. This is based on the fact that dendritic growth is a competitive process where the fastest-growing dendrites are those oriented in the direction of maximum heat transfer, or greatest thermal gradient. Therefore, the dominant dendrite orientation at a given location should be an indicator of the local thermal gradient, and subsequently pool shape. However, it should be noted that some dendrites can extend for tens of millimeters into the ingot and may, as a result, provide a false indicator of the local pool shape. Dendrite orientation is therefore most relevant to melt pool shape near the base of the dendrite or at its lowest point in the ingot.

To compare the hand-measured pool shapes to the automated dendrite processing results, the closest points on the dendrite map were chosen next to the pool profiles, as shown in Figure 11 for slab 2. Slab 2 was chosen for this comparison because it had the best microstructure of the four slabs for manually tracing pool profiles. The top image in Figure 11 shows the overlaid pool profiles (red) on the primary dendrite lines (blue). The bottom image shows the pool profiles (red lines), and the closest dendrite locations used for comparison (blue points).

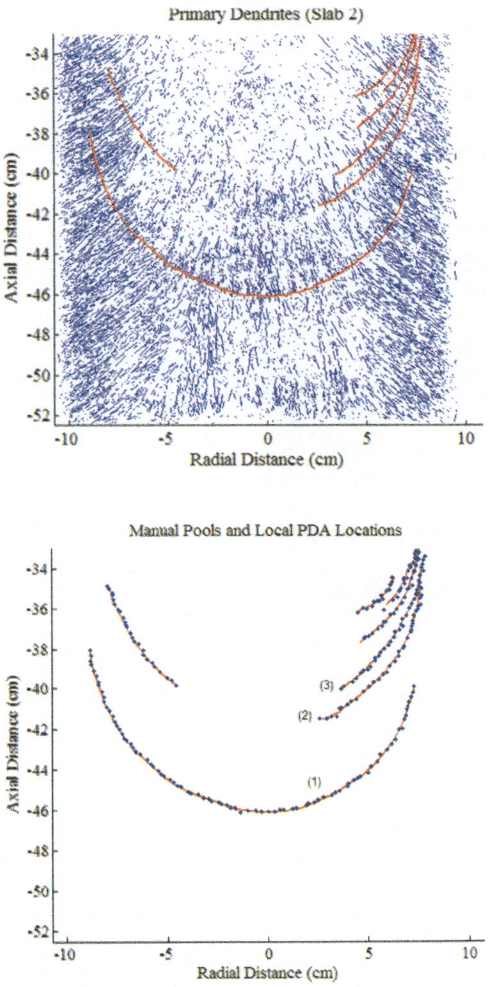

Figure 11. Slab 2 pool profiles (top), and selected dendrite locations (bottom)

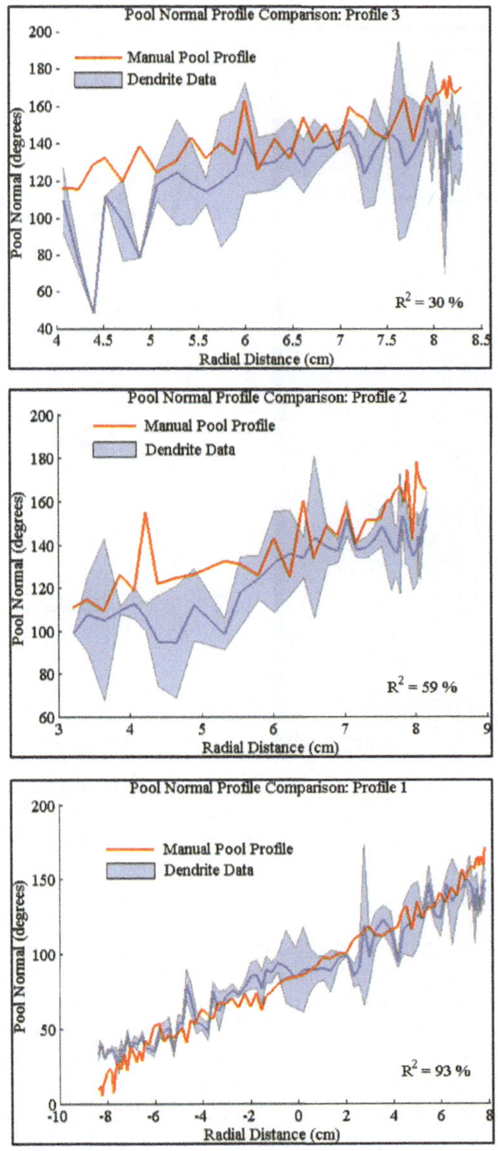

Figure 12. VAR pool profile normal comparisons (slab 2 region) (The shaded region indicates the local standard deviation of the primary dendrite orientations within a 2 mm radius)

The manual pool-shape curves were converted into pool normal *vs.* position by calculating the local slope of the pool ($m = \Delta y/\Delta x$) and converting it into a normal pool angle through the equation $\theta = \tan^{-1}(-1/m)$. A distance-weighted average of all dendrite orientations within a 2 mm radius of each manual pool coordinate was used for comparison. Results of this comparison are shown in Figure 12 for three of the pools indicated in Figure 11. The dendrite-based pool profiles correlate very well with manually measured profiles in some cases, with a coefficient of determination (R^2) value up to 93%. But, in other locations the correlations are relatively poor, with R^2 values below 50%. These initial comparisons indicate that image analysis based upon primary dendrite orientation is a promising technique for the measurement of historical melt pool profiles in VAR ingots. This technique has the distinct advantage over manual measurement in that it can be used throughout the ingot, not just where tree ring or similar structures are visible.

The hand-measured and dendrite-based pool profiles were also compared to the BAR pool profile predictions. The BAR pool profiles were estimated by interpolating where the temperature equaled the liquidus temperature for Ni Alloy 718 in the 2D data set at a given time step. This interpolation of the temperature produced a faceted pool profile because of the gridding used in computations. To overcome this issue, data was smoothed using a mean (moving average) filter prior to comparison with the manual pool measurements (which were also smoothed for comparison). Results of this comparison are shown in Figure 13. At early times the BAR simulation tends to have a melt pool with a flatter base than shown in the manual measurements. However, the pool profiles match reasonably well in the middle region of the ingot (-30 cm to -20 cm).

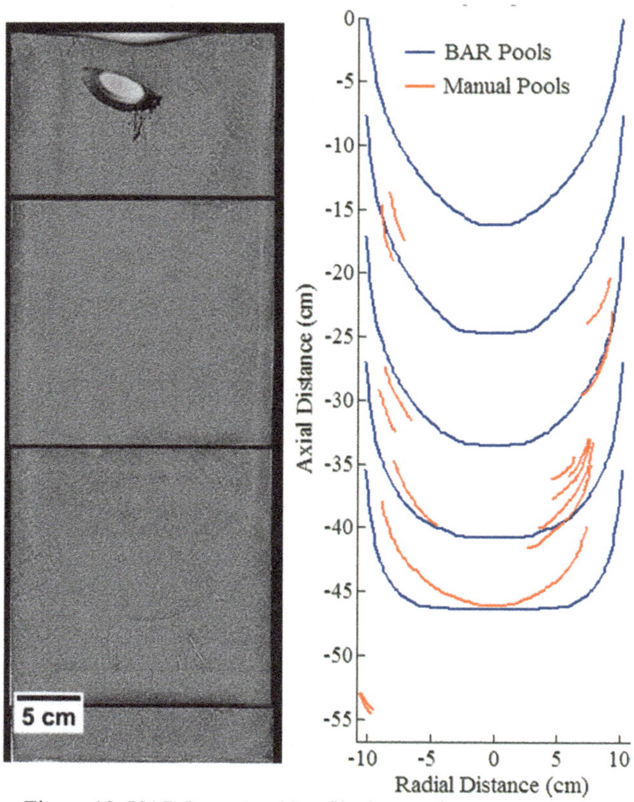

Figure 13. VAR Ingot Pool Profile Comparisons: (left) VAR composite photograph, (right) manual and BAR pool comparisons

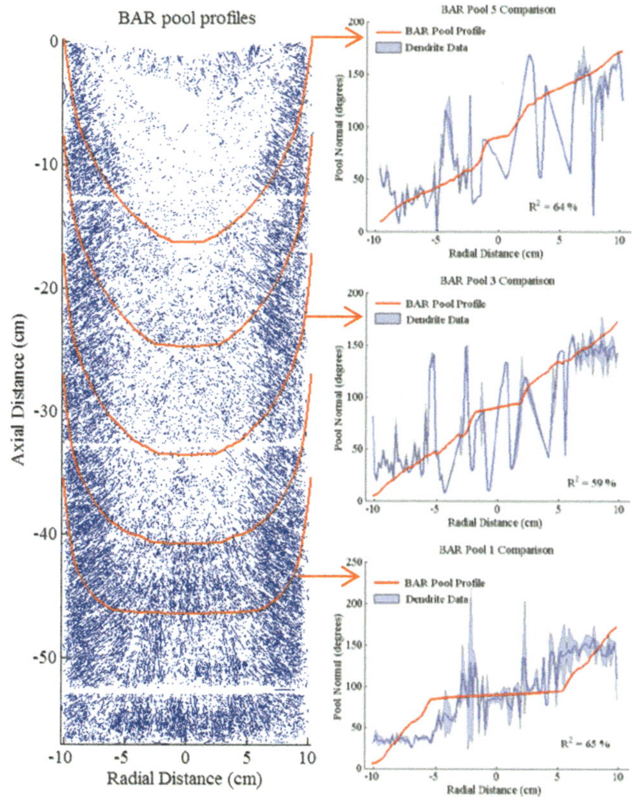

Figure 14. BAR pools and dendrite orientation comparisons

The BAR simulation results were also compared to the dendrite orientation map, using the same procedure shown in Figure 11 and Figure 12. The results of this comparison are shown in Figure 14. The disagreement of the BAR profiles with the local primary dendrite orientation data is similar to that with the manual pool shape measurements. There can be significant scatter in the dendrite data, due in part to the equiaxed solidification structure in the center of the ingot, as shown in Figure 3 and Figure 9.

Solidification Rate Analysis

In addition to pool shape estimates, the dendritic solidification structure can also be used to estimate the local solidification rate, based on existing empirical relationships with the primary and secondary dendrite arm spacing (PDAS, SDAS). To measure the secondary dendrite arm spacing, an offset line was created next to each primary dendrite. The offset line sets the region to be searched for secondary dendrites near the primary dendrite arm. For the initial analysis, an offset of 5 pixels (64 μm) was used. The corresponding pixels in the CLAHE-processed image were then used as data points to generate intensity *vs.* position curves, as illustrated in Figure 15.

The location of secondary dendrite arms was determined by using a local-maximum filter with a window size of 7 pixels (89 μm). The spacing was calculated by taking the mean value of the difference between the resulting maxima locations. As in the primary dendrite analysis, the standard deviation was also calculated for each line, and data that had a relative error (standard deviation / mean) greater than 50% was ignored. The resulting Voronoi diagram for secondary dendrite arm spacing (SDAS) is shown in Figure 16. Unlike the primary dendrite orientation shown in Figure 9, overall trends are less obvious for the SDAS map. This suggests a rather uniform solidification rate throughout the ingot.

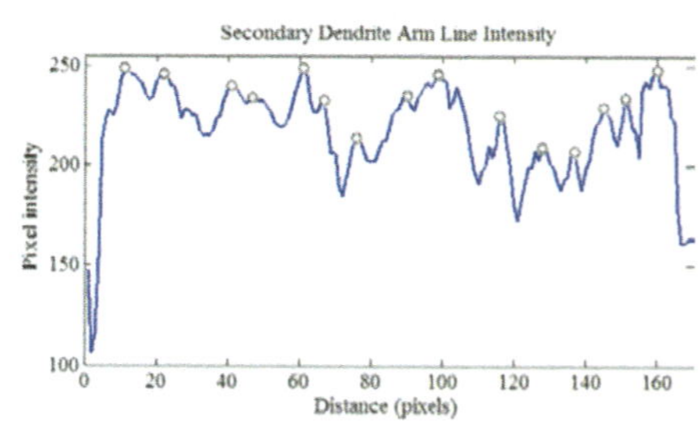

Figure 15. Secondary Dendrite Arm Spacing (SDAS) analysis

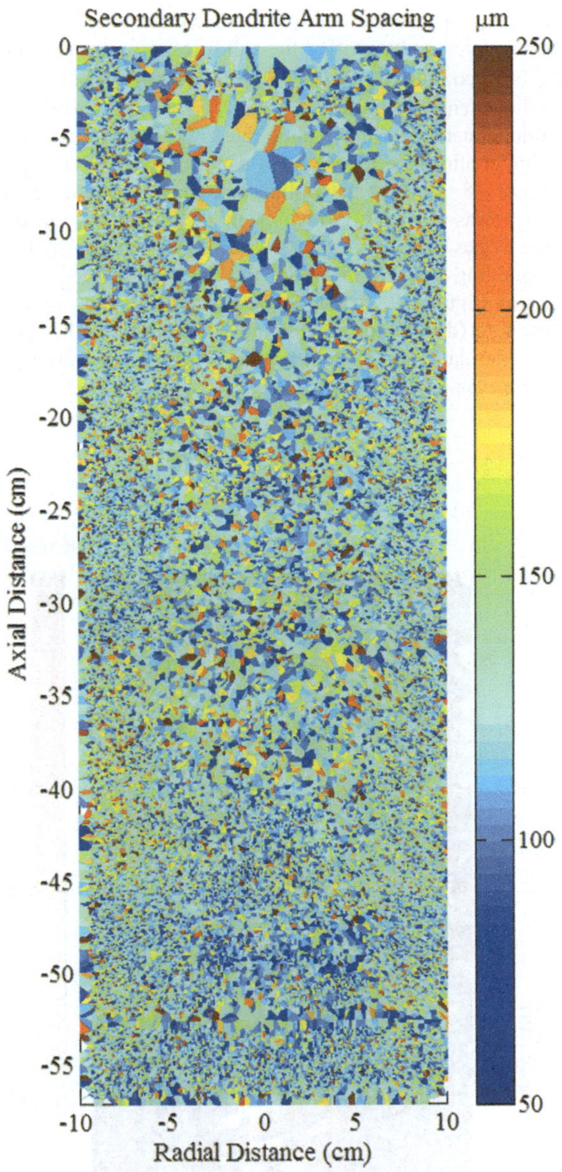

Figure 16. VAR ingot SDAS map. The scale indicates the secondary dendrite arm spacing value, from 50 to 250 µm

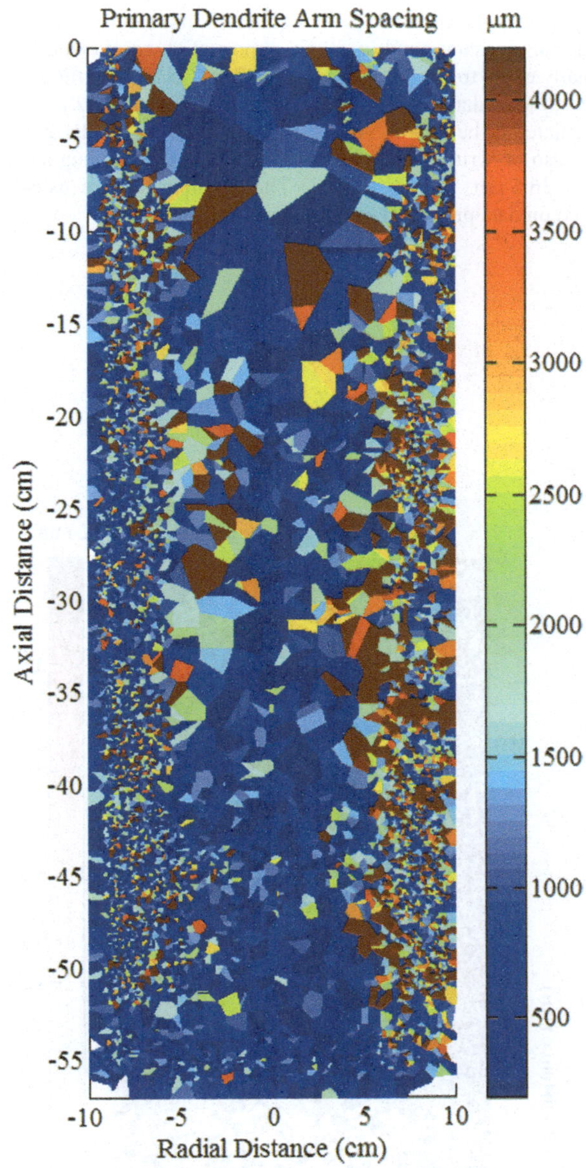

Figure 17. VAR Ingot PDAS map. The scale indicates the primary dendrite arm spacing value, from 0 to 4.2 mm

The primary dendrite arm spacing (PDAS) can also be used to estimate the local solidification rate. The PDAS map was calculated by iterating through each primary dendrite and finding nearby dendrites within a certain range of distances (0.062-12.7 mm) and relative angles (±10 degrees). The distance to the closest dendrite was then used as the PDAS. The resulting Voronoi diagram is shown in Figure 17. The coarse mesh is due to a limited number of recognized dendrites of similar orientation, but this might be improved upon through different image processing approaches.

A notable observation of the PDAS data shown in Figure 17 (and more clearly in Figure 19) is that the central "column" of decreased PDAS is not consistently centered throughout the height of the ingot. Rather, it wanders back and forth several times as the ingot "grows" upward. This may indicate a dynamic, asymmetric pool shape due to a number of causes (arc rotation, cooling asymmetries, etc.), and while this is not a new concept for VAR processes, it is interesting to have supporting quantitative evidence in the solidification microstructure.

Numerous correlations have been created between dendrite arm spacing and solidification rate. The general form is given below in Equation 1, where λ_2 is the secondary dendrite arm spacing, t is the solidification time (time for a given location to go from the liquidus to solidus temperature), G is the thermal gradient across the mushy zone (K/m), v is the liquidus front velocity (m/s), ξ is a material constant, and n is an exponential coefficient that is usually equal to 1/3 [9]. For Nickel Alloy 718, the coefficients for the secondary dendrite arm spacing λ_2 were calculated to be ξ_2 = 6.54 μm, and n_2 = 0.43. A similar equation can also be written for the primary dendrite arm spacing λ_1, with ξ_1 = 36.5 μm, and n_1 = 0.33. These numbers were calculated based on data presented by Bouse and Mihalisin [21].

$$\lambda_2 = \xi t^{n} = \xi (Gv)^{-n} \tag{1}$$

$$\dot{T} = \Delta T \left(\frac{\lambda_2}{\xi}\right)^{\frac{-1}{n}} \tag{2}$$

These equations can be used to solve for the average cooling rate in the mushy zone by writing the solidification time as a function of the local cooling rate $\dot{T}$ and the difference between the liquidus and solidus temperature ΔT (76K for Ni Alloy 718), as $t = \Delta T/\dot{T}$. The final equation for the cooling rate is given as Equation 2. The resulting cooling rate for the ingot using the SDAS data is shown in Figure 18. The mean cooling rate for the ingot using this approach was 6 K/min, which is in the range of conventional cooling rates for Ni Alloy 718 VAR ingots [21]. The corresponding Voronoi image for the PDAS-based cooling rate is shown in Figure 19. It should be noted that some of the low cooling rate (dark blue) regions may correspond with regions of poor image data, and therefore poor dendrite recognition during image processing.

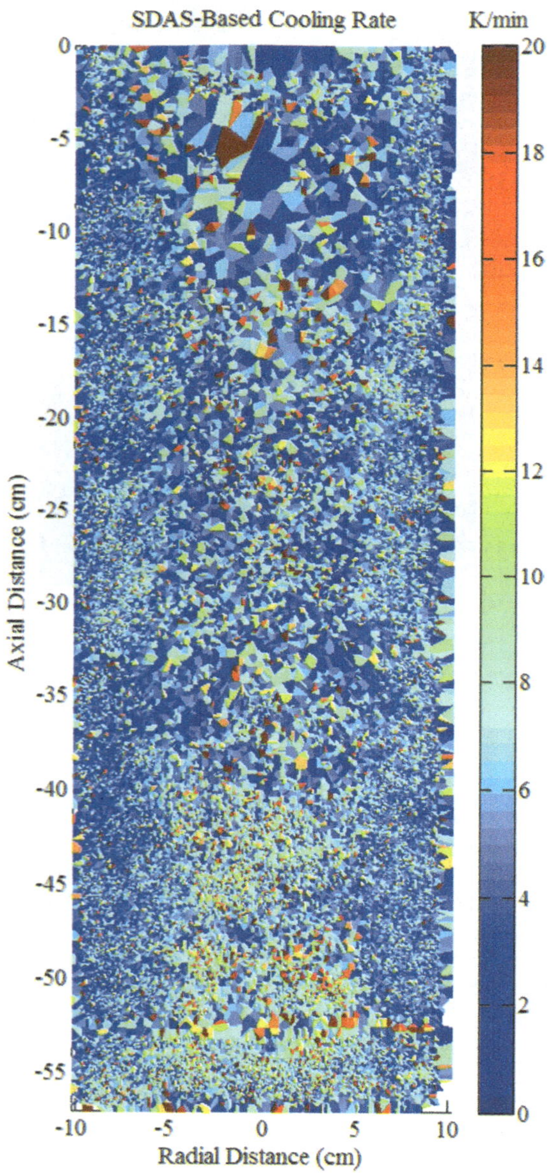

Figure 18. Cooling rate estimated using SDAS measurements

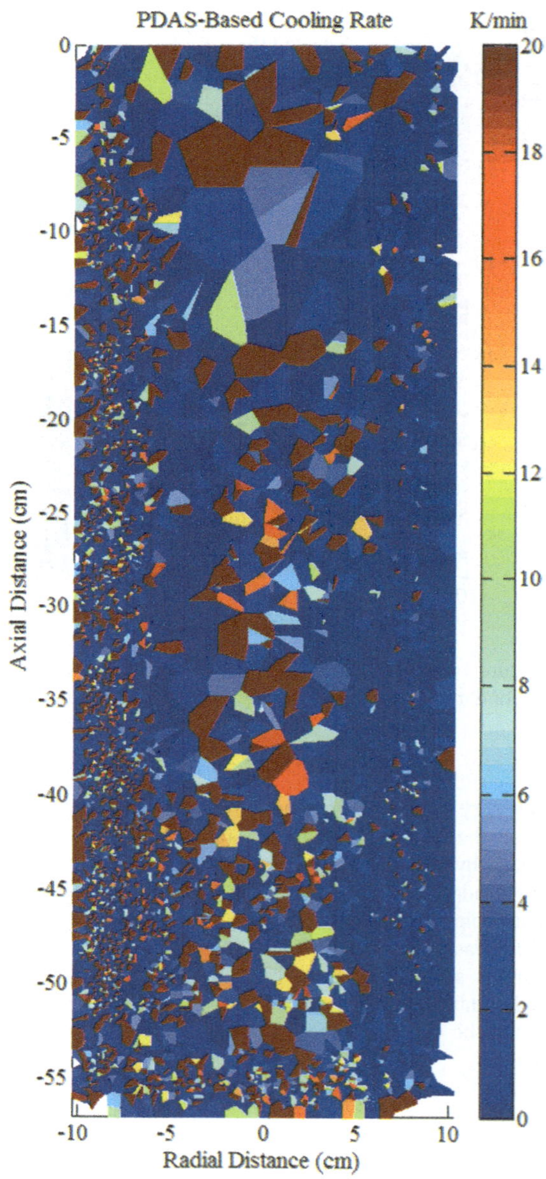

Figure 19. Cooling rate estimated using PDAS measurements

The cooling rates were subsequently compared to data calculated using BAR. This was done by taking the temperature output of BAR, and calculating the time at each location spent between the liquidus and solidus temperatures. The resulting average mushy zone cooling rate for the ingot is shown in Figure 20. The large BAR-predicted cooling rate at the ingot bottom is not immediately evident in the SDAS and PDAS maps, though the general trend of increased cooling rate near the centerline is present.

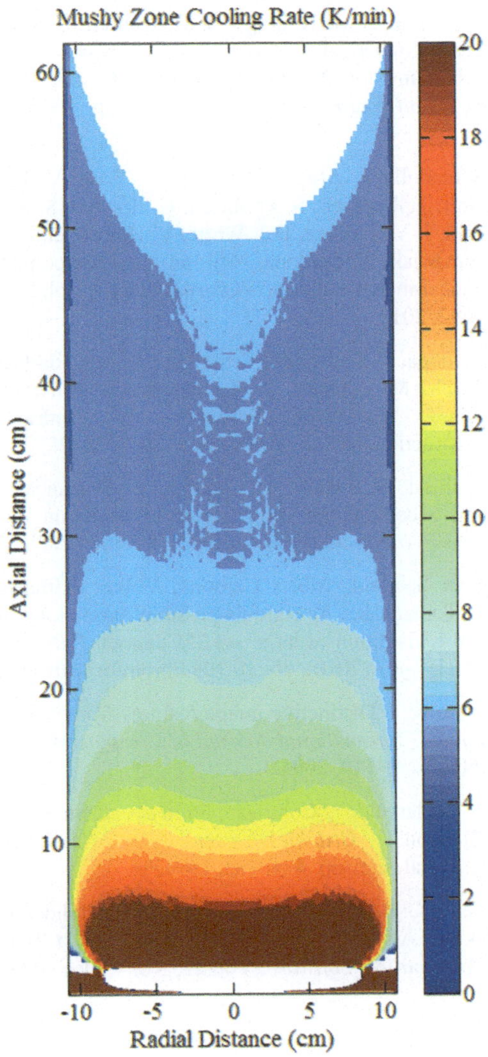

Figure 20. Average mushy zone cooling rate, calculated from BAR results

In addition to the average mushy zone cooling rate, the cooling rate next to the liquidus curve was also calculated, with the results shown in Figure 21. These results were obtained by calculating the mean cooling rate at a given location from the time that location initially reaches the liquidus temperature, plus one minute (equal to 12 time steps in the BAR results). These results are not necessarily useful for comparing against dendrite arm spacing, but they can serve as indicators for regions where freckle channels and tree rings may be present. Specifically, regions with rapid fluctuations in driving current (shown in Figure 2) tend to result in temperature perturbations in the melt pool (shown in Figure 21), which can lead to the aforementioned solidification defects [2, 4, 5]. Comparing these melt pool perturbations to the ingot microstructure data will be the subject of future work.

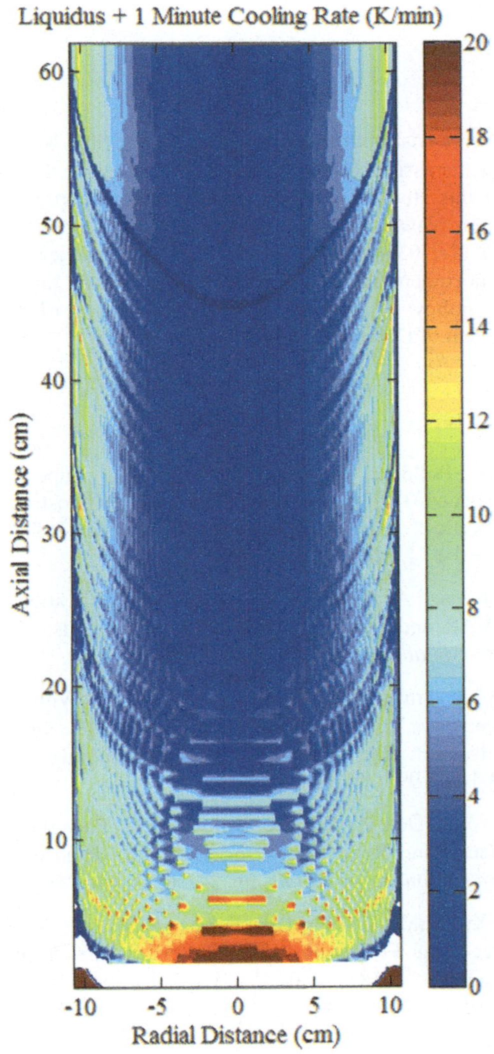

Figure 21. Cooling rate for 1 minute after liquidus, calculated from BAR results

Conclusions

Analysis of the Alloy 718 VAR ingot revealed that BAR simulations tended to predict wider pools at early times than pools calculated using either manual measurements or dendrite orientation maps. However, BAR pool shapes agree reasonably well with the manual measurements during the middle of the melt cycle. It is likely that differences early in the melt cycle are due to ill-defined thermal boundary conditions at the mold walls. However, phenomena such as asymmetric pool shapes indicated by dendrite arm spacing maps are currently beyond the capability of axisymmetric codes like BAR to predict.

Dendrite orientation and spacing maps are a potentially useful method to estimate pool profiles and solidification rates, though there can be significant scatter in the data. Additional image and data processing techniques are expected to yield more consistent results in the future. Future work will also include comparison of observed solidification defects with calculated melt pool perturbations.

Acknowledgements

The authors wish to thank David G. Evans (SMC, New Hartford, NY) for his assistance with etching and preparing the specimens used in this study. Funding for this work was provided by the Office of Naval Research (ONR) through contracts number N00014-11-1-0366 and N00014-12-1-0811. The views expressed in this document are those of the authors and do not reflect the official policy or position of the Department of the Navy, Department of Defense, or the U.S. Government.

References

1 T.M. Pollock, S. Tin, "Nickel-Based Superalloys for Advanced Turbine Engines: Chemistry, Microstructure, and Properties," *Journal of Propulsion and Power*, Vol. 22, No. 2, pp. 361-374, March-April 2006.

2 K.O. Yu, J.A. Domingue, "Control of Solidification Structure in VAR and ESR Processed Alloy 718 Ingots," *Superalloy 718 – Metallurgy and Applications*, TMS, 1989.

3 L.A. Jackman, G.E. Maurer, S. Widge, "White spots in Superalloys," *Superalloys 718, 625, 706 and Various Derivatives*, The Minerals, Metals & Materials Society, pp. 153-166, 1994.

4 J.A. Van Den Avyle, J.A. Brooks, A.C. Powell, "Reducing Defects in Remelting Processes for High-Performance Alloys," *Journal of Metals*, pp. 22-26, March 1998.

5 X. Xu, R.M. Ward, M.H. Jacobs, P.D. Lee, M. McLean, "Tree-Ring Formation during Vacuum Arc Remelting of INCONEL 718: Part I. Experimental Investigation," *Metallurgical and Materials Transactions A*, Vol. 33A, pp. 1795-1804, June 2002.

6 F.J. Zanner, R.L. Williamson, R.P. Harrison, H.D. Flanders, R.D. Thompson, W.C. Szeto, "Vacuum Arc Remelting of Alloy 718," *Superalloy 718 – Metallurgy and Applications*, TMS, 1989.

7 L.A. Bertram, P.R. Schunk, S.N. Kempka, F. Spadafora, and R. Minisandram, "The macroscale simulation of remelting processes," *Journal of Metals*, 50(3):18–21, 1998.

8 L. Nastac, "A Multiscale Transient Modeling Approach for Predicting the Solidification Structure in VAR-Processed Alloy 718 Ingots," *Metallurgical and Materials Transactions B*, September 2012.

9 M. Rappaz, W.J. Boettinger, "On Dendritic Solidification of Multicomponent Alloys with Unequal Liquid Diffusion Coefficients," *Acta mater.*, Vol. 47, No. 11, pp. 3205-3219, 1999.

10 P. Auburtin, T. Wang, S.L. Cockcroft, A. Mitchell, "Freckle Formation and Freckle Criterion in Superalloy Castings," *Metallurgical and Materials Transactions B*, Vol. 31B, pp. 801-811, August 2000.

11 M.M. Franke, R.M. Hilbinger, C.H. Konrad, U. Glatzel, R.F. Singer, "Numerical Determination of Secondary Dendrite Arm Spacing for IN738LC Investment Castings," *Materials Science and Engineering A*, Vol. 42A, pp. 1847-1853, July 2011.

12 J. Schindelin, I. Arganda-Carreras, E. Frise, V. Kaynig, M. Longair, T. Pietzsch, S. Preibisch, C. Rueden, S. Saalfeld, B. Schmid, J.-Y. Tinevez, D.J. White, V. Hartenstein, K. Eliceiri, P. Tomancak, A. Cardona, "Fiji: an open-source platform for biological-image analysis," *Nature Methods*, Vol. 9, No. 7, pp. 676-682, 2012.

13 A. Cardona, S. Saalfeld, J. Schindelin, I. Arganda-Carreras, S. Preibisch, M. Longair, P. Tomancak, V. Hartenstein, R.J. Douglas, "TrakEM2 Software for Neural Circuit Reconstruction," *PLoS ONE* 7(6): e38011, 2012..

14 S. Saalfeld, R. Fetter, A. Cardona, P. Tomancak, "Elastic volume reconstruction from series of ultra-thin microscopy sections," *Nature Methods*, doi:10.1038/nmeth.2072, 2012.

15 Stephan Saalfeld, Albert Cardona, Volker Hartenstein, and Pavel Tomancak, "As-rigid-as-possible mosaicking and serial section registration of large ssTEM datasets," *Bioinformatics*, 26(12), i57-i63 (2010), doi:10.1093/bioinformatics/btq219.

16 D. G. Lowe, "Distinctive image features from scale-invariant keypoints," *International Journal of Computer Vision*, Vol. 2, No. 60, pp. 91-110, 2004.

17 K. Zuiderveld, "Contrast limited adaptive histogram equalization", *Graphics gems IV*, Academic Press Professional, Inc., pp. 474–485, 1994.

18 Barber, C. B., D.P. Dobkin, and H.T. Huhdanpaa, "The Quickhull Algorithm for Convex Hulls," *ACM Transactions on Mathematical Software*, Vol. 22, No. 4, Dec. 1996, p. 469-483.

19 X. Wang, R.M. Ward, M.H. Jacobs, M.D. Barratt, "Effect of Variation in Process Parameters on the Formation of Freckles in INCONEL 718 by Vacuum Arc Remelting," *Metallurgical and Materials Transactions A*, Vol. 39A, pp. 2981-2989, 2008.

20 A. Mitchell, "Solidification in remelting processes," *Materials Science and Engineering A*, Vol. 413-414, pp. 10-18, 2005.

21 G.K. Bouse and J.R. Mihalisin, *Superalloys, Supercomposites and Superceramics*, J.K. Tien, T., Caulfield, ed.s, Academic Press Inc., London, pp. 99-148, 1989.

Aluminum Processing

Liquid Metal Processing & Casting 2013

Proceedings of the 2013 International Symposium on Liquid Metal Processing & Casting
Edited by: Matthew J. M. Krane, Alain Jardy, Rodney L. Williamson, and Joseph J. Beaman
TMS (The Minerals, Metals & Materials Society), 2013

IMPROVEMENT OF MECHANICAL PROPERTIES OF HPDC A356 ALLOY THROUGH MELT QUENCHING PROCESS

Shouxun Ji, Wenchao Yang, Bo Jiang and Z. Fan
Brunel Centre for Advanced Solidification Technology (BCAST), Brunel University, Uxbridge, Middlesex UB8 3PH, United Kingdom

Keywords: High pressure die casting, melt quenching, nucleation, microstructural evolution

Abstract

Melt quenched high pressure die casting (MQ-HPDC) is a new die casting process developed recently for improving the casting quality of the conventional high pressure die casting (HPDC) process. In the MQ-HPDC process, an alloy melt with a specified dose and superheat is quenched by directly pouring the alloy melt into a preheated metallic container. The thermal mass and preheating temperature of the container is selected so that the alloy melt is quenched just below the alloy liquidus and heterogeneous nucleation takes place during the melt quenching. The quenched alloy melt is then fed immediately into the shot sleeve for component casting. In this paper we present the MQ-HPDC process and the resultant microstructures and mechanical properties of a MQ-HPDC A356 alloy.

Introduction

The cold chamber high pressure die casting (HPDC) process has been widely used for manufacturing shaped castings of light alloys due to its high efficiency, high consistency and low further machining, low cost and versatility in component shapes. However, this process still faces several technological challenges. For example, die castings have relatively poor mechanical performance due to porosity and other casting defects and the formation of a large solid lump in the shot sleeve before die filling. Major effort has been made over many years to overcome these challenges, using semisolid metal (SSM) processing [1,2], vacuum die casting [3,4] and squeeze casting [5,6].

Recently, we have developed a melt quenched high pressure die casting (MQ-HPDC) process to overcome the problems associated with conventional HPDC process, in particular, to reduce the casting defects and the formation of the large solid lump in the shot sleeve. This paper presents the MQ-HPDC process and the microstructure and mechanical properties of A356 aluminum alloy processed with the MQ-HPDC process.

The MQ-HPDC process

The MQ-HPDC process (Figure. 1) consists of three sequential steps: melt quenching, melt dosing and high pressure die casting. This process originated from the concept that large dendrites and solidification in the shot sleeve could be reduced by quenching the melt in a metallic container before being transferred into the shot sleeve. For a given alloy melt with a specified superheat and dose, the thermal mass and the preheating temperature of the metallic container can be selected so that the alloy melt can be quenched to a temperature close to the alloy liquidus for enhancing heterogeneous nucleation without significant crystal growth. The quenched alloy melt has the following characteristics:

(a) a temperature just a few degree below the alloy liquidus;
(b) an almost isothermal temperature due to the severe convection caused by melt pouring;
(c) copious nuclei due to significantly enhanced heterogeneous nucleation during melt quenching.

If the nuclei are uniformly distributed throughout the entire volume of the melt, the growth of such nuclei ensures a relatively uniform melt temperature in the shot sleeve before filling the die cavity. This will prevent the formation of the large dendrites and the solidification in the shot sleeve will facilitate die filling and will promote a more uniform solidification after die filling, and therefore reduce the formation of casting defects.

The MQ-HPDC process requires the use of a number of metallic containers. The number required for processing is determined by the specified melt dose, cycle time and container thermal mass. The container will be cooled naturally to the temperature which equals the preheating temperature described previously for a single container.

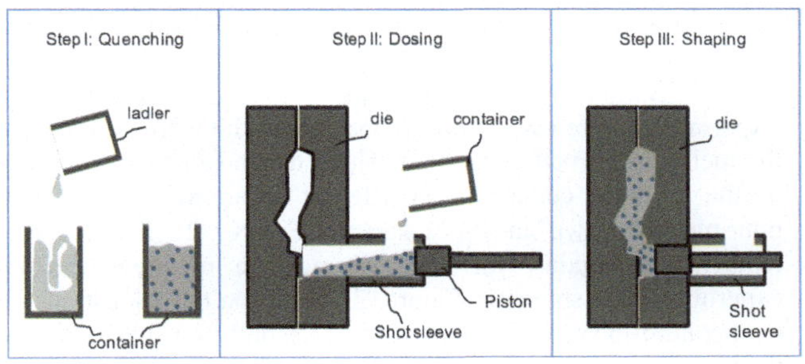

Figure 1: Schematic illustration of the melt quenched high pressure die casting (MQ-HPDC) process.

Experimental

Commercial grade ingots of A356 alloy were used to demonstrate the MQ-HPDC process. The A356 ingots were melted at 710°C in a clay-graphite crucible using an electric resistance furnace. The container used was made of stainless steel with dimensions (in mm) as shown in the inset in Figure 2. The container was preheated to the specified temperature in the furnace. The die casting was carried out using a 4500kN cold chamber die casting machine. The melt was manually dosed into the shot sleeve followed by the standard HPDC process. In the MQ-HPDC process, the melt was poured at 710°C into the preheated container, and then transferred immediately to the shot sleeve followed by a full automatic HPDC cycle. The HPDC die produced three 6.35mm diameter tensile test bars and three square bars (2x6.3mm^2, 3x6.3mm^2, 5x6.3mm^2) in each shot. The die was heated by the circulation of mineral oil at 250°C.

Tensile tests were conducted following ASTM standard B557, using an Instron 5500 Universal Electromechanical Testing Systems with a ±50kN load cell. All tensile tests were carried out at ambient temperature with a 2mm/min crosshead speed. The specimen for microstructural examination was cut from the middle of the tensile test bar. Each specimen was examined using a Zeiss optical microscope with quantitative metallography using an AxioVision 4.3 Quantimet digital image analysis system.

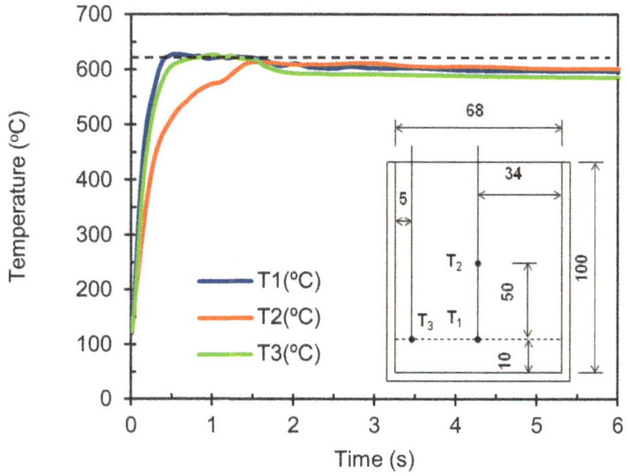

Figure 2: Temperature recorded by three thermo-couples in the stainless steel container during quenching of the A356 melt. The insert shows the dimensions of the container in mm and the positions of the thermo-couples in the container. The temperature of A356 melt was 650°C. The initial container temperature was 200°C.The liquidus temperature of A356 is 615°C.

Results

Experiments were carried out to understand the thermal history of the melt during melt quenching. Three thermocouples were fixed in the metallic container at different locations prior to melt pouring as shown in the insert in Figure. 2. The recorded temperatures against the time during one of the quenching experiments are shown in Figure. 2. At all the three locations the temperature increased in a fraction of second to a maximum and then leveled off to a temperature just below the alloy liquidus (615°C). The slightly slower reaction of the thermal couple T2 is mainly caused by the delayed emersion of T2 in the alloy melt. In consideration of the above time delay, it should be noted that the recorded temperatures reached the maximum value within 1s and achieved an almost isothermal temperature within 2s. It needs to emphasis that the initial temperature includes the responding time of the thermocouple. However, from the time to reach the maximum temperature and the pouring temperature, it is possible to estimate the average cooling rate of the melt during melt quenching. The results showed that the cooling rate during the melt quenching was in the order of 10^2K/s, which is comparable with that achieved in the shot sleeve during HPDC process. As discussed in ref. [7], the cooling rate in the shot sleeve of die caster is at a level of 100 K/s. The similar level of cooling rate will provide similar undercooling for microstructural evolution. It needs to emphasis that the operation from pouring liquid into the container and then to the shot sleeve can results in the temperature variation because of the heat loss and the multiple process steps. However, the process time of melt quenching is very short and the whole process can be automatically achieved during production. Therefore, the benefits are obvious.

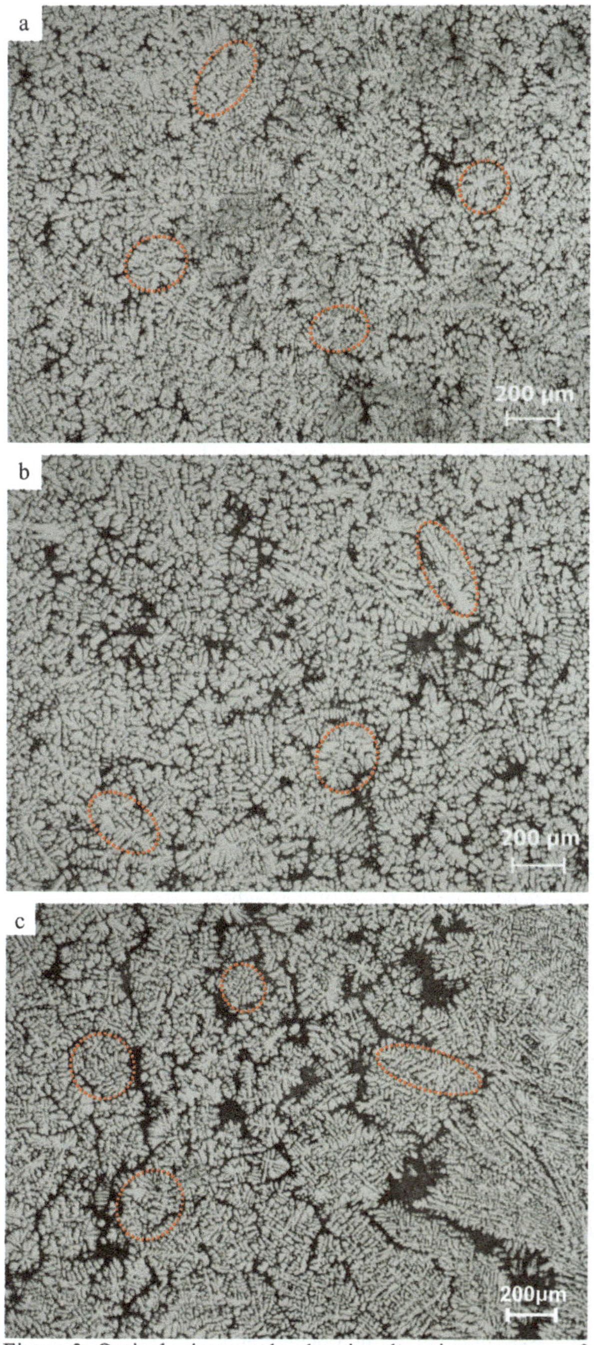

Figure 3: Optical micrographs showing the microstructure of A356 alloy biscuit (ϕ70mm) after HPDC process on a cross section 15mm from the surface in touch with the piston, (a) top of the section, (b) middle of the section, and (c) bottom of the section.

The microstructure of the biscuits (the material solidified in the shot sleeve) was examined to understand the primary phase formation in the shot sleeve prior to die filling. During the conventional HPDC process, superheated alloy melt (about 700°C) is directly poured into the shot sleeve that is at a temperature between 200°C and 400°C. Formation of a solid lump at the bottom of the shot sleeve prior to die filling is inevitable. Such a solid piece is pushed and deformed by the moving piston during the die filling, resulting in severe cracking in the solid.

Such cracked solid is usually found at the bottom of the biscuit (Figure 3c) and occasionally also found at the top of the biscuit. Except the cracked solid in the biscuit, the rest of the biscuits (middle and the top section) exhibit a coarse dendrite structure as shown in Figure 3a&b).

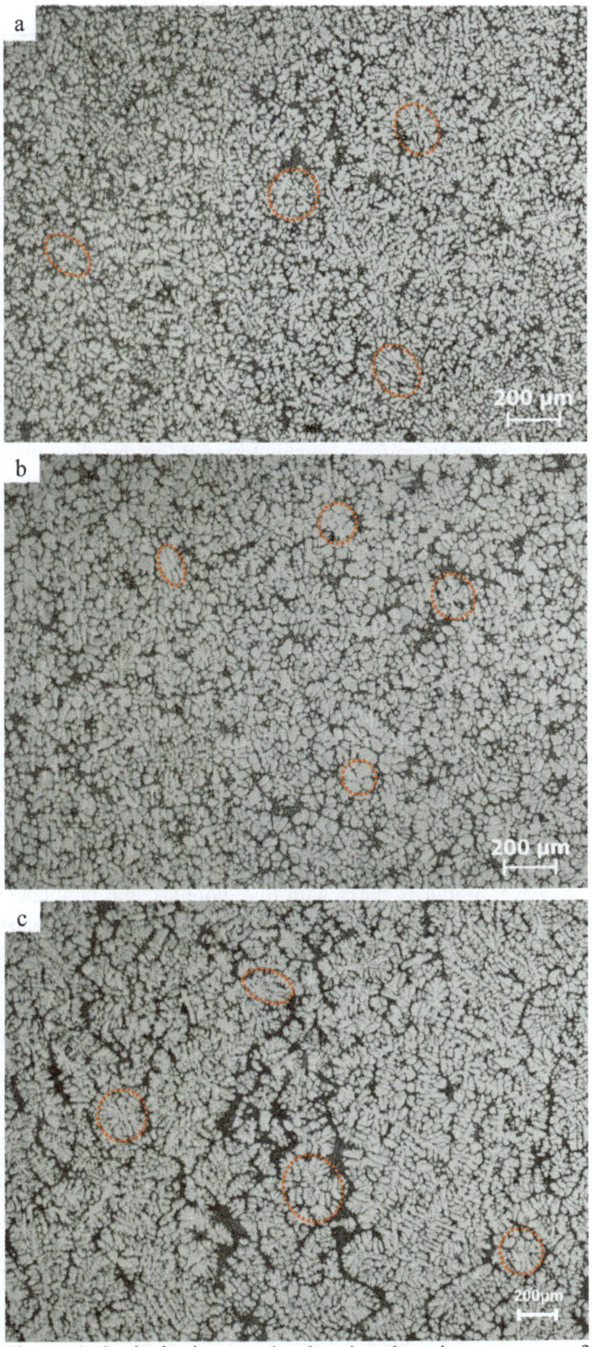

Figure 4: Optical micrographs showing the microstructure of A356 alloy biscuit (ϕ70mm) after MQ-HPDC process on a cross section 15mm from the surface in touch with the piston, (a) top of the section, (b) middle of the section and (c) bottom of the section.

However, although with the same condition, the biscuit produced by the MQ-HPDC process has a different microstructure: (1) the cracked solid region (Figure. 4a) is very much reduced or perhaps even eliminated; (2) no cracked solid was found at the top of the biscuit; (3) the dendrite size was considerably smaller (Figure. 4a&b). This demonstrates that melt quenching can significantly enhance heterogeneous nucleation (smaller dendrite size) and reduce/eliminate solid formation in the shot sleeve.

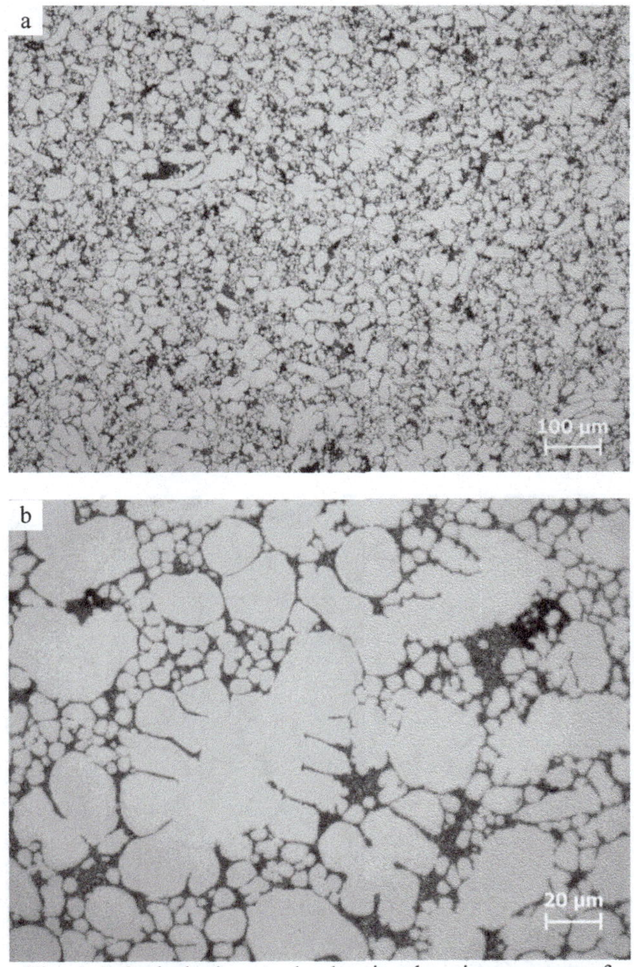

Figure 5: Optical micrographs showing the microstructure of A356 alloy (ϕ6.35mm bar) produced by HPDC process showing (a) the overall microstructure and (b) the morphology of primary α-Al phase.

The microstructures of A356 Al-alloy processed by the conventional HPDC process the MQ-HPDC and are shown in Figures 5 and 6, respectively. The HPDC sample shows fragmented dendrites with an elongated morphology and up to 100μm in size (Figure. 5a), fine spheroids of 8μm in average size (Figure. 5b) and substantial amount of irregular shaped cracks resultant from hot tearing during the final solidification in the die cavity (Figures 5a&b). However, the microstructure of the MQ-HPDC sample is considerably different from that of the HPDC sample. Figure 6 shows that the primary α-Al phase was characterised by the co-existence of two types of globular particles. The relatively large primary α-Al globules (Figure. 6a) with an average size of 35μm were fragmented from the rosettes formed in the shot sleeve, while the fine α-Al globules (Figure. 6b) with an average size of 7.5μm were formed in the die cavity. The MQ-HPDC samples were much less cracked than the HPDC samples, suggesting that there was much less hot tearing in the MQ-HPDC process. For A356 alloy, the measured average porosity level was 0.95% for the HPDC castings and 0.27% for the MQ-HPDC samples. This indicates that MQ-HPDC could produce castings with lower porosity and higher integration.

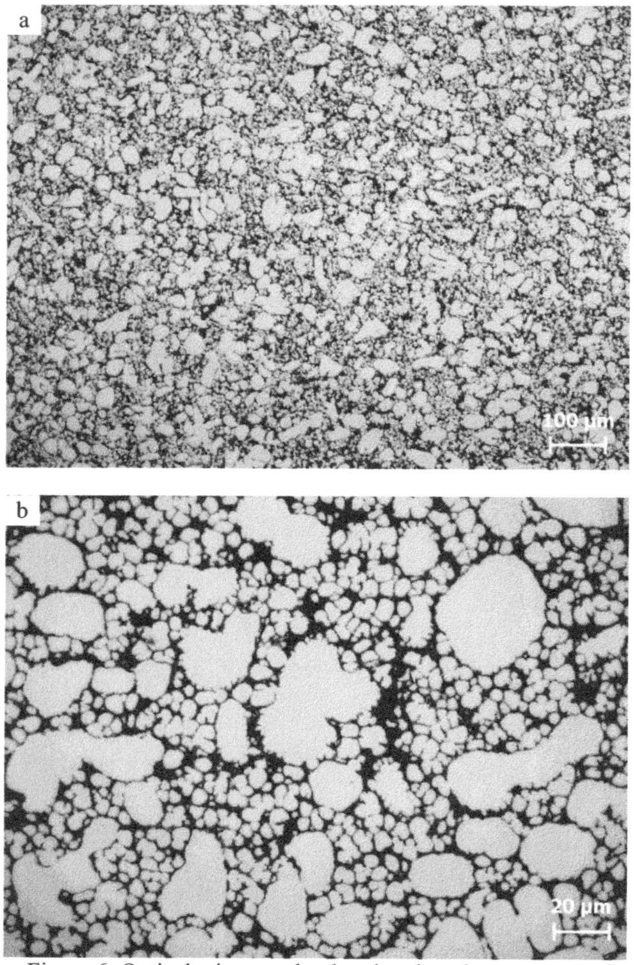

Figure 6: Optical micrographs showing the microstructure of A356 alloy (ϕ6.35mm bar) produced by MQ-HPDC process showing (a) the overall microstructure and (b) the morphology of primary α-Al phase.

The mechanical properties of A356 alloy processed by both MQ-HPDC and HPDC processes are shown in Table 1. Although both yield strength and ultimate tensile strength (UTS) show little difference between the two groups of samples, elongation of the MQ-HPDC samples was significantly improved. The average elongation was increased from 10.1% for HPDC samples to 12.6% for MQ-HPDC samples, a 25% increase.

Table 1: Tensile properties of A356 alloy produced by HPDC and MQ-HPDC processes.

	Yield strength (MPa)	Ultimate tensile strength (MPa)	Elongation (%)
HPDC	120.2±3.97	262.1±2.66	10.1±1.05
MQ-HPDC	120.6±3.56	262.3±2.93	12.6±1.16

Discussion

The unique feature of the MQ-HPDC process is melt quenching prior to conventional HPDC process. The high cooling rate achieved by pouring superheated alloy melt into a relatively cold metallic container results in a large thermal undercooling, which allows an increased number of existing solid particles in the melt to nucleate. However, the growth of such nuclei will be restricted in the container due to lack of time and cooling capacity. The fast growth of the nuclei in the alloy melt in the shot sleeve can generate latent heat, which acts as a temperature regulator to ensure a relatively uniform temperature field in the alloy melt, in comparison of the situation in conventional HPDC process. The simultaneous growth of increased number of primary grains results in a relatively small growth of each primary grain and therefore large dendrites are eliminated, which improve the die filling and result in a reduction of casting defects and improve the mechanical properties.

Summary

A melt quenched high pressure die casting (MQ-HPDC) process has been developed to improve casting quality. In this process, melt quenching by purpose designed metallic container is used to enhance heterogeneous nucleation; and the latent heat released by the growth of the increased number of nuclei in the shot sleeve is used to achieve a relatively uniform melt temperature. This helps to reduce the formation of large dendrites and solidification in the shot sleeve, produces slurry with fine globules, and facilitates the die filling process, reduces casting defects and improves casting quality.

References

[1] M. C. Flemings, "Behavior of metal alloys in the semi-solid state", *Metall. Mater. Trans. A* 22A (1991) 957-81.

[2] Z. Fan, "Semisolid Metal Processing", *Inter. Mater. Review*, 47(2002) 49-85.

[3] X.P. Niua, B.H. Hua, I. Pinwilla, H. Lib, "Vacuum assisted high pressure die casting of aluminum alloys", *J. Mater. Proc. Tech.* 105(2000)119–127.

[4] E.S Kim, K.H Lee, Y.H Moon, "A feasibility study of the partial squeeze and vacuum die casting process", *J. Mater. Proc. Tech.* 105 (2000) 42–48.

[5] A. Maleki, B. Niroumand, A. Shafyei, "Effects of squeeze casting parameters on density, macrostructure and hardness of LM13 alloy", *Mater. Sci. Eng. A*, 428 (2006) 135–140.

[6] M.R. Ghomashchi, A Vikhrov, "Squeeze casting-an overview", *J. Mater. Proc. Tech.*, 101 (2000) 1–9.

[7] S.Ji, Y. Wang, D. Watson, Z. Fan, "Microstructural Evolution and Solidification Behavior of Al-Mg-Si Alloy in High-Pressure Die Casting". *Metall. Mater. Trans. A* 44A (2013)3185-97.

Proceedings of the 2013 International Symposium on Liquid Metal Processing & Casting
Edited by: Matthew J. M. Krane, Alain Jardy, Rodney L. Williamson, and Joseph J. Beaman
TMS (The Minerals, Metals & Materials Society), 2013

REVIEW OF HOT TEARING STUDIES IN Al ALLOYS DURING DIRECT CHILL CASTING

Ashok Kumar Nallathambi, Pavan Kumar Penumakala, Eckehard Specht
Otto von Guericke University Magdeburg, 39106 Magdeburg, Germany

Keywords: Solidification, Shrinkage and deformation, Liquid feeding, Mechanical behavior, Hot tear.

Abstract

DC casting is a well established casting method for producing aluminum and magnesium alloys. One of the major defects frequently encountered in DC casting ingots is hot tearing which originates during the phase transition region. Previous research works reported that the copper content significantly lengthen the solidification interval and further certain critical concentration makes the alloy highly prone to hot cracks (so-called delta curve). In this work, the existing thermal, mechanical, and thermo mechanical hot tearing criteria are extensively reviewed. The critical solid fraction (0.9-0.99) is more focused where the ductility of the material changes drastically. The strength of the solid skeleton, dendritic network, feeding possibility, viscosity of the liquid, alloying element segregation, and cooling profiles are the important parameters considered for the integration of all the existing criteria.

Introduction

A hot tear is a fracture formed during solidification. Shrinkage, hindered contraction and lack of feeding in the mushy zone are the main phenomena that cause hot tear initiation [1]. The complex mechanisms acting during the solidification of metals makes the study of hot tearing a difficult task. Several mechanisms of hot tearing are proposed by various researchers. The criteria can be classified into three main categories based on non-mechanical aspects such as feeding behavior [2-4], based only on mechanical aspects [5-6], and a combination of both non-mechanical and mechanical aspects [1],[7].

To implement the above suggested criteria in the real time castings, the formation of solid dendrites and their growth in mushy zone needs to be studied. Based on the permeability of the solid network in the mushy zone, the transition zone from liquid to solid can be divided into four stages as: Mass feeding, in which both liquid and solid are free to move. Inter dendritic feeding, in which the liquid has to flow through the dendritic network and the permeability of the network is still large enough. Inter dendritic separation, in which the liquid network becomes fragmented and pore formation or hot tearing may occur. With increasing solid fraction, liquid is isolated in pockets or immobilized by surface tension. When the permeability of the solid network becomes too small for the liquid to flow, further thermal contraction of the solid will cause pore formation or hot tearing [1]. The mathematical description of the above mechanisms and related hot tearing criteria are briefly explained in this article.

Non Mechanical Criteria

Feurer's Criteria

This criterion focuses mainly on the liquid feeding and shrinkage during solidification. This approach considers that hot tearing occurs due to the lack of liquid feeding, due to the difficulty of the fluid flow through the mushy zone as a permeable medium competing with solidification shrinkage. Feurer considers two terms, SPV and SRG that indicates maximum volumetric flow rate (feeding term) through a dendritic network and volumetric solidification shrinkage, respectively. The maximum volumetric flow rate per unit volume through a dendritic network can be formulated as

$$SPV = \frac{{f_l}^2 {\lambda_2}^2 P_d}{24\pi c^3 \mu L^2} \quad (1)$$

where f_l is the volume fraction of liquid, λ_2 is secondary dendritic arm spacing, P_s is the effective feeding pressure, μ is viscosity of the liquid and L is the length of coherent mushy zone. The volumetric solidification shrinkage due to the density difference between solid and liquid phases can be calculated as [2],

$$SRG = \frac{\partial lnV}{\partial t} = -\frac{1}{\bar{\rho}}\frac{\partial \bar{\rho}}{\partial t} \quad (2)$$

where $\bar{\rho}$ is the mixture density of the mushy. The Feurer criteria says that hot tearing is impossible if SPV >SRG.

Clyne and Davies Criteria

The hot tearing criteria proposed by Clyne and Davies [3] is based on the theory that during the last stage of freezing, the liquid has the difficulty to move freely in between the strong dendrite network, so that the strain applied during this stage cannot be accommodated by mass feeding. This last stage of solidification is considered as the most susceptible zone for hot cracks. The cracking susceptibility coefficient is defined as the ratio between the vulnerable time period (hot tearing susceptibility) t_R and time available for stress relief process (mass feeding and liquid feeding) t_V

$$HCS = \frac{t_V}{t_R} = \frac{t_{99} - t_{90}}{t_{90} - t_{40}} \quad (3)$$

where t_{99} is the time when the volume fraction of solid f_s is 0.99; t_{90} is the time when f_s is 0.90; t_{40} is the time when f_s is 0.40

Katgerman's Criteria

Katgerman's criteria [4] is derived based on the theoretical considerations of Clyne and Davies and Feurer. The model can be applied to binary and commercial alloys. A hot tearing index is defined as follows:

$$HCS = \frac{t_{99}-t_{cr}}{t_{cr}-t_{40}} \tag{4}$$

where t_{99} is the time at the solid fraction $f_s = 0.99$; t_{40} is the time at the solid fraction $f_s = 0.40$ and t_{cr} is the time to reach a critical point where liquid feeding is inadequate . This critical time is determined using Feurer's criteria and it is the time when SPV=SRG.

Mechanical Criteria

Magnin's Criteria

The criteria proposed by Magnin [5], is based on the strain experienced during solidification in relation to the fracture strain of the mush. This criterion is also focused on the last stages of solidification, where liquid feeding is inadequate. The hot cracking sensitivity, HCS, is taken as the ratio between the circumferential plastic strain ε_c, at solidus temperature and the experimentally determined fracture strain, ε_{cr}, close to the solidus temperature. Thus, the sensitivity can be calculated as

$$HCS = \frac{\varepsilon_c}{\varepsilon_{cr}} \tag{5}$$

If HCS is higher than one, a crack will develop. Therefore, the model can be used to predict hot tearing both qualitatively and quantitatively. While implementing this criterion in real time castings, the circumferential plastic strain can be found from the Thermomechanical simulation of castings.

Prokhorov's Criteria

Prokhorov [8] formulated a mechanical criterion, in which the hot cracking sensitivity is determined by the shrinkage strain rate $\dot{\varepsilon}_{shr}$ and apparent strain rate $\dot{\varepsilon}_{app}$ in the mush in relation to the fracture strain rate of the mush. In this approach, effects of the surrounding configuration are accounted by the apparent strain rate. It is assumed that during solidification, alloys pass through a low-ductility temperature range called the brittle temperature range (BTR). The brittle temperature range is from the coherence temperature until the solidus temperature. The minimum fracture strain rate in this range is $\dot{\varepsilon}_{min}$. A hot tear will form in the solidified body, if

$$\dot{\varepsilon}_{shr} + \dot{\varepsilon}_{app} \geq \dot{\varepsilon}_{min} \tag{6}$$

The above equation is more sensitive to constitutive behavior of the mush, which is generally not precisely known for alloys.

Lahaie criteria

The criterion also focuses on the liquid flow in between the dendritic network. Considering the last stages of solidification, when macroscopic flow of the liquid impossible and liquid film is constrained in between the dendrites, Lahaie [9] proposed a hot tearing model based on the liquid film rupture. Due to the deformation of the solid skeleton, the tensile stresses are induced in the liquid films and liquid will break when the stresses reach critical. The strain, the viscosity of the inter granular fluid, the solid fraction, the isothermal compressibility of the fluid, the surface tension of the liquid, the limiting liquid-film thickness for viscous flow and microstructure are the various parameters that influence the liquid film rupture. The semisolid body at above 0.9 solid fraction is assumed in the form of regular hexagons with thin films of liquid distributed along the solid boundaries, as depicted in Fig 1. The behavior of the liquid films in the hexagonal structure before and deformation is studied in developing hot tearing criteria.

In the undeformed state, it is assumed that the inter granular liquid film is uniformly distributed between the hexagonal grains. When the strain is applied to the solid body, the liquid films start flowing from the inclined channels into horizontal channels. At certain maximum strain ε_{max}, the inclined channels will be completely flowed into horizontal channels. If the strain is applied beyond ε_{max}, and it is assumed that the solid grains are rigid, the only way to preserve compatibility during continued straining is by dilatation of the fluid in the direction of the applied deformation. The liquid film will start to develop the fracture and hot tear forms. During the viscous flow of inter granular liquid from inclined channels to horizontal channels, the strength of the solid is given as [9],

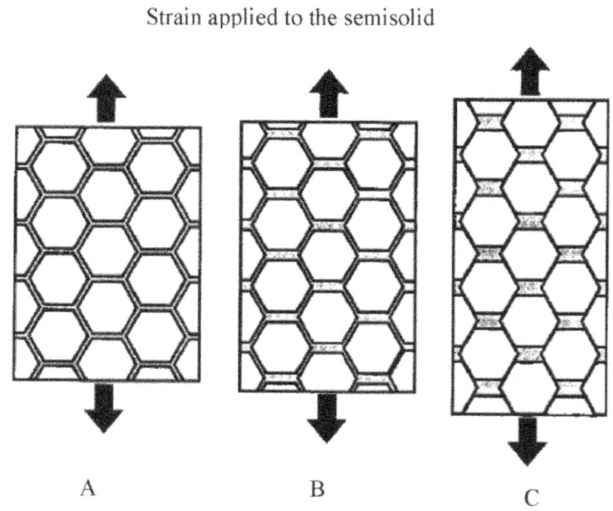

Figure 1. The deformation of a semi solid body due to applied strain. (A) Hexagonal structure with liquid films uniformly distributed (B) Hexagonal structure with liquid films flowing from inclined to horizontal channels (C) Hexagonal structure with liquid films distributed along horizontal channels.(Picture taken from [9])

$$\sigma = \frac{\mu\dot{\varepsilon}}{9}\left(\frac{f_s{}^m}{1-f_s{}^m}\right)^3 \left[\left(1-\frac{1}{2}\left(\frac{f_s{}^m}{1-f_s{}^m}\right)\varepsilon\right)^{-3} + 2\left(1+\left(\frac{f_s{}^m}{1-f_s{}^m}\right)\varepsilon\right)^{-3}\right] \tag{7}$$

where the micro structural parameter m is equal to 1/2 for columnar and 1/3 for Equiaxed grains. The mechanical response of the semisolid body depends on the viscosity of the inter granular liquid μ, applied strain rate $\dot{\varepsilon}$, the accumulated strain ε, and the solid fraction f_s. However, the above equation is valid until the accumulated strain reaches $\varepsilon_{max} = 2\left(\frac{1-f_s{}^m}{f_s{}^m}\right)$. Beyond this maximum strain, the strength of the mush is calculated using the relation,

$$\sigma = \frac{2f_s{}^m K_T{}^{-1}\varepsilon}{3[1+(\varepsilon_{max}-1)f_s{}^m]} \quad (8)$$

where K_T is the isothermal compressibility of the liquid. The complete mechanical response of the mush during the transition from local viscous flow of the intergranular liquid to dilatation of constrained liquid films can be found by the combination of Eq 7 and Eq.8. The total stress of a semisolid body should not exceed the fracture initiation stress. The fracture initiation stress σ_i can be expressed in terms of solid fraction and the accumulated strain as

$$\sigma_i = \frac{4\gamma_l}{3h}\left(1+\left(\frac{f_s{}^m}{1-f_s{}^m}\right)\varepsilon\right)^{-1} \quad (9)$$

The criteria proposes that the hot tear (physically the fracture in the liquid films) will appear, if the stress in semisolid network exceeds the fracture initiation stress.

Criteria Combining Both Non Mechanical and Mechanical Aspects

The coupling between non mechanical aspects such as liquid feeding deep in the mushy zone and the mechanical aspects such as strain and strain rate of the mushy is modeled by several authors with the objective of predicting defect formation. They propose the hot tearing criteria in terms of the strain or strain rate in the mushy zone at which feeding is stopped.

RDG criteria

The RDG (Rappaz-Drezet-Gremaud) criterion [7] focuses on the liquid metal feeding in between the coherent dendrites in the mushy zone. The liquid must be fed to accommodate both solidification shrinkage and mechanical deformation of the solid phase. The liquid feeding is very easy during the early stages of the mush. However, deep in the mushy zone, where the solid network is strongly connected, the liquid feeding is difficult and hot tear may develop at critical regions. The liquid pressure drop in the mush is introduced as a critical factor for the hot tear development. The pressure drop is the difference between metallostatic and local pressure in the mush and depends on the location in the mushy region. Hot tear will develop if the local pressure falls below the cavitaion pressure. Both shrinkage and deformation of the solid contributes to the local liquid pressure drop. The shrinkage and deformation contributions to the pressure drop and are calculated using the mass balance performed over the mushy region. As shown in fig 2, in a reference frame attached to the isotherms, the mass balance can be written as

$$\frac{\partial(f_l\rho_l v_{l,x})}{\partial x} + \frac{\partial(f_s\rho_s v_{l,y})}{\partial y} = \vartheta_T\left[\frac{\partial(f_l\rho_l)}{\partial x} + \frac{\partial(f_s\rho_s)}{\partial x}\right] \quad (10)$$

The deformation rate of solid along the y direction $(\dot{\varepsilon} = \frac{\partial v_{s,y}}{\partial xy})$ and the solidification shrinkage $(\beta = \frac{\rho_s}{\rho_l} - 1)$ are substituted in the above equation and integrating, the expression for the velocity of liquid any point of the mushy zone can be expressed as

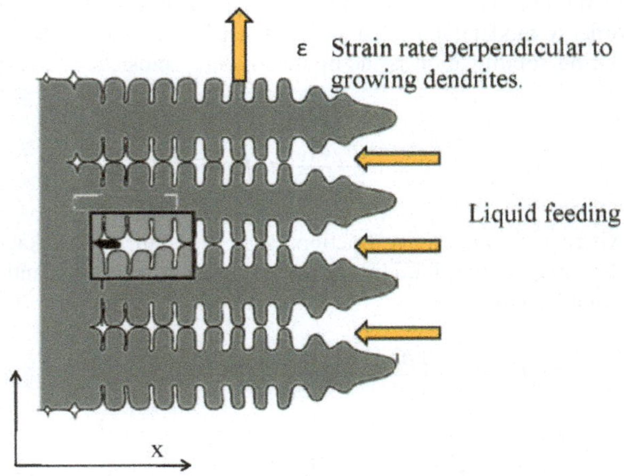

Figure 2. Schematic of the columnar dendrites (Picture taken from [7])

$$f_l v_{l,x} = -(1+\beta)E(x) - v_T\beta f_l \quad (11)$$

with $E(x) = \int f_s\, \dot{\varepsilon} dx$. The liquid velocity can be related to the pressure gradient in the liquid using Darcy equation,

$$f_l v_{l,x} = -\frac{K}{\mu}\frac{dP}{dx} \quad (12)$$

where μ is the viscosity of the liquid, K is the permeability of the mushy zone which is determined as

$$K = -\frac{\lambda_2{}^2}{180}\frac{(1-f_s)^3}{f_s{}^2} \quad (13)$$

Substituting these relations in Eq.11, and integrating over the entire mushy, the shrinkage contribution ΔP_{sh} and the deformation contribution ΔP_ε to the pressure drop are calculated as

$$\Delta P_\varepsilon = -\frac{180}{\lambda_2{}^2}\frac{(1+\beta)\mu}{G}\int_{T_s}^{T_L}\frac{E(T)f_s{}^2}{(1-f_s)^3}dT \quad (14)$$

$$\Delta P_{sh} = -\frac{180}{\lambda_2{}^2}\frac{v_T\beta\mu}{G}\int_{T_s}^{T_L}\frac{f_s{}^2}{(1-f_s)^2}dT \quad (15)$$

The total depression pressure $\Delta P = \Delta P_\varepsilon + \Delta P_{sh}$, is a measure for the hot cracking sensitivity. The hot tear will form if $\Delta P \geq \Delta P_c$, whereΔP_c is the critical depression pressure taken as 2 kPa. The Λ shaped curves, typical of hot tearing, is well reproduced by this criteria, when applied to a variable-concentration aluminum-copper alloy. The susceptibility of the hot tearing increases with addition of copper to a maximum value and the reduces with further addition of copper.

Suyitno Approach

Suyitno modified the above approach and proposed that, the lack of feeding results in the formation of cavity. The cavity formed becomes a hot tear when a critical dimension is reached: otherwise, porosity will result. Thus the model combines the macroscopic cooling process with microscopic porosity prediction. The flow behavior of the semi solid is included in order to model the mechanical response of semi solid. The feeding during solidification is incorporated using transient mass balance as

$$\frac{-\rho_s}{\rho_l}\frac{\partial f_v}{\partial t} = fr + fe \tag{16}$$

where *fv* is the cavity fraction, *fr* is the sum of shrinkage and deformation contributions, *fe* is the feeding rate. Both *fr* and *fe* are calculated as

$$fr = -\left(\frac{\rho_s}{\rho_l} - 1\right)\frac{\partial f_l}{\partial t} + \left(\frac{\rho_s}{\rho_l}\right)\dot{\varepsilon} \tag{17}$$

$$fe = K\left(\frac{P_s}{\mu^2}\right) \tag{18}$$

where K is the permeability of the mush, P_s is the effective feeding pressure, μ is the viscosity of the liquid and L is the length of the porous network. If $fr < fe$, feeding will be sufficient, the flow rate is equal to the shrinkage and deformation rate and no cavity or hot tear will form. If $fr > fe$, the feeding is insufficient and a cavity will form. Once the conditions for the cavity formation are fulfilled, the cavity length a is calculated based on the cavity volume fraction *fv* and the diameter of the cavity as

$$a = C_1\left(\frac{3}{2\pi} f_v V_{char}\right)^{\frac{1}{3}} \tag{19}$$

where C_1 is constant that depends on the grain size and V_{char} is the characteristic volume of the local geometry. According to Griffith approach, the critical cavity length that propagates as hot tear can be calculated as

$$a_{crit} = 4\gamma_s \frac{E}{\pi\sigma^2} \tag{20}$$

The hot cracking sensitivity sensitivity (HCS) is defined by the ratio, $HCS = a_{crit}/a$ and qualitatively if HCS is greater than one, the hot crack will develop.

Constitutive Behavior of the Mush

From the above discussion, it can be inferred that the mechanical response of a semisolid body to an applied strain rate is important in relation to the hot tearing. Several authors, presented about the constitutive behavior of the semi solid based on the experimental observations. RDG criteria consider the deformation of the semisolid is considered only in the direction perpendicular to the dendrites growth direction. MHamdi et.al.[10] provided a method for solving the fully coupled problem taking into account the complex rheology of the mushy alloy. From the physical standpoint, this approach is close to the RDG criterion but it allows for a more accurate description of the strains development in a cast part during an actual process. This approach models two-phase mushy zone as a viscoplastic porous medium saturated with liquid. Many studies [11] have previously shown that as-cast micro structural features such as grain size and shape, porosity, inter metallics and the configuration of the liquid add to the complexity of semi-solid deformation. To date, the application of microstructure modeling to predict semi-solid constitutive behavior has not been possible, due to the difficulty in realistically characterizing the semisolid microstructure. However, in work by Mathier et al [12] this difficulty was addressed through the application of a granular model using the Voronoi tessellation technique to model the solidification of globular grains. Phillion [13] then applied this Voronoi tessellation technique to demonstrate the feasibility of applying microstructure deformation models to semi-solid materials.

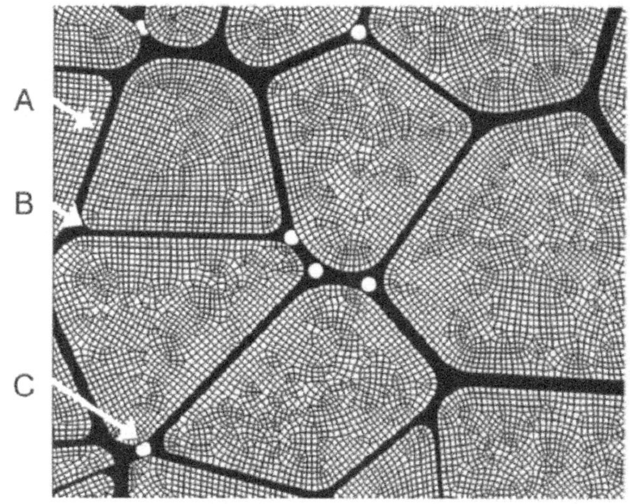

Figure 3. Semi solid micro structure represented by Voronoi tessellation. (A) Grain boundary liquid; (B) Rounded grain vertices; (C) Triple-junction porosity.(Picture taken from [14])

Phillion [14] proposed a model for constitutive relation which include the effects of the key micro structure features such as solid fraction, grain size and the porosity fraction.

$$\sigma = K_p(f_s\,\sigma_s)(\varepsilon_p + \varepsilon_0)^n(1 - \frac{f_p}{1-f_s}) \tag{21}$$

where K_p is the porosity coefficient, ε_0 is the initial strain, *fp* and *fs* are the volume fractions of the pore and solid respectively. The pore volume fraction *fp* can be calculated from the suyitno approach. This relationship is valid for high fraction solid, where the solid phase plays a major role in the deformation properties. Furthermore, this relationship includes the relevant effects of grain size and fraction porosity on the stress strain predictions. In terms of their effect on constitutive behavior, fraction solid is most important, followed by grain size and then fraction porosity. The integration approach of Phillion's material model with suyitno approach can better predict the hot tear sensitivity of the mush.

Summary

Different existing thermal, mechanical, and thermo mechanical hot tearing criteria are extensively reviewed. The mechanisms of interdendritic liquid feeding to accommodate the shrinkage and thermal contraction of the semi solid are explained in brief. The thermal contraction of the mushy depends on the constitutive behavior of the mushy, which is precisely not known. The constitutive behavior which is based on the fraction porosity and grain size can be easily implemented in FEM to simulate the real process. The fraction porosity can be found from the suytino approach, the grain size can be calculated from the local cooling rate at the growth of dendrites.

References

[1] S. W. and L.Katgerman, "Hot tearing criteria evaluation for direct-chill casting of an al-4.5 pct cu alloy," *Metallurgical and Materials Transactions A*, vol. 36A, pp. 1537–1546, 2005.

[2] U.Feurer, "Quality control of engineering alloys and the role of metals science.," in *Delft University of Technology, Delft, The Netherlands*, 1977.

[3] Clyne and T. Davies in *Proceedings of the Conference on Solidification and Casting of metals. Metals Society, London,*, 1979.

[4] L.Katgerman, "Mathematical model for hot tearing of aluminum alloys during dc casting," *Journal of Metals*, vol. 34, pp. 46–49, 1982.

[5] B.Magnin, L. Katgerman, and B. Hannart in *Modeling of Casting Welding and Advanced Solidification Processes VII.*, 1995.

[6] M.Braccini, C. Martin, and M.Suery in *Modeling of casting Welding and Advanced Solidification Processes IX.*, 2002.

[7] M.Rappaz, J.M.Drezet, and M.Gremaud, "A new hot-tearing criterion," *Metallurgical and Materials Transactions A*, vol. 30A, pp. 449–455,1999.

[8] N.N.Prokhorov *Russ.Castings Prod*, vol. 2, pp. 172–175, 1962.

[9] D. Lahaie and M. Bouchard, "Physical modeling of the deformation mechanisms of semisolid bodies and a mechanical criterion for hot tearing," *Metallurgical and Materials Transactions B*, vol. 32B, pp. 697–705, 2001.

[10] M.MHamdi, A.Mo, and C.L.Martin, "Twophase modeling directed towards hot tearing formation in aluminium direct chill casting," *Metallurgical and Materials Transactions A*, vol. 33,pp. 2081–2093, 2002.

[11] D. Eskin, Suyitno, and L.Katgerman, "Mechanical properties in the semi-solid state and hot tearing of aluminium alloys," *Progress in Materials Science*, vol. 49, pp. 629–711, 2004.

[12] V. Mathier, A.Jacot, and M.Rappaz, "Coalescence of Equiaxed grains during solidification," *Modeling and Simulation in Materials science and Engineering*, vol. 12, pp. 479–490, 2004.

[13] A.B.Phillion, S.L.Cockcroft, and P.D.Lee, "A three-phase simulation of the effect of microstructural features on semi-solid tensile deformation," *Acta Materialia*, vol. 56, pp. 4328– 43338, 2008.

[14] A.B.Phillion, S.L.Cockcroft, and P.D.Lee, "Predicting the constituitive behavior of semi solids via a direct finite element solution:application to AA5182," *Modelling and Simulation in Materials science and Engineering*, vol. 17, pp. 1–15, 2009.

Proceedings of the 2013 International Symposium on Liquid Metal Processing & Casting
Edited by: Matthew J. M. Krane, Alain Jardy, Rodney L. Williamson, and Joseph J. Beaman
TMS (The Minerals, Metals & Materials Society), 2013

EFFECT OF SONOTRODE DESIGN ON SIMULTANEOUS GRAIN REFINEMENT AND DEGASSING OF Al ALLOYS BY ULTRASOUND

Jeong Il Youn[1], Young Ki Lee[1], Kee Joo Jung[1], Bong Jae Choi[1], and Young Jig Kim[1]
[1]Sungkyunkwan University, School of Advanced Materials Science and Engineering,
300 Cheoncheon dong, Jangan gu, Suwon, 440-746, Korea

Keywords: Ultrasound, Grain refinement, Degassing, Sonotrode

Abstract

Grain refinement and degassing of Al alloys are the fatal processes to keep the high quality castings. The most widely used process for the grain refinement and the degassing is the addition of inoculants and the rotary degassing, respectively. However, these processes affect each other and make low efficiency of each process, in addition, these increase the cost. The application of the ultrasound for grain refinement and degassing has been studied to modify the processes, but the ultrasound could be only applied in special casting condition, such as solidification or in the semi-liquid state, and the effect of ultrasound in this state seriously decreases because the fluidity of melt extremely dropped. In this study, new process using ultrasound, named as NUMT (Nucleation Ultrasonic Melt Treatment), is suggested as the melt treatment for grain refinement and degassing simultaneously in melt. NUMT process could be established by the special design of the sonotrode made by Ti, and it makes supply the nuclei in Al alloys melt continuously and efficiently. The grain size and the content of hydrogen drop extremely by NUMT.

Introduction

The study to apply the ultrasound in casting process has been studied for the last a decade [1,2]. The application of ultrasound in casting process has been focused on the degassing and the grain refinement, and especially the grain refinement has been mainly studied on the point of view the mechanism that the dispersion of the broken particles of dendrite resulting from the ultrasound injection during solidification or the state which is the coexistence of solid and liquid phase of metal [3]. The process which should be only applied in special casting conditions, as solidification or in the semi-liquid state, could not extend to the general casting process, such as sand casting, permanent mold casting and die casting. The grain refinement and the degassing by the ultrasound injection have been represented as the very effective processes, and mechanisms for the grain refinement under ultrasound injection have been proposed but they have been still arguable [4~6]. The effectiveness which is the most important basis to adapt in industry will be lost, and in addition the degassing by ultrasound never performs well because the fluidity of melt extremely dropped. Based on this research back ground, we have studied to modify the process which can apply in normal casting process, and finally established the process by the new design of the sonotrode. The process named as NUMT (Nucleation Ultrasonic Melt Treatment) is that the ultrasound is injected into Al alloy melt with the special sonotrode. The most important phenomenon of ultrasound injection in liquid is the cavitation effect and the acoustic streaming, but these effects could not have been used in the previous study because of the restriction on the flow caused by the injection condition which only could be the state of the semisolid [7~10]. But, the ultrasound injection into full melt makes it possible to use these unique effects completely and then a lot of new ideas can be suggested as the new process. From this point of view, NUMT could be the most effective process because the process can do the melt treatment for grain refinement and degassing simultaneously from hypo-eutectic to hyper-eutectic Al-Si alloys without the concerns on the alloy composition by using the cavitation and acoustic streaming effect. Fig. 1 shows the mechanism of NUMT, especially, the grain refinement mechanism. NUMT process makes good use of the cavitation conversely to supply Ti substrates in the melt that can act as nucleation sites by the adaption of the sonotrode made of pure Ti for the grain refinement. This paper involves the effect of NUMT process on grain refinement and degassing of aluminum alloy the mechanism based on the analysis of chemical composition and the microstructure of the alloy.

Experimental Procedure

Fig. 2 shows the experimental system for NUMT which can generate and inject the ultrasound in metal melt, for Al casting. The frequency of the system is 20 kHz and the maximum power output is 1900 W/cm^2. The system is consisted of three parts, ultrasound generator, sonotrode and cooler. The ultrasound generator is the piezoelectric device laminated three layer lead zirconate titanate (PZT) as the Bolt-Clamped Langevin Ultrasonic

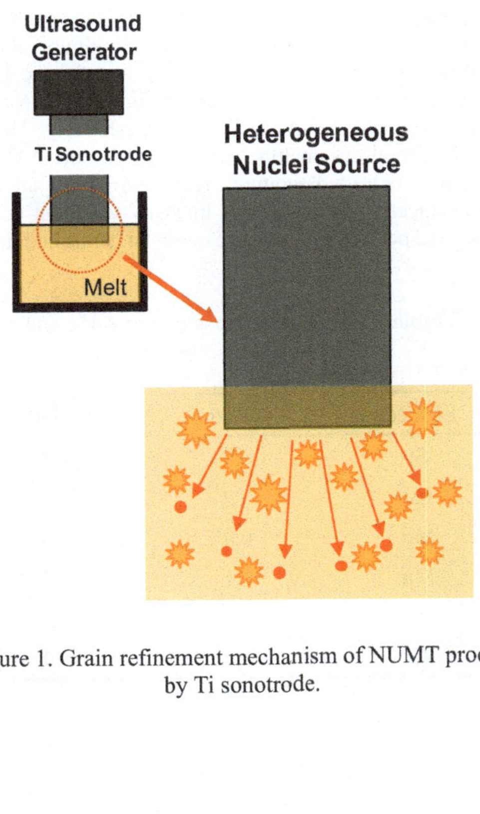

Figure 1. Grain refinement mechanism of NUMT process by Ti sonotrode.

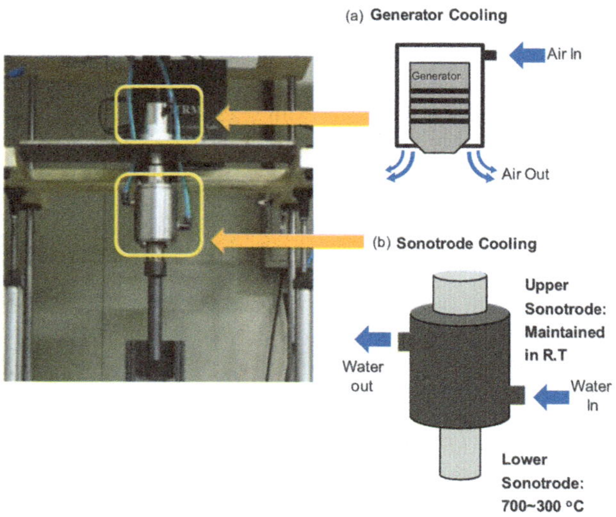

Figure 2. Photograph of NUMT system: (a) air cooling for ultrasound generator and (b) water jacket for prevention of conducting heat of the sonotrode.

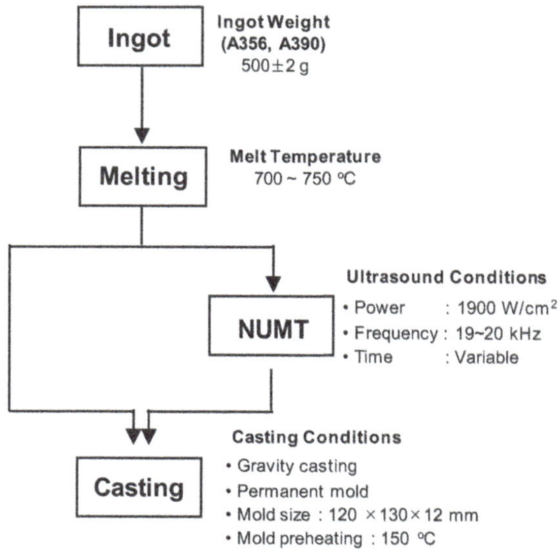

Figure 3. Experimental procedure for NUMT of A356 and A390 aluminum alloy.

Transducer Type (BLT). The PZT piezoelectric lost its property when it is heated to the Curie temperature, therefore the ultrasound generator must keep on cooling not to be heated by the proper method. This system adapted a water jacket and forced air cooling. The water cooling jacket was set up at the booster sonotrode not to transfer the heat through conduct, and the PZT was cooled by forced air cooling supplied with an air compressor as shown in Fig. 2.

The experimental alloys are A356 and A390 aluminum alloy and the chemical compositions are shown in Table I. A356 exhibits good castability, high strength and elongation and considerable ductility, and A390 exhibits the low coefficient of thermal expansion, high hardness and good wear resistance. These alloys are the representative alloys of casting aluminum alloys. The about 500 g alloy ingot was melted by the electric resistance furnace with using an graphite crucible and the A356 alloy was melted at 700 °C and A390 was 750 °C, respectively. The furnace temperature for alloys melting was varied as the NUMT time because the melt temperature increased with an increase ultrasound injection time by high intensity vibration energy. As the results, the pouring temperature could be controlled uniformly.

Table I. Chemical composition of A356 and A390 aluminum alloy. (wt%)

Element	A356	A390
Si	6.9	16.9
Cu	0.2	4.3
Mg	0.4	0.5
Mn	0.1	0.1
Fe	0.1	0.1
Al	Bal.	Bal.

The frequency of the ultrasonic generator is 20 kHz at 25 °C and about 19 kHz at 700 °C, and the ultrasonic power up to 1900 W/cm^2. The experimental procedure shows in Fig. 3. The sonotrode made by Ti was used to inject the ultrasound into alloy melt as the design which makes have the resonant frequency of 20 kHz at 25 °C as shown in Fig. 4 after simulation. When the temperature of the aluminum melt was reached at the experimental condition, the sonotrode was immersed into the melt about 20 mm in depth. The sonotrode was heated for 5 minutes at just above the melt before dipping into melt, and also held in melt for 5 minutes to maintain the melt temperature before NUMT. The melt after the NUMT as each experimental target is poured into the steel mold coated as boron-nitride. After cooling to the room temperature in air, the specimens were prepared to observe the macro and microstructure at each condition. The micro and macro structure of A356 was observed after etched by 0.1 % HF

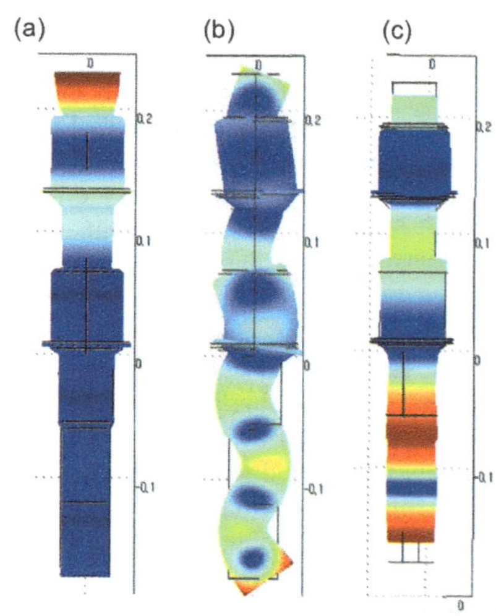

Figure 4. Simulation results of sonotrode vibration mode with frequency: (a) extinction at 17.6 kHz, (b) distortion at 19.2 kHz and (c) normal at 20.0 kHz

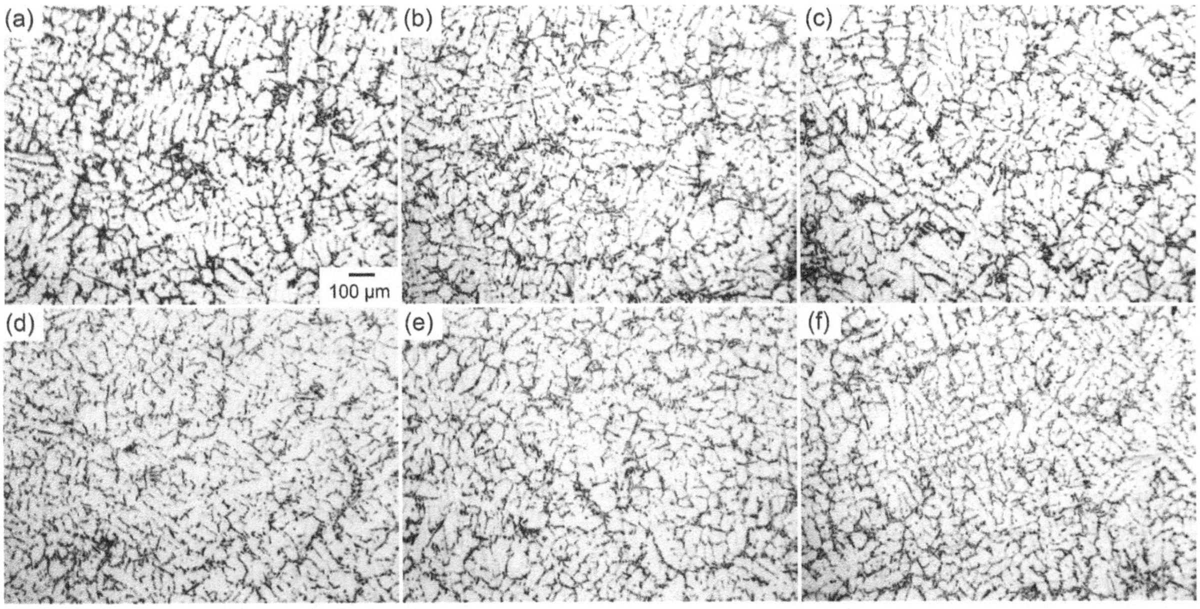

Figure 5. Microstructure of A356 with NUMT time: (a) without NUMT, (b) NUMT for 30 sec, (c) 90 sec, (d) 180 sec, (e) 300 sec, and (f) 600 sec.

and Poulten's reagent (HCl 12 ml + HNO_3 6 ml + HF 1 ml + water 1 ml), respectively. The microstructure of A390 was mainly observed with a perspective on the variation of the size of the primary Si without the special chemical etching. The grain and primary Si size of A356 and A390 were measured by the image analysis software, Image-Pro Plus (Media Cybernetics Inc., U.S.A), to characterize the effect of NUMT on microstructure.

Results and Discussion

Fig. 5 shows the microstructure of A356 alloy after NUMT for 30, 90, 180, 300, and 600 sec and as-cast without NUMT. It can be seen that the microstructure consisted of matrix α Al phase and eutectic Si phase. The microstructure of A356 alloy cast without NUMT, Fig. 5 (a), α phase was coarse dendrite, whose length of

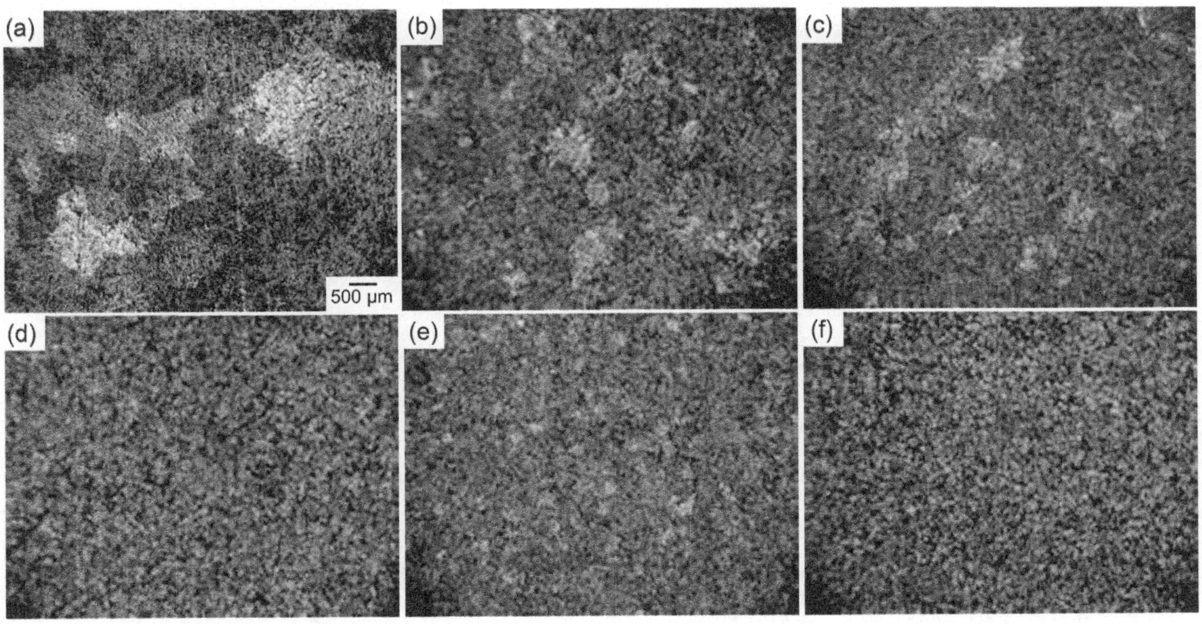

Figure 6. Macrostructure of A356 with NUMT time: (a) without NUMT, (b) NUMT for 30 sec, (c) 90 sec, (d) 180 sec, (e) 300 sec, and (f) 600 sec.

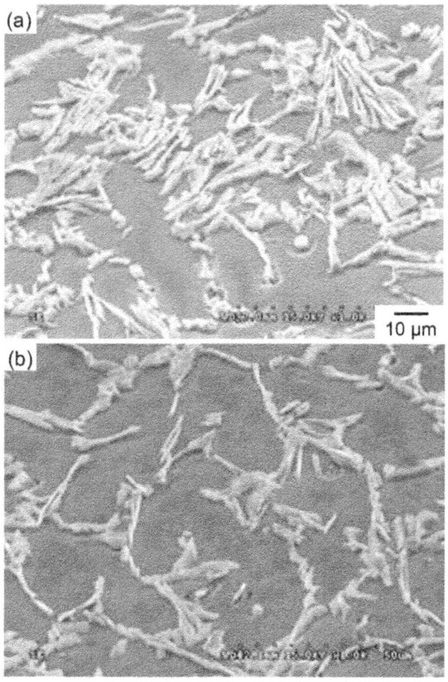

Figure 7. Eutectic Si morphology of A356 alloy (a) without and (b) with NUMT for 300 sec.

secondary arm was about 300~400 µm and width was about 100 µm. The eutectic Si phase was acicular type, because of the absence of a chemical refiner, sodium, strontium and antimony for eutectic modification. A356 with NUMT showed the microstructure of very fine eutectic Si interspersed with fine globular primary aluminum grains with increasing NUMT time as shown in Fig. 7. The huge dendrite arms were changed to the very globular primary aluminum grains with increasing NUMT time as small and rosette type structure with NUMT. The macro etching was carried out to observe the grain clearly and measure the size. The result shown in Fig. 6 suggests that NUMT is very effective to obtain very fine grains of A356 alloy in a short time. The grain size was decreased to a half by NUMT for just 30 sec, and the more fine structure has shown with the more NUMT time.

The effect of ultrasound on the microstructure in A390 alloy is very clear as shown in Fig. 8. This figure shows the microstructure of A390 alloy after NUMT for 30, 90, 180, 300, and 600 sec in alloy melt and as-cast without NUMT. The primary Si of the microstructure of A390 alloy without NUMT was very coarse and non-uniform as shown in Fig. 8 (a). In addition, the primary Si was conglomerated in some area in Al alloy matrix. The average length of primary Si was about 150 µm. As compared with this microstructure, the primary Si of A390 alloy with NUMT was very small and dispersed uniformly in Al alloy matrix without adding other chemical refiners as shown in Fig. 8 (b)~(f). The effects of NUMT of the alloy melt on microstructure refinement of primary Si were increased with NUMT time. The apparent effect on the refinement of the primary Si was appeared to be injected for 90 sec. The optimized injection time for refinement of primary Si was 300 seconds at these experimental conditions and the primary Si size was about 30 µm, and the size was similar after then time. The modification of eutectic Si phase in A390 cannot be established by the only thesis figures, however, it is very obvious that the distribution of eutectic Si is changed to the very uniform with increasing of the NUMT time.

The grain size of A356 and primary Si size was measured by the image analysis to characterize NUMT effect as the size with time, and the results was represented at Fig. 9. The grain size of A356

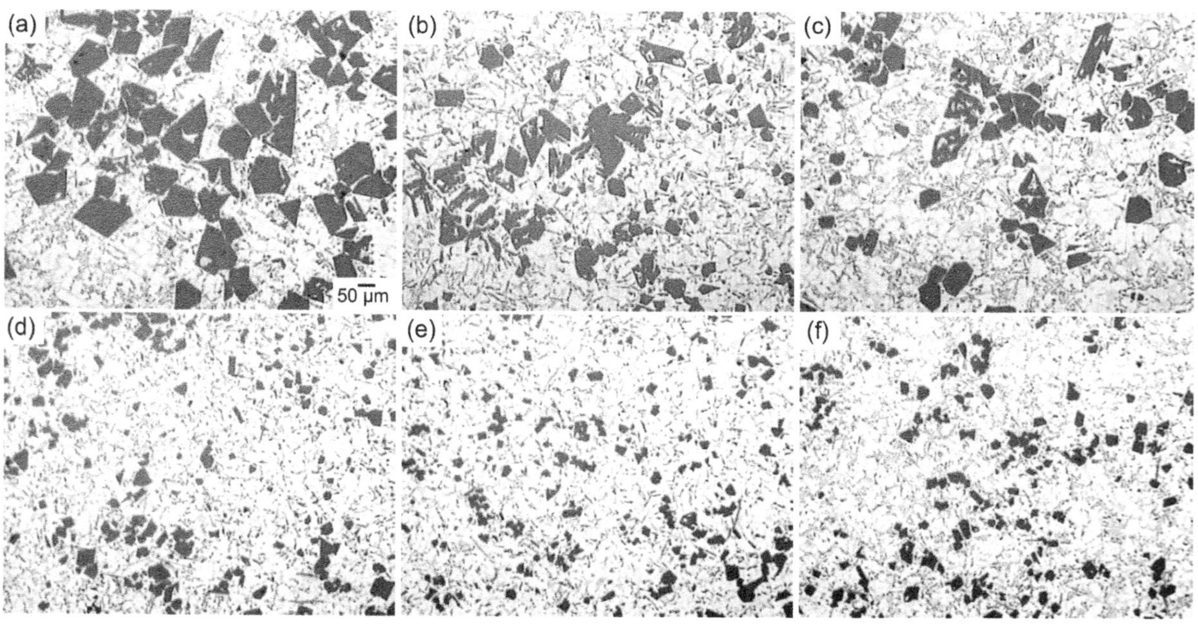

Figure 8. Microstructure of A390 with NUMT time: (a) without NUMT, (b) NUMT for 30 sec, (c) 90 sec, (d) 180 sec, (e) 300 sec, and (f) 600 sec.

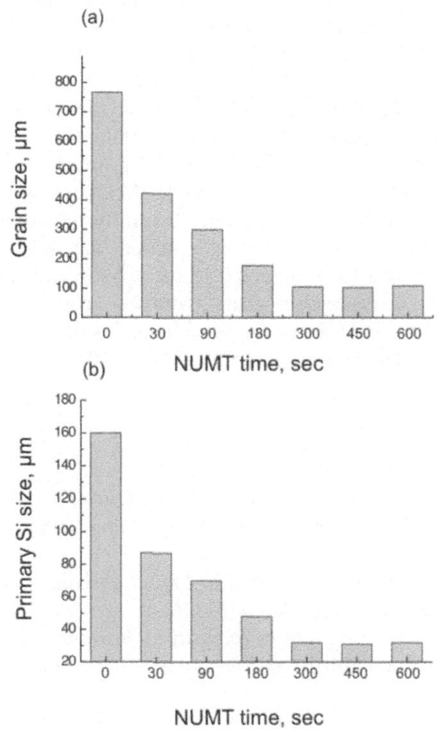

Figure 9. (a) Grain size of A356 and (b) primary Si size of A390 alloy as NUMT time.

The grain size of A356 and primary Si size was measured by the image analysis to characterize NUMT effect as the size with time, and the results was represented at Fig. 9. The grain size of A356 without NUMT was about 760 µm, however, the size decreased with increasing NUMT time and reached to about 100 µm after NUMT for 300 seconds. The size was maintained after the time in spite of increasing of NUMT time. Fig. 9 (b) shows the primary Si size of A390 alloy with NUMT time. The result was similar with A356. The primary Si size of about 180 µm which was shown without NUMT became smaller with increasing NUMT time to about 30 µm by NUMT for 300 seconds too and kept the size.

The mechanism of the grain refinement of A356 and A390 alloys by NUMT is based on the role of the Ti injected in melt by the process. Ti content in melt was analyzed by inductively coupled plasma mass spectrometry at each NUMT conditions of A356 and A390. Ti content increases with NUMT time from 0.08 wt% for 30 sec to 0.31 wt% for 600 sec of NUMT time linearly. The effect of Ti on grain refinement of A356 is very clear. The size of grains is decreased with NUMT time, but the size does not decreased more after NUMT for 300 seconds. The similar result was exhibited at A390, the size of the primary Si of A390 maintains after 300 seconds. This means that there is the minimum time for NUMT on grain refinement, and there is no more effect on that though the more NUMT time exceed the minimum time. This result is related to Ti added to melt in the middle of NUMT through the sonotrode. Grain size is increased with Ti content, however, there is also the minimum Ti content not to affect any more the decreasing of the size. The minimum Ti content for grain refinement was 0.25 wt% and 0.30 wt% at A356 and A390, respectively. This result suggest that optimum NUMT condition

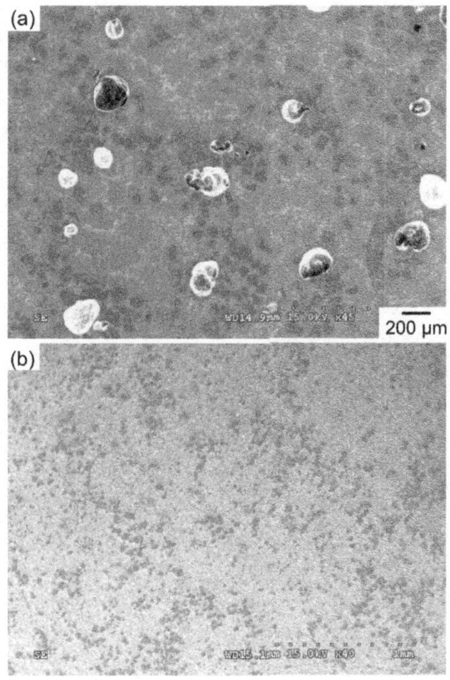

Figure 10. Removal of porosities of A390 alloy by NUMT degassing : (a) without and (b) with NUMT for 15 sec.

for the grain refinement is directly related the Ti content and the degree of grain refinement by NUMT can be controlled by Ti content, than the time. The existence of the minimum content of Ti not to affect any more the grain refinement could be explained by the formation of the intermetallic phase of Ti-Al-Si. The extra Ti cannot be participated the nucleation process as α Al could be easily consumed as the formation of $(AlSi)_3Ti$ or $Ti(AlSi)_2$. This means that there is also maximum Ti content on the grain refinement.

Degassing by NUMT was carried out in A356 and A390 melt. The alloys were melted and then held at predetermined temperatures for a half of an hour before degassing, and the hydrogen content was measured. The hydrogen content before degassing was 0.32 ml and 0.15 ml per aluminum 100 g in A356 and A390, respectively. A356 melt had higher content at about two times than A390. This is because that the Si in A390 makes reduce the solubility of the hydrogen in melt. The degassing by NUMT was carried out as time for 15, 45, 75, 105, and 150 sec at each new melt, without any degassing treatment, and the hydrogen content was measured by a device for the direct measurement of hydrogen concentration (FOSECO Inc., U.K), respectively. Fig. 10 shows the macro and micro porosities on castings. The all macro and micro porosities clearly removed from the castings by NUMT for just 15 seconds. The initial hydrogen concentration was 0.32 ml and 0.15 ml at A356 and A390, respectively. With NUMT, the hydrogen contents decreased sharply with increasing process time in just few seconds and then reached a plateau hydrogen content, which corresponds to the steady-state hydrogen concentration. The results shown in Fig. 11 suggest that NUMT time for degassing of aluminum melt was extremely short.

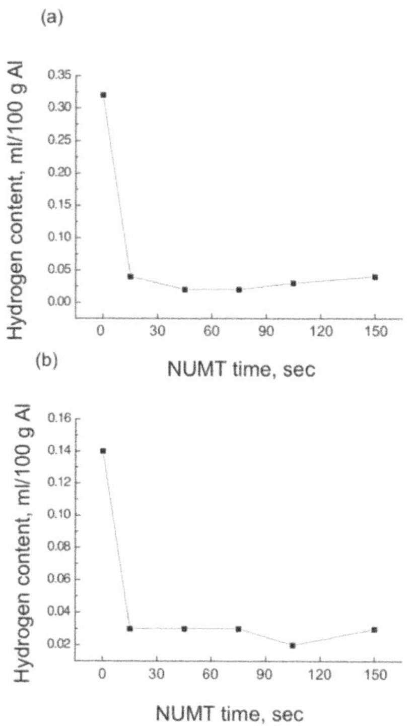

Figure 11. Hydrogen content in melt with NUMT time: (a) A356 and (b) A390.

It is supposed that the high effectiveness of NUMT degassing results from the pressure of the cavities produced in negative pressure [11]. The bubbles produced by blowing of other gas have the positive pressure, and want to enlarge for the expansion of their volume. The size of bubbles is decided by the balance of pressure between blowing gas and surrounding liquid. Therefore, degassing by the gas bubbling, usually using in foundries, is not effective because hydrogen is removed by the adhesion on the surface of bubbles, or just pushed on the melt surface. In contrary, the bubbles produced by cavitation effect have negative pressure as mentioned above. The pressure of cavitation bubbles is lower than surrounding melt, and hydrogen very easily diffuses into the cavitation bubble. This means that the gathering hydrogen atoms change to hydrogen molecular which has some volume, and floating to melt surface by itself. These reasons make increase the efficiency of NUMT degassing.

Summary

This research starts from the questions; how can we use ultrasound for the grain refinement and degassing of Al alloys in melt, simultaneously? For making this answer, we suggested the mechanism by heterogeneous nuclei, named as Nucleation Ultrasonic Melt Treatment. NUMT starts on the basis of the model, and setting up the system for verifying the concept. The system could be worked by the sonotrode made by Ti, and it was designed to transfer the ultrasound energy without loss and supply the nuclei. The effect of NUMT was normalized by the grain size and hydrogen content at various experimental conditions, and these data could be the guideline of the process. In A356, the huge dendrite arms were changed to the very small and rosette type structure with NUMT. The grain size was decreased to a half by NUMT for just dozens of seconds, and the more fine structure has shown with the more process time. The grain size of A356 without NUMT was about 760 μm, however, the size decreased with increasing the time and reached to about 100 μm after NUMT for 300 seconds. The size was maintained after the time in spite of increasing of time. In A390, the primary Si size of about 180 μm which was shown without NUMT became also smaller with increasing the time to about 30 μm by the NUMT for 300 seconds too and kept the size. When the initial hydrogen concentration was 0.32 ml and 0.15 ml at A356 and A390 without degassing, respectively, the hydrogen contents decreased dramatically to 0.02 ml/100 g Al with increasing NUMT time in a few seconds and then reached a plateau of hydrogen content.

Reference

[1] G.I. Eskin, *Ultrasonic Treatment of Light Alloy Melts* (Gordon and Breach Science Publishers, Amsterdam, Netherlands, 1998), 135-166.

[2] O.V. Abramov, *Ultrasound in Liquid and Solid Metals* (CRC press, Florida, U.S.A, 1994), 273-326.

[3] O.V. Abramov, *High Intensity Ultrasonics* (Gordon and Breach Science Publishers, Amsterdam, Netherlands, 1998), 587-656.

[4] G.I. Eskin, G.S. Makarov and Y.R. Pimenov, "Effect of ultrasonic processing of molten metal on structure formation and improvement of properties of high-strength Al-Zn-Mg-Cu-Zr alloys," *Adv. Perform. Mater*, 2 (1995), 43-50.

[5] K. Kocatepe and C.F. Burdett, "Effect of low frequency vibration on macro and micro structures of LM6 alloys," *J. Mater. Sci*, 35 (2000), 3327-3335.

[6] J. Li and T. Momono, "Effect of ultrasonic output power on refining the crystal structures of ingots and its experimental simulation," *J. Mater. Sci. Technol*, 21 (1) (2005), 47-52.

[7] X. Jian, T.T. Meek and Q. Han, "Refinement of eutectic silicon phase of aluminum A356 alloy using high-intensity ultrasonic vibration," *Scr. Mater*, 54 (5) (2006), 893-896.

[8] Y. Osawa, et al, "Effect of Ultrasonic Vibration Pretreatment on Microstructural Evolution and Mechanical Properties of Extruded AZ91 Alloy," *Mater. Trans*, 49 (5) (2008), 972-975.

[9] W. Khalifa, Y. Tsunekawa and M. Okumiya, "Effect of Ultrasonic Melt-Treatment on the Eutectic Silicon and Iron Intermetallic Phases in Al-Si Cast Alloys," *Mater. Sci. Forum*, 638 (2010), 431-436.

[10] M. Abdel-Reihim and W. Reif, " Effect of ultrasonic vibration on the solidifications of alloys containing different microstructures," *METALL*, 38 (2) (1984), 130-132.

[11] A.R.N. Meidani and M. Hasan, "A study of hydrogen bubble growth during ultrasonic degassing of Al–Cu alloy melts," *J. Mat. Pro. Tech*, 147 (3) (2004), 311-320.

Miscellaneous

Liquid Metal Processing & Casting 2013

Proceedings of the 2013 International Symposium on Liquid Metal Processing & Casting
Edited by: Matthew J. M. Krane, Alain Jardy, Rodney L. Williamson, and Joseph J. Beaman
TMS (The Minerals, Metals & Materials Society), 2013

LIQUID METAL ENGINEERING BY APPLICATION OF INTENSIVE MELT SHEARING

Jayesh Patel[1], Yubo Zuo[1], Zhongyun Fan[1]
[1]The EPSRC Centre – LiME, BCAST, Brunel University, London, UB8 3PH, United Kingdom

Keywords: Melt shearing, Grain refinement, Degassing, DC casting, Molten metal pump

Abstract

In all casting processes, liquid metal treatment is an essential step in order to produce high quality cast products. A new liquid metal treatment technology has been developed which comprises of a rotor/stator set-up that delivers high shear rate to the liquid melt. It generates macro-flow in a volume of melt for distributive mixing and intensive shearing for dispersive mixing. The high shear device exhibits significantly enhanced kinetics for phase transformations, uniform dispersion, distribution and size reduction of solid particles and gas bubbles, improved homogenisation of chemical composition and temperature fields and also forced wetting of usually difficult-to-wet solid particles in the liquid metal. Hence, it can benefit various casting processes to produce high quality cast products with refined microstructure and enhanced mechanical properties. Here, we report an overview on the application of the new high shear technology to the processing of light metal alloys.

Introduction

Alloys of light metals such as Aluminum and Magnesium are widely used in various industrial sectors mainly automotive for the purpose of weight saving [1]. Melt quality is crucial for both continuous and shape casting of these light alloys. Gas, oxides and other inclusions in the melt usually deteriorate the quality of the casting products. Liquid metal treatment prior to solidification processing is crucial for ensuring the high quality of the cast products regardless of which casting process is used. The existing methods for liquid metal treatment mainly include melt filtering, mechanical stirring by an impeller, electromagnetic stirring, and rotary degassing. However, such processes are time consuming, energy intensive and high cost. Recently, melt conditioning by intensive melt shearing in a twin-screw machine was studied and used to treat both liquid and semi-solid light alloys [2-4]. Improved microstructure and mechanical properties of both wrought and cast light alloys were achieved by the application of this technology [5-9]. Based on the similar principle, a new technology [10] but with simpler equipment has been developed for treating liquid metals using a high shear unit. The new technology uses a simple rotor-stator unit to provide intensive melt shearing, which disperses effectively the harmful inclusions into fine particles to enhance nucleation during the subsequent solidification processing. Experimental results have demonstrated that the high shear unit can be used for general melt treatment, physical grain refinement, degassing and preparation of metal matrix composites and semisolid slurries [11]. In this paper we offer an overview of the high shear device and its application in processing light alloys.

The New High Shear Device

Intensive melt shearing for melt treatment can be achieved in a rotor-stator unit [10]. A schematic of the new rotor/stator high shear unit is shown in Fig. 1. This unit comprises of a set of rotor and stator attached to an electrical motor with a speed control. During its operation, the motor passes the power to the rotor by the shaft and drives the rotor to rotate and shear the liquid metal in the gap between the rotor and the stator and also in the openings of the stator. The rotation speed can be in the order of 5000-15000 rpm providing a shear rate of up to $10^5 s^{-1}$. The new high shear device provides macro-flow in a volume of melt for distributive mixing and, intensive shearing near the tip of the device for dispersive mixing. The main advantages of the high shear device include significantly enhanced kinetics for chemical reactions or phase transformations, uniform; dispersion, distribution and size reduction of solid particles & gas bubbles, improved homogenisation of chemical composition and temperature fields and also forced wetting of usually difficult-to-wet solid particles in the liquid metal. Therefore, the high shear device can be used for; physical grain refinement by dispersing naturally occurring oxides, for degassing of Aluminium melts, for the preparation of metal matrix composites and also for preparation of semi-solid slurries. Hence, the newly developed high shear technology can be applied to benefit various industrial casting processes such as sand casting, low & high pressure die casting, twin-roll casting and direct chill casting.

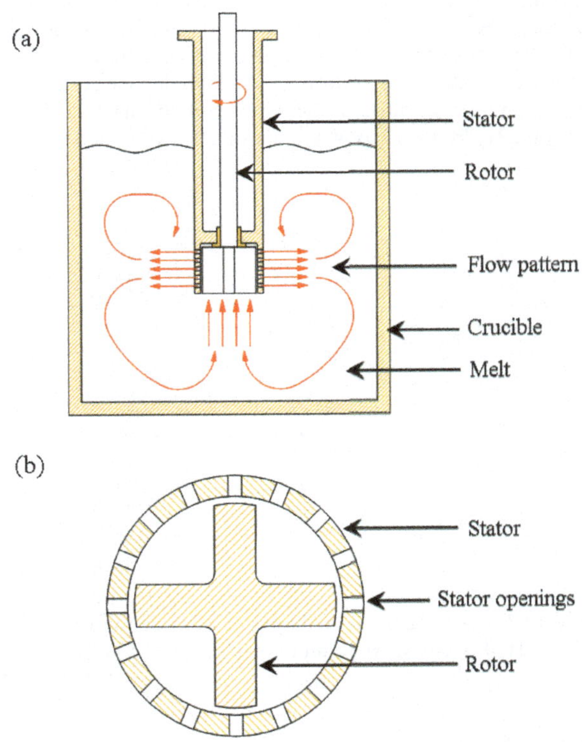

Figure 1: The new rotor-stator high shear device for conditioning of liquid melts by application of intensive melt shearing, (a) inside a crucible of melt illustrating the macroscopic flow pattern and (b) bottom view of the high shear device.

Applications

Melt Conditioning

The new rotor/stator type high shear device can be inserted in a crucible with molten metal, Fig. 1, and apply intensive shearing to disperse particulates that already exist in the melt and due to the sucking action and macro-flow, uniform temperature and chemical composition can be achieved before pouring the melt into a mould to produce a casting. This can be used for simply conditioning liquid melts, preparing semi-solid slurries of alloys and for preparing metal matrix composites where uniform dispersion and distribution of particles like SiC, Alumina etc.., is required to produce better quality castings [11].

Degassing

The high shear device can also be applied to benefit degassing of Aluminum alloy melts in addition to, just shearing in a crucible. A schematic illustration of this is shown in Fig. 2. Since the rotor is spinning at high speed, this creates a centrifugal force that displaces the melt upwards in a pumping action, followed by squeezing the same volume of melt through the stator holes. Conventional degassing of Aluminium melts is done by the rotary degassing method where inert Argon gas is supplied into the melt through a hollow graphite rotor which spins at relatively low speed with the aim of distributing the Argon bubbles evenly throughout the volume of the melt. By replacing the graphite rotor with the new rotor/stator type high shear device, each Argon bubble is dispersed into many tiny bubbles, hence increasing the overall surface area of the bubbles in the melt which in turn improves the degassing efficiency. The degassing efficiency is improved by decreasing the size of the Argon bubbles and by increasing the number density of these bubbles, and at the same time, minimizing disturbance of the melt surface. Large bubbles cause turbulence during their release through the surface of the melt, and at that point some Hydrogen is known to dissolve back into the melt.

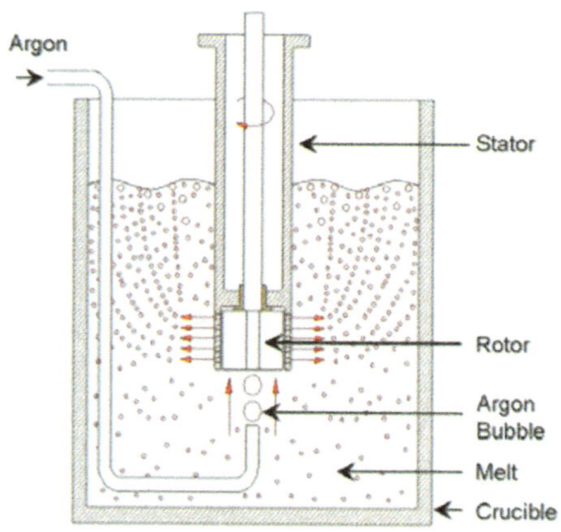

Figure 2: A schematic of the high shear device for degassing (Hydrogen) of Aluminium melts with Argon gas.

Direct Chill Casting

Direct chill casting is the major technology for providing billets or slabs of light alloys for the production of wrought products. However, the billets or slabs made by conventional DC casting process often have coarse and non-uniform microstructures, severe chemical segregation, porosity and hot tearing, which not only burden the downstream thermo-mechanical processing but also have a negative influence on the mechanical properties of the final products. The new rotor-stator high shear device can be applied to conventional DC casting for production of high quality billets and slabs. The combination of the high shear device and DC casting is termed as; Melt conditioned DC casting, the MC-DC casting process [12, 13]. This is schematically illustrated in Fig. 3. In the MC-DC process, the high shear rotor/stator unit is submerged in the sump of a conventional DC casting mould to provide intensive melt shearing during casting. The high shear device operates at a rotation speed of ~6000rpm depending on the size of the mould, providing a shear rate up to $10^5 s^{-1}$. During operation, liquid metal is sucked into the high shear device, is subjected to intensive shearing in the gap between the rotor and stator and is forced to escape from the holes on the stator. The escaping liquid forms a unique macroscopic flow pattern in the sump, as shown in Fig. 3. Due to the sucking action of the high shear device the macroscopic flow in the lower part of the sump is much more intensive than that in the upper part of the sump. Due to this unique melt flow patterns, uniform temperature and chemical composition is maintained throughout the casting process.

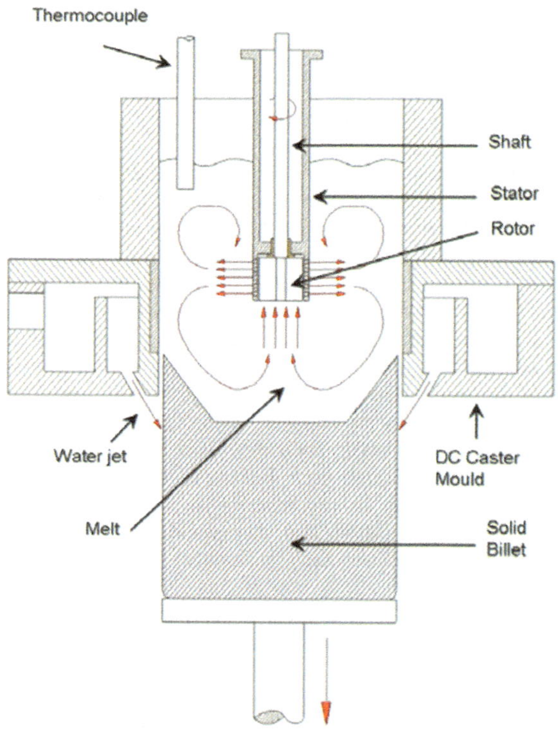

Figure 3: Schematic illustration of the melt conditioned direct chill casting (MC-DC) process, with the high shear device submerged in the sump of a conventional DC mould, also showing the macroscopic melt flow pattern generated by the high shear device.

High Shear Pump

Pumping of molten metal is sometimes necessary in order to quickly transfer melt from a melting furnace to a holding furnace with minimal exposure to air and to avoid unnecessary turbulence due to pick-up of inclusions and also to minimize oxidation. The new high shear device can be utilized as a pump for transfer of melt to a holding furnace or directly to a casting facility such as the 'shot sleeve' in case of high pressure die-casting process or directly to the hot-top in case of the direct chill casting process. A schematic illustration of the multi-stage high shear pump is shown in Fig. 4. The high shear device not only pumps the molten metal but also applies intensive melt shearing at the same time in order to condition the melt. In the multi-stage high shear unit, when the rotor is running at high speed, a volume of melt is sucked upwards into the first stator chamber, then forced out through the stator holes, then sucked back into the second stator chamber and so on, until it is discharged completely in a fully sheared state, out of the high shear pump. The intensively sheared melt can be pumped out of the furnace to feed an insulated laundering system or simply collected and poured for casting. The discharge rate or flow rate can be controlled by the speed of the rotor. There will be a minimum speed required for the sucking and pumping, after which, increasing the speed would directly increase the flow rate. In comparison with the single stage high shear device, Fig. 1, the multi-stage high shear pump ensures that only intensively sheared melt is discharged. In the case of a single stage high shear mixer placed in a large crucible and volume of melt, there is a chance of some melt not been fully sheared. This multi-stage high shear pump can also be allowed to discharge back into the furnace in case of large quantities of melt to ensure that all the melt in the crucible goes through the shearing stages at some point or multiple times. The discharge pipe can be submerged below the surface of the melt so as to avoid unnecessary turbulence on the melt surface to minimise the creation of new oxides films.

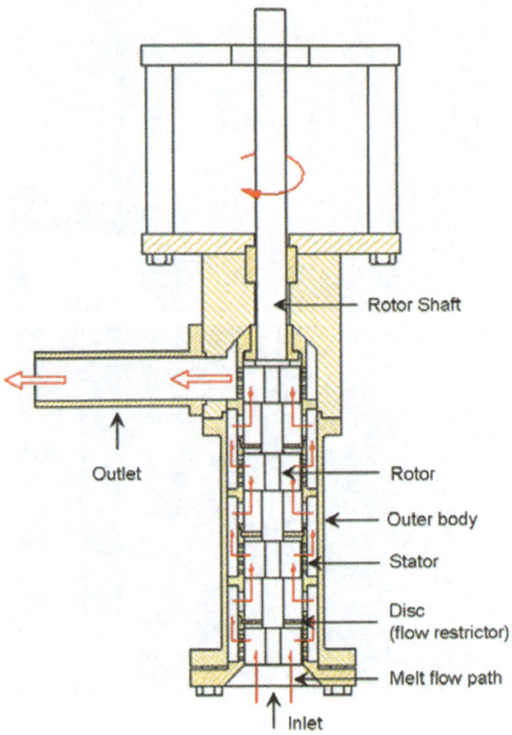

Figure 4: Schematic illustration of the multi-stage high shear pump for supply of fully sheared melt.

Figure 5: Microstructures of AZ91D magnesium alloy cast (a) at 650°C without shearing, (b) at 650°C after shearing, (c) at 605°C without shearing and (d) at 605°C after shearing.

Grain Refinement by Intensive Melt Shearing

In order to demonstrate the effect of intensive melt shearing on the grain refinement of Mg-alloys, AZ91D alloy was sheared at two different temperatures, 650°C and 605°C, and then cast in a TP-1 standard mould [14]. The TP-1 mould was prepared under standard conditions for grain size observation with a constant cooling rate of ~3.5K/s. Reference samples without shearing were also cast at the same temperatures, respectively. The resulting microstructures are shown in Fig. 5. It is clearly observed, that with intensive melt shearing, the microstructure becomes more uniform and the grain size is significantly reduced at both casting temperatures. When measured, the as-cast grain size was reduced from 550μm to 170μm when cast at 650°C, and from 180μm to 90μm when cast at 605°C. It is interesting to note that the grain size of Mg-alloys has a strong dependence on melt superheat and that such superheat effect is suppressed significantly by intensive melt shearing. In the case of Aluminium alloys, similar grain refinement is achieved by application of intensive melt shearing prior to solidification, as shown in Fig. 6.

AZ91D magnesium alloy was also experimented for the effect of intensive shearing in the semi-solid state together with high pressure die-casting process. This resulted in a fine and uniform semi-solid microstructure with the α-Mg grains uniformly distributed with a very narrow size range together with a substantial improvement in tensile elongation. Preparation of metal matrix composites was also experimented with AZ91D magnesium alloy with 9μm diameter SiC particles [9]. Intensive melt shearing improved the wetability and resulted in a uniform distribution of these particles.

Degassing by Intensive Melt Shearing

Dissolved hydrogen gas in Aluminium alloys is detrimental to the mechanical properties of the casting products, due to the fact that during solidification this gas is released which results in porosity within the casting. The hydrogen gas concentration in aluminium alloys is assessed by calculating the density index Di [15, 16], from the measured densities using a water balance. The density index Di is defined as: Di = (Da – Dv) / Da, where Di is the density index, Da is the density of the sample solidified in air, and Dv is the density of sample solidified under partial vacuum using the reduced pressure test (RPT) method, [17, 18]. Fig. 7 shows the density and density index of A7075 Aluminium alloy with and without degassing by the new high shear device. The inert Argon gas that was introduced into the melt at the bottom of the rotor/stator high shear device (as shown in Fig. 2), was set at flow rate of 1 liters/min. After 60s of degassing and intense shearing, the density index was significantly reduced from 13% to less than 0.2%, which indicates that the level of dissolved hydrogen in the alloy can be effectively reduced within a very short period of time by using the new rotor/stator type high shear device. The sectioned samples from the RPT test showed that both the number and size of the porosity were remarkably reduced after intensive melt shearing. Compared with the conventional rotary degassing process, this new shearing technology showed significantly improved degassing efficiency. This is mainly due to the improvement of the uniformity of the inert gas bubbles and the reduction of the bubble size and in consequence the total area of gas-liquid interface. This is one of the important advantages over the conventional degassing process. Another point of degassing by this technology is that, the liquid flow near the liquid surface and the vortex of the liquid are very weak, which can avoid severe turbulence near the liquid surface and, consequently, prevent the entrapment of contaminants and further dissolution of Hydrogen gas from the atmosphere. Isothermal holding of the melt after degassing using the new high shear device still resulted in low density index values for up to 60-90 minutes of holding.

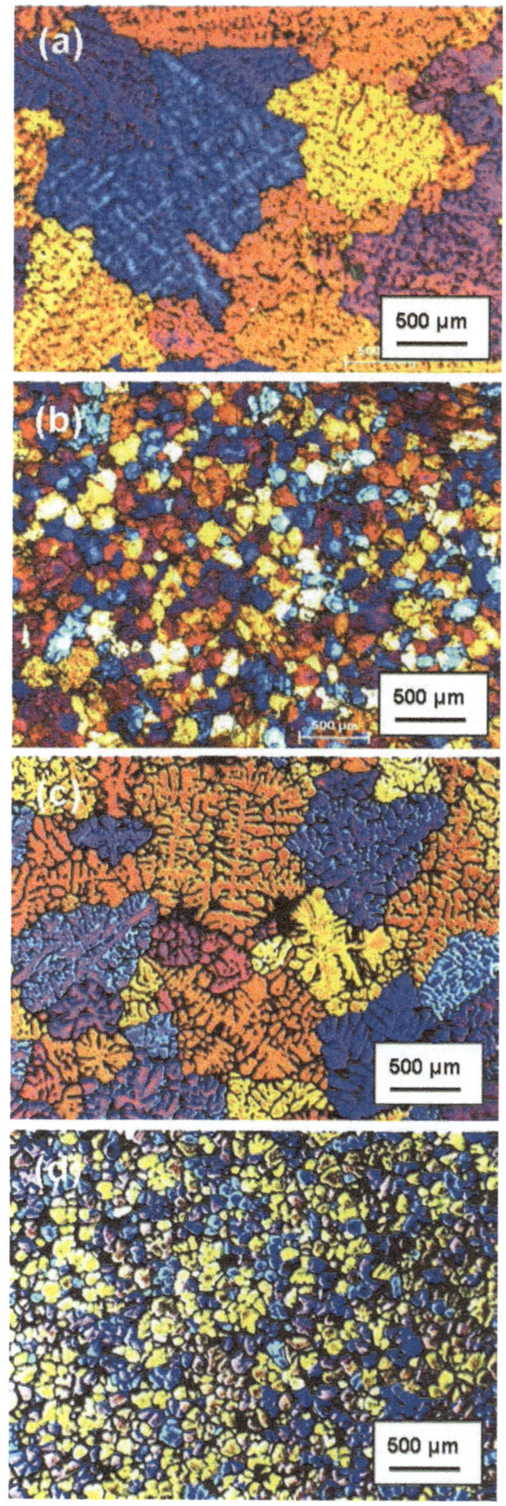

Figure 6: Microstructures of Aluminium alloys (a) A6082 alloy cast at 730°C without shearing, (b) after shearing and (c) A7075 alloy cast at 700°C without shearing, (d) after shearing.

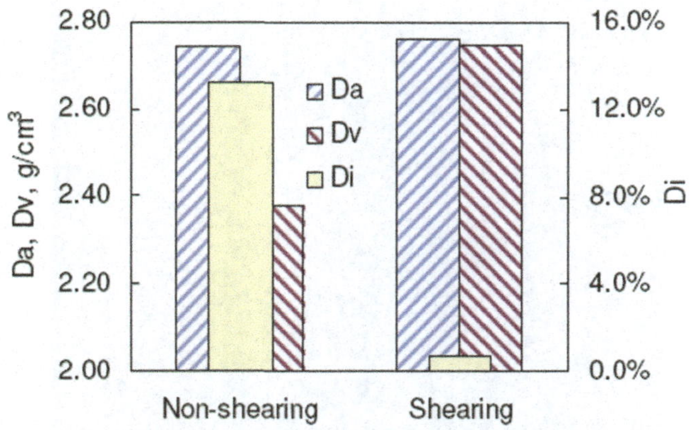

Figure 7: Comparison of density and density index before and after degassing of A7075 Aluminium alloy melt, by the new rotor/stator high shear device with Argon gas.

Direct Chill Casting with Intensive Melt Shearing

The melt conditioned direct chill casting process (MC-DC) is shown in Fig. 3 where the new rotor/stator type high shear device is submerged in the sump of a conventional DC mould to provide intensive melt shearing during casting. In order to demonstrate the effect of *in-situ* intensive melt shearing on the microstructure of the DC cast billet, AZ91D magnesium alloy was cast by the MC-DC casting process by switching on and off the high shear device. The high shear device was submerged in the sump of the DC caster, and was switched-off during the first half of the billet production and then switched-on during the second half of the billet production. Fig. 8 shows the microstructure of the DC cast AZ91D alloy with the transition from the section cast with the high shear device off (bottom) to the section with the high shear device on (top). Without intensive melt shearing, the DC cast billet shows a coarse dendritic microstructure, while with intensive melt shearing it turns sharply into a fine and equiaxed microstructure. From the transition upwards, the billet maintains the same fine and uniform microstructure. It is clear from the result, that during the DC casting process, the application of intensive melt shearing has a significant grain refining effect and can improve the uniformity of the microstructure in the as-cast billets. The DC cast billet of AZ31 magnesium alloy with and without melt conditioning was also analysed for the chemical composition across the cross section of the billet from edge to edge. The resulting difference in chemical composition of this billet is shown in Fig.9. The chemical analysis was done by spark emission spectroscopy. The melt conditioned (MC-DC) billet shows much more uniform chemical compositions from the surface towards the centre of the billet compared with those of the conventional DC cast billets. This is mainly attributed to the achievement of uniform microstructure through out the billet including from the centre to the edge. The macro-etched section of the AZ31 magnesium alloy billet is shown in Fig. 10, in comparison with conventionally DC cast and MC-DC cast billets. In order to understand the grain refining mechanisms by intensive melt shearing, temperature measurements during DC casting were carried out by attaching thermocouples to the head of the high shear device. Based on the direct measurement of the sump depth with stainless steel rods, during the casting process, the bottom of the high shear device is placed about 40mm above the liquid/solid interface. Without melt shearing, when the high shear device is off, the melt temperature in the sump is quite non-uniform; the further the distance from the solidification front the higher the melt temperature. However, with the start of the high shear device, there is a sharp change of melt temperature distribution in the sump. Intensive melt shearing in the sump makes the temperature curve become closer together, resulting in a more or less uniform melt temperature in the sump. The temperature curves also show that after application of the intensive shearing, the melt temperature becomes very close to the liquidus temperature of the alloy being cast, eventually reaching a plateau which is slightly lower that the liquidus temperature. This is deemed to have an important contribution to the microstructural grain refinement and reduced or eliminated chemical segregation achieved in the MC-DC casting process. A similar set of results have also been achieved with Aluminum alloys [13], such as 2xxx, 6xxx and 7xxx series wrought alloys and including some selected cast alloys as well. Microstructural grain refinement and reduced or eliminated chemical segregation is also achieved in the case of MC-DC casting of these Aluminium alloys.

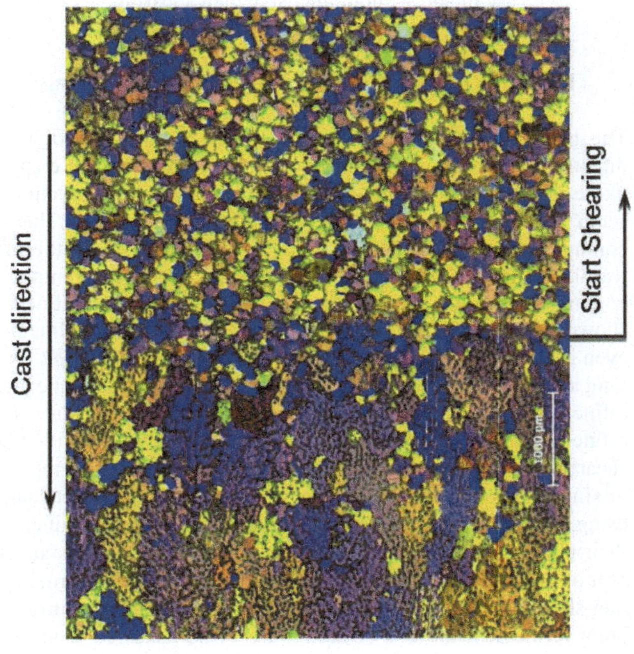

Figure 8: Microstructure of DC cast AZ91D magnesium alloy billet with and without intensive melt shearing, showing the transition from coarse dendritic grain structure without intensive melt shearing (high shear device OFF), to a fine and equiaxed grain structure with application of intensive melt shearing (high shear device ON).

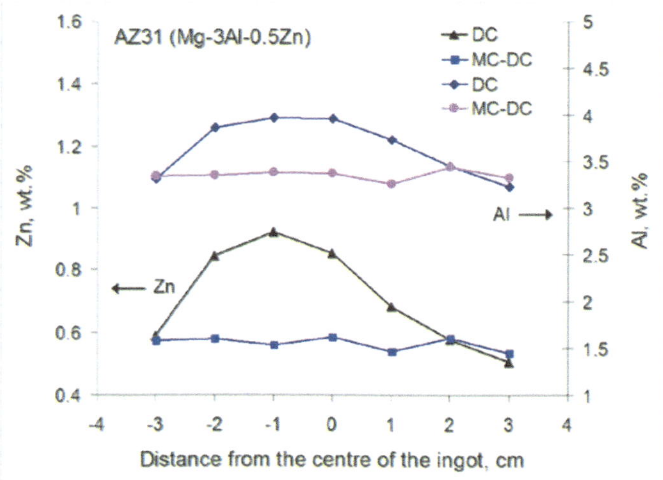

Figure 9: Comparison of chemical composition from the centre towards the edge of the AZ31 Magnesium alloy DC billet cast with and without intensive melt shearing.

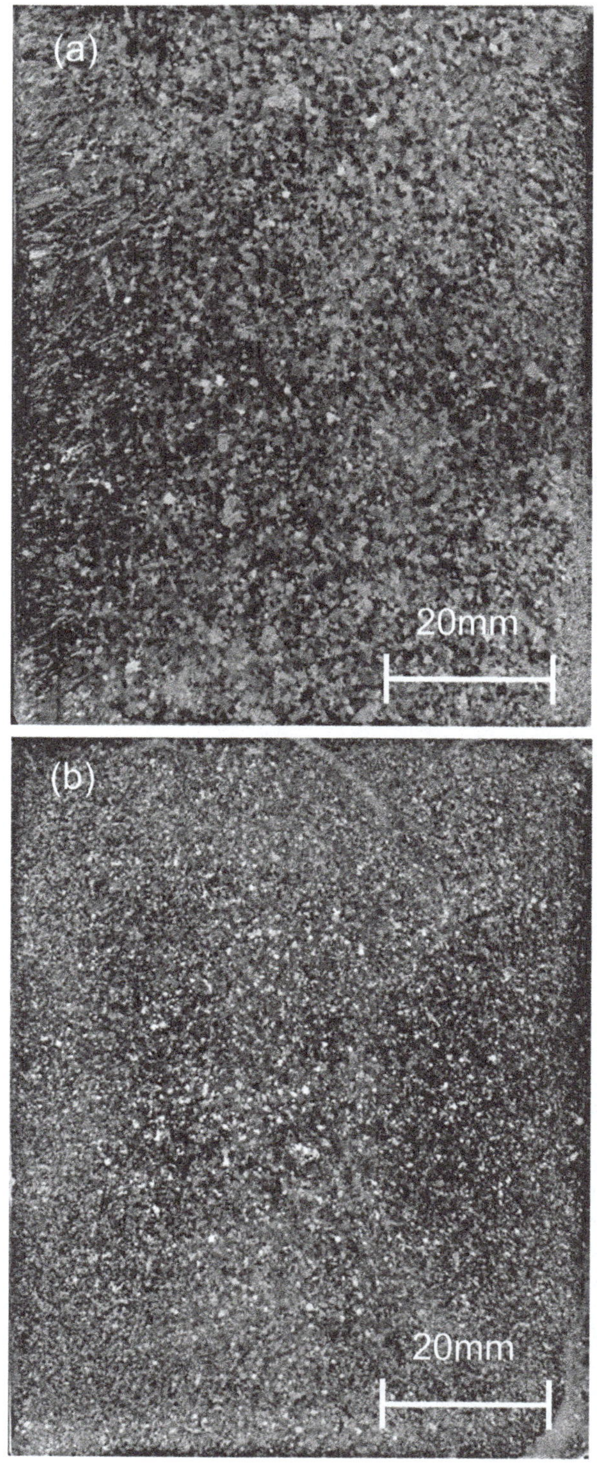

Figure 10: Macrographs of vertically sectioned billets of AZ31 Magnesium alloy; (a) Conventional DC cast and (b) Melt conditioned DC cast.

Recycling of Alloy Scrap by Intensive Melt Shearing

The high shear pump can be utilized for conditioning of melts directly inside a furnace. Since the intensive shearing can disperse oxide films into individual particles, this can benefit recycling of scrap metals. A 10kg melt of AZ91D alloy was sheared continuously at 660°C using the high shear pump where the melt was let to discharge back into the furnace. After 20 minutes of continuous shearing the entire volume of melt was grain refined, as shown in Fig. 11. It is still possible to achieve grain refinement even if the fully sheared melt is mixed with non-sheared melt, as long as there are enough dispersed oxide particles to facilitate grain refinement. An experiment was performed to show that grain refinement is still achieved by mixing sheared melt with non-sheared melt of AZ91D magnesium alloy. After shearing and mixing the two melts in different quantities, samples were taken using the TP-1 standard mould. The samples were examined for their microstructure and grain size measurements were taken. The plot of grain size corresponding to the mixture of sheared melt and non-sheared melt is given in Fig. 12. The sheared melt controls the grain refinement up to a 50% mixture of non-sheared melt, after which the grain size starts to become coarser.

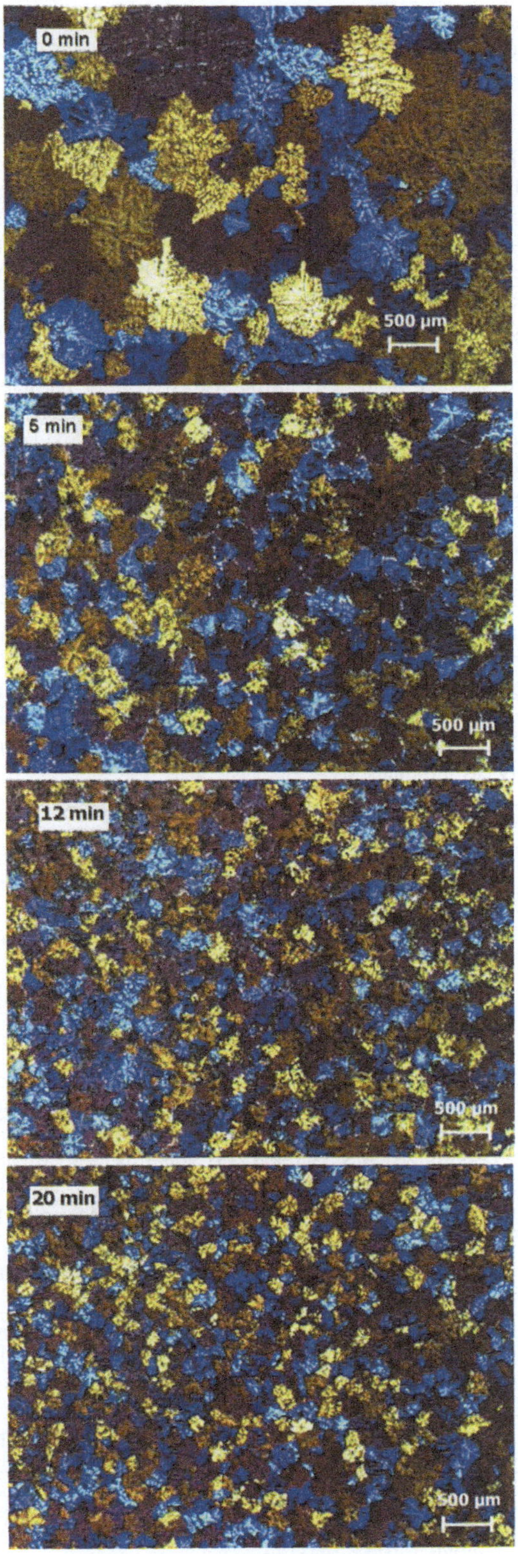

Figure 11: A 10kg melt of AZ91D Magnesium alloy was conditioned using the high shear pump at 660°C, and TP-1 samples were cast after; 0mins, 5mins, 12mins and 20mins.

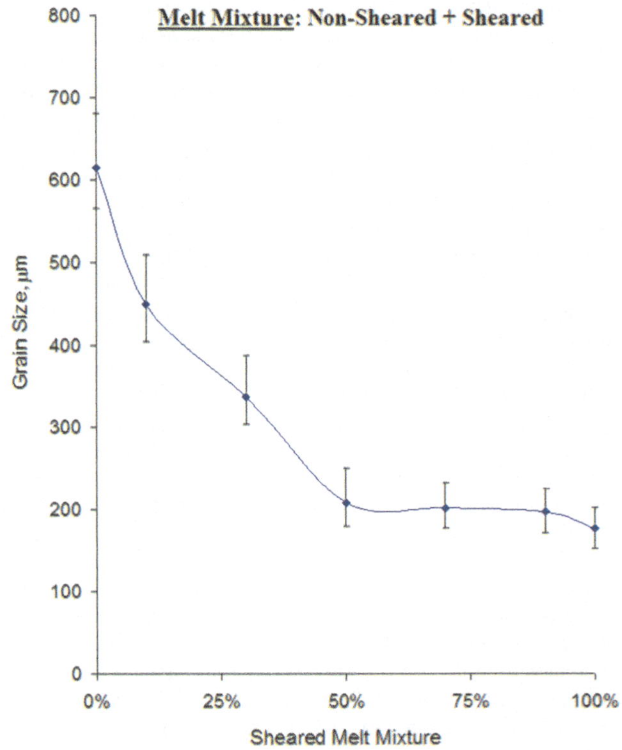

Figure 12: Plot of grain size corresponding to the mixture of fully sheared melt with non-sheared melt of AZ91D Magnesium alloy.

Solidification Mechanisms

Liquid melts of Aluminium and Magnesium alloys inevitably contain oxide in forms of clusters or films with a non-uniform distribution. Such oxides may act as substrates for nucleation but are not effective for grain refinement due to their poor wetability and low number density. However, such oxide clusters and films can be dispersed effectively by intensive melt shearing. It has been confirmed that the well dispersed oxide particles can be as effective nucleation sites [2]. The high shear device can not only provide intensive melt shearing in the gap between the rotor and the stator and in the openings of the stator, which can disperse oxide clusters or films to individual particles, but also forced convection in the melt, which can distribute particles uniformly in the entire volume of the melt. Intensive melt shearing is also believed to be helpful for improving wetability between the liquid alloy and the dispersed oxide particles. It is therefore concluded that the intensive melt shearing can enhance nucleation and make significant contribution to the formation of fine and uniform microstructures. It has been well demonstrated that intensive melt shearing prior to solidification processing is capable of significant grain refinement of Al and Mg alloys, and this grain refining effect has been attributed to the enhanced heterogeneous nucleation on the dispersed oxide particles. However, the effect of *in situ* melt shearing during solidification processing on grain refinement needs to be discussed further. The macroscopic flow pattern that is generated by the rapidly escaping liquid from the openings on the stator provides a distributive mixing action which homogenizes the temperature and chemical composition of the melt. As a consequence, the melt temperature and chemical composition will be uniform as long as the high shear is being applied by the rotor/stator device. In addition, this macroscopic melt flow will be stronger near the tip of the stator and weaker near the surface of the melt. This will prevent the occurrence of gas entrapment from the melt surface. In the specific case of DC casting, where *in situ* melt shearing is applied, there is usually a built-up of solute elements at the advancing solidification front. The forced convection generated by the high shear device will suck the solute elements at the solidification front and distribute them uniformly in the sump. Such uniform fields of both temperature (slightly below liquidus) and chemical composition will create a condition which promotes heterogeneous nucleation and equiaxed growth. In the MC-DC casting process, firstly, the intensive melt shearing generates more potential nucleating particles by dispersing the already existing oxide films. The anticipated increase in particle number density by intensive melt shearing can be as high as 3 orders of magnitude [19]. It is well known that increased nucleation events will lead to grain refinement. Secondly, the uniform temperature field in the sump can increase the survival rate of the existing nuclei by avoiding their dissolution in the hotter melt which is higher than the liquidus temperature of the alloy being cast. This would also additionally contribute to the formation of fine grains. Thirdly, there is a possibility that intensive melt shearing promotes fragmentation of dendrite arms that have already formed at the liquid/solid interface, since the melt temperature in the sump is maintained below the liquidus of the alloy being cast. Hence, the dendrite arm fragmentation phenomena may also contribute partially to the microstructural grain refinement achieved in the MC-DC cast billets, with the anticipation that its contribution will be much less than that from the enhanced heterogeneous nucleation. Besides equiaxed solidification and grain refinement, it should be noted that the MC-DC process exhibits a number of other factors; a grain refined microstructure without the addition of chemical grain refiners and therefore avoiding any adverse effects of these grain refiner additions, eliminates or significantly reduces macro-segregation which in turn facilitates down-stream thermal mechanical processing of the billets such as extrusion and heat treatments, mitigates the harmful effects of oxides and other inclusions by substantially reducing their physical size, reduces the tendency to hot tearing and the formation of gas porosity in the billets and also that it provides an opportunity for DC casting Al and Mg alloys which are difficult or even impossible to DC cast by conventional methods. With regards to enhanced heterogeneous nucleation by the dispersion of potent oxide particles, the application of intensive melt shearing can facilitate recycling of scrap Al and Mg alloys. Al and Mg scrap is usually discarded due to their high content of oxides and other inclusions. Since there are no external chemical additions required to achieve grain refinement by intensive melt shearing, it should be noted that almost any kind of melt can be conditioned in order to produce better quality recycled Al and Mg alloy castings. In addition, the intensive melt shearing would disperse and make a huge number of potent oxide particles, of which only a small fraction of them will get an opportunity to participate during solidification. Hence, after intensive shearing of a melt which contains an increased amount of oxide clusters and films than usual, there will be surplus amount of potent oxide particles left over in the melt upon solidification. It is believed, that these surplus potent oxide particles can get a second chance to participate in the nucleation event, when the already solidified casting or ingot is re-melted and cast again. Grain refinement can still be achieved just by re-melting the once sheared ingot without applying intensive melt shearing again. For the same reason, an intensively sheared melt can also be mixed with primary melt that has not been intensively sheared and still achieve grain refinement upon solidification.

Summary

Liquid metal engineering by application of intensive melt shearing can be achieved by the newly developed Rotor/Stator high shear technology to produce high quality cast products. This new Rotor/Stator type high shear device provides distributive and dispersive mixing and exhibits; significantly enhanced kinetics for phase transformations, uniform dispersion, distribution and size reduction of solid particles and gas bubbles, improved homogenisation of chemical composition and temperature fields, forced wetting of usually difficult-to-wet solid particles in the liquid metal and most importantly, provides physical grain refinement by dispersing naturally occurring oxide particles. Therefore, the new Rotor/Stator type high shear device can be applied; to condition melts of Aluminium and Magnesium light alloys, for efficiently degassing melts of Aluminium alloys from dissolved Hydrogen gas, to benefit conventional casting processes such as DC casting, sand casting, low & high pressure die casting, twin roll casting and also ingot casting with regards to recycling and re-melting of scrap. In addition, the new Rotor/Stator high shear concept can also be combined into a melt conditioning transfer pump for molten metals of Al and Mg alloys and also other low melting point alloys in general, which can be combined with various conventional casting processes to produce high quality cast products.

References

[1] J. Polmear; Light Alloys, Metallurgy of the light metals, 3rd edition, 1995.

[2] Z. Fan, Y. Wang, M. Xia and S. Arumuganathar; Acta Mater., 57, (2009), 4891-4901.

[3] Z. Fan, M. Xia, H. Zhang, G. Liu, J.B. Patel, Z. Bian, I. Bayandorian, Y. Wang, H.T. Li and G.M. Scamans ; Int. J. Cast Met. Res., 22, (2009), 103-107.

[4] Z. Fan, Y. Wang, Z.F. Zhang, M. Xia, H.T. Li, J. Xu, L. Granasy and G.M. Scamans; Int. J. Cast Met. Res., 22, (2009), 318-322.

[5] S. Tzamtzis, N.S. Barekar, J.B. Patel, N. Hari Babu and Z. Fan, Composites A, 40 (2009), 144–151.

[6] J.B. Patel, Y.Q. Liu, G. Shao and Z. Fan; Mat. Sci. Eng. A, Vol. 476, (2008), 341-349.

[7] J.B. Patel, H.W. Zhang, G. Liu and Z. Fan; 11th Int. Conf. Aluminium Alloys (ICAA11), 2008, pp.510-516.

[8] Y.B. Zuo, M.X. Xia, S.M. Liang, Y. Wang, G.M. Scamans and Z. Fan; Mater. Sci. and Technol., 27, (2011), 101-107.

[9] Y.B. Zuo, H. Li, M. Xia, B. Jiang, G.M. Scamans and Z. Fan; Scripta Mater., 64, (2011), 209–212.

[10] Z. Fan, Y. B. Zuo, B. Jiang; Apparatus and method for liquid metals treatment, Application No.1015498.7, 2010, UK Patent.

[11] Z. Fan, Y.B. Zuo and B. Jiang; Materials Science Forum, Vol.690, (2011), 141-144.

[12] J.B. Patel, A.K. Prasada Rao, B. Jiang, Y.B. Zuo and Z. Fan; 9th Int. Conf. Magnesium Alloys and their Applications, Vancouver, BC, Canada, 2012, pp.731-736.

[13] S. Jones, A.K. Prasada Rao, J.B. Patel, G.M. Scamans, and Z. Fan; 13th Int. Conf. Aluminium Alloys (ICAA13), Pittsburgh, 2012, pp. 91-96.

[14] Standard test procedure for aluminium alloy grain refiners (TP-1), The Aluminum Association, Washington, DC, 1990.
[15] D. Dispinar, J. Campbell; Int. J. Cast Met. Res., 17 (5), (2004), 280–286.
[16] D. Dispinar, J. Campbell; Int. J. Cast Met. Res., 17 (5), (2004), 287–294.
[17] S. fox and J. Campbell, Scripta Mater., 43 (2000), 881.
[18] M. Samuel and F.H. Samuel, Metal. Trans. A, 24, (1993), 1857.
[19] H. Men, B. Jiang and Z. Fan, Acta Mater., Vol.58, (2010), 6526.

Proceedings of the 2013 International Symposium on Liquid Metal Processing & Casting
Edited by: Matthew J. M. Krane, Alain Jardy, Rodney L. Williamson, and Joseph J. Beaman
TMS (The Minerals, Metals & Materials Society), 2013

A SIMPLE EXPERIMENTAL APPARATUS FOR TESTING CORE AND CLAD ALLOY COMBINATIONS FOR POTENTIAL FUSION™ CASTING APPLICATIONS

Massimo Di Ciano, Mary A. Wells, David C. Weckman
University of Waterloo Department of Mechanical and Mechatronics Engineering; 200 University Avenue West; Waterloo, Ontario, N2L 3G1, Canada

Keywords: DC Casting, Oxidation, Dissolution, Solidification

Abstract

Novelis Inc. recently developed and patented a unique Direct Chill (DC) casting process whereby two different alloys can be cast simultaneously, producing a laminated aluminum ingot, known as Fusion™[1] casting. In this process, a high quality metallurgical bond is formed between the two simultaneously cast alloys; the mechanism of how this bond formation occurs between the two alloys is not clearly understood. Since direct observation of wetting and interface formation in the Fusion casting process is difficult to obtain in practice, due to temperature and proximity issues, an analog test was designed to mimic the processes on the laboratory scale. The apparatus provides a means of controlling the basic variables thought to influence bond formation in Fusion casting, namely: i) the relative temperatures of the two alloys in contact, ii) the sample contact time, and iii) surface oxidation. The results of two test series will be presented, highlighting the relative importance these variables have on as-cast interface quality.

Introduction to the Fusion DC Casting Process

The invention of laminated aluminum sheet metal has its roots in the US air force. In the 1920's, as high strength aluminum (2xxx series) gained traction in the manufacture of wing skins, concerns regarding their susceptibility to inter-granular corrosion in sea-water environments rose. Research at Alcoa [1]-[2] to alleviate this problem resulted in the invention of the famous Alclad sheet metal. Alclad sheet metal has a core which consists of high strength aluminum, which is metallurgically bonded to an alloy with high corrosion resistance. In the early patent literature [2], the Alclad material was said to be made by casting the core aluminum alloy against solid clad material (in the forms of sheets presumably), and then using a combination of heat treatment and rolling to improve the bond between the two alloys. Modern production of Alclad is done predominately by roll bonding, i.e., forgoing the casting step in the above description; however, successful roll bonding requires meticulous preparation of the mating surfaces prior to rolling, which adds to process costs [3]. Since the 1920's, other markets have immerged for aluminum clad sheet.

An equally important class of aluminum clad products are brazing sheet metals. A brazing sheet consists of a low melting point aluminum alloy clad onto a formable and moderate strength aluminum core. These sheet metals are used in the fabrication of aluminum parts which require brazing processes, such as an automobile heat exchangers. The low melting point alloy is typically an Al-Si brazing alloy, whereas the core alloy is typically a 3xxx series alloy; however, various 5xxx, 6xxx, and 7xxx series alloys can also be successfully brazed together [4]. Again, roll bonding is the predominate route for fabricating brazing sheet stock.

Novelis Inc. utilizes a Direct Chill (DC) casting process, whereby two alloys can be cast simultaneously, to produce clad ingots. This process is known as Fusion casting [5]. This process offers two distinct advantages over current aluminum clad sheet processing routes: i) the ingots produced via Fusion casting can be rolled down to thin-gauge aluminum 'clad' sheet without the meticulous cleaning procedures or high rolling reductions required in traditional roll-bonding production of clad sheet and ii) since the core/clad metallurgical, oxide-free bond is formed during the ingot casting step, a scalping step is saved on one side of each ingot, thus reducing scrap volume and increasing overall process efficiency. Accordingly, other techniques have been proposed in recent years for laminated ingot production [6]-[15].

The Fusion casting process is a direct descendant of traditional DC continuous casting technology. Starting with a generic DC casting setup, a solid divider is inserted into the mould which acts as an inner mould. The divider is typically a water-cooled metal chill plate, hence the name chill bar, see Figure 1. The role of this device is to isolate the liquid metal input (feed) streams. A design specific starting block, (details not shown), effectively separates the two alloys at the start of the casting process. To start the casting, the higher melting point alloy (based on the liquidus temperature) is poured into one region of the mould, thus beginning the solidification process. At a predefined time, the lower melting alloy is poured into the adjacent side of the mould, and begins to solidify independently from the first alloy. The starting block is withdrawn shortly after, and the two streams gradually come into contact and begin the process of bond formation. The ingot is constantly withdrawn with the end result producing a laminate ingot. In general, bond quality will depend on the process conditions during casting, with optimal conditions dependent on the alloys being cast.

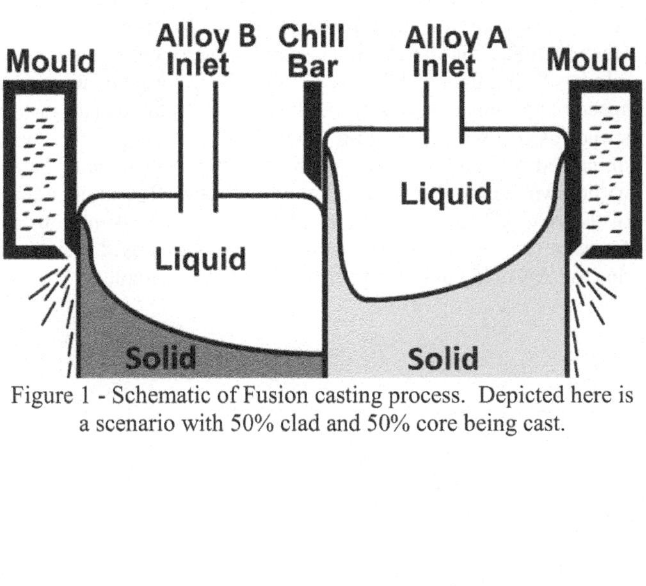

Figure 1 - Schematic of Fusion casting process. Depicted here is a scenario with 50% clad and 50% core being cast.

[1] Fusion is a trademark of Novelis Inc.

To produce a quality composite ingot, the interface between the two alloys must be free of metallurgical defects such as porosity or oxide films. Furthermore, mixing of the two alloys during solidification must be minimized since the goal is to achieve an ingot having two distinct regions (inner and outer region) that vary in alloy composition. The structure of the resulting metallurgical interface between the core and clad can also play an important role in the subsequent processing steps and the final properties of the clad alloy. 'Why', and 'under what conditions' does the Fusion casting process yield quality, oxide-free core/clad interfaces are some of the key intellectual questions to fully understand the bond formation at the interface mechanistically. Since Fusion casting is used to commercially produce aluminum brazing sheet, specifically AA3003 core with AA4045 clad, it has been chosen as the candidate alloy system to study.

Apparatus Details and Testing

As direct observation of wetting and interface formation in the Fusion casting process is difficult to obtain in practice, due to temperature and proximity issues, an analog test was designed which would mimic the wetting and interface process on the laboratory scale. In our approach a unique test apparatus to study this phenomena was designed and built at the University of Waterloo. In the test apparatus, a hollowed out cylindrical AA3003 is preheated to a desired temperature and then immersed into a melt of A4045 at a predefined speed, thus creating an interface between the AA3003-core/AA4045-clad alloys. The apparatus provides a means to independently control the following basic variables: sample and melt temperatures, sample immersion speed, and solid-liquid contact time. Additionally, the ambient atmosphere surrounding the sample and the melt can be carefully maintained using this apparatus, to either limit or promote the degree of surface oxidation on the AA3003 sample and AA4045 melt. The apparatus consists of the following major assemblies: i) a two-zone tube furnace, ii) a stainless steel vessel, iii) a sample positioning system, iii) a sample cooling system, and iv) an ambient gas control system. A cross sectional schematic of the two-zone furnace assembly used to heat and cool the AA3003 specimen and AA4045 melt is shown in Figure 2.

In test series 1, samples were cleaned in alcohol, etched, de-smutted, dried with compressed air, and allowed to stay in a dry box for 24 hours prior to testing. AA4045 was first melted in the bottom portion of the furnace, and its temperature allowed to stabilize. The melt was then skimmed prior to inserting the AA3003 sample into the furnace for sample heat-up and testing. Heating times for the AA3003 samples were approximately 40 minutes. Throughout both series of tests, a constant immersion speed of 1.5 mm/s was used, which is similar to the typical casting speed used in other casting trials [16]. Test samples were immersed into the melt to a depth of ~30mm then immediately withdrawn and cooled with a stream of Argon. The temperatures of the AA3003 sample and the AA4045 melt were kept constant for each test. Furthermore, the test temperatures used were such that the AA4045 could not re-melt the AA3003 samples.

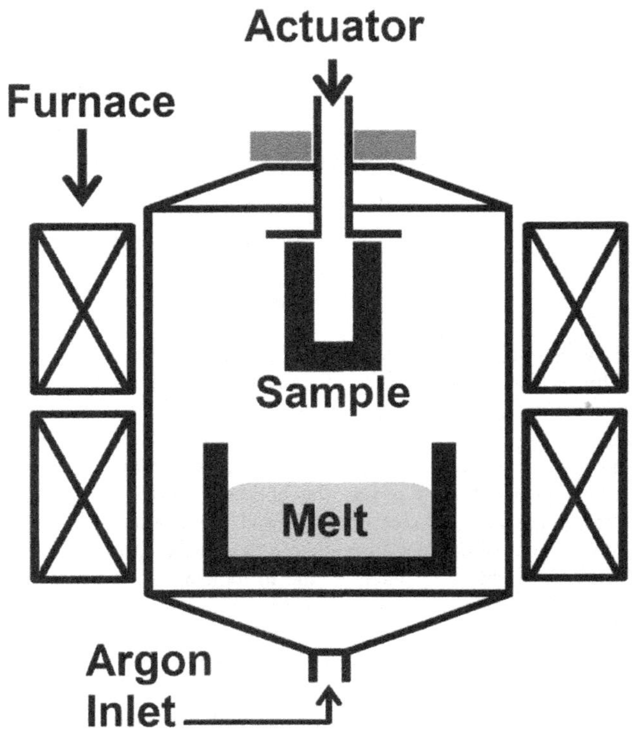

Figure 2 - Schematic of the analog test apparatus.

AA4045 melting and AA3003 sample heating was initiated using an air atmosphere inside the furnace, however; the furnace was specifically purged with Argon at pre-defined times during sample heating. Thus, the degree of surface oxidation present on the AA3003 sample and on the AA4045 melt could be varied. Calculations were also performed to determine the extent of oxidation on the AA3003 samples and on the AA4045 melt prior to immersion of the test specimens, see Figure 3. Oxidation calculations were made using the parabolic rate constants given for AA3004 in [17], and by scaling them to discount the effect of Mg on overall oxidation. Although oxidation of aluminum may not be purely parabolic at high temperatures, see [18]-[21] for example, however this was a good first approximation.

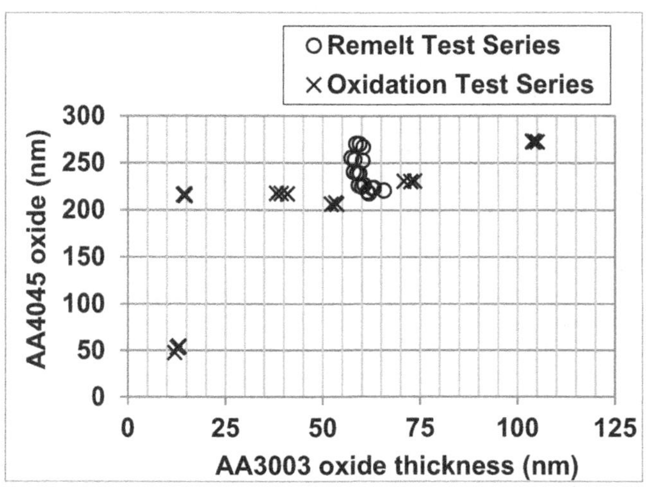

Figure 3 - Calculated sample oxide layer thickness, AA3003, melt oxide layer thickness, AA4045 for both test series.

In test series 2, the AA4045 melt temperature was progressively increased from test to test, while the AA3003 sample temperature was kept constant. This allowed us to study the effect of AA3003 surface re-melting on interface formation. For these tests, the furnace atmosphere was air.

Fusion Casting Trials

Additionally, AA3003/AA4045 ingots were Fusion cast [16] at the Novelis' research centre in Kingston, Ontario. Ingot cross sections were 15" by 6", and were comprised of 50% core and 50% clad material. Casting times were over 5 minutes in length, long enough to reach a steady-state condition [21]. Metallurgical analysis of the interface was performed on the ingots for the purpose of comparing with the analog test results.

Analog Test Results

It was possible to examine and categorize the exterior surfaces of the AA3003 samples after the tests were completed. Qualitatively, the result surfaces could be grouped into one of four categories, one example of each category is shown in Figure 4. The samples shown in this figure had all been immersed to a depth of around ~35 mm. The depth at which the AA4045 melt metal level reaches on the samples can be clearly distinguished in the sample shown in Figure 4a, whereas this is less noticeable in the remaining samples. This sample was part of the oxide tests series, with an estimated AA3003 surface oxide of 15 nm. Above the metal immersion level, the AA3003 remains untouched throughout the test, notice its surface has a dull lustre. Below the metal immersion line, the sample surface appears shiny. Here, the AA4045 melt has interacted with the sample. As can be ascertained by the slight reduction in AA3003 sample diameter below the AA4045 melt metal level, the AA3003 was slightly dissolved by the AA4045 during the time in which it was in contact with the AA4045 melt. Additionally, the shiny surface was very uniform over this sample, with little or no regions below the AA4045 metal level showing bare patches of un-reacted AA3003.

The sample shown in Figure 4b was immersed under the same conditions as that of Figure 4a, however; with this sample the estimated AA3003 surface oxide was 105 nm. This was achieved by exposing the sample to air during the initial stages of sample heating and flushing the furnace with Argon at a pre-defined point in time. Comparing the surface obtained with these test conditions to that shown in Figure 4a, there are two noticeable differences. Firstly, there exists many bare portions on the AA3003 sample surface. These bare spots are below the AA3003 sample immersion depth, meaning they would have technically been in contact with AA4045 but did not interact with the metal. In general, it was observed that as the degree of AA3003 surface oxidation increased, the amount of bare area found on the AA3003 samples after immersion increased. Additionally, it is hard to decipher the final AA3003 sample immersion depth for the sample shown in Figure 4b, as the metal level line is not continuous.

The sample shown in Figure 4c was one from one of the re-melt test series samples. Here, the tests were done in an air atmosphere. By varying the temperature of the AA4045 melt, the effect of AA3003 sample re-melting was assessed. For this particular sample, the AA3003 temperature was 903 K, while the AA4045 melt temperature was 913 K. Thus, re-melting was not expected, as the equilibrium solidus and liquidus temperatures for AA3003 are 915 K and 929 K. The external surface of the AA3003 after immersion into the AA4045 melt are starkly different than those of Figure 4a and 4b. It was clear that the AA3003 sample, while being immersed to a depth of ~35 mm into the AA4045 melt, had little to no interaction with the AA4045. The majority of the AA3003 appeared unaltered after sample immersion, with the exception of small metal fragments of AA4045 metal shown as the darker gray coloured areas in Figure 4c. These AA4045 fragments however were superficially adherent to the AA3003, and in fact were easily removed with the brush of a finger. This behavior, i.e., minimal interaction of AA3003 and AA4045, was consistent for all tests conducted in air where the AA4045 melt temperature was not great enough to induce AA3003 sample re-melting.

When the AA4045 melt temperature was raised to a high enough temperature such that it could re-melt the AA3003 sample, the exterior surfaces of the sample looked similar to that shown in Figure 4d. In this test, the AA3003 temperature was 903 K, while the AA4045 melt temperature was 953 K. Again, these tests samples were immersed to a depth of ~ 35 mm. Upon immersion of the AA3003 sample into the hot melt, AA3003 re-melting ensued. As the re-melting process requires heat transfer, there is a time lag required for re-melting to begin. Thus, the bottom portion of the AA3003 samples, having been exposed to the AA4045 melt for the longest period of time (in relation to a point higher up on the AA3003 sample surface) exhibits the most re-melting. The portion of the AA3003 sample which was in contact with the AA4045 melt for a short period of time showed little signs of re-melting. Similar to Figure 4b, the AA4045 metal immersion level is not clearly discernible, indicating negligible interaction between the AA3003 sample and the AA4045 melt in the region where the contact time was short. The portion of the AA3003 which appeared to have undergone re-melting did have regions where the re-melt surface was shiny, similar to the luster shown in Figure 4a where AA3003 dissolution has taken place. Unlike Figure 4a however; this shiny surface was not continuous and uniform over the re-melted portion of the samples. Additionally, the texture of the re-melted regions appeared rough, indicating a non-continuous re-melting front.

Qualitative ranking of the AA4045/AA3003 sample interaction, based on uniformity, by test conditions was (from best to worse): tests done in Argon, test done in Argon with some air exposure (oxidation during sample heating), tests done in air with AA3003 sample re-melting, tests done in air with no AA3003 sample re-melting.

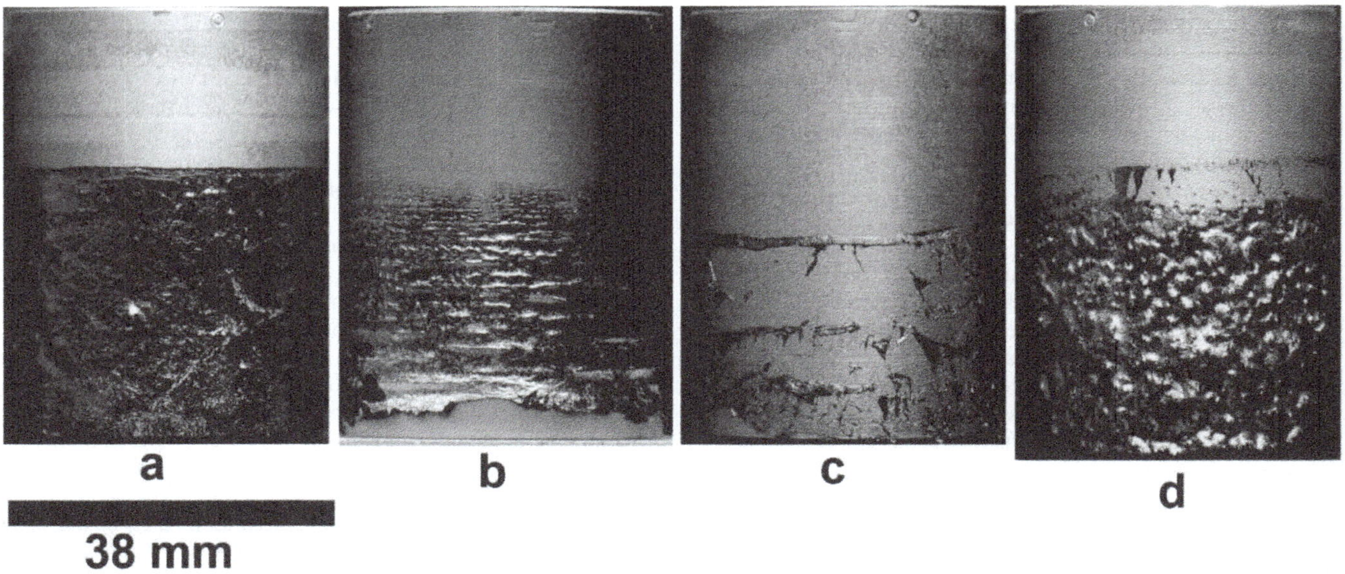

Figure 4 - Exterior surfaces of AA3003 tests samples after immersion in AA4045 melt, a) tested in Argon with low level of AA3003 surface oxidation, b) tested in Argon but AA3003 sample exposed to air for prolonged time during sample heating, c) tested in air and AA4045 temperature 640°C d) tested in air and AA4045 temperature 680°C.

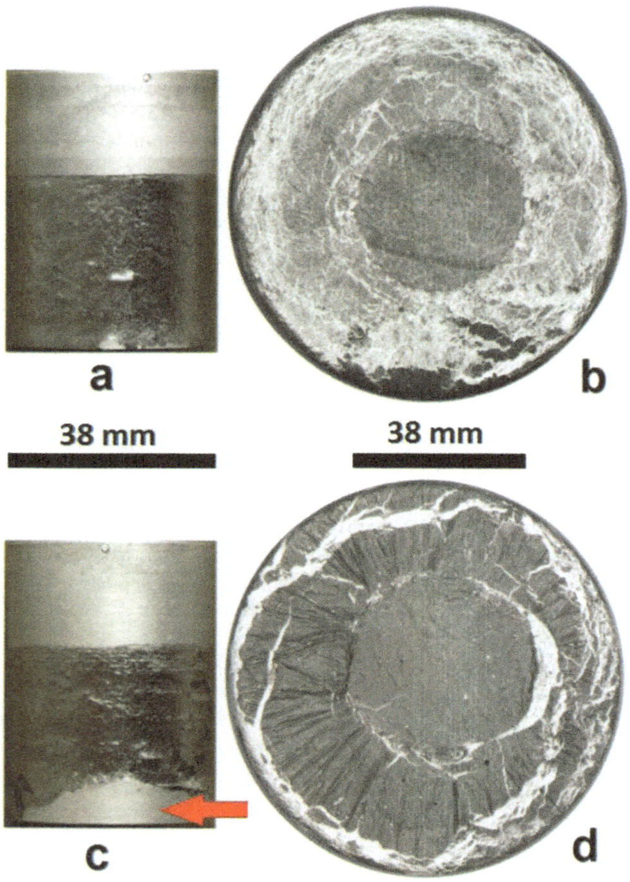

Figure 5 - Effect of AA4045 melt oxidation, a) appearance of AA3003 exterior after immersion into a lightly oxidized AA4045 melt, b) top view of a lightly oxidized AA4045 melt after melt re-solidification, c) appearance of AA3003 exterior after immersion into a heavily oxidized AA4045 melt, notice dry patch (see arrow) along the side of AA3003 sample, and d) top view of a heavily oxidized AA4045 melt after melt re-solidification.

The Oxide Ring

For all tests which were performed, the bottom surfaces of the AA3003 sample was consistently unaffected by immersion into the AA4045 melt (the horizontal portion of the sample to first come into contact with the AA4045 melt). It was also observed that as AA4045 oxidation increased (either by higher AA4045 melt temperatures, or by longer exposure times to air), bare patches were found along the sides of immersed AA3003 samples. These patches were always located near the bottom of the sample (i.e., the first vertical portion of the AA3003 sample to come into contact with the AA4045 melt), see Figure 5a and Figure 5c. Upon visual examination of the AA4045 melts after solidification and cool down, there was an obvious ring pattern found on what would have been the free surface of the AA4045 melt. This can be seen in Figure 5b, for AA4045 which was lightly oxidized, i.e., held in an Argon atmosphere, and in Figure 5d, for AAA4045 melt which had been exposed to atmospheric air. This ring corresponds to the point at which the AA3003 contacted the AA4045 melt upon immersion, and subsequently broke through whatever oxide layer had been grown on the AA4045 melt during the sample heating period. Notice the difference in oxide film colour, and fracturing patterns.

Optical Microscopy of Interfaces

Various test samples were also sectioned longitudinally along there centerline, and standard metallography was performed to assess the AA3003/AA4045 interfaces. Samples from the oxidation test series were typical of those shown in Figure 6. In Figure 6a, the sample shown had minimal exposure to air, and the expected amount of oxidation on the AA3003 sample surface was 15 nm. In this micrograph, the AA3003 material appears dark grey, and has a fine grained microstructure which is barely discernible at this magnification. The lighter toned, and higher contrast material shown in Figure 6a is AA4045 metal. In this example, one clearly sees that the AA4045 liquid penetrated and dissolved some of the original AA3003 sample, leaving behind a sample with a slightly reduced diameter. In Figure 6a, there is a

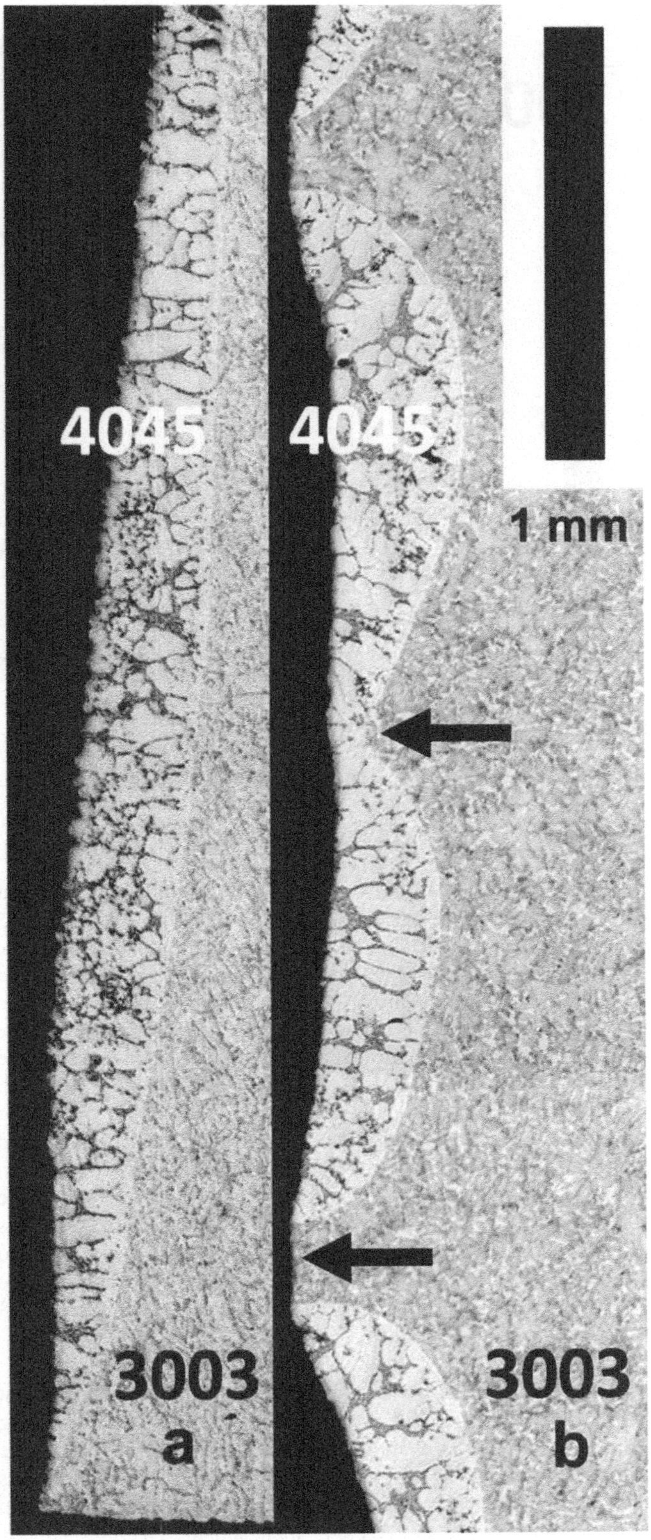

Figure 6 - Longitudinal section (parallel to test immersion direction) of analog test samples conducted in Argon, a) AA3003 surface oxide approximately 15 nm, b) AA3003 surface oxide approximately 55 nm.

clear and continuous interface between the AA3003 sample and the AA4045 material. It is also worth noting that alpha aluminum grains appear to be forming at the AA3003/AA4045 interface, with the direction of the grains perpendicular to that of the interface. This is similar in behavior to previous dissolution studies [22]-[24]. Additionally, the bottom of the sample, which corresponds to the bottom surface shown Figure 6a, is clean of AA4045 and shows no sign of interaction with the AA4045 melt (as mentioned in the previous subsection).

In comparison with Figure 6a, the sample shown in Figure 6b was exposed to air for a longer period of time during sample heating, and had an expected AA3003 oxide layer thickness of 55 nm. This figure shows two bare spots along the left vertical edge of the micrograph, where AA4045 had clearly not penetrated the AA3003 sample. This corresponds to the bare spots depicted in Figure 4b. Adjacent to these bare spots are regions where the AA4045 has again penetrated the AA3003 sample and dissolved a portion of it. In contrast to the sample shown in Figure 6a, the AA3003/AA4045 interface is not continuous and not a straight curve. In addition to regions of bare spots, there are also cusps along the interface. A good example of this is shown along the left edge of this micrograph, about half way up the sample. Here a cusp is clearly shown along the AA3003/AA4045 interface.

A higher magnification image of the AA3003/AA4045 interface is shown in Figure 7. Here the AA3003 material appears on the left hand side of the micrograph, and the AA4045 material on the right hand side. It can be observed that in addition to the alpha aluminum grains which protrude perpendicularly from the AA3003/AA4045 interface, there are also some intermetallic particles present near the interface. As the AA4045 dissolves the AA3003 material, one would expect Mn enrichment of the AA4045 liquid. Upon re-solidification of the AA4045 liquid, this may manifest in additional intermetallics being formed in and around the interface, which would not normally be present in the AA4045 alloy material.

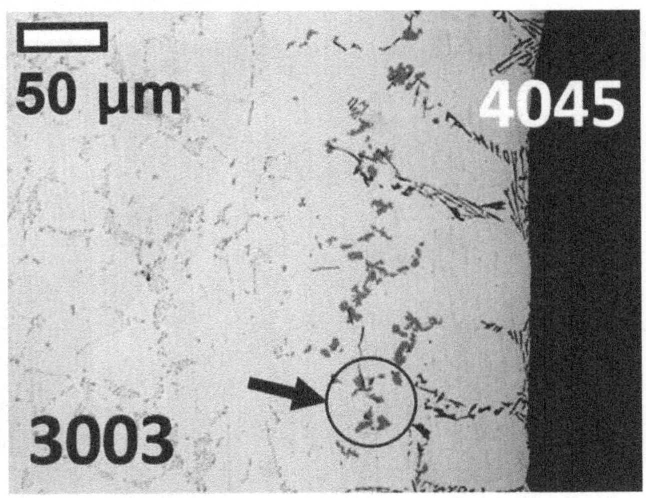

Figure 5 - Longitudinal section of analog test sample. Arrow indicates intermetallics near interface.

Fusion Casting Results

Similar to the analog tests, Fusion cast ingots were sectioned (both transverse and longitudinal directions), and optical metallography was performed. Two longitudinal sections of AA3003/AA4045 Fusion cast ingot are shown in Figure 8 and Figure 9. The sample shown in Figure 8 was taken from the top of the ingot, i.e. it was the last portion of the ingot to be cast.

Hence, this material represents the final point of contact between the AA4045 metal and the AA3003 metal during casting. What looks like a frozen metal meniscus can be seen near the top centre region of the micrograph. In this micrograph the AA3003 appears as the fine grained light gray structure, and the AA4045 appears as a region of large primary aluminum dendrites surrounding by a dark grey eutectic mixture of aluminum and silicon. Here we see that the AA4045 metal has penetrated/dissolved a significant portion of the AA3003 metal surface it has come into contact with. A rough estimate puts the dissolution depth on the order of 1 mm. There are also clear similarities between the analog samples, see Figure 6b, and the Fusion sample shown in Figure 8. Both interfaces are not straight, and both show a cusp partway along the interface. Additionally, primary aluminum grains growing along the interface, in a direction perpendicular to the interface are evident in both Figure 8 and Figure 9.

This primary aluminum grain growth is clearly depicted in Figure 9, at higher magnification. Some other observations to note: the size and spacing of the alpha aluminum grains appear finer in the Fusion cast interfaces when compared with the analog test interfaces. This may be due to the high solidification rates encountered in Fusion DC casting, in comparison to the relatively slow cool down rates experienced by the analog test samples. Also, there are intermetallics present in the AA4045 metal portion of the Fusion cast ingot, which are not present in standard AA4045 cast material, see Figure 9. Here, a series of Chinese Script shaped intermetallic particles are visible, which upon comparison with literature [25] and SEM EDX analysis [16], may potentially be $Al_{15}(FeMn)_3Si_2$ particles. This is another sign that dissolution of AA3003 takes place at the interface during Fusion casting.

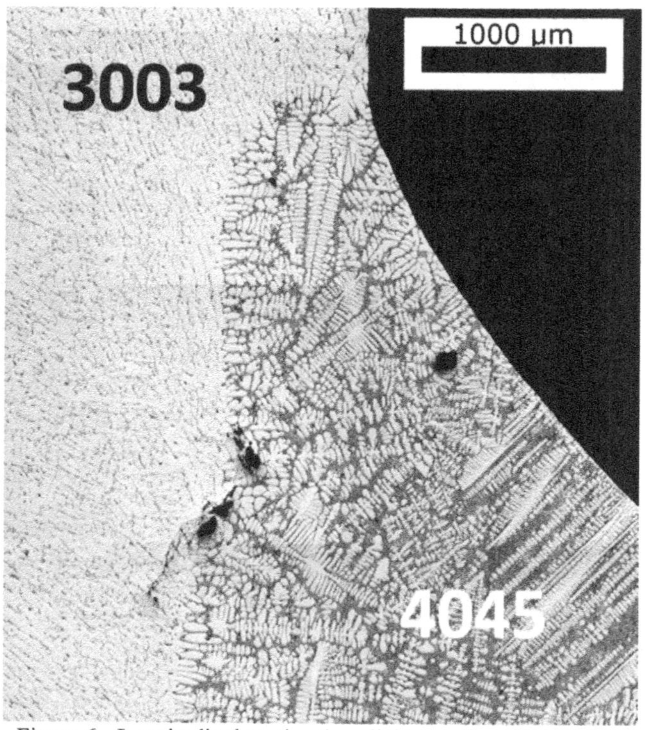

Figure 6 - Longitudinal section (parallel to casting direction) of Fusion cast interface, portion displayed here is the end of the cast with the frozen AA4045 meniscus shown near top centre of figure.

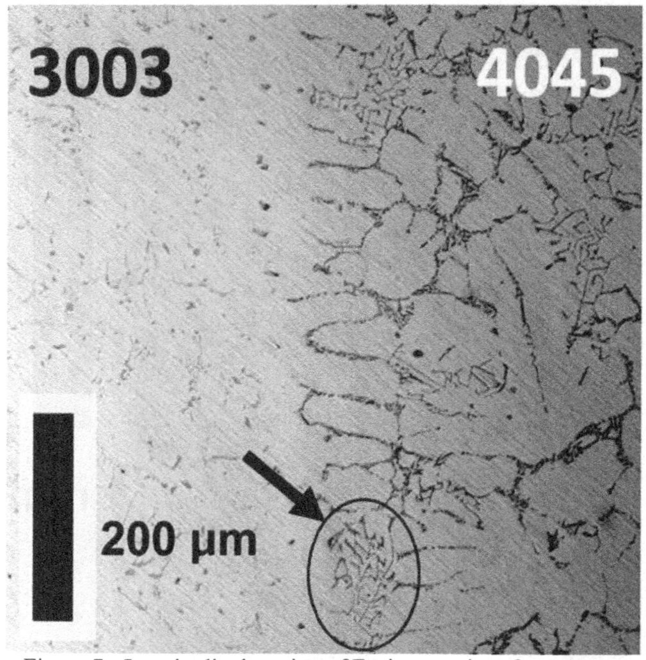

Figure 7 - Longitudinal section of Fusion cast interface. Arrow indicates intermetallics near interface.

Discussion

Oxidation Effect on Interface Formation

As shown in Figure 4a and 4b, there was a clear and marked effect of AA3003 surface oxidation on the final AA3003/AA4045 interface quality. This can be explained by examining the AA3003 sample surface texture. The AA3003 samples used in this study were etched in NaOH prior to testing, which is known to strip any protective alumina oxide layer already present on the AA3003 sample surface. Additionally, it can generate surface defects in the aluminum [26]. This etched, and now defected, surface will re-oxidize unless the partial pressure of oxygen in extremely small. As such, the defects present on the as-etched surface can potentially be sealed by the aluminum re-oxidation process. As the degree of oxidation increases, there is more likelihood these defects become sealed by aluminum oxide. This in turn reduces the number of sites available for AA4045 liquid to penetrate the AA3003 surface oxides upon contact with the AA3003 sample. Thus, less uniform AA4045 and AA3003 dissolution would take place.

It should be noted however, that even when the calculated amount of AA3003 surface oxidation is equivalent, there was a striking difference in results when the furnace atmosphere during sample immersion (into the AA4045 melt) is air or Argon. When tests are done in air and sample re-melting did not occur, there is no interaction between the AA4045 melt and the AA3003 sample. When samples were immersed into the AA4045 melt under an Argon atmosphere, penetration and dissolution of the AA3003 sample readily occurs. This is indicative that the AA4045 oxide, in addition to the AA3003 surface oxide plays a role in interface formation.

Comparison of Interfaces: Analog Tests and Fusion Ingots

The AA3003/AA4045 interface microstructures obtained in analog tests done in Argon atmospheres, see Figure 7 are strikingly similar to those obtained from Fusion cast ingots, Figure 9. This suggests that they share similar interface formation mechanisms. It is postulated that the interface formation mechanism for Fusion cast ingots is largely governed by penetration and dissolution of the AA3003 shell. Additionally, the cusp shown in Figure 6b match those seen in Fusion cast interfaces, see Figure 8. This behavior suggests that the contact time is also an important parameter, as there is a finite time required for bridging of the penetrated sites to occur, via dissolution of the AA3003 metal.

Effect of AA3003 Re-Melting

It was consistently observed that only minimal AA3003/AA4045 interaction took place in the analog tests conducted in air. Re-melting of the AA3003 by the AA4045 liquid did promote bond formation, although it is unclear if re-melting alone is a sufficient pre-requisite for obtaining sound metallurgical bonds during Fusion casting. Recent Fusion casting trials have focused on the role of oxide film growth and movement, in addition to re-melting.

Conclusion

This works suggests that analog testing is a simple method to help elucidate the mechanism of interface formation in Fusion cast ingots. The AA3003/AA4045 interfaces obtained in analog tests qualitatively match those observed from Fusion cast ingots. They both suggest that the mechanism of interface formation involves penetration of the AA3003 surface oxide, followed by dissolution of the AA3003 material by liquid AA4045. How the oxide defects are generated in the real casting process has still not been clearly defined. In the analog test samples, the AA3003 surface oxide morphology has yet to be carefully studied, although it is presumed that local defects are resent in these thermally grown oxide films.

Additionally, AA4045 oxidation has also been shown to play a critical role in interface formation. Observed analog test results suggest that if there is a sufficient AA4045 oxide film present between the AA3003 sample and the AA4045 liquid metal, than it can pose a serious impediment to obtaining a defect free metallurgical bonds between AA3003 and AA4045.

Acknowledgements

The authors would like to acknowledge the efforts of Rosa Ortega Pelayo, Amir Baserinia, Etienne Caron, and the technical staff at the UW department of Mechanical and Mechatronics Engineering. In addition, the support from Novelis Inc., Simon Barker and Mark Gallerneault has been greatly appreciated.

References

[1] E.H. Dix Jr., "'Alclad': a new corrosion resistant aluminum product", National Advisory Committee for Aeronautics, Technical note no. 259, Washington, 1927.

[2] E.H. Dix Jr., "Corrosion-resistant aluminum alloy articles and method of making them the same", US Patent Office, Patent Number 1,856,089, June 28, 1932.

[3] J. Liu et al., "Macro- and micro-surface engineering to improve hot roll bonding of aluminum plate and sheet", *Materials Science and Engineering A*, Vol. 479, 2008, pp. 45-57.

[4] Y. Sugiyama, "Brazing of Aluminum Alloys", *Welding International*, Vol. 3, No. 8, 1989, pp. 700-710.

[5] Mark Douglas Anderson et al., "Method for Casting Composite Ingot", US Patent. US 2005-0011630 A1, January 20, 2005.

[6] Yuan-yuan Li, Jin Fei, and Wei-ping Chen, "Preparation of 2024/3003 Gradient Materials by Semi-Continuous Casting Using Double-Stream-Pouring Technique", *J. Centr. South Univ. Technol.*, Vol. 9, No. 4, Dec 2002, pp 229-234.

[7] Weiwen Zhang et al., "Effect of the depth of the submerged entry nozzle in the mold on heat,flow and solution transport in double-stream-pouring continuous casting", *Journal of Materials Processing Technology*, Vol. 209, 2009, pp 5536–5544.

[8] W.W. Zhang et al., "Effect of plastic deformation on microstructure and hardness of 2024/3003 gradient composite ingot prepared by continuous casting" *Materials Science and Engineering A*, Vol. 505, 2009, pp 120–130.

[9] Yuan-yuan Li et al., "Effect of Deformation Temperature on Microstructures and Properties of 7075-6009 Alloy", *Trans. Nonferrous Met. Soc. China*, Vol. 19, 2009, pp 1037-1043.

[10] R. Nakamura et al., "Roll Caster for the Three-Layer Clad-Strip", *Archives of Materials Science and Engineering*, Vol. 41, issue 2, February 2010, pp 112-120.

[11] Na Li et al., "Technique of Aluminum Alloy Composite by Inversion Casting", *J. Mater. Sci. Technol.*, Vol. 18, No. 2, 2002, pp 187-188.

[12]] E. Takeuchi et al., "Novel continuous casting process for clad steel slabs with level dc magnetic field", *Ironmaking and Steelmaking*, Vol. 24, No. 3, 1997, pp 257-263.

[13] H. Harada et al., "MHD Analysis in Hydrodynamic Casting Process of Clad Steel Slabs", *Applied Mathematical Modelling*, Vol. 22, 1998, pp 873-882,

[14] T.M. Wang et al., "Study on Continuous Casting of Cladding Aluminum Alloys with Electromagnetic Brake", *Materials Research Innovations*, Vol. 14, no. 4, 2010, pp 271-276.

[15] R.J. Kilmer and J.L. Kirby, "Simultaneous Multi-Alloy Casting", US Patent 7,611,778 B2, November 3, 2009.

[16] Rosa E. Ortega Pelayo, "Direct Chill and Fusion Casting of Aluminum Alloys", MASc Thesis, University of Waterloo, 2012.

[17] Jorge A.S. Tenório and Denise C.R. Espinosa, "High-Temperature Oxidation of Al-Mg Alloys", *Oxidation of Metals*, Vol. 53, No. 3-4, 2000, pp 361-372.

[18] W. Smeltzer, "Oxidation of Aluminum in the Temperature Range 400°C-600°C", *Journal of the Electrochemical Society*, Vol. 103, No. 4, 1956, pp 209-214.

[19] A.F. Beck et al., "The Kinetics of the Oxidation of Al in Oxygen at High Temperature", *Corrosion Science*, Vol. 7, 1967, pp 1-22.

[20] A.J. Brock, M.J. Pryor, "The Kinetics of the Oxidation of Aluminum-Copper Alloys in Oxygen at High Temperature", *Corrosion Science*, Vol. 13, 1973, pp 199-227.

[21] B.Q. Li, "A micro/macro model for fluid flow evolution and microstructure formation in solidification processes", *International Journal of Heat and Mass Transfer*, Vol. 38, No.13, 1995, pp 2367-2381.

[22] C.M. Craighead, E.W. Cawthorne, and R.I. Jaffee, "Solution Rate of Solid Aluminum in Molten Al-Si Alloy", *Journal of Metals*, Vol. 7, 1955, pp 81-87.

[23] A.J. Wall, and D.R. Milner, "Wetting and Spreading Phenomena in a Vacuum", *Journal of the Institute of Metals*, Vol. 90, 1961-62, pp 394-402.

[24] B. Dutta and M. Rettenmayr, "An Experimental Investigation on the Kinetics of Solute Driven Remelting", *Metallurgical and Materials Transactions A*, Vol. 31A, 2000, pp 2713-2720.

[25] L. Bäckerud, E. Krol, and J. Tamminen, Solidification Characteristics of Aluminium Alloys, Vol. 1: Wrought Alloys, SkanAluminium, Olso, Sweden, 1986.

[26] S. Adhikari, J. Lee, and K.R. Herbert, "Formation of Aluminum Hydride during Alkaline Dissolution of Aluminum", *Journal of The Electrochemical Society*, Vol. 155, No. 1, 2008, pp. C16-C21.

Proceedings of the 2013 International Symposium on Liquid Metal Processing & Casting
Edited by: Matthew J. M. Krane, Alain Jardy, Rodney L. Williamson, and Joseph J. Beaman
TMS (The Minerals, Metals & Materials Society), 2013

DEVELOPMENT OF A HOT WORKING STEEL BASED ON A CONTROLLED GAS-METAL-REACTION

Roman Ritzenhoff[1], Mohammad Malekipour Gharbi[2]
[1]Head of Quality and Technology, Energietechnik Essen GmbH, Westendstrasse 15, 45143 Essen
[2]Head of Quality Department Retaining Rings, Energietechnik Essen GmbH, Westendstrasse 15, 45143 Essen

Keywords: High Nitrogen Steels, P-ESR melting, nitrogen uptake by gas-metal-reaction, heat treatment, processing

Abstract

As a result of cost sensitiveness, the demand on hot working steels with advanced characteristics and properties are ascending. We have used a controlled gas-metal-reaction in a P-ESR furnace to produce high quality hot working steel. These types of materials are also known as High Nitrogen Steels (HNS). An overview of the development in a pressurized induction furnace to the final industrial scale using P-ESR will be provided. Different heat treatment strategies are conducted and their effect on mechanical properties is investigated.

Introduction

Nitrogen as an alloying element has been known and used in technical applications since the 1940s, initially under the premise for nickel substitution in stainless grades.
Nitrogen in low alloy steels is undesirable due to the formation of brittle nitrides. However, the use of nitrogen in high alloy steels has an array of advantages that makes it appear interesting as an alloying element. In references one find this sufficiently researched, so that in this situation only the most important points need to be summarized [1-5]:

- Significant increase of strength by keeping ductility acceptable
- Improve of corrosion resistance
- Increase high temperature tensile strength
- Extended / stabilized austenite form
- No formation of tension induced martensite induced by high cold working rates
- Avoids / delays the precipitation of inter-metallic phases

HNS-Alloy (High Nitrogen Steels) as specific material group is characterized through an interesting material profile. The nitrogen effect is positive on the mechanical properties, in particular to the offset yield strength.

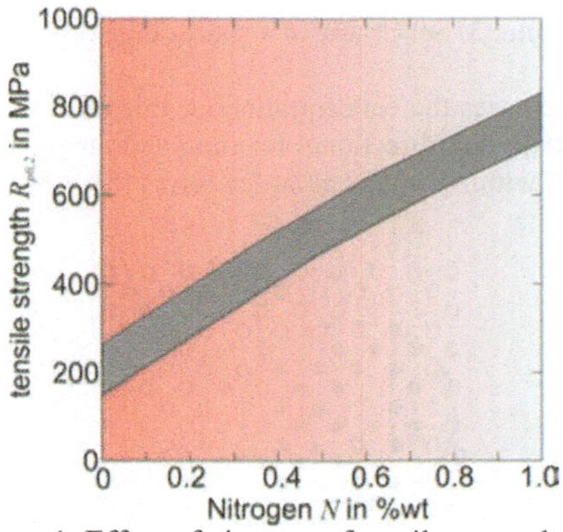

Figure 1. Effect of nitrogen of tensile strength [3]

Figure 1 and 2 clarify the increase in the tensile strength with the increasing nitrogen content and accordingly a comparison overview over the increase in tensile strength and the effect of different elements.

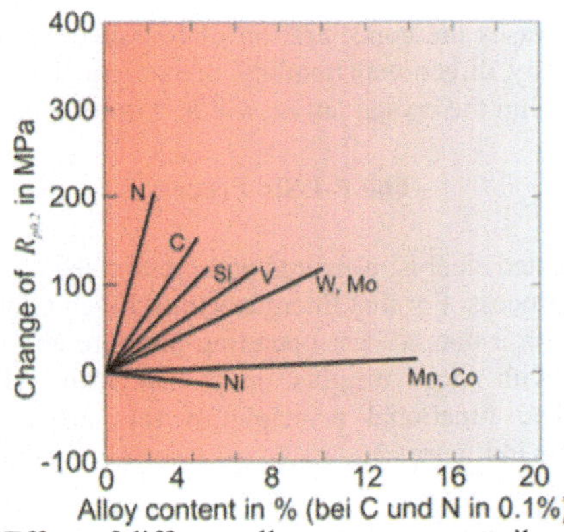

Figure 2. Effect of different alloy content on tensile strength [3]

The increase of the corrosion resistance leads back to the support of the short range order of Cr-atoms. While carbon supports the formation of Cr-atoms clusters on the basis of its electron configuration, nitrogen provides for a uniform allocation of Cr-atoms in a grid so that the risk of a $M_{23}C_6$-formation is reduced. A cluster is to be realized as local accumulation of approx. 100 atoms (see Figures 3 sand 4).

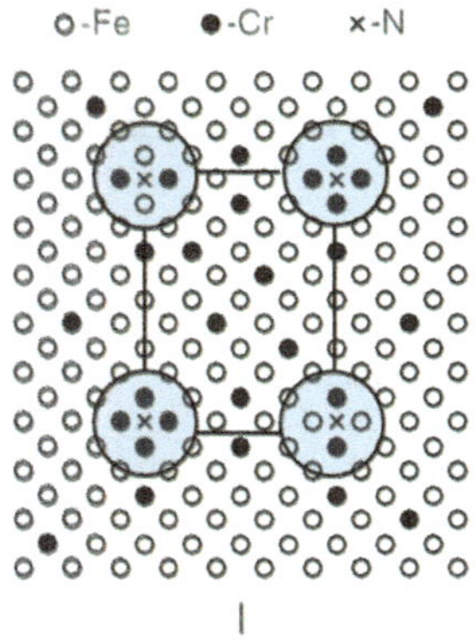

Figure 3. Schematic of a short range order [5]

Nitrogen increases the concentration on free electros in austenitic steels. Thereby non-directional bonding and an equal distribution of the atoms within the crystal lattice forms [5].

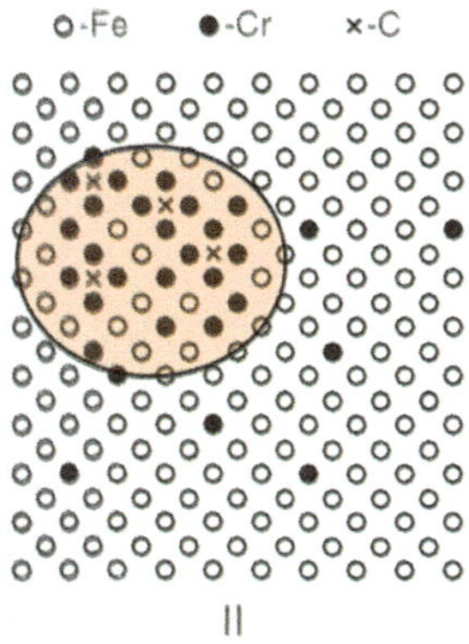

Figure 4. Schematic of a cluster formation [4]

Carbon decreases the concentration of free electros in austenitic steels. Thereby directional bonding of non-equal distribution of the atoms within the crystal lattice will be formed [5].

The P-ESR-Process

The investigated steel is re-melted and pressurized according with the P-ESR-process. For this Energietechnik Essen GmbH operates a furnace with a max 40 bar operating pressure and can achieve production with ingot weights up to 20t with a diameter of 1030mm. The functional principle of this process is shown schematically in Figures 5.

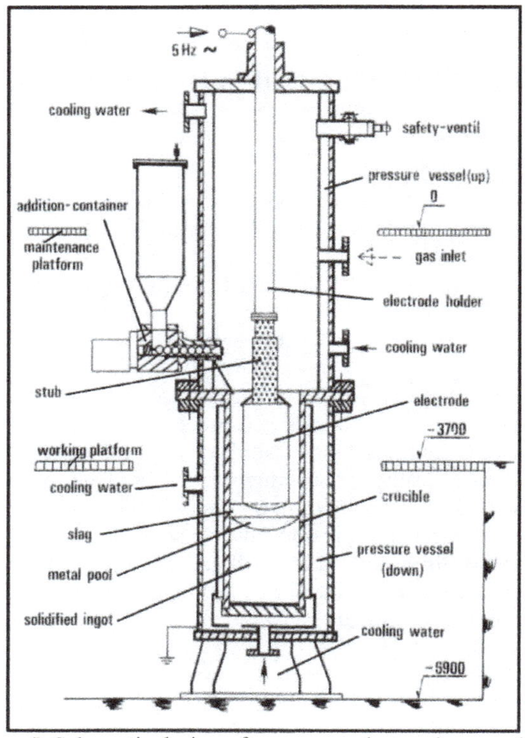

Figure 5. Schematic design of a pressure electro slag remelting furnace (P-ESR) [7]

Basically, the P-ESR process is a conventional remelting facility that works in a pressure tank. The process is designed to meet both, an ESR refining and nitrogen pick up. The metallurgical approach is similar to a standard ESR-process, i.e. refining, low segregation, no porosity or shrinkage, defined microstructure and solidification.

The physical fundamentals of nitrogen pick up are specified over the Sievert's square root law; accordingly the nitrogen solubility is a function of pressure and temperature:

$$[\%N] = k \cdot \sqrt{p_{N2}} \qquad (1)$$

With P_{N2}: Nitrogen partial pressure over melt in bar, k: Material constant (temperature and alloy dependent).

In real systems, the actual solubility is additionally codetermined through the alloy composition. Thermodynamic activities are used to describe the effect of the individual elements.

$$[\%N]_{Fe-X} = \frac{[\%N]_{Fe}}{f_N^X} \cdot \sqrt{p_{N2}} \qquad (2)$$

With $[\%N]_{Fe-X}$: Nitrogen solubility in multi-component systems, $[\%N]_{Fe}$= 0,044% (equilibrium constant in pure Fe at 1600 °C and 1 bar). The activity coefficient f is thereby defined as:

$$\log f_N^X = e_N^X [\%X] \qquad (3)$$

With e_N^X : interaction coefficient, [%X]: Concentration of the elements X in %.

It is obvious that specific elements increase the nitrogen solubility (e.g. manganese), while others reduce the solubility (e.g. Silicum). This has not only an impact on the nitrogen pick up at remelting but also on the precipitation of inter-metallic phases in the solid state. Table 1 summarizes the effect of different elements on nitrogen solubility.

Table 1. Activity coefficients of several elements with effect on the nitrogen solubility in steel at 1 bar.

Element	Coefficient eN	Remarks
C	+ 0.125	Reduces the N-Solubility
Si	+ 0.065	
Ni	+ 0.01	
W	- 0.0015	Increases the N-Solubility
Mo	- 0.01	
Mn	- 0.02	
Cr	- 0.045	
V	- 0.11	
Nb	- 0.06	
Ti	- 0.053	

The nitrogen solubility is accordingly larger in austenitic as in ferrite or martensitic steels.

The nitrogen pick up can occur through the gas phase and from a solid nitrogen carrier, respectively. The choice of a solid body nitrogen pick up medium is down to the following boundary conditions:

- Nitrogen partial pressure: high enough to allow a dissociation at ~ 40 bar
- Characteristics of the slag or flux are not allowed to change (e.g. electrical conductivity, metallurgical properties, etc.)

In practice the standard Si_3N_4 and in some special cases CrN use to pick up nitrogen. Table 2 provides a comparison about advantages and disadvantages of Si_3N_4 and gaseous nitrogen.

Table 2. Advantages and disadvantages different nitrating mediums

Nitrating medium	Advantage	Disadvantage
Si_3N_4	- Nontoxic - Ease of operation and storage - Ease of dissociation	- Very abrasive (joints and gaskets, valves) - Kinetic of dissociation of N^{3-}-ion must be regarded. - Non-continuous allowance on slag - Silicon transfer in melt
N_2 – Gas	- Continuous allowance possible - Simple regulation over the pressure - High equal distribution in ingot - Appropriate for Si-critical steel grades	- slag composition very important - Sievert´s law at high pressure not ideally achieved. - Diffusions conditions in system slag-metal must be known.

The selection of the slag takes place after metallurgical consideration and depends on the alloy. Above all, the slag composition has importance for the nitrogen pick up of the steel.

Alloy Development

The developed alloy consists of the following nominal chemical composition, see Table 3.

Table 3. decomposition of developed alloys

C	Si	Cr	Co	Ni	Mo	Nb/Ta	V	N
0.4	< 0.8	5.0	0.5	1.0	2.5	0.6	0.2	Def.

The nitrogen content is above the solubility limit at atmospheric pressure, i.e. pressurized melting must apply. Due to the defined nitrogen content in combination with a limitation in Si, the use of Si_3N_4 as a nitrating agent was not feasible. Alternatively, nitrogen gas was used. To evaluate the equilibrium and most important boundary conditions, a pressurized induction furnace was used and various trials have been carried out prior to industrial scale melting. Figure 6 and Figure 7 provide an overview of the furnace [6].

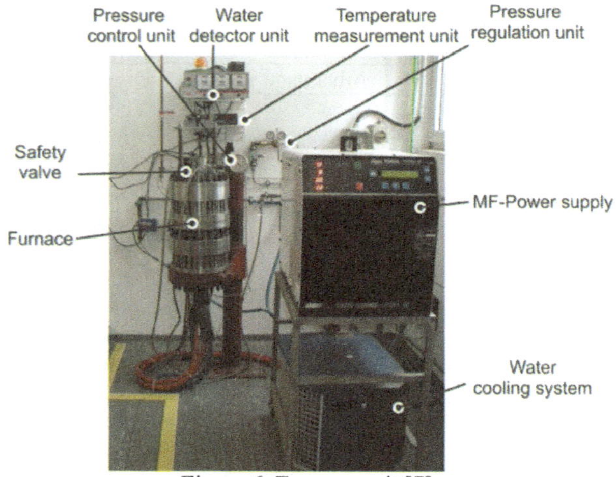

Figure 6. Furnace unit [7]

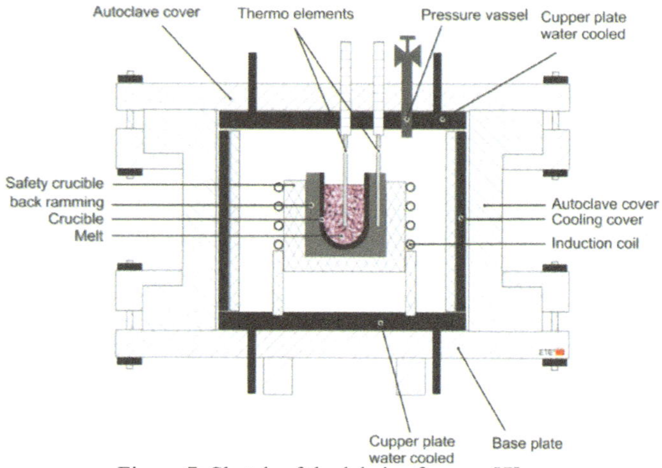

Figure 7. Sketch of the labsize furnace [7]

As seen in Figure 8, there is a clear relation between pressure, time and resulting nitrogen content. The captured data have been used for upscaling the process; however one should consider that P-ESR melting is different to a pressurized induction furnace. This will be explained in more detail within the next section.

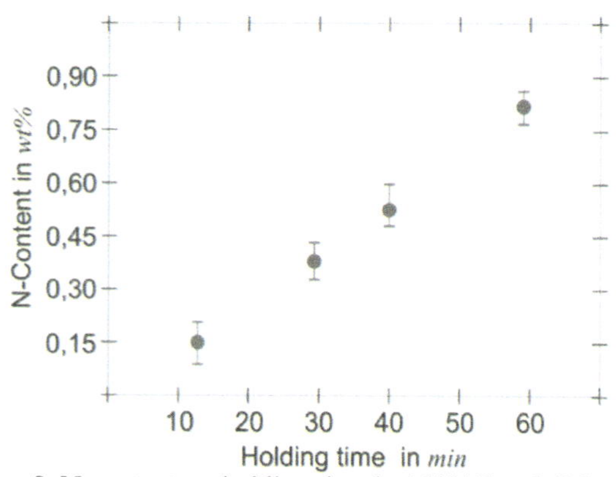

Figure 8. N-content vs. holding time in 1600°C and 40 bar [6]

Industrial Scale Melting

The electrodes have been manufactured using an induction furnace with limited secondary metallurgy, i.e. virgin materials have been used to make up the charge. The cast ingot was fully ground prior to melting to avoid any potential impurities. The P-ESR melting process was carried out as per the standard procedures without any issues. Due to the unknown crack sensitiveness, the ingot size was limited to some extend and the material was directly soft annealed after stripping (see Figure 9).

Figure 9. P-ESR Ingot [7]

Heat Treatment and Mechanical Properties

Based on the existing experiences on High Nitrogen Steels and its heat treatments, a study was carried out to determine the relevant heat treatment parameters. The chemistry of this new alloy was plugged into JMatPro (simulation software) to get a first indication on the most relevant key issues (see Figure 10).

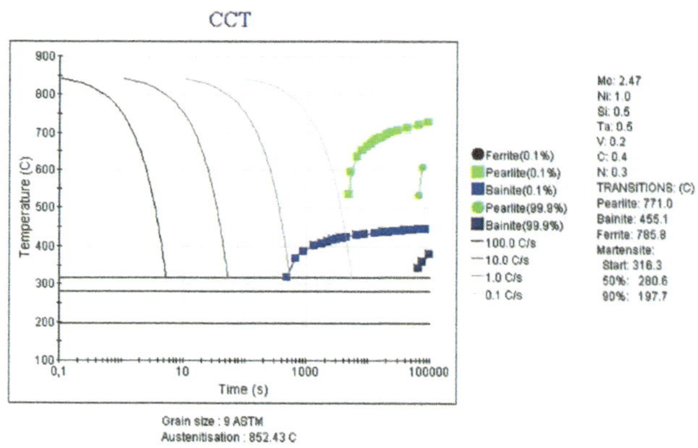

Figure 10. Analytical investigation on developed alloy with JMatPro

The hardness was checked for special aging temperature, the results are shown in Figure 11.

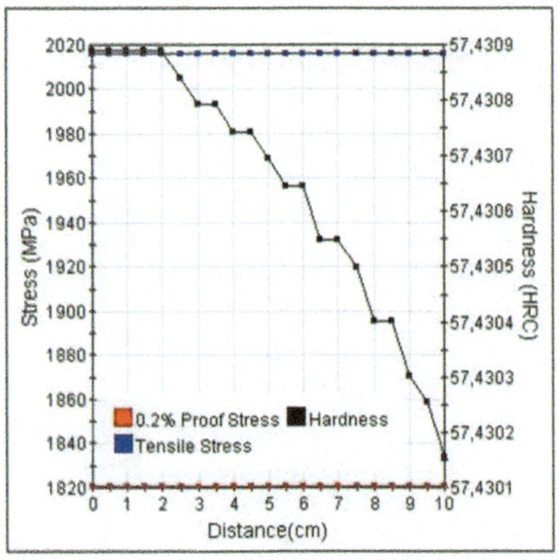

Figure 11. Analytical investigation on developed alloy with JMatPro

To evaluate the warm strength, compression tests at 600 °C (cylinder shape sample: diameter of 10 mm and length of 15mm) were carried out on a hydraulic press. As seen on Figure 11, the force-displacement diagram shows an impressive strength at this temperature. Calculating the strength from the point of plasticity (approx. 100 kN), the resulting compressive strength is around 1900 MPa at 600 °C. Tensile test have not been finished when writing this report.

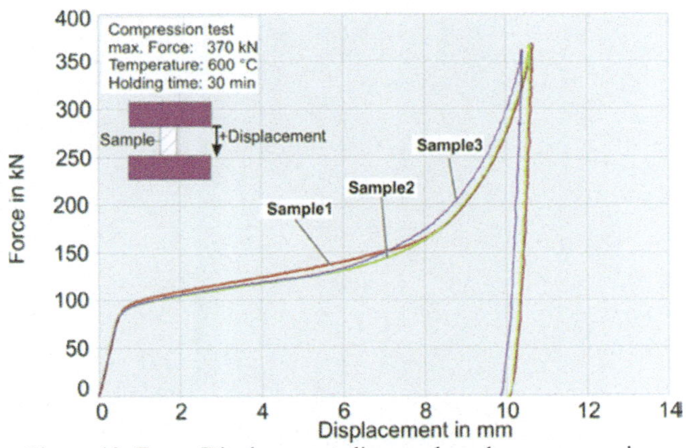

Figure 12. Force-Displacement diagram based on compression test

Summary

New high nitrogen steel was developed using P-ESR melting using a controlled gas-metal-reaction. The alloy confirms a high warm strength which makes it interesting as an engineering material for closed dies, forging mandrels, piercers and high temperature fasteners.

References

[1] Dailly, R.; Hendry, A.: The Effect of Nitrogen on the mechanical behavior of cold-worked austenitic stainless steel rod, S.427-435, High Nitrogen Steels 1998 Conference proceedings, TransTechPublications Ltd, Switzerland, 1998.

[2] Taillard, R.; Vanderschaeve, F.; Foct, J.; Mechanical Behaviour of aged and not prestrained high nitrogen austenitic stainless steels, S.413-420, High Nitrogen Steels 1998 Conference proceedings, TransTechPublications Ltd, Switzerland, 1998.

[3] N.N.: Untersuchungen zur wirtschaftlichen Warmumformung neuer hoch stickstofflegierter nichtrostender Stähle in Abhängigkeit vom Stickstoffgehalt, des Oberflächenzustandes und der Ofenatmosphäre, final rep. AiF project No. 3888N/II, 2007.

[4] Berns H., Gavriljuk V., and Shanina B.: Adv. engg. mater. (2008) No. 12, p. 1083.

[5] Berns H.: Stickstoffmartensit, Grundlage und Anwendung, HTM Härter.-Techn. Mit., Ausgabe 1/2000, Bd. 55, Hansa Verlag, p. 10.

[6] Ritzenhoff, R., Hahn, A., Malekipour Gharbi, M.; Development of High Nitrogen Steels: the new labsize pressurized induction furnace, 5th International Congress on the Science and Technology of Steelmaking 2012, Dresden, Oct. 1-3 ICS 2012.

[7] Energietechnik Essen GmbH, company archive, 2012.

Proceedings of the 2013 International Symposium on Liquid Metal Processing & Casting
Edited by: Matthew J. M. Krane, Alain Jardy, Rodney L. Williamson, and Joseph J. Beaman
TMS (The Minerals, Metals & Materials Society), 2013

NITROGEN CONTROL IN VIM MELTS

P.D. Jablonski and J. A. Hawk
National Energy Technology Laboratory, 1450 Queen Ave. SW, Albany, OR 97321

Keywords: High nitrogen steel, CALPHAD, melt addition.

Abstract

NETL has developed a design and control philosophy for the addition of nitrogen to austenitic and ferritic steels. The design approach uses CALPHAD as the centerpiece to predict the level to which nitrogen is soluble in both the melt and the solid. Applications of this technique have revealed regions of "exclusion" in which the alloy, while within specification limits of prescribed, cannot be made by conventional melt processing. Furthermore, other investigations have found that substantial retrograde solubility of nitrogen exists, which can become problematic during subsequent melt processing and/or other finishing operations such as welding. Additionally, the CALPHAD method has been used to adjust primary melt conditions. To that end, nitrogen additions have been made using chrome nitride, silicon nitride, high-nitrogen ferrochrome as well as nitrogen gas. The advantages and disadvantages of each approach will be discussed and NETL experience in this area will be summarized with respect to steel structure.

Background and Introduction

Alloying with nitrogen has a long history and offers many advantages (1-7). At levels below 0.4 wt%, nitrogen has been used in stainless steels for more than 60 years. The history of the use of nitrogen has been interesting. Initially, nitrogen was used primarily to stabilize the austenite phase in austenitic stainless steels. In small quantities it is a potent solid solution strengthener. It also reduces the tendency of the steel to transform to martensite. In the last thirty years, nitrogen has been used to replace carbon, such as in AISI 304 LN and 316 LN grades, in order to produce austenitic stainless steels with strengths similar to carbon steel but with less susceptibility to sensitization. Mostly nitrogen is used as primary as a solid solution strengthening agent and it is very effective. More recently high nitrogen stainless steels, i.e., with nitrogen levels exceeding 1 wt%, have been produced (3).

With respect to pure iron, nitrogen has low solubility, only about 0.045 wt% at 1600°C and atmospheric pressure. Nitrogen solubility can be increased, however, by alloying and/or increasing the N_2 gas pressure above the melt. Additions of chromium and manganese increase nitrogen solubility, whereas nickel lowers solubility in austenitic alloys. For example, AISI 200 series and Nitronic austenitic stainless steels, based on Fe-Cr-Mn-Ni-N system, were designed for increased nitrogen solubility through alloying. Even with alloying the nitrogen concentration in cast and wrought Fe-Cr-Mn-N stainless steels is typically less than 0.4 wt% (4, 5).

Nitrogen solubility in liquid iron generally follows Sievert's law. The solubility is proportional to the square root of the N_2 gas pressure over the melt. Higher pressures and/or nitrogen levels cause deviation from Sievert's law and increases in nitrogen concentration with pressure occur to a power less than 0.5 (5). Designing an alloying using certain elements enhances the solubility of nitrogen in the melt. For example, elements such as titanium, zirconium, vanadium and niobium, have a positive influence on nitrogen content. However, these elements also have a strong tendency to form nitrides, some of which persist in the liquid well above typical liquidus temperatures. Chromium also increases the solubility of nitrogen in the liquid iron but it does so with a lesser tendency to form nitrides.

From a mechanical behavior perspective nitrogen is added to austenitic stainless steels to increase the yield strength (6). For example, in AISI 304 stainless steel, an almost linear relationship exists between nitrogen wt% and yield strength. In heat resistant ferritic/martensitic steels, nitrogen is added to form a specific matrix precipitate strengthening phase, designated MX (7). The MX phase in its pure form is VN. However, the complex nature of the ferritic/martensitic heat resistant steels is such that the metal portion (M) can include niobium, tantalum, titanium, molybdenum and chromium in addition to vanadium. The interstitial portion (X) of the precipitate can include carbon and boron in addition to nitrogen. From a strengthening perspective MX is coherent with the BCC matrix and coarsens slowly. As such it offers long-term obstacles to dislocation motion during creep.

Alloy Design Considerations

Alloy design is crucial to the success of high nitrogen steels. In general, it is easier to avoid gas formation during melt processing of a high nitrogen austenitic steel versus a high nitrogen ferritic steel since nitrogen is an austenite stabilizer and preferentially partitions to the austenite phase during solidification. The ferrite phase typically has a much lower solubility for nitrogen in comparison to the liquid. Thus, an alloy that solidifies by forming a BCC phase will enrich the liquid with nitrogen as solidification progresses. It is interesting to note that even commercial alloys have regions of "exclusion" in which a level of nitrogen—within alloy specification—cannot be achieved since the alloy preferentially forms nitrogen gas during solidification. Usually this is brought on by a combination of high nitrogen aim, low aim for nitrogen stabilizing elements such as Ni, Mn and Cr and high aim for nitrogen destabilizing elements such as Si (in some cases). Caution is advised however, in classifying an element as strictly nitrogen "stabilizing" or "destabilizing". Complex interactions with other alloy additions as well as the overall phase stability of the alloy need to be taken into account.

NETL's approach utilizes the CALPHAD method using the commercial software ThermoCalc. Usually this region of exclusion within the alloy specification is avoided in the melt shop by heat formulation. However, when a melt shop is unfamiliar with an alloy they may be unaware of this region.

Thus, it is imperative that the alloy specification be thoroughly investigated before formulation and melting. Alloys that form gas in the early stages of solidification, or which form a greater amount, are the most difficult to deal with. Another problematic phenomenon is retrograde solubility. Usually solubility of an element increases as temperature is increased since the entropic contribution to the Gibb's energy also increases. The solubility of gasses can decrease with rising temperature, however, sometimes quite severely. This can have a detrimental effect on weld chemistry for example since temperature control is not as good. It also has implications for process and temperature control during melt operations.

Examples of both a region of "exclusion" within the specification as well as retrograde solubility can be demonstrated by examining type S24100 austenitic stainless steel which is a low cost (low Ni) austenitic steel often used for shafting, bolting, etc., when high corrosion resistance and non-magnetic materials are required. The alloy specification is as follows: C-0.15 max, Mn 11-14, Si-1.0 max, Cr 16.5-19.5, Ni 0.5-2.5, N 0.2-0.45, balance Fe (all values in wt%). Setting C=0.05 and Si=0 (for simplicity) and the rest of the alloy constituents to the specification minimum, a temperature versus weight percent nitrogen diagram can be calculated (Figure 1). This figure demonstrates retrograde solubility for nitrogen. The maximum solubility of nitrogen in the liquid for this alloy is about 0.4 wt% at about 1430°C, which is a eutectic point (and is below the specification maximum for nitrogen in this alloy). At 1600°C the nitrogen solubility is reduced to about 0.3 wt%; thus, as temperature increases the solubility of nitrogen in the melt decreases (retrograde solubility). Clearly, it would not be possible to make a 0.45 wt% N heat of this alloy with the other elements adjusted towards the lower range of their specification. It also is not possible to make a low nitrogen heat under these conditions.

Consider the solidification sequence at for nitrogen at 0.2 wt% (the specification minimum). The liquid starts to solidify as BCC at 1448.2°C; gas first appears at 1406.3°C; FCC begins to form at 1380.6°C; the solidus is reached at 1377.9°C; and the extinction of gas phase formation occurs below this at 1368.2°C (Figure 1). Under these conditions the alloy solidifies with roughly 85 wt% BCC containing 0.11 wt% N and 15 wt% FCC containing 0.41 wt% N. Thus, it is estimated that the nitrogen level achievable is about 0.16 wt% at best. This is mainly due to solidifying the majority of the liquid as BCC with the inherently low solubility of N in the BCC phase. Higher levels of austenite stabilizing elements such as C and Si would help avoid gas formation at the lowest N specification level (other elements remaining near the lower range of the specification). Not surprisingly, C has the greatest effect, being the better austenite stabilizer in comparison to Si. However, with the transition elements set low substantial BCC phase still forms during solidification. With higher C and Si, some of the BCC transforms to FCC during solidification which increases the overall solubility of N in the solid. Thus, it is best to consider how alloy additions affect phase stability which in turn establishes the possible level of nitrogen rather than focus on nitrogen solubility. However, this scenario is strictly an equilibrium consideration and it relies on the transformed BCC (into FCC) to absorb nitrogen by solid state diffusion during solidification to achieve the higher solubility (which is questionable).

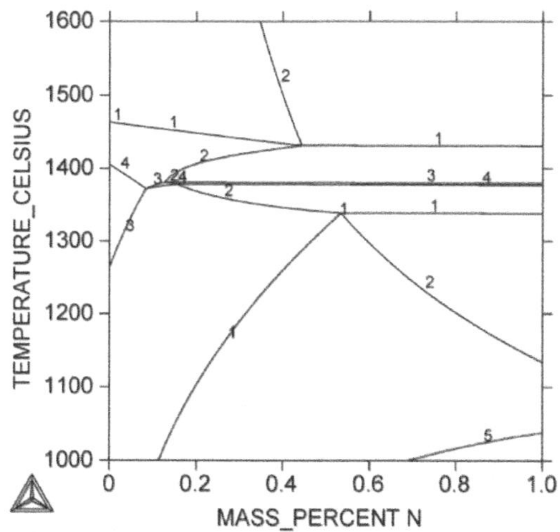

Figure 1: As the above solidifies (N=0.2 wt%): + BCC (1448.2°C), + gas (1406.3°C) + FCC (1380.6°C) – liquid (1377.9°C) – gas (1368.2°C). The ThermoCalc predicted phases are - 1: BCC, 2: Gas, 3: FCC, 4: Liquid.

Method of Nitrogen Addition

Nitrogen can be added to a melt by additions of nitrogen compounds such as CrN or Si_3N_4, additions of high nitrogen master alloys such as high nitrogen ferro-chrome or by the exposure to a controlled nitrogen partial pressure in the gas. Air, of course, consists mostly of nitrogen and could be considered to have a controlled nitrogen partial pressure. However, since the majority balance gas is oxygen, melting in air can cause the formation of a slag layer which inhibits the uptake of nitrogen from the atmosphere. In the lab scale furnaces at NETL, which range in capacity from a few pounds to a few hundred pounds, the strategy utilized is to melt under a controlled atmosphere containing a known amount of nitrogen and to also utilize a solid nitrogen addition. Thus, the nitrogen level is achieved by the solid addition and maintained by the gas. To further complicate matters, most of the melts at NETL are made with virgin stock of high purity components. The impact of this is that the melt chemistry continually changes as each component is melted in turn. This means that the melt will have a changing ability to absorb nitrogen as the melt progresses. One way to combat this is to make the solid addition and the cover gas addition after the full melt is established. This is not always possible, however. Also, the solid addition can locally super saturate the melt with nitrogen leading to uncontrolled boiling and nitrogen evolution. An alternative is to load the solid addition dispersed in the lower portion of the melt stock within the crucible along with high nitrogen stabilizing elements such as Cr and Mn. With respect to "solid" nitrogen additions, the high nitrogen ferro-chrome is perhaps the most preferred since it has a low melting point and dissolves easily into the melt while the Si_3N_4 is much less preferred since it appears to dissolve as a peritectic reaction which can be quite slow. The concentration of higher nitrogen solubility elements eases the absorption of the solid nitrogen into the melt. A critical component to the success of this strategy is the level of nitrogen overpressure.

Each element comprising the melt will have a thermodynamic activity at a given set of conditions (temperature, chemistry, etc.) including nitrogen. A convenient way to calculate the activity of nitrogen is to utilize any of the commercially available thermodynamic software packages such as ThermoCalc. The goal, of course, is to equalize the activity of nitrogen in the melt with the activity of nitrogen in the gas ($a_{N_2,gas}$). If the gas is ideal, a reasonable estimate of activity is the nitrogen gas pressure divided by the standard conditions pressure (typically one atmosphere). Thus, if the desired activity were 0.1, then the required nitrogen gas pressure would be 76 torr. It turns out that such low pressures, by themselves, are insufficient to maintain the nitrogen in the melt in alloys that solidify either partially or completely as ferritics. This is because the BCC crystal structure has a much lower solubility for nitrogen. Thus, during solidification these alloys expel nitrogen into the liquid and quickly supersaturate the remaining liquid, causing interdendritic gas bubble formation. Unfortunately, it is typical for nitrogen bearing ferritics to have nitrogen melt activities on the order of 0.01-0.10; thus, relatively low partial pressure of nitrogen is called for. To combat the potential for gas evolution during solidification additional gas pressure of an inert gas such as argon has been employed. Total pressure, on the order of a few hundred torr, appears to be adequate. Austenitic alloys, with their typically higher nitrogen aims, will have target nitrogen gas activities on the order of 0.3-0.9, and thus, require a nitrogen partial pressure of a few hundred torr eliminating the need for additional inert gas pressure.

Experimental Results

In this section examples of austenitic and ferritic/martensitic alloys will be given. The first example is an experimental high nitrogen austenitic alloy with a composition of Fe- 0.4 C- 5.95 Mn- 0.26 Si- 21.26 Cr- 6.1 Ni- 2.0 Mo (wt%) with the intended use for high wear applications. A temperature versus weight percent N plot for this alloy is shown in Figure 2. This composition shows the retrograde solubility described earlier. The desired nitrogen content for this alloy was 0.50 wt%; however, the calculated maximum nitrogen solubility in the liquid is 0.49 wt% at the liquidus temperature (1380°C). Upon reevaluation, a nitrogen content of 0.45 wt% was deemed acceptable from a materials point of view, but the desire for "as high as possible" remained. Thus, a design target of 0.47 wt% was selected. With this level of nitrogen, gas bubbles may form in the melt at temperatures above about 1407°C; however, even with a useful superheat of ~50°C (1430°C) the nitrogen solubility in the liquid is 0.455, which is still reasonable.

With the base composition and target nitrogen taken into account, the nitrogen activity in the gas phase can be calculated at the liquidus ($a_{N_2,gas}$(at T = 1380°C) = 0.905). This can be used to calculate the partial pressure of nitrogen required to achieve this chemistry: $P_{N_2} = P*a_{N_2,gas} = 760*0.905 = 687$ torr. Thus, rounding up, a partial pressure of 700 torr of high purity nitrogen was used when melting the constituents which constituted this alloy. A high nitrogen ferro-chrome master alloy was also used. This was placed near the bottom of the charge dispersed within the Mn, Ni, C and some of the Cr. Layered on top of this was a compact of ~1/3 of the iron followed by a compact of Cr, Mo and Si followed by two additional compacts of the remaining Fe charge. After melting, the nitrogen wt% was found to be 0.47, which is reasonably close to the desired level of 0.50 and very near the design level.

Some closing comments are necessary. This alloy has some features which make it limited in its practicality, and thus, its usefulness: First, the retrograde nitrogen solubility and high nitrogen content makes repeatable chemistry results in doubt at this level of nitrogen. An aim of 0.3 would be much more practical. Second, this alloy would not be field weldable since the nitrogen content raises the activity above atmospheric nitrogen. These two conspire against the practical use of the alloy. However, it does serve as a useful example (if not an extreme case) of this technique.

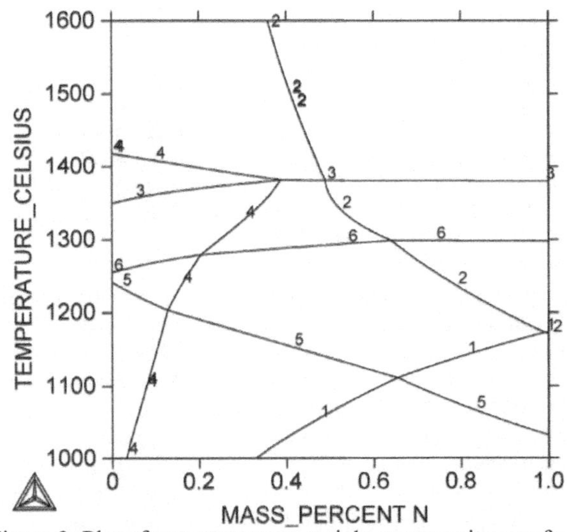

Figure 2: Plot of temperature vs. weight percent nitrogen for an alloy with a composition of Fe-0.4 C-5.95 Mn-0.26 Si-21.26 Cr-6.1 Ni-2.0 Mo (wt%). The ThermoCalc predicted phases are - 1: HCP, 2: Gas, 3: FCC, 4: BCC, 5: $M_{23}C_6$, 6: Liquid.

This next example is an experimental high nitrogen austenitic steel with a composition of Fe- 0.1 C- 8 Mn- 0.5 Si- 21 Cr- 7 Ni- 3 Cu- 0.3 Mo- 1 W- 0.8 Nb (wt%) with the intended use for high strength applications. A temperature versus weight percent N plot for this alloy is shown in Figure 3. This alloy also shows retrograde solubility for nitrogen with maximum nitrogen solubility in the liquid of about 0.5 wt% at about 1368°C. The nitrogen solubility decreases to about 0.34 wt% at 1600°C. A reasonable nitrogen aim for this nominal composition would be below 0.45 wt% nitrogen. An alloy was subsequently designed with an aim of 0.3 wt% nitrogen. This alloy had a liquidus of about 1375°C and an $a_{N_2,gas}$(at T = 1375°C) = 0.257, which equates to a nitrogen partial pressure of 195 torr. This alloy was melted with nitrogen partial pressure of 320 torr. This partial pressure is high enough to produce a nitrogen content of about 0.37 wt%, if held for an extended time at the liquidus. However, the melt was poured as soon as the desired temperature was reached (typically less than 25 minutes). As a result, minimal nitrogen pickup from the gas was observed. Several heats of this formulation were made using these conditions with the resulting nitrogen contents ranging from 0.284 to 0.334 wt%.

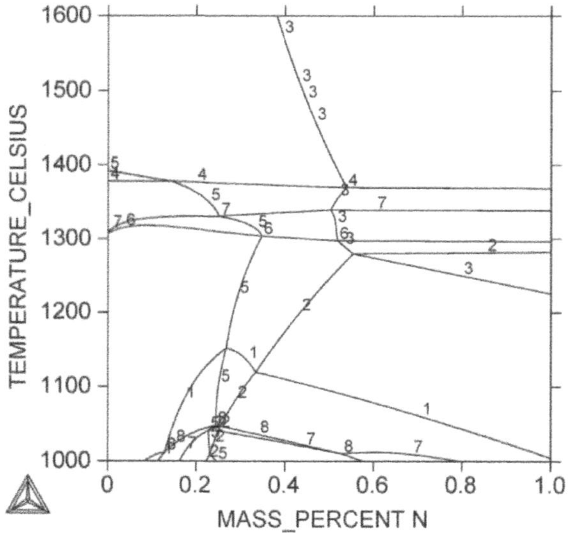

Figure 3: Plot of temperature vs. weight percent nitrogen for an alloy with a composition of Fe-0.1 C-8 Mn-0.5 Si-21 Cr-7 Ni-3 Cu-0.3 Mo-1 W-0.8 Nb (wt%). The ThermoCalc predicted phases are - 1: Z-phase, 2: HCP, 3: Gas, 4: FCC, 5: BCC, 6: Liquid, 7: MC, 8: $M_{23}C_6$.

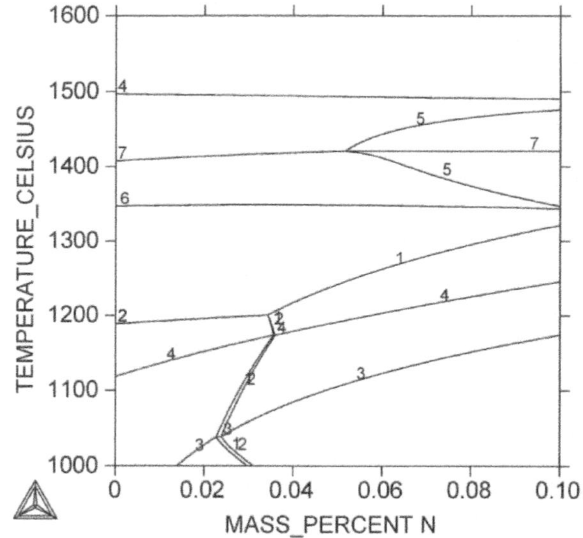

Figure 4: Plot of temperature versus weight percent nitrogen for an alloy with a composition of Fe-0.13C-10.6Cr-0.41Mo-0.20V-0.88Cu-0.69Mn-0.23Si-0.006Nb-0.06Ni-1.97W-0.001B (wt%). The ThermoCalc predicted phases are - 1: BN_HP4, 2: M_2B_TETR, 3: MC, 4: BCC, 5: Gas, 6: Liquid, 7: FCC.

This next alloy example is an experimental ferritic/martensitic steel with possible applications for power plant components. The composition is Fe-0.13C-10.6Cr-0.41Mo-0.20V-0.88Cu-0.69Mn-0.23Si-0.006Nb-0.06Ni-1.97W-0.001B with a nitrogen aim of 0.045 (wt%). A plot of temperature versus weight percent nitrogen for this alloy is shown in Figure 4. Note that the range of nitrogen composition is an order of magnitude lower than that of Figures 1-3 because of the much lower nitrogen solubility in ferritic alloys. Below 1600°C no gas/liquid phase field is observed through 0.1 wt% nitrogen both at and above the liquidus. Between the liquidus and solidus, the maximum nitrogen solubility with gas exclusion is about 0.052 wt% at about 1418°C. There is higher nitrogen solubility both above and below this temperature. Considering the solidification of an alloy with 0.045 wt% nitrogen, solidification starts from the liquid as a BCC phase when the liquidus is reached at 1494°C. At this point the nitrogen activity in the gas phase is $a_{N_2,gas}$(at T = 1494°C) = 8.92×10^{-2}. The alloy continues to solidify as BCC until 1420°C at which point it switches to FCC, which continues through the completion of solidification. Concurrent with this solidification path is the transformation of about half of the original BCC phase to the FCC phase structure (equilibrium solidification). This increase in FCC phase is the reason gas phase formation is pushed to higher nitrogen contents as shown in Figure 4. This alloy was melted using a nitrogen partial pressure equal to 70 torr with the balance argon for a total overall pressure of 200 torr. The resulting nitrogen content was 0.044 wt% with no signs of gas formation.

This next example is another ferritic/martensitic steel, alloy P92, an advanced commercial alloy which is used for power plant piping. The compositional range of this alloy is C 0.07-0.13, Cr 8.5-9.5, Mo 0.3-0.6, V 0.15-0.25, Mn 0.3-0.6, Si 0.5 max, Nb 0.04-0.09, Ni 0.4 max, W 1.5-2, B 0.001-0.006 and nitrogen from 0.03-0.07 (balance Fe, all wt%). A plot of temperature versus weight percent nitrogen for this alloy with nominal (middle) values for all the elements is shown in Figure 5. Upon cursory inspection, this looks very much like Figure 4; however, it appears that the maximum nitrogen solubility with gas exclusion is about 0.04 wt% N, which is much lower than the previous example and results in a large region of exclusion within the specification. Exploration within the alloy specification only provided marginal improvement in this limited solubility. Increasing the C, Mn, Si, Cr, W, V, Nb, B and Ni contents gave a slight increase in nitrogen solubility, while increasing Mo slightly decreased the solubility, giving a predicted maximum nitrogen content of about 0.047 wt% while simultaneously avoiding gas formation. Thus the alloy, as specified, has a substantial region of exclusion with respect to nitrogen content. A heat of this alloy using the nominal composition was formulated with a nitrogen aim of 0.039 wt%. The liquidus for this alloy is about 1506°C at which point the nitrogen activity in the gas phase is $a_{N_2,gas}$(at T = 1506°C) = 8.95 $\times 10^{-2}$. This alloy was melted using a partial pressure of 70 torr nitrogen with the balance argon for a total overall pressure of 200 torr. The total melt time was approximately 45 minutes. The resulting chemistry gave 0.037 wt% N with no signs of gas formation.

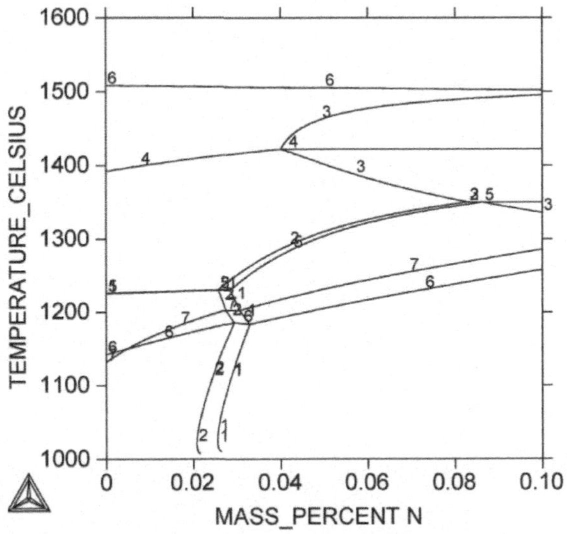

Figure 5: Plot of temperature vs. weight percent nitrogen for an alloy with a composition of Fe-0.07-0.13C-8.5-9.5Cr-0.3-0.6Mo-0.15-0.25V-0.3-0.6Mn-0.5Si-0.04-0.09Nb-0.4Ni-1.5-2W-0.001-0.006B (wt%). The ThermoCalc predicted phases are - 1: M_2B_TETR, 2: BN_HP4, 3: Gas, 4: FCC, 5: Liquid, 6: BCC, 7: MC.

Concluding Remarks

The practical considerations regarding the formulation and melting of high nitrogen steels has been presented. Often times alloys will have regions of exclusion in which certain chemistries cannot be made by melt processing due to gas formation either in the melt or during solidification. Some alloys exhibit retrograde solubility of nitrogen. These tend to be austenitic alloys with their inherently higher nitrogen aims. Nitrogen can be added to the melt through additions of nitrogen compounds such as CrN or Si_3N_4, additions of high nitrogen master alloys such as high nitrogen ferro-chrome or by the exposure to a controlled nitrogen partial pressure in the gas. NETL formulates nitrogen containing alloys with the addition of nitrogen containing solids along with melting under a controlled nitrogen partial pressure. The philosophy is one of establish and maintain with respect to nitrogen content. The desired nitrogen partial pressure is estimated by calculating the activity of nitrogen in the gas at the liquidus temperature ($a_{N_2,gas}(T_L)$). The nitrogen partial pressure for austenitic alloys tends to be on the order of a few hundred torr while that of ferritic alloys tends to be a few tens of torr. This nitrogen gas pressure alone appears to be adequate while the low pressure required in ferritics appears to benefit from added inert gas pressure to raise the total gas pressure to a few hundred torr. Example calculations and practical melt experiences are provided.

Disclaimer

References

1. R.P. Reed, Nitrogen in Austenitic Stainless Steels, J. Met., Vol. 41, Issue 3, 1989, p.16.
2. M.O. Speidel and R.M. Pedrazzoli, High Nitrogen Stainless Steels in Chloride Solutions, Mater. Perform., Vol. 31, No. 9, 1992, p.59.
3. J.W. Simmons, Overview: High-Nitrogen Alloying of Stainless Steels, Mater. Sci. Eng., Vol. 207, Issue 2, 1996, p. 159.
4. R.D. Pehlke and J.F. Elliott, Solubility of Nitrogen in Liquid Iron Alloys: 1. Thermodynamics, Trans. AIME, Vol. 218, Dec. 1960, p. 1088.
5. A.H. Satir-Kolorz and H.K. Feichtinger, On the Solubility of Nitrogen in Liquid Iron and Steel Alloys Using Elevated Pressure, Z. Metallks., Vol. 82, No. 9, 1991, p. 689.
6. K.J. Irvine, D.T. Llewellyn and F.B. Pickering, High-Strength Austenitic Stainless Steels, J. Iron Steel Inst., Vol. 199, 1961, p. 153.
7. F.B. Pickering, Historical Development and Microstructure of High Chromium Ferritic Steels for High Temperature Applications, in Microstructural Development and Stability In High Chromium Ferritic Power Plant Steels, A. Strang and D.J. Gooch, Eds., The Institute of Materials, 1997, The University Press, Cambridge, UK, p. 1.

Proceedings of the 2013 International Symposium on Liquid Metal Processing & Casting
Edited by: Matthew J. M. Krane, Alain Jardy, Rodney L. Williamson, and Joseph J. Beaman
TMS (The Minerals, Metals & Materials Society), 2013

CHARACTERISATION AND MODELLING OF MICROSEGREGATION IN LOW CARBON CONTINUOUSLY CAST STEEL SLAB

Dayue Zhang, Martin Strangwood
School of Metallurgy and Materials, University of Birmingham, Edgbaston, Birmingham, B15 2TT, UK

Keywords: Microalloyed steel, Microsegregation, Clyne-Kurz model

Abstract

Segregation during casting of microalloyed steels can affect subsequent processing and final properties. In optimising composition – processing combinations an accurate prediction of the composition and distribution of solute-rich and solute-depleted regions is needed. A series of low carbon (0.06 - 0.14 wt % C) as-cast steel slabs have been investigated using statistically significant SEM-EDS mapping to characterise the compositions of solute-rich and solute-depleted regions as well as their size and spacing. A Clyne-Kurz model has been applied to these alloys (which all involve the peritectic formation of austenite during their solidification) using Thermo-Calc predictions of transformation temperatures and compositions. Initial modelling using carbon re-distribution predicted solute-rich fractions within 3 % of experimentally-determined values. Measured composition agreement with Clyne-Kurz and Thermo-Calc predictions was less good. The microalloying elements such as Nb seemed to have little effect on the prediction of solute-rich fractions. This paper will report the progress in modelling and characterisation of segregation in continuously cast steel slabs.

Introduction

Microalloyed structural steels are widely used in many major steel market sectors (e.g.automotive and aerospace industries) due to their combination of high strength and good toughness. This arises from a combination of small grain size and distribution of fine precipitates. Nb, Ti and V are the most commonly used microalloying elements, which form carbide, nitride and/or carbonitride precipitates along with C and/or N. These microalloying additions show varying tendencies to segregate to the interdendritic regions during initial casting of the steel, which can result in inhomogeneous composition, structure and properties in the final product [1, 2]. The variation in [Nb] and pinning particle contents between the solute-rich and solute-depleted regions can lead to variation in critical grain growth temperatures and form bimodal grain structure (abnormally large grains present in a matrix of fine grains) during reheating[3, 4]. Reduced toughness and increased scatter, due to the bimodal grain structure, has been found in some Nb-microalloyed thermo-mechanically controlled rolled (TMCR) plates consisting of banded coarse and fine grains [2].

Accurately understanding and simulating the micro-structural development during solidification of Nb microalloyed steel calls for the solution of a combination of microsegregation and thermodynamic models. Investigators have employed various types of microsegregation models for calculating the solute content from the primary solidified phase rejected during soldification into the interdendritic liquid [5, 6, 7]. The solidification of low carbon microalloyed steel is modified by the occurrence of the peritectic transformation involving solidification as and solute redistribution from both delta ferrite and austenite. Several studies have been reported in the open literature, although there is less agreement between these and experimental results [8, 9, 10]. In addition, commercially available thermodynamic software packages (e.g. Thermo-Calc and MT-DATA) have been used to predict micro- and macro-segregation in alloy systems [9, 11]. Davis and Strangwood have predicted the microsegregation of microalloying elements in continuously cast steels using Thermo-Calc, but this has not been extended to volume fractions of solute-rich or solute-depleted regions [11].

In modeling the structure and properties of hot rolled microalloyed steels then it is necessary to account for the amount, size and composition of solute-rich and solute-depleted regions prior to considering reheating and hot deformation processes.
This paper reports progress in determining spatial composition variations in as-cast material.

Modeling Approach

Micosegregation during steel solidification has been predicted by many models with different assumptions and simplifications [7, 12, 13]. Typically, most of microsegregation models consist of an expression for solute concentration in terms of solid fraction between the neighbouring dendrite arms. There are several common assumptions to all quantitative treatments: equilibrium at the liquid-solid interface, no significant undercooling before solid nucleation and constant partition coefficient of solute element. The two basic limiting models are the lever-rule and the Scheil equation. The former is an equilibrium solidification model, which assumes complete mixing in the solid and liquid phases. At a temperature T^*, the composition of solid at the solid-liquid interface, C_S, as the function of fraction solidified, f_s, is

$$C_S = \frac{kC_0}{1-(1-k)f_S} \tag{1}$$

where k is the equilibrium partition coefficient, and C_0 is the bulk composition. In term of the system temperature, the solid fraction may be written as

$$f_S = \left(\frac{1}{1-k}\right)\left(\frac{T_L - T^*}{T_f - T^*}\right) \tag{2}$$

where T_f is the melting temperature of pure material, pure iron for steel (1536 °C), T_L is the liquidus temperature.

Because solid-state diffusion is slow, especially for the larger solute element atoms such as Nb, the lever-rule model is not accurate during solidification [7]. The Scheil equation deals with non-equilibrium conditions, assuming complete mixing in the liquid, no diffusion in the solid and local equilibrium at the solid-liquid interface. In the Scheil model, the following quantitative expression is applied:

$$C_S^* = kC_0(1-f_S)^{(k-1)} \tag{3}$$

or, expressing the fraction of solid as a function of temperature

$$f_S = 1-\left(\frac{T_L - T}{T_f - T}\right)^{1/(k-1)} \tag{4}$$

However, it is apparent that the Scheil equation inadequately estimates the final solute composition concentration, because the Scheil equation generates an infinite value at f_s=1. This model is only used for very rapid solidification process such as laser welding with the cooling rates over 10^2 °C/s [14]. In order to predict microsegregation during solidification accurately, the finite solute diffusion in solid must be considered. Brody and Flemings [15] have proposed a model that assumes finite back – diffusion in the solid and infinite diffusion in the liquid, based on one-dimensional solute redistribution. By assuming that the local interface advance velocity V decreased parabolically with increasing time, the final equation for a binary solution is:

$$C_S^* = kC_0[1+f_S(2\alpha k-1)]^{(1-k)/(2\alpha k-1)} \tag{5}$$

$$f_S = \left(\frac{1}{1-2\alpha k}\right)\left[1-\left(\frac{T_f - T}{T_f - T_L}\right)^{(1-2\alpha k)/(k-1)}\right] \tag{6}$$

$$\alpha = \frac{D_S t_f}{X^2} \tag{7}$$

where α is a Fourier number for the solute element. D_S is the diffusion coefficient of the solute element in the solid phase, t_f is the local solidification time, and X is the length scale of the microsegregation, usually regarded as equal to half of the secondary dendrite arm spacing, λ_{SDAS}.

The model is not accurate when α is very large, since even mass conservation cannot be satisfied. A number of modifications have been developed to address this limitation [16, 17, 18]. Among them, the Clyne-Kurz model [16] seems to be more popular. This modification makes the model physically reasonable by replacing α by the function $\Omega(\alpha)$ as follows:

$$\Omega(\alpha) = \alpha\left(1-\exp\left(-\frac{1}{\alpha}\right)\right)-\frac{1}{2}\exp\left(-\frac{1}{2\alpha}\right) \tag{8}$$

$$C_S^* = kC_0[1+f_S(2\Omega k-1)]^{(1-k)/(2\Omega k-1)} \tag{9}$$

$$f_S = \left(\frac{1}{1-2\Omega k}\right)\left[1-\left(\frac{T_f - T}{T_f - T_L}\right)^{(1-2\Omega k)/(k-1)}\right] \tag{10}$$

The equations mentioned above are all semiempirical results, but they can be very useful in predicting the microsegregation if applied carefully [6, 7, 17].

As noted above, the solidification of a range of microalloyed steels involves the peritectic reaction. For these steels the development of microstructure can be summarized as in Figure 1.

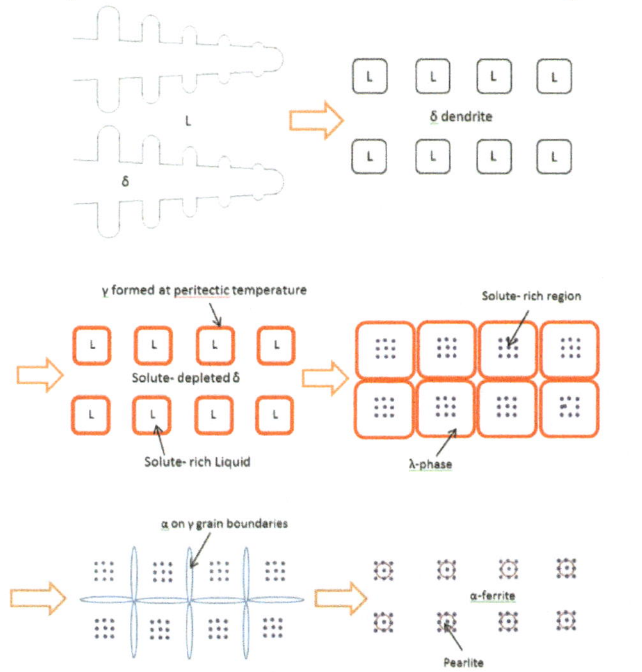

Figure 1: Schematic diagram of solute-rich and solute-depleted areas formation on solidification of the continuous cast steel slab.

For continuously cast steel, δ dendrites first form inside the liquid. The solute elements are rejected by the solid into the interdendritic liquid, so that the δ phase has a low alloy element content and the liquid in the interdendritic space is enriched in alloy elements. Initially the interdendritic regions are interconnected, but as the fraction of solid increases, these get pinched off as isolated pockets of solute-rich liquid. Cooling to the peritectic temperature allows austenite to nucleate at the δ/liquid interface before growing into both the δ phase and liquid. Diffusion in austenite is much slower than diffusion in δ phase; diffusion of interstitial elements (C,N) in the delta phase is about one order of magnitude faster than in the γ-phase. Diffusion of the substitutional elements such as Si, Mn, V in the delta phase is approximately 2 or 3 (especially Nb) orders of magnitude greater than that in the γ-phase at the peritectic temperature. Due to the low diffusion rate in the γ phase, the solute is unable to move between the δ phase and liquid and thus the compositions of the solute-rich and solute-depleted regions is largely fixed. With decreasing temperature, the δ-ferrite and liquid fully transform into γ-phase and the solute rich regions locate in the centre of austenite. When the temperature falls below Ar_3, α-ferrite forms on the γ grain boundaries and grows into the surrounding austenite. Pearlite finally forms in the solute-rich region containing higher alloying element content.

If the hypothesis summarized above holds, then the fraction of solute-depleted material should correspond to the fraction solidified at the peritectic temperature. The composition at the boundary between solute-rich and solute-depleted material should also correlate to the compositions of the solid phases at the peritectic temperature.

Thus a Clyne-Kurz model has been used to predict the fraction and composition of solute-rich and solute-depleted material using the peritectic temperature predicted from Thermo-Calc software. The latter was also used to predict solute compositions at that temperature. These predictions are compared with experimental characterization of as-cast steels.

Experimental Procedure

As-continuously cast low-carbon microalloyed steel slabs, table I, containing of a range of alloying elements were investigated. The structural and ship building steels were supplied by Shougang Steel Company, China as 220 mm thick slabs, nitrogen levels were not specified, but these will not significantly affect segregation or solidification behavior. Slab 1 was supplied by Tata Steel, UK as a 290 mm thick, 1800 mm wide slab. Longitudinal sections were mounted in conductive bakelite, polished to a 0.05 OPA suspension polish and etched in 2%-nital. All samples were taken at quarter thickness. Optical observations were conducted on a Zeiss Akioskop-2 microscope using AxioVision 4.6.3 image capture and analysis software to quantify microstructural feature, such as the secondary dendrite arm spacing (SDAS). Chemical characterization of dendrite segregation was performed on polished, unetched and lightly etched samples using a Jeol 7000 SEM with an Oxford instruments EDS system. Statistical sampling using 200 or 300 point SEM-EDS composition measurements were carried out on each steel using a square array of points, Figure 2. The EDS step size (the distance between two consecutive measurements) was larger than the average secondary dendrite arm spacing of the cast samples. Several sections per sample were analysed. The number of grid points and the grid spacing of the investigated three steels are listed in Table II. These randomly sampled electron microanalysis data were ranked using a Flemings – Gungor (F-G) approach [19, 20]. The F-G sorting is a statistically significant and accurate means of quantifying the distribution of alloying elements and all elements are sorted independently into ascending or descending order according to their segregation direction regardless of sampling location. This led to smooth and continuous concentration profiles from which changes in slope were used to distinguish variations in the solidification behavior.

Results and Discussion

Predicted solidification sequence

Using the bulk compositions (Table I), the equilibrium solidification sequences were predicted using Thermo-Calc. The Ship building and Slab 1 grades were predicted to solidify via the sequence:

$$L \rightarrow L + \delta \rightarrow L + \delta + \gamma$$

The Structural steel was predicted to solidify fully as δ-ferrite, but at only 1 K above the temperature at which austenite should appear. Hence, with non-equilibrium cooling, all three should be expected to comply with the hypothesis presented above for formation of solute-rich and solute-depleted regions involving the peritectic reaction. The Structural steel is predicted to complete solidification as a (δ+γ) mix, whilst the other two steels are predicted to complete solidification as single-phase austenite. Before these differences occur the solute-rich and solute-depleted regions should have been established. The predicted temperatures for onset of the peritectic reaction are 1479, 1482 and 1483 °C respectively for Structural steel, Ship-building steel and Slab 1.

Microstructure of as-cast slabs

The Clyne-Kurz approach requires input of a solidification rate, which was determined from the secondary dendrite arm spacing at the quarter thickness position of the slab (all experimental results came from this position). The microstructures of the as-continuously cast steel slabs, Figure 3, all exhibited a ferrite – pearlite structure. In the Ship building and Slab 1 samples there was a clear ferritic dendrite structure and the SDAS values was readily obtained from mesurements of centre-to-centre spacings of the interdendritic pearlite colonies. The Structural steel, however, had a much lower pearlite content, which made the dendritic structure less easy to define, leading to unacceptable errors in the SDAS values determined optically.

Table I: Major alloying element contents (wt %) of the as-cast slabs

Steel	C	Si	Mn	Al	V	Nb	Other
Structural	0.06	0.30	1.52	0.0358	-	0.052	-
Ship building	0.14	0.38	1.35	0.039	0.003	0.002	0.01 Ni, 0.01Cu,0.004 Ti, 0.001 Mo
Slab 1	0.10	0.31	1.42	0.046	0.052	0.045	0.52 Ni, 0.002 Ti, 0.008 N

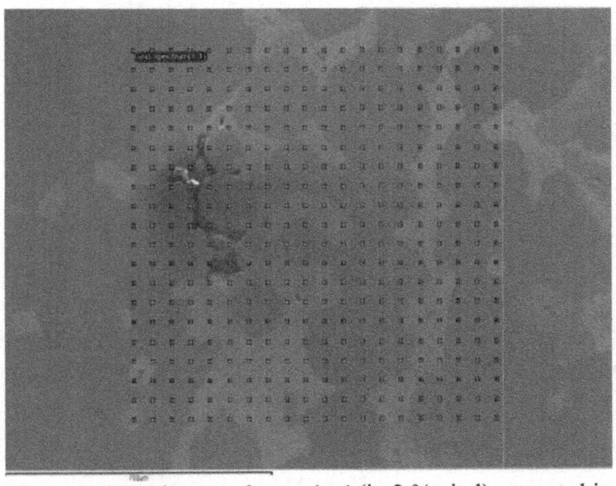

Figure 2: SEM image of an etched (in 2 % nital) as-cast ship building steel with overlaid sampling square grid points and dendritic (A) and interdendritic (B) areas identified.

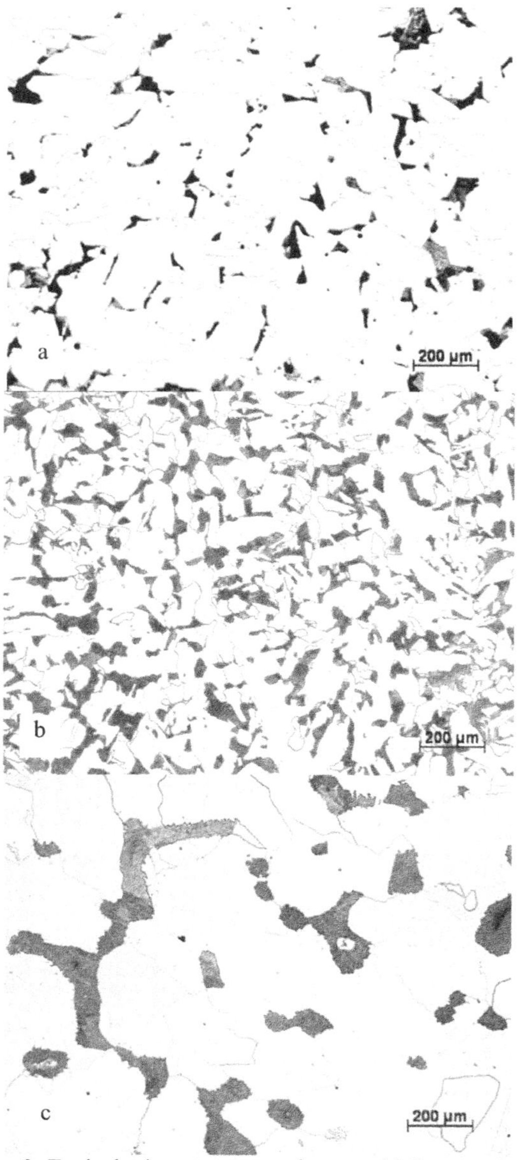

Figure 3: Typical microstructures of as-cast (a) Structural steel, (b) Ship building steel and (c) Slab 1 steel at the quarter thickness positions

Mathematical Modelling of Solidification Based on Clyne-Kurz Model

Assumptions

This present model was based on the Clyne-Kurz microsegregation model and Thermo-Calc software. Carbon diffusion is regarded as one-dimensional (normal to the solid-liquid interface). The other main assumptions adopted for the formulation of the mathematical model are:

i. The cooling rate is constant for the solidification temperature range.
ii. The growth of δ phase is controlled by carbon diffusion.
iii. The condition of equilibrium is obtained at the δ and liquid interface.
iv. The carbon partition coefficient in the δ phase is assumed constant.
v. Due to the high diffusivity of carbon atoms during this temperature range, there is little undercooling from the equilibrium temperature for the peritectic reaction [6]. So the equilibrium condition was assumed at the peritectic temperature.

SDAS and Cooling Rate

In the Clyne-Kurz model, the secondary dendrite arm spacing λ_{SDAS} is an important parameter as it has a great effect on the development of micro-segregation during solidification. An empirical relationship between SDAS and cooling rate was used based on equation 11, which has previously been fitted to experimental data [7]:

$$\lambda_{SDAS} = (169.1 - 720.9 \cdot (\%C)) \cdot C_R^{-0.4935} \quad \text{for } 0<(\%C)\leq 0.15$$
$$= 143.9 \cdot C_R^{-0.3616} \cdot (\%C)^{(0.5501-1.996(\%C))} \quad \text{for } 0.15<(\%C) \tag{11}$$

where C_R is the cooling rate (°C /s) and $(\%C)$ is the carbon content (wt % C)

The mean pearlite colony spacing of Ship building and Slab 1 steels, which corresponds to the secondary dendrite arm spacing were measured. From this, the cooling rates of Ship building steel and Slab 1 steel were calculated by equation (11). Because the Structural and Ship building steels were cast using the same caster with the same slab dimensions by Shougang, their cooling rates on solidification should be very similar. As the error in measuring SDAS from pearlite colonies was high, the cooling rate has been assumed to be the same for these two steels so that the SDAS of Structural steel was calculated by equation (11). The cooling rates and SDASs of all investigated steels were listed in Table II. Table II also includes the spacing of the square grid used for statistical SEM-EDS analysis as that needs to be greater than the SDAS for F-G ranking of the results.

Table II: Characteristic Second Dendrite Arm Spacing (SDAS) of Structural, Ship building and Slab1 steels and experimental parameters used for SEM-EDS grid scanning and modeling parameters predicted by Thermo-Calc software.

Steel	Structural	Ship building	Slab 1
Cooling rate (°C/s)	0.64	0.64	0.18
λ_{SDAS} (μm)	157	85	225
Number of F-G grid points	200	300	300
Grid point spacing (μm)	400	150	400
ΔT_S (K)	41.2	51.3	47.7
t_f (s)	64	80	265
Peritectic T (°C)	1481.4	1483.5	1485.9

Parameters Predicted by Thermo-Calc Software Considering the Effect of Microalloying Elements

For the constant cooling rate assumed in the present work, the local solidification time (t_f) simplifies to

$$t_f = \frac{\Delta T_S}{C_R} \tag{12}$$

where $\Delta T_S = T_L - T_S$ is the solidification temperature range, T_S is the solidus temperature. ΔT_S and t_f (s) of all investigated steels were listed in Table II, along with the equilibrium predicted peritectic temperature. In the case of the Structural steel this was taken from extrapolation of the δ-ferrite and austenite liquidus values. The volume fraction of δ ferrite as a function of temperature was simulated in order to determine the fraction of δ ferrite present when austenite was first predicted to form. Integrating equation (10) and the parameters in Table II; the δ phase fraction f_S which corresponded to the solute-depleted fractions were calculated at the predicted peritectic temperature and the corresponding solute-rich fractions were listed in Table III. Also, the predicted solute-rich fractions were greater than the pearlite area fractions, suggesting that pearlite colonies cannot represent all the solute-rich areas and some parts of the ferrite probably have higher alloying element contents. The Structural steel is predicted to contain a smaller amount of solute-rich material than the Ship building and Slab 1 steels. This is consistent with the lower carbon content (0.06%) increasing the liquidus temperature whilst lowering peritectic temperature. The δ-ferrite develops over a greater temperature range before λ-phase forms so that more partition occurs. The major difference between the Ship building and Slab 1 grades was the cooling rate, which gives little change in predicted solute-rich fraction consistent with the very rapid diffusion of C at these temperatures.

Experimental Profiles

Results for the F-G array of SEM-EDS spectra have so far been analysed to quantify the major alloying element analysis (Mn, Si and Ni). WDS will be undertaken to characterise the microalloying element contents. The ranking of the EDS results for these elements all showed similar trends, as shown by the [Mn] data in Figure 4. The compositional profiles of Mn in the Structural, Ship building and Slab 1 steels all showed a similar shape; the middle parts of the Mn compositional profiles were approximate straight lines and at the end of the lines the slopes of the curves started to change and increase until 100% solid fraction. The change of slope could be associated with a change in the degree of mixing in the liquid, e.g. as interdendritic regions get 'pinched off' or by a change effective partition coefficient. The latter would be consistent with the first appearance of austenite. Slabs with different SDASs would be expected to give 'pinch off' at different fractions solidified. The similarity of the plots for the Ship building and Slab 1 steels would support the change in slope is associated with austenite appearance. The change of slope was thus defined as the critical position of λ-phase forming at the interface of δ-ferrite and liquid, which separates the solute-rich and solute-depleted regions. These solute separation points are marked by the intersection points of the red dotted lines in Figure 4. At this point, λ-phase substitutes for δ-ferrite growing into liquid and also growing back into already-solidified δ-ferrite. The back diffusion of Mn atoms in λ-phase is much slower than that in δ-ferrite which would not allow the change in slope of the Mn compositional profile to be removed by back diffusion.

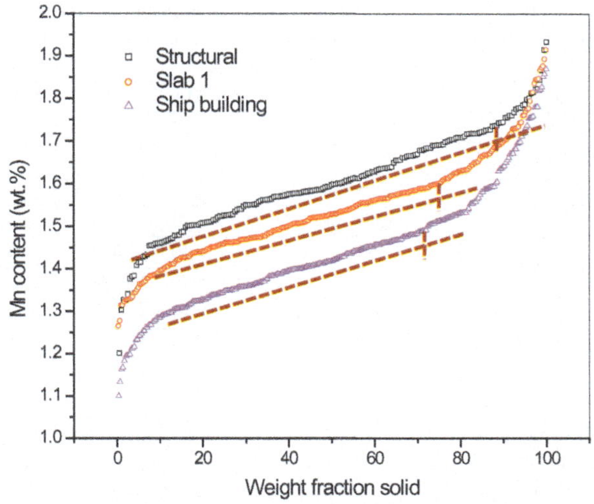

Figure 4: Mn content as a function of solid fraction in Structural, Ship building and Slab 1 steels.

Table III: Comparison of predicted and experimental solute rich fractions (%) and measured Mn contents at solute separating points and pearlite area fractions (%)

Steel		Structural	Ship building	Slab 1
Solute-rich fraction (%)	C-K prediction	10	25	26
	T-C prediction	0	20	16
	F-G measurement	12	28	24
Pearlite area fraction (%)		8.3	22.4	17.7

Measured solute-rich fractions in Table III agreed well with the C-K predictions within 2-3 %, whereas the liquid fractions, just above the peritectic temperature (basis of Thermo-Calc prediction) gives poor agreement. Thus, the Clyne-Kurz approach gives good correlation for the volume fraction of solute-rich material for this range of compositions and casting route.

Composition Comparison

The interfaces of δ-ferrite and liquid, δ-ferrite and λ-phase, γ-phase and liquid were assumed to be under equilibrium condition. Using equation (9), the elemental concentrations in δ-ferrite at the interface as a function of solid fraction and the boundary equilibrium conditions, the interface concentrations of solute elements in the δ-ferrite, λ-phase and liquid at peritectic temperature were calculated. These calculations treated the steels as a series of binary alloys with no mutual interaction effects between the alloying components and are listed in Table IV. The elements C, Ni and Mn ($k^{\delta/L} < k^{\lambda/L}$) are austenite-forming elements which can stabilize the austenitic structure and thus their contents in γ-phase were in between the values for δ-ferrite and liquid, allowing atom diffusion from liquid to δ-ferrite through γ-phase. Elements such as P, S, Nb, V, Ti are ferrite-forming elements tending to destabilize austenite resulting in lower contents in γ-phase than those in δ-ferrite and liquid, which prevented the further transportation from solute-rich region to solute-depleted regions.

Table IV: Clyne-Kurz prediction of interface elements concentratio at peritectic temperature of Structural, Ship building and Slab 1 steels

Steel		C	Si	Mn	Nb	Ni
Structural	δ	0.045	0.32	1.70	0.050	-
	λ	0.087	0.32	1.84	0.037	-
	L	0.295	0.52	2.59	0.274	-
Ship building	δ	0.068	0.36	1.30	0.001	0.010
	λ	0.131	0.35	1.40	0.001	0.011
	L	0.445	0.58	1.98	0.008	0.013
Slab 1	δ	0.047	0.30	1.39	0.033	0.516
	λ	0.092	0.29	1.50	0.024	0.583
	L	0.311	0.48	2.11	0.180	0.658

Measured Mn contents at solute separation points from the F-G sorting of the SEM-EDS data were used as a quantitative measure of the composition prediction, both C-K and T-C. The C-K prediction ignores interactions between different alloying elements, whereas the T-C prediction should deal with these, but at equilibrium. The predictions and measured [Mn] at the phase separation points in the three steels were listed in Table V.

Table V: Comparison of measured and predicted [Mn] contents (wt %) in austenite at the phase separation point (measured) and at the peritectic temperature (predicted)

Steel	Structural	Ship building	Slab 1
C-K	1.84	1.40	1.50
T-C	1.63	1.32	1.42
Measured	1.73	1.49	1.61

The agreement between measured and predicted composition values is less good than for the volume fraction of solute-rich material. In addition to the differences being greater, there is also no monotonic trend for the discrepancy, e.g. for the Structural steel the measured value falls between those for C-K and T-C predictions, whilst for the other two steels the measured values exceed both sets of predicted values. This would suggest that there are strong inter-alloying element effects that are concentration-dependent. This would be consistent with the trends shown in Table V for overall carbon level. A full quantitative prediction requires modification of the effective partition coefficients, based on the interactions accounted for by solution thermodynamic models.

Conclusions

1. The combination of a Clyne-Kurz-type mathematical model and Thermo-Calc has been used as a first step in modelling segregation behaviour in three continuously-cast microalloying steel slabs. Due to the low diffusivity of solute substitutional elements in γ-phase, the layer of γ-phase formed at the interface of δ-ferrite and liquid was used to separate the solute-rich and solute-depleted regions. Predicted solute-rich fractions are in agreement with experimental results measured by statistically significant SEM-EDS mapping (F-G sorting).
2. The interface concentrations of solute elements in the δ-ferrite, λ-phase and liquid at peritectic temperature were predicted by present model. The predicted interface Mn contents in λ-phase are lower than measured Mn contents at the solute separating point. The predicted composition need to be further verified.
3. The present model only took into account carbon diffusion. Microalloying elements such as Nb seemed to have little influence on the prediction of the solute-rich and solute-depleted fractions.

Acknowledgements

The authors would like to thank the School of Metallurgy and Materials at the University of Birmingham for providing the facilities required for the research and Tata Steel and Shougang for providing samples.

References

1. Davis, C.L. and Strangwood, M., *Preliminary Study of the Inhomogeneous Precipitate Distributions in Nb-Microalloyed Plate Steels.* J. Mater. Sci., 2002. 37(6): p. 1083-1090.

2. Wu, S.J. and Davis, C.L., *Effect of Duplex Ferrite Grain Size Distribution on Local Fracture Stresses of Nb-Microalloyed Steels.* Mater. Sci. Eng. A-Struct. Mater. Prop. Microstruct. Process., 2004. 387: p. 456-460.

3. Chakrabarti, D., Davis, C., and Strangwood, M., *Development of Bimodal Grain Structures in Nb-Containing High-Strength Low-Alloy Steels During Slab Reheating.* Metall. Mater. Trans. A-Phys. Metall. Mater. Sci., 2008. 39A(8): p. 1963-1977.

4. Chakrabarti, D., Davis, C.L., and Strangwood, M., *Characterisation of Bimodal Grain Structures and Their Dependence on Inhomogeneous Precipitate Distribution During Casting.* Materials Science Forum, 2005. 500-501: p. 613-620.

5. Choudhary, S.K. and Ghosh, A., *Mathematical Model for Prediction of Composition of Inclusions Formed During Solidification of Liquid Steel.* ISIJ Int., 2009. 49(12): p. 1819-1827.

6. Won, Y.M., et al., *Effect of Cooling Rate on Zst, Lit and Zdt of Carbon Steels near Melting Point.* ISIJ Int., 1998. 38(10): p. 1093-1099.

7. Won, Y.M. and Thomas, B.G., *Simple Model of Microsegregation During Solidification of Steels.* Metall. Mater. Trans. A-Phys. Metall. Mater. Sci., 2001. 32(7): p. 1755-1767.

8. Matsumiya, T., *Mathematical Analyses of Segregations and Chemical Compositional Changes of Nonmetallic Inclusions During Solidification of Steels.* Mater. Trans. JIM, 1992. 33(9): p. 783-794.

9. Natsume, Y., Shimamoto, M., and Ishida, H., *Numerical Modeling of Microsegregation for Fe-Base Multicomponent Alloys with Peritectic Transformation Coupled with Thermodynamic Calculations.* ISIJ Int., 2010. 50(12): p. 1867-1874.

10. Ueshima, Y., et al., *Analysis of Solute Distribution in Dendrites of Carbon-Steel with Delta-Gamma-Transformation During Solidification.* Metallurgical Transactions B-Process Metallurgy, 1986. 17(4): p. 845-859.

11. Davis, C.L. and Strangwood, M., *Segregation Behaviour in Nb Microalloyed Steels.* Mater. Sci. Technol., 2009. 25(9): p. 1126-1133.

12. Voller, V.R. and Beckermann, C., *A Unified Model of Microsegregation and Coarsening.* Metall. Mater. Trans. A-Phys. Metall. Mater. Sci., 1999. 30(8): p. 2183-2189.

13. Wang, C.Y. and Beckermann, C., *A Unified Solute Diffusion-Model for Columnar and Equiaxed Dendritic Alloy Solidification.* Mater. Sci. Eng. A-Struct. Mater. Prop. Microstruct. Process., 1993. 171(1-2): p. 199-211.

14. Sarreal, J.A. and Abbaschian, G.J., *The Effect of Solidification Rate on Microsegregation.* Metallurgical Transactions a-Physical Metallurgy and Materials Science, 1986. 17(11): p. 2063-2073.

15. Brody, H.D. and Flemings, M.C., *Solute Redistribution in Dendritic Solidification.* Transactions of the Metallurgical Society of Aime, 1966. 236(5): p. 615-&.

16. Clyne, T.W. and Kurz, W., *Solute Redistribution During Solidification with Rapid Solid-State Diffusion.* Metallurgical Transactions a-Physical Metallurgy and Materials Science, 1981. 12(6): p. 965-971.

17. Kobayashi, S., *Solute Redistribution During Solidification with Diffusion in Solid-Phase — a Theoretical Analysis.* Journal of Crystal Growth, 1988. 88(1): p. 87-96.

18. Ohnaka, I., *Mathematical Analysis of Solute Redistribution During Solidification with Diffusion in the Solid Phase.* Transactions of the Iron and Steel Institute of Japan, 1986. 26(12): p. 1045-1051.

19. Flemings, M.C., et al., *Microsegregation in Iron-Base Alloys.* Journal of the Iron and Steel Institute, 1970. 208: p. 371-&.

20. Gungor, M.N., *A Statistically Significant Experimental-Technique for Investigating Microsegregation in Cast Alloys.* Metallurgical Transactions a-Physical Metallurgy and Materials Science, 1989. 20(11): p. 2529-2533.

Proceedings of the 2013 International Symposium on Liquid Metal Processing & Casting
Edited by: Matthew J. M. Krane, Alain Jardy, Rodney L. Williamson, and Joseph J. Beaman
TMS (The Minerals, Metals & Materials Society), 2013

THE PRACTICAL APPLICATION OF MINOR ELEMENT CONTROL IN SMALL SCALE MELTS

P.D. Jablonski and J. A. Hawk
National Energy Technology Laboratory, 1450 Queen Ave. SW, Albany, OR 97321

Keywords: Tramp elements, Sulfur, vacuum induction melting.

Abstract

Many times small scale melts are made for the purposes of alloy development, component evaluation, or simply economic concerns when "commercial" alloys are unavailable in small quantities. Not only is it critical that the major alloy elements meet the desired levels, but "tramp" elements or trace element additions must also be controlled. Control of tramp and trace elements on the small scale is often done differently simply due to the scale of the melt or the equipment available. In this paper several approaches will be presented that have been used at NETL in manufacturing alloys for in-house research, including, for example, vacuum refining. Also, the relative effectiveness of various gettering elements will be explored. The successes achieved as well as the failures of the various approaches will be discussed in terms of thermodynamic and kinetic considerations. The presentation will conclude with practical alloy examples.

Background and Introduction

The control of tramp elements has been a concern for the foundry man for ages. The foundational control efforts are by the selection of high purity melt stocks. However, either by lack of availability or by extreme costs the melt formulation often times includes unwanted tramp elements such as oxygen, carbon and sulfur, for example. In nickel alloys, sulfur can be detrimental even at low levels. Sulfur is known to strongly concentrate at the grain boundaries (1). Sulfur enrichment at grain boundaries can be detrimental to oxide scale adhesion (2, 3). Furthermore, sulfur can have an adverse effect on grain boundary ductility (4) and has been found to embrittle the gamma/gamma prime interface (5). Finally, sulfur can form unwanted phases. In the case of alloys containing carbon and niobium, for example, it has been shown that M_2CS phases can form. These carbo-sulfide phases are low melting and embrittling phases (6, 7), which can be problematic during hot working and other subsequent operations.

Sulfur levels in iron and nickel-iron alloys have been reduced by induction melting in fused lime crucibles (8), while unwanted metalloid elements were removed by vacuum distillation (9). These metalloid elements, Bi, Sb, Sn, and As, along with Pb, also tend to concentrate at the grain boundaries similar to sulfur. The reduction of unwanted elements through vacuum distillation can be described by the Langmuir loss equation:

$$L = P\left(\frac{M}{RT}\right)^{0.5} \quad (1)$$

where L = *Langmuir Loss [g/(cm²s)], P = partial pressure of pure element at temperature [Pa = g/(cm.s²)], RT = gas constant times absolute temperature [erg/mole = g.cm²/mole], and M = molecular weight of element [g/mole].*

The time required to remove a particular element from the melt is given by:

$$t(T,C) = \frac{W}{LA} \quad (2)$$

where t = *time to lose a particular element from the melt [s]; a function of T and C, T = absolute temperature [K], C = mass fraction of element in melt, W = mass of element in the melt (taken as the melt weigh times the weight fraction) [g]* and A = *surface area of the melt (taken as the cross-sectional area of the crucible) [cm²].*

Combining equations 1 and 2, and accounting for the fact that in alloys it is the partial pressure of the component in solution, gives:

$$t(T,C) = \frac{W}{A}\left(\frac{RT}{M}\right)^{0.5}\left(\frac{1}{aP}\right) \quad (3)$$

where a = *activity of element in the melt* and P = *partial pressure of pure element at temperature [Pa = g/(cm.s²)].*

Experimental Approach

For the sulfur distillation experiments, high purity starting materials were used to formulate each heat with a target composition of Ni- 22 Cr (wt%). Approximately 6800 g of raw materials were used with the Cr formed into two compacts. The nickel was cut from 70 mm diameter bar stock. Each alloy was induction melted and held for the prescribed time and temperature. Three different superheat temperatures were employed: T_L + 50°C, T_L + 100°C, and T_L + 150°C. These melts were held for 20 minutes (T_L is the liquidus temperature). The effect of hold time was explored with a superheat of T_L + 100°C, i.e., hold times at this superheat were 10, 20 and 60 minutes. In experiments where a getter was used, it was either added to the melt once the prescribed temperature was reached or included with the charge materials. Each melt run was held for a certain time/temperature. These melts were poured into 70 mm diameter steel molds. After casting, a 2 mm thick slice was sectioned from each ingot approximately 40mm from the as-cast top. These slices were used for chemical analysis. The metal chemistries were determined by x-ray fluorescence (XRF) using a Rigaku ZSX Primus II, with NIST traceable standards (reported values are accurate to 0.01 wt%). Carbon and sulfur chemistries were determined with a LECO CS444LS using NIST certified standards (reported values are accurate to 10 ppm for C and 1 ppm for S). Oxygen and nitrogen chemistries were determined with a LECO TC436AR using NIST certified standards (reported values are accurate to 1 ppm). Trace element chemistries were determined with either the Varian 820 MS with precision to 0.02 ppm at 1 ppm, or the Varian Vista-Pro with precision to 0.001 ppm at 0.1 ppm and 0.1 ppm at 5 ppm again using NIST certified standards (reported values are accurate to 1 ppm).

Results

To make use of equation 3, the activity of S in a liquid metal melt pool was calculated with ThermoCalc employing the melt hold temperature and alloy formulation with the SSUB4 database appended onto the TCFe6 database. This, along with typical starting sulfur contents (0.01 wt%), the crucible diameter (9.5 cm) and the partial pressure of the pure element were the variables needed to make predictions. Figure 1 shows the sulfur loss time versus temperature (°C) for Ni-22Cr alloy with 100 ppm S which shows increasing time is required as temperature decreases. The loss rate is also a function of composition and decreases substantially with decreasing S content based on equation 3. Figure 2 shows the sulfur loss time versus sulfur content (at T_L + 100°C). This figure shows that the loss time increases substantially below about 20 ppm S based on equation 3. Both these figures predict that holding a melt for a few seconds to a few minutes at reasonable superheats should drop the sulfur content down to only a few ppm (Figures 1, 2).

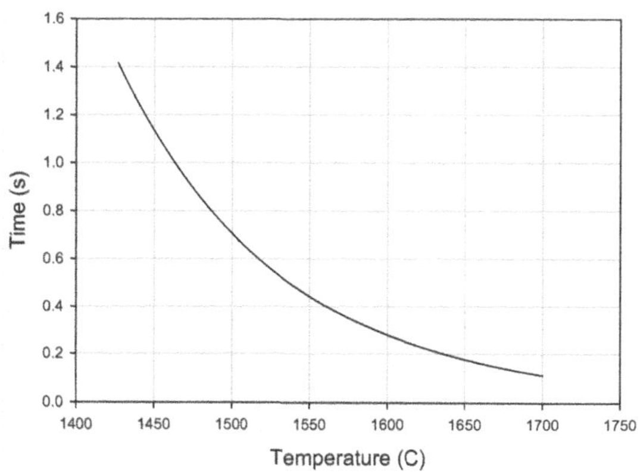

Figure 1: Schematic representation of sulfur loss rate versus temperature (°C) for Ni-22Cr alloy with 100 ppm S.

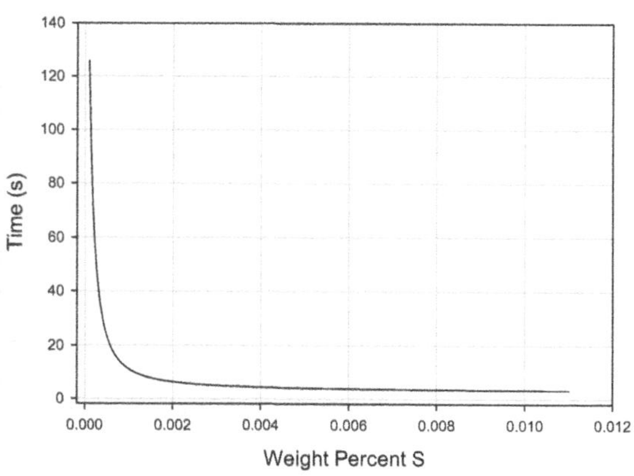

Figure 2: Schematic representation of the time it takes to reduce S level in a nominal Ni-22Cr alloy with 100 ppm S and T_L + 100°C.

To evaluate S removal with vacuum experimentally, Ni-22Cr melts were made employing the hold times and superheats described previously. The Ni remelt stock comes with a S content of 1-2 ppm. The Cr stock comes with 133 ppm S and was compacted into two "pucks". Thus, based on the input values, levels of roughly 30 ppm S are expected in the Ni-22Cr alloy after melting and solidification. However, historically the S level is closer to 100 ppm under normal melt conditions. The additional S was traced to a zirconia wash that was used to protect the mold. A low sulfur wash was employed for subsequent experiments.

The experimental conditions and resulting S content in Ni-22Cr alloys is shown in Table I. Three superheats (50, 100 and 150°C) with a 20 min hold were employed as well as three hold times (10, 20 and 60 min) for a total of 5 conditions. The first heat was melted in a small VIM furnace. Due to a furnace excursion during a subsequent melt, this furnace was unavailable for the remaining experiments. Thus, the first melt conditions were repeated in a second (larger) VIM furnace (although it employed the same size crucible). This same larger furnace was used for all subsequent melts of this series. These experiments show that reducing S to levels much below 40 ppm was not possible using the standard VIM melt practice and that Equation 3 is not a very good predictor for S distillation. Not surprisingly, longer melt hold times and/or higher superheat temperature led to more Cr loss (although not significant). The more extreme conditions also led to lower carbon contents, presumably by CO boil rather than vapor loss of C. Since the goal was to reduce S to as low a level as possible, preferably 10 ppm or lower, the effect of chemical additions to "getter" S (i.e., tie up and remove) was next evaluated.

Table I: Experimental melt conditions and subsequent chemistries.

Experiment	Cr (w/o)	S (ppm)	C (ppm)	O (ppm)	N (ppm)
50°C/20 min	21.74	50	28	45	61
50°C/20 min	21.74	43	43	173	46
150°C/20 min	21.57	43	<10	59	27
100°C/60 min	21.67	43	16	90	12
100°C/20 min	21.67	39	12	38	17
100°C/10 min	21.63	41	18	42	22

Besides using a vacuum to distill sulfur it may also be possible to remove sulfur by "gettering" with an element that has high S affinity. Such elements include Group II elements Mg, and Ca, the lanthanides, including La and Ce as well as Y. The following will be a brief presentation of additional thermodynamic considerations. Once again, the base alloy chemistry will be a simple binary of Ni-22 Cr. The addition of 0.1 atomic percent of the following sulfur active elements from groups IIA and IIIA of the periodic table (Mg, Ca, Y, La, and Ce) were considered at 1477°C, a temperature at which the base material is fully molten. The thermodynamic results are shown in Figure 3 as a plot of W(S) (weight fraction sulfur) versus BPW (weight fraction phases). Thus, a value of 10^{-6} W(S) is 1 ppm S by weight. Below about 0.325 wt% S the only sulfides that remain are those formed from Groups IIA and IIIA elements. The amounts of each that form can then be used to create a ranking as to their effectiveness in gettering S: La > Ce > Y > Ca > Mg. The critical sulfur level, below which the respective sulfides no longer form (when all these elements are present), can then be used to create yet another ranking: Ca > La > Y > Mg> Ce. Thus, La was favored by one measure while Ca was favored by another.

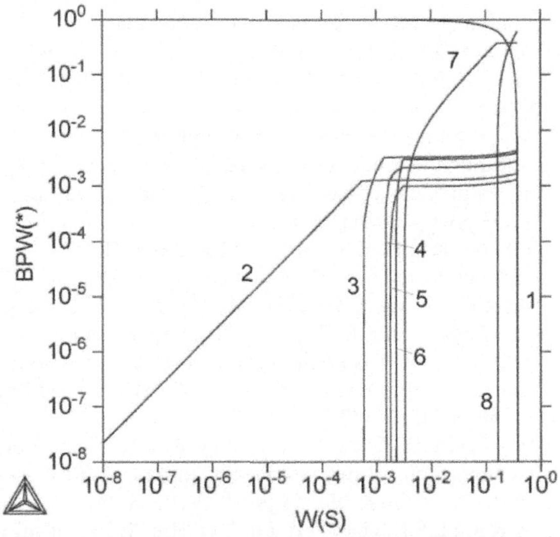

Figure 3. The thermodynamic results of W(S) (weight fraction sulfur) versus BPW (weight fraction phases) are shown in the plot above. The phase identities are as follows: 1: liquid, 2: CaS, 3: La_2S_3, 4: YS, 5: MgS, 6: CeS, 7: CrS, 8: NiS.

Since handling elemental Ca can be somewhat problematic the combination of Mg, Y, La, and Ce (i.e., no Ca) was evaluated. Interestingly, La appeared to rank highest using both measures, forming compounds at the lowest S levels evaluated and forming the highest weight fraction (Figure 4).

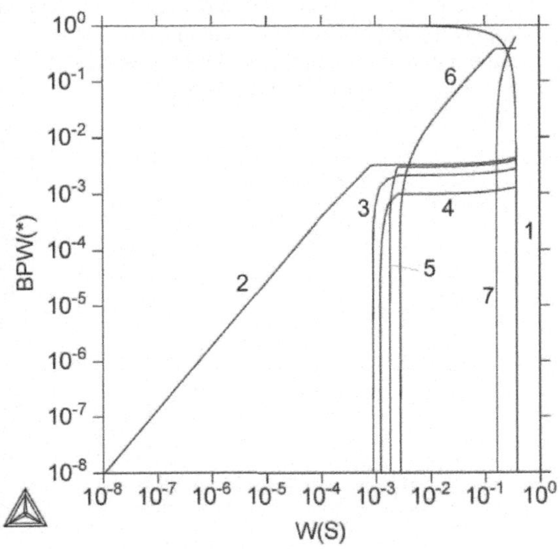

Figure 4. The thermodynamic results of W(S) (weight fraction sulfur) versus BPW (weight fraction phases) are shown in the plot above. The phase identities are as follows: 1: liquid, 2: La_2S_3, 3: YS, 4: MgS, 5: CeS, 6: CrS, 7: NiS.

To evaluate S removal with melt additions of sulfur-active elements, additions were made to Ni-Cr melts. The same Ni and Cr remelt feedstocks used in the vacuum distillation experiments were used in these experiments. All were added after a full molten pool was established with a temperature of T_L + 50-100°C. One experiment with Mg was run with the addition contained within the Cr compact. Magnesium and Y were added as elemental additions. Additions of master alloys of La (NiCrLa; 32.3 wt% La), Y (NiCrY; 10 wt% Y), and Mg (NiMg; 14.4 wt% Mg) were also made. The master alloys were useful in several ways. In the cases of Mg, La, and Y, alloying increased the density of the addition easing its incorporation into the melt rather than allowing it to sit on the top of the melt pool allowing it to oxidize rather than react with the melt. It was also felt that the NiMg master alloy would have a lower Mg vapor pressure (lower Mg activity). The added benefit of the Y master alloy is that it had a greatly reduced melting temperature, thereby easing its incorporation into the melt. All experimental melts were held at temperature for 10 minutes following the master alloy addition. Table II contains the experimental details and the results of these melts.

Table II: The experimental conditions and resulting S and residual active element contents in Ni-XCr (where X = 22, 30, 40 & 50) alloys are shown below. Each melt was made under vacuum with very low leak rates.

Experiment	S (ppm)	Retained (ppm)
Ni-22Cr		
2.0 g Mg	54	10 Mg
25.6 g IncoMg 1	59	110 Mg
2.0 g Mg in ea. Cr	31	18 Mg
26.5 g NiCrLa	4	102 La
7.1 g Y	23	247 Y
23.3 g NiCrY	14	246 Y
Ni-30Cr		
35.1 g NiCrLa	5	106 La
Ni-40Cr		
46.2 g NiCrLa	33	106 La
Ni-50Cr		
35.1 g NiCrLa	9	227 La
25.4 g NiCrLa	67	61 La
35.1 g NiCrLa	7	118 La

These experiments showed that the best results were obtained with the addition of a La containing master alloy. Additional observations included: Mg tends to vaporize, which may reduce its effectiveness. Use of the NiMg master alloy appeared to reduce the Mg vapor pressure (i.e., less Mg loss). Incorporating the Mg into the Cr compacts may have helped to reduce S, however, this may also be attributable to the increase in Mg addition (2x). Additions of Y (elemental or master alloy) did not prove as effective in reducing S and there was the secondary problem of retaining more of the S active element (the preference is that the S be removed and no other additions remain).

Interestingly, addition of the Y master alloy to more complex alloys was very effective in reducing S, so more work in this area may be warranted. It may be that Y reacts with S in the Ni-Cr binaries forming compounds but these remain in the melt (which leads to the higher retained Y values). In the more complex alloys the YS that forms in the melt may be modified by the other alloy constituents, which in turn modifies the tendency of the sulfide to wet or otherwise stick to the crucible. Further work is also needed to verify this. Some variation in the S results was observed with the NiCrLa master alloy which may be attributed to the level of superheat at the time the addition is made. It was observed that lower superheat temperature resulted in less sulfur removal. Again this might be attributable to reduced reaction kinetics or perhaps less than adequate removal of the sulfide particles.

The sulfur reduction strategy has been applied to other model alloys. These alloys are nominally Ni-22Cr-XMo-0.05C with X=0, 1, and 8 wt%. These alloys were made by utilizing some of the low sulfur heats shown in Table II, along with additional amounts of Cr, Mo and C to bring the chemistry up to the desired levels. An effort was also made to eliminate the retained La. This was achieved by making a NiO addition. This addition was strategically placed in the lower level of the crucible while the C addition was placed at the upper level since C can easily reduce NiO. The resulting chemistries of the three heats are given in Table III. As can be seen, S remained low and the La was also removed. Thus, the main objectives were achieved.

Table III: Chemistries of experimental Ni-22Cr-XMo-0.05C alloys are given below. Cr, Mo, and C are given in weight percent while N, O, S and La are given in ppm.

	Cr	Mo	C	N	O	S	La
Ni22Cr0Mo	22.18	0.016	0.04	49	8	11	21
Ni22Cr1Mo	22.04	0.99	0.05	54	14	4	3
Ni22Cr8Mo	22.04	7.77	0.05	64	11	11	3

The NiCrLa master alloy is effective in reducing sulfur contents in Ni-Cr melts. Based on calculations ESR, with the proper selection of slag components (e.g., Ca) can be used as an alternative way to remove S and provide low S melt stock for VIM melting. Further exploration of these methods is planned in the future.

Disclaimer

"This report was prepared as an account of work sponsored by an agency of the United States Government. Neither the United States Government nor any agency thereof, nor any of their employees, makes any warranty, express or implied, or assumes any legal liability or responsibility for the accuracy, completeness, or usefulness of any information, apparatus, product, or process disclosed, or represents that its use would not infringe privately owned rights. Reference herein to any specific commercial product, process, or service by trade name, trademark, manufacturer, or otherwise does not necessarily constitute or imply its endorsement, recommendation, or favoring by the United States Government or any agency thereof. The views and opinions of authors expressed herein do not necessarily state or reflect those of the United States Government or any agency thereof."

References

1. S. Floreen, J.H. Westbrook, "Grain boundary segregation and the grain size dependence of strength of nickel-sulfur alloys", Acta Met 17, pp1175-1181, (1969).
2. M.A. Smith, W.E. Frazier, B.A. Pregger, "effect of sulfur on the cyclic oxidation behavior of single crystalline, nickel-base superalloy", Mat Sci and Eng A203 pp383-398, (1995).
3. J.A. Nychka, D. R. Clarke, and Gerald H. Meier, "Spallation and transient oxide growth on PWA 1484 superalloy", Mat Sci and Eng A 490 pp359-368 (2008).
4. A.J. Ramirez and J.C Lippold, "High temperature behavior of Ni-base weld metal part II – insight into the mechanism for ductility dip cracking", Mat Sci and Eng A 380 pp245-258 (2004).
5. K. Chen, L.R. Zhao, J.S. Tse, "Sulfur embrittlement on gamma/gamma prime interface of Ni-base single crystal superalloys", Acta Mat 51 pp 1079-1086 (2003).
6. W.R. Sun, S.R. Guo, D.Z. Lu, Z.O. Hu, "Effect of sulfur on the solidification and segregation in Inconel 718 alloy", Mat Lett 31 pp195-200 (1997).
7. J. Dong, M. Zhang, Xishan Xie, R.G. Thompson, "interfacial segregation and cosegregation behavior in a nickel-base alloy 718", Mat Sci and Eng A 328 pp8-13 (2002).
8. P.V. Strushchenko, B.N. Putimttsev, Yu. A. Gratsianov, and A.G. Ryl'Nikova, "Vacuum induction furnace refining of iron and iron-nickel alloys in a fused lime crucible", Steel in the USSR Volume: 7, Issue: 4, pp221-224 (1977).
9. B.G. Chernov, P.Y. Ageev, "Behaviour of lead bismuth antimony tin and arsnic in iron and nickel in melts in vacuum induction furnace", STAL IN ENGLISH-USSR Issue: 11 pp939-1004 (1968).

Proceedings of the 2013 International Symposium on Liquid Metal Processing & Casting
Edited by: Matthew J. M. Krane, Alain Jardy, Rodney L. Williamson, and Joseph J. Beaman
TMS (The Minerals, Metals & Materials Society), 2013

A TWO-DIMENSIONAL INFILTRATION DYNAMICS MODEL OF C-SiC COMPOSITES

K. Iqbal [1]*, J. Sha[1] and A. Maqsood[2]
[1]State Key Laboratory of Structural Analyses for Industrial Equipment, School of Aeronautics and Astronautics, Dalian University of Technology, 116024 Dalian, China
[2]Centre for Emerging Sciences, Engineering & Technology, Islamabad, Pakistan

Keywords: Composites, Silicon infiltration

Abstract

This article presents the mathematical modeling of Si infiltration material in a porous carbon fiber preforms. Capillary infiltration kinetics in the limit of both interface and diffusion control for situations where the capillary radius decreases with time, and the contact angle was assumed constant during infiltration were studied. One major manufacturing process of the carbon fiber reinforced SiC composite materials is the liquid silicon infiltration (LSI) of a porous carbon fiber preforms to form C-SiC composites. The results show that the large-radius (r_0 = 5 μm) capillary yields faster infiltration kinetics and longer penetration lengths than the small-diameter (r_0 = 1 μm).

Introduction

Ceramic matrix carbon fibre (CMC) reinforced composites today are being developed for the application in high-performance structures [1]. In addition to, high-performance ceramics are gaining increasing attention because (i) the good oxidation resistance of the ceramic matrix makes the composites attractive for high-temperature applications, and (ii) continuously decreasing price of carbon fibers makes the use of carbon fiber reinforced composites economically feasible [2, 3]. C-SiC composites are used for nose tip of reusable space vehicles, leading edges of hypersonic vehicles, jet-vanes for thrust vectoring and brake disc of high speed automobiles [1-3]. One major manufacturing process of the carbon fiber reinforced SiC composite materials is the liquid silicon infiltration (LSI) of a porous carbon fiber preforms to form C-SiC composites. The processing by LSI consists on the infiltration of a porous carbon fiber preforms with molten silicon using a conventional furnace. Carbon and silicon react to form SiC at temperatures in the range of 1450-1650 °C, and dwell time between 1-3h under vacuum [4]. Understanding of infiltration dynamics of molten silicon in carbon fiber preforms is essential for the development of LSI based C-SiC composites. Direct observation of silicon infiltration in carbon fiber preforms is not possible; because of high temperature and vacuum conditions. Therefore mathematical modeling is the only way to get close results with the experimental values.
This article presents mathematically modeling of the capillary infiltration kinetics in the limit of both interface and diffusion control for situations where the capillary radius decreases with time, and the contact angle was assumed constant during infiltration. In this system, pore shrinkage caused due to the deposition of a product phase of a higher specific volume alters the resulting interface [4].

Analysis

In this section, model considerations will be presented to analyse the reactive wetting in the liquid-silicon/solid-carbon system. A two-dimensional schematic view of the infiltration dynamics model of C-SiC composites is shown in Fig. 1. We let σ is the surface tension, r is the pore radius, g the gravity, ρ is the fluid density, μ is the fluid viscosity, and h is the infiltration height. The differential equation of motion for capillary rise under quasi-steady state can be written as:

$$h\frac{dh}{dt} = \frac{\rho r^2}{8\mu}\left(\frac{2\sigma\cos\theta}{\rho r} - gh\right) \tag{1}$$

Interface-limited capillary rise

As mentioned earlier, the pore sizes shrink with time as solid SiC is formed, and the reaction kinetics can be expressed by the following relationship:

$$r(t) = r_0 - kt \tag{2}$$

Here r_0 is the initial capillary radius, k is the reaction-rate constant, and t is the time. Substituting for r(t) in eq. [1] with g = 0 and θ = θ_∞, gives

$$\frac{4\mu}{\sigma}hdh = \left(r_0 - kt\right)\cos\theta_\infty dt \tag{3}$$

Taking the integration in eq. [3], gives

$$\frac{4\mu}{\sigma}h^2 = \cos\theta_\infty\left(2r_0t - kt^2\right) \tag{4}$$

Diffusion-limited capillary rise

For the diffusion-limited capillary rise, the capillary radius decreases parabolically with time by the following relationship:

$$r(t) = r_0 - m\sqrt{t} \tag{5}$$

where m is a parabolic-reaction rate constant, and r_0 is the initial capillary radius at t = 0. Substituting for r(t) in eq. [1] with g = 0 and θ = θ_∞, gives

$$\frac{4\mu}{\sigma}hdh = \left(r_0 - m\sqrt{t}\right)\cos\theta_\infty dt \tag{6}$$

Taking the integration in eq. [6], gives

$$\frac{6\mu}{\sigma} h^2 = \cos\theta_\infty \left(3r_0 t - 2mt^{\frac{3}{2}}\right) \tag{7}$$

Results and Discussion

According to eqs. [4] and [7], infiltration length and penetration length of porous carbon fiber preforms of two different initial radii (r_0 = 1 μm and r_0 =5 μm) have been done and analyzed (Figures 2). The experimental Si infiltration heights were calculated [4] at 1650 °C for different durations ranging from 6 to 180 s (Table 1). Figures 2 show the theoretical Si infiltration profiles in capillaries of initial radii, (r_0 = 1 μm in (a) and (b) and r_0 = 5 μm in (c) and (d)); based on (i) interface-limited (Eq. 4), and (ii) diffusion-limited capillary rise (Eq. 7) with constant $\theta = \theta_\infty$ and variable r. The values of k, m, θ_∞, σ, and μ were used in the simulation for the Si/C couples at 1480 °C, and reported in [4, 5].

Conclusions

Capillary infiltration kinetics under the limiting cases of interface and diffusion control has been derived for the reactive-infiltration phenomenon. Understanding of infiltration dynamics of molten silicon in carbon fiber preforms is essential for the development of LSI based C-SiC composites. A new feature of the model is the inclusion of the pore shrinkage caused due to the deposition of a product phase. The results show that the large-radius (r_0 = 5 μm) capillary yields faster infiltration kinetics and longer penetration lengths than the small-diameter (r_0 = 1 μm).

References

[1] W. Krenkel, and F. Berndt, "C/C-SiC composites for space applications and advanced friction systems," Materials Science and Engineering A, 412 (2005), 177-181.

[2] H.A. El-Hija, W. Krenkel, and S. Hugel, "Development of C/C-SiC brake pads for high-performance elevators," International Journal of Applied Ceramic Technology, 2 (2005), 105-113.

[3] J.P. Wang, J.M. Qian, and Z.H. Jin, "Microstructure of C/C composites prepared by chemical vapor infiltration method with vaporized kerosene as a precursor," Materials Science and Engineering A, 419 (2006), 162-167.

[4] S. Kumar, A. Kumar, R. Devi, A. Shukla, and A.K. Gupta, "Capillary infiltration studies of liquids into 3D-stitched C-C performs Part B: Kinetics of silicon infiltraion," Journal of the European Ceramic Society, 29 (2009), 2651-2655.

[5] A.G. Odeshi, H. Mucha, and B. Wielage, "Manufacture and characterization of a low cost carbon fibre reinforced C/SiC dual matrix composite," Carbon, 44 (2006), 1994-2001.

Table I. The experimental Si infiltration heights by C-C preforms

S. No.	Infiltration time, s	Infiltration height, m
1	6	0.032
2	18	0.066
3	24	0.070
4	30	0.072
5	36	0.074
6	48	0.088
7	90	0.118
8	108	0.121
9	138	0.126
10	180	0.148

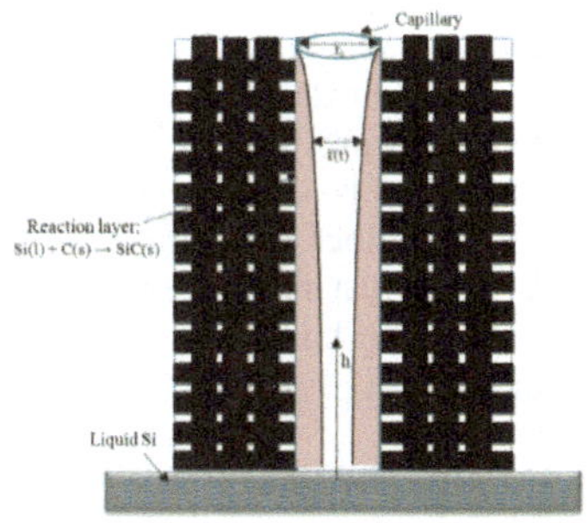

Figure 1. A two-dimensional schematic view of the infiltration dynamics model of C-SiC composites.

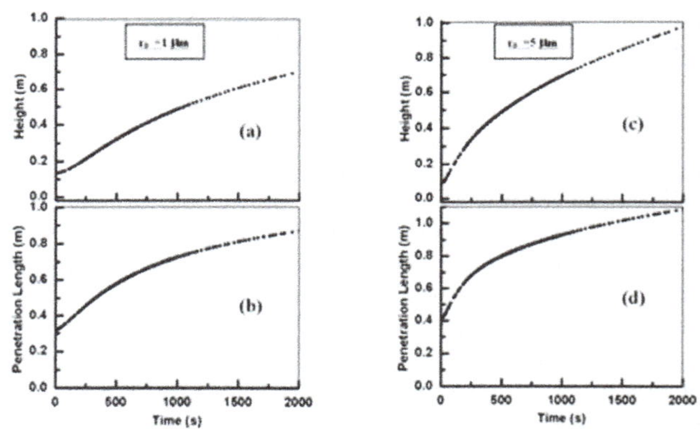

Figure 2. Theoretical Si infiltration profiles in capillaries of initial radii, (r_0 = 1 μm in (a) and (b) and r_0 = 5 μm in (c) and (d)); based on (i) interface-limited (Eq. 4), and (ii) diffusion-limited capillary rise (Eq. 7) with constant $\theta = \theta_\infty$ and variable r.

Proceedings of the 2013 International Symposium on Liquid Metal Processing & Casting
Edited by: Matthew J. M. Krane, Alain Jardy, Rodney L. Williamson, and Joseph J. Beaman
TMS (The Minerals, Metals & Materials Society), 2013

INVESTIGATION OF PECULIARITIES OF DEEP DEOXIDIZED STEEL INTERACTION WITH REFRACTORY LINING OF LADLE DURING THE MANUFACTURING OF LARGE FORGE INGOTS WITH MASS UP TO 415 TON

M.V. Efimov[1], A.A. Selyutin[1], P.M. Yavtushenko[1], A.A. Troyanskyy[2], A.D. Ryabtsev[2], V.V. Pashynskyi[2]
[1]PJSC "Energomashspetsstal", Kramatorsk, Donetsk Region, Ukraine, 84306
[2]State High Educational Institution, Donetsk National Technical University, Artyoma str. 58, Donetsk, Ukraine, 83001

Keywords: deep deoxidized steel, nonmetallic inclusions, exogenous inclusions, slag – metal – refractory system, refractory destroying

Abstract

In the modern power engineering machinebuilding the high demands are made for the quality of forged semiproducts for producing of different parts. It is possible to guarantee their high quality only at conditions of metallurgical defects absence in initial ingots before forging.

In result of fulfilled investigations of deep deoxidized steel quality for producing of large-size ingots, the evidence of secondary contamination of steel by exogenous nonmetallic inclusions were obtained. In order to determine the source of contamination, the chemical and phase composition of inclusions were investigated. It is established that they are formed in result of complex interaction in "slag – metal – refractory" system. The mechanism of system component interaction at three levels of ladle which corresponds to contact of refractory lining mostly with slag, metal or slag-metal was studied. With using of petrography analysis and scanning electron microscopy techniques the peculiarities of refractory destroying and metal contamination were shown.

Introduction

Currently, one of the conditions of the stable development of power generation industry is providing of new generation units with high reliability in operation. Therefore, the modern power engineering is requiring increased quality for original ingots (forgings) for producing of modern equipment components. It is possible to guarantee the high quality of forgings through the absence of metallurgical defects in the initial ingot. To do this, it is necessary to get pure on harmful impurities steel with dispersed and uniformly distributed non-metallic phase. The quality of material, a complete resource of service characteristics of parts are formed greatly during melting and casting. This may be provided by the application of modern metallurgical engineering and technology[1-3].

This approach is realized in new productive route to provide a high level of metallurgical warranties of ingot in one processing module, using the productive unit structure, which is universal. This approach to the production is implemented in electric steelmaking shop of Public Joint Stock Company "Energomashspetsstal" (PJSC "EMSS") during melting of different grades of steel, which is poured into large ingots weighing up to 415 tons for the production of turbines shaft, rolls for rolling mills and other large products [4-6]. To implement this approach, it was necessary to transform the production of large steel ingots to a modern technological system on PJSC "EMSS". But new technology has many features, which may have the influence on final content and distribution of non metallic phase. Current work is devoted to analysis of developed productive route efficiency in terms of cast metal purity from nonmetallic inclusions.

Multifunctional Modular Production Scheme for Large Size Ingots Production

The modern multifunctional modular production scheme was taken as the basis for technical modernization of PJSC "EMSS", which is based on the application of integrated production units, allowing the sequence of heating, refining, deoxidizing, alloying the steel in the ladle. In general, production scheme include:

The melting of carbon or alloyed semi-product in electric arc furnaces EAF-50t and EAF-100t, including the melting of charge materials, metal dephosphorization and its heating to a temperature of pouring.

Slag-free pouring of semi-product in teeming ladle with filling to the last one de-oxidants and slag materials (if necessary, fulfillment of the operations of alloying metal is possible) in combination with steel steering by blowing of inert gas via elements, mounted in the bottom of the ladle.

Secondary treatment technology in the ladle furnace, comprising:

- arc heating of melt in the ladle in order to compensate the loss of heat during processing of the melt at atmospheric pressure and under vacuum;
- a continuous steering of the melt during all process steps of secondary treatment in order to equalize the steel temperature and the intensify the mass transfer processes during desulfurization, diffusion and sediment reduction, alloying and degassing;
- -injection under the slag of powder materials in a stream of gas through uncooled expended lance to adjust the carbon content;
- feeding into the metal the reagents in the form of a powder wire with wire feeder for deep steel deoxidation and modification of non-metallic inclusions.

Vacuum processing of deoxidized metal under the refining slag in chamber type degasser, with the possibility of additional alloying and adjusting the chemical composition (about 60% of the steel subjected to evacuation to reduce the hydrogen content of less than 1.5-2.0 ppm and 30% - to decrease the content of non-metallic inclusions).

Vacuum oxygen decarburization;

Pouring of steel into molds (including under the vacuum in the vacuum chamber) to obtain the required quality of surface and internal structure of a large ingot while reducing the consumption of the metal, as well as to achieve economies of molds, pallets and refractories;

Large ingot solidification and its cooling in the mold, stripping of ingot and transportation for later processing.

The Technology of Production of Forging Ingots with Weight 415 Tons

The melting of steel is fulfilled in electric arc steelmaking furnaces of traditional design EAF 50t and EAF 100t, where deep dephosphorization is realized. After the mass fracture of phosphor becomes not more than 0.004%, oxidizing slag is removed. During tapping deoxidizing is fulfilled by aluminium additions and heat insulation slag mixtures are introduced in ladle.

Collection of metal for 415 ton ingot pouring is realized by consequent tapping of heats from EAF 50t furnace via transporting ladle to the teeming ladle and from EAF100t - in teeming ladle directly. Ladle used at the LFU ladle is the same as the degassing ladle and the teeming ladle Teeming ladle is installed on special stand, after that the metal from transporting ladle is transferred to it through the damper. Teeming ladle, filled by metal, is transported to the out-of-furnace treatment facility. To collect necessary quantity of metal, metal in the every ladle after teeming from the furnace placed subsequently in LFU and degasser. The consequent pouring of steel was carried out from four teeming ladles in vertical mould under a vacuum through a tundish with protection of stream by argon. The minimal residence time of molten metal in the teeming ladle was .... and maximal - .....

Treatment of steel at ladle-furnace unit is fulfilled under highly basic deoxidized refining slag in order to deep desulfuration, compensation of heat losses during treatment, correction of chemical composition and providing of overheating of metal before the degassing. The composition of the ladle treatment of metal of electro steelmaking shop includes ladle furnace and vacuum chamber

It is necessary to consider a substantial increase in the residence time of melting in steel ladles. This is for two reasons:

1. Purity of steel for harmful impurities, dispersed and uniformly distributed structure of the non-metallic phase ensures rather long time of ladle treatment. The average duration of treatment in the Ladle-Furnace Unit (LFU) is 141 and 88 minutes, and Vacuum Degasser (VD) - 57 and 47 min, respectively, for 100 and 50-ton ladles. The average processing time of melting in aggregates of ladle metallurgy is about 198 and 135 minutes for the 100 and the 50-ton ladles, respectively.

2. Low productivity of arc steelmaking furnaces leads to increase of storage time for casting of metal ingots, weighing more than 100 tons. The average waiting time between the tapping and the start of treatment in the LFU is for the 100 and 50-ton steel ladle 84 and 51 minutes, and between the LFU and VD - 21 and 10 min, respectively. The overall average waiting time - 105 and 61 minutes for the 100 and the 50-ton ladles, respectively.

High velocity of desulfuration is provided by low oxidizing of melt from the first minutes of treatment [4]. Velocity and degree of desulfuration of melt increases with intensity of argon blowing through the aperture plugs [5].

Parallel to heating and desulfuration of the steel, the correction of alloying elements content is fulfilled. Adding of ferroalloys is made at increased argon consumption. Maximal portion of ferroalloys is not more than 200 kg. After ferroalloys addition and (if necessary) carburizing, the averaging argon blowing is fulfilled for period not less than 5 minutes after that the probe of steel for chemical composition determination is picked up.

Exposure of metal in ladle from finishing of treatment in LFU to beginning of degassing usually does not exceed 30 min.

After obtaining of necessary chemical composition and temperature, the steel teeming ladle with metal transferred for degassing. To provide the good degassing the refining slag is partially removed

Degassing of metal is realized in two stages. At first step, the liquid metal in the ladle is subjected to treatment in chamber degasser. Second, finishing step is fulfilled during the pouring of metal in mould, placed in vacuum chamber.

Pouring of metal after treatment in degasser is fulfilled in vacuum in special chamber too. For casting of ingots with 415 ton by mass. the special mould was designed and manufactured. It provides the good formation and crystallization of ingot.

Ingot pouring is fulfilled from the top at vacuum less than 1.5 mBar through the intermediate stopper type tundish with mullite – corundum lining. Corundum barrels with diameter 55 mm are used and velocity of pouring in this case consists 7.0 – 7.5 ton/min. The process of pouring is monitored through the special looking windows in the lid of vacuum chamber. The parts of casting mould are heated before the assembling to the temperature more than 100 °C. In bottom plate of mould the cast metal insert was mounted. In order to decrease the liquation processes in ingot the multipouring technology was used. It based on the principle of different metal composition concerning C and Mo obtaining in different ladles. To decrease the liquation, the low content of such elements as P (0.009%), S (0.005%) and Cu (0.19%) was obtained. After pouring, the top part of ingot was warm-kept by additions of heat insulating mixtures in quantity 2 kg per ton of ingot.

Investigation of Non-Metallic Inclusions in Large-Mass Ingots

One of the important questions of the technology is quality inspection. To do this, the laboratory was organized at the plant and all types of research are conducted: from the chemical analysis to non-destructive testing.

Taking into account large size of ingots, three variants of samples selection were used:

- sampling of liquid metal at different steps of liquid metal treatment;
- sampling from specially designed points of ingots or forged semiproducts;
- sampling from the points, where defects were found by ultrasonic inspection.

For evaluation of inclusion content, the methods of optical microscopy and SEM microscopy were used. Scanning electron microscope JSM-6490LV (JEOL, Japan) was equipped by:

-energy dispersive spectrometer INCA Penta FETx3 (Oxford Instruments, England),

- wave spectrometer INCA Wave (Oxford Instruments, England)

- detector of backscattered electron diffraction HKL (Oxford Instruments, England).

Quantitative determination of contamination by non-metallic inclusions was conducted with the use of software package "Videotest" (programmatic module «Metal 1.2») in the automatic and manual modes.

Procedure of ultrasonic inspection was performed on forged billet. Specimens for investigation of non-metallic phase were sampled from the specially designed points of billet or from the sections, where inadmissible defects were found by ultrasonic inspection technique. Analysis has shown that most informative results were obtained from samples, selected on the basis of ultrasonic inspection results., because in these points the maximal relative content of different nonmetallic inclusions was found. .

Preliminary analysis has shown that at average concentration of S and P less than 0.005 – 0.007% and deep deoxidizing provide very low volume fraction of inclusion (less than 0.05% by vol.). Typical manganese sulphide of spherical shape is shown on Figure 1. It was detected single sulphides only, usually not in every field of view. Average size of such type of inclusion did not exceed 10 microns.

Picture, mentioned above, is typical for stable technology. At these conditions the main fracture of inclusions was represented by single inclusions with sporadic location on micro-, meso-, and macro levels. Mostly the formation of such inclusion deals with sporadic deviation of manufacturing process. Statistical analysis has shown that in practice of PJSC "EMSS" this defect formed very rarely.

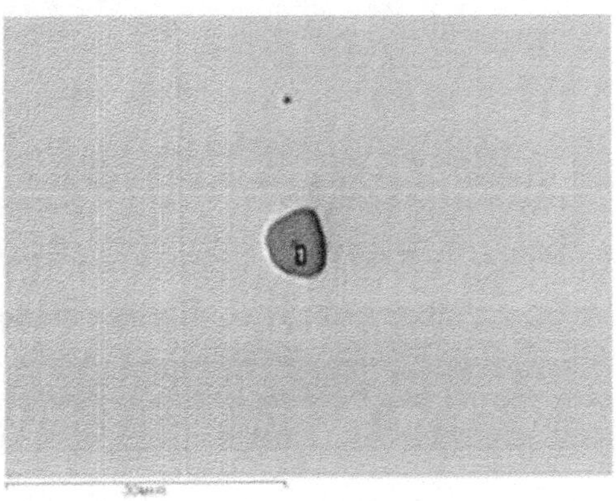

Figure 1. The general SEM view of manganese sulphide inclusion (1) in investigated steel

But simultaneously with endogenous manganese sulphides, another variant of inclusions was found. Typical their view is shown on Figure 2.

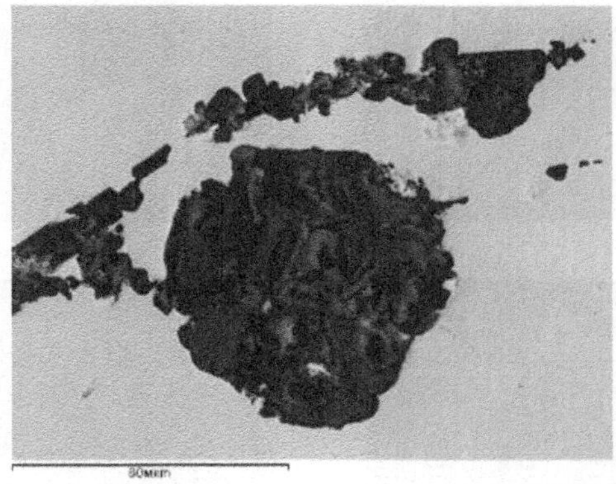

Figure 2. The general SEM view of combined exogenous inclusion in investigated steel

Main difference deals with morphology and chemical composition of these inclusions. They are non homogenous and have relatively large size and complex nonuniform composition. Distribution of chemical elements in complex inclusion from previous figure is shown on Figure 3. As it is seen from the Figure 3, the inclusions are oxide phases primarily of aluminum, calcium and silicon. In addition, the clusters of manganese sulphide and magnesium compounds are present in inclusions. The difference between the inclusions of round and elongated morphology is different quantitative ratio of the individual components. It can be assumed that the formation of rounded inclusions is due to coagulation of aluminates and silicates under the influence of calcium compounds in the compact conglomerates and flotation of inclusions in the liquid metal, while the elongated chain of inclusions formed in the interdendritic spacing during crystallization.

Compounds of calcium are not included in the composition of the steel and can get into the metal from the slag phase or from refractory lining. Oxides of aluminum and silicon may have an endogenous origin (formed as a result of the processes of deoxidation) and exogenous (result of refractory destroying). In the in favor of this hypothesis is the large size of the particles (especially aluminum oxide), and the presence in a non-metallic phase of magnesium oxide, which can get into metal only from the refractories.

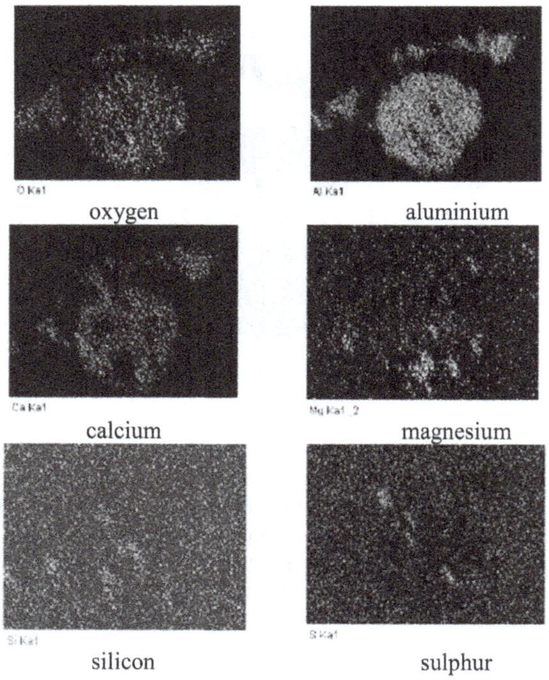

Figure 3. Distribution of chemical elements in complex inclusion (see Figure 2) (X-ray microspectral analysis)

The morphology of inclusions of such type usually is irregular (Figure 4). It is the additional evidence of their exogenous origin.

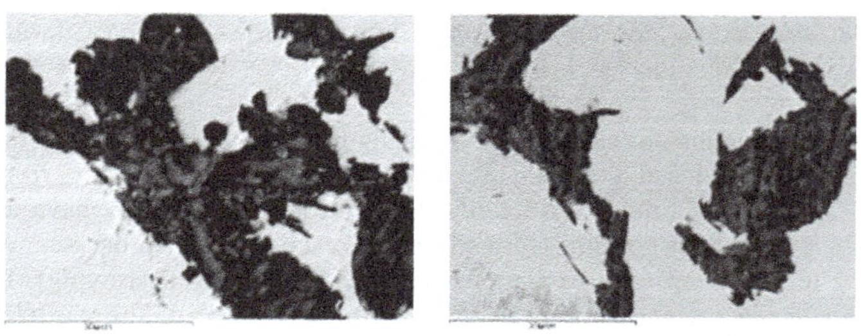

Figure 4. Typical morphology of combined exogenous inclusion

However, formation of endogenous and exogenous particles always occurs in the production of metal. Therefore the additional investigations are necessary to find main mechanism. In order to do this, the investigation of structure and chemical composition of pouring ladle refractory lining was fulfilled.

Interaction of Refractory Lining with Deep Deoxidized Molten Metal

The samples were taken from the refractory lining of the pouring ladle after full working cycle before repair of lining. The samples of refractory were selected from different areas of ladle lining: the sample from the zone, which was mostly in contact with slag (upper part of ladle), with slag and metal alternating during the pouring (middle part), mostly with metal (bottom part), as well as a control sample. The petrography section view of control sample is shown on Figure 5.

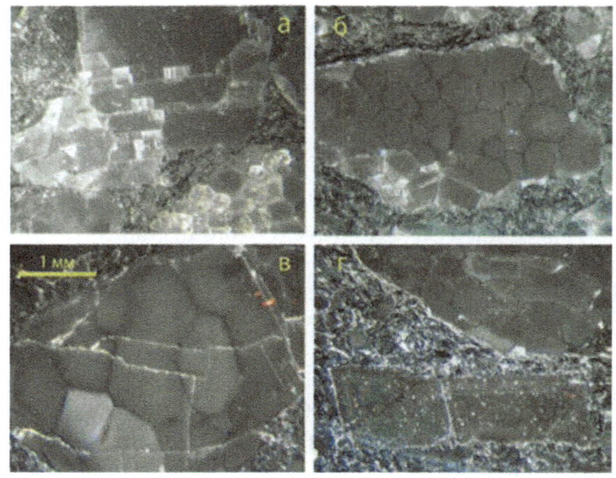

Figure 5. Structure of control sample

Structure of refractory from slag, slag – metal and metal zones are shown on Figures 6 – 8 respectively.

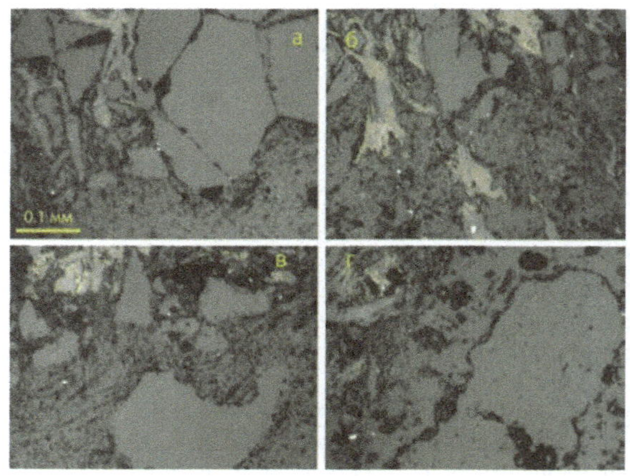

Figure 6. Structure of sample from slag zone

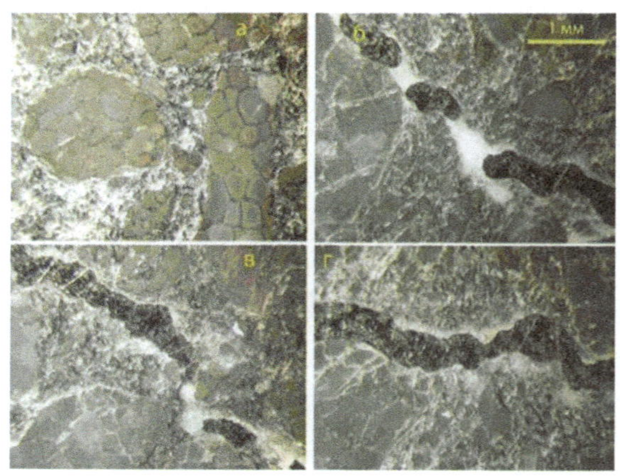

Figure 7. Structure of sample from metal - slag zone

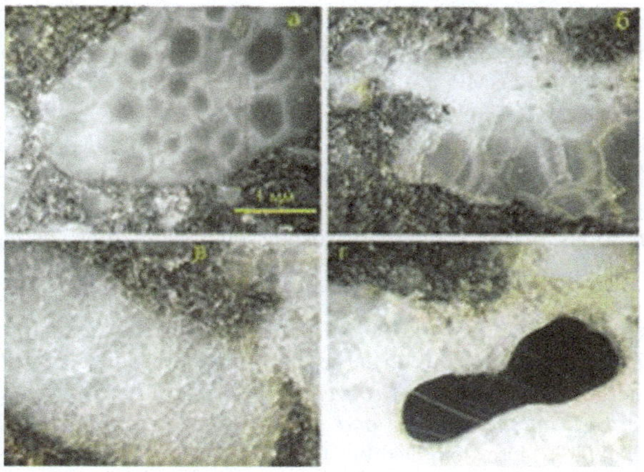

Figure 8. Structure of sample from metal zone

Analysis of general structure of refractory has shown that structure transformations in refractory take place during the exploitation. Degree of transformation development depends on the total time of refractory contact with molten metal. Thus, at Figure 6 we can see that structure transformation in contact with slag is minimal. With increasing duration of contact with molten metal the transformation becomes more developed (compare Figure 7 and Figure 8).

In zones, where contact with metal takes place, main transformation deals with formation of cracks and filling of them by liquid metal. Penetration of metal in crack (Figure 7) is accompanied by a halo of refractory conversion of replacing minerals of binder by cryptocrystalline aggregates.

In a sample from the metal belt (Figure 8) replacement of binder by cryptocrystalline mineral aggregate and subsequent recrystallization, develops in cracks and in the surrounding area. Transformed zone occupies the large areas.

The results of X-ray microspectral analysis of chemical elements distribution in initial refractory are shown on Figure 9 and table 1.

As can be seen from table 1, binder consists mostly of silicon and calcium oxide (points 2 – 6), whereas grains are the magnesium oxide (points 1,7). In result of interaction with deep deoxidized metal, the destroying of binder may take place (Figure 10) and dissolution of magnesium oxide grains in binder develops.

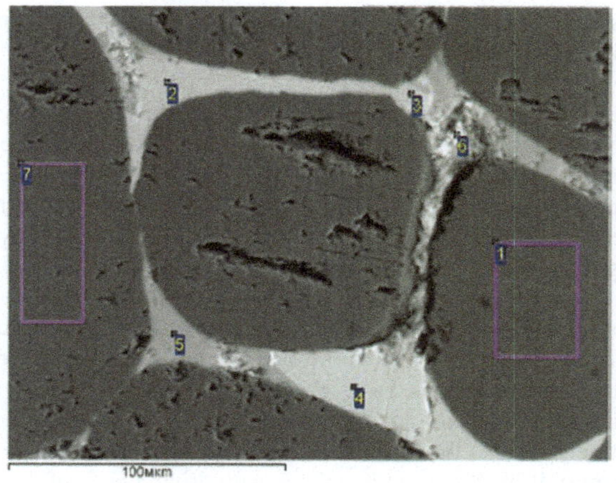

Figure 9. The general SEM view of refractory in initial state and locations of points for analysis

Table 1. Results of chemical analysis (points – on Figure 9)

Point	C	O	Mg	Al	Si	Ca
	Content of element, % wt.					
1	22.84	35.12	40.99	0.08	0.03	0.02
2	23.44	29.54	4.87	0.03	13.06	28.36
3	25.24	29.35	6.43	0.05	12.49	25.99
4	21.97	30.26	5.35	0.07	13.07	28.68
5	24.31	31.66	10.99	0.10	13.14	19.14
6	35.65	23.05	1.47	0.08	2.28	26.42
7	22.51	35.44	40.99	0.13	0.01	0.05

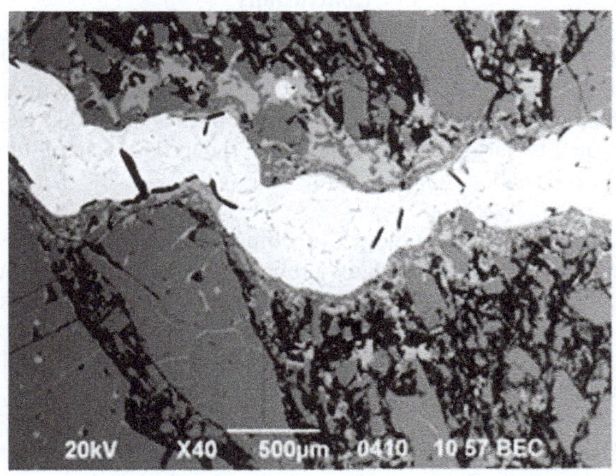

Figure 10. Penetration of molten metal in refractory through the binder areas.

This process may cause the destroying of refractory material on pieces of microsize, which may form the exogenous inclusions, shown on Figure 2 – Figure 4.

Discussion of Obtained Results

It is shown that for stable technology at low concentration of S and P and deep deoxidizing very low volume fraction of inclusion (less than 0.05% by vol.) was detected. Inclusions has the endogeneous nature and are represented by spherical manganese sulphides. Average size of inclusion did not exceed 10 microns.

Mostly the formation of such inclusion deals with sporadic deviation of manufacturing process.

Investigation confirms that in the points, where defects were found by ultrasonic inspection, most of inclusion are formed by complex oxides of type $(Al,Ca,Mg,Si)_mO_n$ that have the exogenous origin.. In favour on this assumption, the form of defects talks, namely areas of irregular shape, filled by the conglomerates of particles of different form and composition.

Another argument in behalf on the exogenous origin of including is their spatial localization in the volume of ingot. For the endogenous inclusions more characteristic is enriching of axial area of ingot, while in this case most inclusions are located relatively close to the surface of ingot (billet) (at the distance 70 – 120 mm). Probably it can be related with driving back of the large inclusion, getting in mould, from a surface to the center during rapid crystallization of skin area with subsequent their mechanical «fixing» between growing columnar crystals.

Due to difference of surface tension on boundary of non-metallic inclusions with metal and slag, coagulation of them with slag droplets takes place. But it is very important that during investigation the presence of inclusions, formed by pure slag particles, have not been found. It means that technology of tapping and pouring prevents the entering of large portion of slag in metal and only interaction of slag droplets with products of refractory erosion causes the formation of slag – inclusion aggregates. Therefore to prevent the formation of such type of defect, it is necessary to use the refractory with improved properties and increased resistance to the action of molten metal.

Conclusions

1. The analysis showed that the production capacity of PJSC "Energomashspetsstal" can solve actual scientific and engineering problems and implement modern approaches in production of high-quality large forging ingots for the production of turbine shafts and backup rolls for rolling mills.

2. The developed technology of melting, secondary treatment and casting of large forging ingots provides the high degree of purity of steel on compounds of sulphur and phosphorus and the oxides of aluminium and mixture of complex oxides of type $(Al,Ca,Mg,Si)_mO_n$ become a basic non-metallic phase in large mass ingots.

3. The most probable source of their formation is destruction of refractory lining of furnaces and ladles; however it is possible to assume, that part of them formed in the process of deoxidizing and modifying of melt due to coagulation of aluminates and silicates under the influence of calcium compounds in the compact conglomerates and flotation of inclusions in the liquid metal

References

1. Dub V.S., Makarychevf E.V., Markov I.I. "Large ingot – current state amd future" *Electrometallurgy*, 1999. № 5.- P.22-30.
2. Dyudkin D.A., Bat S.Yu., Grinberg S.E., Marintsev S.N. *Production of steel at ladle – furnace complex*. Donetsk: South-West, 2003, 62 p.
3.Farouk Sidikki, Kodak A.V., Kasyan G.I., Yavtushenko P.M, Popik N.I. "Modernization of out of furnace technology of steel treatment in ESMP JSC "MMZ "Istil (Ukraine)" *Metal and casting of Ukraine*, 2006. №1. 19 p.
4. Efimov M.V., Selyutin A.A., Yavtushenko P.M., Safonov V.M., Troyanskyy A.A., Ryabtsev A.D. "Modern technology of large ingots production for energy generation machinebuilding in conditions of JSC "Energomashspetsstal" *Metal and casting of Ukraine*, 2012. №9. P. 3-7.
5. Efimov M.V., Panov V.V., Kolomoec A.N., Lobanov A.I., Pashynskyi V.V., Snizko O.A., Ryabtsev A.D. "Investigation of non-metallic inclusions in large-mass ingots, produced by PJSC "Energomashspetsstal" *Proceeding of the 5th International Congress on the Science and Technology of Steelmaking 2012 "ICS 2012"*, Dresden, October 1-3/ 2012. – P. 152.
6. Efimov M.V, Selyutin A.A., Kapustin V.V., Troyanskyy A.A., Ryabtsev A.D. "Producing of 355 ton ingot for forging in the conditions of PJSC "Energomashspetsstal" *Proceeding of the 5th International Congress on the Science and Technology of Steelmaking 2012 "ICS 2012"*, Dresden, October 1-3/ 2012. – P. 148.

Proceedings of the 2013 International Symposium on Liquid Metal Processing & Casting
Edited by: Matthew J. M. Krane, Alain Jardy, Rodney L. Williamson, and Joseph J. Beaman
TMS (The Minerals, Metals & Materials Society), 2013

CHARACTERIZATION OF THE INTERFACIAL REACTION BETWEEN TITANIUM CASTINGS AND ALUMINA MOLD CONTAINING THE ALPHA-CASE COMPOUNDS

Lee Seul[1], Bong-Jae Choi[1], Jeong-IL Youn[1] and Young-Jig Kim[1]
[1]School of Advanced Materials Science & Engineering, Sungkyunwan University,
300 Chunchun-dong, Jangan-gu, Suwon Gyenggi-do, 440-746, Republic of Korea

Keywords: Titanium, Alpha-case, Investment casting, Mold, Reaction phase

Abstract

The interfacial reactions of titanium with investment molds containing alpha-case compound systems were studied. Especially, the alumina molds that are containing titania which is one of the alpha-case compounds were used as a mold material for the alpha-case control. Micro Vickers-hardness and microstructure was analyzed for evaluation of reactivity of titanium with mold material. It was found that the thickness of the alpha-case was effectively reduced, indicating that the titania in the mold can increase the inertness of the alumina mold. The microstructural and compositional characterizations of molds were also analyzed for confirming effect of alpha-case control mechanism. The composition of the mold surface turns titania into oxygen-deficient titania which act as effective barrier on the mold element diffusion.

Therefore, the based on the thickness of the alpha-case and the maximum hardness of titanium castings, alumina mold containing titania may be the best choice for titanium investment casting.

Introduction

Titanium has a high specific strength with a low density that is only about 60 percents of that of steel or nickel-base superalloy. Therefore titanium has been widely used for chemical, electric power and aero space industries as major metal material by taking advantage of their characteristics. [1-3] In addition, cast pure titanium has been increasingly used to make fixed and removable prostheses because of its outstanding biocompatibility and light weight compared to conventional dental alloys. [4] Also, it was widely recognized that investment casting is the best option for producing complex net or near net shape titanium components with a minimum waste material in view of the poor machinability and workability of these alloys. [5, 6]

However, liquid titanium is well known as a universal solvent, because violent reactions with gasses, liquids, and solids result in contamination of the titanium. Furthermore titanium is extremely reactive to ceramics used in conventional investment mold at high temperatures, resulting in interfacial reaction layers. The result of this interaction is an oxygen and metallic element enriched surface layer known as the alpha-case. [7, 8] In the case of reaction between titanium and alumina, the alpha-case layer, consisting of TiO_2 (interstitial compound) and Ti_3Al (substitutional compound), was formed. [3, 7] In general, alpha-case can lower the fatigue strength and tensile ductility because of larger microcracks in the alpha-case. Therefore this alpha-case, which is hard and brittle because of its high oxygen content, is usually removed from the castings by chemical milling. However, removal process is waste producing, expensive, and limits the complexity and detail that can be achieved in a casting. [3, 6, 7]

This study was to investigate the alpha-case on the pure titanium castings using an alumina based mold that are containing alpha-case compound (interstitial and substitutional compound) for alpha-case control. Finally, alpha-case control mechanism was verified at the interfacial reactions between titanium and alumina based mold containing alpha-case composition during casting.

Experimental Procedure

The wax patterns for the examination of metal/mold reactions were made by pouring molten wax into a simple cylindrical silicon rubber mold (Ø 15 mm×30 mm). Subsequently, the patterns coated with Al_2O_3, Al_2O_3+10wt%Ti, Al_2O_3+10wt%Ti_5Si_3 and Al_2O_3+10wt%Ti-Vac slurry, which was mixing with colloidal silica (15nm, 30 wt% SiO_2) and powder (purity of 99%, 45 μm). The shell molds were dried under the condition of 21±1 °C and 50±10 % humidity for 4 h. The dipping, stuccoing with the Al_2O_3 (purity of 99%, 1-0.5 mm) procedures were repeated two times. After the primary layer coating, the patterns were coated with the back-up layers by the chamotte. The dewaxing of the shell mold was carried out at around 150 °C and 0.5 MPa in an autoclave. Finally, the Al_2O_3, Al_2O_3+10Ti, Al_2O_3+10Ti_5Si_3 shell molds were fired at 950 °C for 2 h at atmosphere condition. And in the case of Al_2O_3+10Ti-Vac mold was fired at 950 °C for 2h at vacuum condition (fewer than 70 mtorr). The shell mold burnout conditions are given in Table I.

Table I: The alumina based mold burnout conditions.

Oxide	Al_2O_3 : Ti or Ti_5Si_3 (wt%)	Binder	Burnout Condition
Al_2O_3	-	colloidal silica	1223 K /2 h Atmosphere
Al_2O_3+10Ti	9:1		
Al_2O_3+10Ti_5Si_3			
Al_2O_3+10Ti-Vac			1223 K /2 h 70 mtorr

The casting was progressed using plasma arc melting furnace for preventing any contamination from refractory crucibles. A titanium (ASTM B-348, grade 2) rod of 120g was melted and casted into the alumina based mold under argon atmosphere. And the surface of titanium castings were observed optical microscope (Olympus PME3) and electron prove microanalysis (EPMA, JEOL EPMA JXA-8900R) for the identification of alpha-case composition. Furthermore, the micro Vickers hardness (Mitutoyo MVK-H2) was used to determine the thickness of the alpha-case. Also, Microstructural and compositional characterization of molds were analyzed by scanning electron microscopy (SEM, JEOL

JSM-7600F), X-ray diffraction (XRD, M18XHF-SRA) and electron probe microanalyses (EPMA).

Result and Discussion

Analysis of Mold Composition

Figure 1 shows the XRD spectra results of the alumina based mold. Firstly, TiO_2 and Al_2O_3 were found in the Al_2O_3+10Ti mold. This result was signified that the Ti which was added in the mold was change into the TiO_2. And that is because of the reaction of titanium with the atmospheric oxygen produced TiO_2 during the burnout process. Also, P. Kofstad et al. [9] reported that titanium is oxidized through the recrystallization to TiO_2 at temperature over 950 °C. Secondary the reaction phase in the Al_2O_3+10Ti_5Si_3 mold was not formed and that was implied that there were no reactions between the Al_2O_3 and Ti_5Si_3 in the mold. Finally, the Al_2O_3 +10Ti-Vac mold was composed of Ti_2O, Ti_5Si_3 and Al_2O_3, indicating that titanium reacted with small amount of oxygen in the vacuum chamber and also, colloidal silica. And the results of the thermodynamic calculation based the XRD result at 1200 K were shown along with the reaction equations below. [10, 11]

$$2Ti(s) + O_2(g) \rightarrow 2TiO_2(s) \quad (1)$$
$$\Delta G = -598.51 \text{ kJ/mol}$$

$$3SiO_2(s) + 23Ti(s) \rightarrow Ti_5Si_3(s) + 6Ti_2O(s) \quad (2)$$
$$\Delta G = -1352.74 \text{ kJ/mol}$$

$$2Ti(s) + 1/2O_2(g) \rightarrow Ti_2O(s) \quad (3)$$
$$\Delta G = -475.00 \text{ kJ/mol}$$

ΔG has negative value for the formation of TiO_2, Ti_2O and Ti_5Si_3 in the Al_2O_3+10Ti, Al_2O_3+10Ti-Vac mold. The comparison result XRD and thermodynamic calculation, the reaction (1), (2) and (3) confirm a spontaneous reaction

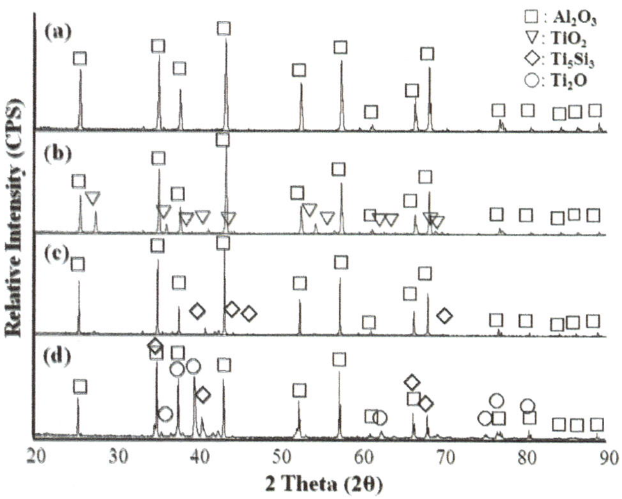

Figure 1: XRD spectra of oxide burnout at 950 °C for, 2h (a) Al_2O_3, (b) Al_2O_3+10Ti, (c) Al_2O_3+10Ti_5Si_3 (d) Al_2O_3+ 10Ti-Vac

Titanium Reactivity with Alumina Based Molds

Figure 2 illustrates the titanium castings microstructure of the regions below the surfaces of titanium castings made with alumina based molds. The titanium casting was located at the left side. There appeared a quite serious reaction region between the titanium and the Al_2O_3, Al_2O_3+10Ti_5Si_3, Al_2O_3+10Ti-Vac molds. And the morphology of alpha-case was lamellar structure. However, there is no visible reaction layer in the titanium surface using Al_2O_3+10Ti mold (Casting surface was similar to the titanium matrix). The result of micro Vickers hardness in titanium castings from the outer to inner surfaces are shown in Figure 3 The titanium made in Al_2O_3 molds are similar harden layer to the Al_2O_3+10Ti_5Si_3 and Al_2O_3+10Ti-Vac mold. And the thickness of the reaction layers was approximately 350 μm. The castings from the Al_2O_3+10Ti mold had an extremely thinner harden layer about 50 μm than the other molds. In Figure 4, it can be observed that the alpha-case composition on the casting surface with respect to the aluminum, silicon, oxygen. The oxygen was detected throughout, from the surface to the inner region for all the castings. The aluminum, silicon and oxygen diffusion coefficient in titanium matrix were determined from the rate of increases of the interface width. The aluminum, silicon and oxygen diffusion coefficient were 1.4×10^{-9} cm^2/s (at 1256-1523 K), 6.3×10^{-5} cm^2/s (1223-1423 K), and 4.4×10^{-7} cm^2/s (923-1073 K), respectively. [12, 13]

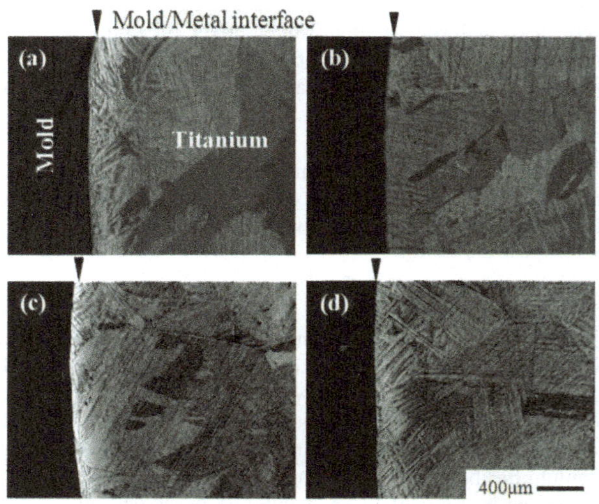

Figure 2: Metal/mold reactions of titanium with (a) Al_2O_3, (b) Al_2O_3+10Ti, (c) Al_2O_3+10Ti_5Si_3 (d) Al_2O_3+10Ti-Vac

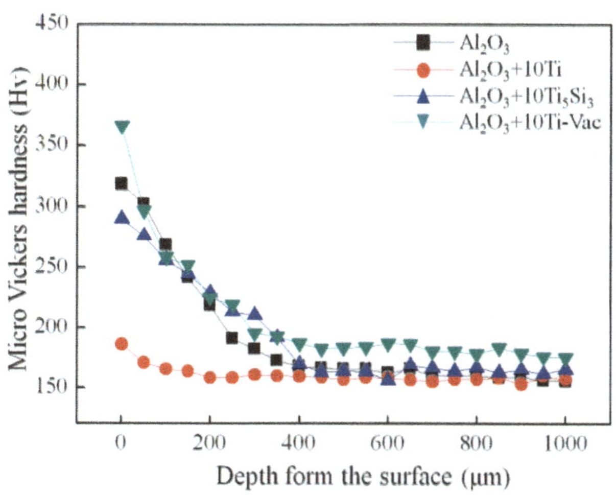

Figure 3: Micro-Vickers hardness profile of titanium with (a) Al_2O_3, (b) Al_2O_3+10Ti, (c) Al_2O_3+10Ti_5Si_3 (d) Al_2O_3+10Ti-Vac

The titanium castings made with Al_2O_3 mold can easily observe that the Al and Si were severely dissolved in the alpha-case. C. Frueh et al. indicated the Silica, used as a binder for the mold, drives the metal mold reaction regardless of the refractory used. [7] And Saha and Misra signified the formation of the Ti_5Si_3 at the

Table II: EPMA elemental spot analysis of Al_2O_3+10Ti-Vac mold surface of before and after the casting.

Mold	Before casting /After casting	Position	Ti	O	Al	Si	Total
Al_2O_3+10Ti -Vac	Before casting	A (Al_2O_3+SiO_2)	0.01	56.86	27.28	15.86	100
		B (Ti_2O)	70.84	28.71	0.33	0.12	100
	After casting	C (AlO)	0.04	44.47	55.22	0.40	100

casting surface during the casting of titanium in zircon sand molds. [6] Also, S. Y. Sung et.al suggested that the alpha-case is TiO_2 and Ti_3Al between titanium and Al_2O_3. [14] The interstitial diffusion of oxygen and substitutional aluminum dissolved form mold material. The high temperature during casting procedure induced decomposition of the Al_2O_3 mold and substitutional diffusion of aluminum and silicon of colloidal silica into titanium castings. Also, the castings made with Al_2O_3+10Ti_5Si_3 and Al_2O_3+10Ti-Vac have not silicon layer but a thick aluminum layer. In this results mean that the Ti_5Si_3 in the mold side could not control the aluminum and oxygen diffusion but silicon diffusion. Also, the diffusion of oxygen from the Al_2O_3+10Ti-Vac mold to titanium castings had no effect. However, in drop castings made in Al_2O_3+10Ti molds, very thin aluminum layer was found on the metal mold interface. It means that the diffusion length of metal and oxygen effectively decrease by act of TiO_2.

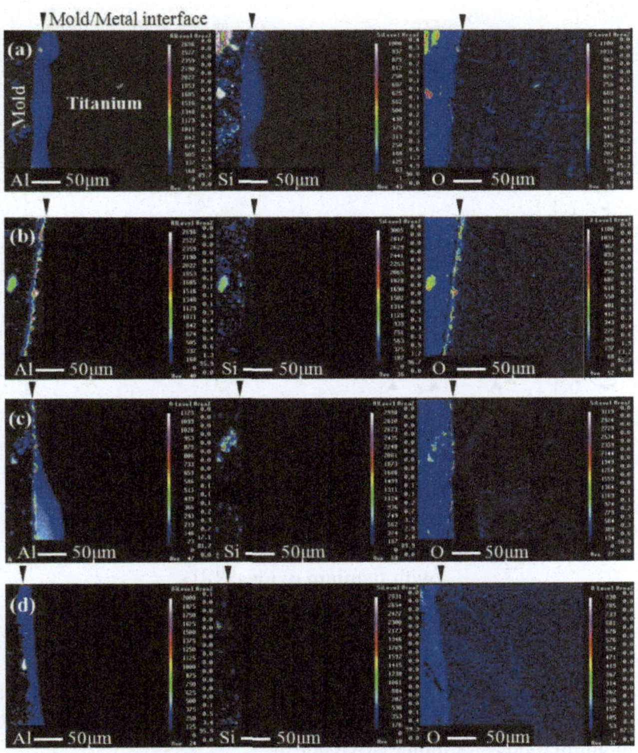

Figure 4: Comparison of elemental mapping image of oxygen, aluminum and silicon in titanium castings in (a) Al_2O_3, (b) Al_2O_3+10Ti, (c) Al_2O_3+10Ti_5Si_3 and (d) Al_2O_3+10Ti-Vac

Therefore the oxygen diffusion length is thicker than aluminum and silicon in titanium matrix. Consequently, castings made in Al_2O_3 mold, the Ti_3Al, Ti_5Si_3 and TiO_2 and can be formed by the aluminum, silicon and oxygen diffusion in the high concentration element layer. And also, the silicon layers at the titanium/Al_2O_3+10Ti-Vac or Al_2O_3+10Ti_5Si_3 surface was not visible because of the Ti_5Si_3. However, the castings in Al_2O_3+10Ti mold are apparent that the benefits of metal and oxygen diffusion resistance increase.

The Role of Reaction Phase on the Metal/Mold Reaction

Effect of Ti_2O and Ti_5Si_3 on the Metal/Mold Reaction The diffusion of oxygen and aluminum from the Al_2O_3+10Ti-Vac mold into the titanium castings took place during the casting process. The mold surface was analyzed to investigate the diffusion behavior of the Al_2O_3+10Ti-Vac molds. As shown in Figure 5, spherical Ti_2O and plate-like Al_2O_3 primary coating layer, which was build the network with colloidal silica existed in

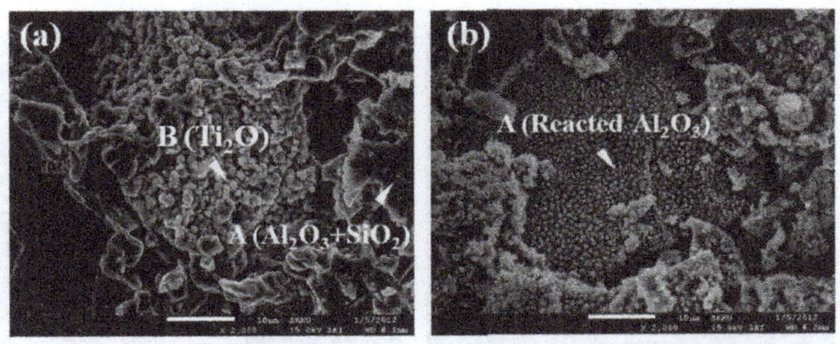

Figure 5: Mold surface change of Al_2O_3+10Ti-Vac mold (a) before the casting and (b) after the casting.

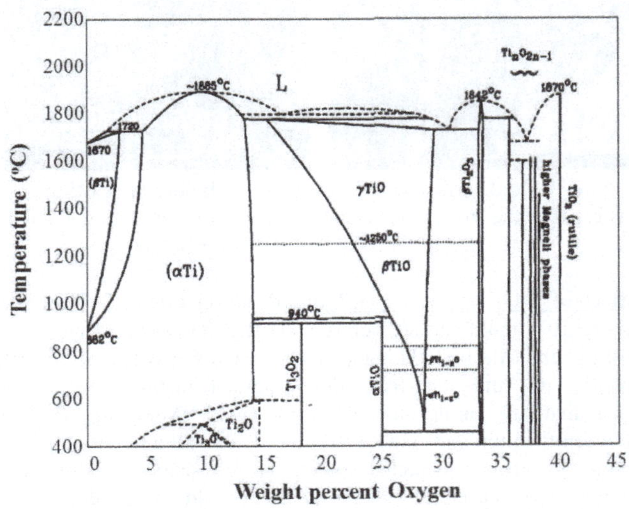

Figure 6: Binary phase diagram of Ti-O. [15]

before the casting Al_2O_3+10Ti-Vac mold surface. However, because of the interfacial reaction between titanium and mold, most of spherical Ti_2O was hard to observe after the casting mold surface and the only coarse alumina grit can be observed (Table II). These results of interaction can be explained using phase diagram that was illustrated at figure 6. The metastable Ti_2O in the Al_2O_3+10Ti-Vac mold was transformed into stable alpha-titanium (14wt% oxygen solubility limit) dissolved a significant amount of oxygen during the casting temperature. As the transformed stable alpha-titanium reacted with titanium melts, most of alpha-titanium swept away or melted inward the titanium melts and then the coarse alumina grit exposed at mold surface.

Table III: EPMA elemental spot analysis of Al_2O_3, Al_2O_3+10Ti mold surface of before and after the casting.

Mold	Before casting /After casting	Position	Ti	O	Al	Si	Total
Al_2O_3	Before casting	A (Al_2O_3+SiO_2)	0.01	52.95	34.62	12.42	100
	After casting	B (AlO)	0.19	49.24	50.38	0.19	100
Al_2O_3+10Ti	Before casting	C (Al_2O_3+SiO_2)	0.14	53.6	32.9	13.36	100
		D (TiO_2)	26.32	67.51	5.09	1.08	100
	After casting	E (AlO)	0.82	49.84	48.55	0.80	100
		F (TiO)	55.29	41.55	2.37	0.8	100

Furthermore, the silicon diffusion was effectively reduced because of the Ti_5Si_3 in the Al_2O_3+10Ti-Vac and Al_2O_3+10Ti_5Si_3 mold. The Ti_5Si_3 had a higher binding energy (1095 ± 40 kJ/mol) than the silica (769 ± 12 kJ/mol) causes the highest melting temperature. [16, 17] Therefore, the Ti_5Si_3 was more stable than silica during the casting process so the diffusion of silicon was effectively prevented. However, the Ti_5Si_3 cannot prevent the oxygen and aluminum diffusion. Consequently the alpha-case was composed of metal element and oxygen.

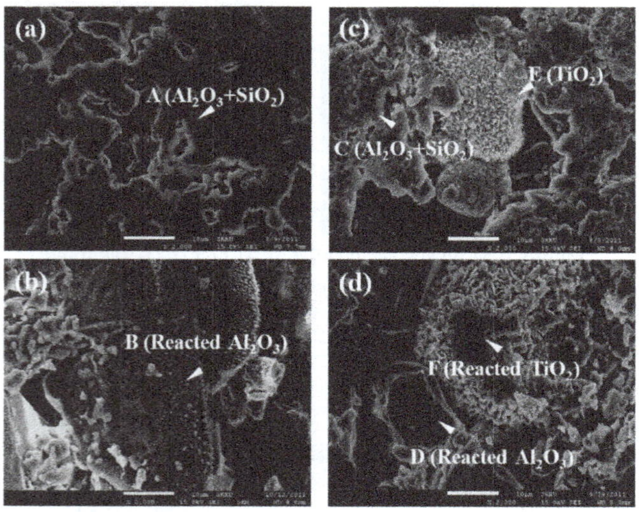

Figure 7: Mold surface change of (a) Al_2O_3 before the casting, (b) Al_2O_3 after the casting, (c) Al_2O_3+10Ti before the casting and (b) Al_2O_3+10Ti after the casting

Effect of TiO_2 on the Metal/Mold Reaction Figure 7 show the Al_2O_3+10Ti mold surface before and after the casting. Before the casting, the Al_2O_3+10Ti mold surface was consisted of prickly sphere TiO_2 and plate-like morphology of Al_2O_3 primary coat layer that are similar to the Al_2O_3 mold. After the reaction between titanium and Al_2O_3+10Ti mold, a small amount of Al_2O_3 primary coat was detected, moreover the shape of TiO_2 was change. The morphology of Al_2O_3+10Ti mold change during the casting process was plate-like surface TiO_2 and polygonal Al_2O_3. Because of the decomposition of the Al_2O_3+10Ti mold and diffusion of oxygen into the titanium melts, the chemical composition altered oxygen deficient-TiO_2 and oxygen deficient-Al_2O_3 that oxygen and metal element were in the ratio of one to one (table III). Figure 8 illustrated the estimate the oxygen solubility limits in liquid titanium at various temperatures when titanium was in contact with its oxides. It was shown that oxygen solubility limit was restricted when the ratio of oxygen on the titanium decreased. [18] Therefore, the oxygen deficient-TiO_2 in the mold side could be restricting oxygen diffusion, also, the titanium oxide layer on the casting surface act as a diffusion barrier during the casting process owing to the decline of concentration gradient between titanium and titanium oxide in the Al_2O_3+10Ti mold.

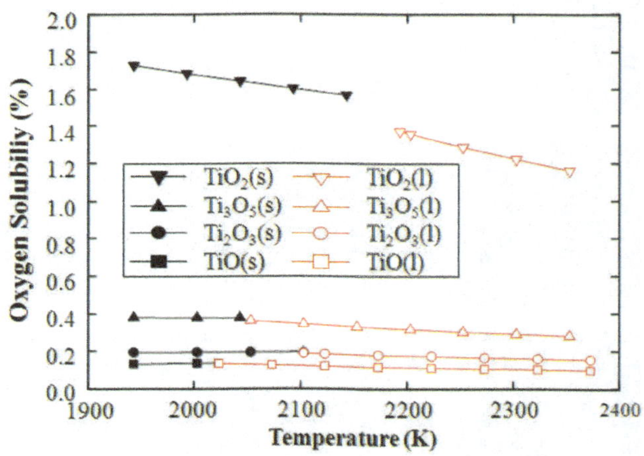

Figure 8: Solubility limit of oxygen in liquid titanium in contact with titanium oxides [18]

Conclusion

Interfacial reaction was investigated between titanium and alumina based mold that are containing alpha-case compound (Ti_5Si_3 and titanium oxide) for alpha-case control.

1. The surface hardened layer of the casting made with an Al_2O_3 mold was about 350 μm thick, and the aluminum, silicon and oxygen rich layer were located along the casting surface, because the metal element and oxygen diffused to the titanium melts. Moreover, the oxygen was faster diffused than the metal element since; the oxygen had a high diffusion coefficient.

2. The silicon concentration was reduced to Al_2O_3+10Ti, Al_2O_3+10Ti_5Si_3 in the titanium castings but, the oxygen and aluminum concentration were not decreased. Because the Ti_5Si_3 was very stable during the casting procedure, therefore, the control of silicon diffusion was effective.

3. The mold stability of oxides could be obtained by determination of microhardness profile and EPMA result in cast titanium. And the most stable mold was Al_2O_3+10Ti, because the TiO_2 was act as a diffusion barrior by reducing oxygen solubility.

Reference

[1] C.C. Hung, P.L. Lai, C.C. Tsai, T.K. Huang and Y.Y. Liao, " Pure titanium casting into titanium-modified calcia-based and magnesia-based investment molds," Mater. Sci. Eng. 454-455 (2007), 178–-182.

[2] W. J. Boettinger, M. E. Williams, S. R. Coriell, U. R. Kattenr, and B.A. Mueller, "Alpha Case Thickness Modeling in Investment Castings," *Metall. Mater. Trans.* 31 (B) (2000), 1419-1427

[3] S. Jones, C. Yuan, "Advances in shell moulding for investment casting," J. *Mater. Process. Tech.* 135 (2003), 258–265

[4] Matthew J. Donachie, Jr. *Titanium a technical guide 2nd*, 2000

[5] Y. W. Chang and C. C. Lin, "Compositional Dependence of Phase Formation Mechanisms at the Interface Between Titanium and Calcia-Stabilized Zirconia at 1550°C," *J. Am. Ceram. Soc.* 93 (2010), 3893-3901.

[6] R. L. Saha, R. D. K. Misra, "Formation of low-melting eutectic at the metal-mould interface during titanium casting in zircon sand moulds," *J. Mater. Sci Letters.* 10 (1991), 1318-1319.

[7] C. Frueh, D. R. Poirier and M.C. Maguire, "The Effect of Silica-Containing Binders on the Titanium/Face Coat Reaction," *Metall. Mater. Trans.* 28 (B) (1997), 919-916.

[8] R. L. Saha, T. K. Nandy, R. D. K. Misra, and K. T. Jacob, "On the Evaluation of Stability of Rare Earth Oxides as Face Coats for Investment Casting of Titanium," *Metall. Mater. Trans.* 21 (B) (1990), 559-556.

[9] P. Kofstad, P. B. Anderson and O. J. Krudtaa, "Oxidation of titanium in the temperature range 800–1200°C," *J. less-common metals.* 3 (1961), 89-97

[10] I. Barin, Thermochemical Data of Pure Substances, 1994

[11] L. D. Teng, F.M. Wang, W.C. Li, "Thermodynamics and microstructure of Ti–ZrO2 metal–ceramic functionally graded materials," *Mater. Sci. Eng*, 293 (A) (2000), 130–136

[12] Z. Liu and G. Welsch, "Literature Survey on Diffusivities of Oxygen, Aluminum, and Vanadium in Alpha Titanium, Beta Titanium, and in Rutile," *Metall. Trans.* 19 (A) (1988) 1121-1125.

[13] E. A. Branded and G. B. Brook, Smithells Metals Reference Book, 7th ed., Butterworth-Heinemann Ltd., UK (1992).

[14] S. Y. Sung, Y. J. Kim, "Alpha-case formation mechanism on titanium investment castings," *Mater. Sci. Eng*, 405 (A) (2005), 173-177

[15] ASM Handbook 3, (The materials information company 1992)

[16] E. A. Branded and G. B. Brook, Smithells Metals Reference Book, 7th ed., Butterworth-Heinemann Ltd., UK (1992)

[17] Oleg N. Senkov, Daniel B. Miracle, Sergey A. Firstov, Metallic Materials with High Structural Efficiency, (2004), 287–308

[18] L. N. Belyanchikov, "Thermodynamics of Titanium-Based Melts: II. Oxygen in Liquid Titanium," *Metall.,* 2010 (2010), 1156-1163.

Proceedings of the 2013 International Symposium on Liquid Metal Processing & Casting
Edited by: Matthew J. M. Krane, Alain Jardy, Rodney L. Williamson, and Joseph J. Beaman
TMS (The Minerals, Metals & Materials Society), 2013

THEORETICAL DESIGN OF CONTINUOUS CASTING PROCESS USING SEMI ANALYTICAL METHOD

Pavan Kumar Penumakala, Ashok Kumar Nallathambi, Eckehard Specht
Otto von Guericke University Magdeburg, 39106 Magdeburg, Germany

Keywords: Solidification, Temperature compensation, Semi analytical study, Nondimensionalised form.

Abstract

A semi analytical study is presented for the solidification of liquid metals in the form of ingots or billets in a theoretical manner. A 2-D heat transfer with solidification effects are considered in non dimensional form and solved using numerical methods. The surface of the ingot is assumed to be maintained at constant temperature which is practically reasonable. This assumption also delivers the kind of cooling profile which has to be applied in reality. Fourier and Phase change numbers, mass flow rate, specific heat, conductivity, and ingot geometry are the influencing parameters which governs the total metallurgical length of the plant. At the end of solidification, the ingot must undergo temperature compensation before rolling. Therefore, the influence of temperature distribution on the compensation length is presented in non-dimensional form. The novelty of the study is that the extreme two cases such as plate and square considered here will provide many design guidelines.

Introduction

Continuous casting is a well established technique for the production of steel billets and ingots and successfully used by many industries across the world. For example, in a continuous casting of steel alloys, heat is removed in primary mold cooling and in the secondary water cooling zones as shown in Fig 1. The liquid metal starts to solidify and small layers of solid steel will be formed along the walls of the mold. After the mold exit, the solidified strand with liquid core will be bent horizontally and the cooling water is continuously sprayed on the surface of the strand. The liquid core rejects the heat to the cooling water and solidifies completely after a certain time. The entire heat transfer process is transient in nature and it is important to know the time required for the complete solidification. However, the heat removal rate and the corresponding heat flux and/or heat transfer coefficient in the different zones is difficult to find and can be obtained from the costly industrial trails.

It is attempted to develop a general model which finds the time required for complete solidification based on the approximation of constant surface temperature. The model is valid for most of the metals ranging from copper to steel alloys, that are to be cast by continuous casting and the approximation of constant surface temperature is valid. The model also finds the influence of cross section on the time of solidification. The infinitely long plate, square and rectangular cross sections are studied and compared.

Continuous casting of major alloys is integrated with hot rolling and the ingots after the complete solidification are directly sent to hot rolling mills. At the moment of complete solidification, the ingots has solidus temperature at the core and a different temperature at the surface. But the rolling requires a uniform temperature through the ingot and they must be heat compensated for some time to achieve this. The typical temperature curves are shown in figure 2.

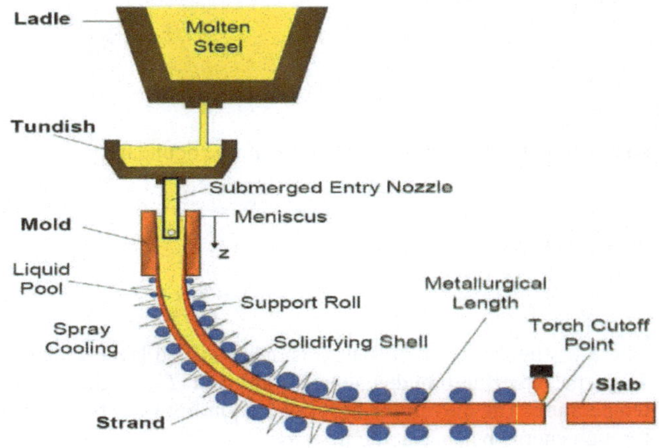

Figure 1: Schematic representation of Continuous Casting of Steels [1]

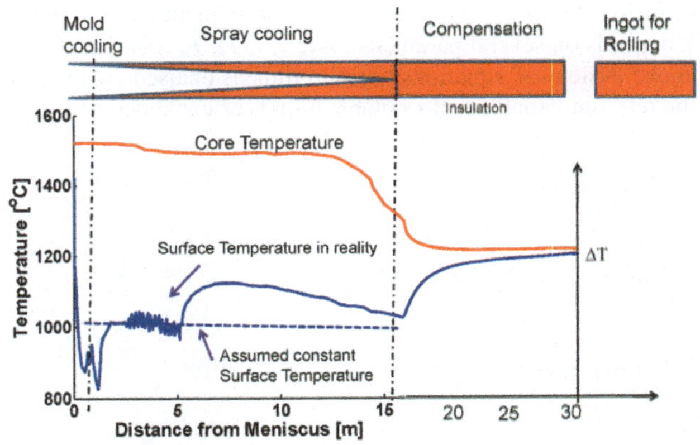

Figure 2: Temperature profiles of an ingot during the Continuous Casting Process [2]

A model is developed to find the time required for this compensation. This model is also generalized in a way that it studies the influence of cross section on the compensation time. In this way, this study attempts to find the minimum time required for the complete solidification and for the compensation of differently sized ingots. The following sections explain about the analytical and numerical procedures.

Analytical Solution for Infinite Slab

Consider a domain that contains fully molten metal at liquidus temperature. In order to solidify, the cooling is applied at the surface by maintaining it at constant temperature which is well below the liquidus temperature. A thin solidified layer forms at the surface and its thickness increases with time. The heat transfer is assumed to be one dimensional and solidification front which separates the liquid and solid will move along the positive x direction. The Fig.3 shows the solidified layer and the corresponding temperature profiles at different times t_1 , t_2 and t_f. The heat conduction in the solidified layer is governed by Fourier differential equation of the form

$$\rho c \frac{\partial T}{\partial t} = \frac{\partial^2 T}{\partial x^2} \tag{1}$$

The initial condition is

$$T(t = 0) = T_f \tag{2}$$

and the boundary conditions are, at the phase boundary

$$T(x = \delta) = T_f \tag{3}$$

on the surface

$$T(x = 0) = T_w \tag{4}$$

where s is thickness of the slab, δ is the solidified layer thickness, T_f is the freezing/liquidus temperature and T_w is the surface temperature. The movement of solidification front can be obtained through energy balance at the front as

$$\lambda \frac{\partial T}{\partial x}(x = \delta) = \Delta h.\rho.\frac{dT}{dt} \tag{5}$$

where λ is thermal conductivity, ρ is density, Δh is the latent heat of fusion. From the above equations, it is evident that the solution T depends on several parameters like x, t, T_f, T_w, Δh, λ and c. The above system of equations are non-dimensionalsed by arranging the relevant variables into suitable groups as explained below.

1. Dimensionless temperature $\qquad \theta = \frac{T-T_W}{T_f-T_W}$

2. Dimensionless space coordinate $\qquad X = \frac{x}{s/2}$

3. Fourier Number $\qquad Fo = \frac{\alpha t}{(s/2)^2}$

4. Dimensionless solidification thickness $\qquad \Delta = \frac{\delta}{(s/2)}$

5. Phase change Number $\qquad Ph = \frac{c(T_f-T_W)}{\Delta h}$

The appropriate substitution of the above dimensionless numbers in the differential equation Eq.1 results in [3]

$$\frac{\partial \theta}{\partial Fo} = \frac{\partial^2 \theta}{\partial X^2} \tag{6}$$

with initial and boundary conditions modified as

$$\theta(Fo = 0) = 1 \tag{7}$$

$$\theta(X = 0) = 1 \tag{8}$$

$$\theta(X = 1) = 1 \tag{9}$$

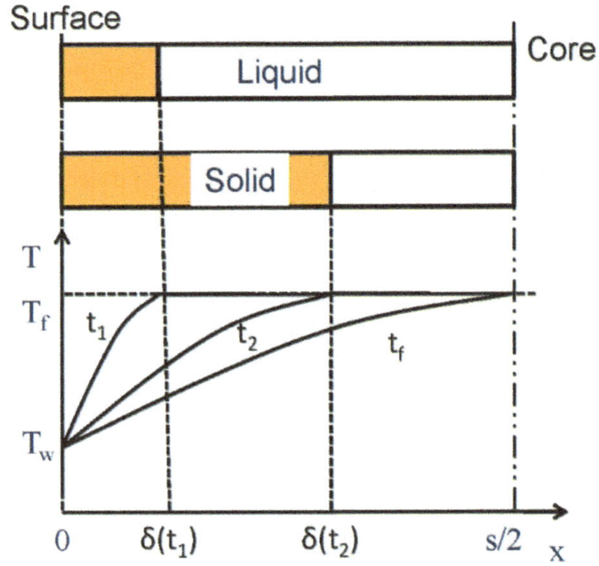

Figure 3: Local temperature profiles during solidification

and the energy balance at the phase boundary modified as

$$\frac{\partial \theta}{\partial X}(X = \Delta) = \frac{1}{Ph}\frac{d\delta}{dFo} \tag{10}$$

The dimensionless solidification thickness (Δ) is thus a function of Fourier number (Fo) and Phase change number (Ph). The analytical solution of Eq. 6 by imposing the energy balance at the phase boundary gives the relation between Δ, Fo and Ph as

$$\frac{\Delta\sqrt{\pi}}{2\sqrt{Fo}}.e^{\frac{\Delta^2}{4Fo}}\,\mathrm{erf}\left(\frac{\Delta}{2\sqrt{Fo}}\right) = Ph \tag{11}$$

The analytical solution is explained briefly in [3]. For the complete solidification, Δ=1 and Eq.11 becomes,

$$\frac{1\sqrt{\pi}}{2\sqrt{Fo_f}}.e^{\frac{1}{4Fo_f}}\,\mathrm{erf}\left(\frac{1}{2\sqrt{Fo_f}}\right) = Ph \tag{12}$$

where Fo_f is the Fourier number corresponding to the complete solidification.

$$Fo_f = Fo(\Delta = 1) \tag{13}$$

This relation Eq.12 is depicted in Fig. 4. For a given Ph and a complete solidification Δ=1, Fo_f is calculated first. The time required for the complete solidification is calculated from the definition of Fo_f as

$$t_f = Fo_f \frac{(s/2)^2}{a} \tag{14}$$

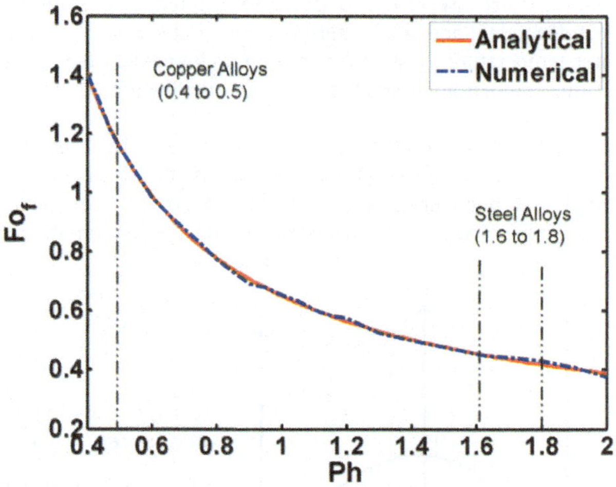

Figure 4: Fourier number vs Phase change number for infinite slab

This time can be converted to casting length with the use of casting speed V_c as,

$$L_f = t_f V_c \tag{15}$$

The casting speed can be calculated from the mass flow rate $\dot{M}$ as

$$V_c = \frac{\dot{M}}{\rho B s} \tag{16}$$

with B and s are the width and thickness of the plate. By substituting the expressions for the time and casting speed, the length required for complete solidification can be expressed as

$$L_f = \frac{c\dot{M}}{4\lambda} \frac{s}{B} Fo_f \tag{17}$$

and assuming b as the ratio of width to thickness,

$$L_f = \frac{c\dot{M}}{4\lambda} \frac{Fo_f(Ph,b)}{b} \tag{18}$$

From the above equation, it is evident that the solidification length (L_f) depends on the ratio of Fo_f and b. However, Fo_f itself is a function of b. For the semi infinite plate $b = \infty$ and L_f can be expresses as

$$L_f = \frac{c\dot{M}}{4\lambda} Fo_f(Ph) \tag{19}$$

From the Eq.19, it can be inferred that the solidification length is proportional to mass flow, Fourier number and inversely proportional to thermal conductivity. The change of initial melt temperature, wall temperature, specific heat and latent heat changes the Ph value. Once Phase change number (Ph) and degree of solidification (Δ) are known, the solidification length is calculated using the above relations. In this way, the solution is generalized for different kind of metal alloys.

Analytical Solution for Finite Slab

Generally in the casting process, the ingots are either rectangular or square in shape. The heat conduction is typically two dimensional in these bodies and the differential equation that governs the heat transfer becomes [4],

$$\rho c \frac{\partial T}{\partial t} = \frac{\partial^2 T}{\partial x^2} + \frac{\partial^2 T}{\partial y^2} \tag{20}$$

Now, the space dimensions are non dimensionalised with $X = x/(B/2)$, $Y = y/(s/2)$ and the above equation becomes

$$\frac{\partial \theta}{\partial Fo} = \frac{\partial^2 \theta}{\partial X^2} + \frac{s^2}{B^2} \frac{\partial^2 \theta}{\partial Y^2} \tag{21}$$

The analytical solution for Eq.21 is a difficult task and the imposition of solidification condition makes the problem more complex. The governing equations are solved numerically for the finite slabs. In the numerical solution, the solidification front needs to be determined in each time step to impose boundary condition Eq.5. Again this is a difficult task and in this study, the fixed grid method is used which considers both liquid and solid as a mixture in single computation cell. The solidification and the corresponding latent heat release are incorporated by increasing the specific heat.

$$c = c + \Delta h \frac{\partial f}{\partial T} \tag{22}$$

where f is the liquid fraction. By modifying the specific heat and non dimensionalising the Eq 1, the final form of the governing equation for 1D becomes,

$$\left(1 + \frac{1}{Ph} \frac{\partial f}{\partial \theta}\right) \frac{\partial \theta}{\partial Fo} = \frac{\partial^2 \theta}{\partial X^2} \tag{23}$$

Similarly for the 2D problems, Eq 20 is modified as

$$\left(1 + \frac{1}{Ph} \frac{\partial f}{\partial \theta}\right) \frac{\partial \theta}{\partial Fo} = \frac{\partial^2 \theta}{\partial X^2} + \frac{s^2}{B^2} \frac{\partial^2 \theta}{\partial Y^2} \tag{24}$$

For Two dimensional geometries, the solidification length is thus a function of solidification thickness (δ), Fo, Ph and the geometric parameter b as presented in Eq.18. The numerical solutions are obtained by using COMSOL Multiphysics 4.3a. To check the accuracy of numerical method, the nondimensionalised governing equation for one dimensional geometry Eq.23 is solved numerically and compared with the analytical solution given by Eq.12. The relation between Ph and Fo_f is analyzed. First the Ph is fixed with a certain value and the governing equation is solved for Fo_f . The Ph value is varied from 0.4 to 2 and in each case, Fo_f is recorded for complete solidification with degree of solidification equals to one. Fig.4 shows the variation of Fo_f with Ph and there is a good agreement between analytical and numerical solutions.

Generally the copper alloys has specific heat in the range of 380 kJ/kg.K , latent heat in the range of 205kJ/kg and initial liquidus temperature of 1120°C. For an assumed constant surface temperature of 1000°C, the Ph number for most of the copper alloys varies from 0.4 to 0.5. The steel and its alloys has specific heat in the range of 700 kJ/kg.K , latent heat in the range of 272 kJ/kg and initial liquidus temperature of 1550°C. For an assumed constant surface temperature of 900°C, the Ph number for most of

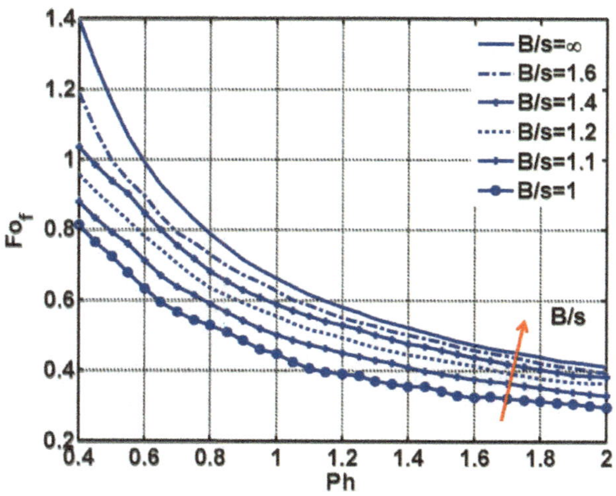

Figure 5: Fourier number vs Phase change number for various values of b

the steel alloys varies from 1.6 to 1.8. The method can be easily extended to low or high Ph numbers if needed. However, too low Ph numbers can produce the numerical challenges.

The developed numerical method is used to solve the governing equation Eq 24 for 2D geometries. In each trail, first the geometric parameter b and Ph is fixed and the Fo_f required for complete solidification is estimated. Now with the same value of b, similar to the one dimensional case, Ph is varied from 0.4 to 2 and the corresponding Fo_f is recorded in each trails. After this b value is changed from 1 to 2 in steps and the similar trails has been carried out. The Fig 5 shows the variation of Fo_f with Ph for various b values.

The value of b=1 represents the case of square cross section. As the value of b increases, one can observe that the Fo_f vs Ph curve reaches to 1D plate solution. Almost at the value of b=1.8, the solution of 2D equals to the 1D. This means that the two dimensional cross sections with width greater than 1.8 times of thickness can be considered as plate and the heat transfer is purely one dimensional.

In the continuous casting problems, it is important to know the minimum length required for the complete solidification. For a given cross section of the ingot to be cast, b value is known. Ph can be calculated from the alloy properties and surface temperature. By the use of b and Ph, Fo is estimated from the Fig 5. From this Fo_f value, the corresponding solidification time is calculated using Eq.14. The final solidification of continuously cast alloys depends purely on the geometric ratio b rather than the actual ingot dimensions.

Temperature Compensation

As explained in the previous sections, ingot has different temperature at core and surface after the solidification. In order to achieve constant temperature throughout the ingot, the surface is insulated for some time such that the surface cannot exchange heat with the ambient. The heat transfer takes place from the core to surface because of the temperature difference. This process is time dependent and stops after both core and surface reaching the same temperature. Now, it is important to know the time required for this compensation and the corresponding length required.

This transient problem can be solved by using the nondimensionalised Fourier differential of the form Eq.21. The initial and boundary conditions that govern the compensation problem are clearly explained in Fig.6. The distribution of

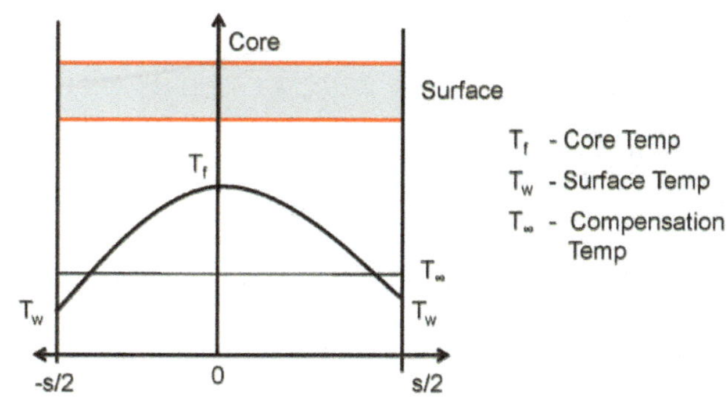

Figure 6: Initial temperature Profile inside the body before the compensation [5]

temperature inside the ingot is assumed to be linear. The equation is solved numerically by keeping all the boundaries insulated at all times. Here, a new parameter φ is defined as the ratio of the temperature difference between core and surface after compensation to the maximum initial temperature difference. The degree of compensation φ can be expressed as

$$\varphi = \frac{\Delta T}{(T_f - T_w)} \quad (25)$$

where ΔT is the temperature difference between core and surface after compensation. It is often allowed to have a some difference between core and surface temperatures. For this reason φ can vary from 0 to 0.2. The φ value of 0.2 represents that the temperature difference between core and surface after compensation is 20 percent of the maximum initial temperature difference. By following the same procedure explained in solidification, first b is fixed and then the curve of compensation Fo_c vs φ is estimated. The same curve is estimated for various b values and the variation of compensation Fo_c with b is plotted in Fig.7. The compensation length increases with the mass flow rate and b. It can be inferred from the above that the time of compensation of continuously cast alloys depends on b. The b value rather than the actual ingot dimensions plays an important role in the compensation also. As b increases, the compensation Fourier number and thus the compensation time is increased.

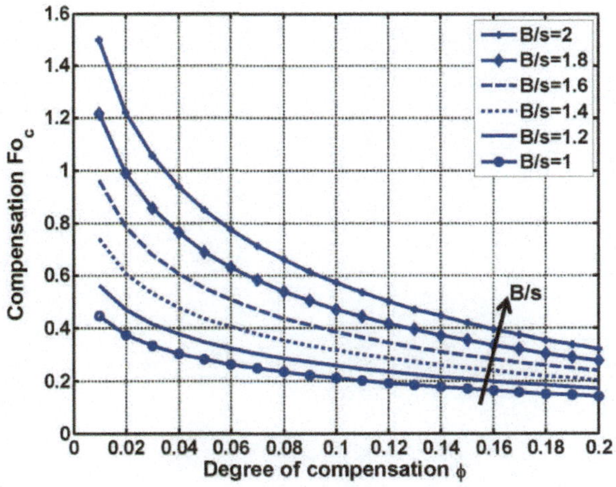

Figure 7: Variation of compensation Fo with φ for arious b values

Effect of Casting Speed

For a fixed b value, the casting speed depends on mass flow rate. From Eq. 19, it can be seen that the solidification length depends on the material properties and mass flow rate for a fixed b value. For a known values of b ratio and Ph, the freezing Fourier number Fo_f can be calculated from Fig.5. This Fo_f *value* can be used to find the solidification length using Eq.19.

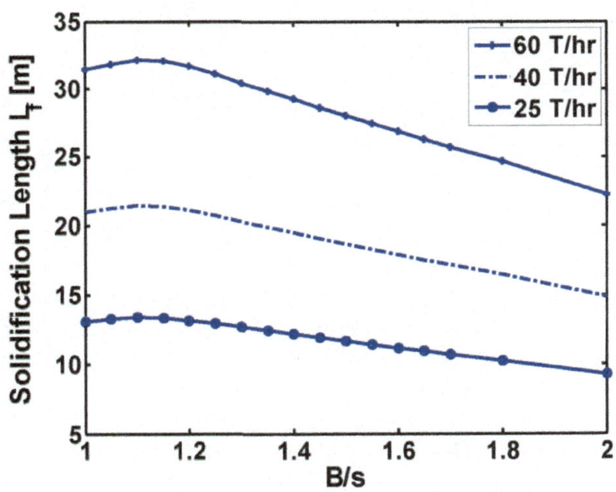

Figure 8: Solidification length variation with b for steel

Fig.8 shows the variation of solidification length with b and mass flow. The solidification length increase with increase in mass flow rate(or casting speed). The solidification length increases with increase in b until 1.1 and then decreases gradually. This is evident from the Eq.18, that the solidification length depends on the ratio of Fo_f to b, and Fo_f in turn is a function of b. For the solidification length point of view, the higher b is recommendable. During the billet casting, generally the cross section shapes are nearly square with b as one and for thin slab casting b is very high.

But in contrast to the solidification length, the compensation length always increases with increase in b. The compensation length for different mass flow rates and the various b values is plotted in Fig.9. During the square billet casting compensation length is less and for thin slabs compensation length is more.

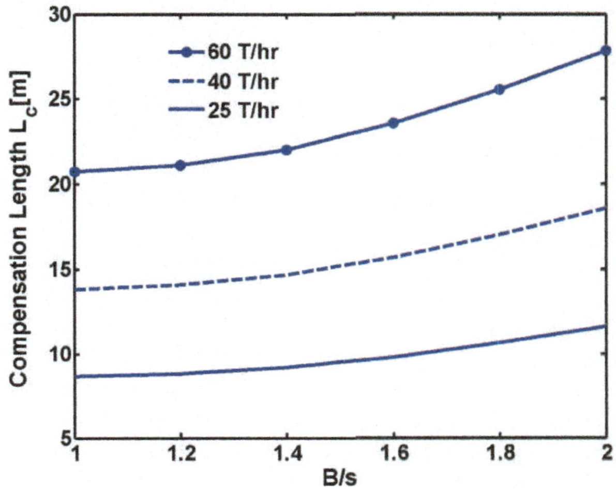

Figure 9: Compensation Length for various B/s for φ=0.1 for steel

Summary

An analytical model for one dimensional solidification is presented for the constant surface boundary condition. A numerical model for both one dimensional and two dimensional heat transfer and solidification is presented. The procedure to find the time and the length required for complete solidification is explained. The model is valid for the continuous casting of alloys ranging from copper to steel. The influence of ingot cross section is explained in detail. It is found that the width to thickness ratio plays a key role in finding the solidification and compensation length.

References

[1] X.Zhou, "Heat transfer during spray water cooling using steady experiments," Master's thesis, University of Illinois at Urbana-Champaign,,2009.

[2] H.Wang, G.Li, Y.Lei, Y.Zaho, Q.Dai, and J.Wang, "Mathematical heat transfer model research for the improvement of continuous casting slab temperature," *ISIJ International*, vol. 45,pp. 1291–1296, 2005.

[3] Incropera, Dewitt, Bergmann, and Lavine, *Fundamentals of Heat and Mass Transfer*. Wiley Publishing Inc, 2006.

[4] H.S.Carslaw and J.C.Jaeger, *Conduction of Heat in Solids*. Oxford University Press, 1959.

[5] E.Specht and R.Alt, "Temperature compensation in heated solids," *Process Metallurgy*, vol. 11, pp. 569–575, 1990.

Proceedings of the 2013 International Symposium on Liquid Metal Processing & Casting
Edited by: Matthew J. M. Krane, Alain Jardy, Rodney L. Williamson, and Joseph J. Beaman
TMS (The Minerals, Metals & Materials Society), 2013

PHOSPHORUS PARTITIONING DURING EAF REFINING OF DRI BASED STEEL

Mohammed A. Tayeb[1,2], Richard Fruehan [1], Seetharaman Sridhar[3]
[1]*Carnegie Mellon University, 5000 Forbes Ave., Pittsburgh, PA, 15213, USA*
[2]*Sabic Technology and Innovation, P.O.BOX 11669, Jubail Industrial City, 31961, Saudi Arabia*
[3] *University of Warwick, Warwick Manufacturing Group, Coventry CV4 7AL, UK*

Keywords: Phosphorus, Partitioning, EAF, DRI, Slag, Steelmaking

Abstract

Flexibility in raw material has prompted interest in better control of phosphorus (P) in molten steel. While significant work has been done on the partition of phosphorus to slags with chemistries relevant to those in the Basic Oxygen Furnace (BOF), there is little reported work on slag chemistries corresponding to that in Electric Arc Furnace (EAF) when direct reduced iron (DRI) is used. In the current phosphorus equilibration between molten Fe-P alloys and slags relevant to EAF-DRI manufacturing is carried out and the thermodynamic limit is compared to existing partition correlations developed for BOF. It is found that the existing correlations cannot be extended to EAF-DRI conditions and a new correlation is presented. The correlation based on thermodynamic data measured in the laboratory is validated with EAF plant data.

Introduction

Over the past decade, steelmakers have been facing two major obstacles: the elevation in raw materials prices and the requirements for manufacturing clean steel with very low amounts of impurities of phosphorus, nitrogen, and sulfur. Phosphorous (P), which is a non-metallic residual element, is un-desired and should be controlled. It is largely originated from the gangue or metal in the DRI. In fact, the DRI contains significantly higher levels of phosphorus as compared to scrap and the content is expected to increase as cheaper sources of raw materials are sought. Iron ore price has increased dramatically to five times what it was in 2006 and this encourages the use of lower grade iron ore that contains higher levels of P in both EAF and integrated routes of steelmaking. With the requirement of steels with low phosphorus content such as in thin sheets, deep drawn applications, automobile exteriors and all high uniform deformability applications and with the known detrimental effects of phosphorus in steel which includes various forms of embrittlement that reduce the toughness and ductility, a better understanding of thermodynamics and kinetics of the slag/metal reactions that govern phosphorus removal is required. [1-7]

Phosphorus in Steelmaking

The oxidation of phosphorus (dephosphorization) occurs according to the following equation written in ionic form as [4]:

$$\underline{P}+2.5(FeO)+1.5(O^{2-})=(PO_4^{3-})+2.5Fe \qquad (1)$$

Or as molecular form:

$$\underline{P}+2.5(FeO)=(PO_{2.5})+2.5Fe \qquad (2)$$

Equilibrium constant of equation 1 can be defined as:

$$K_1=\frac{a_{PO_4^{3-}}}{a_{\underline{P}}P_{O_2}^{5/4}a_{O^{2-}}^{3/2}}=C\frac{(\%P)\gamma_{PO_4^{3-}}}{[\%P](T.Fe)^{5/2}\gamma_{FeO}^{5/2}a_{O^{2-}}^{3/2}} \qquad (3)$$

$$=Ck_p\frac{\gamma_{PO_4^{3-}}}{\gamma_{FeO}^{5/2}a_{O^{2-}}^{3/2}}=C'k_p$$

From equations 3, two terms can be defined:

The phosphorus partition ratio:

$$L_P=\frac{(\%P)}{[\%P]} \qquad (4)$$

And the apparent equilibrium constant:

$$k_P=\frac{L_P}{(T.Fe)^{5/2}} \qquad (5)$$

In the above equations:
The underlined represents the content in the metal while the parentheses represent the content in the slag
a: activity in the metal or in the slag
P: partial pressure
γ: activity coefficient
T.Fe: total iron in the slag

To evaluate the interaction between different parameters, it is useful to use multivariate analysis. This analysis is used to evaluate a statistical dephosphorization equation. The slag-metal system of EAF is sensitive to several parameters and is an extremely dynamic system. The slag chemistry and furnace temperature should be known to model the dephosphorization chemical reactions. The slag chemistry can change with oxidation of metallic charge materials during melting, reduction of FeO from slag, addition of CaO and MgO, and slag flushing. [8]

Equilibrium phosphorus distribution correlations had been investigated in the past by many researchers [9-21] and phosphorus distribution equations had been developed. Briefly, some correlations will now be reviewed:

Equation by Healy [14]:

$$Log(\frac{L_P}{Fe_t^{2.5}}) = 0.08(\%CaO) + \frac{22350}{T} - 16 \tag{6}$$

The equation developed by Healy in 1970 related phosphorus partition ratio to the CaO content in the slag and temperature. Many researchers [12, 18] have concluded the inaccuracy of this correlation.

Equation by Suito et al. [19]:

$$Log(\frac{L_P}{Fe_t^{2.5}}) = 0.072[(\%CaO) + 0.3(\%MgO) + 0.6(\%P_2O_5) + 0.6(\%MnO)] + \frac{11570}{T} - 10.52 \tag{7}$$

Equation by Ide and Fruehan [15]:

$$Log(\frac{L_P}{Fe_t^{2.5}}) = 0.072[(\%CaO) + 0.15(\%MgO) + 0.6(\%P_2O_5) + 0.6(\%MnO)] + \frac{11570}{T} - 10.52 \tag{8}$$

In equation 8 above, re-evaluation of MgO coefficient of equation 7 was made by using CaO crucibles in order to ensure that the slag had variable MgO contents.

In a recent work by Assis et al. [9], change of coefficients of equation 7 was made and the following revised correlation was introduced:

$$Log(\frac{L_P}{Fe_t^{2.5}}) = 0.068[(\%CaO) + 0.42(\%MgO) + 1.16(\%P_2O_5) + 0.2(\%MnO)] + \frac{11570}{T} - 10.52 \tag{9}$$

While the work on developing relations such as the aforementioned ones is quite extensive, it is not clear which and to what extent they represent thermodynamic equilibrium or not. It is also not clear to what extent the role of metal droplets plays in industrial processes. Finally, despite the extensive work on phosphorus reactions in steelmaking, limited data exists in the literature that treats and tests low basicity and FeO contents of 25-50% slags that are relevant to DRI based EAF. All the correlations developed by Suito et al. [19] and the revised ones by Ide and Fruehan [15] and by Assis et al. [9] were mainly developed for high range basicity (CaO/SiO_2). Therefore, the goal of this work is to evaluate phosphorus equilibrium between metal and slag in conditions similar to DRI based EAF slags.

Evaluation of Correlations

As mentioned earlier, there are many correlations for phosphorus equilibrium. To test the validity of equilibrium equation 7, all the data in the literature [10, 11, 18, 19] are used to get the plot of Figure 1, which represents log k_p versus coefficients of equation 7.

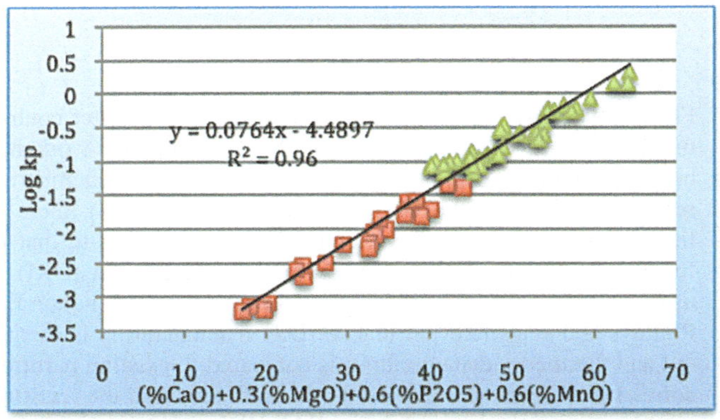

Figure 1. Log k_p vs. coefficients of equation 7, T=1650°C. The triangles represent Basu's et al. [10, 11] data; the squares represent Suito's et al. [18, 19] data.

Overall, the data fits reasonably well with the line. The slight difference in the slope from the equilibrium equation 7 might be due to the fact that the very low basicity ($CaO/SiO_2<0.5$) and/or very high FeO (>60%) data have been omitted.

On the other side, looking at the target chemical compositions relevant to DRI-EAF as summarized in Table I, Figure 2 is plotted. It is obvious that the scatter is more and the slope is higher compared to Figure 2. The R^2 value in addition is less. Because the plot is in a log scale, even with the closest point to the line, which represents the equilibrium equation, the predicted and the measured phosphorus in the metal vary significantly. The equilibrium equations 7, 8, and 9, therefore, are directed to BOF slags.

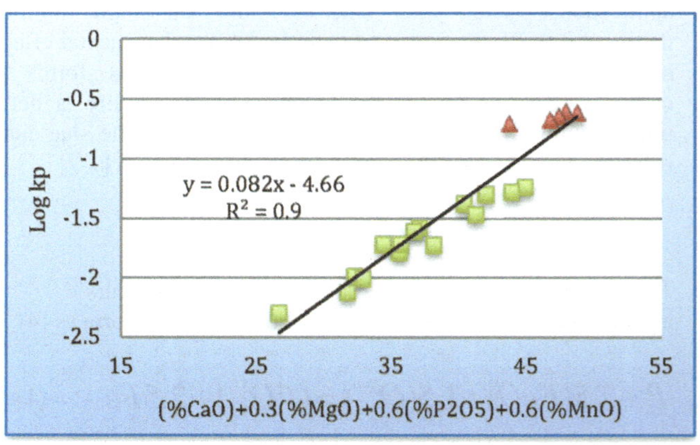

Figure 2. Log kp vs. coefficients of equation 7, T=1600oC, CaO/SiO2: 0.9-2.1 and FeO: 22-52%. The triangles represent Basu's et al. [10, 11] data; the squares represent Suito's et al. [18, 19] data.

Table I. DRI-EAF target basicity and iron oxide content

	CaO/SiO_2	FeO
DRI-EAF slags	0.9 to 2.1	22 to 52

Experimental Procedure

The experimental equipment and procedures are similar to that described elsewhere [9, 17]. Therefore, only a brief description follows.

Designing the Chemical Compositions

This work concentrates on basicity ranges from 1.0 – 2.1 and FeO concentrations of 25-40% on the slag system $CaO\text{-}SiO_2\text{-}MgO\text{-}FeO\text{-}P_2O_5$.

Equilibration Time

The time required to reach equilibrium between the slag and the metal varies and depends on compositions and temperature. It can be noted that a time between 3 and 10 hours was reported as sufficient for equilibration [9-11, 13, 15, 16, 18, 19, 22, 23]. To ensure that equilibrium between the metal and the slag is achieved, 10 hours is used in the present work as equilibrium time. The present work has approached equilibrium from both sides i.e. slag to metal and metal to slag. For equilibrium, the two approaches should give the same phosphorus partition ratio values.

Materials Preparation

Master Alloy Approximately 600 g of alloy was prepared by adding 0.05 wt.% phosphorus in the form of iron phosphide Fe_3P to the iron chips (99.95 wt.%). The total mix was put in MgO crucible in a resistance box furnace in high purity Ar gas and heated up to 1600 °C.

Master slag A total of around 300 g was prepared. Initially, CaO, SiO_2 and MgO dried in air at 900°C for 8 h in a resistance box furnace. Then, the products were mixed and heated for 4 h at 1600 °C in Ar gas. When the mixture was at 1600 °C, the furnace was opened and the melt was quenched in water to ensure homogenized mixing. Then, the solid was crushed to powder and decarburized at 1200°C for 10 h. The main objective of pre-melting the slag is ensuring a fully liquid slag at the actual experiments. In all experiments, FactSage [24] was used to ensure that the chemistry corresponds to a fully liquid slag at 1600 °C and the FeO content was modified accordingly. To confirm that the slag was liquid, it was heated to 1600°C and imaged through a Confocal Laser Scanning Microscope (CLSM). Three slag samples were prepared with slag basicity of 1.0 and 2.0 and it was observed that the slag was liquid before reaching 1600°C. A horizontal tube furnace manufactured by CM Furnaces with $MoSi_2$ heating elements was used for the equilibrium experiments. The schematic is shown in Figure 3.

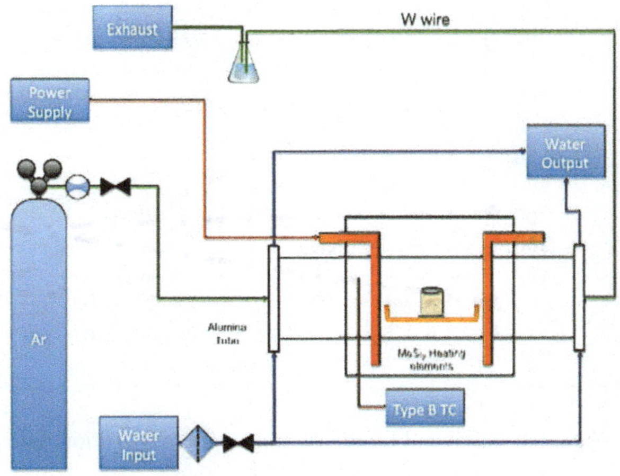

Figure 3. Furnace Setup Schematic

Experiment's Description

Equilibration was approached from both sides, i.e. phosphorus transfers from metal to slag (dephosphorization) and from slag to metal (phosphorization). 12 g of metal and 6 grams of slag were used. The steel phase was added to the bottom of the crucible. Then, the weighted mixed slag was placed at the top with an iron disk separator.

Each experiment lasts around 2 days. The controller was programmed for 15 hours heating until it reaches 1600 °C. The holding (equilibrium time) was set to be 10 hours. A quickly drop of a temperature to 1300 °C was done. Finally, a slow cooling rate was set for 14 hours. The samples were removed when room temperature was reached. The metal and slag was then separated and sent for chemical analysis.

Results and Discussion

To modify the equilibrium equation to suit the DRI-EAF slag compositions, EAF database should be established. From Figure 2 previously shown, we could conclude that the number of points in the literature that represents the DRI-EAF target compositions is not enough and more laboratory experiments should be performed. Accordingly, fourteen laboratory experiments were performed in this study with basicity ranging from 1.0 – 2.1 and FeO contents ranging between 25-40 wt.%. Figure 4 shows the initial and equilibrium phosphorus partition ratio for two experiments. 10 hours hold time does appear to ensure thermodynamic equilibrium.

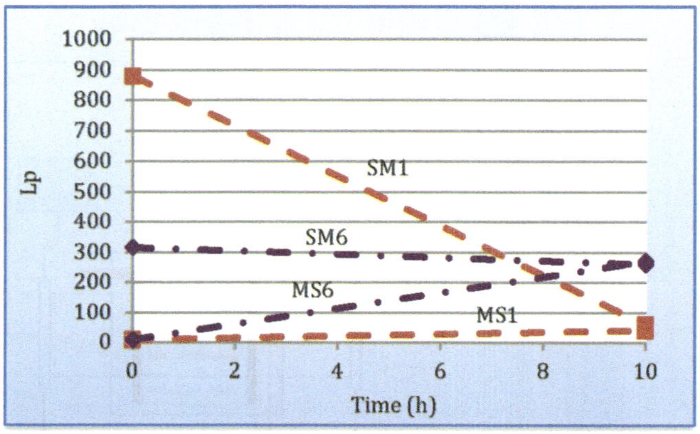

Figure 4. Initial and Equilibrium phosphorus partition ratios for experiments 1, and 6.

Using the new data with all the existing data in the literature with the target slag chemical compositions, a better equilibrium equation 10 is formulated. Figure 5 represents the new plot.

$$Log(\frac{L_P}{Fe_t^{2.5}}) = 0.061[(\%CaO) + 0.2(\%MgO) + 0.54(\%P_2O_5) - 0.7(\%MnO)] + \frac{11150}{T} - 10.52 \quad (10)$$

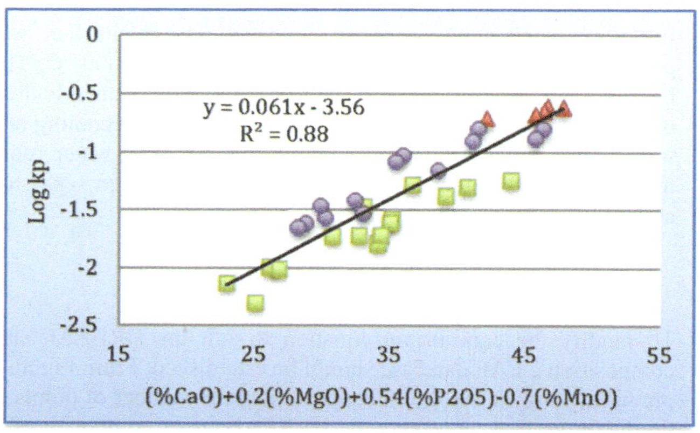

Figure 5. Log k_p vs. coefficients of the present study equation 10, T=1600°C. The triangles represent Basu's et al. [10, 11] data; the squares represent Suito's et al. [18, 19] data; the circles represent current work data.

R square in the plot shown in Figure 5 shows that with the inclusion of the new data, the scatter is greater. In present work, P_2O_5 content was fixed at roughly 1 wt.% whereas Basu's et al. [10, 11] experiments were performed with 3-8 wt. % P_2O_5 and Suito's et al. [18, 19] experiments varied the P_2O_5 content between 0.7-5 wt.%.

It can be noticed that the current work data tends to indicate higher phosphorus distribution ratios than what is predicted. In other words, there appears to be more phosphorus in the slag and less in the metal for a given slag composition. Suito and his group [18, 19] used 3-5 hours as equilibration time depending on the slag compositions while in the present work, 10 hours was used as equilibrium time. If we select two points with similar chemical compositions as shown in Table II, it can be seen that (i) at lower basicity, the current data and Suito's data have similar phosphorus distribution ratio whereas (ii) at higher basicity the data differs by a factor of 3. The cause for this is not known at this time. More experiments and closer analysis of the data are being performed. The analytical uncertainties at very low phosphorus contents may partially account for the scatter of the data. The phosphorus in the slag may present in the form of PO_4^{3-} or $P_2O_7^{4-}$ ions. The assumption that the phosphorus in the slag presents as only PO_4^{3-} ions might in addition account for the deviation.

Table II. Two selected points from Suito et al. [18, 19] and present data

CaO	MgO	P_2O_5	Lp	
30.8	14.8	0.971	35	Suito's data
30	14.08	0.93	41	Current work data
36.9	11.6	0.722	63	Suito's data
38.1	10.44	1.15	173	Current work data

Industrial Data

Hadeed Company [25] provided the authors with data from 14 heats from an Electric Arc Furnace with low basicity and FeO contents of less than 40 wt.%. In addition to these 14 points, there is open data in the literature [26] with 24 points EAF data: basicity of less than 2.5 and FeO contents range from 27 to 50 wt.%. A third set of scrap-based EAF plant data recently provided from an international steel producer was also analyzed. Using these data, the newly developed equation 10 is tested. The results are shown in Figure 6. Although, the slope is higher than the one in equation 10, R square is 0.86, which is reasonably good. Therefore, the new equation is considered reasonable in predicting phosphorus partition ratio in conditions similar to EAF real plant practice. However, if we plot the plant data with the equilibrium equations 7 and 10 as shown in Figure 7, obviously it is clear that equation 10 is better than equation 7 in predicting the phosphorus partition ratio. It should be noted, however, that the plant data consistently falls below the equilibrium line possibly indicating equilibrium is not achieved in the EAF.

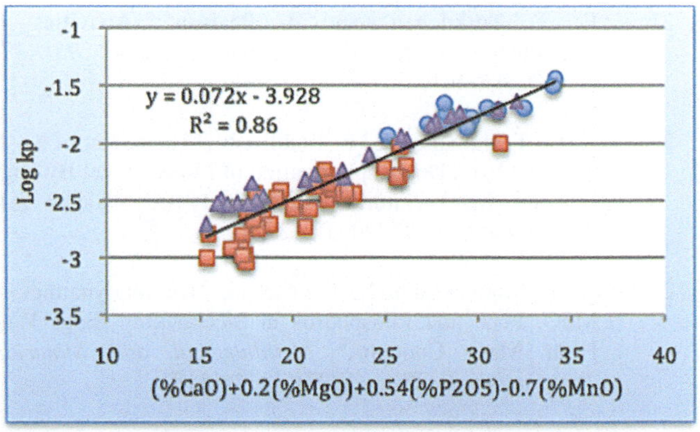

Figure 6. Log k_p vs. coefficients of the present study equation 10 for EAF plant data. The triangles represent EAF data [26]; the circles represent Hadeed data [25]; the squares represent the international steel producer data.

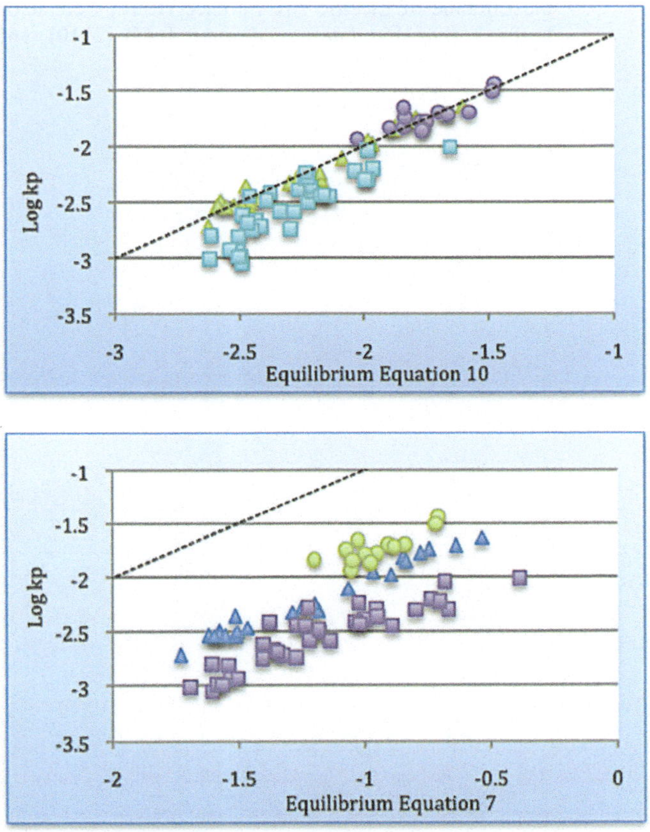

Figure 7. Log k_p vs. equilibrium equations 7 and 10 (the dash line). The triangles represent EAF data [26]; the circles represent Hadeed data [25]; the squares represent international steel producer data.

Conclusion

1. The Presented paper focused on evaluation of existing correlations on dephosphorization effectiveness of EAF slags when using DRI. A new correlation is developed based on laboratory experiments and plant data.
2. In equilibration experiments, 10 hours was used as holding time and appears to ensure that thermodynamic equilibrium was achieved. In addition, equilibrium was approached by transferring phosphorus from slag to metal and vice versa and observing the same phosphorus partition ratio values is a good test of equilibrium.
3. Many factors contribute to predict phosphorus partition ratio and their effects are overlapping. Accordingly, linear equation is produced that take into account several parameters including CaO, MgO, P_2O_5, and temperature. The equation is developed based on laboratory data. It is validated with plant data, which showed a good fit to the developed equation with R Square of more than 0.85. It showed scatter from the equilibrium equation, though.
4. For low basicity, phosphorus partition ratios for laboratory and plant data are very close and this may explain why the correlation works reasonably well with the plant data.
5. Current work data predicts higher phosphorus partition ratio than either the existing data in the literature or the plant data especially for higher basicity. More work and analysis are being performed to explain the differences between present work and existing data.
6. The literature lacks experiments data with Al_2O_3 addition. It is an essential part of EAF slags unlike BOF. This is a focus of an ongoing work by the authors of this paper. Upon more data is being available, new correlation will be developed for EAF slags with addition of Al_2O_3 and will be tested with EAF plant data.

Acknowledgements

The authors wish to express their thanks to the Center of Iron and Steelmaking Research (CISR) at Carnegie Mellon University and Saudi Arabia Basic Industrial Company (SABIC) for their financial support of this study. The authors also thank SABIC for their financial support of Tayeb, M. during his study at Carnegie Mellon University. In addition, Tayeb, M. would like to thank Mr. Andre Assis for the valuable discussions.

References

1. Index Mundi. www.indexmundi.com.

2. C. L. Briant, and S. K. Banerji, *Intergranular Fracture in Ferrous Alloys in Nonaggressive Environments* (Academic Press, 1983), 21-58.

3. J. R. Low et al., "Alloy and Impurity Effects on Temper Brittleness of Steel," *Trans Aime*, 242 (1968), 14-24.

4. R. J. Fruehan, *The Making, Shaping and Treating of Steel* (Pittsburgh, Pa: The AISE Steel Foundation, 1998), 1-134.

5. S. Andreson, "Educated Use of DRI/HBI Improves EAF Energy Efficiency and Yield and Downstream Operating Results," Midrex Technologies Inc., (2002).

6. T. A. Boom, D. R. Fosnacht, and D. M. Haezebrouck, "The Influence of Phosphorus on the Properties of Sheet Steel Products and Methods Used to Control Steel Phosphorus Levels in Steel Product Manufacturing - Part I," *Iron and Steelmaker*, 17(9) (1990), 35-41.

7. M. A. Tayeb, "Introduction to EAF Steelmaking," Course presented to HADEED steel plant employees, (2009).

8. C. Manning, "Behavior of Phosphorus in DRI/HBI During Electric Furnace Steelmaking" (Ph.D. thesis, Carnegie Mellon University, 2000).

9. A. Assis, R. Fruehan, and S. Sridhar, "Phosphorus Equilibrium Between Liquid Iron and CaO-SiO_2-MgO-FeO Slags," Proceedings from AISTech, Atlanta, GA, (2012), 861-870.

10. S. Basu, A. K. Lahiri, and S. Seetharaman, "Phosphorus Partition Between Liquid Steel and CaO-SiO_2-P_2O_5-MgO Slag Containing Low FeO," *Metallurgical and Materials Transactions B*, 38(3) (2007), 357-366.

11. S. Basu, A. K. Lahiri, and S. Seetharaman, "Phosphorus Partition Between Liquid Steel and CaO-SiO_2-FeO_x-P_2O_5-MgO Slag Containing 15 to 25 Pct FeO," *Metallurgical and Materials Transactions B*, 38(4) (2007), 623-630.

12. S. Basu, A. K. Lahiri, and S. Seetharaman, "A Model for Activity Coefficient of P_2O_5 in BOF Slag and Phosphorus Distribution Between Liquid Steel and Slag," *ISIJ International*, 47(8) (2007), 1236-1238.

13. H. Ishii, and R. J. Fruehan, "Dephosphorisation Equilibria Between Liquid Iron and Highly Basic CaO-Based Slags Saturated With MgO," *ISS Transactions*, (1997), 47-54.

14. G. W. Healy, "A New Look At Phosphorus Distribution," *JISI*, 208(7) (1970), 664-668.

15. K. Ide, and R. J. Fruehan, "Evaluation of Phosphorus Reaction Equilibrium in Steelmaking," *Iron & steelmaker*, 27(12) (2000), 65-70.

16. C. M. Lee, and R. J. Fruehan, "Phosphorus Equilibrium Between Hot Metal and Slag," *Ironmaking & steelmaking*, 32(6) (2005), 503-508.

17. M. A. Tayeb, R. Fruehan, and S. Sridhar, "Equilibrium Between Liquid Iron and EAF Slags," Proceedings from 5th International Congress on the Science and Technology of Steelmaking, Dresden, Germany, (2012).

18. H. Suito, R. Inoue, and M. I. Takada, "Phosphorus Distribution Between Liquid Iron and MgO Saturated Slags of the System CaO-MgO-FeO_x-SiO_2," *Trans. Iron Steel Inst. Jpn.*, 21(4) (1981), 250-259.

19. H. Suito, and R. Inoue, "Phosphorus Distribution Between MgO-Saturated CaO-Fe_tO-SiO_2-P_2O_5-MnO Slags and Liquid Iron," *Transactions of the Iron and Steel Institute of Japan*, 24(1) (1984), 40-46.

20. H. Suito, and R. Inoue, "Thermodynamic Assessment of Hot Metal and Steel Dephosphorization With MnO-Containing BOF Slags," *ISIJ international*, 35(3) (1995), 258-265.

21. E. T. Turkdogan, and J. Pearson, "Activities of Constituents of Iron and Steelmaking Slags," *JISI*, 175 (1953), 398-401.

22. G. Li, T. Hamano, and F. Tsukihashi, "The Effect of Na_2O and Al_2O_3 on Dephosphorization of Molten Steel By High Basicity MgO Saturated CaO-FeO_x-SiO_2 Slag," *ISIJ International*, 45(1) (2005), 12-18.

23. A. T. Morales, and R. J. Fruehan, "Thermodynamics of MnO, FeO, and Phosphorus in Steelmaking Slags With High MnO Contents," *Metallurgical and Materials Transactions B*, 28(6) (1997), 1111-1118.

24. "FactSage™ (Version 6.4)," (1976-2013).

25. "Saudi Iron & Steel Company," (1979-2013).

26. T. Borovsky, J. Kijac, and M. Domovec, "The Slag Composition Influence on the Dephosphorization and on the Lifetime of Electric Arc Furnace Hearth Referactory Lining," *Acta Metallurgica Slovaca*, 16(3) (2010), 165-171.

AUTHOR INDEX
Proceedings of the 2013 International Symposium on Liquid Metal Processing and Casting

SUBJECT INDEX
Proceedings of the 2013 International Symposium on Liquid Metal Processing and Casting

The manufacturer's authorised representative in the EU is Springer Nature Customer Service Centre GmbH, Europaplatz 3, 69115 Heidelberg, Germany. If you have any concerns regarding our products, please contact ProductSafety@springernature.com

Printed and bound by CPI Group (UK) Ltd, Croydon, CR0 4YY
01/10/2026
02225886-0002